Hydrological Sciences in the Anthropocene

IAHS Scientific Assembly 2022

Proceedings of the International Association of Hydrological Sciences (PIAHS) Volume 385, 2024

Montpellier, France
29 May – 3 June 2022

ISBN: 979-8-3313-0727-1

Printed from e-media with permission by:

Curran Associates, Inc.
57 Morehouse Lane
Red Hook, NY 12571

Some format issues inherent in the e-media version may also appear in this print version.

International Association of Hydrological Sciences
Centre for Ecology and Hydrology
Wallingford, Oxfordshire OX10 8BB
United Kingdom

Phone: 441 491 692 442
Fax: 441 491 692 448

http://iahs.info/

Additional copies of this publication are available from:

Curran Associates, Inc.
57 Morehouse Lane
Red Hook, NY 12571 USA
Phone: 845-758-0400
Fax: 845-758-2633
Email: curran@proceedings.com
Web: www.proceedings.com

TABLE OF CONTENTS

Proc. IAHS, 385, 1–4, 2024
https://doi.org/10.5194/piahs-385-1-2024

Open Access
PIAHS

The hydrological system as a living organism

Hubert H. G. Savenije

Water Resources Department, Delft University of Technology, Delft, the Netherlands

Correspondence: Hubert H. G. Savenije (h.h.g.savenije@tudelft.nl)

Received: 15 June 2022 – Accepted: 18 November 2022 – Published: 18 April 2024

Abstract. Hydrology is the bloodstream of the terrestrial system. The terrestrial system is alive, with the ecosystem as its active agent. The ecosystem optimises its survival within the constraints of energy, water, climate and nutrients. The key variables that the ecosystem can modify are the controls on fluxes and storages in the hydrological system, such as: the capacities of preferential flow paths (preferential infiltration, recharge and subsurface drainage); and the storage capacities in the root zone, wetlands, canopy and ground surface. It can also, through evolution, adjust the efficiency of carbon sequestration and moisture uptake. Some of these adjustments can be made fast, particularly rootzone storage capacity, infiltration capacity, vegetation density and species composition. These system components are important controls on hydrological processes that in hydrological models are generally considered static and are determined by calibration on climatic drivers of the past. This leads to hydrological models that are dead and incapable to react to change, whereas the hydrological system is alive and will adjust. The physical law driving this evolutionary process is the second law of thermodynamics with the Carnot limit as its constraint. This physical limit allows optimisation techniques to explore the reaction of the hydrological system and its components to change in climatic drivers. This implies a new direction in the theory of hydrology, required to deal with change and addressing the Unsolved Problems in Hydrology.

Keywords. UPH 19; ecosystem-based hydrology; new theory

1 Introduction

Physical processes in the atmosphere, the hydrology, ecosystems, and geology are interconnected, influencing each other with numerous feedbacks at a wide range of temporal scales. This was realised as early as the 18th century by Alexander von Humboldt who did empirical research in many parts of the world and came to the conclusion that all these processes were connected (Wulf, 2015). In present times, in the science of hydrology, specialisation and reductionism has forced us towards fragmentation and focus on laboratory processes instead of system-wide and interdisciplinary research. This has led to wide applicability of laboratory scale-based methods that have been erroneously upscaled to catchment level. The most well-known examples are the Richards and Darcy equations which, supported by ever stronger computational power, are applied at scales for which they are not fit. They may give reasonable answers, but not for the right reasons, demanding numerous calibration parameters at different scales. At catchment-scale, these concepts are flawed.

The main problem with these lab-scale-based methods is that they neglect preferential flow patterns that are dominant in nature. Wherever water moves through a medium, fractal-type patterns appear. These are omnipresent in nature: in the veins of leaves, in branches and root systems, in rills of overland flow, in alluvial vans, in seepage of water, in soil infiltration, in sub-surface drainage patterns, in the shape of river networks and in the veins of our own body. Although hard to observe, these drainage patterns are also present in the groundwater system, which is the cause of the mismatch between the travel times of dissolved substances and the water in predominant groundwater models.

These patterns are formed by different processes at different time scales. They evolve over time and are not static. The speed of pattern formation in geo-morphological processes depends on the erodibility and consistency of the medium through which the water flows and often has long time scales. Biological processes of pattern formation are much faster.

But although these processes are dominant, main stream hydrological models don't take them into account.

Instead, traditional models: (1) split-up the Earth in small fragments, destroying patterns, preferential pathways and system behaviour; (2) consider the substrate and ecosystem static; (3) are dead and don't consider the ecosystem as an active agent which can adjust to change; (4) are unnecessarily complex; and (5) rely heavily on calibration. This goes against what we observe in nature and there is an urgent need for a new hydrological theory that considers the hydrological system as a living organism (Savenije and Hrachowitz, 2017). Catchments are alive!

2 Towards a new theory of hydrology

The ecosystem, as the most active agent in the natural system, manages the water. It does this for a simple evolutionary purpose: to optimise survival. If an ecosystem had not managed its water resources well, it would simply no longer be there. People, who also manage the water system, obey similar laws of survival, by trial and error. But in the head waters of a catchment, the ecosystem is dominant and has a high potential for adaptation under existing climatic, landscape and geological constraints. With the ecosystem as the active water manager, the hydrological system is alive and able to adjust to changing climatic or human-induced circumstances (e.g. Nijzink et al., 2016). But although the ecosystem-atmosphere-hydrology-geology interaction is very complex, surprisingly, hydrological laws we observe in nature, such as the linear reservoir, resistance to flow, Budyko's curve, Darcy's equation, etc., are often described by simple equations (e.g. Dooge, 1986, 1997; McDonnell et al., 2007). How can this be explained?

The only physical law that has a unique direction in time is the second law of thermodynamics, which functions at all scales, and which states that entropy can only increase. The Earth is a "dissipative structure" that exchanges low entropy for high entropy. It does so at Maximum Power, close to the "Carnot limit" of a dissipative engine (Kleidon, 2016). The ecosystem that has the potential to evolve, appears to operate close to the Carnot limit, the most efficient way in which free energy can be converted into work. It appears that maximizing the Power of a natural process often leads to surprisingly "Simple Laws" (Kleidon et al., 2013).

Hydrology is the bloodstream of the ecosystem. There are several ways in which the ecosystem manipulates the hydrological system. The canopy interception not only provides immediate feedback to the atmosphere maintaining atmospheric moisture, it also channels water to dominant dripping points and stemflow, directing the water to places where infiltration is facilitated and preventing surface runoff. Through preferential patterns this infiltration recharges the root zone moisture and percolation to the groundwater. If there is too much water, then the excess moisture is evacuated through preferential sub-surface drainage patterns that are formed under the root zone. In addition, the ecosystem creates sufficient storage in the root zone to overcome critical period of drought. Hence, the ecosystem stores and partitions the precipitation to its advantage and in doing so determines the parameter values of hydrological models.

The root zone storage capacity is a crucial parameter in all hydrological models, which is normally calibrated, but which can be predicted by a simple water balance method similar to the way in which engineers size a reservoir (Gao et al., 2014a). Considering hydrological systems as a living, evolving ecosystem holds the key to independent determination of model parameters, such as the root zone storage capacity (De Boer-Euser et al., 2016; Wang-Erlandsson et al., 2016), and its spatial distribution (Gao et al., 2019), but can potentially also be applied to interception, infiltration and drainage capacity.

3 How should we model catchments?

The essence of hydrological modelling is to translate what we observe in the actual environment to the model space. From our real-life observations, we formulate a perceptual model, followed by a conceptual model and a model structure, and finally mathematical equations that quantify these processes (Beven, 2011). Each of these steps is equally important, but – in the process – we should take the following in consideration:

1. Accounting for landscape in the model structure. The landscape reflects different formation processes that have developed over longer time scales, but which have different dominant runoff processes, such as surface runoff, sub-surface runoff, groundwater recharge, groundwater seepage, freeze-melt cycles, etc. (e.g.: Savenije, 2010; Gharari et al., 2011; Gao et al., 2014b).

2. Accounting for the geology in relation to model structure, defining model structure and slow groundwater processes (Fenicia et al., 2014).

3. Accounting for land use, since different land covers will have different phenology and evaporation, storage and runoff characteristics.

4. Considering the ecosystem as the active agent that manipulates the hydrological processes to optimise its survival (Savenije and Hrachowitz, 2017).

5. Using the 2nd law of thermodynamics to try to find the laws and parameters that govern the energy transitions, by maximizing the power of transferring free energy to work (the Carnot limit).

Although the last item is a deeper item for research, the first four are very practical and can be applied readily (Mao and

Liu, 2019; Bouaziz et al., 2020). This ecosystem-centred approach has the potential to lead to more robust equations with physical parameters, which reduce the need for calibration and enhance our capacity to predict hydrological behaviour in ungauged basins.

4 Tipping points and the limits of ecosystems to adjust

A good illustration of how ecosystems adapt to climatic drivers and of transition points that occur when an ecosystem is no longer capable of meeting atmospheric conditions is provided by Singh et al. (2020). In a number of transects through Africa and South America they showed how ecosystems adapt their root zone storage capacity to the occurrence of drought and dry spells, indicating the point where evergreen rainforest gradually switches to less dense (often deciduous) forest, such as the African Miombo forest, until a maximum root zone storage is achieved, after which a transition occurs to savannah and dry land vegetation. These tipping points are crucial because they are often strengthened by positive feedbacks, such as in the transition of rainwater harvesting tiger bush to barren land.

5 How to deal with human interference

One may question whether the ecosystem-centred approach also applies to parts of the world that are strongly influenced by human interventions. Clearly in hydrological modelling it is required to take into account land use. If the use is a monoculture, whether covered with forest of agricultural crops, then it is important to distinguish between annual or multi-annual crops. A multi-annual single-species crop will obey the same survival rules as a natural ecosystem, although adjustment to change is different. In a landcover with a wide range of species, adjustments can occur not only by species adaptation but also by invasion of more fit species, changing the composition of species. In all cases it is important to take into account the ecosystem traits. Evergreen forest will have to store enough water to deal with dry spells or dry seasons, catering for a dry period with a longer return period for survival (e.g. once in 20–30 years). Deciduous forests have an additional option. They can shed their leaves earlier than normal in case rains stop early in the wet season; this happens for instance in the African Miombo forest. In the Miombo, some trees can even tap into the deeper groundwater. The return period can thus be shorter than for evergreen forest (e.g. once in 10–20 years). Grasslands have even more options. They can go dormant, stopping evaporation until the next rainy season starts. As a result, they cater for within season droughts with a return period of about once in 1–2 years. Single year crops act in much the same way as grass land, catering only for within year dry spells. Finally, irrigated crops react again differently, whereby the human being aims at a crop survival of about once in 5 years, depending on the value of the crop. In urban areas, one should look at the vegetation cover and apply the approach to the vegetated part only.

Of course, it still requires creativity of the modeller. The essence is the different frame of mind, putting the ecosystem at the centre of the approach; popularly said: "try to think like an ecosystem". Ask yourself the question "if I were the ecosystem, what would I do to improve my survival and reproduction?".

6 Conclusion

UPH #19 of the 23 unsolved problems in hydrology reads: "How can hydrological models be adapted to be able to extrapolate to changing conditions, including changing vegetation dynamics?" (Blöschl et al., 2019). This paper offers a promising new venue for addressing this question. Considering the hydrological system as a system that can adapt itself to changing climatic and human circumstances, while trying to optimise its conditions for survival, is a way to discover parameter values that agree with an unknown future. Instead of confronting a calibrated model, tuned to the past (i.e. a dead model), to changing atmospheric drivers, one should use a living and adaptable model that corresponds with these changing circumstances. Partly that can be done by evolving parameter values, and partly by "exchanging space for time", i.e. by lending the ecosystem characteristics from other parts of the world where these climate conditions already exist (e.g. Bouaziz et al., 2022).

Dealing with a world that is continuously changing is the main challenge of the coming decades (Sivapalan et al., 2012; Montanari et al., 2013) and the ecosystem-centred approach may be a promising way to address this challenge.

Data availability. No data sets were used in this article.

Competing interests. The author has declared that there are no competing interests.

Disclaimer. Publisher's note: Copernicus Publications remains neutral with regard to jurisdictional claims in published maps and institutional affiliations.

Special issue statement. This article is part of the special issue "IAHS2022 – Hydrological sciences in the Anthropocene: Variability and change across space, time, extremes, and interfaces". It is a result of the XIth Scientific Assembly of the International Association of Hydrological Sciences (IAHS 2022), Montpellier, France, 29 May–3 June 2022.

Review statement. This paper was edited by Christophe Cudennec and reviewed by two anonymous referees.

References

Beven, K. J.: Rainfall-runoff modelling: the primer, John Wiley & Sons, 2011.

Blöschl, G., Bierkens, M. F., Chambel, A., et al.: Twenty-three unsolved problems in hydrology (UPH) – a community perspective, Hydrolog. Sci. J., 64, 1141–1158, 2019.

Bouaziz, L. J. E., Steele-Dunne, S. C., Schellekens, J., Weerts, A. H., Stam, J., Sprokkereef, E., Winsemius, H., Savenije, H., and Hrachowitz, M.: Improved understanding of the link between catchment-scale vegetation accessible storage and satellite-derived Soil Water Index, Water Resour. Res., 56, e2019WR026365, https://doi.org/10.1029/2019WR026365, 2020.

Bouaziz, L. J. E., Aalbers, E. E., Weerts, A. H., Hegnauer, M., Buiteveld, H., Lammersen, R., Stam, J., Sprokkereef, E., Savenije, H. H. G., and Hrachowitz, M.: Ecosystem adaptation to climate change: the sensitivity of hydrological predictions to time-dynamic model parameters, Hydrol. Earth Syst. Sci., 26, 1295–1318, https://doi.org/10.5194/hess-26-1295-2022, 2022.

De Boer-Euser, T., McMillan, H. K., Hrachowitz, M., Winsemius, H. C., and Savenije, H. H. G.: Influence of soil and climate on root zone storage capacity, Water Resour. Res., 52, 2009–2024, 2016.

Dooge, J. C. I.: Looking for hydrologic laws, Water Resour. Res., 22, 46S–58S, 1986.

Dooge, J. C. I.: Searching for simplicity in hydrology, Surv. Geophys., 18, 511–534, 1997.

Fenicia F., Kavetski, D., Savenije, H. H. G., Clark, M. P., Schoups, G., Pfister, L., and Freer, J.: Catchment properties, function, and conceptual model representation: is there a correspondence?, Hydrol. Process., 28, 2451–2467, 2014.

Gao, H., Hrachowitz, M., Schymanski, S. J., Fenicia, F., Sriwongsitanon, N., and Savenije, H. H. G.: Climate controls how ecosystems size the root zone storage capacity at catchment scale, Geophys. Res. Lett., 41, 7916–7923, 2014a.

Gao, H., Hrachowitz, M., Fenicia, F., Gharari, S., and Savenije, H. H. G.: Testing the realism of a topography-driven model (FLEX-Topo) in the nested catchments of the Upper Heihe, China, Hydrol. Earth Syst. Sci., 18, 1895–1915, https://doi.org/10.5194/hess-18-1895-2014, 2014b.

Gao, H., Birkel, C., Hrachowitz, M., Tetzlaff, D., Soulsby, C., and Savenije, H. H. G.: A simple topography-driven and calibration-free runoff generation module, Hydrol. Earth Syst. Sci., 23, 787–809, https://doi.org/10.5194/hess-23-787-2019, 2019.

Gharari, S., Hrachowitz, M., Fenicia, F., and Savenije, H. H. G.: Hydrological landscape classification: investigating the performance of HAND based landscape classifications in a central European meso-scale catchment, Hydrol. Earth Syst. Sci., 15, 3275–3291, https://doi.org/10.5194/hess-15-3275-2011, 2011.

Kleidon, A.: Thermodynamic foundations of the Earth system, Cambridge University Press, Cambridge, 2016.

Kleidon, A., Zehe, E., Ehret, U., and Scherer, U.: Thermodynamics, maximum power, and the dynamics of preferential river flow structures at the continental scale, Hydrol. Earth Syst. Sci., 17, 225–251, https://doi.org/10.5194/hess-17-225-2013, 2013.

Mao, G. and Liu, J.: WAYS v1: a hydrological model for root zone water storage simulation on a global scale, Geosci. Model Dev., 12, 5267–5289, https://doi.org/10.5194/gmd-12-5267-2019, 2019.

McDonnell, J. J., Sivapalan, M., Vaché, K., et al.: Moving beyond heterogeneity and process complexity: A new vision for watershed hydrology, Water Resour. Res., 43, W07301, 2007.

Montanari, A., Young, G., Savenije, H. H. G., et al.: "Panta Rhei, Everything Flows": Change in hydrology and society; The IAHS Scientific Decade 2013–2022, Hydrolog. Sci. J., 58, 1256–1275, 2013.

Nijzink, R., Hutton, C., Pechlivanidis, I., Capell, R., Arheimer, B., Freer, J., Han, D., Wagener, T., McGuire, K., Savenije, H., and Hrachowitz, M.: The evolution of root-zone moisture capacities after deforestation: a step towards hydrological predictions under change?, Hydrol. Earth Syst. Sci., 20, 4775–4799, https://doi.org/10.5194/hess-20-4775-2016, 2016.

Savenije, H. H. G.: HESS Opinions "Topography driven conceptual modelling (FLEX-Topo)", Hydrol. Earth Syst. Sci., 14, 2681–2692, https://doi.org/10.5194/hess-14-2681-2010, 2010.

Savenije, H. H. G. and Hrachowitz, M.: HESS Opinions "Catchments as meta-organisms – a new blueprint for hydrological modelling", Hydrol. Earth Syst. Sci., 21, 1107–1116, https://doi.org/10.5194/hess-21-1107-2017, 2017.

Singh, C., Wang-Erlandsson, L., Fetzer, I., Rockström, J., and Van der Ent, R., Rootzone storage capacity reveals drought coping strategies along rainforest-savanna transitions, Environ. Res. Lett., 15, 124021, https://doi.org/10.1088/1748-9326/abc377, 2020.

Sivapalan, M., Savenije, H. H. G., and Bloeschl, G.: Socio-hydrology: A new science of people and water, Hydrol. Process., 26, 1270–1276, 2012.

Wang-Erlandsson, L., Bastiaanssen, W. G. M., Gao, H., Jägermeyr, J., Senay, G. B., van Dijk, A. I. J. M., Guerschman, J. P., Keys, P. W., Gordon, L. J., and Savenije, H. H. G.: Global root zone storage capacity from satellite-based evaporation, Hydrol. Earth Syst. Sci., 20, 1459–1481, https://doi.org/10.5194/hess-20-1459-2016, 2016.

Wulf, A.: The Invention of Nature. Alexander von Humboldt's New World, John Murray Publ., London, 2015.

Proc. IAHS, 385, 5–10, 2024
https://doi.org/10.5194/piahs-385-5-2024

Open Access PIAHS

Analysis of high-resolution rain records in FVG, northeastern Italy

Elisa Arnone[1], **Dario Treppiedi**[2], **and Leonardo Valerio Noto**[2]

[1]Dipartimento Politecnico di Ingegneria e Architettura, University of Udine, Udine, 33100, Italy
[2]Dipartimento di Ingegneria, University of Palermo, Palermo, 90128, Italy

Correspondence: Elisa Arnone (elisa.arnone@uniud.it)

Received: 29 May 2022 – Accepted: 9 January 2023 – Published: 18 April 2024

Abstract. The Friuli Venezia Giulia (FVG) region, northeastern Italy, records the heaviest precipitation annual totals of the country. The region counts on a dense ground-station network constituted by 2 main rain-gauges networks, whose sampling frequency has been progressively increased from 60 up to 1 min step. In this work, we propose a comprehensive analysis of the available dataset of continuous series at high temporal resolution (i.e. 60, 30, 5 and 1 min) to verify whether trends in very short rainfalls are underway. We adopt the quantile regression (QR) method which allows to detect changes in the tails of the rainfall distributions and to screen the whole rainfall time series. At this aim, we first introduce a method to check and correct the continuous series by removing the suspicious outliers, based on references values. Significant increasing trends at 5 % of significant level have been detected on some of the analysed stations.

Keywords. Field observation; new monitoring; high-resolution records; UPH 1; SDG 13.3.

1 Introduction

Rainfall observations are among the most important input data for the hydrological and geomorphological modelling. Depending on the physical process being simulated, an appropriate spatial and temporal resolution of data must be used. Temporal variability of rainfall is one of the main source of uncertainty in hydrological simulation and flood forecasting (Huang et al., 2019). Specifically, temporal variability becomes important when dealing with urban hydrology (Berne et al., 2004), very small mountain catchments, rainfall-driven geomorphic processes such as erosion, debris flow or shallow landslides. In fact, from a modelling perspective, it has been demonstrated that model performance as well as values of the calibrated parameters can significantly vary the at different temporal rain data resolutions (Wang and Takase, 2009).

From a statistical perspective, many works in literature analyse time series of rainfall and rainfall-derived indices in order to assess past variability and changes (Caporali et al., 2021 and references therein). Trends in precipitation regimes are commonly assessed in terms of total precipitation or wet days. Whereas with regard to the extreme precipitations, most of the studies investigate changes on annual maxima at 1 h duration or greater (e.g. Arnone et al., 2013). Only a few researches focus on sub-hourly durations (e.g., Treppiedi et al., 2021). The reason is mainly due to the scarce availability of sufficiently long series at such durations. Detecting trend methods such as the widely used non-parametric Mann–Kendall (hereinafter MK) test can be applied to series of annual maxima that are length enough. An alternative method is based on the quantile regression (hereinafter QR) analysis (Koenker, 2005), which performs a linear regression on the continuous data time-series, taking into account those values greater than a selected quantile. The method is more flexible than the MK (Treppiedi et al., 2021, Villarini et al., 2011), however, the drawback is the necessity of continuous data.

The region of Friuli Venezia Giulia (FVG), in the northeastern Italy, records the heaviest precipitation annual totals in the country. Due to the complex orography and other prone factors, effects of both prolonged and extreme precipitation is particularly damaging (Arnone et al., 2023). Additionally, the mountain basins of the Alpine area are characterized by very short time of concentrations, thus short-extreme events are of interest, and rainfall analyses at sub-daily scale could not be appropriate. The region is equipped with rain gauges of recent technology, capable of recording data at a frequency up to 1 min. However, two issues may arise when dealing with such series: (i) quality of the dataset in terms of errors, misreading, equipment malfunction (Huang et al., 2019) and (ii) the length of the time series not sufficiently long for statistical analyses.

To address the above mentioned issues, in this work, we propose: (i) a comprehensive analysis of the dataset at high temporal resolution (i.e. 60, 30, 5 and 1 min) available for the FVG region; (ii) a quality check of the continuous series at 5 min together with a correction method. Finally, in order to verify whether trends in very short rainfall durations are underway, the QR method at different thresholds and durations is applied to the corrected series of a selected locations.

2 Study area and dataset

Friuli Venezia Giulia (FVG) is characterized by a complex orography, which includes the plain area of the northern Adriatic Sea (southern side), and the pre-alpine and alpine ridges (Julian and Carnic) on the northern side. The orography controls the mean annual precipitation distribution across the region that ranges from about 950 mm across the plain and coastal area to about 3150 mm across the Prealps (Fig. 1). The maximum total rainfall is recorded in the Julian Prealps, between NE Italy and NW Slovenia. Daily rainfall precipitation may exceed 500 mm in a return period of 20–30 years (Ceshia et al., 1991). In general, the climate is classified as a humid continental climate (i.e., Dfb according to the Köppen climate classification), with warm summer and coldest month with below zero temperatures.

The ground rain-gauges network of the FVG region consists of more than 200 ground stations, which are all managed by the regional Civil Protection Agency. The entire network is composed mainly by two different series of rain-gauge networks (Fig. 1): (i) the network originally managed by the Hydrographic Operative Unit that is now based on CAE technology and that consists of 162 rain-gauges (hereinafter CAE network); (ii) the network built with the SIAP-Micros hardware technology, which is directly managed by the agency OSMER ARPA-FVG (OSservatorio MEteorologico Regionale-Agenzia Regionale per la Protezione dell'Ambiente) and consists of 46 rain-gauges (hereinafter SIAP-Micros network).

The recording frequency rate varies from 60 to 30, 5 and 1 min, depending on the station and temporal horizon (see Sect. 4.1). However, validated and certified data are guaranteed by the agency only for resolutions equal and greater than 1 h. All data can be retrieved from the OMNIA platform developed by OSMER ARPA-FVG (http://omnia.meteo.fvg.it/, last access: 13 January 2023).

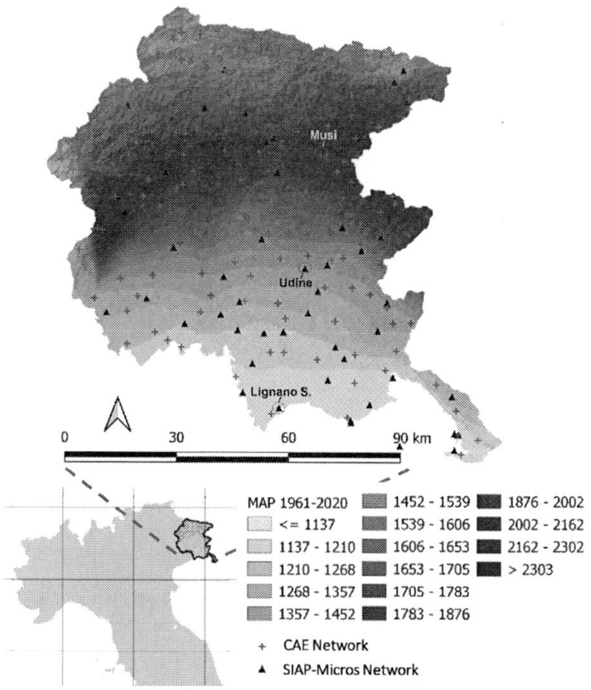

Figure 1. Mean Annual Precipitation (MAP) of Friuli Venezia Giulia region (northeastern Italy) and distribution of the two main rain-gauges networks, i.e. CAE and SIAP-Micros.

3 Methodology

3.1 Trend detection method

The Quantile Regression (QR) method (Koenker, 2005) is an extension of the standard linear regression model that allows to perform a regression on different quantiles instead of only on the mean. While simpler non-parametric tests as the Mann-Kendall (MK) are usually applied to rainfall-derived specific data sets, e.g. the annual maxima, the advantage of the QR is that it provides a more complete view of the statistical properties of a sample, by inspecting at different quantile levels the tail of the rainfall distribution. Differently than the simple linear regression in which the coefficients are evaluated through the least square minimization problem, the QR model is based on a minimization of the sum of the weighted absolute value of a difference between the ith observation (y_i) and the τth quantile line $y(\tau,x) = \beta_0(\tau) + \beta_1(\tau)x$ at x_i (Treppiedi et al., 2021):

$$\min_{y(\tau,x)} \left(\begin{array}{c} \sum\limits_{y_i \geq y(\tau,x_i)} \tau \, |y_i - (\beta_0(\tau) + \beta_1(\tau)x_i)| + \\ \sum\limits_{y_i < y(\tau,x_i)} (1-\tau) \, |y_i - (\beta_0(\tau) + \beta_1(\tau)x_i)| \end{array} \right) \quad (1)$$

The intercept, β_0, and the slope, β_1 thus depend on the quantile level τ, for any $0<\tau<1$. Points above the quantile line are weighted by τ, while those below the quantile line are weighted by $(1-\tau)$, thus meaning that the greater the considered quantile, the more relevant are points above the quantile line in the evaluation of the slope and intercept of the regression line. The trends significance can be evaluated by means of the Student's t, which is used here to reject the null hypothesis (slope of the quantile line is equal to zero) with a significance level of 5 % and 10 %.

As previously mentioned, differently than the MK test, the QR screens the whole rainfall time series and the trend result could be significantly affected by the presence of outliers in the series, which could be due to errors in recording data.

3.2 Data validation

Rainfall data are first check for consistency in terms of unphysical measurements based on realistic ranges (Dal Gesso et al., 2019). Specifically, the application of the QR method requires a verification mainly oriented to the identification, in the series at high resolution recording rate, of suspicious outliers or recurring values. At this aim, we developed a very simple method of data "check and cleaning" which is based on the comparison with the verified series provided by the agency.

Given a series at sub-hourly recording frequency rate, fr, this is first aggregated at the resolution of the validated series, i.e. 60 min. Let us name as $A(\text{fr})$ the aggregated series from the original recorded data and as $R(60)$ the validated series at 60 min considered here as reference and as not affected by errors. The two series are compared at each temporal step, and the following conditions could apply:

1. Perfect agreement between the two values, i.e. $A(\text{fr}) = R(60)$;
2. *No-data* value either in $A(\text{fr})$ or in $R(60)$ series;
3. Mismatching between $A(\text{fr})$ and $R(60)$ values, with $A(\text{fr}) > R(60)$; let us label this error as *Mismatch I*;
4. Mismatching between $A(\text{fr})$ and $R(60)$ values, with $A(\text{fr}) < R(60)$; let us label this error as *Mismatch II*.

No-data values (condition 2) are discarded. Values of *Mismatch I* (condition 3) must be considered as certain errors given that they do not find confirmation in the reference series and, above all, they could significantly affect the results of the QR at high quantile levels. Values of *Mismatch II* (condition 4) provide an underestimation of the confirmed values at 60 min. However, including them in the final series would not invalidate the analysis of the tail of rainfall distribution.

The identified outliers and errors are thus discarded from the aggregated series, and then the series at the original frequency rate is cleaned-up. Consistency in space through the measurements of the nearest neighbour stations is not checked-up. However, this is done for the validation of the reference series (ARPA-FVG, 2021).

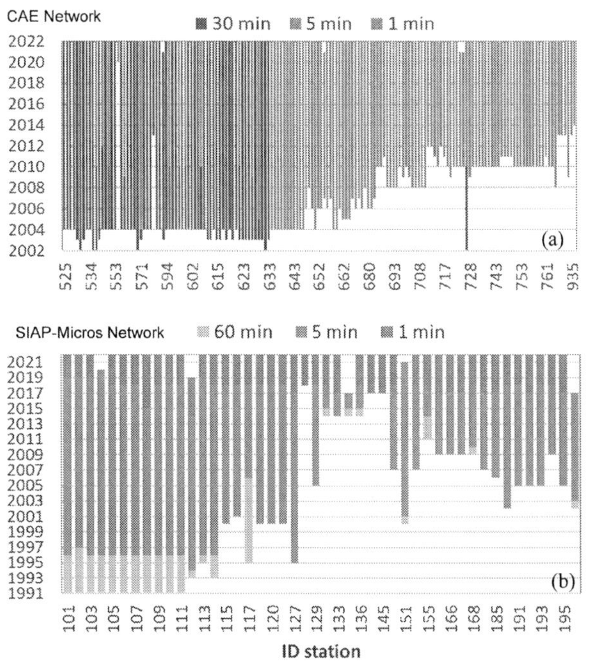

Figure 2. Temporal horizon and recording frequency rate of the working rain-gauges for the CAE network **(a)** and the SIAP-Micros network **(b)**.

4 Results

4.1 Dataset screening

Figure 2 shows the temporal horizon and the recording frequency rate of the working rain-gauges for the two networks. The CAE network records continuously mostly from 2004, with a frequency rate of 30 min for most of the rain-gauges (Fig. 2a, blue bars). The recording frequency has been progressively increased up to 1 min around the 2010s for both new gauges or existed ones. Up to today, the 68 % of the rain-gauges records at 1 min and the 4 % has a continuous record length of at least almost 20 years (2002–2022).

The SIAP-Micros rain-gauges have been installed starting from the 1990s at an initial recording frequency of 1 h (Fig. 2b). The resolution has been increased up initially to 5 min (around the 1997) and more recently ($\sim$ 2018) to 1 min. Then, in this case, the continuous series with longest recording length are at a rate of 5 min, with an average of 22 years for the oldest rain-gauges (30 % of the network).

The data validation method described in Sect. 3.2 has been tested on the time series with fr = 5 min of the SIAP-Micros network, which have the longest sampling records. As an example, Fig. 3 shows the results for the ID station #106 lo-

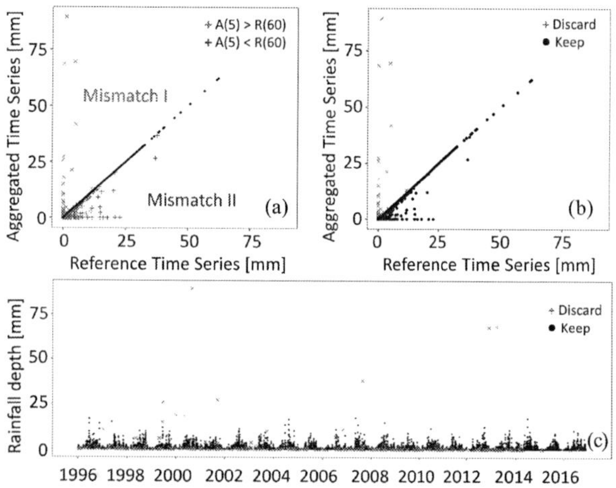

Figure 3. Comparison between reference series at 1 h, $R(60)$ and aggregated series at 1 h from the recorded data at 5 min, $A(5)$: in **(a)** the types of mismatches, in **(b)** the discarded and kept values and in **(c)** the cleaned series a 5 min step.

cated in Udine, the second biggest city of FVG (Fig. 1). From Fig. 3a it can be noticed that the aggregated series $A(5)$ highlights some "spikes", i.e. values much greater than the references, $R(60)$, up to ~ 80 mm, which does not find confirmation into the reference series. All the red values are thus discarded (Fig. 3b), for a total of about 0.54 % of the recorded data. Figure 3c shows the corrected series at 5 min step and it is highlighted how the correction method allowed to discard the suspicious outliers' values ($\gtrsim 60$ mm/5 min) that could invalidate the quantile regression analysis.

4.2 Trend detection

The quantile regression has been applied to the corrected 5 min time series for three quantile levels, i.e. 0.90, 0.95 and 0.99. Additionally, aggregated series at durations from 10 min to 1 h at 10 min step and of 1, 3, 6, 12 and 24 h have been assessed, for a total of 11 durations.

For the sake of brevity, the results are here reported for three selected stations representative of three different significant locations, i.e. the previous mentioned Udine (#106), Lignano Sabbiadoro (hereinafter Lignano S., #118), which is located along the coastline (Fig. 1) and Musi (#190), located in the eastern part of the Prealps ridges and which records the maximum annual rainfall total in the region (Fig. 1). Table 1 reports the significance of the trend detecting test. It can be noticed that:

- Udine shows a significant decreasing trend at 5 % for precipitations of 5 and 10 min (0.95 and 0.99 quantiles); an increasing trend at 5 % is detected only at 20 min and 0.9 quantile. Considering a significance level of 10 %, and increasing trend is detected also for precipitations of 24 h (0.9 quantile).

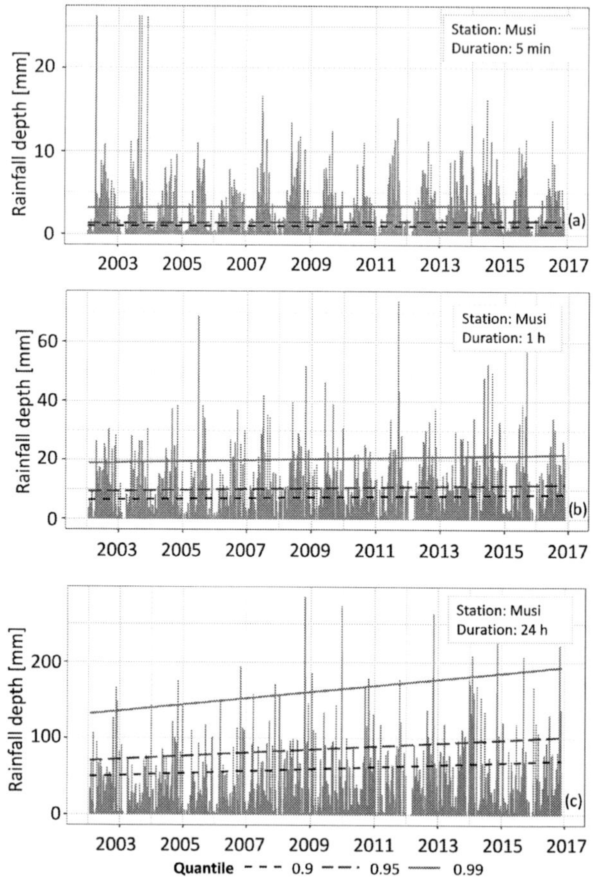

Figure 4. Quantile regression lines for quantiles 0.9, 0.95 and 0.99, at the durations of 5 min **(a)**, 1 h **(b)** and 24 h **(c)** for the station located in MUSI (#190).

- Lignano S. shows significant increasing trends at 5 % for almost all the durations and quantile levels.
- Musi shows increasing significant for all the durations and 0.9 and 0.95 quantiles, as well. The most significant trend at 5 % and for the highest quantile (0.99) is detected only for durations of 10 min, 12 and 24 h.

For the sake of example, Fig. 4 reports the time series of Musi station with the corresponding regression lines for the above highlighted durations. It is notable the significant increase in precipitation at 24 h, especially at 0.99 quantile.

5 Conclusions

We proposed a comprehensive analysis of the rainfall dataset available for the Friuli Venezia Giulia (FVG), northeastern Italy. The rain-gauges that compose the two main networks are characterized by a sampling frequency rate that varies from 60 to 30, 5 min and 1 min, depending on the historical period and on the station. We checked the consistency

Table 1. Significant trends at 5 % and 10 % significance levels obtained for three selected stations (i.e., Udine, Lignano S. and Musi) at 11 durations (D) and 3 quantiles ($q_1 = 0.9$; $q_2 = 0.95$; $q_3 = 0.99$).

D	Station								
	Udine (#106)			Lignano S. (#118)			Musi (#190)		
	q_1	q_2	q_3	q_1	q_2	q_3	q_1	q_2	q_3
5 min	.	−	−	.	.	+	+	+	○
10 min	−	−	−	.	+	+	+	+	+
20 min	+	∇	∇	+	+	+	+	+	.
30 min	.	.	.	+	+	+	+	+	.
40 min	.	.	.	+	+	+	+	+	.
50 min	.	.	.	+	+	+	+	+	○
1 h	.	.	∇	+	+	+	+	+	.
3 h	.	.	.	+	+	+	+	+	.
6 h	.	○	.	+	+	+	+	+	○
12 h	.	.	.	+	+	+	○	+	+
24 h	○	.	.	+	+	+	+	+	+

+/− Increasing/decreasing significant at 5 %.
○/∇ Increasing/decreasing significant at 10 %.
. No trend.

of the continuous series at 5 min in order to verify the presence of significant trends through the Quantile Regression (QR) method. The method allows to explore the upper tail of the probability distribution function of the data if very high quantiles are considered. For this reason, we first corrected the original data to remove eventual suspicious outliers, by comparing the series with the validated series at 1 h provided by the local agency that manages the data. The results of the trend detection highlight the presence of significant changes in both short and long precipitations mainly in the station along the coastline and in the Prealps.

Future works aim at further validate the method of data correction and discussing the presence of trend across the entire region, focusing on eventual spatial correlation.

Code availability. Codes used in this work can be asked to authors upon motivated request.

Data availability. Data used in this article can be found, on request, at the website of omnia platform by ARPA-OSMER http://omnia.meteo.fvg.it/ (last access: 16 January 2023).

Author contributions. EA is the first and corresponding author. All the authors contributed to the study conception and design. EA collected the data; DT and EA performed the analysis and LVN supervised the work. First draft of the paper was written by EA and all authors commented on previous versions of the manuscript. All authors read and approved the final manuscript.

Competing interests. The contact author has declared that none of the authors has any competing interests.

Disclaimer. Publisher's note: Copernicus Publications remains neutral with regard to jurisdictional claims in published maps and institutional affiliations.

Special issue statement. This article is part of the special issue "IAHS2022 – Hydrological sciences in the Anthropocene: Variability and change across space, time, extremes, and interfaces". It is a result of the XIth Scientific Assembly of the International Association of Hydrological Sciences (IAHS 2022), Montpellier, France, 29 May–3 June 2022.

Review statement. This paper was edited by Christophe Cudennec and reviewed by two anonymous referees.

References

Agenzia Regionale per la Protezione e per l'Ambiente del Friuli Venezia Giulia (ARPA-FVG): Return periods of annual maxima series, Technical Report, https://www.meteo.fvg.it/clima/clima_fvg/02_documenti_descrittivi, 2021.

Arnone, E., Pumo, D., Viola, F., Noto, L. V., and La Loggia, G.: Rainfall statistics changes in Sicily, Hydrol. Earth Syst. Sci., 17, 2449–2458, https://doi.org/10.5194/hess-17-2449-2013, 2013.

Arnone, E., Zoratti, V., Formetta, G., Bosa, S., and Petti, M.: Predicting peakflows in mountain river basins and data-scarce areas: a case study in northeastern Italy, Hydrolog. Sci. J., accepted, https://doi.org/10.1080/02626667.2022.2162408, 2023.

Berne, A., Delrieu, G., Creutin, J.-D., and Obled, C.: Temporal and spatial resolution of rainfall measurements required for urban hydrology, J. Hydrol., 299, 166–179, 2004.

Caporali, E., Lompi, M., Pacetti, T., Chiarello, V., and Fatichi, S.: A review of studies on observed precipitation trends in Italy, Int. J. Climatol., 41, E1–25, https://doi.org/10.1002/joc.6741, 2021.

Ceschia, M., Micheletti, S., and Carniel, R.: Rainfall over Friuli Venezia Giulia: High Amounts and Strong Geographical Gradients, Theor. Appl. Climatol., 43, 175–180, 1991.

Dal Gesso, S., Arnone, E., Venturini, M., Cucchi, M., and Petitta, M.: SOON: The Station Observation Outlier finder, in: Managing Water Resources for a Sustainable Future, edited by: Garrote L., Tsakiris G., Tsihrintzis V.A., Vangelis H., and Tigkas D., Proc. of the 11th World Congress of EWRA, Madrid, Spain, 25–29 June 2019, ISBN 978-618-84419-0-3, 2019.

Huang, Y., Bárdossy, A., and Zhang, K.: Sensitivity of hydrological models to temporal and spatial resolutions of rainfall data, Hydrol. Earth Syst. Sci., 23, 2647–2663, https://doi.org/10.5194/hess-23-2647-2019, 2019.

Koenker, R.: Quantile Regression, Cambridge University Press, Cambridge, ISBN 9780511754098, https://doi.org/10.1017/CBO9780511754098, 2005.

Treppiedi, D., Cipolla, G., Francipane, A., and Noto, L. V.: Detecting precipitation trend using a multiscale approach based on quantile regression over a Mediterranean area, Int. J. Climatol., 41, 5938–5955, 2021.

Villarini, G., Smith, J. A., Baeck, M. L., Vitolo, R., Stephenson, D. B., and Krajewski, W. F.: On the frequency of heavy rainfall for the Midwest of the United States, J. Hydrol., 400, 103–120, 2011.

Wang, B. H. and Takase, K: Effects of temporal resolution on hydrological model parameters and its impact on prediction of river discharge, Hydrol. Sci. J., 54, 886–898, https://doi.org/10.1623/hysj.54.5.886, 2009.

Proc. IAHS, 385, 11–16, 2024
https://doi.org/10.5194/piahs-385-11-2024

Open Access PIAHS

The performance of rainwater harvesting systems in the context of deep uncertainties

Gabriela Cristina Ribeiro Pacheco[1,2] **and Conceição de Maria Albuquerque Alves**[1]

[1]Environmental Technology and Water Resources Post-Graduation Program (PTARH), University of Brasília, Brasília, 70910-900, Brazil
[2]Federal Institute of Goiás (IFG), Goiânia, 74055-110, Brazil

Correspondence: Gabriela Cristina Ribeiro Pacheco (gabrielacrpacheco@gmail.com)

Received: 29 May 2022 – Revised: 21 November 2022 – Accepted: 9 January 2023 – Published: 18 April 2024

Abstract. Rainwater harvesting systems (RHS) are a relevant alternative of water supply in urban areas with increasing water demand and limited water availability. But these systems depend on several parameters that present uncertainties as well-characterized uncertainties whose probability functions are known and deep uncertainty factors that doesn't have analytical representation of their variability. This study evaluates the influence of water demand, tariff and discount rate (deep uncertain factors) on the feasibility of RHS for different scenarios of uncertainties. The systems were evaluated using the following performance criteria: Satisfied Demand, Reliability, Percentage of Rainwater Harvesting, Net Present Value, Net Present Value Volume and Benefit Cost Rate. We simulated the RHS performance for sixteen system configurations, comprised of eight categories of residential buildings according to representative water consumption (ranging from 4.748 to 44.673 m^3 per month) and two typical catchment areas for each of the eight groups of demands (ranging from 60 to 400 m^2) in the city of Rio Verde located in the central of Brazil. Each system was evaluated under the context of 1000 States of the World (SOWs) defined using the Latin Hypercube Sampling (LHS) method (in the case of the deep uncertainty factors) and bootstrapping resampling (for precipitation). Results showed slight difference on performance criteria among precipitation scenarios, maybe due to the fact that the synthetic rainfall series preserved the pattern and the total rainfall volume among the series which is reasonable for the location. However, the water tariff and discount rate showed a significant influence in the performance criteria confirming the relevance of deep uncertainty factors in the evaluation of RHS performances.

Keywords. Water conservation; rainwater harvesting; deep uncertainties; UPH 21; SDG 6

1 Introduction

Climate, social and economic uncertainties pose greater challenges to water resources security in urban areas that already face limited water supply and urban sprawl and increasing water demand (AWWA, 2019). The universal access to water requires public policies adapted to local reality, in addition to supply management and water conservation practices such as Rainwater Harvesting Systems (RHS).

RHS consist of an ancient water supply technique in which rainfall is captured in the household, stored and used in non-potable or even potable uses in communities where there is limited water supply service. The system requires a collect surface area (roof or floor), a storage tank and a water quality control according to the final use of the collected rainwater. Due to the increase in the water demand associated to limited water availability, this technique has been rescued in some communities.

RHS have potential to save from 20 % to more than 50 % of the amount of water needed to supply household demands (Ghisi and Ferreira, 2007; Ennenbach et al., 2017; Cáceres et al., 2019; Custódio and Ghisi, 2019; Hofman-Caris et al., 2019). However, these benefits are variable with rainfall regime, local economic aspects and the RHS features. In order to get the best performance of a RHS one should consider on site feature demands and climate variability, as well as other uncertainties that could influence the performance of the system. The variability of the local precipitation regime has been proved to impact the performance of RHS especially due to climate change (Youn et al., 2012; Lash et al., 2014; Haque et al., 2016).

The variability in the rainfall regime is usually referred as well characterized uncertainties. However, there are other parameters that might have some impact in the performance of RHS and do not have their variability represented by any analytical expression. These parameters are defined in the literature as deep uncertainty factors (Knight, 1921; Kwakkel et al. 2016). There is no consensus of a conceptual model that portrays the long-term future of these factors, the probability distribution that describes a variable over time and how to evaluate the results and the importance of each one (Lempert et al., 2003).

In rainwater harvesting systems, water demand, tariff, discount rate and rate of increase in operation and maintenance costs may be considered as deep uncertainty factors (DU factors). The impact of deep uncertainty factors on system viability must be evaluated as soon as their inclusion in the assessment of management strategies reduces proposed measures vulnerabilities (Lempert, 2002; Trindade et al., 2019).

A recent National Water and Sanitation Law included the rainwater harvesting system as an important alternative to improve the access to clean and safe water in the country (Brazil, 2020). This study evaluates the influence of deep uncertainties factors on rainwater harvesting systems feasibility in the Rio Verde city in Central West region of Brazil. This city has socio and economic characteristics that represent most of the Brazilian medium size cities. Additionally, most of the households buildings and parcels have space to incorporate a RHS.

2 Method

The influence of DU factors and precipitation variability on RHS performance were evaluated in the context of different States of the World (SOWs). 1000 scenarios of rainfall were created and associated with 1000 set of deep uncertainty factors generated using Latin Hypercube Sampling (LHS) method. Table 1 presents the lower and upper limits for each DU factor defined based on the variability of historical values of these parameters.

30-years daily rainfall series were generated through bootstrapping resampling from the 24-years observed daily rainfall serie. The hydrologic year in the simulation started in October, at the beginning of the wet season and in order to avoid the discontinuity between December and January in different years and to preserve event order in the bootstrap.

In addition, eight demand values were adopted corresponding to the consumption ranges of the normal residential category defined by the Water Utility. For each water demand level, we defined two sizes of typical catchment areas, according to the characteristics of the most frequent buildings in the city and the water consumption in residences. Each of the 16 RHS were evaluated for all 1000 SOWs of precipitation associated to 1000 vectors of DU factors, totalizing 16 000 scenarios.

2.1 The computational experiment

The methodology included four major steps: (1) the selection of the storage reservoir capacity; (2) the definition of the state of the world scenarios; (3) the economic evaluation of the RHS performance; (4) the computation of the selected RHS performance criteria. The steps were performed for each of the 16 000 scenarios.

The size of the reservoir in the RHS was computed according to a mass balance simulation, indicated in Eqs. (1) and (2), using the YAS method (Yield After Spillage – $\theta = 0$). Three system performance metrics were computed for each selected storage volume, according to Eqs. (3), (4) and (5). The indicator Satisfied Demand (SD) corresponds to the percentage of overall demand met by the RHS, the Reliability (REL) refers to the number of days that the demand is fully met and the Rainwater Harvested (RH) is the percentage of rainwater volume that is actually consumed in the household.

$$\mathrm{Vc}_t = \min \begin{cases} D_t \\ \theta \cdot \mathrm{Vac}_t + Vr_{t-1} \end{cases} \tag{1}$$

$$\mathrm{Vr}_t = \min \begin{cases} \mathrm{Vac}_t + \mathrm{Vr}_{t-1} - \mathrm{Vc}_t \\ \mathrm{Vrinf}_t - (1-\theta) \cdot \mathrm{Vc}_t \end{cases} \tag{2}$$

$$\mathrm{SD} = \sum\nolimits_{t=1}^{T} \frac{\mathrm{Vc}_t}{D_t} \tag{3}$$

$$\mathrm{REL} = \frac{\sum_{t=1}^{T} t\,(se\,\mathrm{Vc}_t = D_t)}{T} \tag{4}$$

$$\mathrm{RH} = \frac{\sum_{t=1}^{T} \mathrm{Vc}_t}{\sum_{t=1}^{T} \mathrm{Vac}_t} \tag{5}$$

Vc: consumed volume in the day (m^3), Vr: volume of water in the reservoir in the day (m^3), Vac: volume available for capture in the day (m^3), D: daily demand in the day (m^3), Vrinf: inferior reservoir volume, θ: moment of water use ($0 \leq \theta \leq 1$) in this study $\theta = 0$, SD: satisfied demand (%), REL: reliability (%), RH: rainwater harvested (%), t: day, T: simulation time (days).

In a second stage, the reservoirs economic evaluation in the considered SOWs was carried out, the systems cash flow and the economic indicators were calculated: Benefit Cost Rate (BCR), Net Present Value (NPV) and Net Present Value Per Volume Consumed (NPVV), as Eqs. (6), (7) and (8).

$$\mathrm{BCR} = \frac{\sum_{m=1}^{T} \frac{\mathrm{Fc}_m}{(1+(i \cdot \mathrm{tvd}))^m}}{\mathrm{Inv}} \tag{6}$$

Table 1. DU factors and their range of variability.

Deep uncertainty factors	Inferior limit	Superior limit
Increase on operation and maintenance costs	1.03	1.2
Discount rate	0.7	1.60
Water and sewage tariff for the residential category	1.00	1.25

$$\mathrm{NPV} = \sum\nolimits_{m=1}^{T} \frac{\mathrm{Fc}_m}{(1+(i \cdot \mathrm{tvd}))^m} - \mathrm{Inv} \quad (7)$$

$$\mathrm{NPVV} = \frac{\mathrm{NPV}}{\sum_{t=1}^{T} \mathrm{Vc}_t} \quad (8)$$

Fct: cash flow (USD), i: monthly discount rate (% per month), tvd: discount rate variation rate (%), m: month, Inv: initial investment (USD), BCR: Benefit-cost Rate, NPV: Net Present Value, NPVV: Net present value per volume consumed.

Finally, the model determined the indicators for the selected storage reservoir volume (reservoirs volumes that presented the highest NPV value). A set of minimum performance criteria was defined and the number of scenarios that satisfied this condition was computed.

2.2 Data considered in the analysis

The study was carried out for the municipality of Rio Verde. The 24-year historical series of rainfall data (starting in 1 October 1996) were taken from the National Institute of Meteorology (INMET).

Table 2 shows the sixteen configurations of households according to monthly demand and typical catchment area of the buildings defined in this study. Each household ID represents a water demand level associated to a size of roof area. We considered two typical catchment areas according to a visual search (using Geographic Information Systems) in remote sensing maps of the Rio Verde city. These demand values were established from the average monthly consumption values for each of the eight consumption ranges of typical household according to data from the water utility. Once the RHS supply water for non-potable demand, the RHS demand for each household was assumed to be 53 % of the average monthly demand according to the study of Sant'Ana (2011) for a city with similar climate.

The economic evaluation of the RHS was based on the December 2020 values of the National System for Research on Civil Construction Costs and Indexes (SINAPI) and the cost of 18 storage tanks volumes commercially available, pump, upper reservoir, filter and accessories. The analysis for the operation and maintenance costs included recommendations from the Brazilian Technical Regulations NBR 15527 (ABNT, 2019).

Table 2. Setting of typical households in terms of monthly water demand and roof (catchment) areas.

Household ID	Water demand per household (m^3 per month)	Catchment area (m^2)
1 and 2	5484	60 and 100
3 and 4	12 949	60 and 100
5 and 6	17 808	100 and 200
7 and 8	22 759	150 and 200
9 and 10	27 806	200 and 300
11 and 12	34 431	300 and 350
13 and 14	44 668	350 and 400
15 and 16	67 860	350 and 400

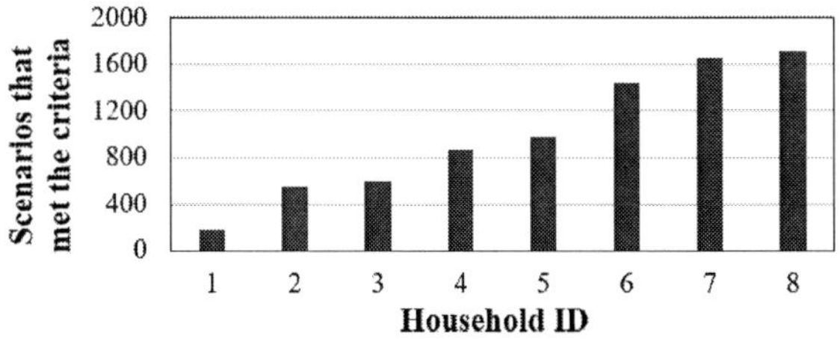

Figure 1. Number of scenarios that met the criteria by household ID.

We considered the local water utility (Saneago) 2021 tariff structure in the revenues computation. The company also charges a fixed minimum cost of USD 2.45 (USD 1 = BRL 5.55) and a tariff of sewage, corresponding to 100 % of the water tariff.

The discount rate was based on the SELIC (basic interest rate of the Brazil economy) for November 2021, 7.65 % per year (0.6162 % per month) (Central Bank, 2021). For the RHS tanks sizing a surface runoff rate of 0.8 was adopted and an initial discharge of 2 mm.

3 Results Analysis and Discussion

The performance criteria for the different scenarios and the number of scenarios that meet the defined performance criteria ($\mathrm{NPV} \geq 0$, $\mathrm{NPVV} \geq 0$, $\mathrm{BCR} \geq 0\,\%$, $\mathrm{SD} \geq 50\,\%$, $\mathrm{REL} \geq 50\,\%$ and $\mathrm{RH} \geq 30\,\%$) was evaluated as indicated in Fig. 2. The lines indicate each scenario and connect the values of the performance criteria (that have different units) al-

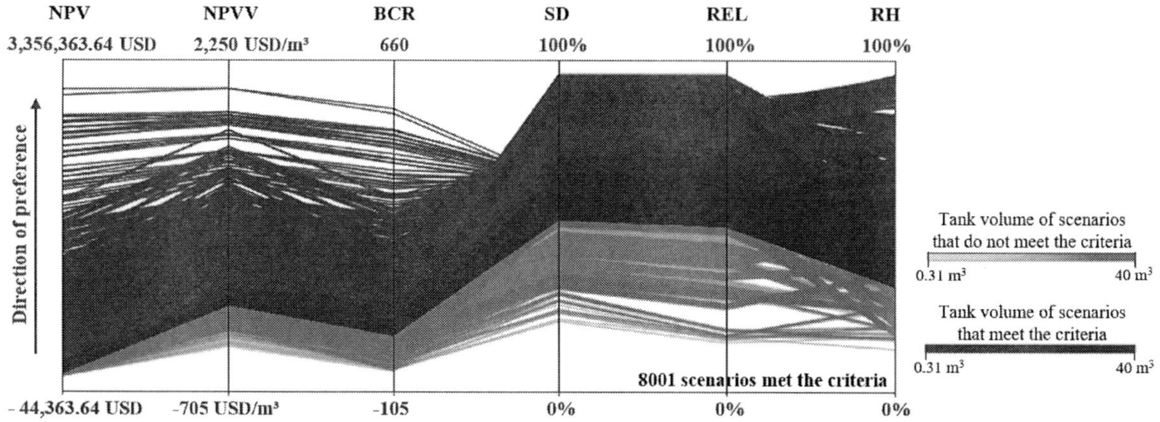

Figure 2. Comparative analysis of systems performance indicating the tank volume and the criteria found for the 16 000 scenarios, the scenarios that met the criteria and its direction of preference.

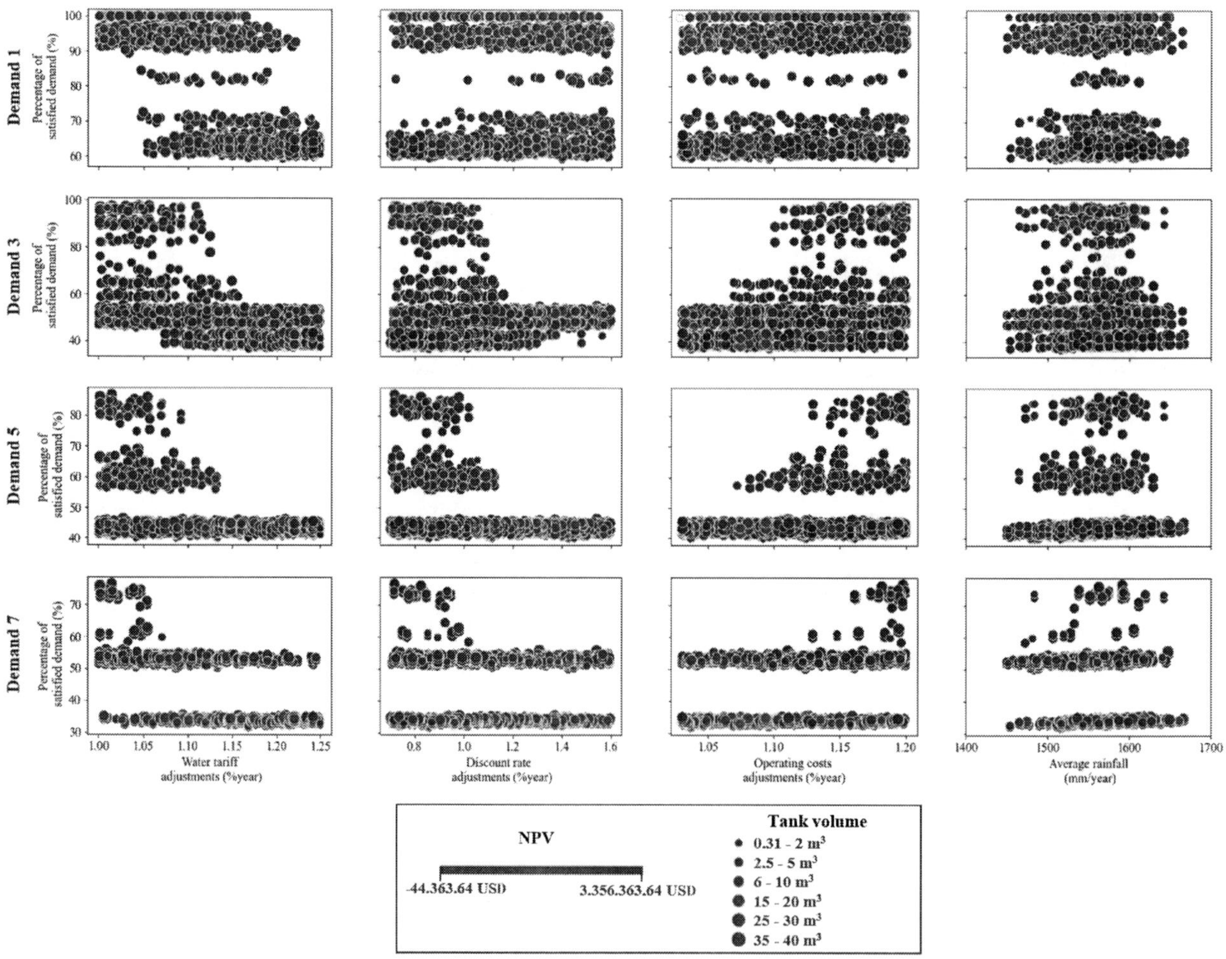

Figure 3. Sensitive analysis of deep uncertainty factors.

lowing the performance comparison among scenarios. The direction of preference indicates the direction of the goal expected for each performance criteria. From the 16 000 scenarios, 8.001 met the criteria. It is observed that the SD and REL values are similar for most of the scenarios.

To assess how the results vary for different consumption ranges, the number of scenarios that meet the criteria for different consumption ranges were observed as indicated in Fig. 1. For households with lower demand, the number of scenarios that meet the criteria is reduced. There is a tendency for scenarios with lower demand for rainwater to present higher values of SD and REL. However, the economic indicators and the RH are lower in scenarios of higher demand values. This fact demonstrates that buildings with higher demands tends to present lower technical indicators (SD, REL) and higher economic indicators (NPV, NPVV) under the context of climate and deep uncertainties.

In Fig. 3 is illustrated the variation in the adopted tanks, NPV and SD with deep uncertainty factors for four of the eight categories of demands. The highest NPV values occur for the lowest discount rate and the highest water tariff. As already noted, the highest values of NPV occur for the lowest values of SD, including a reduction in the maximum values of this indicator as demand increases. It is also noted that larger storage volumes are associated with lower SD values. For households 5, 6, 9, 10, 13 and 14, there were only scenarios with high SD values for low water tariff, discount rate and high values of operational costs.

The variability of NPV and SD was evaluated for the different rainfall series and there is no clear influence of the precipitation regime in NPV and SD once the analysis considers the same location and the method to generate the precipitation series do not capture future variability.

4 Conclusions

This work evaluated the performance of different RHS configuration for the town of Rio Verde under the context of uncertainties related to climate and deep uncertainties. Sixteen categories of household were defined according to water demand and catchment area to capture rainwater. The RHS were considered to supply non-potable water uses. Initial results indicate the influence of deep uncertainties such as the water tariff and the discount rate, while the influence of operating costs was not noticed. The highest values of NPV occurred for the lowest values of discount rate and for the highest water tariffs.

The impact of these two parameters was more relevant than the impact of demand, catchment area and the rainfall regime, proving the importance of incorporating deep uncertainties in the evaluation of RHS.

Considering the performance criteria defined in the methodology, more than half of the 16 000 scenarios for the RHS in Rio Verde city were favourable. It is noteworthy that 50 % of service and reliability is compatible with RHS since the buildings have another source of water supply. The RHS functions as an alternative of water supply in these places, especially for non-potable uses. In any case, it is necessary to evaluate uncertainties impacts in sites with significantly different rainfall regimes, tariff structures and demand patterns.

Code availability. The code is available at GitHub: https://github.com/gabrielacrpacheco/Rainwater-harvesting-feasibility (last access: 25 January 2023; https://doi.org/10.5281/zenodo.7566355, Pacheco and Alves, 2023).

Data availability. Rainfall data are available at http://bdmep.inmet.gov.br/ (last access: 5 February 2021, login required; INMET, 2021). Demand and water tariff data are in the water utility data base and may be available by request to the authors by email.

Author contributions. All authors contributed to the study conception and design. The first draft was written by GCRP and CdMAA reviewed and improved it several times.

Competing interests. The contact author has declared that neither of the authors has any competing interests.

Disclaimer. Publisher's note: Copernicus Publications remains neutral with regard to jurisdictional claims in published maps and institutional affiliations.

Special issue statement. This article is part of the special issue "IAHS2022 – Hydrological sciences in the Anthropocene: Past and future of open, inclusive, innovative, and society-interfacing approaches". It is a result of the XIth Scientific Assembly of the International Association of Hydrological Sciences (IAHS 2022), Montpellier, France, 29 May–3 June 2022.

Acknowledgements. The authors acknowledge the financial support given by FAP-DF, University of Brasília Graduate School and Federal Institute of Goiás (IFG).

Financial support. This research has been supported by FAP-DF, University of Brasília Graduate School and Federal Institute of Goiás (IFG).

Review statement. This paper was edited by Christophe Cudennec and reviewed by two anonymous referees.

References

ABNT: NBR 15527: "Aproveitamento de água de chuva de coberturas para fins não potáveis – Requisitos", Brazilian Association of Technical Standards, 2nd edn., Rio de Janeiro, ISBN 978-85-07-07999-6, 2019.

AWWA: State of the water industry report, https://www.awwa.org/Portals/0/AWWA/ETS/Resources/2019_STATE%20OF%20THE%20WATER%20INDUSTRY_post.pdf (last access: 3 January 2020), 2019.

Brazil: Lei n. 14.026, de 15 de julho de 2020, http://www.planalto.gov.br/ccivil_03/_Ato2019-2022/2020/Lei/L14026.htm, last access: 27 July 2020.

Cáceres, P. S., Ramos, S. R., and Sant'ana, D. R.: Potencial de redução da exploração dos recursos hídricos pelo aproveitamento de água pluvial em edificações residenciais no Distrito Federal, Cadernos de Arquitetura e Urbanismo – Paranoá 23, 11–19, https://doi.org/10.18830/issn.1679-0944.n23.2019.02, 2019.

Central Bank: Taxas de juros básicas - Histórico, https://www3.bcb.gov.br/sgspub/consultarvalores/consultarValoresSeries.do?method=get, last access: 15 December 2021.

Custódio, D. A. and Ghisi, A.: Assessing the Potential for Potable Water Savings in the Residential Sector of a City: A Case Study of Joinville City, Water, 11, 1–18, https://doi.org/10.3390/w11102074, 2019.

Ennenbach, M. W., Larrauri, P. C., and Lall, U.: County-scale rainwater harvesting feasibility in the United States: climate, collection area, density, and reuse considerations, J. Am. Water Resour. As., 54, 1–20, https://doi.org/10.1111/1752-1688.12607, 2017.

Ghisi, E. and Fereira, D. F.: Potential for potable water savings by using rainwater and greywater in a multi-storey residential building in southern Brazil, Build. Environ., 42, 2512–2522, https://doi.org/10.1016/j.buildenv.2006.07.019, 2007.

Haque, M. M., Rahman, A., and Samali, B.: Evaluation of climate change impacts on rainwater harvesting, J. Clean. Prod., 137, 60–69, https://doi.org/10.1016/j.jclepro.2016.07.038, 2016.

Hofman-Caris, R., Bertelkamp, C., Waal, L., Brand, T. V. D., Hofman, J., Aa, R. V. D., and Hoek, J. P. V.: Rainwater Harvesting for DrinkingWater Production: A Sustainable and Cost-Effective Solution in The Netherlands? Water, 11, 511, https://doi.org/10.3390/w11030511, 2019.

INMET: Rainfall data, Banco de Dados Meteorológicos, http://bdmep.inmet.gov.br/, last access: 5 February 2021.

Knight, F. H.: Risk, Uncertainty and Profit, University of Illinois at Urbana-Champaign's Academy for Entrepreneurial Leadership Historical Research Reference in Entrepreneurship, https://ssrn.com/abstract=1496192 (last access: 2 December 2021), 1921.

Kwakkel, J. H., Haasnoot, M., and Walker, W. E.: Comparing robust decision-making and dynamic adaptive policy pathways for model-based decision support under deep uncertainty, Environ. Model. Softw., 86, 168–183, https://doi.org/10.1016/j.envsoft.2016.09.017, 2016.

Lash, D., Ward, S., Kershaw, T., Butler, D., and Eames, M.: Robust rainwater harvesting: probabilistic tank sizing for climate change adaptation, J. Water Clim. Change, 5, 526–539, https://doi.org/10.2166/wcc.2014.080, 2014.

Lempert, R. J.: A new decision sciences for complex systems, P. Natl. Acad. Sci. USA, 99, 7309–7313, https://doi.org/10.1073/pnas.082081699, 2002.

Lempert, R. J., Popper, S. W., and Bankes, S. C.: Shaping the next one hundred years: new methods for quantitative, long-term policy analysis, MR-1626-RPC, RAND, Santa Monica, CA, ISBN 0-8330-3275-5, 2003.

Pacheco, G. C. R. and Alves, C. d. M. A.: Rainwater harvesting feasibility considering deep uncertainties, Zenodo [code], https://doi.org/10.5281/zenodo.7566355, 2023.

Sant'Ana, D. R.: A socio-technical study of water consumption and water conservation in Brazilian dwellings, PhD thesis, Oxford Brookes University, 2011.

Trindade, B. C., Reed, P. M., and Characklis, G. W.: Deeply uncertain pathways? Integrated multi-city regional water supply infrastructure investment and portfolio management, Adv. Water Resour., 134, 103442, https://doi.org/10.1016/j.advwatres.2019.103442, 2019.

Youn, S., Chung, E., Kang, W. G., and Sung, J. H.: Probabilistic estimation of the storage capacity of a rainwater harvesting system considering climate change, Resour. Conserv. Rec., 65, 136–144, https://doi.org/10.1016/j.resconrec.2012.05.005, 2012.

Proc. IAHS, 385, 17–23, 2024
https://doi.org/10.5194/piahs-385-17-2024

Isotopic signature of precipitation in the semi-arid region of Ceará, Northeastern Brazil

Marjorie B. Kreis[1,2,3,4], **Jean-Denis Taupin**[3], **Nicolas Patris**[3], **and Eduardo S. P. R. Martins**[4]

[1]University of Montpellier, Montpellier, France
[2]Dep. of Civil Engineering, Federal Univ. of Ceará, Fortaleza, Brazil
[3]HSM, Univ Montpellier, CNRS, IRD, Montpellier, France
[4]Research Institute of Meteorology and Water Resources (FUNCEME), Fortaleza CE, Brazil

Correspondence: Marjorie B. Kreis (marjorie.kreis@hotmail.fr)

Received: 23 May 2022 – Revised: 11 January 2023 – Accepted: 11 January 2023 – Published: 18 April 2024

Abstract. The study of stable isotopes in the precipitation of Ceará was conducted on monthly rainfall samples between 2011 and 2019 and on daily rainfall samples between 2018 and 2019 in order to support the global knowledge of hydrological processes in semi-arid regions given the scarcity of such data in the Northeast of Brazil. Results showed that the frequency of rainfall sampling in semi-arid regions may induce an isotopic bias, due to the partial evaporation from the water collectors during monthly storage causing isotopic fractionation. In the absence of adequate data processing, this bias may lead to a wrong definition of the Local Meteoric Water Line (LMWL), which can have major consequences on the understanding of groundwater recharge processes. The processing of the monthly isotopic data (elimination of the evaporated data) and the analysis of daily data highlighted that the LMWL of Quixeramobim is characterized by a y-intercept value greater than $+12$ ‰, indicating that rainwater is composed of vapours from marine origin, but also from continental recycling. Analysis of $\delta^{18}O$ monthly variations in relation to climatic factors suggested the importance of convective systems and relative humidity in the control of the isotopic composition of precipitation in a semi-arid tropical region.

Keywords. SGD 6; new monitoring

1 Introduction

The study of the stable isotopic composition of precipitation is an important tool for characterizing the groundwater input signal (Terzer et al., 2013; Gastmans et al., 2017). Indeed, the content analysis of stable isotopes (^{18}O, ^{2}H) in the different compartments of the water cycle and the construction of local isotopic "meteoric water lines" constitute a powerful integrative approach to understand the origin and organization of groundwater flows within aquifers, and to identify potential relationships between surface water and groundwater (Sukhija et al., 2006; Terzer et al., 2013; Santoni et al., 2016; Hao et al., 2018; Ahmed et al., 2019; Babaye et al., 2019). Moreover, stable isotopes in precipitation provide elements for understanding climatic factors that affect the precipitation regimes (Gastmans et al., 2017). To date, the isotopic characterization of rainwater in the semiarid region of Northeastern Brazil was only addressed by a few studies (Gastmans et al., 2021). Moreover, the isotopic precipitation data available were mainly linked to the establishment of the Global Network of Isotopes in Precipitation (GNIP), which started in the 1960s and was paralyzed in the 1990s for several reasons, despite the extensive use of these data to understand the hydrological processes and to identify the mechanisms of recharge and salinization in Northeastern aquifers (Gastmans et al., 2021). This study aims, therefore, to improve and update the characterization of the isotopic signal of rainfall in the semiarid region of Ceará in order to support the understanding of rainfall isotope variability and aquifer systems functioning in semi-arid regions.

2 Study area

The study zone is located at approximately 180 km from the Atlantic Ocean coast, in the rural municipality of Quixeramobim (5°11′40″ S, 39°18′42″ W), in the state of Ceará in Northeastern Brazil. The local biome is caatinga. The climate of the study area is semi-arid and characterized by an irregular distribution of rainfall, high evapotranspiration (ET) rates and recurring periods of drought (Burte, 2008; Martins and Júnior, 2017; Pontes Filho et al., 2020). The wet season extends from December to July, but almost 75 % of the precipitation is concentrated in four months of the year, from February to May. These four rainy months are related to the influence of the Inter-Tropical Convergence Zone (ITCZ), where the northeast and southeast trades flow together. The ITCZ, in the north of the northeast of Brazil, is characterised by strong upward motion and heavy rainfall (FUNCEME, 2018). The mean annual rainfall in Quixeramobim, calculated from the Thiessen polygon method for the 1961–2021 period, is around 709 mm, with values ranging from 218 to 1432 mm. For the same period, the mean annual potential evapotranspiration (PET) is around 1775 mm, which leads to a high annual water deficit (> 1000 mm yr^{-1}). In Quixeramobim, the last important drought dates from 2012–2019, with the 2012–2016 period corresponding to the most severe 5-year drought recorded in the systematic record (40 % decrease in annual rainfall). The temperature of the study area has low thermal amplitude, with monthly mean temperatures ranging between 25.7 and 28.5 °C and an annual mean of 27.2 °C. The relative humidity presents an annual average of 61.0 %, with monthly mean values varying from 41.6 % in July to 85.9 % in February (data from INMET Station no. 82586).

3 Material and methods

3.1 Rainwater sampling

A monthly monitoring of the stable isotope composition of rainwater was operated in Quixeramobim between January 2011 and July 2019 from three experimental rainfall stations (Inmet-Quixeramobim, Algodões-ZéNobre and Vista Alegre stations), amounting to a total of 172 samples. This monthly sampling was performed on the 1st day of the month (±48 h) between January and July (period concentrating 95 % of precipitation). Moreover, a daily monitoring of the isotopic composition of rainwater was implemented in the study region, between March 2018 and October 2019, from two conventional rainfall stations (Radar and Riacho Verde stations). Daily sampling was carried out every day of the year around 07:00 local time (UTC−3), for rainfall events greater than 1 mm d^{-1}, accumulating a total of 212 samples. The water collector of the rainfall stations was stored inside an isothermal box given the impossibility of burying the collection system. Furthermore, they were lined with aluminum foil to minimize the effects of evaporation, especially during the monthly storage. Coordinates and supplementary information of the five rainfall stations are given in Table 1.

3.2 Water stable isotopes analyses

Rainwater samples were taken from the water collector and packaged in 20 ml transparent plastic bottles. Bottles were hermetically sealed and stored in a refrigerator until transport to the laboratory. Water stable isotopes (^{18}O, ^{2}H) were measured on an Isoprime mass spectrometer at the LAMA laboratory of Hydrosciences Montpellier in France. Oxygen-18 composition was measured with the conventional CO_2-H_2O equilibrium dual inlet technique, with a precision of ±0.08 ‰. Hydrogen-2 was measured by a continuous flow method after reduction of water on Cr powder at 1050 °C, with a precision of ±0.8 ‰. Results are expressed in δ (‰) on the V-SMOW scale. Deuterium excess (d_{exc}) was calculated according to the Dansgaard definition ($d_{exc} = \delta^2H - 8 \times \delta^{18}O$).

4 Results and interpretation

4.1 Determination of rainfall characteristic isotopic values

4.1.1 Daily timescale

Rainwater samples taken from the daily monitoring over the 2018-2019 period in Quixeramobim showed $\delta^{18}O$ and δ^2H median values of −2.06 ‰ and −3.3 ‰, respectively. As in other dry tropical regions with strong seasonal contrasts (Taupin et al., 1997), the daily monitoring revealed an important variability of the isotopic composition of rainfall with values ranging from −12.97 ‰ to +2.27 ‰ for $\delta^{18}O$ and from −98.5 ‰ to +20.2 ‰ for δ^2H. The most depleted isotopic values (with $\delta^{18}O < -10.92$ ‰) were found in April 2018 on the two rainfall stations (Radar and Riacho Verde) and correspond to the last 2 d of a heavy precipitation event (4 to 5 consecutive rainy days accumulating a total of ≈ 12 % of annual rainfall, with 50 mm of rain on the penultimate day). Deuterium excess (d_{exc}) calculated from this dataset evidenced values ranging from −2.9 ‰ to +22.1 ‰ with a median value of +13.7 ‰. Only 13 % of the dataset showed evaporated values, with $d_{exc} < +10$ ‰, corresponding to minor rainfall events (< 10 mm) and an undersaturation of the atmosphere.

4.1.2 Monthly timescale

Analysis of the 172 samples of the monthly monitoring in the classical plot of $\delta^{18}O$ vs. δ^2H showed a highly evaporated isotopic line compared to the Global Meteoric Water Line (GMWL), with a line slope of 6.41 ($r^2 = 0.90$). Furthermore, more than 51 % of the data show d_{exc} values of less

Table 1. Coordinates and information of the rainfall stations (Kreis et al., 2020).

	Monthly monitoring			Daily monitoring	
Station Name	Inmet-Quixeramobim	Algodões-ZéNobre	Vista Alegre	Radar	Riacho Verde
Station ID	–	–	–	766	814
Station type	Experimental			Conventional	
Municipality	Quixeramobim				
Latitude (WGS84)	−5.17	−5.38	−5.34	−5.07	−5.36
Longitude (WGS84)	−39.29	−39.58	−39.40	−39.27	−39.60
Altitude	208 m	289 m	206 m	663 m	278 m
Distance to the coast	172 km	212 km	195 km	165 km	212 km

than +10 ‰ regardless of the amount of monthly precipitation (d_{exc} median value of +9.8 ‰). In order to determine whether the evaporated samples of the monthly monitoring are due to partial evaporation of the droplets during rainfall or to evaporation problems during monthly water storage, the $\delta^{18}O$ values of the monthly samples were compared to the monthly isotopic signal calculated by weighting the daily $\delta^{18}O$ by precipitation. The comparison of the isotopic data was made for a same period (2018–2019) and between two stations with similar characteristics in terms of distance from the coast and altitude (Algodões-ZéNobre and Riacho Verde stations). This comparison highlighted that the data from the monthly monitoring present values generally more enriched and evaporated compared to the monthly amount-weighted data from the daily monitoring (Fig. 1). This means that the evaporated values of the monthly samples (with $d_{exc} < +10$ ‰) are not only related to partial evaporation of precipitating water droplets, but to an evaporation phenomenon causing isotopic fractionation in the water collector during monthly storage. The demonstration of this isotopic bias in the monthly data is something that could change the fundamental understanding of crystalline aquifer recharge processes in Ceará and NEB, since shallow groundwater generally reflects the long-term amount-weighted mean isotopic composition of precipitation (Chang et al., 2020) and that the majority of isotopic studies in NEB are made from monthly data. Despite the existence of this isotopic bias, the relative variations of the monthly monitoring are consistent with those of the daily monitoring (Fig. 1). On a monthly scale, the data from the daily monitoring in Quixeramobim highlighted the seasonal variability of isotopic composition of precipitation, with monthly concentrations more depleted during the wettest months (from February to May, Fig. 1), as it was evidenced in Fortaleza (Da Silva Nobre et al., 2019).

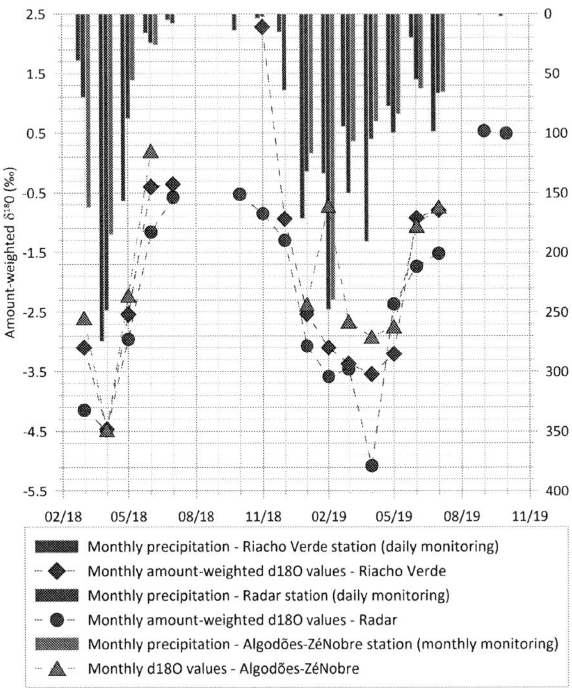

Figure 1. Seasonal variability of monthly isotopic composition of precipitation obtained from the daily monitoring for the 2018–2019 period and comparison with the isotopic data obtained from the monthly monitoring network (Algodões-ZéNobre station). Monthly means of daily $\delta^{18}O$ values are weighted by the amount of precipitation.

4.1.3 Annual timescale

The amount-weighted annual mean isotopic composition of precipitation, determined from daily data that does not exhibit isotopic bias, is −3.21 ‰ for $\delta^{18}O$ and −12.0 ‰ for $\delta^{2}H$ ($n = 212$). More depleted values (0.45 ‰ $< \Delta\delta^{18}O <$ 0.58 ‰) are observed on the Radar sta-

Table 2. Amount-weighted annual mean isotopic composition of precipitation determined from daily monitoring data.

	Riacho Verde		Radar		Mean
Altitude	278 m		663 m		–
Distance to the coast	212 km		165 km		–
	Amount-weighted mean $\delta^{18}O$	Rainfall	Amount-weighted mean $\delta^{18}O$	Rainfall	Amount-weighted mean $\delta^{18}O$
2018 (incomplete year)	−3.19 ‰	507 mm	−3.64 ‰	525 mm	−3.42 ‰
2019 (complete year)	−2.80 ‰	858 mm	−3.38 ‰	792 mm	−3.08 ‰
2018–2019	−2.95 ‰	–	−3.48 ‰	–	−3.21 ‰

Table 3. Annual amount-weighted mean $\delta^{18}O$ values obtained from the monthly network stations between 2011 and 2019. Rainfall data from INMET station no. 82586. Values of $\delta^{18}O$ of the monthly monitoring suffer from a slight bias, so they should not be used in terms of absolute values.

Year (from January to July)	Annual amount-weighted mean $\delta^{18}O$ (‰)	d_{exc} mean (‰)	"Annual" rainfall (from January to July) of the INMET Station no. 82586 (mm)
2011	−2.24	5.1	939
2012	−0.91	7.9	299
2013	−1.57	6.3	513
2014	−1.48	9.6	431
2015	−1.73	9.0	416
2016	+0.33	7.5	389
2017	−2.03	11.8	653
2018	−2.75	10.5	688
2019	−2.28	12.8	616

tion compared to the Riacho Verde station (Table 2). This depletion of heavy isotopes may be mostly linked to the elevation difference between the two stations (the other environmental parameters being approximately similar for both stations on the annual scale). There is a certain interannual variability of the rainfall isotopic signal. Indeed, the 2011–2019 monthly monitoring evidenced enriched amount-weighted annual isotopic values during the 2012–2016 drought compared to the other years (Table 3).

4.2 Definition of the Local Meteoric Water Line

The presence of an isotopic bias in the samples of the monthly monitoring required the processing of the data before correctly calculating the LMWL. Therefore, all the data with $d_{exc} < +10$ ‰ were deleted from the monthly database (51 % of the data). Then, the plotting of the monthly isotopic data between 2011 and 2019 on a $\delta^{18}O$ vs. $\delta^{2}H$ graph (Fig. 2) allowed to obtain a preliminary LMWL of Quixeramobim, characterized by the following equation Eq. (1):

$$\delta^{2}H\ (‰) = 8.06 \times \delta^{18}O\ (‰) + 12.6 \quad (r^2 = 97\,\%) \tag{1}$$

Evaporated samples from the daily monitoring (13 % of the data) were also not included in the construction of the LMWL. Then, the preliminary LMWL of Quixeramobim, obtained from the daily monitoring realized between March 2018 and October 2019 (Fig. 2), is defined by the following equation Eq. (2):

$$\delta^{2}H\ (‰) = 8.18 \times \delta^{18}O\ (‰) + 14.5 \quad (r^2 = 98\,\%) \tag{2}$$

The slope of the LMWL of Quixeramobim of both Eqs. (1) and (2) is consistent with that of the GMWL. However, the y-intercept value is greater than +12 ‰ in both datasets despite the proximity of the study zone to the Atlantic coast. Nonetheless, data obtained in the study region from the unevaporated precipitation record available in Moura (2013) or in the IAEA-GNIP data of Fortaleza (Ceará, Brazil) also showed a y-intercept value > 10 ‰, of respectively +12.3 ‰ ($r^2 = 98$ %, $n = 8$, 2011–2012 period) and +12.7 ‰ ($r^2 = 97$ %, $n = 50$, 1965–1984 period). Moreover, the isotopic monitoring of daily rainfall highlighted the high d-excess values observed in the rainwater (between +10.1 ‰ and +22.1 ‰ for the non-evaporated data) and allowed, therefore, to confirm the LMWL derived from the monthly data. Consequently, the high y-intercept value of the LWML of Quixeramobim indicates that rainwater is composed of moisture from marine origin (from the Atlantic, considering the proximity of the study zone to the coast), but also from continental recycling. As a consequence, d-excess value < +12 ‰ in groundwater evidences the presence of water that has undergone evaporation, and could reveal indirect infiltration processes. This observation has major consequences in the understanding of aquifer recharge processes in NEB.

4.3 Influence of local or regional parameters

The study of the temporal variability of the rainfall isotopic composition at the daily scale showed that the isotopic sig-

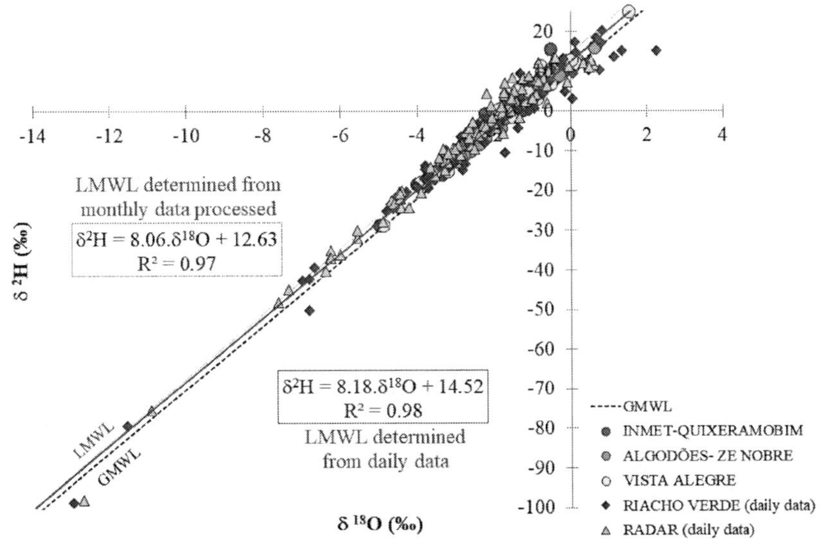

Figure 2. Local Meteoric Water Line (LMWL) of Quixeramobim determined from the processing of monthly data between 2011 and 2019 ($n = 83$) and daily data between 2018 and 2019 ($n = 184$).

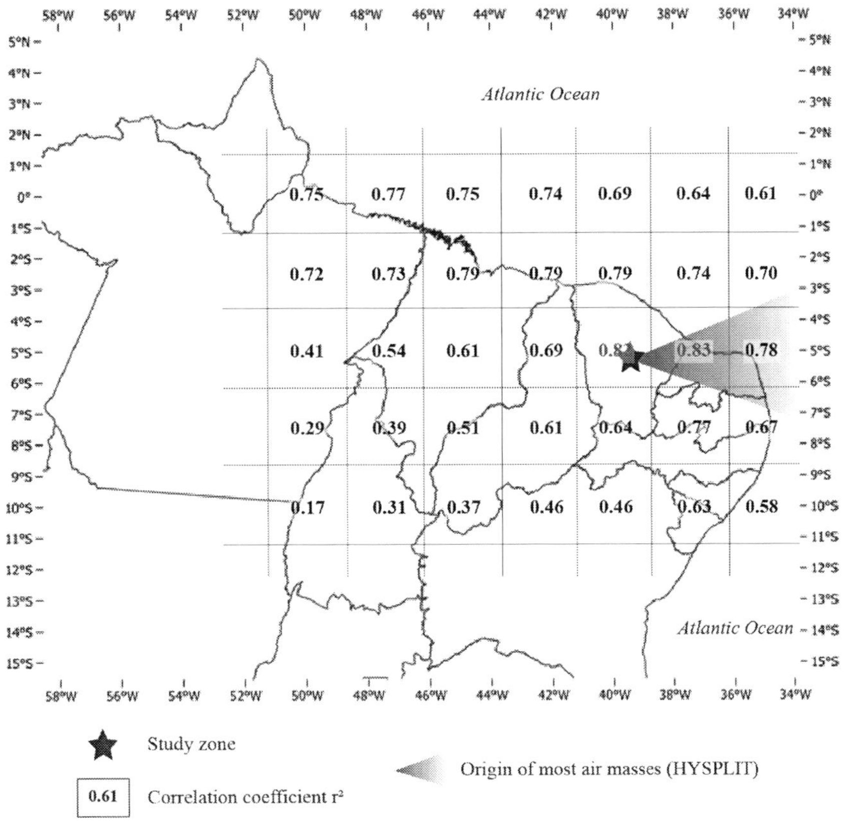

Figure 3. Correlation coefficient r^2 between monthly $\delta^{18}O$ (obtained from daily data) and OLR in Northeast Brazil.

nal of the Quixeramobim's precipitation is not related to the local parameters, such as the relative humidity ($r^2 < 0.4$), the temperature ($r^2 < 0.2$) or the rainfall amount ($r^2 < 0.1$). Nonetheless, at the monthly scale (obtained from the daily monitoring), better correlations are found with the relative humidity ($r^2 < 0.8$) and the rainfall amount ($r^2 < 0.6$). In terms of regional parameters, the simulation over the year 2019 of the origin of air masses at the daily scale (47 events), tracked using HYSPLIT at several altitudes (1000, 3000 or 5000 m) over 6 last days (2500 to 4000 km before reaching the isotopic station), highlighted that most of vapours originate from the Atlantic Ocean at the East of the study zone between 5° N and 15° S (33 events). Therefore, variations in the isotopic composition of precipitation cannot be explained by changes in the moisture origin. However, the convective activity of the study region could be partly responsible for the monthly isotopic variations, considering the good correlation between $\delta^{18}O$ and the Outgoing Longwave Radiation (OLR) data ($r^2 < 0.8$; Fig. 3). OLR define also the main moisture direction with higher correlations at the east of the study zone highlighted with retro trajectories from HYSPLIT.

5 Conclusion

This isotopic study, based on the analysis of monthly rainfall samples between 2011 and 2019 (172 samples) and daily rainfall samples between 2018 and 2019 (212 samples), allowed to improve the characterization of the isotopic signal of precipitation in the semiarid region of Ceará. Results highlighted an important variability of the $\delta^{18}O$ and δ^2H values with daily values ranging from −12.97 ‰ to +2.27 ‰ and from −98.5 ‰ to +20,2 ‰ respectively. The annual amount-weighted mean isotopic composition of precipitation is −3.21 ‰ for $\delta^{18}O$ and −12.0 ‰ for δ^2H. However, a seasonal variability of the isotopic composition of precipitation was observed at the monthly scale, with monthly heavy isotope concentrations more depleted during the rainiest months (from February to May). Comparisons between monthly and daily data pointed out that the frequency of rainfall sampling may induce an isotopic bias and lead to a wrong definition of the LMWL in the absence of adequate data processing, which could affect the understanding of aquifer recharge processes. In this study, the selection of non-evaporated isotopic data allowed to define the preliminary LMWL of Quixeramobim by the equation $\delta^2H\ (‰) = 8.06 \times \delta^{18}O\ (‰) + 12.6$ ($r^2 = 97\ \%$), which indicates that rainwater is not only composed of moisture from direct marine origin, but also from continental recycling despite the proximity of the study zone to the Atlantic coast. At last, analyses of the isotopic temporal variations showed that the isotopic signal of precipitation is not linked to the amount effect at the daily scale, even if a better correlation is observed at the monthly scale. Nonetheless, analyses evidenced the importance of convective systems and relative humidity in the control of the isotopic composition of the Quixeramobim's precipitation.

Code availability. This study used the software HYSPLIT Atmospheric Transport and Dispersion Modeling System of the NOAA Air Resources Laboratory (https://www.ready.noaa.gov/HYSPLIT.php, last access: 26 April 2022; Stein et al., 2015) to simulate the origin of air masses.

Data availability. The research data presented in this study can be found in Kreis (2021), accessible at: https://www.theses.fr/2021MONTG031 (last access: 23 May 2022). The isotopic data of Fortaleza are from the GNIP database of IAEA, and can be found in IAEA/WMO (2022; https://nucleus.iaea.org/wiser, last access: 23 May 2022, login required).

Author contributions. MBK carried out the investigations and interpretations of this study under the scientific supervision of JDT and ESPRM. NP performed the stable isotope analyses. MBK prepared the manuscript with contributions from all co-authors.

Competing interests. The contact author has declared that none of the authors has any competing interests.

Disclaimer. Publisher's note: Copernicus Publications remains neutral with regard to jurisdictional claims in published maps and institutional affiliations.

Special issue statement. This article is part of the special issue "IAHS2022 – Hydrological sciences in the Anthropocene: Variability and change across space, time, extremes, and interfaces". It is a result of the XIth Scientific Assembly of the International Association of Hydrological Sciences (IAHS 2022), Montpellier, France, 29 May–3 June 2022.

Acknowledgements. This work is part of the multidisciplinary French-Brazilian cooperation between IRD and FUNCEME. The authors gratefully acknowledge the NOAA Air Resources Laboratory (ARL) for the provision of the HYSPLIT transport and dispersion model (https://www.ready.noaa.gov, last access: 26 April 2022) used in this publication.

Financial support. This research has been supported by the LAMA laboratory and the FUNCEME.

Review statement. This paper was edited by Christophe Cudennec and reviewed by two anonymous referees.

References

Ahmed, A. A., Shabana, A. R., and Saleh, A. A.: Using hydrochemical and isotopic data to determine sources of recharge and groundwater evolution in arid region from Eastern Desert, Egypt, J. Afr. Earth Sci., 151, 36–46, https://doi.org/10.1016/j.jafrearsci.2018.11.024, 2019.

Babaye, M. S. A., Orban, P., Ousmane, B., Favreau, G., Brouyère, S., and Dassargues, A.: Characterization of recharge mechanisms in a Precambrian basement aquifer in semi-arid south-west Niger, Hydrogeol. J., 27, 475–491, https://doi.org/10.1007/s10040-018-1799-x, 2019.

Burte, J.: Os pequenos aquiferos aluviais nas areas cristalinas semiaridas: Funcionamento e estrategias de gestao. Estudo de caso no Nordeste brasileiro, PhD Thesis, University of Montpellier, https://theses.hal.science/tel-00399714 (last access: 5 July 2021), 2008.

Chang, H. K., Gonçalves, R. D., Aggarwal, P. K., Stradioto, M. R., Hespanhol, E. C., Sturchio, N. C., and Araguas, L. J.: Groundwater isotope ratios reflect convective and stratiform (paleo) precipitation fractions in Brazil, J. Hydrol., 585, 124801, https://doi.org/10.1016/j.jhydrol.2020.124801, 2020.

Da Silva Nobre, M. E., Gomes, D. F., Vasconcelos, S. M. S., de Lima Neto, I. O., Ferreira, S. K., and de Ponti Souza, D.: Estudo hidroquímico-ambiental do aquífero aluvionar do baixo Jáguaribe, Itaiçaba–Ceará, Revista do Instituto Geológico, 39, 77–92, https://doi.org/10.33958/revig.v39i3.601, 2019.

FUNCEME: Para entender melhor a previsão climática para a estação chuvosa no Ceará e glossário de termos meteorológicos, FUNCEME, Fortaleza, 23 pp., http://www.funceme.br/?page_id=9153 (last access: 5 July 2021), 2018.

Gastmans, D., Santos, V., Galhardi, J. A., Gromboni, J. F., Batista, L. V., Miotlinski, K., and Govone, J. S.: Controls over spatial and seasonal variations on isotopic composition of the precipitation along the central and eastern portion of Brazil, Isot. Environ. Healt. S., 53, 518–538, https://doi.org/10.1080/10256016.2017.1305376, 2017.

Gastmans, D., Garpelli, L. N., dos Santos, V., de Lima, C., Quaggio, C. S., Santarosa, L. V., and Kirchheim, R. E.: Contribuição dos isótopos estáveis da água (H e O) no conhecimento dos aquíferos brasileiros: estado da arte e perspectivas futuras, Derbyana, 42, https://doi.org/10.14295/derb.v42.734, 2021.

Hao, S., Li, F., Li, Y., Gu, C., Zhang, Q., Qiao, Y., and Zhu, N.: Stable isotope evidence for identifying the recharge mechanisms of precipitation, surface water, and groundwater in the Ebinur Lake basin, Sci. Total Environ., 657, 1041–1050, https://doi.org/10.1016/j.scitotenv.2018.12.102, 2018.

IAEA/WMO: Global Network of Isotopes in Precipitation, The GNIP Database, https://nucleus.iaea.org/wiser, last access: 23 May 2022.

Kreis, M.: Origine et dynamique de salinité des aquifères fracturés cristallins de la région semi-aride du Ceará, Brésil, PhD Thesis, University of Montpellier, Federal University of Ceará, NNT: 2021MONTG03, https://www.theses.fr/2021MONTG031 (last access: 23 May 2022), 2021.

Kreis, M., Taupin, J. D., Patris, N., and Martins, E. S. P. R.: Isotopic characterisation and dating of groundwater recharge mechanisms in crystalline fractured aquifers: example of the semi-arid Banabuiú watershed (Brazil), Isot. Environ. Healt. S., 56, 418–430, https://doi.org/10.1080/10256016.2020.1797275, 2020.

Martins, E. S. P. R. and Júnior, F. D. C. V.: O clima da Região Nordeste entre 2009 e 2017: Monitoramento e previsão, Parcerias Estratégicas, 22, 63–79, 2017.

Moura, I. B. M. D.: Estudos das águas da bacia hidrográfica do Rio Banabuiú no trecho entre Quixeramobim e Banabuiú, Ceará, Brasil, PhD Thesis, Federal university of Fortaleza, https://rigeo.cprm.gov.br/handle/doc/1188 (last access: 5 July 2021), 2013.

Pontes Filho, J. D., Souza Filho, F. D. A., Martins, E. S. P. R., and Studart, T. M. D. C.: CopulaBased Multivariate Frequency Analysis of the 2012–2018 Drought in Northeast Brazil, Water, 12, 834, https://doi.org/10.3390/w12030834, 2020.

Santoni, S., Huneau, F., Garel, E., Vergnaud-Ayraud, V., Labasque, T., Aquilina, L., and Celle-Jeanton, H.: Residence time, mineralization processes and groundwater origin within a carbonate coastal aquifer with a thick unsaturated zone, J. Hydrol., 540, 50–63, https://doi.org/10.1016/j.jhydrol.2016.06.001, 2016.

Stein, A. F., Draxler, R. R., Rolph, G. D., Stunder, B. J. B., Cohen, M. D., and Ngan, F.: NOAA's HYSPLIT Atmospheric Transport and Dispersion Modeling System, B. Am. Meteorol. Soc., 96, 2059–2077, https://doi.org/10.1175/BAMS-D-14-00110.1, 2015 (code available at: https://www.ready.noaa.gov/HYSPLIT.php, last access: 26 April 2022).

Sukhija, B. S., Reddy, D. V., Nagabhushanam, P., Bhattacharya, S. K., Jani, R. A., and Kumar, D.: Characterisation of recharge processes and groundwater flow mechanisms in weathered-fractured granites of Hyderabad (India) using isotopes, Hydrogeol. J., 14, 663–674, https://doi.org/10.1007/s10040-005-0461-6, 2006.

Taupin, J. D., Gallaire, R., and Arnaud, Y.: Analyses isotopiques et chimiques des précipitations sahéliennes de la région de Niamey au Niger: implications, IAHS-AISH P., 244, 151–162, https://www.documentation.ird.fr/hor/fdi:010015696 (last access: 5 July 2021), 1997.

Terzer, S., Wassenaar, L. I., Araguás-Araguás, L. J., and Aggarwal, P. K.: Global isoscapes for δ^{18}O and δ^{2}H in precipitation: improved prediction using regionalized climatic regression models, Hydrol. Earth Syst. Sci., 17, 4713–4728, https://doi.org/10.5194/hess-17-4713-2013, 2013.

Proc. IAHS, 385, 25–29, 2024
https://doi.org/10.5194/piahs-385-25-2024

Open Access PIAHS

Estimation of expected annual damage, EAD

Dan Rosbjerg

Department of Environmental and Resource Engineering, Technical University of Denmark, Kongens Lyngby, 2700, Denmark

Correspondence: Dan Rosbjerg (daro@dtu.dk)

Received: 11 May 2022 – Revised: 2 January 2023 – Accepted: 9 January 2023 – Published: 18 April 2024

Abstract. The sensitivity of expected annual damage, EAD, is analytically analysed by applying a log-linear relation between return periods and corresponding damages. It is found that the smallest return period for damage should be estimated as precisely as possible, that the percentage uncertainty in the damage estimate is transformed into the same percentage uncertainty in the EAD estimate, and that it is possible to extrapolate beyond the largest return period with corresponding damage assessment. The precision of the estimate of EAD is investigated in detail in the case of only few available data, and it is found that two different methods for numerical integration may result in strongly diverging results. By applying a piecewise log-linear damage function, it is shown that the log-linear model provides a trustworthy estimate of EAD, also in the case of few available data. Finally, the modifications needed in the special case of threshold exceedance data instead of annual maxima data are presented.

Keywords. UPH 21; SDG 13; Modelling; Pluvial flooding; Estimation techniques

1 Introduction

In the future, large urban drainage structures in Denmark will be designed based on economic criteria instead of design based on a fixed return period (SVK, 2017). The societal value of the structure will be evaluated as the reduction in net present value of future damages less the net present value of investments and maintenance costs. Implicitly, it has been assumed that EAD (the expected annual damage) can be trustworthy determined. This, however, may strongly depend on the estimation method and the available data. Analytical estimation based on a log-linear relation between return period and corresponding damage (Olsen et al., 2015; Rosbjerg, 2017) is here further developed by introducing a stepwise integration. The method is evaluated by analysing two cases with, respectively, ample and sparse data availability. It is concluded that use of the analytical model is superior to straightforward numerical integration. Both annual maximum data and threshold exceedance data (known as peak over threshold series, POT, or partial duration series, PDS) are considered.

2 Theory

If the distribution function for the annual maximum X is denoted $F_X(x) = P\{X \le x\}$, the T-year event x_T is given by

$$F_X(x_T) = 1 - \frac{1}{T_X} = 1 - p_X \quad (1)$$

Consequently

$$x_T = F_X^{-1}\left(1 - \frac{1}{T_X}\right); \quad T_X \ge 1 \quad (2)$$

The damage caused by the T_X year event is denoted by $D_X(T_X)$ and the density function of X by $f_X(x) = \mathrm{d}F_X(x)/\mathrm{d}x$. Accordingly, EAD can be determined by

$$\mathrm{EAD} = \int_0^\infty D_X(x_T)\, f_X(x_T)\,\mathrm{d}x_T = \int_1^\infty \frac{D_X(T_X)}{T_X^2}\mathrm{d}T_X \quad (3)$$

or

$$\mathrm{EAD} = \int_0^1 D_p(p_X)\,\mathrm{d}p_X; \quad D_p(p_X) = D_X\left(\frac{1}{p_X}\right) \quad (4)$$

In practise, the integrations in Eqs. (3) and (4) are carried out inside the limits defined by available data.

In the case of threshold exceedance data with k exceedances per year in average and the distribution function for threshold exceedances Y denoted by $G_Y(y)$, we get

$$G_Y(y_T) = 1 - \frac{1}{k\,T_Y} = 1 - p_Y \tag{5}$$

Consequently

$$y_T = G_Y^{-1}\left(1 - \frac{1}{k\,T_Y}\right); \quad T_Y \geq \frac{1}{k} \tag{6}$$

Hereby EAD becomes

$$\mathrm{EAD} = \int_{1/k}^{\infty} \frac{D_Y(T_Y)}{T_Y^2}\mathrm{d}T_Y \tag{7}$$

or

$$\mathrm{EAD} = k\int_0^1 D_p(p_Y)\mathrm{d}p_Y; \quad D_p(p_Y) = D_Y\left(\frac{1}{k\,p_Y}\right) \tag{8}$$

Experience has shown and confirmed by the present analysis that a log-linear relation between return period and damage can be a good approximation.

$$D(T) = a\ln T + b \tag{9}$$

where it may be assumed that the equation is valid also for $T -> \infty$. If $D(T) = 0$ for $T = T_s$, then by integration from T_s to ∞ we get

$$\mathrm{EAD} = \frac{a}{T_s}\left(1 + \ln\frac{T_s}{T_0}\right); \quad T_0 = \exp\left(-\frac{b}{a}\right) \tag{10}$$

Integration between two arbitrary return periods T_i and T_j greater than T_s leads to the EAD-contribution

$$\mathrm{EAD} = \frac{a}{T_i}\left(1 + \ln\frac{T_i}{T_0}\right) - \frac{a}{T_j}\left(1 + \ln\frac{T_j}{T_0}\right) \tag{11}$$

If the log-linear relation is valid until $T = T_c$ and then constant for $T_c < T < \infty$, the contribution to EAD from the constant part is

$$\mathrm{EAD} = \frac{a}{T_0} - \frac{a}{T_c}\left(1 + \ln\frac{T_c}{T_0}\right) + \frac{D(T_c)}{T_c} \tag{12}$$

3 Case A: Good data coverage

First, some initial sensitivity analyses are carried out based on annual maxima data with corresponding damage costs in a case with good data coverage, see Table 1. Implicitly, all costs are in mill. DKK. Case A refers to an idealised situation, where the data closely corresponds to the log-linear assumption.

Table 1. Case A – Damage cost as function of return period.

T	$D(T)$	$\ln T$
1	0	0
3	70	1.099
8	130	2.079
15	180	2.708
30	260	3.401
60	280	4.094
150	375	5.011
300	450	5.704

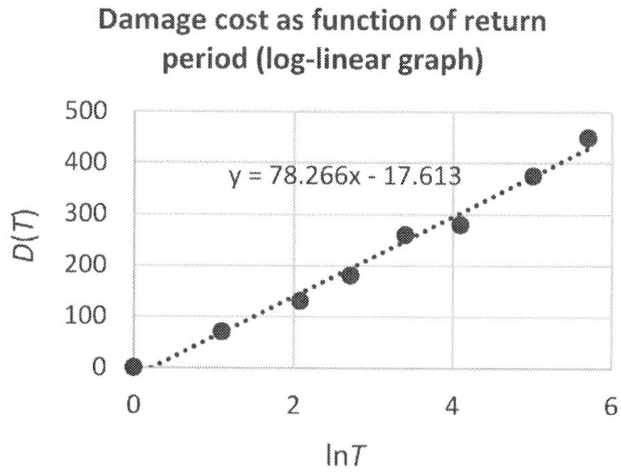

Figure 1. Case A – Cost function in log-linear plotting.

Table 2. Case A – Sensitivity of EAD.

Basic solution $T_s = 1$	62.5
$T_s = 2$	57.5
$D(T) + 10\,\%$	68.2
$D(T) - 10\%$	56.2
Without extrapolation	60.8
Numerical integration over T	61.7
Numerical integration over p	70.0
$D(T)$ constant above $T = 500$	62.3

Table 3. Case B – Damage cost as function of return period.

T	$D(T)$	$\ln T$
1.6	0	0.470
5	0.0009	1.609
20	20.5	2.996
50	49.5	3.912
100	65.5	4.605

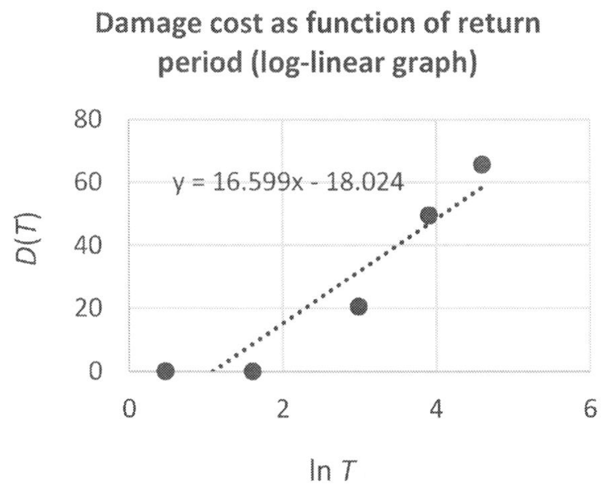

Figure 2. Case B – Cost function in log-linear potting.

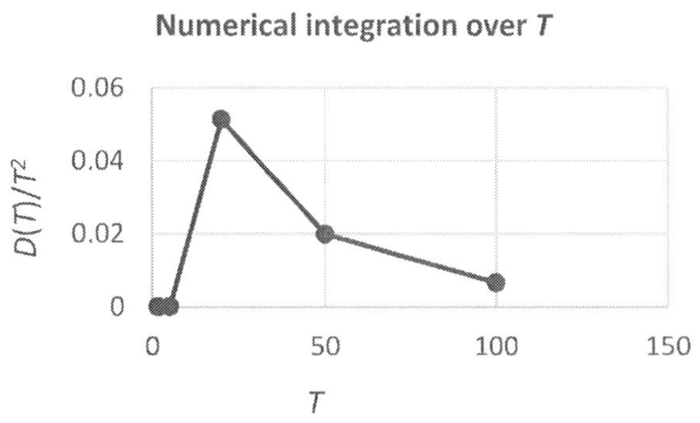

Figure 3. Case B – Numerical integration over T (area below the curve).

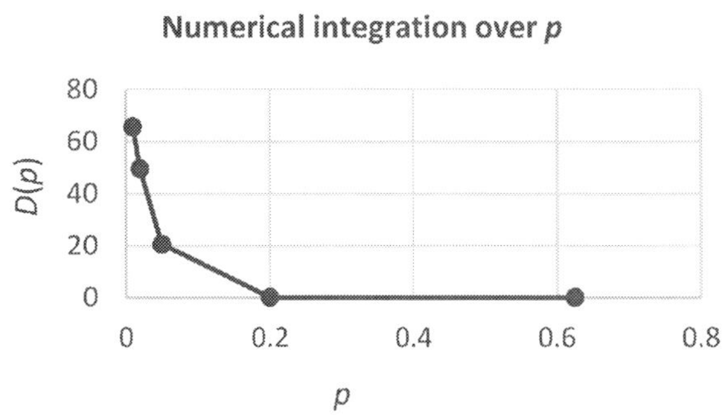

Figure 4. Case B – Numerical integration over p (area below the curve).

Inserting the regression parameters acquired from Fig. 1 into Eq. (10) results in the basic result shown in Table 2. The sensitivity of the result is investigated by varying input values one by one as indicated in Table 2. The numerical integrations are carried out based on Eq. (3).

Some differences, but not substantial, are noticed between the obtained methods. An uncertainty of 10 % in the damage cost is seen to cause an uncertainty in EAD also of 10 %. Generally, the different results are of the same order of magnitude.

4 Case B: Sparse data coverage; numerical vs. analytical integration

Corresponding values obtained from Rambøll (2021) of annual maxima return periods and damage costs are shown in Table 3, and regression parameters obtained from a log-linear plot can be seen in Fig. 2. The limited number of cost estimates reflects the fact that it is quite labour intensive to estimate the costs. The figure reveals that the log-linear model provides a poor fit. However, by omitting the first data pair, which is justifiable since the resulting EAD value does not change at all when applying usual numerical integration, the log-linear fit is substantially improved. Having now only four data points justifies the characterization "sparse".

In Figs. 3 and 4, the curves for numerical integration over, respectively, T and p are shown. Contrary to Case A, the integration results are here strongly differing. The area under the curve in Fig. 3 is found to be EAD = 2.11, whereas the area under the curve in Fig. 4 is calculated to EAD = 3.16 with no obvious explanation for the difference other than a strong non-linearity being present in both integrands. It makes no difference whether the first point in Table 3 is included or excluded.

The above results and corresponding results using the log-linear model combined with analytical integration are shown in Table 4. Suffix w means calculation without extrapolation, suffix e indicates calculation with extrapolation and suffix t the sum. Both one, two and pointwise log-linear approximations using Eq. (11) are applied.

The results show that the log-linear model combined with analytical integration is generally applicable, also in the case of a sparse data set, where straightforward numerical integration appears inadequate. It is also seen that extrapolation can provide a significant supplement to the EAD value obtained by integration inside the interval covered by data.

Table 4. Case B – Numerical vs. analytical integration.

Integration method	EAD_w	EAD_e	EAD_t
Numerical integration over T	2.11		
Numerical integration over p	3.16		
Analytical integration without $T = 1.6$ using one log-linear approximation	2.97	0.86	3.83
Analytical integration without $T = 1.6$ using two log-linear approximations	2.72	0.95	3.67
Analytical integration using pairs of data points	2.74	0.90	3.64

Table 5. Case C – Damage cost as function of return period.

T	$D(T)$	$\ln T$
0.33333	0	−1.099
0.7	15	−0.357
1	30	0
3	70	1.099
8	130	2.079
15	180	2.708
30	260	3.401
60	280	4.094
150	375	5.011
300	450	5.704

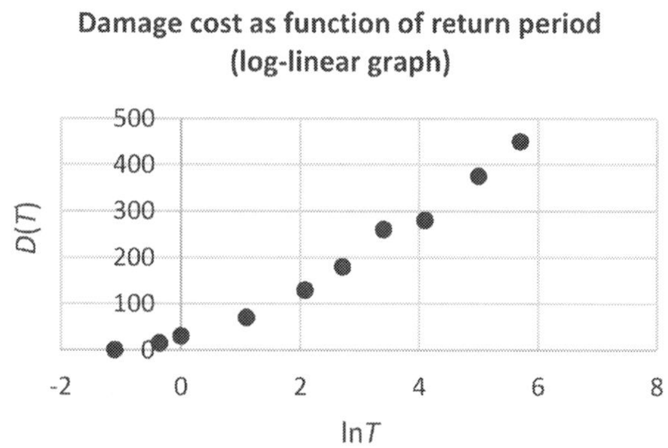

Figure 5. Case C – Cost function in log-linear plotting.

5 Case C: Estimation of EAD with exceedance data

Here almost the same example as in case A is used, however slightly modified for $T = 1$ and supplemented with two data pairs for $T < 1$, see Table 5. The averages number of exceedances per year is $k = 3$. The dataset is depicted in a log-linear graph in Fig. 5.

The results of the different integration methods are shown in Table 6. As in case A, we see only moderate divergence between the different integration methods. Integration over T gives the largest value with integration over p being slightly below. However, in comparison with case A it is found that adding damage costs for return periods below $T = 1$ implies a significant increase in the EAD estimate.

Table 6. Case C – Integration using exceedance data.

Exceedance data	EAD_w	EAD_e	EAD_t
Numerical integration over T	106.4		
Numerical integration over p	101.1		
Analytical integration using pairs of data points	95.7	1.8	97.5

6 Conclusions

By analysing three cases, the following experience has been obtained.

Sensitivity

- It is important to know the return period for which damages get started, as frequent but small damages contribute significantly to EAD.
- Percentwise uncertainty in damage costs implies the same percentwise uncertainty in the estimate of EAD.
- The added supplement to EAD obtained by extrapolation can be significant.

Numerical integration

- Using numerical integration may result in diverging results when integrating over, respectively, T and p.
- In the case of a sparse data set, and for small values of T, the difference may be substantial.

Analytical integration

- Using a log-linear model combined with analytical integration is a flexible and convenient tool. A piecewise application down to data pairs is easily performed. The method provides trustworthy results and can be extrapolated to infinity, contrary to ordinary numerical integration.

Code availability. The applied software is developed by the author and not publicly accessible.

Data availability. The data in Case A and Case C are artificially created by the author. The data in Case B are publicly available in Rambøll (2021; https://www2.mst.dk/Udgiv/publikationer/2018/manual.pdf, last access: 21 January 2023).

Competing interests. The author has declared that there are no competing interests.

Disclaimer. Publisher's note: Copernicus Publications remains neutral with regard to jurisdictional claims in published maps and institutional affiliations.

Special issue statement. This article is part of the special issue "IAHS2022 – Hydrological sciences in the Anthropocene: Variability and change across space, time, extremes, and interfaces". It is a result of the XIth Scientific Assembly of the International Association of Hydrological Sciences (IAHS 2022), Montpellier, France, 29 May–3 June 2022.

Review statement. This paper was edited by Christophe Cudennec and reviewed by two anonymous referees.

References

Olsen, A. S., Zhou, Q, Linde, J. J., and Arnbjerg-Nielsen, K.: Comparing methods of calculating expected annual damage in urban pluvial flood risk assessments, Water, 7, 255–270, https://doi.org/10.3390/w7010255, 2015.

Rambøll: Manual for the application of Skrift 31, Miljøstyrelsen, https://www2.mst.dk/Udgiv/publikationer/2018/manual.pdf (last access: 21 January 2023), 2021 (in Danish).

Rosbjerg, D.: Optimal adaptation to extreme rainfalls in current and future climate, Water Resour. Res., 53, 535–543, https://doi.org/10.1002/2016WR019718, 2017.

SVK: Methods for determination of the service level for rainwater on the surface, IDA Spildevandskomitéen, https://spildevandskomiteen.dk/skrift-31/ (last access: 21 January 2023), 2017 (in Danish).

Proc. IAHS, 385, 31–37, 2024
https://doi.org/10.5194/piahs-385-31-2024

Open Access PIAHS

Streamflow simulation using the event-based SCS-MS/LR model in the Hitiaa catchment, French Polynesia

Garance Tanguy[1], **Christophe Bouvier**[2], **and Lydie Sichoix**[1]

[1]Laboratoire GEPASUD, Université de la Polynésie française, BP 6570, 98702 FAA'A, Tahiti, French Polynesia

[2]Institut de Recherche pour le Développement (IRD), HydroSciences Montpellier, UMR 5151 CNRS-IRD-UM, 34000 Montpellier, France

Correspondence: Garance Tanguy (garance.tanguy24@gmail.com)

Received: 26 May 2022 – Revised: 15 November 2022 – Accepted: 9 January 2023 – Published: 18 April 2024

Abstract. The volcanic tropical Island of Tahiti is prone to heavy rainfalls and flash floods which regularly generate severe damages. Hydrological response on the island is highly variable from one catchment to the other, in terms of runoff production as well as peak flow amplitude. Infrastructures designers there require an operational method able to calculate the design flood metrics on any Tahitian catchment, taking into account the variability of their response. This study applies a parsimonious distributed event-based conceptual model to the observed rainfall-runoff events of 9 small catchments ($< 6\,\text{km}^2$). A modified version of the Mishra and Singh Soil Conservation Service (SCS-MS) runoff model, associated with the Lag-and-Route (LR) model was applied to series of 9 to 176 rainfall-runoff events for each catchment that occurred between 2004 and 2021. Two dominant parameters are to be considered for the SCS-MS model: the size of the soil reservoir (Si) and the initial water content (M) in this soil reservoir. SCS-MS was found to perform better than SCS-CN and to fit better the asymptotic behaviour of the rainfall-runoff relationship. Si could be fixed to 2500 mm for all catchments except Nymphea where Si had to be lowered to 80 mm, because of the high density of urbanization over this catchment. M parameter was set variable from one event to the other as the initial condition of the model. The Nash criterion was calculated for each event and their median value ranged from 0.39 to 0.79 across the catchments, proving the flexibility of the model. A clear relationship could be fitted between the median M and Antecedent Precipitation Index (API), both calculated for each catchment. Further testing of this model on similar tropical volcanic context should be undertaken to verify its flexibility and to consolidate the calibration of its parameters.

Keywords. UPH 19; PUB; Modelling; Flash floods; Urbanization; Conceptual model; Calibration

1 Introduction

Flash floods have been recognized as the main natural risk in the Island of Tahiti, where high urbanization rate increases the vulnerability of people, homes and infrastructures (Lamaury et al., 2021). There is a strong need for a flood predetermination method to assist engineers in the design of hydraulic infrastructure.

Flash floods result from the combined effects of extreme rainfall events and the steep morphology of the catchments (Wotling, 1999). However, their relative influences on the hydrological response and the variability of these influences across the island have not yet been captured by a rainfall-runoff model.

Wotling (1999) and Pheulpin (2016) verified the ability of event-based distributed conceptual models to reproduce the hydrological response of three western and one southern catchment. The low number of studied catchments didn't allow them to explain the spatial variability of hydrological response, nor to relate the model parameters to measurable characteristics of the catchment such as geomorphology, soil moisture, altitude or exposure to trade winds.

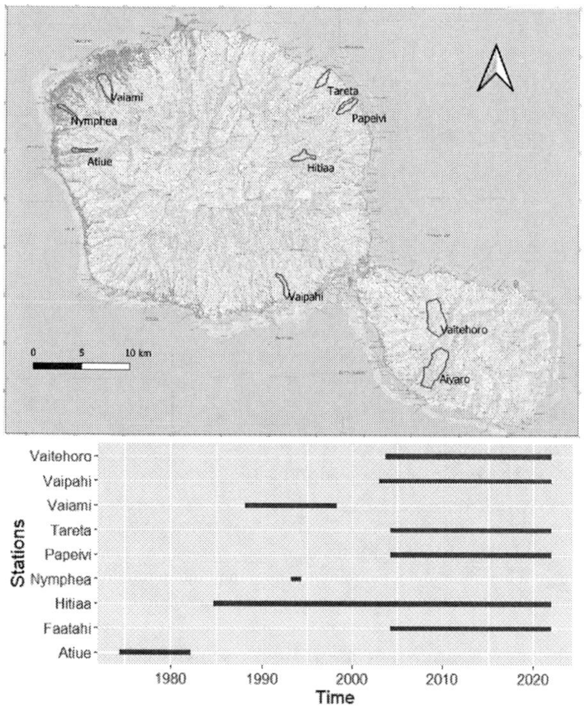

Figure 1. Location of the gauged catchments chosen to calibrate the rainfall-runoff model, and record periods of the associated discharge stations. IGN map available at (https://www.data.gouv.fr/fr/datasets/r/3aad9e99-50f8-4bb1-8b27-2cf3b19301af, last access: 20 November 2021).

The main objective of this study is to select and apply a rainfall-runoff model that is both robust to the hydrological variability in Tahiti and ready to be regionalized. The chosen model should be parsimonious to facilitate its transfer to ungauged catchments, and its parameters are to be related to physical features of the catchments.

2 Study site and data

2.1 The island of Tahiti

Tahiti is a volcanic island belonging to the Society Archipelago in the middle of the Pacific Ocean. Tahiti is the biggest – area of 1042 km^2 –, the highest – Mount Orohena culminates at 2241 m – and the most populous – 192 760 inhabitants in 2017 – island of French Polynesia. The Island is subject to a southern tropical climate, controlled by northeasterly trade winds and the South Pacific Convergence Zone. Rainfall pattern is very heterogeneous within the island, and it is also the case for extreme events. Wotling (1999) and Pheulpin (2016) modelled extreme rainfall events from topography and exposure to trade winds.

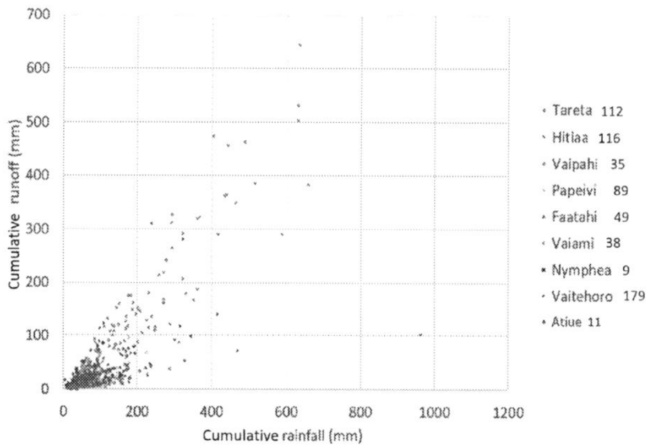

Figure 2. Spatial variability of hydrologic behaviours: Cumulative rainfall and runoff of events extracted from observation data in the studied catchments. Number of events of each series is indicated next to the name of the catchment.

2.2 Studied catchments

Over the 33 catchments that are or have been gauged in Tahiti, nine catchments were chosen to test and calibrate the rainfall-runoff model. Record periods are indicated in Fig. 1.

They were selected for the quality of the data and because they are small-sized catchments, with area lower than 6 km^2. Thus their area is considered small enough to neglect the variability of physio-climatic variables within them, so that each of them can be representative of a specific environment, at least in terms of altitude and rainfall input.

Locations of the catchments are selected in order to capture the spatial variability of rainfall exposure within the island. Both leeward and windward sides of the main island – Tahiti Nui – are considered, as well as the peninsula – Tahiti Iti.

Each catchment is instrumented with one or two rain gauges and the rainfall is interpolated by the Thiessen method. For each catchment, series of events are extracted at a 5min timestep from the rainfall and flow records. An event is extracted if it occurs after a duration of at least 3 h with rainfall intensity below 0.3 mm in 5 min. Cumulative rainfall and runoff must exceed thresholds that are adjusted from one catchment to the other so that the number of collected events is enough. The direct runoff volume for each event is then calculated by subtracting the base flow from the total instantaneous flow. Resulting runoff coefficients highlight the variability of the hydrological response among the selected catchments (Fig. 2).

3 Rainfall runoff model

The Mishra et al. (2003) Soil Conservation Service (SCS-MS) runoff model was modified and associated with the Lag-and-Route (LR) model to simulate the complete flood hydrographs. Note the model operates for each cell of a regular grid mesh. Each cell produces an elementary hydrograph, and the addition of all the elementary hydrographs leads to the complete flood hydrograph.

3.1 SCS-CN and SCS-MS model

Original SCS is based on the following equations:

$$P = \mathrm{Ia} + F + Q \tag{1}$$

$$\frac{Q}{P - \mathrm{Ia}} = \frac{F}{S} \tag{2}$$

$$\mathrm{Ia} = \lambda S \tag{3}$$

with P standing for total precipitation, "Ia" for initial abstraction, F for cumulative infiltration, Q for direct runoff, S for potential maximum retention.

Combined together, Eqs. (1), (2) and (3) lead to:

$$Q = \frac{(P - \mathrm{Ia})^2}{P - \mathrm{Ia} + S} \tag{4}$$

Mishra and Singh (2003), introduced the concept of equality between the runoff coefficient C and the degree of saturation Sr. Sr being expressed as:

$$Sr = \frac{F}{S + M} \tag{5}$$

with M the initial soil moisture, and C as:

$$C = \frac{Q}{P} \tag{6}$$

Equation (2) of the SCS becomes:

$$\frac{Q}{P - \mathrm{Ia}} = \frac{F + M}{S + M} \tag{7}$$

The total soil reservoir capacity Si was introduced as:

$$\mathrm{Si} = S + M \tag{8}$$

Equation (4) of the SCS is rewritten for the SCS-MS as follows:

$$Q = \frac{(P - \mathrm{Ia})(P - \mathrm{Ia} + M)}{P - \mathrm{Ia} + S + M} \tag{9}$$

Equation (4) was then differentiated as a function of time to calculate the direct runoff at each timestep. To adapt them to the dynamic flow calculation, the SCS-MS equations have been modified in the same way as in the work of Coustau et al. (2012):

- To take into account the potential decrease of the runoff coefficient between floods, the cumulative rainfall P (t)

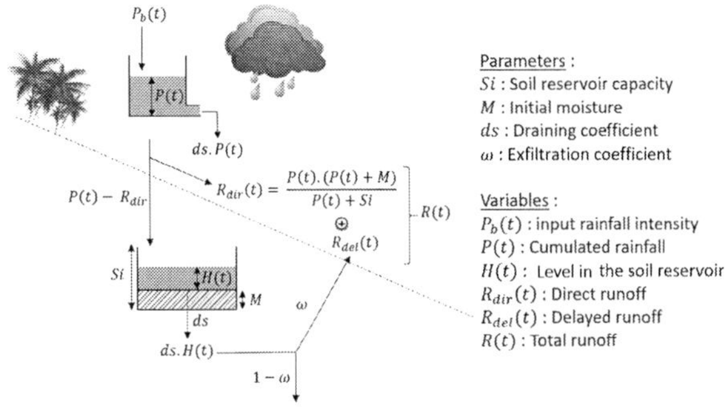

Figure 3. Conceptual diagram of the modified SCS-MS production function.

is considered as the level of a reservoir of infinite capacity to which an emptying is applied:

$$\frac{\mathrm{d}P\,(\mathrm{t})}{\mathrm{d}t} = P_\mathrm{b}\,(t) - \mathrm{d}s.P\,(\mathrm{t}) \tag{10}$$

where ds $[\mathrm{T}^{-1}]$ is the linear decay coefficient and P_b (t) is the input rainfall intensity.

- Soil drainage is represented by the application of a drain modulated by the same coefficient ds:

$$\frac{d\,H\,(\mathrm{t})}{\mathrm{d}t} = P_\mathrm{b}\,(\mathrm{t}) - R_\mathrm{dir}\,(\mathrm{t}) - \mathrm{d}s.H\,(\mathrm{t}) \tag{11}$$

where R_dir (t) $[\mathrm{L\,T}^{-1}]$ is the direct runoff intensity.

- Delayed flow R_del is introduced by recovering a fraction ω of the soil discharge at the outlet:

$$R_{del}\,(\mathrm{t}) = \omega.\mathrm{d}s.H\,(\mathrm{t}) \tag{12}$$

where ω is a dimensionless exfiltration coefficient.

5 parameters are associated to SCS-MS: Si, Ia, M, ds and ω.

3.2 Lag and Route model

The Lag-and-Route model (Maidment, 1993) routes the runoff production of each cell m to the outlet of the catchment.

According to the production function, each cell produces an elementary hydrograph at each time t,equation can be written as:

$$\begin{cases} Q_\mathrm{m}\,(\mathrm{t}) = 0 \ \text{ if } \ t < t_\mathrm{o} + T_\mathrm{m} \\ Q_\mathrm{m}\,(\mathrm{t}) = \frac{R(t_\mathrm{o})}{K_\mathrm{m}} \cdot \exp\left(-\frac{(t-(t_\mathrm{o}+T_\mathrm{m}))}{K_\mathrm{m}}\right).A \quad \text{otherwise} \end{cases} \tag{13}$$

where $R(t_\mathrm{o})$ is the runoff produced by the m cell at time t_o, T_m is the propagation time and K_m is the diffusion time. They are linked by the K_0 [–] constant as follows:

$$K_\mathrm{m} = K_0 \cdot T_\mathrm{m} \tag{14}$$

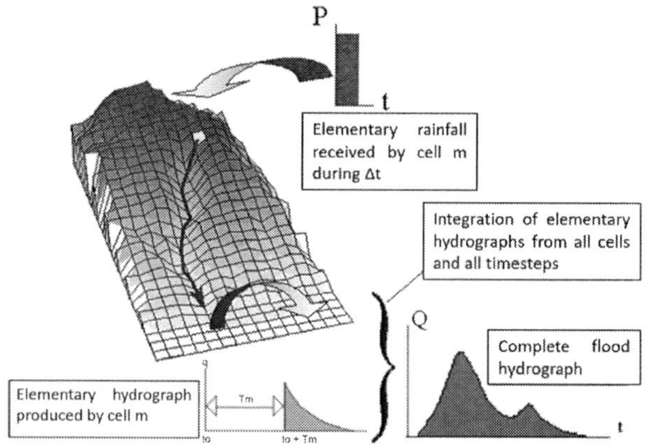

Figure 4. Lag-and-Route transfer function (modified from Maidment, 1993).

T_m is the travel time from the m cell to the catchment outlet. It can be expressed by:

$$T_\mathrm{m} = \sum_{k=1}^{n} \frac{l}{V_0} \tag{15}$$

where n is the number of cells of length l between the cell m and the outlet and V_0 is linked to the velocity of water transfer. V_0 and K_0 are the input parameters for the transfer function.

4 Model calibration

SCS-MS/LR was calibrated on each catchment from the series of observed rainfall-runoff events. 5 parameters must be calibrated for the production function (Si, Ia, M, ω and ds) and 2 parameters for the transfer function (V_0 and K_0). K_0 was arbitrarily fixed to 0.7 for all events and all catchments because highly related to V_0 (Coustau et al., 2012; Nguyen and Bouvier, 2019). Ia was fixed for all events and all catchments from graphical fit of the theoretical balance sheet equation (Eq. 9) on observations of event cumulative rainfall and runoff (Fig. 2). Si, ds and ω were first calibrated by trial and error and sensibility analysis over each series of events paying attention to the median value of the NSE criterion (Nash and Sutcliffe Efficiency, Nash and Sutcliffe, 1970). Explored ranges of values are from 0.01 to 0.6 for ω, from 0.5 to 2 $\mathrm{m\,s^{-1}}$ for ds and from 10 to 3000 mm for Si, based on previous applications of the SCS model (Coustau et al., 2012; Nguyen and Bouvier, 2019). Then these three parameters were uniformed for all catchments when possible in order to reduce the risk of equifinality.

Parameter M stands for the initial condition of the model, and should vary from one event to the other. According to its physical meaning, V_0 should also be set variable from one event to the other. These two state parameters should be related to initial conditions like the level of soil humidity at the beginning of each event. For each catchment, such initialization relationships were built between optimized values of M and V_0 on one hand, and measured baseflow Q_0 at the beginning of each event on the other hand. Q_0 is used here as a proxy for Antecedent Moisture Conditions.

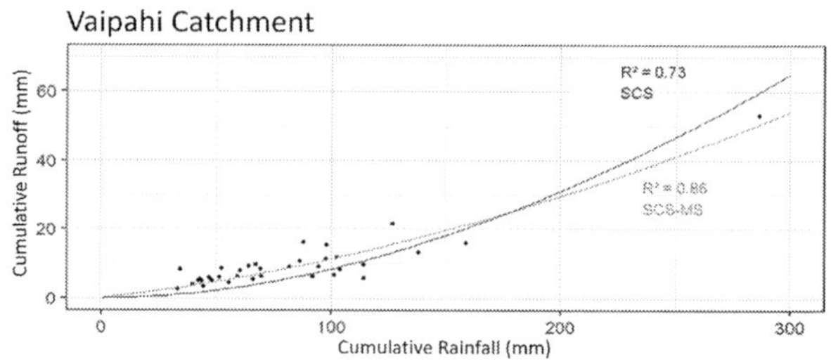

Figure 5. Comparative fits of the SCS and SCS-MS models on event cumulative rainfall and runoff observed in the Vaipahii catchment.

Parameters M and V_0 were simultaneously optimized for each event, using the SIMPLEX method (Nelder and Mead, 1965) and NSE as efficiency criterion. SIMPLEX explores the parameter space and converges to a pair of final parameters (M, V_0) either after 300 iterations or when the difference in NSE between two successive iterations is less than 10^{-9}.

5 Results

5.1 Improvement of the SCS-MS model compared to the SCS

Before calibrating the SCS-MS/LR model in dynamics, its balance equation (Eq. 9) is adjusted for each catchment on the observed event rainfall and runoff totals. The objective is to calibrate Ia parameter and to verify that the SCS-MS can adapt to the diversity of hydrological behaviours that can be observed in Tahiti. Compared to the SCS equation (Eq. 4), SCS-MS fits better the whole range of observed rainfall-runoff events, especially for catchments with the lowest observed runoff coefficient like Vaipahi. For these catchments, SCS has a tendency to under-estimate the low flow events when SCS-MS crosses the scatterplot at its centre (Fig. 5).

With Ia parameter greater than 0, balance equations of both models proved to be inadequate to fit observations, especially for catchments with observed runoff coefficients below 0.4. Ia parameter was then set to 0 for both models.

5.2 Parameter calibration and model efficiency

To measure the efficiency of SCS-MS/LR on each catchment, NSE was calculated for each event, and then the median of the NSE was calculated over each series of events.

SCS-MS/LR model showed little sensitivity to the parameters ds, ω and Si. It was therefore possible to find uniformed values of these parameters over the range of catch-

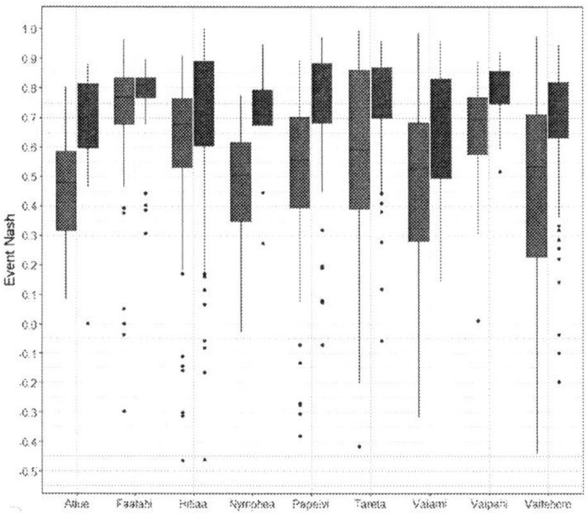

Figure 6. Distribution of Nash coefficients for each catchment, with M and Vo being optimized **(a)** and fixed to median values of optimized sets **(b)**.

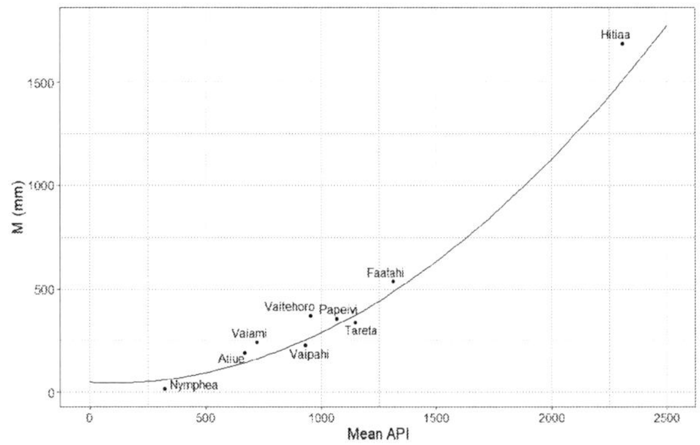

Figure 7. Adjustment of a relationship between the median M parameter calculated for the series of event of each catchment, and the mean API (Antecedent Precipitation Index), also calculated for each event series.

ments. ds and ω were set to 1 d^{-1} and 0.2 respectively, and Si could be set to 2500 mm for all basins except Nymphea. For Nymphea, the optimal value of Si is 80 mm, and setting the NSE to 2500 degrades the median of NSE criterion from 0.77 to 0.08. For the other catchments, setting Si at 2500 mm leads to negligible degradation of median NSE, which remains above 0.73 (Fig. 6).

Initialization relationships between Q_0 and optimized values of V_0 and M parameters were found not robust enough to be retained: R^2 coefficients of the adjusted relationships were found lower than 0.1 for 6 catchments out of 9 for M, and lower than 0.1 for all catchments for V_0. The model is therefore unable to reproduce the temporal variability of the hydrological response, but it remains relevant for characterize the median hydrological behavior of each catchment. For each catchment, M and Vo were therefore set to the medians of the optimised sets. Median M varies between 8 and 1710 mm depending on the catchment, and Vo varies between 0.9 and 1.6 m s^{-1}. These two parameters are not correlated, which is reassuring with regard to the risk of equifinality.

Median NSE values ranged from 0.73 to 0.87 across the catchments when M and V_0 optimized and from 0.39 to 0.79 with M and V_0 fixed to the median of optimizes values (Fig. 6).

Setting Vo and M to a single value for each basin lowers the NSE criteria because the model no longer takes into account the temporal variability of the hydrological response. The poor performance of the relationships between these two parameters and the AMC proxy can be explained by the existence of spatial uncertainties in rainfall that interfere with the optimisation of M and Vo. Rainfall patterns in Tahiti are highly variable in space (Benoit et al., 2022) and a high level of uncertainties is associated with values of spatial rainfall averages. These uncertainties are reflected in the hydrological model on the optimization of the initial condition until blurring the relationship between the initial condition and a proxy for AMC (Tanguy et al., 2023).

5.3 Towards the regionalization of the model parameters

After calibrating M and Vo separately for each catchment, their spatial variability, attempts were made to explain their spatial variability. The median values of M were linked to a proxy of AMC in order to explain its spatial variability across catchments.

Relationship between median M and median Q_0 was not relevant, but a clear relationship was identified between the median M and the mean Antecedent Precipitation Index (API) calculated separately for each catchment (Fig. 7). API was formulated by Fedora and Beschta (1989) as follows:

$$\mathrm{API_t} = (\mathrm{API}_{t-\Delta_t}\mathrm{K}) + P_{\Delta t} \quad (16)$$

K is a recession constant that we fixed to 0.85. We calculated API at a daily time step for the rainfall time series of each catchment, before extracting the daily API corresponding to the day before the beginning of each rainfall event.

A hypothesis to explain the low Si value in the Nymphea catchment lies in the difference of land use. The land use type of each catchment was estimated from satellite images of the island (Google maps). All catchments are mainly natural with peri urban areas smaller than 20 % of the total catchment surface, except Nymphea which is the only catchment to be entirely urbanized. Reducing the size Si of the soil reservoir for urban basins is coherent since artificial surfaces are mostly impermeable. Surface runoff is favoured compared to natural basins where an increased production capacity allows more infiltration, storage and subsurface runoff.

6 Conclusions

The modified version of SCS-MS/LR was calibrated and tested on series of observed rainfall-runoff events from 9 Tahitian catchments. V_0 and M could not be initialized from AMC proxy, and had to be fixed for each catchment. The model no longer restitute the temporal variability of rainfall-runoff response, but is still relevant to provide the average behavior of each catchment. Evaluated by the median values of NSE calculated for each catchment, the model efficiency is affected by the approximation made on V_0 and M but remains above 0.39. This model fulfills the flexibility required to restitute the hydrological spatial variability in Tahiti and its small number of parameters makes it a good candidate for regionalization.

About the regionalization of the model parameters, Si can be set according to the land use type. The median M is significantly related to the average API, which is a proxy for the average moisture state of each catchment. This study presents soil moisture as a key variable to regionalize rainfall-runoff models. However, the regionalization is not completed because V_0 spatial variations still remain unexplained, and because the model has not been validated.

This model and the relationship found between M and mean API are promising for event rainfall-runoff modelling in tropical volcanic context. However, it could be applied to other catchments in similar context in order to consolidate the calibration of its parameters and to clarify the regionalization relationships. The impact of land use type on the Si parameter could especially be specified.

Code availability. Data was treated using VISHYR module of ATHYS environment (http://www.athys-soft.org/, last access: 15 November 2022; HSM, 2022). The hydrological model was implemented with the MERCEDES module of ATHYS environment (http://www.athys-soft.org/, last access: 15 November 2022; HSM, 2022). Figures were made using Matplotlib version 3.3.3 (Caswell et al., 2020) available under the Matplotlib license at https://matplotlib.org/ (last access: 15 November 2022; Matplotlib, 2022), and ggplot2 (Wickham, 2016) at https://ggplot2.tidyverse.org (last access: 15 November 2022).

Data availability. Hydrometeorological data is available upon request from Groupement d'Etudes et de Gestion du Domaine Public de Polynésie Française (secretariat@equipement.gov.pf) and Météo France (contact.polynesie-francaise@meteo.fr).

Author contributions. LS was the principal investigator of the project. CB, GT and LS conceived together the framework of the study. GT analyzed the data, performed the numerical simulations and prepared the graphical outputs. CB contributed to the critical interpretations of the results, sharing his deep knowledge on hydrological modelling. GT wrote the manuscript in consultation with CB and LS.

Competing interests. The authors declare that they have no conflict of interest.

Disclaimer. Publisher's note: Copernicus Publications remains neutral with regard to jurisdictional claims in published maps and institutional affiliations.

Special issue statement. This article is part of the special issue "IAHS2022 – Hydrological sciences in the Anthropocene: Variability and change across space, time, extremes, and interfaces". It is a result of the XIth Scientific Assembly of the International Association of Hydrological Sciences (IAHS 2022), Montpellier, France, 29 May–3 June 2022.

Acknowledgements. This project is also sponsored by a local private company named Société Polynésienne de l'Eau, de l'Electricité et de Déchets (SPEED). The authors are grateful to the French Weather Agency (Direction Interrégionale en Polynésie française – Météo France) and to the Polynesian public service named Direction de l'Equipement (Groupement d'Etudes et de Gestion du Domaine Public de Polynésie Française – GEGDP) in particular for providing the rainfall and water level gauge dataset of the Island of Tahiti. The authors are very grateful to the Hydropower company, Engie-EDT-Marama Nui, for logistic support during fieldwork on the Hitiaa catchment.

Financial support. This work is supported by the Government of French Polynesia – Ministère de la Recherche through the project E-CRQEST (grant no. 05832 MED 2019-08-26). This study is also linked with the initiated project ERHYTM, Contrat de Projets Etat/Polynésie française (grant no. 7938/MSR/REC 2015-12-05), focusing on flood hazards over Tahiti and Moorea.

Review statement. This paper was edited by Christophe Cudennec and reviewed by two anonymous referees.

References

Benoit, L., Sichoix, L., Nugent, A. D., Lucas, M. P., and Giambelluca, T. W.: Stochastic daily rainfall generation on tropical islands with complex topography, Hydrol. Earth Syst. Sci., 26, 2113–2129, https://doi.org/10.5194/hess-26-2113-2022, 2022.

Caswell, T. A., Droettboom, M., Lee, A., Hunter, J., Firing, E., Stansby, D., Klymak, J., Hoffmann, T., Sales de Andrade, E., Varoquaux, N., Hedegaard Nielsen, J., Root, B., Elson, P., May, R., Dale, D., Lee, J.-J., Seppänen, J. K., McDougall, D., Straw, A., and Katins, J.: matplotlib/matplotlib v3.1.3 (v3.1.3), Zenodo [code], https://doi.org/10.5281/zenodo.3633844, 2020.

Coustau, M., Bouvier, C., Borrell-Estupina, V., and Jourde, H.: Flood modelling with a distributed event-based parsimonious rainfall-runoff model: case of the karstic Lez river catchment, Nat. Hazards Earth Syst. Sci., 12, 1119–1133, https://doi.org/10.5194/nhess-12-1119-2012, 2012.

Fedora, M. A. and Beschta, R. L.: Storm runoff simulation using an antecedent precipitation index (API) model, J. Hydrol., 112, 121–133, 1989.

HSM: ATelier HYdrologique Spatialisé, HSM [data set], http://www.athys-soft.org/, last access: 15 November 2022.

Lamaury, Y., Jessin, J., Heinzlef, C., and Serre, D.: Operationalizing urban resilience to floods in Island Territories – Application in Punaauia, French Polynesia, Water, 13, 337, https://doi.org/10.3390/w13030337, 2021.

Maidment, D. R.: Handbook of Hydrology, edited by: David, R., Maidment, ISBN 0-07-039732-5, McGraw-Hill, 1424 pp., 1993.

Matplotlib: Matplotlib: Visualization with Python, Matplotlib [data set], https://matplotlib.org/, last access: 15 November 2022.

McBride, M.: New books–Handbook of Hydrology, edited by: Maidment, D. R., Ground Water, 32, 331, 1994.

Mishra, S. K., Singh, V. P., Sansalone, J. J., and Aravamuthan, V.: A modified SCS-CN method: characterization and testing. Water Resources Management, 17, 37-68, 2003.

Nash, J. E. and Sutcliffe, J. V.: River flow forecasting through conceptual models part I – A discussion of principles, J. Hydrol., 10, 282–290, 1970.

Nelder, J. A. and Mead, R.: A simplex method for function minimization, The Comput. J., 7, 308–313, 1965.

Nguyen, S. and Bouvier, C.: Flood modelling using the distributed event-based SCS-LR model in the Mediterranean Réal Colobrier catchment, Hydrol. Sci. J., 64, 1351–1369, 2019.

Tanguy, G., Sichoix, L., Bouvier, C., and Leblois, E.: Sensitivity of an event-based rainfall-runoff model to spatial uncertainty in precipitation : a case study of the Hitiaa catchment, Island of Tahiti, submitted, 2023.

Pheulpin, L., Recking, A., Sichoix, L., and Barriot, J. P.: Extreme floods regionalisation in the tropical island of Tahiti, French Polynesiam, in: E3S Web of Conferencesm Vol. 7, p. 01014, EDP Sciences, https://doi.org/10.1051/e3sconf/20160701014, 2016.

Wickham, H.: ggplot2: Elegant Graphics for Data Analysis, Springer-Verlag New York, ISBN 978-3-319-24277-4, https://ggplot2.tidyverse.org, ggplot2 [code], 2016.

Wotling, G.: Caractérisation et modélisation de l'aléa hydrologique à Tahiti, Montpellier (FRA), Montpellier, USTL, IRD, 355 p., multigr. (Mémoires Géosciences-Montpellier, 18), Th. Géosciences, Mécanique, Génie Mécanique, Génie Civil, USTL, Montpellier, 20 December 1999, 2000.

Proc. IAHS, 385, 39–45, 2024
https://doi.org/10.5194/piahs-385-39-2024

Open Access PIAHS

Using of hydrological model and geospatial tool to assess climate change impact on the hydropower potential of the White Bandama watershed in Côte d'Ivoire (West Africa)

Koffi Claude Alain Kouadio[1,2], Siélé Silué[1], Ernest Amoussou[3], Kouakou Lazare Kouassi[4], Arona Diedhiou[5,2], Talnan Jean Honoré Coulibaly[6], Salomon Obahoundjé[2], Sacré Regis Didi[4], and Houebagnon Saint Jean Coulibaly[6]

[1]Faculty of Biological Sciences, University of Peleforo Gon Coulibaly, Korhogo, Côte d'Ivoire
[2]African Centre of Excellence on Climate Change, Biodiversity and Sustainable Agriculture, Abidjan, Côte d'Ivoire
[3]Department of Geography and Territory Management, University of Parakou, Parakou, Benin
[4]Laboratory of Environmental Sciences and Technologies, University of Jean Lorougnon Guédé, Daloa, Côte d'Ivoire
[5]Laboratory of Transfer Studies in Hydrology and Environment, University of Grenoble Alpes, Grenoble, France
[6]Laboratory of Geosciences and Environment, University of Nangui Abrogoua, Abidjan, Côte d'Ivoire

Correspondence: Koffi Claude Alain Kouadio (kclaudealain@gmail.com)

Received: 26 May 2022 – Revised: 18 January 2023 – Accepted: 22 January 2023 – Published: 18 April 2024

Abstract. This study was carried out in the White Bandama watershed (WBW) in Côte d'Ivoire (West Africa). The objective is to assess the impacts of future climate change (CC) on the hydropower potential (HPP) of the WBW. The methodology is based on coupling the SWAT (Soil and Water Assessment Tool) hydrological model with the Geographic Information System (GIS) QGIS to assess HPP on streams and evaluate the impacts of future CC on HPP of the watershed. Historical and climate projection data (precipitation, minimum and maximum temperature) for a set of three Regional Climate Models (RCM) from CORDEX-AFRICA (CCCma-CanRCM4, CCLM4-8-17 and REMO 2009) under RCP 4.5 were used. The biases of the ensemble mean were corrected by the Delta-change method. The relative change of streamflow discharge and HPP was assessed as the relative difference between the projection periods (2041–2070 and 2071–2100) and the reference period (1976–2005). The results showed a total of 22 future hydropower potential sites in the watershed. These sites were identified, geolocated and classified according to their potential capacity of generation in 82 % as small (1–25 MW potential capacity), 9 % as medium (25–100 MW potential capacity) and 9 % as large (more than 100 MW potential capacity) hydropower. The climate models' ensemble projected an upward trend for both the annual mean discharge of rivers and HPP of the WBW according to RCP 4.5 for the periods 2041–2070 and 2071–2100. On the annual cycle, the months of August and September will record the highest monthly mean flows between 150 and 200 $m^3\,s^{-1}$ while the months from November to April will record low monthly mean flows in the WBW.

Keywords. UPH 19; SDG 7; hydropower potential modelling; geospatial analysis; climate change impact; Côte d'Ivoire

1 Introduction

Hydroelectricity (HE) represents the largest source of renewable electricity in the world with a global production of more than 4250 TWh of electricity (IHA, 2022). In West Africa, HE contributes about 40 % of total electricity production (Barbier et al., 2009). In Côte d'Ivoire, HE is the second source of energy production after thermal energy and can contribute to the production of clean electricity with low Greenhouse Gas (GHG) emissions (Kouadio et al., 2020; IEA, 2020). HE could be an essential energy of the future to mitigate climate change (CC). However, there is a duality relationship between hydroelectricity and climate (Kouadio et al., 2022). Although contributing to CC mitigation through its low GHG emission, hydroelectricity can be affected by climate due to changes in the hydrological cycle (Berga, 2016). HE is dependent on climatic parameters (precipitation, temperature, insolation, potential evapotranspiration [PET], etc.) (De Souza Dias et al., 2018). These parameters influence the availability of water resource, the hydropower potential (HPP) and therefore the HE. Previous studies by Amoussou et al. (2012), Kouame et al. (2019) and Kouadio et al. (2020, 2022) have shown that CC can influence the intensity and duration of precipitation as well as river flows. It can also vary the PET. Furthermore, rainfall deficits cause runoff deficits (Savane et al., 2002) and impacts on HE (Kouadio et al., 2020). In the literature, numerous scientific works, in particular authors such as Hamududu and Killingtveit (2012), Grijsen (2014) and, Turner et al. (2017) focused, on the one hand, on the potential impacts of CC on water resources/on hydrology and, on the other hand, on hydropower. All these studies have endeavored to demonstrate that CC has impacts on both water resources and hydroelectric production. Analyzing the impacts of CC on hydrology and HPP in different watersheds in Africa, Yira et al. (2021) and Nonki et al. (2021) proposed a methodology based on the use of a hydrological model and climate model outputs. Thus, this study is a contribution to the establishment of a methodology using hydrological modelling coupled with geospatial analysis to assess the impacts of CC on the HPP in the White Bandama watershed (WBW) in Côte d'Ivoire. The purpose of this paper is to adopt a hydrological model combined with a Geographic Information System (GIS) to assess the potential impacts of future CC on the HPP of the White Bandama River watershed. This article is structured as follows: after the Introduction, which presents the background and objectives of this study, the Materials and Methods section describes the materials and methods used, the main results and discussion are presented in the Results and Discussion section. This manuscript ends with the Conclusion section.

2 Material and methods

2.1 Study area

This study was conducted in the WBW with an area of approximately 32 400 km^2 (Côte d'Ivoire) (Fig. 1). The WBW extends from the center to the north of Côte d'Ivoire. Its geographic coordinates in the WGS 84 reference vary between 5° and 6°30′ in longitude and between 6°30′ and 10°30′ in latitude. The White Bandama River has its source in the northern part of Côte d'Ivoire between Boundiali and Korhogo, at an altitude of 480 m. The WBW is under the influence of two climates: the Baoulean climate or equatorial climate or attenuated transition with annual precipitations varying on average between 1000 and 1600 mm yr^{-1} and the sudanian climate or tropical climate of transition with annual precipitations of unimodal distribution varying between 1000 and 1200 mm yr^{-1} (Goula et al., 2007). The population living on the WBW is estimated $\sim$ 3 000 000 inhabitants (Ivorian Institute of Statistics, 2014). In 1972, one of the major socio-economic assets of Côte d'Ivoire, the Kossou hydroelectric dam was built in this basin, with an installed capacity of 174 MW.

2.2 Description, data and configuration of the SWAT model

In this study, hydrological modelling was carried out with the SWAT (Soil and Water Assessment Tool) model (https://swat.tamu.edu/software/, last access: 17 May 2022) coupled with the QGIS interface (https://download.qgis.org, last access: 17 May 2022). The choice of this model is justified by a wide field of applications and use in several relatively recent studies. In addition, coupling facilitates access to variables and parameters. It offers a set of GIS tools for the development, execution and editing of hydrological inputs, and promotes the management of raster, vector and alphanumeric data. SWAT is an agro-hydrological model developed by researchers at the Agricultural Research Service (ARS) of the United States Department of Agriculture (USDA) in the 1990s. It is a physically based, spatially distributed, basin-scale hydrological model developed to simulate the impact of land use and management practices on water quantity and quality (surface water and groundwater). A detailed description of the model can be found in the work of Neitsch et al. (2011) and Arnold et al. (2012). The SWAT model simulates the water cycle through the hydrological Eq. (1):

$$SW_t = SW_0 + \sum_{i=1}^{t} \left(R_i - Q_{si} - E_i - W_i - Q_{qwi}\right) \tag{1}$$

With: SW_t: final soil water content (mm), SW_0: initial soil water content on day (mm), t: time (day), R_i: precipitation of day i (mm), Q_{si}: surface runoff on day i (mm), E_i: potential evapotranspiration on day i (mm), W_i: percolation or

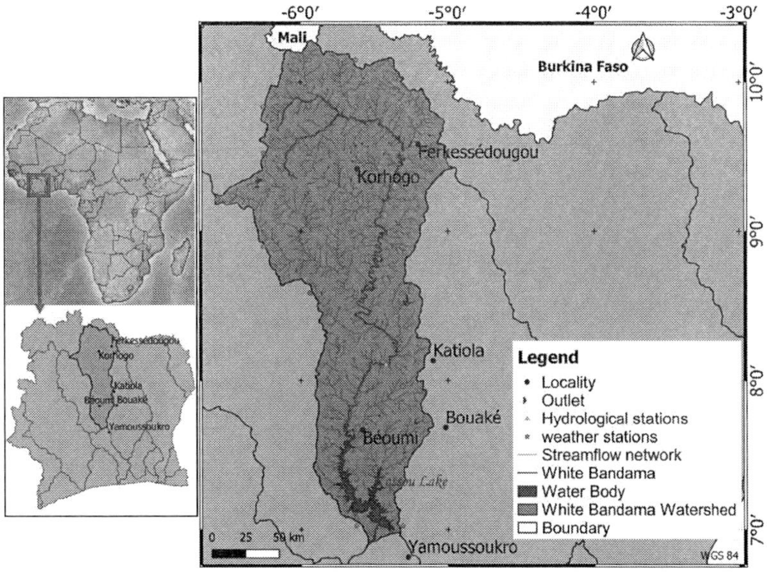

Figure 1. Location of the White Bandama Watershed (WBW).

infiltration of day i (mm), $Q_{\mathrm{qw}i}$: low water flow or quantity of water returning to the soil of day i (mm).

Several data were needed to configure the model and simulate the flows. The satellite images were obtained by the NASA (National Aeronautics and Space Administration) and NGA (National Geospatial-Intelligence Agency) radar-surveying mission in 2000. These satellite images are 30 m spatial resolution and are available on the United States Geological Survey (USGS) website https://earthexplorer.usgs.gov (last access: 10 December 2021). In addition, daily reanalysis data of precipitation, temperature (minimum and maximum), relative humidity, solar radiation, wind speed available for a period of 34 years (1980 to 2013) with a resolution of 38 km were used. They are available at https://globalweather.tamu.edu/ (last access: 10 December 2021; Saha et al., 2014). These reanalysis data collected at a daily time step were aggregated in monthly time step before the simulation of the SWAT model at the monthly time step. The daily hydrometric data observed were collected from the Ivorian direction of human hydraulics (DHH) for the stations of Marabadiassa (1980 to 2013) and Badikaha (1982 to 1989). The model setup also required soil and land cover data in raster form. For this purpose, a soil map of the WBW with a spatial resolution of 1 km provided by the Food and Agriculture Organization of the United Nations (FAO) was used. The WBW land cover map from 2006, with a spatial resolution of 300 m, was obtained from the ESA website (http://due.esrin.esa.int/page_globcover.php, last access: 10 December 2021; Leroy et al., 2006). PET was computed using the Penman-Monteith equation. The variable storage method developed by Williams (1969) was chosen for the discharge simulation. Runoff was computed using the Curve Number (CN) method. The simulation period was defined from 1980 to 2013 (34 years) with a warming period of 2 years (1980–1981). The configured and calibrated model was used to assess the potential impacts of CC. The model was calibrated from 1982 to 1985 (4 years), and validated from 1986 to 1989 (4 years) at a monthly time step by the coefficient of Nash-Sutcliffe (NS) and the coefficient of determination (R^2).

2.3 Assessment of future climate impacts on hydropower potential

For the assessment of CC impacts, the outputs (precipitation, minimum and maximum temperature) of three CORDEX-AFRICA climate models (0.44° resolution) were used: CCCma-CanRCM4, CCLM4-8-17 and REMO 2009. These data were obtained for the RCP 4.5 scenario for the reference period P0 (1976–2005) and the projection periods P1 and P2 respectively (2041–2070) and (2071–2100). In this study, the methodological approach for evaluating the impacts of CC on the HPP is based on a process similar to that used by De Oliveira et al. (2017). Bias correction of the dataset mean of the three climate models was performed by the Delta-Change method (Teutschbein and Seibert, 2012). The HPP was assessed over the reference periods P0 and projection periods P1 and P2 of the RCP 4.5 by Eq. (2) of Maher and Smith (2001).

$$\mathrm{HPP} = \rho g Q \cdot h \tag{2}$$

With: HPP: hydropower potential (W), ρ: density of water ($1000\,\mathrm{kg\,m^{-3}}$), g: acceleration due to gravity ($\mathrm{m\,s^{-2}}$):

$9.81\,\mathrm{m\,s^{-2}}$, Q: flow in ($\mathrm{m^3\,s^{-1}}$) on the section of the network, h: vertical drop height in meters (m).

The relative change (Δ) characterized by the fluctuations makes it possible to assess the potential CC impacts over the projection periods (2041–2070 and 2071–2100) by Eq. (3).

$$\Delta = \left(\frac{\overline{X}_{\mathrm{proj}} - \overline{X}_{\mathrm{ref}}}{\overline{X}_{\mathrm{ref}}}\right) \times 100 \quad (3)$$

With: Δ: Relative change or fluctuation, $\overline{X}_{\mathrm{proj}}$: Average value of the variable calculated on the projection period $\overline{X}_{\mathrm{ref}}$: Average value of the variable calculated on the reference period.

3 Results and discussion

3.1 Model calibration and validation

The calibration and validation of the SWAT model on the WBW are shown in Fig. 2a and b. The results are illustrated here for the Badikaha hydrological station. The Nash-Sutcliffe (NS) and determination (R^2) coefficients obtained (NS and $R^2 > 0.6$) indicate that there is a very good correlation between the simulated and observed flows. The whole correlation over the calibration period (NS = 0.83 and $R^2 = 0.839$) is much better than that of the validation period (NS = 0.70 and $R^2 = 0.772$). Furthermore, the model underestimated flood peaks. It would be a limit of the SWAT model.

3.2 Evaluation of the HPP in the WBW

The calibrated and validated model was used for the evaluation of the HPP in the basin. Figure 3 presents potentially exploitable hydroelectric sites and their geospatial distribution in the basin. Twenty-two sites have been identified, geolocated and classified according to their potential production capacity at 82 % small (potential capacity 1–25 MW), 9 % medium (potential capacity 25–100 MW) and 9 % large (more of 100 MW) hydropower (Table 1). The results show that the WBW has a good potential in small HE. This would be a major asset in the decentralization of electrification in this basin.

Table 1. Distribution of HE sites by type.

SW	No. of sites	Generation capacity per HE class (MW)	HE type	Distribution
60	2	[1–25[	Small HE	
101	3	[1–25[	Small HE	
118	3	[1–25[	Small HE	18 sites or 82 %
119	2	[1–25[	Small HE	
123	6	[1–25[	Small HE	
134	2	[1–25[	Small HE	
120	2	[25–100[	Medium HE	2 sites or 9 %
171	2	> 100	Large HE	2 sites or 9 %
	22			

3.3 Impacts of future CC on the HPP

The analysis of the impacts of future CC was carried out on the HPP of the 22 sites spread over eight Sub-Watersheds (SW 60, 101, 118, 119, 120, 123, 134 and 171). The relative changes in monthly average flow and HPP for SWs 101 and 171 are shown in Figs. 4 and 5. Compared to the reference period P0, all the climate models project a general upward trend in flows on the eight SWs for the periods P1 and P2 according to the RCP 4.5 scenario. On an annual cycle, flows will increase from July to October. This increase in flows over this period is estimated at a rate of ∼ 35 % and 53 % respectively on P1 and P2 for SW 101, and ∼ 54 % and 65 % respectively on P1 and P2 for SW 171. Like the discharges, the models project an increase in HPP on all SWs compared to the reference period P0. Furthermore, the results show that SW 171 will register the largest HPP with a monthly average during the months of August and September oscillating between 400 and 500 MW according to the RCP 4.5 scenario. The SW101 will record the lowest HPP with a monthly average maximum of less than 18 MW during these same months.

4 Discussion

The analysis of CC impacts on the HPP used outputs from a set of three CORDEX- Africa climate models (ensemble mean). According to Pandey et al. (2019), the use of a set of models would contribute to the reduction of uncertainties in climate projections. The results of the climate impact analysis project a general upward trend in monthly mean streamflow and HPP on the WBW according to RCP 4.5. This increase in flows on P1 and P2 could be explained, on the one hand, by a resumption of rainfall and, on the other hand, by a degradation of the vegetation cover in the medium and long term. This situation would allow strong flow to the detriment of infiltration, as Amoussou et al. (2012). The degradation or reduction of the vegetation cover would favor strong surface runoff to the detriment of infiltration. In addition, the work of Kling et al. (2016), based on the reference (1998–2014) and future (2046–2065) periods, estimate that an increase in precipitation in West Africa would be of the order of 1 % to 6.8 %. Moreover, according to the fifth report of the GIEC (2014), an increase in precipitation until 2100 is projected according to the scenario considered. However, these projections remain marked by many uncertainties. For this reason, in this study it would be interesting to consider the general trend marked by a future increase in flows and HPP over the basin for P1 and P2 compared to the reference P0.

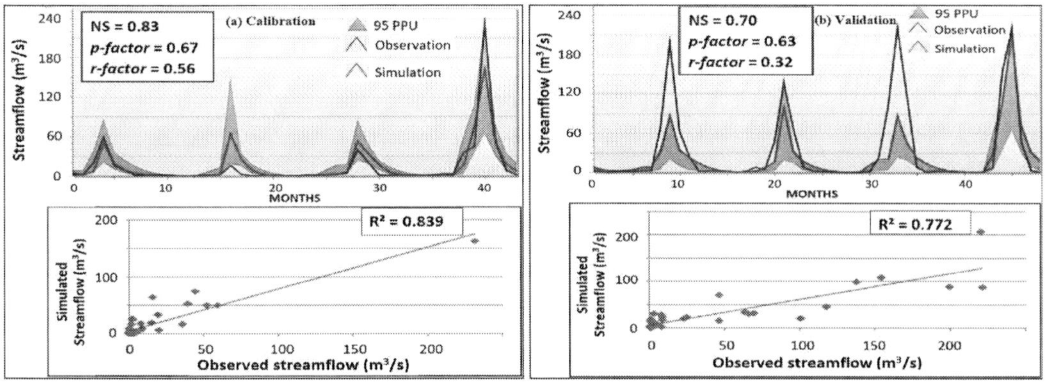

Figure 2. **(a)** Calibration of the SWAT model of the WBW from 1982 to 1985 (Badikaha station) and **(b)** Validation of the WBW SWAT model from 1986 to 1989 (Badikaha station). * p-factor: percentage of observed data framed by 95 % prediction uncertainty (95PPU). * r-factor: thickness of the 95PPU strip.

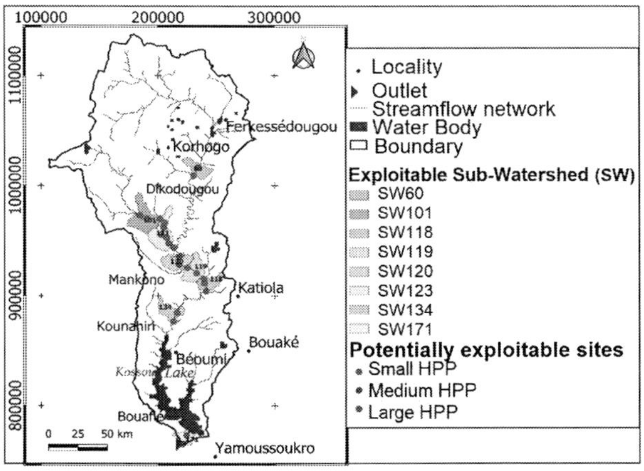

Figure 3. Geospatial distribution of the 22 hydroelectric sites in the WBW.

In the work of Kouadio et al. (2020), hydrological dynamics was analyzed under CC in the WBW. This study has shown that the Kossou dam plays a role in reducing floods in the watershed. Moreover, the analysis of monthly flows has demonstrated that the flood flow incoming to the Kossou Lake was skimmed ~ 93.7 %

5 Conclusion

The objective of this study was to assess the potential impacts of CC on the HPP of the WBW in Côte d'Ivoire. For this purpose, the SWAT hydrological model combined with geospatial analysis was applied. The potential impacts of CC on the monthly flow and HPP of the basin were analyzed over the projection periods P1 and P2 compared to the reference period P0, under the RCP 4.5 scenario. The analysis of these impacts also focused on eight (8) SWs with twenty-two (22) HE sites identified (SW 60, 101, 118, 119, 120, 123, 134 and 171). In comparison with the P0 reference period, a general upward trend in flows and HPP on the WBW was observed. On the annual cycle, the models have projected that the months of August and September will register the highest flows between 150 and 200 $m^3 s^{-1}$ with a high HPP while the months of November through April will register the lowest flows with a low HPP. In addition, an increase in flows and HPP on the 8 SWs for periods P1 and P2 compared to P0 is projected. The models also showed that the SW 171 would have the largest HPP with a monthly average during the months of August and September oscillating between 400 and 500 MW. In addition, SW 101 will record the lowest HPP with an average monthly maximum of less than 18 MW during these same months. This general increase in flows could however increase the problems of siltation, eutrophication and the risk of flooding in certain areas and for certain activities in this basin. To this end, additional studies should be carried out to better assess these consequences on the basin.

Data availability. The data used in this study is publicly available. The reanalysis data are freely available at https://globalweather.tamu.edu/ (SWAT, 2021), and the satellite data are available on the USGS website (https://earthexplorer.usgs.gov, USGS, 2021) and the ESA website (http://due.esrin.esa.int/page_globcover.php, ESA, 2021). However, the observed data comes from national service databases available upon request.

Author contributions. Conceptualization: KCAK, SS, EA, and AD; Data analysis: KCAK, EA, and AD; Methodology: KCAK, AD, and TJHC; Validation, KCAK, and AD; Drafting of preliminary manuscript: KCAK, SS, and AD; Revision and writing of the final manuscript: KCAK, SS, EA, KLK, AD, TJHC, HSJC, SRD,

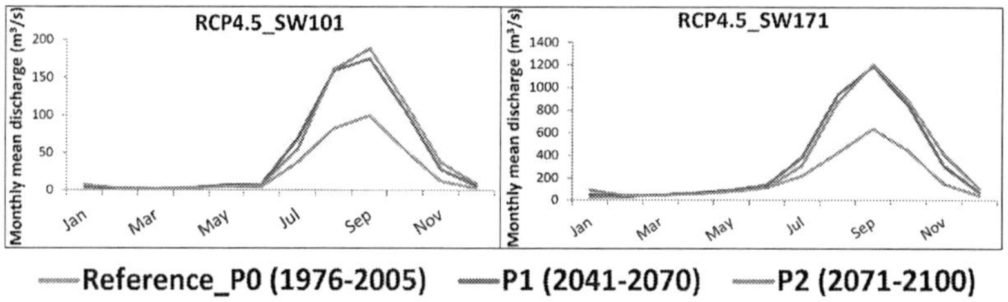

Figure 4. Projected monthly mean discharges of SW101 and 171 under RCP 4.5 of the ensemble mean of the three RCM.

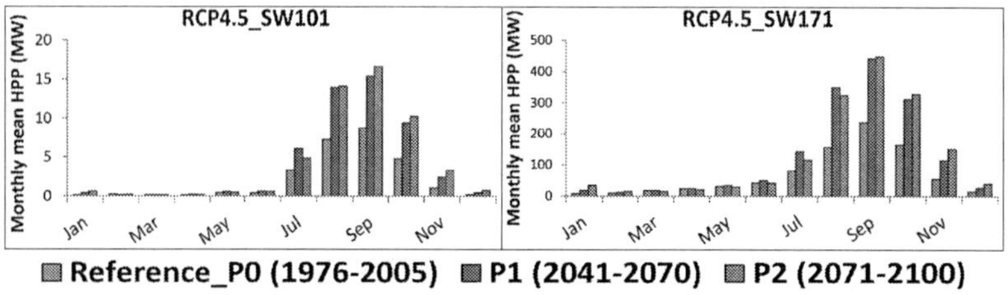

Figure 5. Projected monthly mean HPP of SW 101 and 171 under RCP 4.5 of the ensemble mean of the three RCM.

and SO. The final version of the article was approved by all authors: KCAK, SS, EA, KLK, AD, TJHC, SO, SRD, and HSJC.

Competing interests. The contact author has declared that none of the authors has any competing interests.

Disclaimer. Publisher's note: Copernicus Publications remains neutral with regard to jurisdictional claims in published maps and institutional affiliations.

Special issue statement. This article is part of the special issue "IAHS2022 – Hydrological sciences in the Anthropocene: Variability and change across space, time, extremes, and interfaces". It is a result of the XIth Scientific Assembly of the International Association of Hydrological Sciences (IAHS 2022), Montpellier, France, 29 May–3 June 2022.

Acknowledgements. The authors would like to thank the organizing committee of the XIth Scientific Assembly of the International Association of Hydrological Sciences in Montpellier 2022. We also thank IAHS through the SYSTA grant for funding the first author to attend this conference and for giving the opportunity to publish this paper. Thanks also to Jean-Marie Kileshye-Onema and the anonymous reviewer who contributed to improving the quality of this paper.

Review statement. This paper was edited by Christophe Cudennec and reviewed by Jean-Marie Kileshye-Onema and one anonymous referee.

References

Amoussou, E., Camberlin, P., and Mahé, G.: Impact de la variabilité climatique et du barrage Nangbéto sur l'hydrologie du système Mono-Couffo (Afrique de l'Ouest), Hydrolog. Sci. J., 57, 805–817, https://doi.org/10.1080/02626667.2011.643799, 2012.

Arnold, J. G., Moriasi, D. N., Gassman, P. W., Abbaspour, K. C., White, M. J., Srinivasan, R., Santhi, C., Harmel, R. D., Van Griensven, A., Van Liew, M. W., Kannan, N., and Jha, M. K.: SWAT: Model use, calibration, and validation, T. ASABE, 55, 1491–1508, 2012.

Barbier, B., Yacouba, H., Maïga, A. H., Mahé, G., and Paturel, J.-E.: Le retour des grands investissements hydrauliques en Afrique de l'Ouest: les perspectives et les enjeux, Geocarrefour, 84, 31–41, https://doi.org/10.4000/geocarrefour.7205, 2009.

Berga, L.: The Role of Hydropower in Climate Change Mitigation and Adaptation: A Review, Engineering, 2, 313–318, https://doi.org/10.1016/J.ENG.2016.03.004, 2016.

De Oliveira, V. A., de Mello, C. R., Viola, M. R., and Srinivasan, R.: Assessment of climate change impacts on streamflow and hydropower potential in the headwater region of the Grande river basin, Southeastern Brazil, Int. J. Climatol., 37, 5005–5023, https://doi.org/10.1002/joc.5138, 2017.

De Souza Dias, V., Pereira da Luz, M., Medero, G. M., and Tarley Ferreira Nascimento, D.: An Overview of Hydropower Reservoirs in Brazil: Current Situation, Future Perspectives and Impacts of Climate Change, Water, 10, 592, https://doi.org/10.3390/w10050592, 2018.

ESA (European Space Agency): GlobCover, http://due.esrin.esa.int/page_globcover.php, last access: 10 December 2021.

GIEC: AR5 Climate Change 2014: Impacts, Adaptation, and Vulnerability. Part B: Regional Aspects, Contribution of Working Group II to the Fifth Assessment Report of the Intergovernmental Panel on Climate Change, edited by: Barros, V. R., Field, C. B., Dokken, D. J., Mastrandrea, M. D., Mach, K. J., Bilir, T.

E., Chatterjee, M., Ebi, K. L., Estrada, Y. O., Genova, R. C., Girma, B., Kissel, E. S., Levy, A. N., MacCracken, S., Mastrandrea, P. R., and White, L. L., Cambridge University Press, Cambridge, United Kingdom and New York, NY, USA, 688 pp., https://www.ipcc.ch/report/ar5/syr/ (last access: 10 January 2022), 2014.

Goula, B. T. A., Konan, B., Brou, Y. T., Issiaka, S., Fadika, V., and Srohourou, B.: Estimation des pluies exceptionnelles journalières en zone tropicale: cas de la Côte d'Ivoire par comparaison des lois lognormale et de Gumbel, Hydrolog. Sci. J., 52, 49–67, https://doi.org/10.1623/hysj.52.1.49, 2007.

Grijsen, J.: Understanding the impact of climate change on hydropower: the case of Cameroon, World Bank Report No. 87913, Washington, DC, Africa Energy Practice, IBRD/IDA, https://documents1.worldbank.org/curated/en/243651468010867538/pdf/879130ESW0P1140Box385106B00PUBLIC0.pdf (last access: 3 May 2022), 2014.

Hamududu, B. and Killingtveit, A.: Assessing climate change impacts on global hydropower, Energies, 5, 305–322, https://doi.org/10.3390/en5020305, 2012.

IEA: Statistics report, Renewables Information: Overview, https://www.iea.org/reports/renewables-information-overview (last access: 19 May 2022), 2020.

IHA: Hydropower Status Report, Sector trends and insights, https://www.hydropower.org/status-report, last access: 5 October 2022.

Ivorian Institute of Statistics: Report of the population statistics from the general population and housing census in Côte d'Ivoire, https://www.ins.ci/documents/RGPH2014_expo_dg.pdf (last access: 3 November 2021), 2014.

Kling, H., Stanzel, P., and Fuchs, M.: Regional assessment of the hydropower potential of rivers in West Africa, Enrgy. Proced., 97, 286–293, https://doi.org/10.1016/J.EGYPRO.2016.10.002, 2016.

Kouadio, C. A., Kouassi, K. L., Diedhiou, A., Obahoundje, S., Amoussou, E., Kamagate, B., Paturel, J., Coulibaly, T. J. H., Coulibaly, H. S. J. P., Didi, R. S., and Savane, I.: Assessing the Hydropower Potential Using Hydrological Models and Geospatial Tools in the White Bandama Watershed (Côte d'Ivoire, West Africa), Front. Water, 4, 844934, https://doi.org/10.3389/frwa.2022.844934, 2022.

Kouadio, K. C. A., Amoussou, E., Coulibaly, T. J. H., Diedhiou, A., Coulibaly, H. S. J.-P., Didi, R., and Savane, I.: Analysis of hydrological dynamics and hydropower generation in a West African anthropized watershed in a context of climate change, Modeling Earth Systems and Environment, 6, 2197–2214, https://doi.org/10.1007/s40808-020-00836-4, 2020.

Kouame, Y. M., Obahoundje, S., Diedhiou, A., François, B., Amoussou, E., Anquetin, S., Didi, R. S., Kouassi, L. K., N'guessan Bi, V. H., Soro, E. G., and Yao, E. K.: Climate, Land Use and Land Cover Changes in the Bandama Basin (Côte D'Ivoire, West Africa) and Incidences on Hydropower Production of the Kossou Dam, Land, 8, 103, https://doi.org/10.3390/land8070103, 2019.

Leroy, M., Bicheron, P., Brockmann, C., Krämer, U., Miras, B., Huc, M., Ninô, F., Defourny, P., Vancutsem, C., Petit, D., Amberg, V., Berthelt, B., Arino, O., and Ranera, F.: GlobCover: a 300 m global land cover product for 2005 using ENVISAT MERIS time series, In Proceedings of ISPRS Commision VII Mid-Term Symposium: Remote Sensing: from Pixels to Processes, Enschede (NL), 538–542, http://due.esrin.esa.int/page_globcover.php (last access: 10 December 2021), 2006.

Maher, P. and Smith, N.: Pico hydro for village power: A practical manual for schemes up to 5 kW in hilly areas, 2nd Edn., IntermediateTechnology Publications, 5, Nottingham, UK, https://energypedia.info/images/9/92/Ph_manual_s.pdf (last access: 26 February 2021), 2001.

Neitsch, S. L., Arnold, J. G., Kiniry, J. R., and Williams, J. R.: Soil and Water Assessment Tool Theoretical Documentation Version 2009, USDA-ARS GRASSLAND Publication, Texas Water Resources Institute Technical Report No. 406, 77843-2118, Texas, College Station, https://swat.tamu.edu/media/99192/swat2009-theory.pdf (last access: 26 February 2021), 2011.

Nonki, R. M., Lenouo, A., Tchawoua, C., Lennard, C. J., and Amoussou, E.: Impact of climate change on hydropower potential of the Lagdo dam, Benue River Basin, Northern Cameroon, Proc. IAHS, 384, 337–342, https://doi.org/10.5194/piahs-384-337-2021, 2021.

Pandey, B. K., Khare, D., Kawasaki, A., and Mishra, P. K.: Climate Change Impact Assessment on Blue and Green Water by Coupling of Representative CMIP5 Climate Models with Physical Based Hydrological Model, Water Resour. Manage., 33, 141–158, https://doi.org/10.1007/s11269-018-2093-3, 2019.

Saha, S., Moorthi, S., Wu, X., Wang, J., Nadiga, S., Tripp, P., Behringer, D., Hou, Y.-T., Chuang, H., Iredell, M., Ek, M., Meng, J., Yang, R., Mendez, M.P., van den Dool, H., Zhang, Q., Wang, W., Chen, M., and Becker, E.: The NCEP Climate Forecast System Version 2, J. Climate, 27, 2185–2208, https://doi.org/10.1175/JCLI-D-12-00823.1, 2014.

Savane, I., Coulibaly, K. M., and Gioan, P.: Variabilité climatique et ressources en eaux souterraines dans la région semi-montagneuse de Man, Science et changements planétaires/Sécheresse, 12, 231–237, 2002.

SWAT (Soil & Water Assessment Tool): CFSR Global Weather Data for SWAT 1979–2014, https://globalweather.tamu.edu/, last access: 10 December 2021.

Teutschbein, C. and Seibert, J.: Bias correction of regional climate model simulations for hydrological climate-change impact studies: Review and evaluation of different methods, J. Hydrology, 456–457, 12–29, https://doi.org/10.1016/j.jhydrol.2012.05.052, 2012.

Turner, S. W. D., Hejazi, M., Kim, S. H., Clarke, L., and Edmonds, J.: Climate impacts on hydropower and consequences for global electricity supply investment needs, Energy, 141, 2081–2090, https://doi.org/10.1016/j.energy.2017.11.089, 2017.

USGS (United States Geological Survey): EarthExplorer, https://earthexplorer.usgs.gov, last access: 10 December 2021.

Williams, J. R.: Flood routing with variable travel time or variable storage coefficients, T. ASAE, 12, 100–103, 1969.

Yira, Y., Mutsindikwa, T. C., Bossa, A. Y., Hounkpè, J., and Salack, S.: Assessing climate change impact on the hydropower potential of the Bamboi catchment (Black Volta, West Africa), Proc. IAHS, 384, 349–354, https://doi.org/10.5194/piahs-384-349-2021, 2021.

Proc. IAHS, 385, 47–52, 2024
https://doi.org/10.5194/piahs-385-47-2024

Open Access PIAHS

IAHS2022 – Past and future of open, inclusive, innovative, and society-interfacing approaches

Managed groundwater recharge at the farm scale in pre-Saharan Morocco

Yassine Khardi[1,2,3], **Guillaume Lacombe**[1,2], **Benoit Dewandel**[4], **Abdelilah Taky**[1], **Jean-Christophe Maréchal**[4], **Ali Hammani**[1], **and Sami Bouarfa**[5]

[1]Hassan II Agronomic and Veterinary Institute, 6202, Rabat, Morocco
[2]CIRAD, UMR G-EAU, 34398, Montpellier, France
[3]L'institut Agro Montpellier, 34060, Montpellier, France
[4]BRGM Occitanie region, UMR G-EAU, 34000, Montpellier, France
[5]INRAE Occitanie-Montpellier, UMR G-EAU, 34000, Montpellier, France

Correspondence: Yassine Khardi (ykhardi@gmail.com)

Received: 27 May 2022 – Revised: 19 January 2023 – Accepted: 22 January 2023 – Published: 18 April 2024

Abstract. The oases of the pre-Saharan basin of Wadi Ferkla in southeastern Morocco receives low and erratic rainfall (annual average of 141 mm and inter-annual standard deviation of 70 mm). From the 1980s, surface water and groundwater are increasingly used due to the expansion of irrigation, mainly along two wadis, namely Wadis Ferkla and Satt originating in the High-Atlas and the Anti-Atlas Mountains, respectively. Their flows reach the Ferkla's irrigated perimeters only when the volume of the flood events exceed upstream evaporation, withdrawals and riverbed's infiltration. Nowadays, these irrigated perimeters exert significant pressure on groundwater resources, through numerous drillings equipped with pumping systems, most of them being powered by solar energy. This increasing water demand situation incentivizes individual farmers to design and implement innovative techniques to increase water access for their farms. For instance, the spreading of floodwaters – an ancestral and collective irrigation practice in traditional oases – is currently being modernized by individual farmers. The new technique consists in partially diverting flood flows into earthen basins. The stored water either infiltrates to recharge local aquifers, or is pumped for flood irrigation of date palms. An experimental protocol was set up to characterize groundwater recharge below one of these on-farm basins equipped with a recharge well. Barometric probes were installed in the basin, in the recharge well and in neighboring boreholes to automatically monitor water levels. A topographic survey of the monitoring points and of the basin aimed at deriving piezometric levels from water levels measurements and estimating the height-surface-volume curves of the basin. After 7 months of continuous monitoring, 3 flood events were recorded. The establishment of the basin water balance at a fine time-resolution allowed estimating its different components including the infiltration rate influencing groundwater recharge. An analytical modeling of this process was developed to assess its effect on groundwater level variations. This approach aims to contribute to a broader reflection on securing water management in this fragile oasis ecosystem.

Keywords. Water level monitoring; field observations; groundwater recharge; modelling; oasis; pre-Saharan Morocco

1 Introduction

The pre-Saharan Morocco has experienced a significant expansion of pumped agriculture since the 1980s. The pressure on the already scarce groundwater resource has increased considerably. Nowadays, some farms have set up groundwater recharge structures (Todd and Mays, 2004) to increase the supply of irrigation water from wadi floods (Hashemi et al., 2013).

Several authors have evaluated the feasibility of farm-level groundwater recharge (Arshad et al., 2014; Godwin et al., 2022) and the performance of farm-level well recharge structures (Soni et al., 2020).

Worldwide, infiltration rates from groundwater recharge basins vary between 0.3 and $1\,\mathrm{m\,d^{-1}}$. The volumes infiltrated into these basins are between 600 and $14\,000\,\mathrm{m^3\,d^{-1}}$ (Wuilleumier and Seguin, 2008). The capacity of the injection wells fluctuates between 1700 and $6000\,\mathrm{m^3\,d^{-1}}$ (Casanova et al., 2013). It is important to assess the effect of infiltration on groundwater. The analytical solution developed by Hantush (1967) for assessing the growth and decay of the groundwater mound below an infiltration basin is used worldwide and very accurate compared to numerical modelling (Warner et al., 1989; Carleton, 2010). Recently, this solution has been adapted to the case where a pumping or recharge well is located near or inside the recharge area (Dewandel et al., 2021).

In this paper, we assess the infiltration rate of a farm-scale recharge basin from its water balance and we evaluate its impact on the local aquifer. The research site, located in Ferkla (Morocco), consists of a floodwater storage basin built around and above a well formerly used for pumping.

2 Material and methods

2.1 Study site

The farm studied (5°02′01.00″ W, 31°29′40.00″ N) is located in the arid region of Drâa-Tafilalet (Morocco). It is located on the left bank of Wadi Satt which originates in the Anti-Atlas. Its runoff reaches the farm during flood periods only. The average rainfall is around $141\,\mathrm{mm\,yr^{-1}}$ with a standard deviation of around 70 mm (Ait Bouijane station 1961–2018). This farm relies on the quaternary formations of Wadi Satt and pump water from both quaternary and underlying fractured schist-quartzite aquifers. Like all the farms in Ferkla, this farm is irrigated by pumping groundwater throughout the year. It has a basin that allows partial storage of flood water and local recharge of the water table. After filling the basin with a maximum storage area and depth of about $6400\,\mathrm{m^2}$ and 2.8 m, respectively, part of the stored water is pumped during the day using solar energy, and channeled into a pipe to irrigate the date palms by gravity. Some palm trees are planted inside the storage basin.

2.2 Material

In order to assess groundwater recharge from the flood basin and its well, four barometric probes recording surface and groundwater levels every 10 min (Fig. 1) were installed in the basin, in its well and in two boreholes located 30 and 360 m away from the basin. The depth of the well inside the basin does not exceed 12 m. The other boreholes are more than 80 m deep and capture water from the Ordovician quartzite and the shale aquifers. One probe placed in the open air aimed to remove the biases caused by the atmospheric pressure variations from the water level measurements. Measured water levels were converted to piezometric levels based on a topographic survey.

Periodic measurements with a manual piezometer aimed to validate automatic records. A topographic survey of the basin was carried out and processed in ArcGIS to derive the height-volume-surface curve of the basin. Daily precipitation and evapotranspiration data calculated by the Penman-Montheith method ($\mathrm{ET_0}$) were available from the nearest meteorological station (Goulmima, ORMVA-Tf).

2.3 Methods

2.3.1 Calculating the infiltration rate from the basin

Infiltration from the basin and its well were estimated from the simplified water balance equation of the basin Eq. (1) obtained during periods with no pumping, no inflow and no rainfall. During these periods visually identified on the graph depicting temporal variations of the basin water level, its water balance simplifies as follows:

$$\Delta h_{\mathrm{basin}} = -\left(q + E + \mathrm{ET_{palm}}\right) \tag{1}$$

where Δh is the variation of the basin water level (mm) over time, q, E and $\mathrm{ET_{palm}}$ are the infiltration (mm), the evaporation (mm) and the evapotranspiration of the palm trees located in the basin, respectively, over the same period of time.

These computation periods were selected at least 24 h after each flood event to ensure the absence of delayed inflow that can bias the infiltration rate calculations, and during the nighttime when there is no pumping from the basin. Infiltration rates were assessed for different water levels in the basin, over time periods of a few hours and exhibiting near-constant drawdown. E and $\mathrm{ET_{palm}}$ were calculated with Eqs. (2) and (3), respectively:

$$E = k \times E_{\mathrm{tank}} = k \times \left(\frac{\mathrm{ET_0}}{f}\right) \tag{2}$$

where k is the coefficient of the Colorado tank measuring E_{tank}, usually equal to 0.8, f is the Penman reduction factor of $\mathrm{ET_0}$ used to estimate E_{tank}, equal to 0.82 for semi-arid areas (Riou, 1971).

$$\mathrm{ET_{palm}} = k_{\mathrm{c}} \times \mathrm{ET_0} \tag{3}$$

where k_{c} is the crop coefficient of the date palm. It depends on the vegetative stage and varies between 0.8 and 1. In order not to under evaluate $\mathrm{ET_{palm}}$, k_{c} was fixed to 1.

The infiltration rates q calculated over the selected periods were plotted against their corresponding mean basin water level (h_{basin}). Spurious values likely caused by undetected

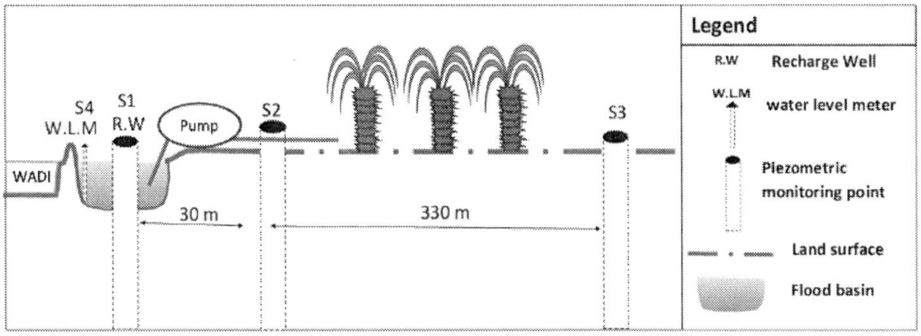

Figure 1. Cross section of the study site and location of the probes S1, S2, S3 and S4.

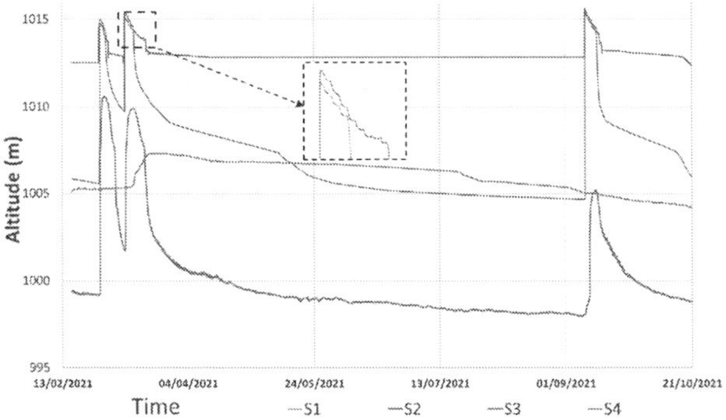

Figure 2. Water level in the basin (S4) and in the piezometric monitoring points (S1, S2, S3).

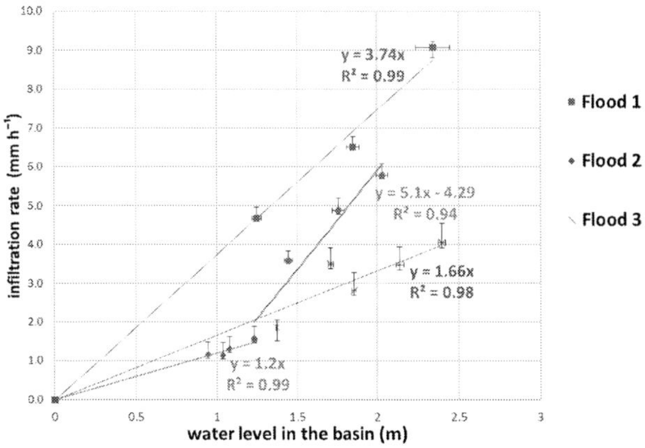

Figure 3. Infiltration rates as a function of water depth in the basin.

pumping or probe dysfunctions were removed from the sample (3 values among 17). Linear regressions with R^2 exceeding 0.94 were fitted to the scatter plots of each recorded flood using Eq. (4):

$$q = a \times h_{\text{basin}} + b \tag{4}$$

Different values for the parameters a and b were obtained for each flood event because of the change in the geometry and permeability of the basin due to siltation. On one of the recorded floods, the infiltration rates were calculated for shallow depths. Thus, two infiltration equations were calculated for this flood, depending on the water level, either lower or greater than 1.4 m.

Infiltration rates of the well located in the basin were computed for water levels lower than those ensuring hydraulic connection with the basin. These infiltration rates were calculated from the well discharge curve. The calculation intervals selected from this curve are the same for all recorded flood events. This velocity is multiplied by the cross-sectional area of the well to derive the seepage rate from the well.

2.3.2 Analytical modelling of groundwater recharge

In order to model the response of the water table below the studied recharged basin, the analytical model developed by Dewandel et al. (2021) which is based on the equations of Hantush (1965 and 1967), was used. The model accounts for the aquifer characteristics (storage coefficient, permeability and aquifer thickness) and for the infiltration flow as a function of time (Eq. 4), to estimate piezometric variations.

For the purpose of the modelling, the geometry of the basin was approximated to a rectangle with the same average area of ponded water. The aquifer thickness used in the model was assigned to 30 m, which corresponds to the saturated quaternary deposit and the top of fractured Ordovician aquifer in the studied region (Ressources en Eau Du MAROC Tome III, 1977). An additional probe located in the vicinity of the wadi 3 km upstream showed no reaction of the piezometry during floods. Therefore, the natural river recharge was assumed to be negligible and not accounted for in the model.

The calibration of the model was performed by adjusting the permeability and storage coefficients values so as to obtain a visually satisfactory match between observed and simulated piezometric levels at S2 and S3.

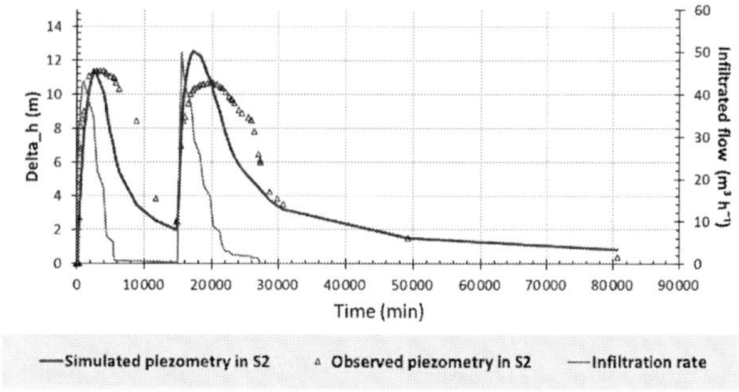

Figure 4. Modelling of groundwater recharge in S2 for floods 1 and 2 ($K = 4.1 \times 10^{-6}\,\mathrm{m\,s^{-1}}$; $S = 3.5 \times 10^{-3}$).

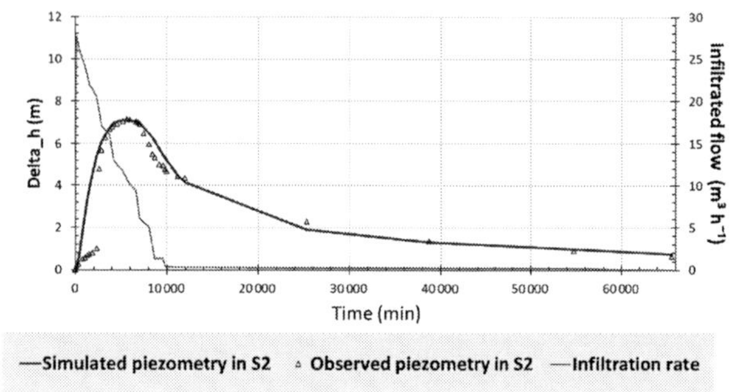

Figure 5. Modelling of groundwater recharge in S2 for flood 3 ($K = 2.2 \times 10^{-6}\,\mathrm{m\,s^{-1}}$; $S = 8.5 \times 10^{-3}$).

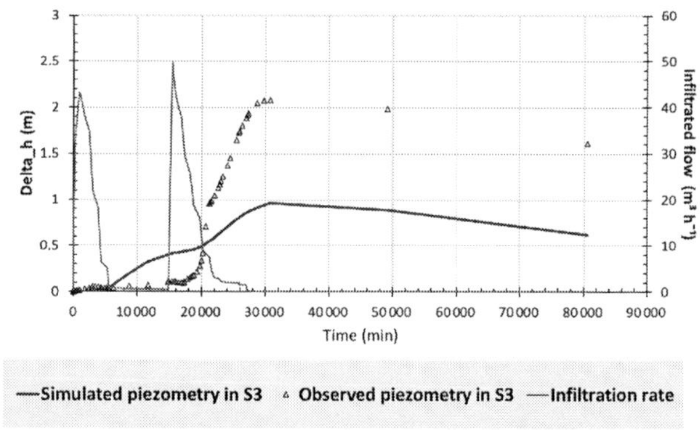

Figure 6. Modelling of groundwater recharge in S3 for floods 1 and 2 ($K = 4.1 \times 10^{-6}\,\mathrm{m\,s^{-1}}$; $S = 3.5 \times 10^{-3}$).

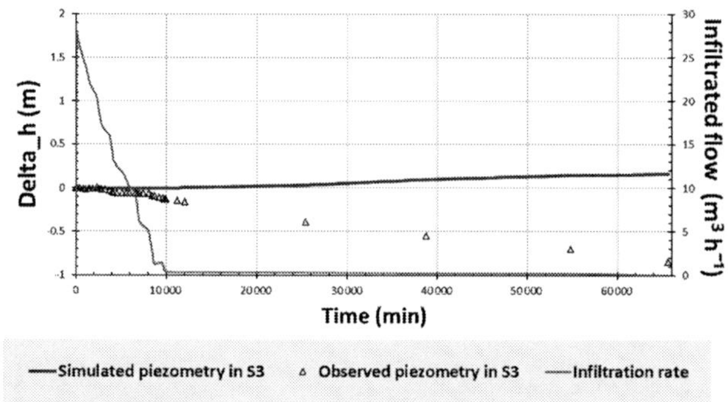

Figure 7. Modelling of groundwater recharge in S3 for the 3rd flood ($K = 2.2 \times 10^{-6}\,\mathrm{m\,s^{-1}}$; $S = 8.5 \times 10^{-3}$).

3 Results and discussion

3.1 Description and analysis of piezometric and limnimetric measurements

During the three recorded flood events (Fig. 2), the water levels in the basin (S4) and in the well inside the basin (S1) initiate a sharp and rapid increase (filling stage), immediately followed by a smoother drop (emptying stage). At the beginning of the emptying stage, S1 and S4 follow similar decrease rates, suggesting that the well and the basin are hydraulically connected. The staircase-like shape of the curves (Fig. 2) reflects the intermittence of pumping that occurs during daytime to irrigate the farm.

After several days of nearly synchronized drawdown of S1 and S4, S1 initiates a much sharper decrease, suggesting that the well disconnects from the basin.

The monitoring points S1 and S2 respond to all filling of the basin. The maximum groundwater level reached during each flood in S2 decreases overtime. In contrast, point S3 did not seem to respond to the first and the third flood, and its response to the second flood is delayed.

3.2 The infiltration rate in the basin

The basin infiltration rates vary between 1 and $9\,\mathrm{mm\,h^{-1}}$. Figure 3 shows that the infiltration rates decrease overtime (from flood 1 to flood 3). This is assumed to result from the silting up of the basin and the local clogging of the bottom. Furthermore, it is found that the rates are proportionally linear to the depth of water in the basin; the infiltration equations from the basin are thus deduced.

The basin has a maximum storage capacity of approximately $6500\,\mathrm{m^3}$, the volumes of water infiltrated during the three events fluctuate between 2100 and $3150\,\mathrm{m^3}$. Moreover, the sum of evaporation and evapotranspiration accounts for 3 %, 7 % and 14 % of the infiltrated volumes for the first, second and third floods, respectively. As for the infiltration flows from the recharge well, they do not exceed $0.7\,\mathrm{m^3\,h^{-1}}$, i.e. less than $17\,\mathrm{m^3\,d^{-1}}$. These values are negligible compared to the infiltration from the basin, which are between 240 and $380\,\mathrm{m^3\,d^{-1}}$. Thus, we are only interested in infiltration from the bottom of the basin in the remainder of this work.

3.3 Calibration of the groundwater recharge model

The model was first fitted at point S2 where all the fills of the basin induced an increase of the observed and modelled water levels. The permeability and the storage coefficient values yielding the best match between the observed data and the simulated values are $K = 4.1 \times 10^{-6}\,\mathrm{m\,s^{-1}}$ and $S = 3.5 \times 10^{-3}$, respectively. The third flood was fitted with a permeability coefficient, $K = 2.2 \times 10^{-6}\,\mathrm{m\,s^{-1}}$, slightly lower than that used for the first two floods (Fig. 5). This slight difference may be due to errors in flow estimation and to the artefact of modelling the basin as a rectangle. These K and S values are typical of free fractured aquifers (Dewandel et al., 2006; Lachassagne et al., 2021).

The calibration values of permeability and storage coefficient obtained in S2 were also used in S3 (Figs. 6 and 7), assuming that the structure of the aquifer is homogenous. However, the model could not reproduce the groundwater level fluctuations observed in S3 during the three floods, even for other K and S values. Thus, we may conclude that there is no direct link between the infiltration basin and S3. The piezometric increase in S3 may be due to the return flow from irrigation (Fig. 6) that occurs around this point.

4 Conclusions

Modelling that takes into account the infiltration flow rates from the basin and does not take into account the natural recharge of the wadi and the recharge by the well has allowed the response at piezometer (S2) to be reproduced reasonably. This modelling also shows that a farther piezometer (S3) does not respond to the pond infiltration, probably due to the absence of a dense enough fractures network to connect the aquifer on a large scale. After each flood, infiltration rates decrease because of the basin siltation. In summary, this recharge basin allows the groundwater to be recharged locally over a radius less than 360 m, which may improve locally water sustainability for agriculture. Also, we show that on this experimental site, the recharge induced by the wadi is weak, likely because of its low-permeable bed, and that the recharge induced by the well inside the basin is weak too, probably due to the partial or total clogging caused by the flood deposit.

Data availability. The data that support the findings of this study are available from the corresponding author upon reasonable request.

Author contributions. YK, GL, AT, and AH. designed the research; YK performed the processing of data; YL, GL, and BD performed the research and the modelling; YK analysed the data. YK, GL, and BD wrote the paper; AH, SB, and JCM provided correction and remarks to the paper.

Competing interests. The contact author has declared that none of the authors has any competing interests.

Disclaimer. Publisher's note: Copernicus Publications remains neutral with regard to jurisdictional claims in published maps and institutional affiliations.

Special issue statement. This article is part of the special issue "IAHS2022 – Hydrological sciences in the Anthropocene: Past and future of open, inclusive, innovative, and society-interfacing approaches". It is a result of the XIth Scientific Assembly of the International Association of Hydrological Sciences (IAHS 2022), Montpellier, France, 29 May–3 June 2022.

Acknowledgements. We thank the Massire project (IFAD, policy number 2000002013) for funding the research and the state agencies ORMVA-Tf, ABH-GZR and ANDZOA for field support and data sharing.

Review statement. This paper was edited by Christophe Cudennec and reviewed by Mohammad Merheb and one anonymous referee.

References

Arshad, M., Guillaume, J. H. A., and Ross, A.: Assessing the Feasibility of Managed Aquifer Recharge for Irrigation under Uncertainty, Water, 6, 2748–2769, https://doi.org/10.3390/w6092748, 2014.

Carleton, G. B.: Simulation of groundwater mounding beneath hypothetical stormwater infiltration basins, US Geol. Surv. Sci. Inv. Rep. 2010-5102, 64 pp., https://doi.org/10.3133/sir20105102, 2010.

Casanova, J., Cagnimel, M., Devau, N., and Pettenati, M.: Recharge artificielle des eaux souterraines: état de l'art et perspectives 227, https://infoterre.brgm.fr/rapports/RP-61821-FR.pdf (last access: 26 May 2022), 2013.

Dewandel, B., Lachassagne, P., Wyns, R., Maréchal, J. C., and Krishnamurthy, N. S.: A generalized 3-D geological and hydrogeological conceptual model of granite aquifers controlled by single or multiphase weathering, Hydro-ecological functioning of the Pang and Lambourn catchments, UK 330, J. Hydrol., 330, 260–284, https://doi.org/10.1016/j.jhydrol.2006.03.026, 2006.

Dewandel, B., Lanini, S., Hakoun, V., Caballero, Y., and Maréchal, J.-C.: Artificial aquifer recharge and pumping: transient analytical solutions for hydraulic head and impact on streamflow rate based on the spatial superposition method, Hydrogeol. J., 29, 1009–1026, https://doi.org/10.1007/s10040-020-02294-9, 2021.

Godwin, I. A., Reba, M. L., Leslie, D. L., Adams, R. F., and Rigby, J. R.: Feasibility of farm-scale infiltration galleries for managed aquifer recharge in an agricultural alluvial aquifer of Northeast Arkansas, Agr. Water Manag., 264, 107531, https://doi.org/10.1016/j.agwat.2022.107531, 2022.

Hantush, M. S.: Growth and decay of groundwater-mounds in response to uniform percolation, Water Resour. Res., 3, 227–234, https://doi.org/10.1029/WR003i001p00227, 1967.

Hashemi, H., Berndtsson, R., Kompani-Zare, M., and Persson, M.: Natural vs. artificial groundwater recharge, quantification through inverse modeling, Hydrol. Earth Syst. Sci., 17, 637–650, https://doi.org/10.5194/hess-17-637-2013, 2013.

Lachassagne, P., Dewandel, B., and Wyns, R.: Review: Hydrogeology of weathered crystalline/hard-rock aquifers – guidelines for the operational survey and management of their groundwater resources, Hydrogeol. J., 29, 2561–2594, https://doi.org/10.1007/s10040-021-02339-7, 2021.

Ressources en Eau Du MAROC Tome III: Hydrogéologie Bassin versant [WWW Document], Scribd, https://fr.scribd.com/document/494028246/Ressources-en-Eau-Du-MAROC-Tome-III (last access: 4 May 2022), 1977.

Riou, C.: Evaporation en bac et évapotranspiration potentielle: Expression de l'évaporation d'une petite nappe d'eau en fonction des données sous abri, Int. Assoc. Sci. Hydrol. Bull., 16, 53–65, https://doi.org/10.1080/02626667109493048, 1971.

Soni, P., Dashora, Y., Maheshwari, B., Dillon, P., Singh, P., and Kumar, A.: Managed Aquifer Recharge at a Farm Level: Evaluating the Performance of Direct Well Recharge Structures, Water, 12, 1069, https://doi.org/10.3390/w12041069, 2020.

Todd, D. K. and Mays, L. W.: Groundwater Hydrology, John Wiley & Sons, 656 pp., ISBN 0471059374, 9780471059370, 2004.

Warner, J., Molden, D., Chehata, M., and Sunada, D. K.: Mathematical analysis of articial recharge from basins 1, Jawra, https://doi.org/10.1111/J.1752-1688.1989.TB03077.X, 1989.

Wuilleumier, A. and Seguin, J. J.: Réalimentation artificielle des aquifères en France, Une synthèse, Rapport final, BRGM/RP-55063-FR, 122 pp., 2008.

Proc. IAHS, 385, 53–58, 2024
https://doi.org/10.5194/piahs-385-53-2024

Open Access PIAHS

Artificial Recharge of the Shallow Alluvial Aquifer as an Adaptation Strategy in the Garonne Valley, France

Nazeer Asmael[1], **Alain Dupuy**[1], **and Bernard Leroy**[2]

[1]EPOC UMR CNRS-Université de Bordeaux-Institut Polytechnique de Bordeaux, ENSEGID-Bordeaux INP, 1 allée Fernand Daguin, 33607 Pessac Cedex, France
[2]SMEAG Syndicat Mixte d'Études et d'Aménagement de la Garonne, 61 rue Pierre Cazeneuve, 31200 Toulouse, France

Correspondence: Nazeer Asmael (nazeer.asmael@ensegid.fr)

Received: 25 May 2022 – Revised: 19 January 2023 – Accepted: 22 January 2023 – Published: 18 April 2024

Abstract. The Garonne River is the primary river in the southwest part of France. The Quaternary alluvial aquifer along this River may represent a substantial water resource, especially for agricultural activities well-developed in the Garonne Valley. However, this shallow aquifer hosts numerous irrigation wells reducing the baseflow during the intensive pumping periods. It is recharged by rainfall, lateral inflow from the hillside (overlying terraces), and the river bed seepage during the flood periods. The aquifer sustains the River during the dry periods. Furthermore, the potential recharge of this aquifer is particularly sensitive to annual climate fluctuation and consequently affects the ecosystems and related socio-economy. Groundwater artificial recharge can be considered an innovative and sustainable nature-based solution. The runoff water from The Techno-Pole Agen – Garonne (TAG) zone is collected in retention basins and is a potential source to recharge the shallow alluvial aquifer. The study aims to model water infiltration and understand the aquifer response and, consequently, the effects on river low flow. Within the framework of this study, 132 wells/boreholes were used in to determine the groundwater level fluctuations and to create its maps. The measurements showed that the artificial recharge increased the groundwater level by more than 1 m close to the retention basin after the rainstorm event. Similarly, a three-dimensional (3D) groundwater model shows a similar magnitude aquifer response to the induced infiltration. Consequently, this model satisfactorily represents the interest of the artificial recharge of the alluvial aquifer and permits predictions about whether the aquifer can maintain the low flow of in the Garonne River. To this end, it was estimated that the infiltrated water would take about 4 months to reach the River, which is an appropriate time to sustain it during the dry periods.

Keywords. Climate change; Alluvial aquifer; Artificial recharge; Groundwater modelling; Water resources management, Garonne valley; France

1 Introduction

After the Loire and Seine, the Garonne River is the third of the large French Atlantic rivers. Its drainage basin has an area of 57 000 km^2 (Lancaster, 2005) and up to Tonneins, the last gauging station uninfluenced by the tidal action, it is calculated to be 50 000 km^2 (Biancamaria et al., 2019). The Garonne Basin is mainly influenced by the Oceanic and Mediterranean climate as well as snowmelt from the mountainous areas.

In the Garonne Valley, water is a crucial element for economic and social development. Both surface water and groundwater resources usually meet the needs of all uses. However, the spatial and temporal distribution of these resources does not always match the demands. Several studies have mentioned how the current and future negative impacts of climate change influence the hydrology of the Garonne River and its surrounding alluvial aquifers (Caballero et al., 2007; Hendrickx and Sauquet, 2013; Agence de l'eau Adour-Garonne, 2014; Guivarch et al., 2015; Le Treut, 2018).

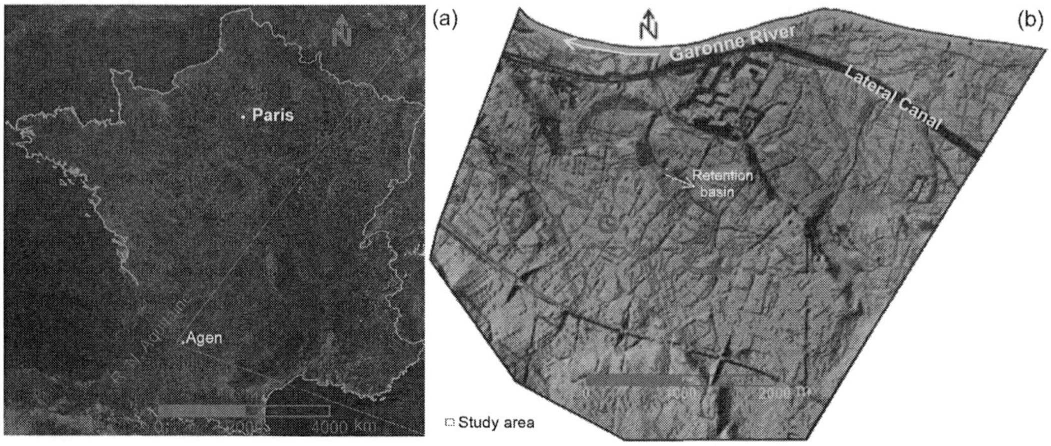

Figure 1. Map of France, prepared by QGIS, free and open source software, showing the location of the study area in Nouvelle Aquitaine Region **(a)** and the topographic units of the study area according to IGN (French national geographic institute) spatial satellite data with a 5 m resolution **(b)**.

Groundwater and surface water resources interact in a range of geological, topographical, and climatic settings, and they have varying degrees of connection (Ivkovic, 2009). The shallow alluvial aquifer system investigated in this work overlies an impermeable layer of Tertiary Molasse sediments and it supplies water to thousands of irrigation wells. According to the location and the degree of incision of the River into the Tertiary Molasse, the Garonne River can temporarily recharge the alluvial aquifer during flood periods. On the other hand, this aquifer drains and sustains the low flow of the River during dry periods. This aquifer drainage takes place mainly in summer and is often the main source for the River during this period (Brugeron et al., 2018). The impacts of climate change on the water resources in the Garonne Valley lead to seeking adaptation strategies, including artificial aquifer recharge.

The Techno-Pole Agen - Garonne (TAG) project is under construction close to the City of Agen southwest France. The project and its surrounding areas (about $20\,\mathrm{km}^2$), where the current major occupation is essentially agricultural activities, are taken as a study area. The alluvial aquifer in this area will be used as a potential source to create a peri-urban, natural, green, and temperate zone in order to combat rising temperatures. The study aims to: (1) determine the potential response of the alluvial aquifer to artificial recharge; (2) identify how this strategy will participate in increasing the baseflow toward the Garonne River during dry periods and (3) recognize the potential of this strategy as a management tool in the Garonne Valley to mitigate the impacts of climate change on water resources.

2 Materials and methods

2.1 Study area

The study area is located in the southwest of France (Fig. 1). The TAG Project is considered as one of the most important project from the economical point of view in the southwest of the region. It is under construction within an area of about 240 ha near Agen. The project zone and surrounding area were taken as a study area ($\approx 20\,\mathrm{km}^2$). This area is characterized by a temperate transitional climate between the Oceanic and Mediterranean regimes. It is characterized in the south and southwest by hillsides, while the floodplain is dominated in the north and northeast. The relief, according to IGN (French national geographic institute) spatial satellite data with a 5 m resolution, ranges from about 83 m in the south toward the hillsides to 32 m in the north near the Garonne River.

The alluvial deposits are dominant in the study area (Danneville, 2016). The low alluvial plain, consisting of gravel and large pebbles intercalate with a sandy matrix, occupies the north and northeast parts. In contrast, the low terrace predominantly contains weathered deposits of silt and sandy-clay. The alluvial deposits are limited at the bottom by the molassic layer. The thickness of these deposits is between 6 to 10 m and decreases towards the hillside, where the silt, clay, and sand cover intercalates with the aeolian sediments of the hills.

The substantial variation in the sedimentological composition of the alluvial deposits induces subsurface heterogeneity leading to an important change in hydraulic properties of the alluvial aquifer over several orders of magnitude in a relatively small area. The hydraulic conductivity values of this aquifer vary from 10^{-2} to $10^{-5}\,\mathrm{m\,s^{-1}}$.

 https://doi.org/10.5194/piahs-385-53-2024

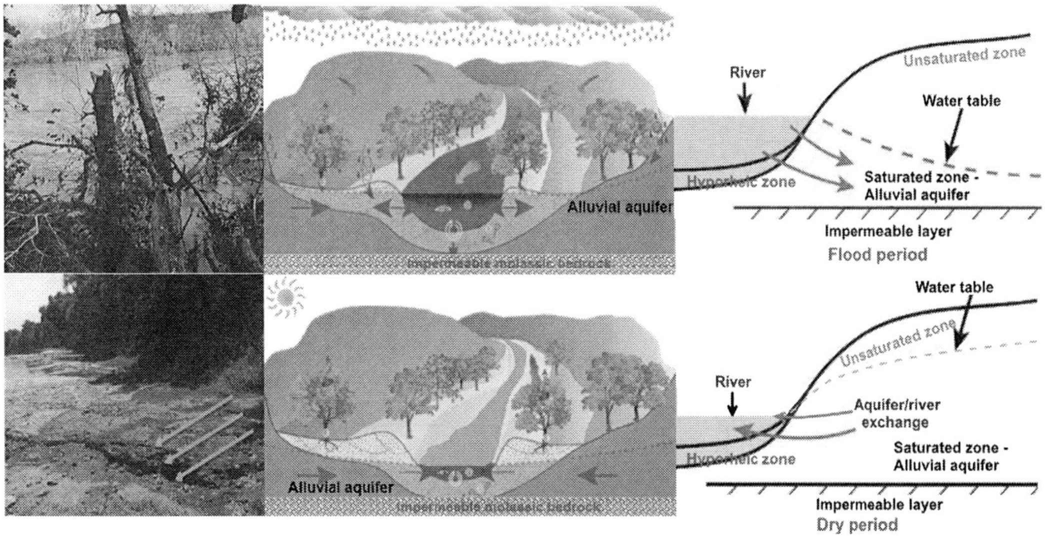

Figure 2. Alluvial aquifer/Garonne River interface and exchange patterns during flood and dry periods, after the Rhône Mediterranean and Corse Water Agency and Queensland Government.

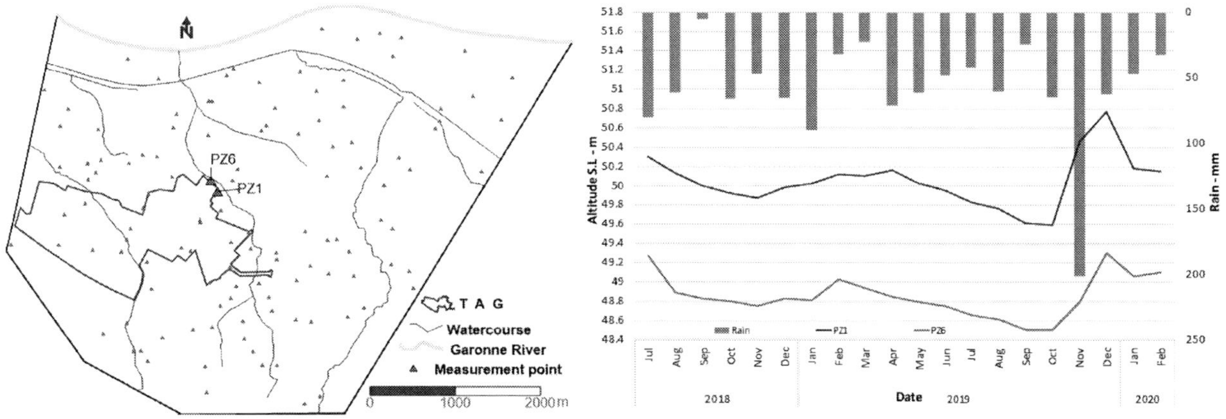

Figure 3. Location maps of observation wells (left) and monthly groundwater level fluctuations in two boreholes located close to the retention basin, as well as monthly rainfall variation measured in Agen station close to the TAG area.

2.2 Methods

2.2.1 The observed relationship between the alluvial aquifer and the Garonne River in the study area

The impermeable molassic rocks that form the riverbed of the Garonne River are characterized by a low altitude compared with the adjacent alluvial aquifer. The alluvial aquifer is characterized by a shallow flushed system that drains into the River during low flow periods when the groundwater level is higher than the river level. Conversely, during the flood periods, when the river level is higher than the groundwater level, the surface water seeps and recharges the aquifer either through the river banks if they are not clogged or through the infiltration when the river water reaches about 2 km further from the River in the flooded plain. However, the exchanges between the aquifer and the River depends on the River's head (Fig. 2).

These exchanges have significant quantitative and qualitative consequences, particularly during low-flow periods when the aquifer contributes to maintain the River's flow and its thermal buffering capacity. The river temperature was measured locally as 29 °C in August 2019 while the temperature of the drained groundwater in the River was 15 °C in the same period.

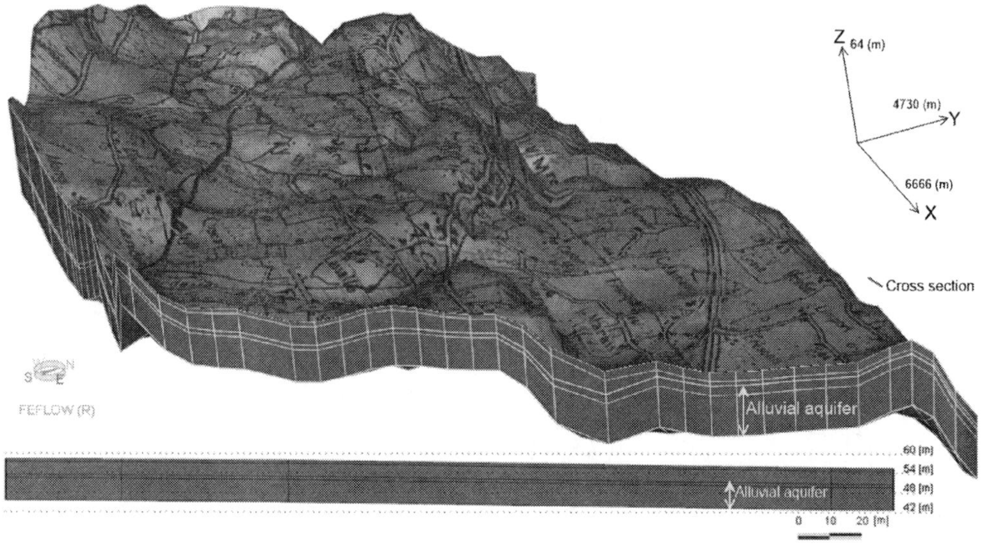

Figure 4. The three-dimensional (3D) representation model of the study area.

2.2.2 Potential of artificial recharge of the alluvial aquifer in the TAG area

Based on the observed hydraulic relationship between the Garonne River and the adjacent alluvial aquifer, the artificial recharge of this aquifer has been initiated.

The runoff water in the TAG area is collected in the retention basins in order to enhance the recharge of the aquifer and its later use. Following the storm event of November 2019, the retention basin has filled up. Consequently, the infiltrated water has created a piezometric dome that has been observed in the two boreholes (PZ_1 and PZ_6) located near the basin where the groundwater level has increased by about 1 m (Fig. 3).

3 Groundwater modeling result

By using FEFLOW (DHI-WASY 2010) and according to IGN (French national geographic institute) spatial satellite data with a 5 m resolution, the two-dimensional (2D) horizontal finite element mesh was generated. After that, the three-dimensional (3D, Fig. 4) slice elevations, layer properties and boundary conditions were defined.

The model domain was discretized into elements and nodes with five model layers. The first layer represents vegetal soil, its thickness changes between 1–3 m. The second layer consists of silt and sandy silt with a thickness changes between 1–7 m. The third layer represents a thin sandy layer with a thickness ranging from 1 to 2 m. The fourth one constitutes the alluvial aquifer (gravels, pebbles, and sandy-clay) housed in the molassic formations with a thickness varying from 6 to 10 m.

The calibrated hydraulic conductivity values with the inverse method in a steady state, in order to simulate the measured hydraulic heads, shows that these values change between 10^{-2} and 10^{-5}. The calibrated model has allowed us to specify the spatial distribution of the aquifer permeability and to visualize the main flow directions in relation to the equipotential lines. These lines show that the infiltration basin is located in an area characterized by relatively high permeability.

The superposition of the measured and calculated groundwater level contours (Fig. 5) for the same period indicates a good agreement. The two sets of piezometric contours, plotted every 2 m, are relatively close to each other over the whole study area.

The calibrated model has defined the initial groundwater level, which is used to run it in the transient regime taking into account the effect of artificial recharge from the retention basin (Fig. 6). A piezometric dome and significant rise in groundwater level can be observed near the retention basin after storm events and, consequently, runoff water infiltration (Fig. 6). Thus, the initial piezometric contour value has moved about 500 m toward a groundwater flow direction.

The results show that the infiltrated water during March-April takes about 4 months to reach the Garonne River which is an appropriate time to sustain the low flow of the river.

4 Conclusions

The application of artificial recharge of the alluvial aquifer in the study area is found to be a significant tool to maintain groundwater level. Groundwater modelling results show how this mechanism can be used to sustain the low flow of the Garonne River. It was estimated that the infiltrated wa-

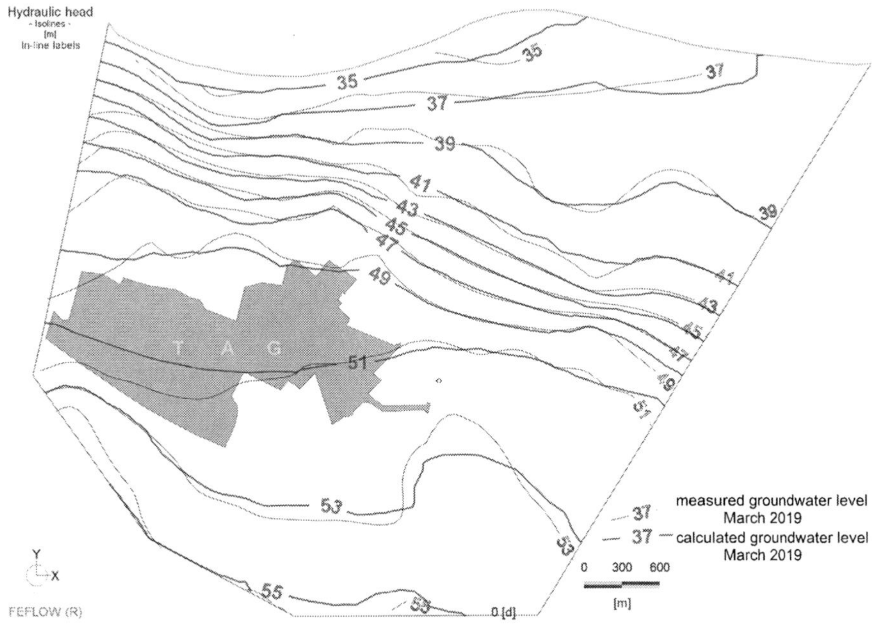

Figure 5. Measured and calculated groundwater level maps, March–2019.

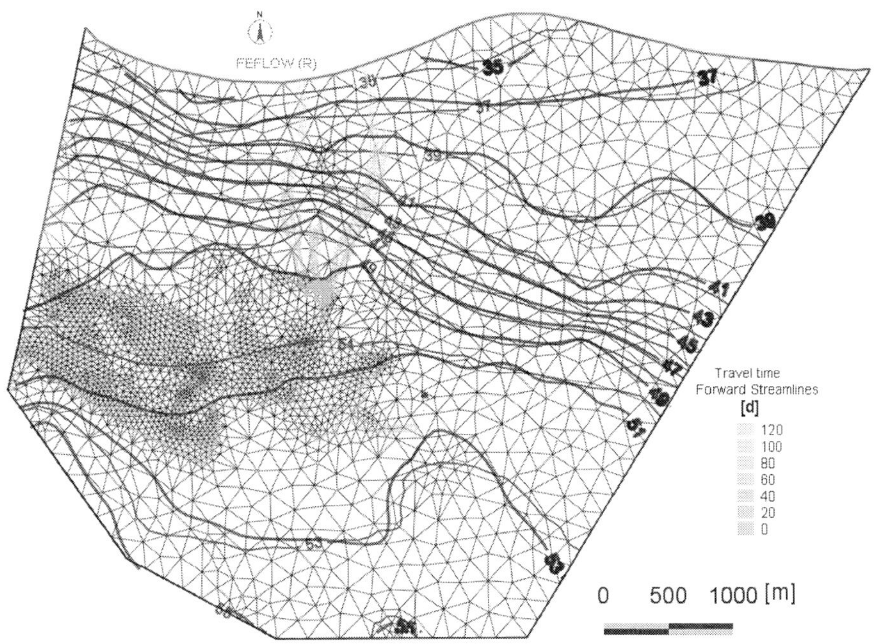

Figure 6. Groundwater flow patterns towards the Garonne River after artificial recharge applied in the TAG area.

ter would take about 4 months to reach the River, which is an appropriate time to sustain it during the dry periods if the recharge takes place between January and April. This pragmatic and potentially eco-responsible management tool can be considered as an adaptation strategy to climate change in the whole Garonne Valley.

Based on the obtained results of this project, three other pilot sites in the Garonne valley were selected to be studied and subsequently evaluate their potential for artificial recharge using the Canal Lateral of the Garonne River as a source. This will help to recharge the Garonne River during the dry period with thermally tempered water and contribute to proper functioning of its environment. However, this methodology is appropriate to preserve the water resources in the study area and to maintain the agricultural activities of the inhabitants, who depend mainly on agricultural production.

Data availability. No data sets were used in this article.

Author contributions. All the authors contributed to the work presented in this paper and commented on the manuscript at all stages. NA, AD and BL designed the methodological framework and revisions. NA and AD performed the simulation and analysed the results. NA prepared the manuscript with contributions of AD and BL.

Competing interests. The contact author has declared that neither of the authors has any competing interests.

Disclaimer. Publisher's note: Copernicus Publications remains neutral with regard to jurisdictional claims in published maps and institutional affiliations.

Special issue statement. This article is part of the special issue "IAHS2022 – Hydrological sciences in the Anthropocene: Variability and change across space, time, extremes, and interfaces". It is a result of the XIth Scientific Assembly of the International Association of Hydrological Sciences (IAHS 2022), Montpellier, France, 29 May–3 June 2022.

Acknowledgements. We thank the ACMG (Climatological Association of the Middle Garonne) for their technical support.

Financial support. This research has been supported by the Nouvelle Aquitaine Region (File no 2017-1R20104) and by SMEAG, Syndicat mixte d'études et d'aménagement de la Garonne (File no CT_2019-068).

Review statement. This paper was edited by Christophe Cudennec and reviewed by two anonymous referees.

References

Agence de l'eau Adour-Garonne: Eau et Changements climatiques en Adour-Garonne, Les enjeux pour la ressource, les usages et les milieux, http://oai.eau-adour-garonne.fr/oai-documents/60721/GED_00000000.pdf (last access: 4 March 2022), 2014.

Biancamaria, S., Mballo, M., Le Moigne, P., Sánchez Pérez, J. M., Espitalier-Noël, G., Grusson, Y., Cakir, R., Häfliger, V., Barathieu, F., Trasmonte, M., Boone, A., Martin, E., and Sauvage, S.: Total water storage variability from GRACE mission and hydrological models for a 50 000 km^2 temperate watershed: the Garonne River basin (France), J. Hydrol., 24, 100609, https://doi.org/10.1016/j.ejrh.2019.100609, 2019.

Brugeron, A., Schomburgk, S., Cabaret, O., Bault, V., Bel, A., Salquebre, D., Fourniguet, G., Lamotte, C., and Parmentier, M.: Synthèse sur la cartographie et la caractérisation des alluvions dans le référentiel hydrogéologique BDLISA, Rapport final BRGM/RP-67533-FR, 2018.

Caballero, Y., Voirin-Morel, S., Habets, F., Noilhan, J., Le Moigne, P., Lehenaff, A., and Boone, A.: Hydrological sensitivity of the Adour-Garonne river basin to climate change, Water Resour. Res., 43, W07448, https://doi.org/10.1029/2005WR004192, 2007.

Danneville, L.: Contribution des eaux souterraines aux débits et à la qualité des eaux de surface – Exemple de la Garonne, de ses sources à la confluence du Tarn, https://tel.archives-ouvertes.fr/tel-01420437 (last access: 18 March 2022), 2016.

DHI-WASY GmbH: FEFLOW 6, Finite elements subsurface flow and transport simulation system, In User's Manual, DHI-WASY GmbH: Berlin, Germany, 2010.

Guivarch, M., Dany, A., Toulzac, N., Guiheux, A., and Le Gall, G.: État Initial du SAGE Vallée de la Garonne, Rapport, Version 6, Février 2015.

Hendrickx, F. and Sauquet, E.: Impact of warming climate on water management for the Ariège River basin (France), Hydrolog. Sci. J., 58, 976–993, https://hal.archives-ouvertes.fr/hal-01117889v2, 2013.

Ivkovic, K. M.: A top-down approach to characterise aquifer-river interaction processes, J. Hydrol., 365, 145–155, https://doi.org/10.1016/j.jhydrol.2008.11.021, 2009.

Lancaster, R. R.: Fluvial evolution of the Garonne River, France: integrating field data with numerical simulations, https://digitalcommons.lsu.edu/gradschool_theses/3351 (last access: 18 March 2022), 2005.

Le Treut, H.: Les impacts du changement climatique en Aquitaine, in: L'air et l'eau – Presses Universitaires de Bordeaux, chapitre 8, https://books.openedition.org/pub/663?lang=fr{#}text (last access: 10 March 2022), 2018.

Queensland Government: Groundwater dependent ecosystem pictorial conceptual model "alluvia": version 1.5, Queensland Government, Brisbane, 2017.

QGIS Development Team (2019): QGIS Geographic Information System, Open Source Geospatial Foundation Project, http://qgis.osgeo.org, last access: 12 November 2019.

Proc. IAHS, 385, 59–64, 2024
https://doi.org/10.5194/piahs-385-59-2024

Open Access
PIAHS

Comparative study from ground-based rain gauges vs. rainfall products at different time steps in the southeast of the Republic of Djibouti

Golab Moussa Omar[1,2], **Jean-Emmanuel Paturel**[1], **Christian Salles**[1], **Gil Mahé**[1], **Mohamed Jalludin**[2], **Frédéric Satgé**[3], **Mohamed Ismail Nour**[4], **and Abdillahi Hassan Hersi**[4]

[1]HSM, Univ. Montpellier, CNRS, IRD, IMT Alès, Montpellier, France
[2]Centre d'Etudes et de Recherches de Djibouti, Institut des Sciences de la Terre, Route de l'aéroport, Djibouti-ville, République de Djibouti
[3]UMR228 ESPACE-DEV, Univ. Montpellier, IRD, Univ. Antilles, Univ. Guyane, Univ. La Réunion, Montpellier, France
[4]Agence Nationale de la Météorologie de Djibouti, Djibouti, République de Djibouti

Correspondence: Golab Moussa Omar (golabmoussa@gmail.com)

Received: 29 May 2022 – Revised: 23 January 2023 – Accepted: 23 January 2023 – Published: 18 April 2024

Abstract. The Republic of Djibouti is a small country in the Horn of Africa and as for most regions of Africa, ground rainfall stations are sparse. This study aims to compare at different time steps (annual, monthly, and daily) 15 rainfall estimation products (P-datasets) to 5 reference ground-based rainfall stations, over the period of 1980–1990. To classify the reliability of these products, several metrics were considered, the Kling Gupta Efficiency (KGE) and the Heidle Skills Scores (HSS at daily time step). For the study area, 5 P-datasets present high KGE values at annual and monthly time step: EWEMBI, GPCC, JRA-55 Adj, MSWEP and WFDEI-GPCC; the best 5 performing products at daily time step are CHIRPS, EWEMBI, JRA-55 Adj, MSWEP and WFDEI-GPCC. To complete this comparative analysis, it will be necessary to perform an analysis at the watershed scale and on a new temporal window.

Keywords. UPH 16; field observation; precipitation-datasets; Republic of Djibouti

1 Introduction

The Republic of Djibouti is a small country (23 200 km^2) located in the Horn of Africa and as for most regions of Africa, ground rainfall stations are sparse and unequally distributed.

To palliate the low density of the rainfall observation network in some areas, there are alternative or complementary products we can find on the Web such as rainfall estimate products (P-Datasets). Since the 1980s, these P-Datasets have blossomed and offer interesting opportunities because, usually the dataset is continuous, and spatially available and, rainfall data are assessed at daily or hourly time steps.

We can define three groups of rainfall estimation products according to the input data and rainfall data retrieval process: those based on spatial information from ground measurement points, those based on reanalysis data derived from physical and dynamic models, and those based on satellite information using passive microwave (PMW) and infrared (IR) information (Le Coz and van de Giesen, 2020; Satgé et al., 2020). Lots of combinations of these three groups exist on the WEB.

The objective of this study is to determine which of these rainfall products is the most reliable compared to ground-based rainfall stations for the south-east of Djibouti. The south-east zone concentrates most of the population and economic exchanges between Djibouti and its neighbouring countries. Therefore, our study focuses on this part of the country This study is the first component of a broader work that aims to improve the knowledge and understanding of the hydrological processes involved in the Ambouli basin. The Ambouli basin is one of the country's largest catchments

(794 km^2) and is of major importance. The Ambouli aquifer is the main source of drinking water for the city of Djibouti. The Ambouli basin. is of particular importance for flood risk of the city of Djibouti.

2 Study area, Rain gauges and P-Datasets

2.1 Study area

The whole territory of the Republic of Djibouti presents an arid tropical climate which is characterized by the irregularity and the weakness of precipitations, by high temperatures during all the year, by the absence of perennial water course and by a very intense evaporation. The climate of the country is characterized by two distinct seasons: a cool season from October to April and a hot season from May to September (Houmed-Gaba, 2009). Due to geology of volcanic origin, the Republic of Djibouti is characterized by a very steep relief which is a succession of massifs, plateaus and plains also represented in the southeastern part (study area).

2.2 Rain gauges

There are 14 rainfall stations in the south-east of the Republic of Djibouti managed by the "Agence Nationale de la Météorologie de Djibouti (ANM)". Out of 14 stations, only 5 have at least 10 months or more of data available per year. Thus, we use a network of 5 rain gauges: Aerodrome, Serpent, Ali-Sabieh, Hol-Hol and Loyada (Fig. 1). The rainfall data from the network cover the period 1951–1990. The rainfall are at monthly time step for 4 rain gauge stations (Serpent, Ali-Sabieh, Hol-Hol and Loyada) and the Djibouti Aerodrome station is at daily time step (over the period 1981–2020). These rainfall data are regulated by the quality control and quality assurance system of the World Meteorological Organization (WMO).

2.3 P-Datasets

Products selection for testing was motivated by their availability over the region of interest and as they are commonly used in operational or research studies focusing on hydrology or agronomy. Table 1 presents temporal and spatial characteristics of the P-Datasets.

3 Methodology

This study assesses the performance of 15 precipitation estimation products at daily, monthly, and annual time step against the ground rainfall stations of the south-eastern zone of the Republic of Djibouti.

The prerequisite for the choice of studied period is a shared time window between P-Datasets and rain gauge data. Since the P-Datasets are only available from the beginning of the 1980's, the common study period is 1980–1990.

The P-Datasets validation is carried out on a spatially aggregated level: point-to-pixel. The grid values of the precipitation estimation products containing each rain gauge station are extracted and pairwise compared.

The performances of P-Datasets were evaluated using the quantitative metrics (KGE and MBE) and categorical indices (POD, FAR, Accuracy, Error and HSS), at annual, seasonal, monthly and daily temporal scale

3.1 Pre-processing (pre-treatment)

As the P-Datasets differ in spatial resolution – ranging from 0.0375° for TAMSAT v.3 to 1° for GPCC v.7 – and for easy comparison, the P-Datasets were resampled to a spatial resolution of 0.1° by 0.1° using bilinear interpolation (Nikulin et al., 2012; Akinsanola et al., 2016; Satgé et al., 2020). Moreover, to address the uniformity in the temporal span, the P-Datasets available at a sub-daily time step were aggregated to a daily time step records over the period of study.

3.2 Metrics

The comparative study includes quantitative and categorical metrics. Although wide range of metrics are available to assess data performance (Wang et al., 2003; Segele et al., 2008; Diro et al., 2009; Ayehu et al., 2018), there is no single one that encapsulates all aspects of interest. For this reason, it is useful to consider several metrics and to understand the type of information or insight they might provide (Akinsanola et al., 2016).

3.2.1 Quantitative metrics

The metrics used for the ground to rainfall products statistical comparison are:

- Kling Gupta Efficiency (KGE) is the Euclidean distance computed using the coordinates of bias ($\beta = \frac{\overline{y}}{\overline{x}}$), standard deviation ($\gamma = \frac{1}{\beta} \times \frac{\sigma_y}{\sigma_x}$) and correlation coefficient $\left(r = \frac{\sum_{i=1}^{n}(x_i-\overline{x})(y_i-\overline{y})}{\sqrt{\sum_{i=1}^{n}(x_i-\overline{x})^2}\sqrt{\sum_{i=1}^{n}(y_i-\overline{y})^2}} \right)$ (Gupta et al., 2009; Clark et al., 2021):

$$\mathrm{KGE} = 1 - \sqrt{(r-1)^2 + (\beta-1)^2 + (\gamma-1)^2} \quad (1)$$

- Mean Bias Eror (MBE) is used to estimate the average bias in the rainfall product, and it provides a good indication of the mean overestimate of predictions. A positive value of MBE means an overestimation:

$$\mathrm{MBE} = \frac{1}{n}\sum_{i=1}^{n} x_i - y_i \quad (2)$$

where x corresponds to the ground reference rainfall and y corresponds to the P-Datasets.

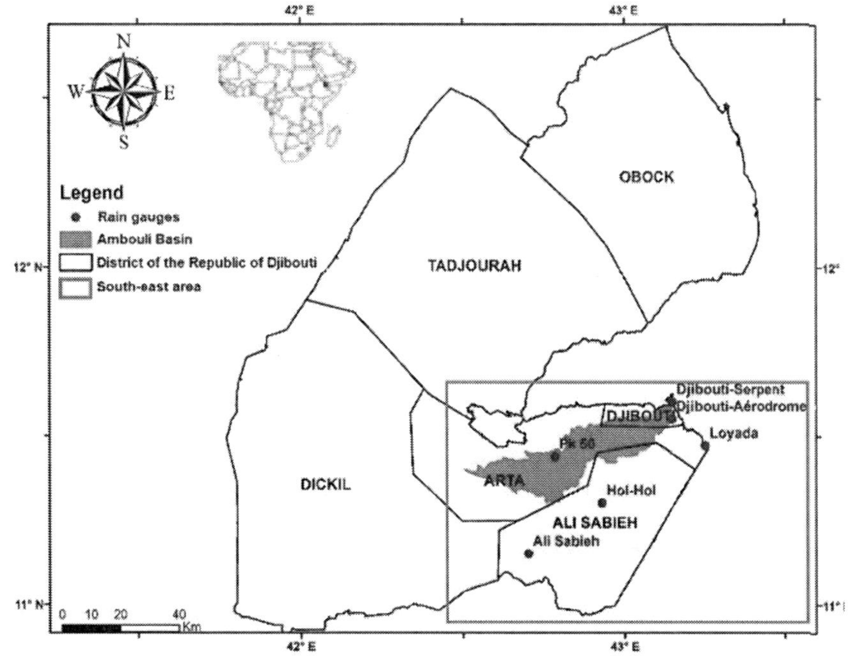

Figure 1. Location of the study area and rain gauge stations.

Table 1. Main characteristics of the P-datasets – S = satellite, R = reanalysis and G = gauge information. Spatial coverage refers to the absolute maximum and minimum latitude with precipitation information.

Acronym	Data	Temporal Coverage	Temporal Resolution	Spatial Coverage	Resolution
CPC v.1	G	from 1979	Daily	Global	0.5°
GPCC v.7	G	1901–2013	Monthly	Global	1°
ERA5	R	from 1979	3 h	60°	0.75°
ARC-2	S, G	from 1983	Daily	Africa	0.1°
PERSIANN-CDR	S, G	1983–2016	Daily	60°	0.25°
TAMSAT	S, G	from 1983	Daily	Africa	0.0375°
EWEMBI	R, G	1979–2013	Daily	Global	0.5°
JRA-55 Adj	R, G	1959–2012	3 h	Global	0.56°
WFDEI CRU	R, G	1979–2016	Daily	Land	0.5°
WFDEI GPCC	R, G	1979–2016	Daily	Land	0.5°
CHIRP v.2	S, R	from 1981	Daily	50°	0.05°
CHIRPS v.2	S, R, G	from 1981	Daily	50°	0.05°
MERRA-2 PT	S, R, G	from 1980	Hourly	Global	0.5°
MERRA-2 PTC	S, R, G	from 1980	Hourly	Global	0.5°
MSWEP v.2.2	S, R, G	from 1979	3 h	Global	0.1°

3.2.2 Categorical metrics

The performances of rainfall estimate products are also analyzed using a, b, c and d categorical indices. Five statistical indicators based on a contingency table (Table 2) are computed:

- Probability of detection (POD):

$$\frac{a}{a+c} \tag{3}$$

- False Alarm Ratio:

$$\mathrm{FAR} = \frac{b}{a+b} \tag{4}$$

Table 2. Contingency table used for the categorical statistical analysis of rainfall.

		Gauge observation	
		Rainfall	No Rainfall
Rainfall	Rainfall	Hits "a"	False Alarms "b"
Products	No Rainfall	Misses "c"	Correct Negatives "d"

– Accuracy:

$$A = \frac{a+d}{a+b+c+d} \quad (5)$$

– Error

$$E = \frac{b+c}{a+b+c+d} \quad (6)$$

– Heidle Skill Scores

$$\mathrm{HSS} = \frac{2\cdot(a\cdot d - b\cdot c)}{[(a+c)\cdot(c+d)+(a+b)\cdot(b+d)]} \quad (7)$$

4 Results and discussion

4.1 Annual Comparisons (Fig. 2)

Strong agreement with the rain gauge data is observed for some of the P-datasets. Five P-Datasets have a high KGE – EWEMBI, GPCC, JRA_55_Adj, MSWEP and WFDEI_GPCC. WFDEI_GPCC is the best P-Datasets whereas MERRA-2 PTC and ERA 5 had the worst performance.

Eight of rainfall estimates products – CPC, ERA5, EWEMBI, GPCC, JRA-55_Adj, MERRA-2_PT, MERRA-2_PTC and WFDEI_GPCC – overestimated rainfall with positive MBE values.

Loyada rain gauge has the lowest KGE except for the rainfall estimation products TAMSAT and WFDEI_CRU. This can be explained either by the location of the station or the digitalization of the data (because it is a manual rain gauge station, when data are collected and then they are digitalized by a technician which can generate an uncertainty.

4.2 Monthly comparisons (Table 3)

Regarding the KGE, all P-Datasets systematically present values within −0.16 and 0.70 which is much more efficient than the annual time step. The top five P-Datasets at monthly time step are the same as those at annual time step: EWEMBI, GPCC, JRA-55_Adj, MSWEP and WFDEI_GPCC. WFDEI_GPCC is always the best among these fives. As for the annual time step, MERRA-2 PTC corresponds to the least reliable P-Dataset.

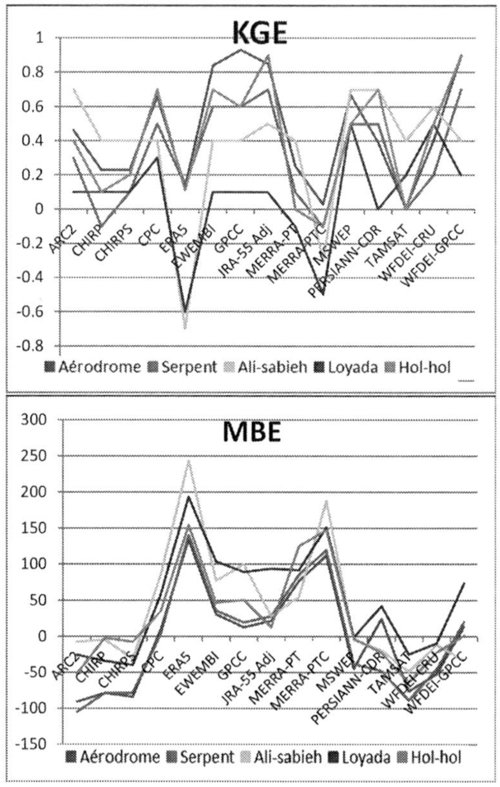

Figure 2. Statistical validation of rainfall products for rain gauge stations in the south-eastern part of the Republic of Djibouti, at annual time step for the period 1980–1990.

Table 3. Monthly statistical indicator.

Rainfall Estimate Product	KGE	MBE (mm)
ARC2	0.49	−4.38
CHIRP	0.116	−3.4
CHIRPS	0.2	−4.01
CPC	0.4	3.88
ERA5	−0.14	14.7
EWEMBI	0.58	4.91
GPCC	0.54	4.67
JRA-55 Adj	0.66	3.08
MERRA-2 PT	0.06	7.23
MERRA-2 PTC	−0.16	11.97
MSWEP	0.65	−1.52
PERSIANN_CDR	0.14	7.76
TAMSAT	0.18	−3.41
WFDEI-CRU	0.44	−2.34
WFDEI-GPCC	0.7	1.27

Table 4. Quantitative (KGE to MBE column) and categorical (POD to HSS column) statistical indicators at daily time step at Djibouti aerodrome station.

P-Dataset	KGE	MBE	POD	FAR	Accuracy	Error	HSS	HSS > 1
Arc2	−0.02	−0.05	0.06	0.93	0.88	0.12	0	0.08
CHIRP	−0.14	0.05	0.83	0.92	0.33	0.67	0.02	0.12
CHIRPS	−0.05	−0.13	0.24	0.75	0.9	0.1	0.19	0.21
CPC	−0.04	0.26	0.45	0.89	0.71	0.29	0.07	0.14
ERA5	−0.22	0.42	0.95	0.92	0.26	0.74	0.03	0.13
EWEMBI	0.19	0.13	0.76	0.9	0.53	0.47	0.07	0.21
JRA-55 ADJ	0.3	0.11	0.72	0.9	0.53	0.47	0.06	0.21
MERRA-2 PT	−0.31	0.39	1	0.93	0.07	0.93	0	0.12
MERRA-2 PTC	−0.2	0.36	0.9	0.93	0.24	0.76	0.01	0.11
MSWEP	0.17	−0.12	0.29	0.9	0.78	0.22	0.06	0.18
PERSIANN-CDR	0.17	0.06	0.38	0.9	0.74	0.26	0.06	0.16
TAMSAT	−0.23	−0.12	0.1	0.92	0.86	0.14	0.01	0.04
WFDEI-CRU	−0.06	0.01	0.71	0.9	0.54	0.46	0.06	0.16
WFDEI-GPCC	0.26	0.09	0.74	0.9	0.55	0.45	0.07	0.21

4.3 Daily comparisons (Table 4)

For the daily time step analyses, we use data from Djibouti Aerodrome station for the period 1981–2005 which is the common period between the station and all the P-Datasets. The GPCC product is not present in the analysis of daily time step data because it has a monthly temporal resolution.

The ability of the P-Datasets to quantify the amount of daily precipitation is relatively low, with most products having negative KGE values (Table 4). Only the best performing products at the annual and monthly time step EWEMBI, JRA-55_Adj, MSWEP v.2.2, and WFDEI_GPCC have positive KGE values. Concerning MBE, values are relatively homogeneous for all products from −0.13 to 0.42 mm. The low value of MBE is due to the low number of rainy days.

Table 4 shows the results of the POD, FAR, Accuracy, Error and HSS indices. Regarding POD, the P-Datasets MERRA-2 (PT and PTC) and ERA5 (the worst performing products at annual and monthly time step) almost do not detect non-rainy days, which explain the almost perfect POD values for these products. EWEMBI, JRA-55 Adj, and WFDEI-GPCC (the best performing products at the annual and monthly time step) show POD values < 0.7. Regarding FAR, we observe relatively low values for the products (FAR > 0.9 except for CHIRPS which presents a FAR equal to 0.75). For the Accuracy and Error indices, the results are moderately satisfactory for most of the products; however, as for the FAR values, the CHIRPS product presents the best Accuracy and Error values, respectively 0.9 and 0.1.

Satgé et al. (2020), defined a non-rainy day using several threshold values ranging from 0 to 25 mm were applied for the P-Datasets. They found that maximum value of HSS was obtained with a threshold value of 1 mm. We consider the same threshold of 1 mm and observe that "HSS > 1 mm" are higher than "HSS" for all the P-Datasets. The best performing P-Datasets in terms of "HSS > 1 mm" are the same as at the annual and monthly time step (EWEMBI, JRA-55 Adj, MSWEP and WFDEI-GPCC) with the addition of CHIRPS.

5 Conclusion

Quantitative and categorical metrics allow us to determine which of the P-Datasets is the most reliable to reproduce the rainfall amounts measured by a 5 rain gauge network regulated by the quality control and quality assurance system of the World Meteorological Organization (WMO) over the south-east of Republic of Djibouti. For the annual and monthly time step, the KGE is used for the product rankings. For the daily time step, we add to KGE a categorical index, the HSS performance index, which allows us to assess the ability of these products to detect rainfall events.

For our study area, five P-Datasets present high KGE values at annual and monthly time step: EWEMBI, GPCC, JRA-55 Adj, MSWEP and WFDEI-GPCC. Therefore, the best performing products at the daily time step are CHIRPS, EWEMBI, JRA-55 Adj, MSWEP and WFDEI-GPCC.

To complete this first study, and to deepen the knowledge concerning the Ambouli basin, it would be necessary to carry out a comparative analysis at the basin scale and use another temporal window to validate the reliability of these products.

Data availability. The data used are from the databases of the Centre d'Etudes et de Recherches de Djibouti (CERD) and the Agence Météorologique Nationale (ANM) are available on request from these institutions. Concerning the data of the rainfall estimation products, they are available in the official sites of the rainfall products mentioned in Table 1.

Author contributions. GMO did the methodology, data collection, data analysis, writing and editing. JEP did the methodology, data analysis, writing/reviewing and editing. CS did the methodology, data analysis, writing and editing. GM did the methodology, data analysis, writing and editing. MJ did the data analysis, review writing and editing. FS did the methodology and data collection. MIN did the data collection. AHH did the data collection.

Competing interests. The contact author has declared that none of the authors has any competing interests.

Disclaimer. Publisher's note: Copernicus Publications remains neutral with regard to jurisdictional claims in published maps and institutional affiliations.

Special issue statement. This article is part of the special issue "IAHS2022 – Hydrological sciences in the Anthropocene: Variability and change across space, time, extremes, and interfaces". It is a result of the XIth Scientific Assembly of the International Association of Hydrological Sciences (IAHS 2022), Montpellier, France, 29 May–3 June 2022.

Review statement. This paper was edited by Christophe Cudennec and reviewed by two anonymous referees.

References

Akinsanola, A. A., Ogunjobi, K. O., Ajayi, V. O., Adefisan, E. A., Omotosho, J. A., and Sanogo, S.: Comparison of five gridded precipitation products at climatological scales over West Africa Comparison of five gridded precipitation products at climatological scales over West Africa, Meteorol. Atmos. Phys., 129, 669–689, https://doi.org/10.1007/s00703-016-0493-6, 2016.

Ayehu, G. T., Tadesse, T., Gessesse, B., and Dinku, T.: Validation of new satellite rainfall products over the Upper Blue Nile Basin, Ethiopia, Atmos. Meas. Tech., 11, 1921–1936, https://doi.org/10.5194/amt-11-1921-2018, 2018.

Clark, M. P., Vogel, R. M., Lamontagne, J. R., Mizukami, N., Knoben, W. J. M., Tang, G., Gharari, S., Freer, J. E., Whitfield, P. H., Shook, K. R., and Papalexiou, S. M.: The abuse of popular performance metrics in hydrologic modeling, Water Resour. Res., 57, e2020WR029001, https://doi.org/10.1029/2020WR029001, 2021.

Diro, G. T., Grimes, D. I. F., Black, E., O'Neill, A., and Pardo-Iguzquiza, E.: Evaluation of reanalysis rainfall estimates over Ethiopia, Int. J. Climatol., 29, 67–78, https://doi.org/10.1002/joc.1699, 2009.

Gupta, H. V., Kling, H., Yilmaz, K. K., and Martinez, G. F.: Decomposition of the mean squared error and NSE performance criteria: Implications for improving hydrological modelling, J. Hydrol., 377, 80–91, https://doi.org/10.1016/j.jhydrol.2009.08.003, 2009.

Houmed-Gaba, A.: Hydrogéologie des milieux volcaniques sous climat aride. Caractérisation sur site expérimental et modélisation numérique de l'aquifère basaltique de Djibouti (Corne de l'Afrique), PhD thesis, University of Poitiers, Poitiers, France, 220 pp., 2009.

Le Coz, C. and van de Giesen, N.: Comparison of rainfall products over sub-saharan africa, J. Hydrometeorol., 21, 553–596, 2020.

Nikulin, G., Jones, C., Giorgi, F., Asrar, G., Buchner, M., Cerezo-Mota, R., Christensen, O. B., Deque, M., Fernandez, J., Hansler, A., van Meijgaard, E., Samuelsson, P., Sylla, M. B., and Sushama, L.: Precipitation climatology in an ensemble of CORDEX-Africa regional climate simulations, J. Climate, 25, 6057–6078, 2012.

Satgé, F., Defrance, D., Sultan, B., Bonnet, M.-P., Seyler, F., Rouché, N., Pierron F., and Paturel, J.-E.: Evaluation of 23 gridded precipitation datasets across West Africa, J. Hydrol., 581, 124412, https://doi.org/10.1016/j.jhydrol.2019.124412, 2020.

Segele, Z. T., Leslie, L. M., and Lamb, P. J.: Evaluation and adaptation of a regional climate model for the Horn of Africa: rainfall climatology and interannual variability, Int. J. Climatol., 29, 47–65, https://doi.org/10.1002/joc.1681, 2008.

Wang, Y., Sen, O. L., and Wang, B.: A highly resolved regional climate model (IPRC-RegCM) and its simulation of the 1998 severe precipitation event over China. Part I: model description and verification of simulation, J. Climate, 16, 1721–1736, 2003.

Proc. IAHS, 385, 65–69, 2024
https://doi.org/10.5194/piahs-385-65-2024

Open Access PIAHS

Regional multi-objective calibration for distributed hydrological modelling: a decision tree based approach

Matteo Pesce[1], Alberto Viglione[1], Jost von Hardenberg[1], Larisa Tarasova[2], Stefano Basso[2,6], Ralf Merz[2], Juraj Parajka[3,4], and Rui Tong[3,4,5]

[1]Department of Environment, Land and Infrastructure Engineering, Politecnico di Torino, Turin, Italy
[2]Department Catchment Hydrology, Helmholtz Centre for Environmental Research, Halle (Saale), Germany
[3]Centre for Water Resource Systems, TU Wien, Vienna, Austria
[4]Institute of Hydraulic Engineering and Water Resources Management, TU Wien, Vienna, Austria
[5]Department of Hydraulic Engineering, Tsinghua University, Beijing, China
[6]Norwegian Institute for Water Research (NIVA), Oslo, Norway

Correspondence: Matteo Pesce (matteo.pesce@polito.it)

Received: 15 June 2022 – Revised: 20 January 2023 – Accepted: 22 January 2023 – Published: 18 April 2024

Abstract. Large scale modelling is becoming increasingly important in hydrology, particularly to characterize and quantify changes in the hydrological regime, whose drivers are typically large-scale phenomena, up to the global scale (e.g., climate change). This can be done with distributed models by estimating spatially consistent model parameters i.e. parameters having a functional relationship with catchment characteristics. In this study we adopt the newly developed PArameter Set Shuffling (PASS) approach, based on a machine learning decision tree algorithm, for the regional calibration of the TUWmodel over North-Western Italy. The method exploits observed patterns of locally calibrated parameters and catchment (climatic and geomorphological) descriptors, to derive functional relationships between the variables. The calibration procedure is performed by including snow cover information, as captured by MODIS datasets, in the model efficiency function. The results show that the PASS regionalization procedure allows to obtain very good regional model efficiencies, without significant loss of performance when moving from training to test catchments and from calibration to verification period, confirming the robustness of the methodology. We also highlight that using snow information in the calibration procedure is helpful to obtain spatially consistent model parameters for this study area. In the spirit of "obtaining good results for the right reasons", this should be a preferred approach when performing the regional calibration of distributed hydrologic models over mountainous regions.

Keywords. UPH 19; distributed hydrological modelling; parameter regionalization

1 Introduction

One of the main challenges of hydrological modelling at the large scale is the estimation of spatially distributed model parameters, which are consistent with climatic and geomorphologic features of the territory, according to a process-based approach (Archfield et al., 2015; Clark et al., 2016; Gupta et al., 2014; Mizukami et al., 2017; Paniconi and Putti, 2015). The goal is to obtain parameter sets for each gridded element or hydrologic unit having the same functional relationship with climatic and geomorphological variables. Many studies dealt with parameters regionalization techniques, in particular within the context of runoff prediction in ungauged basins (e.g., Blöschl et al., 2013; Merz and Blöschl, 2004; Parajka et al., 2013; Seibert, 1999; Troch et al., 2003). An innovative approach was introduced by Merz et al. (2020), with the idea of using machine learning based algorithms to derive relationships between locally calibrated parameters and catchment descriptors, in order to predict parameters in a spatially distributed mode. The procedure, called PArameter Set Shuffling (PASS), provides good results for plain areas like Ger-

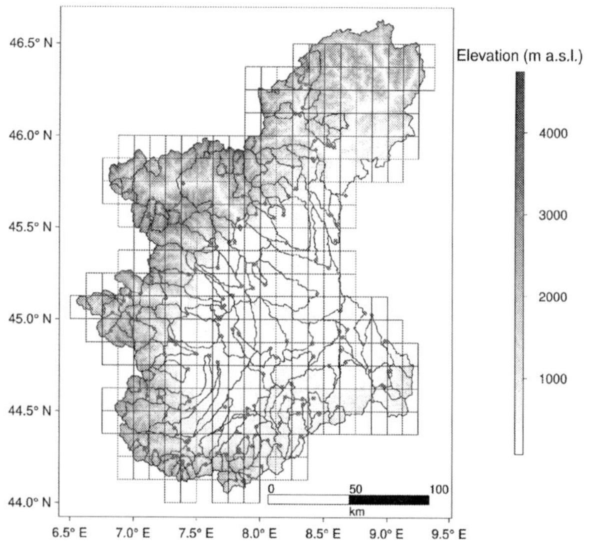

Figure 1. Stream gauges (red points), catchments boundaries (black lines) and pixels with meteorological inputs (blue cells) over Piemonte and Valle d'Aosta. The background represents mean elevation (m a.s.l.).

many but, so far, it has not found application in other regions. This work presents the PASS approach in a more complex environment like the Alpine Region, by integrating snow information in the regionalization procedure.

2 Data and Methods

Data from the stream gauge network managed by the environmental protection agency of the Piemonte Region, in North-Western Italy, are used (Streamflow data, https://www.arpa.piemonte.it/rischinaturali/accesso-ai-dati/annali_meteoidrologici/annali-meteo-idro/banca-dati-idrologica.html, last access: 31 January 2023). The study area is characterized by a range of topographical elevations and a temperate continental climate, but conditions are not uniform in space. In the Alps, a tendency towards more humid conditions prevails, while the central area is characterized by lower annual precipitation and higher temperatures.

Precipitation [mm] and mean air temperature [°C] data are provided by a gridded dataset, obtained by interpolation of daily observations. These observations are taken from a dense network of meteorological stations and cover the period from 1957 up to the present day (Optimal Interpolation (OI) Dataset, http://www.arpa.piemonte.it/rischinaturali/tematismi/clima/confronti-storici/dati/dati.html, last access: 31 January 2023; Arpa Piemonte). The study domain consists in total of 305 pixels of size $0.125° \times 0.125°$ (Fig. 1). Temperature data are used for the calculation of the potential evapotranspiration, by using the Blaney-Criddle equation based on the FAO procedure. A digital elevation model (DEM) at 90 m resolution (Digital elevation model, http://www.earthenv.org/, last access: 31 January 2023) and catchment boundaries for 197 catchments are also considered. Snow cover is mapped by combining MODIS products from Terra (MOD10A1) and Aqua (MYD10AQ) satellites, and in particular daily maps of the Normalized Difference Snow Index (NDSI) at 500 m spatial resolution. NDSI values larger than 40 are identified as snow. The combination is performed to reduce the effect of clouds; pixels which are classified as clouds or missing in Terra (NDSI > 100), are replaced by pixels from Aqua if these are classified as snow covered or snow free. The percentage of snow pixels in each grid cell of Fig. 1 is counted and the value is retained if the percentage of clouds is lower than 60 %. In this study, 79 catchment descriptors are used to infer regionalization rules, divided into 5 categories: climate (precipitation and temperature) statistics, calculated from the OI dataset; the 27 ETCCDI indices (http://etccdi.pacificclimate.org/list_27_indices.shtml, last access: 31 January 2023); morphological descriptors obtained from the digital elevation model (referred to elevation, slope and aspect); land-use classes derived from the Corine Land Cover dataset (2018) and a series of soil characteristics, obtained from the Harmonized World Soil Database.

The relationship between catchment descriptors and model parameters is established using decision trees (rpart R function). The PASS procedure is applied together with the TUWmodel (Appendix of Parajka et al., 2007), which follows the structure of the HBV model (Bergström, 1995). The algorithm used for local calibration is the differential evolution (DEoptim R package), which implements the global optimization of a real-valued function of a real-valued parameter vector (Mullen et al., 2011). We consider as model efficiency function the arithmetic average of two quantities:

– the Kling-Gupta Efficiency (KGE) for discharge:

$$1-\sqrt{(r-1)^2+\left(\frac{\mathrm{CV}_{\mathrm{sim}}}{\mathrm{CV}_{\mathrm{obs}}}-1\right)^2+\left(\frac{\mu_{\mathrm{sim}}}{\mu_{\mathrm{obs}}}-1\right)^2}$$

where r is the linear correlation coefficient between observed and simulated discharge, $\mathrm{CV}_{\mathrm{sim}}/\mathrm{CV}_{\mathrm{obs}}$ is the ratio between the coefficients of variation of simulated and observed discharges, $\mu_{\mathrm{sim}}/\mu_{\mathrm{obs}}$ is the ratio between the means of simulated and observed discharges.

– a Snow Cover Efficiency, which is given by the following formula:

$$\mathrm{SC}=1-\frac{1}{n\cdot N}\sum_{1}^{n}\left(\sum_{1}^{N}|\mathrm{SC}_{\mathrm{sim}}-\mathrm{SC}_{\mathrm{obs}}|\right)$$

where N is the number of days, n is the number of cells, $\mathrm{SC}_{\mathrm{sim}}$ is 1 if the simulated snow water equivalent (SWE) exceeds 1 mm and 0 otherwise, and $\mathrm{SC}_{\mathrm{obs}}$ is 1 if the % of snow pixels within the cell is greater than 10 % and 0 otherwise.

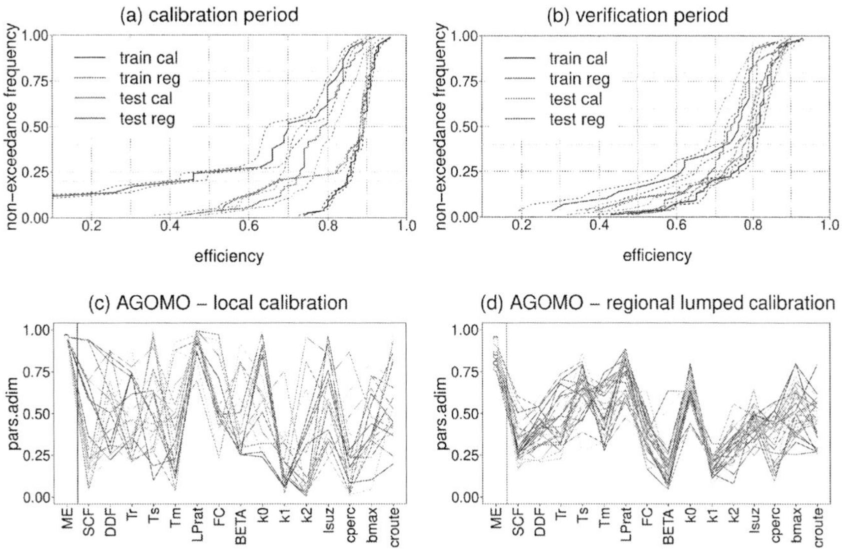

Figure 2. **(a)** Non exceedance cumulative distribution function of model efficiencies ($ME = 0.5 \cdot KGE + 0.5 \cdot SC$) during the calibration period: "train_cal" shows MEs obtained by using locally calibrated lumped parameters for 76 training catchments, "test_cal" shows MEs obtained by using locally calibrated lumped parameters for 28 test catchments, "train_reg" shows MEs obtained by running the model with predicted distributed parameters for 76 training catchments, "test_reg" shows MEs obtained by running the model with predicted distributed parameters for 28 test catchments. **(b)** Non exceedance cumulative distribution function of model efficiencies ($ME = 0.5 \cdot KGE + 0.5 \cdot SC$) during the verification period. The continuous lines represent the mean model efficiency, while the dashed lines represent the 10th and 90th percentiles of model efficiencies for 30 parameter sets. **(c)** 30 normalized parameter sets found by local calibration for the Agogna river catchment at Momo, North-Western Italy. **(d)** 30 normalized parameter sets found by regional calibration for the Agogna river catchment at Momo, North-Western Italy.

The model efficiency is so defined:

$$ME = 0.5 \cdot KGE + 0.5 \cdot SC$$

From the entire database, only stations having discharge data over the period 2000–2020 are considered for the analysis (104 stations). We split the timeseries into a calibration period (2010–2020) and a verification period (2000–2010). We perform the local calibration over the 104 catchments; from this set, only catchments with at least 5 parameter sets having local model efficiency (ME) > 0.75 and having a catchment area $< 1000\,km^2$ are used for the training of PASS, following the assumptions of the method (Merz et al., 2020). These are the so called training catchments. For these sites, the best 30 parameter sets are used for running the regionalization. The remaining catchments, called test catchments, are instead used for validation.

3 Results and discussion

The results for the calibration period (2010–2020) and the verification period (2000–2010) are shown in Fig. 2a and b. Considering the calibration period, while the medians of the mean performances of training and test catchments in local calibration ("train cal" and "test cal", respectively) are almost identical, applying the regionalization procedure, the median of MEs of regionally predicted distributed parameters sets for the training catchments ("train reg") is 0.785, and the median regional efficiency for the test catchments ("test reg") is 0.700.

When moving to the verification period (2000–2010), the local model efficiencies have a substantial decrease. The median of regionally distributed MEs for training catchments decreases from 0.785 to 0.740, while the median of regionally distributed MEs for test catchments increases from 0.700 to 0.760. In general, we observe that the medians of regional MEs for training and test catchments are very close and the regional performance is not significantly degraded when moving from the calibration to the verification period. This confirms the robustness of the methodology in parameter estimation across spatial and temporal scales. In Fig. 2c and d, examples of normalized locally calibrated and regionally lumped calibrated parameter sets for the Agogna River at Momo are shown. Parameters found by regional calibration show a much smaller spread compared to locally calibrated parameters. As found in Merz et al. (2020), also for this Italian dataset regionalization reduces the effect of parameter equifinality.

Observing the maps of distributed model parameters, they appear to be highly consistent with the climatic and geomorphological features of the region; in particular snow param-

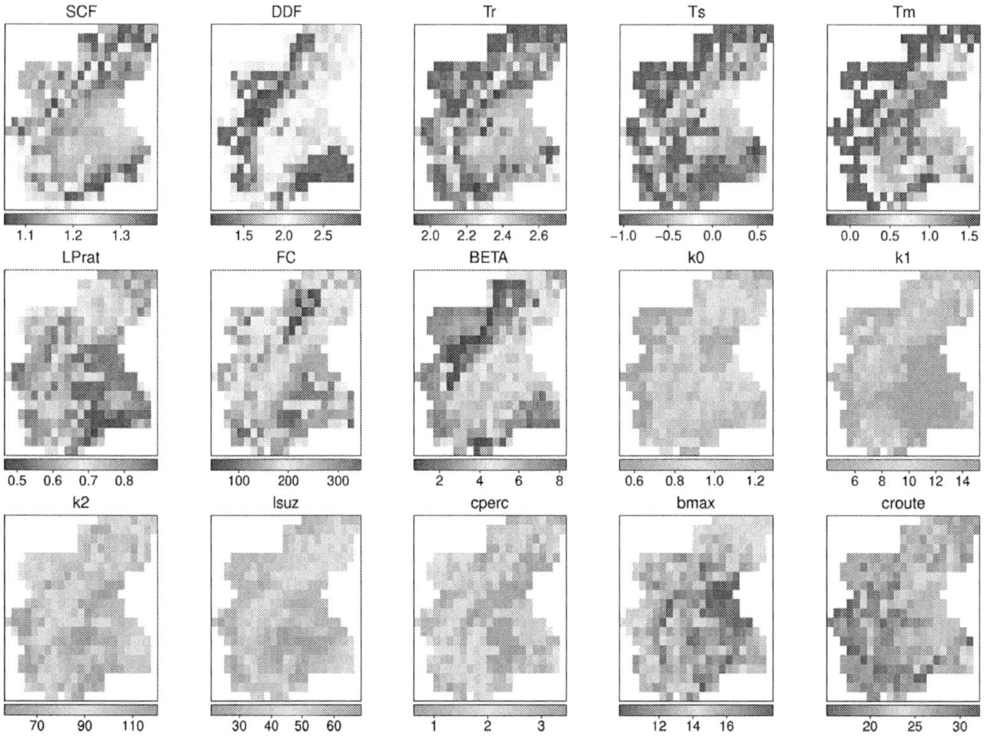

Figure 3. Regional maps of distributed TUWmodel parameters obtained with PASS. The color indicates the mean value of the parameter among 30 regionalizations.

eters (SCF, DDF, Tr, Ts, Tm) seem to be in accordance with the three different geographical areas which characterize the region: the Alpine range on the West and North, the Apennines on the South and the Po River Valley (Fig. 3).

4 Conclusions

The results of this study indicate that the PASS procedure can be efficiently applied over the Alpine Region, with satisfactory results, also by taking advantage of snow cover information for the calibration and validation of the model. The model efficiencies for train catchments don't decrease significantly when moving from calibration to the verification period and actually for test catchments an increase is observed. This confirms the robustness of the procedure in the parameter estimation over different spatial and temporal domains. Overall, the results for this study area are in line with what has already been found in Merz et al. (2020). The spatial distribution of the parameters is consistent with the climatic and geomorphological characteristics of the region and this confirms that parameters are controlled by these variables. Following a process-based approach, the use of snow information can be crucial for a good model representation of the processes. This will allow to get "right results for the right reasons", which is of particular importance in hydrology, especially for large-scale distributed modelling studies which require ungauged streamflow prediction.

Data availability. Precipitation, temperature and daily discharge data are freely available on the ARPA Piemonte website (see Sect. 2 for further information). The other types of data can be provided by the corresponding author upon request.

Author contributions. LT, SB and RM provided suggestions for organizing the code and data; JP and RT prepared the snow data; AV and MP worked on the data; MP wrote the manuscript draft; AV, JvH and LT reviewed and edited the manuscript.

Competing interests. The contact author has declared that none of the authors has any competing interests.

Disclaimer. Publisher's note: Copernicus Publications remains neutral with regard to jurisdictional claims in published maps and institutional affiliations.

Special issue statement. This article is part of the special issue "IAHS2022 – Hydrological sciences in the Anthropocene: Variability and change across space, time, extremes, and interfaces". It is a result of the XIth Scientific Assembly of the International Association of Hydrological Sciences (IAHS 2022), Montpellier, France, 29 May–3 June 2022.

Acknowledgements. We would like to thank the Environmental Agency of Piemonte Region (ARPA Piemonte), in particular the Department of Forecasting Systems and the Department of Natural and Environmental Risks, for providing data used in this study.

Review statement. This paper was edited by Christophe Cudennec and reviewed by two anonymous referees.

References

Archfield, S. A., Clark, M., Arheimer, B., Hay, L. E., McMillan, H., Kiang, J. E., Seibert, J., Hakala, K., Bock, A., Wagener, T., Farmer, W. H., Andreassian, V., Attinger, S., Viglione, A., Knight, R., Markstrom, S., and Over, T.: Accelerating advances in continental domain hydrologic modeling, Water Resour. Res., 51, 10078–10091, https://doi.org/10.1002/2015WR017498, 2015.

Bergström, S.: The HBV model, Computer Models of Watershed Hydrology, edited by: Singh, V. P., Water Resources Publications, Highlands Ranch, Colorado, 443–476, ISBN 9781887201742, 1995.

Blöschl, G., Sivapalan, M., Wagener, T., Savenije, H., and Viglione, A. (Eds.): Runoff prediction in ungauged basins: synthesis across processes, places and scales, Cambridge University Press, 484 pp., ISBN 9781107028180, 2013

Clark, M. P., Schaefli, B., Schymanski, S. J., Samaniego, L., Luce, C. H., Jackson, B. M., Freer, J. E., Arnold, J. R., Moore, R. D., Istanbulluoglu, E., and Ceola, S.: Improving the theoretical underpinnings of process-based hydrologic models, Water Resour. Res., 52, 2350–2365, https://doi.org/10.1002/2015WR017910, 2016.

Gupta, H. V., Perrin, C., Blöschl, G., Montanari, A., Kumar, R., Clark, M., and Andréassian, V.: Large-sample hydrology: a need to balance depth with breadth, Hydrol. Earth Syst. Sci., 18, 463–477, https://doi.org/10.5194/hess-18-463-2014, 2014.

Merz, R. and Blöschl, G.: Regionalisation of catchment model parameters, J. Hydrol., 287, 95–123, https://doi.org/10.1016/j.jhydrol.2003.09.028, 2004.

Merz, R., Tarasova, L., and Basso, S.: Parameter's Controls of Distributed Catchment Models – How Much Information is in Conventional Catchment Descriptors?, Water Resour. Res., 56, e2019WR026008, https://doi.org/10.1029/2019WR026008, 2020.

Mizukami, N., Clark, M. P., Newman, A. J., Wood, A. W., Gutmann, E. D., Nijssen, B., Rakovec, O., and Samaniego, L.: Towards seamless large-domain parameter estimation for hydrologic models, Water Resour. Res., 53, 8020–8040, https://doi.org/10.1002/2017WR020401, 2017.

Mullen, K., Ardia, D., Gil, D., Windover, D., and Cline, J.: DEoptim: an R package for global optimization by differential evolution, J. Stat. Softw., 40, 1–26, https://doi.org/10.18637/jss.v040.i06, 2011.

Paniconi, C. and Putti, M.: Physically based modeling in catchment hydrology at 50: Survey and outlook, Water Resour. Res., 51, 7090–7129, https://doi.org/10.1002/2015WR017780, 2015.

Parajka, J., Merz, R., and Blöschl, G.: Uncertainty and multiple objective calibration in regional water balance modelling: case study in 320 Austrian catchments, Hydrol. Process., 21, 435–446, https://doi.org/10.1002/hyp.6253, 2007.

Parajka, J., Viglione, A., Rogger, M., Salinas, J. L., Sivapalan, M., and Blöschl, G.: Comparative assessment of predictions in ungauged basins – Part 1: Runoff-hydrograph studies, Hydrol. Earth Syst. Sci., 17, 1783–1795, https://doi.org/10.5194/hess-17-1783-2013, 2013.

Seibert, J.: Regionalisation of parameters for a conceptual rainfall-runoff model, Agr. Forest Meteorol., 98, 279–293, https://doi.org/10.1016/S0168-1923(99)00105-7, 1999.

Troch, P. A., Paniconi, C., and van Loon, E. E.: Hillslope-storage Boussinesq model for subsurface flow and variable source areas along complex hillslopes: 1. Formulation and characteristic response, Water Resour. Res., 39, 1316, https://doi.org/10.1029/2002WR001728, 2003.

Proc. IAHS, 385, 71–77, 2024
https://doi.org/10.5194/piahs-385-71-2024

IAHS2022 – Past and future of open, inclusive, innovative, and society-interfacing approaches

Agent-Based Modelling for representing water allocation methodologies in the irrigation system of the Formoso River Basin, Brazil

Déborah Sousa[1], **Conceição Alves**[1], **Fernán Vergara**[2], **Cássio Coelho**[3], **and Célia Ralha**[3]

[1]Department of Civil and Environmental Engineering, University of Brasília, Brasília, 70910-900, Brazil
[2]Department of Environmental Engineering, Federal University of Tocantins, Palmas, 77001-090, Brazil
[3]Department of Computer Science, University of Brasília, Brasília, 70910-900, Brazil

Correspondence: Déborah Sousa (deborahsousa.eng@gmail.com) and Conceição Alves (cmaalves@gmail.com)

Received: 29 May 2022 – Revised: 23 January 2023 – Accepted: 24 January 2023 – Published: 18 April 2024

Abstract. The intersection of water and social systems is strictly linked to society's adjacent political and economic contexts, which adds complexity to water-related issues, such as new demographical, market, and behavioural conditions. Traditional hydrological modelling approaches have often failed to include social aspects and their impacts on hydrological processes when representing human actions as fixed external conditions, thus, neglecting dynamic and heterogeneous behaviours that are intrinsically human. As an alternative, Agent-Based Models (ABMs) can represent the complexity of modelled water resources systems and their interchange with social contexts. The Brazilian Formoso River Basin (FRB) is a human-water system characterized by intense agricultural activity where conflicts among water users are present. Aligned with the principles of socio-hydrology, this study proposes an ABM to represent the interaction of natural and social systems in the Urubu River basin, a sub-basin of the FRB. The model considers farmers and regulatory authorities as agents who act according to the Belief-Desire-Intention (BDI) paradigm. Farmers are heterogeneous decision-makers with distinct collaborative profiles towards water management strategies. Finally, a methodology to extract information on farmers' behaviour towards cooperation from water demand data is presented, which might contribute to the conceptualization of more complex socio-hydrological models.

Keywords. Water allocation; agent-based model; socio-hydrology; conceptualization; UPH 18; SDG 6; Formoso River Basin

1 Introduction

Water is a common-pool resource whose lack of effective regulation may lead to the "tragedy of the commons" demonstrated by Hardin (1968). Thus, the adequate allocation of water among users is imperative. Water allocation has been important to support communities by ensuring water availability for essential activities, such as human consumption, sanitation, food production, and other economic activities. Consequently, changes in society have resulted in new challenges for problems related to water and its allocation (Dinar et al., 1997). In addition, the interface of water systems with political, social and economic contexts adds complexity to water allocation, especially in watersheds characterized by water use conflicts (Carraro et al., 2006).

Particularly, in Brazilian agricultural basins the decrease in water availability, whether due to fluctuations in the natural hydrological regime as a result of climate change or due to the intensification of consumption by users, may increase the frequency of water crisis episodes. Consequently, uncertainty about water supply may generate or intensify conflicts among irrigation users themselves, as well as enhance competition between this and other uses (ANA, 2021), including non-consumptive uses, such as the ecological flow.

Traditional models developed to support decision-making regarding water resources often fail to consider social aspects and their associated heterogeneities in the representation of hydrological processes by considering human actions as fixed boundary conditions (Khan et al., 2017). As an alternative, Agent-Based Models (ABMs) can represent the complexity of systems in numerous environmental applications, such as land use and cover change (Ralha et al., 2013) and urban water supply (Berglund, 2015). Such models are based on a bottom-up approach where system-level outcomes result from agent-level interactions. These agents represent autonomous entities that inhabit a shared environment and follow behavioural rules to achieve their own interests.

In the water resources domain, agent-based modelling follows the concept of the socio-hydrology (SH) discipline (Sivapalan et al., 2012). Socio-hydrology aims to explore and understand the dynamic and co-evolutionary aspects of coupled human-water systems, including the possible generation of emergent behaviours from this two-way interaction. SH is also comprised within the scientific decade 2013–2022 of the International Association of Hydrological Sciences (IAHS), entitled "Panta Rhei – Everything Flows" (Montanari et al., 2013), dedicated to research activities regarding changes in both hydrology and society.

Additionally, measuring and modelling social information from existing data in order to incorporate them into socio-hydrological analysis is still a work in progress, as expressed by the 18th Unsolved Problem in Hydrology (UPH 18): "How can we extract information from available data on human and water systems in order to inform the building process of socio-hydrological models and conceptualisations?" (Blöschl et al., 2019).

Regarding the reasoning of modelled agents, the Belief-Desire-Intention (BDI) is a paradigm based on Bratman (1987)'s human practical reasoning model. The BDI architecture has been used to formalize the behaviour of more elaborate cognitive agents, and thus represent the complexity of human decision-making, including that of farmer agents (Kock, 2008; Liang et al., 2016; Taillandier et al., 2017).

The implementation of coupled socio-hydrological models allowing the integration of water users' decisions, represented by a BDI agent model, into hydrological models may contribute to the evaluation of management strategies including the negotiable water allocation, especially during drought events or dry seasons.

In order to support the management of conflicts over water use, this work presents an agent-based model to represent the effect of the cooperative behaviour of farmers on water availability in the context of a Brazilian agricultural river basin. The model is focused on simulations during the dry season, when water diversion for irrigation is the major use and water levels are low.

2 Methodology

2.1 Case study overview

The proposed socio-hydrological framework has been developed based on the water allocation process in the Formoso River Basin (FRB), sited in the state of Tocantins, north of Brazil. The FRB is an agricultural basin and covers approximately 21 300 km^2 (Fig. 1a). The region receives an average of 1600 mm of precipitation annually. However, the monthly average precipitation ranges from 1–5 mm during the dry season (May–September), to 120–320 mm during the wet season (October–April) (Alves et al., 2016; Fragoso et al., 2021).

The Irrigation District of the Formoso river (DIRF) was established in 1980 and is now the second-largest public irrigation project in Brazil, covering an area of over 20 000 ha (ANA, 2021). Therefore, agricultural activity plays a central role in the local economy and is by far the major water user in the basin. The region's strong seasonality along with its hydrophilic soil allows double-cropping systems to be established, with alternate cultivation of two crops in the same area over an annual cycle, mainly rice and soybean seeds. Rice crops are flood-irrigated, whereas soybean crops are irrigated with subsurface irrigation through channels supplied by stationary pumps (Tocantins, 2007; Vergara et al., 2013; Silva, 2015).

There are 52 agricultural properties that operate a total of 105 active stationary hydraulic pumps in the FRB (GAN, 2022). These pumps divert water from the rivers (Formoso, Urubu, Duerê, or Xavante) for irrigation purposes in the basin, each with a daily average demand of 1620 $L\,s^{-1}$ (IAC, 2018). Some of the properties withdraw water from more than one river.

The cultivation of soybean seeds in the FRB during the dry season represents the critical period concerning water availability in the basin (Silva, 2015). Consequently, the reduction of water availability along with high levels of irrigation water consumption led to streamflow depletion in some stretches of the Formoso and Urubu rivers in June–July 2016, jeopardizing the water security in the region. Judicial interventions were introduced, including the temporary suspension of water use permits and associated pumps operation, resulting in the increase of conflicts among water users (IAC, 2018).

Technical measures were requested by the State to the Federal University of Tocantins (UFT). The GAN (High-Level Management) project is an initiative of several specialists from the Institute of Attention to Cities (IAC), linked to the UFT. The GAN project resulted in the proposal of a series of water management rules and strategies in the 2018–2019 Biennium Plan (IAC, 2018). One of these strategies involved installing, operating, and transmitting water consumption data from irrigation pumps in real-time and continuously every 15 min (https://gan.iacuft.org.br/, last access: 23 January 2023).

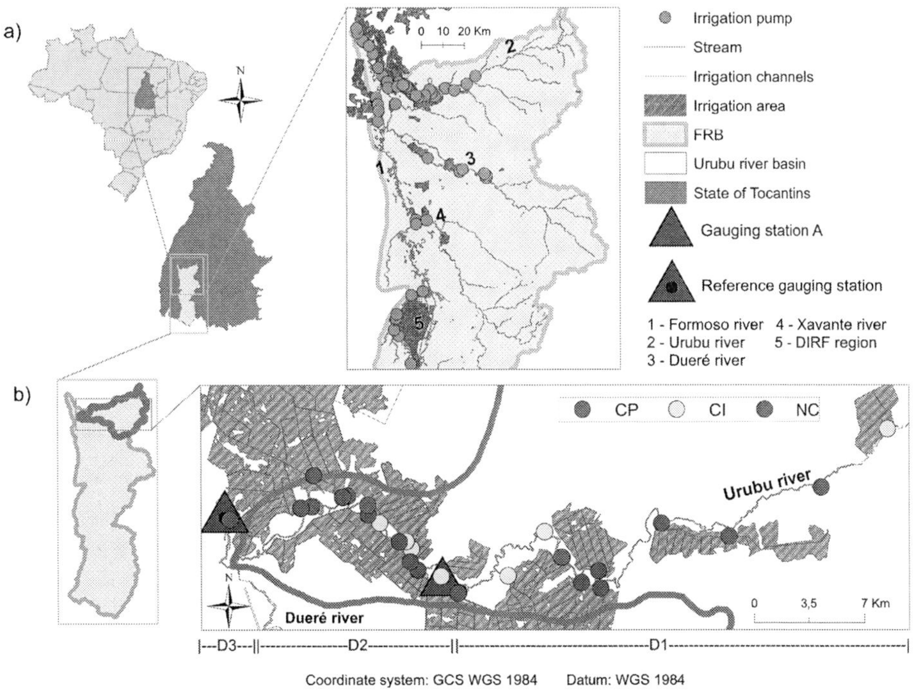

Figure 1. Map and overview of the irrigation network and irrigation area in **(a)** the Formoso River Basin (FRB), and **(b)** the Urubu river basin. Data source: GAN (2022).

2.2 Modelling proposal

Based on the Formoso River basin case, the model presented in this study provides a framework to represent farmers' decisions in agricultural basins towards water use during the dry season. It builds the representation of a socio-hydrological system in order to assess the economic, social, and environmental performance of different configurations of water allocation and management strategies.

Heterogeneity of the decision-making of farmer agents was introduced in the model. Based on reports and interviews with local experts carried out in February 2022, each irrigation pump operation was categorized as collaborative/non-collaborative according to its owner's engagement in the technical recommendations and maintenance indicated by the GAN project. Furthermore, the results of the interviews were confronted with the GAN database of water withdrawals during 2019–2021 describing the farmers' commitment to providing automatic water withdrawal information during soybean season (May–August).

These analyses resulted in the definition of three behaviour profiles towards water use (Table 1). Cooperative-Proactive (CP) behaviour is related to those farmers who agree with the GAN project and also cooperate with frequent data transmission within the GAN system for a certain pump ($\leq 25\,\%$ data transmission failure). Conversely, if a farmer consents to the project, but does not collaborate with consistent data transmission ($> 25\,\%$ transmission failure), their operational behaviour was classified as Cooperative-Ideological (CI). Finally, if a farmer does not collaborate with the project itself nor with the data transmission, they were classified as Non-Cooperative (NC).

An Agent-Based Model (ABM) is proposed to represent the social behaviour and interaction among actors in the basin. The initial conceptualization was outlined using the Tropos methodology for agent-oriented software development (Bresciani et al., 2004).

3 Proposed Agent-Based Model

The proposed ABM was based on the fundamental interrelation between two higher-level entities: Water Users and Water Regulatory entities. Water users' main goal was considered as achieving a welfare state, which is an abstract concept that may have different meanings depending on the user type, e.g. industry users aim to maximize their profit, whereas household water consumption users desire to have their basic sanitary and domestic necessities fulfilled. The present ABM considers water use for agricultural purposes, so the water user agents are farmers.

In general, the main role played by agricultural agents in ABMs is the consumption of water for irrigation purposes. Their attributes may include their geographic location, the size of their property, as well as demographic and socioe-

Table 1. Classification and Distribution of farmer agents' behaviour in the Urubu River basin. Data source: GAN (2022).

Behaviour profile	Agrees with project	Data transmission >75 %	Farmer agents in each category (%) ($n=24$)
CP[a]	Yes	Yes	17 %
CI[b]	Yes	No	29 %
NC[c]	No	No	54 %

[a] Cooperative-Practical. [b] Cooperative-Ideological. [c] Non-Cooperative.

conomic aspects (Kaiser et al., 2020). Since farmer agents are instances of water users, they also aim to reach welfare, which is represented here by fulfilling the objective of generating income through crop production. Each property and associated irrigation pumps were considered as being operated by a single farmer agent.

Regulatory agents, who may be national or local authorities, have as their main objective to provide sustainable water resources to water users through regulatory instruments. In Brazil, ANA (National Water and Sanitation Agency) is the national regulatory entity for water resources that cross more than one state, whereas water resources located within the boundaries of a single state are operated by local regulators. In the specific case of the Formoso river basin, the local regulatory agency is represented by Naturatins, which provides regulatory services for all the state of Tocantins.

Regulatory entities may issue water permits to water users, according to water availability, economic objectives, and environmental protection. Their ability to communicate with other agents is a very relevant attribute. In general, the reasoning of regulatory agents may include knowledge bases on water use permits, previous allocation rules, and historical records of hydrological conditions (Kaiser et al., 2020).

3.1 Results for the Urubu River basin

Following the methodology presented in Sect. 2.2, the cooperative behaviour of the 24 farmer agents whose pumps are located in the Urubu River basin (Fig. 1b) was characterized as: most of them act as Non-Cooperative (54 %), followed by Cooperative-Ideological (29 %) and Cooperative-Proactive (17 %). A summary of this analysis can be seen in Table 1. The Urubu River basin is modeled as representative of the FRB social-hydrological interaction since all three behaviour profiles could be identified there.

Based on the work of Volken (2022), irrigation pumps were gathered in three different demand groups according to their location in the basin: (i) D1, before gauging station A; (ii) D2, after gauging station A and before the Dueré river confluence; (iii) D3, after the Dueré confluence and before the reference gauging station (Fig. 1b).

Table 2. Water use rules for the Urubu River basin. Source: IAC, IAC (2018).

Stream	Gauging Station	Yellow stage[a]	Red stage[b]
Urubu	Reference	3.98 m (1 July)	2.20 m (1 August)

In our model, levels below the ones described, or starting from the date in parentheses, induce the following rules: [a] Water demand decrease. [b] Water withdrawal interruption.

3.2 Agent decision-making process

In the present agent-based modelling exercise, both farmers and the regulatory entity are decision-makers who act according to the Belief-Desire-Intention paradigm (BDI).

The Biennium Plan (IAC, 2018) is a drought mitigation plan developed for the FRB, which contains management measures (water use rules) triggered by predefined dates and water levels in specific sites along the main channels of the basin. According to historical data identified in reference gauging stations, the Biennium Plan issued two different water use rules: (i) Yellow stage is established when water levels go down below the attention threshold level, or on the 1st of July, whichever happens first, initiating a slot rotation in the irrigation network; (ii) red stage is defined when levels go down below the critical threshold level, or on the 1st of August, whichever happens first, inducing the interruption of water withdrawals. The Yellow stage was implemented in the model reducing the original water withdrawal. The rules defined for the Urubu river are provided in Table 2. These rules apply to the pumps (water demand sites) located upstream of the stream's respective reference gauging station.

The local water regulator agent for the Urubu River basin case was modelled as having the single "Regulate" capability, that is, issuing and communicating the Biennium Plan rules to farmers according to the current date and/or observed streamflow levels.

Farmer agents were defined with two capabilities: (i) Irrigate, and (ii) Generate Income, the former being a necessary condition for the latter. The most important decision of the farmer agent is defined at the beginning of the simulation when they decide the total land area they will reserve to grow soybean seeds and consequently sell as crop production at the end of the crop cycle.

In order to accomplish the "Irrigate" desire, farmers take into consideration the following factors: their perception of water supply quantity in the previous wet season, the neighbourhood effect within their demand group (D1, D2 or D3), the Biennium Plan rules on the Urubu river, the current date and stream level, and their potential irrigable area. These are the farmer's beliefs which are applied to their decision of expected irrigated area (EIA) by introducing multiplicative factors (α and β, dimensionless) that either reduce or maintain their total potential irrigable area (PIA) (Eq. 1).

Table 3. Multiplicative factors' definitions and potential values for each behaviour profile.

Factor	Definition	Behaviour profile	Range
α	Previous wet season perception	CP	[0.89–0.95] ∪ 1.0[a]
		NC	1.0
		CI	[0.92–1.0]
β	Influence of neighbours profile	CP	[0.89–0.95]
		NC	[0.9–1.0]
		CI	[0.92–1.0]

[a] First condition of Rule 1.

CP and CI profiles are assumed to have an inclination to reduce their areas more than NC agents. Thus, by combining α and β values (Table 3), CP farmers were considered to reduce up to 20 % (minimum of 11 %), whereas that percentage was defined as 15 % and 10 % for CI and NC farmers, respectively.

$$\text{EIA} = \text{PIA} \times \alpha \times \beta \tag{1}$$

Moreover, a set of behavioural rules were defined (Rules 1–6):

- Rule 1: IF previous wet season ≥ average THEN $\alpha = 1$ ELSE $\alpha \leq 1$ (see Table 3).
- Rule 2: IF current day = 1 July OR water level ≤ attention level THEN reduce water withdrawal
- Rule 3: IF current day = 1 August OR water level ≤ critical level THEN interrupt water withdrawal
- Rule 4: IF current day = 1 July AND water level ≤ attention level THEN reduce water withdrawal
- Rule 5: IF current day = 1 August AND water level ≤ critical level THEN suspend water withdrawal
- Rule 6: IF nNC / (nT) > 0.75 THEN max(β) ELSE IF nCP / (nT) > 0.75 THEN min(β) ELSE β = rnd().

For all three profiles, if their perception of the previous wet season is that it was greater than or equal to the average, then there is no reduction on their expected irrigation area (first condition of Rule 1). On the contrary, a reduction factor is considered for cooperative farmers (Table 3).

The Biennium Plan regulation, as seen from Table 2, depends on the current date and/or observed stream level. CP farmers were considered to entirely comply to this regulation (Rules 2 and 3), whereas CI farmers were assumed as being more resistant to them, needing both level and date criteria to be true (Rules 4 and 5) to collaborate. NC farmers were considered as maintaining their water consumption regardless of the Biennium Plan commands.

Neighbourhood effect was considered based on the overall predominant profile within the respective demand group D_i ($i = 1, 2, 3$) of farmers (Rule 6). If the number of NC farmers (nNC) in the D_i group is greater than 75 % of the total number of farmers in that group (nT), farmers are influenced by non-collaborative attitudes. Thus, for all categories of farmer agents in that group, β is assumed as their respective maximum value for that factor (upper bound values for β from Table 3). On the contrary, if the number of CP farmers (nCP) is greater than 75 %, farmers are stimulated to act cooperatively, and thus β is given by the lower bound values. For other conditions, β is assumed as a random value within the intervals presented in Table 3 according to a uniform distribution.

Finally, in order to achieve the desire of Generating income, at the end of the simulation, the actual Total Irrigated Area (TIA) (Eq. 2) is calculated from the Total Consumed Water (TCW, [volume]) multiplied by a factor that correlates TIA to TCW (AVf, [area volume^{-1}]). Consequently, the total revenue (Rev) is calculated as the TIA ([area]) multiplied by the productivity (p, [mass area^{-1}]) and the soybean market price (smp, [price mass^{-1}]) (Eq. 3).

$$\text{TIA} = \text{TCW} \times \text{AVf} \tag{2}$$

$$\text{Rev} = \text{TIA} \times p \times \text{smp} \tag{3}$$

The neighbourhood influence is being considered in order to evaluate the joint effect of a collective water permit which is a command and control management tool being evaluated by the National Water Agency in Brazil. Enacting new regulations, issuing penalties or incentives, and introducing educational measures may influence users' behaviours (Akhbari and Grigg, 2013). All of these management measures may interfere in the current distribution of behaviours presented in Table 1.

Thus, the introduction of new factors on the conceptualization of farmers' BDI, or the simulation of scenarios with farmers assuming different configurations might help assess the effect of social behaviours of farmers on total water withdrawals and the basin's economy, and, accordingly the importance of initiatives that enhance collaboration among users. Additionally, the integration of the presented ABM into a hydrological model resent might contribute to visualizing the impact of farmers' decisions on the streamflow availability.

4 Conclusions

This paper proposes an Agent-Based Model to represent the decision of farmers toward water use and regulatory rules. The model encompasses the profiles of farmers in the Formoso River basin, an agricultural basin where drought events jeopardized economic activities in the last few years. A framework considering the interaction between a simple regulatory agent and farmers is presented. Based on the BDI architecture, agents act according to specific intentions aligned with their beliefs in order to achieve distinct desires, namely, income generation, and sustainable water resources, in the case of farmers and water regulators, respectively. The update of the presented framework by considering other behaviour configurations for farmers, as well as other profiles for the regulatory agent, may help elucidate important aspects of both regulation and cooperation mechanisms within the basin. Finally, the development of a methodology that evaluates both quantitative and qualitative data to extract information on farmers' behaviour adds to the conceptualization of socio-hydrological models.

Data availability. Some data can be provided by the corresponding authors upon request.

Author contributions. Conceptualization: All authors. Data curation: DS, CA, FV. Formal analysis: DS, CA. Methodology: All authors. Project administration and Resources: FV. Supervision: CA, FV. Validation: FV. Writing – original draft preparation: DS, CA. Writing – review and editing: All authors.

Competing interests. The contact author has declared that none of the authors has any competing interests.

Disclaimer. Publisher's note: Copernicus Publications remains neutral with regard to jurisdictional claims in published maps and institutional affiliations.

Special issue statement. This article is part of the special issue "IAHS2022 – Hydrological sciences in the Anthropocene: Past and future of open, inclusive, innovative, and society-interfacing approaches". It is a result of the XIth Scientific Assembly of the International Association of Hydrological Sciences (IAHS 2022), Montpellier, France, 29 May–3 June 2022.

Acknowledgements. This paper contains part of the research from the Master's Dissertation of the first corresponding author, Déborah Sousa, developed at the University of Brasília, Brazil. Professor Célia Ralha thanks the support received from the Brazilian National Council for Scientific and Technological Development (Conselho Nacional de Desenvolvimento Científico e Tecnológico – CNPq), grant no. 309688/2021-3. The authors are also grateful to the editor and the two anonymous reviewers whose valuable comments helped improve the quality of this paper.

Financial support. This research has been supported by the Coordination for the Improvement of Higher Education Personnel – Brazil (CAPES) – CAPES-ANA-DPB Program (grant no. 88887.144867/2017-00).

Review statement. This paper was edited by Christophe Cudennec and reviewed by two anonymous referees.

References

Akhbari, M. and Grigg, N.: A Framework for an Agent-Based Model to Manage Water Resources Conflicts, Water Resour. Manag., 27, 4039–4052, https://doi.org/10.1007/s11269-013-0394-0, 2013.

Alves, K. C. C. L. F., Viola, M., de Mello, C., Giongo, M., and dos Santos, A.: Distribuição da precipitação mensal, anual e máxima diária anual na bacia hidrográfica do rio Formoso Tocantins, Ambiência, 12, 49–70, https://doi.org/10.5935/ambiencia.2016.01.03, 2016.

ANA: Atlas Irrigação: uso da água na agricultura irrigada, Agência Nacional de Águas (ANA), MDR, Governo Federal, 2 Edn., http://atlasirrigacao.ana.gov.br (last access: 23 January 2023), 2021.

Berglund, E. Z.: Using Agent-Based Modeling for Water Resources Planning and Management, J. Water Resour. Plan. Manag., 141, 04015025, https://doi.org/10.1061/(ASCE)WR.1943-5452.0000544, 2015.

Blöschl, G., Bierkens, M. F., Chambel, A., Cudennec, C., Destouni, G., Fiori, A., Kirchner, J. W., McDonnell, J. J., Savenije, H. H., Sivapalan, M., Stumpp, C., Toth, E., Volpi, E., Carr, G., Lupton, C., Salinas, J., Széles, B., Viglione, A., Aksoy, H., Allen, S. T., Amin, A., Andréassian, V., Arheimer, B., Aryal, S. K., Baker, V., Bardsley, E., Barendrecht, M. H., Bartosova, A., Batelaan, O., Berghuijs, W. R., Beven, K., Blume, T., Bogaard, T., de Amorim, P. B., Böttcher, M. E., Boulet, G., Breinl, K., Brilly, M., Brocca, L., Buytaert, W., Castellarin, A., Castelletti, A., Chen, X., Chen, Y., Chen, Y., Chifflard, P., Claps, P., Clark, M. P., Collins, A. L., Croke, B., Dathe, A., David, P. C., de Barros, F. P. J., de Rooij, G., Baldassarre, G. D., Driscoll, J. M., Duethmann, D., Dwivedi, R., Eris, E., Farmer, W. H., Feiccabrino, J., Ferguson, G., Ferrari, E., Ferraris, S., Fersch, B., Finger, D., Foglia, L., Fowler, K., Gartsman, B., Gascoin, S., Gaume, E., Gelfan, A., Geris, J., Gharari, S., Gleeson, T., Glendell, M., Bevacqua, A. G., González-Dugo, M. P., Grimaldi, S., Gupta, A. B., Guse, B., Han, D., Hannah, D., Harpold, A., Haun, S., Heal, K., Helfricht, K., Herrnegger, M., Hipsey, M., Hlaváčiková, H., Hohmann, C., Holko, L., Hopkinson, C., Hrachowitz, M., Illangasekare, T. H., Inam, A., Innocente, C., Istanbulluoglu, E., Jarihani, B., Kalantari, Z., Kalvans, A., Khanal, S., Khatami, S., Kiesel, J., Kirkby, M., Knoben, W., Kochanek, K., Kohnová, S., Kolechkina, A., Krause, S., Kreamer, D., Kreibich, H., Kunstmann, H., Lange, H., Liberato, M. L. R., Lindquist, E., Link, T., Liu, J., Loucks, D. P., Luce, C., Mahé, G., Makarieva, O., Malard, J., Mashtayeva, S., Maskey, S., Mas-Pla, J., Mavrova-Guirguinova, M., Mazzoleni, M., Mernild, S., Misstear, B. D., Montanari, A., Müller-Thomy, H., Nabizadeh, A., Nardi, F., Neale, C., Nesterova, N., Nurtaev, B., Odongo, V. O., Panda, S., Pande, S., Pang, Z., Papacharalampous, G., Perrin, C., Pfister, L., Pimentel, R., Polo, M. J., Post, D., Sierra, C. P., Ramos, M.-H., Renner, M., Reynolds, J. E., Ridolfi, E., Rigon, R., Riva, M., Robertson, D. E., Rosso, R., Roy, T., Sá, J. H., Salvadori, G., Sandells, M., Schaefli, B., Schumann, A., Scolobig, A., Seibert, J., Servat, E., Shafiei, M., Sharma, A., Sidibe, M., Sidle, R. C., Skaugen, T., Smith, H., Spiessl, S. M., Stein, L., Steinsland, I., Strasser, U., Su, B., Szolgay, J., Tarboton, D., Tauro, F., Thirel, G., Tian, F., Tong, R.,

Tussupova, K., Tyralis, H., Uijlenhoet, R., van Beek, R., van der Ent, R. J., van der Ploeg, M., Loon, A. F. V., van Meerveld, I., van Nooijen, R., van Oel, P. R., Vidal, J.-P., von Freyberg, J., Vorogushyn, S., Wachniew, P., Wade, A. J., Ward, P., Westerberg, I. K., White, C., Wood, E. F., Woods, R., Xu, Z., Yilmaz, K. K., and Zhang, Y.: Twenty-three unsolved problems in hydrology (UPH) – a community perspective, Hydrol. Sci. J., 64, 1141–1158, https://doi.org/10.1080/02626667.2019.1620507, 2019.

Bratman, M.: ntentions, Plans, and Practical Reason, Harvard University Press, https://doi.org/10.2307/2184441, 1987.

Bresciani, P., Perini, A., Giorgini, P., Giunchiglia, F., and Mylopoulos, J.: Tropos: An Agent-Oriented Software Development Methodology, Auton. Agent. Multi.-Ag., 8, 203–236, https://doi.org/10.1023/B:AGNT.0000018806.20944.ef, 2004.

Carraro, C., Marchiori, C., and Sgobbi, A.: Application of Negotiation Theory to Water Issues, SSRN Electron. J., 1, 3641, https://doi.org/10.2139/ssrn.722362, 2006.

Dinar, A., Rosegrant, M. W., and Meinzen-Dick, R.: Water Allocation Mechanisms: Principles and Examples, The World Bank, 1, 1779, https://doi.org/10.1596/1813-9450-1779, 1997.

Fragoso, D., Evangelista, B., Caetano, G., and Cardoso, E.: Análise do regime pluviométrico da região do projeto rio Formoso em Formoso do Araguaia – TO, Agr.-Environ. Sci., 7, 1–14, https://doi.org/10.36725/agries.v7i2.5442, 2021.

GAN: Gestão Alto Nível, instituto de Atenção às Cidades, Universidade Federal do Tocantins [data set], https://gan.iacuft.org.br/ (last access: 23 January 2023), 2022.

Hardin, G.: The Tragedy of the Commons, Science, 162, 1243–1248, https://doi.org/10.1126/science.162.3859.1243, 1968.

IAC: Plano do Biênio 2018–2019, https://central.to.gov.br/download/43038 (last access: 23 January 2023), 2018.

Kaiser, K. E., Flores, A. N., and Hillis, V.: Identifying emergent agent types and effective practices for portability, scalability, and intercomparison in water resource agent-based models, Environ. Model. Softw., 127, 104671, https://doi.org/10.1016/j.envsoft.2020.104671, 2020.

Khan, H. F., Yang, Y. C. E., Xie, H., and Ringler, C.: A coupled modeling framework for sustainable watershed management in transboundary river basins, Hydrol. Earth Syst. Sci., 21, 6275–6288, https://doi.org/10.5194/hess-21-6275-2017, 2017.

Kock, B.: Agent-based models of socio-hydrological systems for exploring the institutional dynamics of water resources conflict, master's Thesis – Massachusetts Institute of Technology, USA, Master Thesis, Massachusetts Institute of Technology, http://hdl.handle.net/1721.1/44199 (last access: 23 January 2023), 2008.

Liang, X., Chen, H., Wang, Y., and Song, S.: Design and application of a CA-BDI model to determine farmers' land-use behavior, SpringerPlus, 1581, https://doi.org/10.1186/s40064-016-3245-7, 2016.

Montanari, A., Young, G., Savenije, H., Hughes, D., Wagener, T., Ren, L., Koutsoyiannis, D., Cudennec, C., Toth, E., Grimaldi, S., Blöschl, G., Sivapalan, M., Beven, K., Gupta, H., Hipsey, M., Schaefli, B., Arheimer, B., Boegh, E., Schymanski, S., Baldassarre, G. D., Yu, B., Hubert, P., Huang, Y., Schumann, A., Post, D., Srinivasan, V., Harman, C., Thompson, S., Rogger, M., Viglione, A., McMillan, H., Characklis, G., Pang, Z., and Belyaev, V.: "Panta Rhei–Everything Flows": Change in hydrology and society – The IAHS Scientific Decade 2013–2022, Hydrol. Sci. J., 58, 1256–1275, https://doi.org/10.1080/02626667.2013.809088, 2013.

Ralha, C. G., Abreu, C. G., Coelho, C. G. C., Zaghetto, A., Macchiavello, B., and Machado, R. B.: A multi-agent model system for land-use change simulation, Environ. Model. Softw., 42, 30–46, https://doi.org/10.1016/j.envsoft.2012.12.003, 2013.

Silva, A. J. R.: Sensoriamento remoto como subsídio para a gestão agrícola: estudo de caso do projeto de irrigação rio Formoso do Araguaia – TO, instituto Federal, Goiás, Brazil, MS thesis, Instituto Federal de Goiás, 2015.

Sivapalan, M., Savenije, H. H. G., and Blöschl, G.: Socio-hydrology: A new science of people and water, Hydrol. Process., 26, 1270–1276, https://doi.org/10.1002/hyp.8426, 2012.

Taillandier, P., Bourgais, M., Caillou, P., Adam, C., and Gaudou, B.: A BDI Agent Architecture for the GAMA Modeling and Simulation Platform, Proceedings of the Multi-Agent Based Simulation XVII—International Workshop, MABS 2016, 3–23, 10399, https://doi.org/10.1007/978-3-319-67477-3_1, 2017.

Tocantins: Plano de Bacia Hidrográfica do Rio Formoso – PBH Rio Formoso, no Estado do Tocantins – Relatório Síntese, Secretaria de Recursos Hídricos e Meio Ambiente, https://central3.to.gov.br/arquivo/269430/ (last access: 23 January 2023), 2007.

Vergara, F. E., dos Reis, F. C., Filho, L. N. L. M., and Rezende, C. D. S. A.: Proposta de vazão de referência Q_{90} para o Rio Formoso na bacia do Araguaia, Engenharia Ambiental: Pesquisa e Tecnologia, 10, 84–102, 2013.

Volken, N. J.: Modelagem hidrológica para análise de balanço hídrico e avaliação de cenários de gestão de demanda agrícola na bacia hidrográfica do rio Urubu – TO, Dissertação de Mestrado, Universidade de Brasília, Brasil, MS thesis, Universidade de Brasília, 2022.

Proc. IAHS, 385, 79–84, 2024
https://doi.org/10.5194/piahs-385-79-2024

Open Access PIAHS

Spatial characterization of drought through CHIRPS and a station-based dataset in the Eastern Mediterranean

Christina Orieschnig[1] **and Yonca Cavus**[2,3,4]

[1]UMR G-EAU, UMR HydroSciences, University of Montpellier, IRD, Institut Agro, Montpellier, France
[2]Department of Civil Engineering, Beykent University, Istanbul, Turkey
[3]Graduate School, Istanbul Technical University, Istanbul, Turkey
[4]Faculty of Environment and Natural Resources, University of Freiburg, Freiburg, Germany

Correspondence: Christina Orieschnig (christina.orieschnig@umontpellier.fr)

Received: 26 May 2022 – Revised: 24 January 2023 – Accepted: 26 January 2023 – Published: 18 April 2024

Abstract. Drought is a natural hazard which occurs in all climatic zones and affects different sectors, such as irrigation, energy, water supply, and ecology. Monitoring and predicting drought are pressing challenges, as drought is becoming more common and severe owing to the impacts of climate change and increased climatic variability. However, in many areas of the globe, the temporal and spatial characterization of droughts and drought severities are hindered by a lack of reliable, locally-measured long-term data and unevenly distributed, erratic meteorological stations. In this situation, remote sensing datasets such as Climate Hazards Group InfraRed Precipitation with Station Data (CHIRPS) can offer valuable insights into long-term developments and the spatial characteristics of droughts. Nonetheless, it is often uncertain to which extent data such as CHIRPS succeeds in representing local dynamics and how this varies between geographical regions and climate zones. In this analysis, we aim to evaluate spatial drought conditions over the Seyhan River basin in Turkey in the Eastern Mediterranean. Using the Standardized Precipitation Index (SPI) as a drought index, the applicability of CHIRPS as a long-term satellite precipitation product for drought monitoring is investigated. We compare two spatial representations of the SPI: one derived on a per-pixel basis from CHIRPS data since 1981, the other based on data from 19 meteorological stations scattered across the basin, which was spatialized using inverse distance weighted interpolation (IDW). Our results offer insights into the relative accuracy of CHIRPS data and avenues towards optimizing the quality of spatial drought characterization.

Keywords. SDG13; Drought monitoring; CHIRPS; Remote sensing; Seyhan River Basin (Turkey)

1 Introduction

Drought is one of the most common natural hazards, especially with the global increase in water demand, and the looming impacts of climate change. As such, combating droughts and mitigating their effects is an important action item in many landmark documents and strategies, such as the Sustainable Development Goals Resulting from a complex constellation of climatological and hydrological factors, a drought typically affects a wide swath of the population, with environmental, social, and economic effects. Where this is the case, drought indices such as the Standardized Precipitation Index (SPI, McKee et al., 1993) can be used to characterize these events and, consequently, to contribute to drought management. However, given the stochastic nature of droughts, a large and reliable dataset is is necessary for such a characterization. In many places of the world, this basis of data is lacking, which renders the calculation of drought indices over larger areas difficult.

Remotely sensed precipitation products such as CHIRPS (Climate Hazards Group InfraRed Precipitation with Station data, Funk et al., 2015) can provide an alternative pathway to a spatialized calculation of drought indices, especially in data sparse regions. However, the applicability and accuracy of these data need to be assessed in multiple ways before they

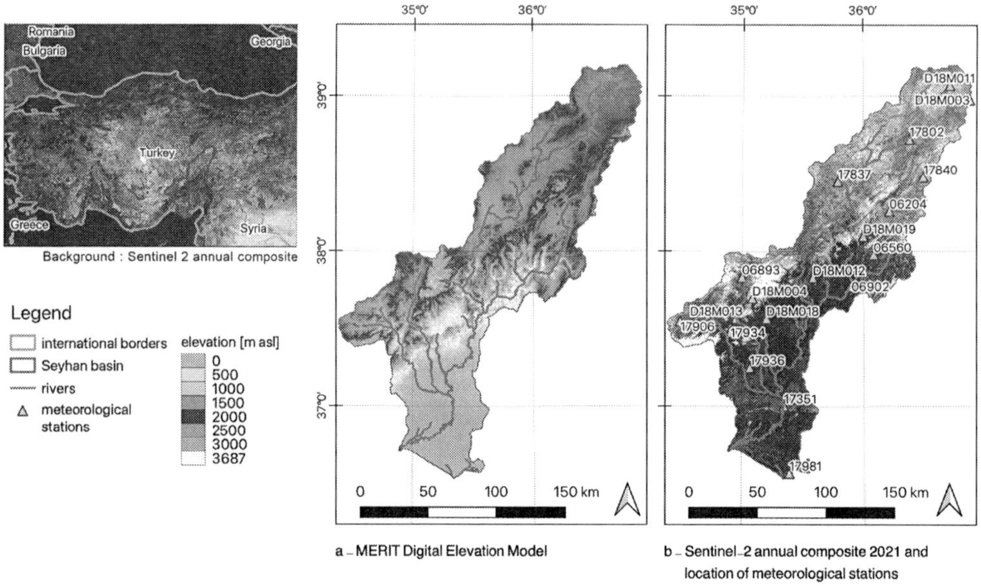

Figure 1. The Seyhan River Basin in Turkey, its elevation, and drainage network.

can be applied with confidence. In this study, we compare the use of CHIRPS for the calculation of the SPI over the Seyhan River Basin against the use of data from nineteen meteorological stations. Inverse distance weighted (IDW) interpolation was used to determine the spatial distribution of precipitation deficits from in-situ data and compare it to CHIRPS-derived maps.

With this study, we aim to respond to the following research questions:

- Is CHIRPS capable of accurately estimating in-situ precipitation in Mediterranean areas such as the Seyhan River Basin?

- Are CHIRPS-based SPI values capable of accurately reflecting the values and distribution of SPI calculation using in-situ data?

- Which additional insights does a spatial SPI raster derived from CHIRPS offer, compared to a spatialized representation of point-based data?

2 Study Area

Located in Southern Turkey, the Seyhan River Basin stretches from the mountains of Central Anatolia to the lowlands along the Mediterranean coast (Fig. 1). The northern part of the river basin is characterized by steep terrain gradients, higher annual precipitation amounts, and lower temperatures. In the southern part of the basin, summers are hot and dry, while winters are warm and rainy. The coastal areas are characterized by a semi-arid Mediterranean climate. The Seyhan Basin stretches across 22 139 km^2, with a north-south extent of approximately 300 km, and an average east-west extent of approximately 75 km. Its elevation ranges from 0 m a.s.l. at the Mediterranean coast to 3687 m a.s.l. 19 meteorological stations are scattered over the river basin. Their distribution is uneven, with a cluster of stations located in the central part of the basin, and very few in the south-western and north-eastern parts of the basin (Fig. 1b). The Seyhan Basin has been the region of interest for a number of hydrological studies, due to its large regional importance for irrigation, flood control, and hydropower generation (Cavus and Aksoy, 2019), which lends a high priority to the characterization of droughts in this area.

3 Methods

3.1 Data

In this study, two different types of data were used: monthly in-situ precipitation data from 19 meteorological stations and CHIRPS precipitation data extracted from the cloud computing platform Google Earth Engine (GEE, Gorelick et al., 2017). A complete list of the meteorological stations whose data was used in this study is provided in Cavus and Aksoy (2019). The stations are located between 22 and 1750 m a.s.l. in altitude and register between 274 and 1006 mm annual precipitation on average. The data used originates from between 1960 and 2016, though the length of observation periods varies between stations, and there are some discontinuities in the precipitation time series at several stations. For instance, station 6560 provides data only between 1986 and 1995 and additionally misses four months of data in this period. In contrast, station 17351 offers a

Table 1. RMSE characteristics for SPI time series, averaged across stations.

RMSE	SPI-1	SPI-3	SPI-6	SPI-9	SPI-12	SPI-24
MIN	0.602	0.597	0.534	0.470	0.432	0.398
AVG	0.765	0.801	0.782	0.779	0.779	0.828
MAX	0.918	1.083	1.007	1.048	1.091	1.269
SD	0.088	0.136	0.133	0.157	0.181	0.263

complete and consistent time series between 1960 and 2016. Overall, there are multiple years for which data from fewer than 19 stations is available for the calculation of SPI values. The data was provided by the Turkish State Meteorological Service (MGM) and the General Directorate of State Hydraulic Works (DSI). Monthly precipitation sums were used for the analysis.

The second data type used was CHIRPS, a long-duration, high-resolution precipitation product based on remotely sensed infrared Cold Cloud Duration (CDD) observations, corrected with local station data where available. It provides data at various temporal resolutions – daily, pentadal, and monthly – for the period since 1981, and at a spatial resolution of 0.05 arcdeg (Funk et al., 2015). For the purposes of this analysis, daily CHIRPS data available via GEE were summed up to calculate monthly precipitation sums for the years between 1981 and 2016.

3.2 Comparison of Data

The in-situ precipitation data was compared to CHIRPS values in three ways. First, the monthly precipitation sums were compared directly to pinpoint any systematic and random deviations between two time series. Second, the SPI was calculated using data from both sources. The SPI was developed by McKee et al. (1993) and is one of the most well-known drought indices. Its principal advantage is that it overcomes issues arising from the use of a non-standardized distribution by first transforming the distribution of the rainfall record to a normal distribution by fitting it to a gamma distribution through an equal-probability transformation. Furthermore, the SPI is advantageous as it can be calculated for different time scales.

In this application, the SPI was calculated at time scales of 1, 3, 6, 9, 12 and 24 months. The reason for this choice in interval is that SPI-1 and -3 are typically used to characterize meteorological drought, SPI-6 and -9 are used to characterize agricultural drought, and SPI-12 and -24 are used to characterize hydrological drought, with effects on reservoirs and groundwater. To compare the results obtained from the two different data sources, the root mean squared error (RMSE) was calculated. To investigate possible sources of bias, RMSE values were then correlated with parameters such as the duration of in-situ time series and the elevation of the station. This analysis was carried out twice: once using the entirety of the available time series from both data sources, and once using only the period for which both types of data are available. The purpose was to exclude bias caused by the length of the data series. A more in-depth comparison of SPIs was also carried out for 1990, one of the most well-known drought years in this river basin (Cavus and Aksoy, 2019).

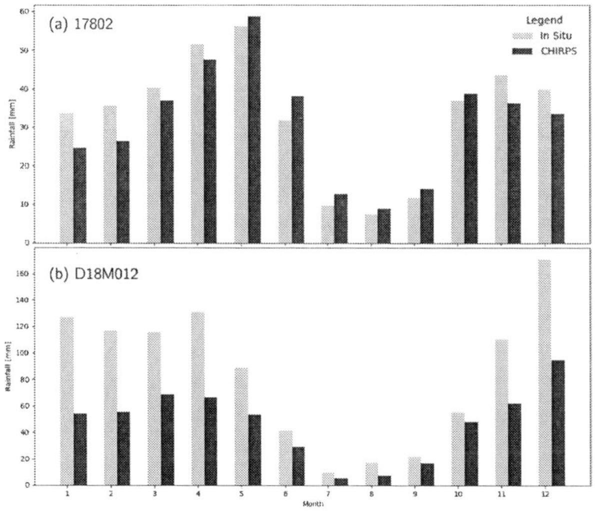

Figure 2. Comparison of monthly total precipitation, measured by in-situ stations (light blue) and by CHIRPS (dark blue) at stations 17802 and D18M012.

Finally, a spatial comparison of data was conducted. This contrasted a pixel-wise calculation of the SPI for the drought year 1990 derived directly from CHIRPS and a spatialized representation of SPI values calculated using data from in-situ stations and inverse distance weighted (IDW) interpolation.

4 Results

4.1 Comparison of Precipitation Data

The results of the comparison of monthly precipitation show that the average RMSE is 39 mm/month with a standard deviation of 17 mm/month. CHIRPS performs best in estimating in-situ rainfall at stations 17802 and 17837 (Fig. 2a) with a RMSE of 14 and 15 mm/month respectively. It performs worst at station D18M012 with a monthly RMSE of 72 mm (Fig. 2b). Generally, CHIRPS overestimates precipitation at the majority of stations. In higher altitudes, however, CHIRPS underestimates precipitation for most of the year, likely due to its limited capability to detect snowfall. However, CHIRPS largely succeeds in reflecting the distribution, if not the exact values, of precipitation.

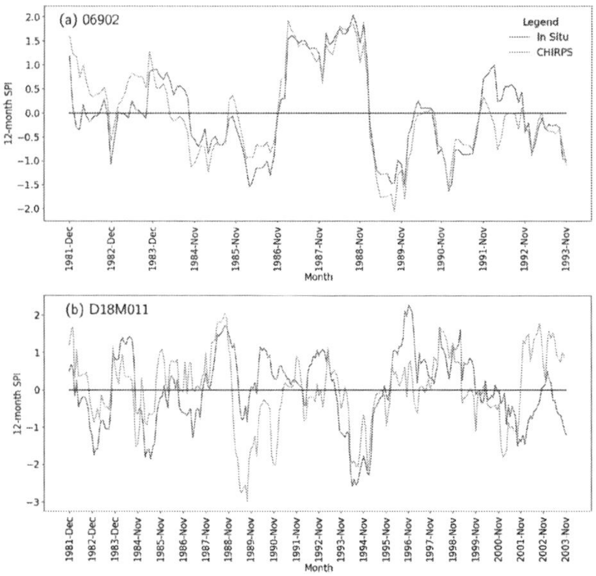

Figure 3. Comparison of SPI-12 values at the best- and worst-performing stations in terms of RMSE.

In addition, there are no significant correlations between the RMSE characterizing the disagreement between CHIRPS and in-situ data in terms of the length of the in-situ time series or of station elevation.

4.2 Comparison of SPI Values

The minimum, average, and maximum RMSE, as well as the standard deviation of each SPI (1–24) were calculated (Table 1). The results show significant deviations between each of the SPI values calculated using CHIRPS and using in-situ data. The longer the period over which the SPI is calculated, the smaller the minimum RMSE, but the larger the maximum RMSE and the standard deviation. The best overall agreement can be found for SPI1 with an average RMSE of 0.765 and a standard deviation of 0.088.

Average RMSE values vary considerably between stations, with stations at which CHIRPS performed well in accurately representing precipitation amounts not necessarily showing the lowest RMSE values. The station whose CHIRPS-derived SPI values most closely match the values calculated using in-situ data is 06902 (RMSE = 0.517). The one with the largest RMSE is D18M011 (RMSE = 1.040). Figure 3 shows the SPI12 for these two stations with the best and worst agreement between SPI values calculated from CHIRPS data and those derived from in-situ data. The graphs show that even for stations where the disagreement between absolute SPI values is large enough to result in high RMSE values, the general temporal dynamics of SPI values are similar.

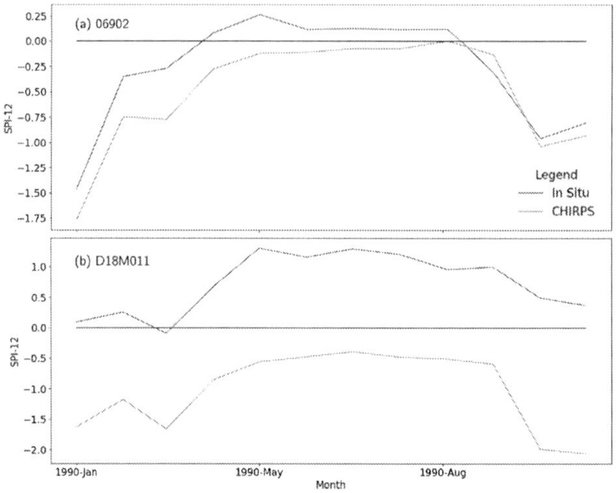

Figure 4. Comparison of SPI-12 values for January 1990.

This parallel behaviour is particularly apparent when taking a closer look at the more in-depth comparison of SPI during the drought year 1990 (Fig. 4). A synchronistic tendency can be observed at both stations. At D18M011, the dynamics of the values calculated from CHIRPS closely mirror those calculated from in-situ data, even though there is a fundamental disagreement over whether the SPI itself is positive or negative. The situation for station 06902 is similar, though the overall difference between the two values is much smaller.

Finally, as with the precipitation data, there is no significant correlation between the RMSE and station elevation or the length of the respective time series. No significant differences were observed depending on whether the full time series was used for each data set, or the concordant time series.

4.3 Comparison of Spatial SPI Distribution

The spatial comparison between a pixel-wise CHIRPS extract of SPI12 values for the drought year 1990 and a raster calculated using IDW interpolation from in-situ data shows widespread agreement (Fig. 5). The interpolation captures the general spatial distribution of SPI, though the absolute values differ slightly from the ones calculated from CHIRPS. One notable deviation can be seen in the north of the case study area, where station 17802 registers a far larger precipitation deficit than CHIRPS. This is reflected in the surrounding pixels, with the interpolated map showing a SPI hotspot, while the map extracted from CHIRPS actually shows lower SPI values than surrounding areas.

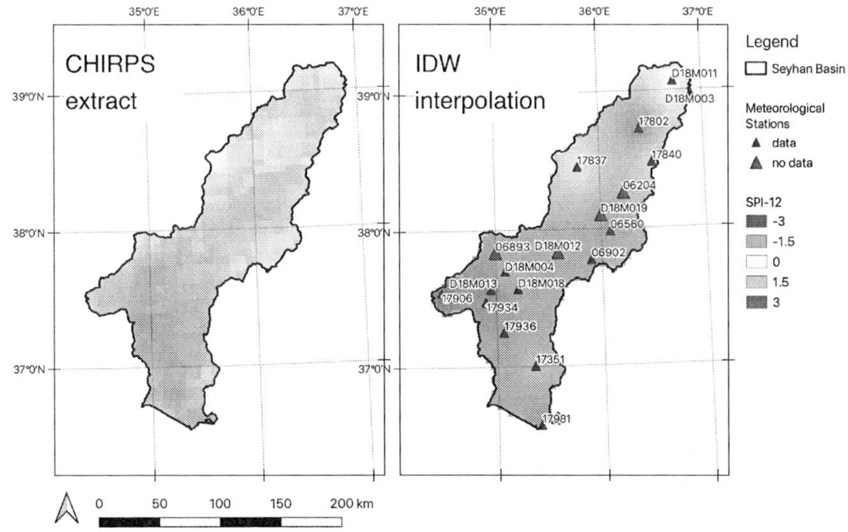

Figure 5. Spatial comparison of SPI-12 for January 1990.

5 Discussion

The analysis of precipitation values shows that the agreement between CHIRPS-derived precipitation values and those measured in-situ varies considerably between stations. Some tendencies that could be observed, such as the trend towards underestimating precipitation in elevated areas that likely experience snowfall, have been observed in previous, targeted studies (e.g. Bai et al., 2018). However, the high individual variations suggest that other variables also contribute towards the deviations, such as local terrain effects and storm paths. This merits further investigation.

With regard to the SPI, the temporal and spatial dynamics derived from CHIRPS and in-situ data exhibit strong parallelisms, but the RMSE between values is considerable for all SPI intervals and all stations. Even where RMSEs are relatively low, there is still occasionally a fundamental disagreement over whether the SPI is in the positive or negative range (Fig. 4b). It must also be highlighted that the stations at which RMSEs are lowest for the precipitation data are not the same that exhibit the lowest RMSEs for the SPI. This indicates that for the calculation of the SPI, the deviance from the observed norm and thus the overall consistency is more important than the absolute agreement between monthly precipitation values.

Finally, it must be noted that even though the precipitation measurements extracted from CHIRPS don't succeed completely in capturing the values of in-situ stations or generating the same SPI, they still hold the merit of providing a consistent time series. This is illustrated well in Fig. 5, which shows that only 15 out of 19 stations provided sufficient data for the calculation of SPI12 at this time point.

6 Conclusion and Outlook

This study has largely answered all three research questions outlined in the introduction. We can assert that the capacity of CHIRPS to estimate in-situ precipitation in the Seyhan Basin varies considerably between stations, with some monthly RMSEs being as low as 14 mm and as high as 72 mm. It can be noted that CHIRPS-derived SPIs are capable of representing the dynamics of the drought index, but not its absolute value – both in terms of spatial and temporal variability. The SPI map extracted from CHIRPS offers little additional information in comparison to the one generated from in-situ information, at least for the year 1990.

In terms of future perspectives, it would be interesting to investigate what role local terrain effects and storm paths play in the deviation of in-situ and CHIRPS values. Furthermore, an interesting prospect is offered by the possibility of adding other remotely sensed precipitation datasets to the comparison, such as data from the Tropical Rainfall Measuring Mission (TRMM).

Code and data availability. The data are available upon request to the corresponding author.

Author contributions. CO extracted the CHIRPS data from Google Earth Engine, performed the statistical comparison between datasets, and wrote the first draft of the paper. YC provided the station-based dataset, contributed to the development of the methodology, and provided extensive revisions for the first and second drafts of the paper.

Competing interests. The contact author has declared that neither of the authors has any competing interests.

Disclaimer. Publisher's note: Copernicus Publications remains neutral with regard to jurisdictional claims in published maps and institutional affiliations.

Special issue statement. This article is part of the special issue "IAHS2022 – Hydrological sciences in the Anthropocene: Variability and change across space, time, extremes, and interfaces". It is a result of the XIth Scientific Assembly of the International Association of Hydrological Sciences (IAHS 2022), Montpellier, France, 29 May–3 June 2022.

Acknowledgements. The second author is supported by the DAAD "Research Grants – Bi-nationally Supervised Doctoral 454 Degrees/Cotutelle" Program in the University of Freiburg, for which is the second author is thankful.

Review statement. This paper was edited by Christophe Cudennec and reviewed by Mohammad Merheb and one anonymous referee.

References

Bai, L., Shi, C., Li, L., Yang, Y., and Wu, J.: Accuracy of CHIRPS Satellite-Rainfall Products over Mainland China, Remote Sens., 10, 362, https://doi.org/10.3390/rs10030362, 2018.

Cavus, Y. and Aksoy, H.: Spatial drought characterization for Seyhan River basin in the Mediterranean region of Turkey, Water, 11, 1331, https://doi.org/10.3390/w11071331, 2019.

Funk, C. C., Peterson, P. J., Landsfeld, M. F., Pedreros, D. H., Verdin, J. P., Rowland, J. D., Romero, B. E., Husak, G. J., Michaelsen, J. C. and Verdin, A. P.: A quasi-global precipitation time series for drought monitoring, US Geological Survey data series, 832, 1–12, https://doi.org/10.3133/ds832, 2015.

Gorelick, N., Hancher, M., Dixon, M., Ilyushchenko, S., Thau, D., and Moore, R.: Google Earth Engine: Planetary-scale geospatial analysis for everyone, Remote Sens. Environ., 202, 18–27, https://doi.org/10.1016/j.rse.2017.06.031, 2017.

McKee, T. B., Doesken, N. J., and Kleist, J.: The relationship of drought frequency and duration to time scales, in: Proceedings of the 8th Conference on Applied Climatology, 17, 179–183, 1993.

Proc. IAHS, 385, 85–89, 2024
https://doi.org/10.5194/piahs-385-85-2024

Uptake by end-users of a PUB approach made available as a Web Service

Tom Loree[1], **Hervé Squividant**[1], **Josette Launay**[1,2], **Alban de Lavenne**[3], **and Christophe Cudennec**[1]

[1]Institut Agro, INRAE, SAS, Rennes, France
[2]CRESEB – Centre de Ressources et d'Expertise Scientifique sur l'Eau de Bretagne, Rennes, France
[3]Université Paris-Saclay, INRAE, UR HYCAR, Antony, France

Correspondence: Christophe Cudennec (cudennec@agrocampus-ouest.fr)

Received: 21 June 2022 – Accepted: 30 January 2023 – Published: 18 April 2024

Abstract. A consolidated geomorphology-based approach for discharge hydrograph Prediction in Ungauged Basins (PUB) through deconvolution of discharge signals from gauged donor catchments, their transposition and their convolution towards target outlets, has been made available online to end-users through a Web Service (Web Processing Service Application Programming Interface – WPS API) for the synoptic peninsular region of Brittany, France. In the spirit of hydrological services, the SIMFEN Web Service allows anyone to execute the hydrological modelling online through OGC® interoperability standards; collection of public hydrometric data; connection to a pre-existing Web Service for geomorphometric analysis; and online visualization. Thanks to tangible indicators, we here show how the users' community is actually active in using the Web Service, i.e. in uptaking the recent science-operational translation of a scientific development into a Service.

Keywords. UPH 7; UPH 21; UPH 22; PUB; Hydrogeomorphology; Science-Solution; Brittany

1 Introduction

In hydrology, scientific knowledge and methodological progress are critical for society in terms of security and sustainable development, polycentric governance and citizen engagement (Young et al., 2015; Ceola et al., 2016; Di Baldassarre et al., 2019; Nardi et al., 2022). Scientific outcomes in terms of methodology have always had a high innovative potential for different end-users including operational hydrometeorological entities, water management institutions, and policy makers. The translation of methods into Services (WMO, 2015) is a major challenge which can be inscribed in the hydro-informatic side of Open Science (UNESCO, 2021; Cudennec et al., 2022b) in terms of data FAIRness and relevance (see Cudennec et al., 2020 on FAIR and SQUARE data; Pecora and Lins, 2020 on e-monitoring the nature of water; Hall et al., 2022), of efficient workflows and chainings (Cudennec et al., 2022a), and of operationalization and assimilation of innovations (Dixon et al., 2022). The scientific work is naturally valued through scientific publications with its own rules and metrics, yet the translation into innovation towards end-users needs to go beyond scientific publications and bibliometrics (Cudennec and Hubert, 2008; Koutsoyiannis et al., 2016; Cudennec et al., 2022b). Here we illustrate how a consolidated modelling approach of the PUB – Prediction in Ungauged Basins type has been translated into a Web Service (Web Processing Service Application Processing Interface – WPS API) and how indicators of use can be considered.

2 SIMFEN: A Web Service to make a PUB modelling approach available

Prediction in Ungauged Basin (PUB) remains an ambitious task, especially when dealing with discharge series, i.e. hydrographs and derived statistics of regime (Blöschl et al., 2013; Hrachowitz et al., 2013) and related signatures (McMillan, 2021). For any gauged basin, Cudennec (2000) proposed the inversion of a geomorphology-based transfer function and so the deconvolution of the measured discharge time series to assess the net rainfall time series. The simu-

Published by Copernicus Publications on behalf of the International Association of Hydrological Sciences.

lated discharge time series at the outlet of a similar target catchment (in particular nested or neighbouring) is then obtained by transposition from the donor catchment, then the convolution with the corresponding transfer function as deduced from its observed geomorphometry and a robust assumption on the streamflow velocity. This geomorphology-based hydrological approach (Cudennec, 2007) has been tested in different well described contexts with pseudo-ungauged strategies: semi-arid Tunisia (Boudhraâ et al., 2009, 2018), temperate oceanic France (de Lavenne et al., 2015, 2016) and pluvio-nival Canadian Québec (Ecrepont et al., 2019). Robustness has been assessed regarding the calibration configuration (Boudhraâ et al., 2018; de Lavenne et al., 2016) and the resolution of the digital elevation model used as a geomorphometric reference (Ecrepont and Cudennec, 2020). Based on these grounds, simulations in truly ungauged catchments have been performed, in particular towards coastal bays (de Lavenne and Cudennec, 2019), which are water quality-related hotspots (Heal et al., 2021; Neville et al., 2022), being fragile fresh/marine water ecosystems where water-conveyed nutrient over-loads can induce eutrophication and green tides.

This consolidated geomorphology-based approach for discharge Prediction in Ungauged Basins (PUB) through deconvolution of discharge signals from gauged donor catchments, their transposition and their convolution towards target outlets, has been made available online to end-users through a Web Processing Service Application Programming Interface (WPS API) for the synoptic peninsular region of Brittany, France. In the spirit of hydrological Services, the SIMFEN WPS API (https://geosas.fr/simfen, last access: 10 March 2023, Dallery et al., 2022) allows anyone to execute the hydrological modelling package transfR (de Lavenne et al., 2021, 2023) online through: (1) Open Geospatial Consortium (OGC®) interoperability standards; (2) collection and use of public hydrometric data; (3) connection to the pre-existing MNTSurf WPS API for geomorphometric analysis (Squividant et al., 2015); (4) visualization using a collaboratively developed Mviewer; and (5) innovative WPS API chaining workflows facilitated by an Open Science spatial data infrastructure (Bera et al., 2015). The ability to model discharge series at any ungauged outlet of the synoptic region is offered to specialists of other disciplines, non-modeller water practitioners and interested citizens to support interdisciplinarity, water monitoring and management, and related science-society-policy debates and actions. Figure 1 illustrates the SIMFEN Web Service through its online interface.

3 Dissemination and indicators of use by end-users

The modelling approach had been used and presented in different catchments across the Brittany region (a coherent hydrographic area of $29\,640\,\mathrm{km}^2$ – Fig. 2a), with corresponding stakeholders over the last decade; but the expertise was sitting with the scientific team, whereas the demand for tailored simulations was increasingly soliciting. The principle of the SIMFEN Web Service was envisaged with key stakeholders, i.e. the Brittany regional government (Région Bretagne) and the corresponding water basin institution (Agence de l'Eau Loire Bretagne) which funded the development. The SIMFEN Web Service was beta-tested with a sample of end-users in 2019 (Dallery et al., 2022) and was then opened to end users late 2019. A public announcement was made via the different communication vehicles of the scientific team, the funders, and the regional CRESEB network for science-policy dialogue in the field of water. Beyond the water-related stakeholders, announcements and demonstrations were made within the regional GeoBretagne community dealing with data sharing. These multichannel advertisements have then been amplified and consolidated via a dedicated webinar presentation on 20 June 2021 (https://www.creseb.fr/cycle-webinaire-hmuc/, last access: 10 March 2023), incorporation in teaching at master level, and a constructive hotline interaction with pro-active end-users.

Figure 2 displays different indicators, of the metric and mapping types, of end-using between 1 January 2020 and 9 February 2023. Assessing these indicators is allowed by the digital monitoring of the Web Service actioning. The ultimate use of the simulations obtained cannot be fully captured by the digital monitoring, and the trial-and-error or expert-exploration iterations cannot be easily detected and accounted for. Yet informal feedbacks from end-users, announced and detected specific analysis works led by water institutions, consulting companies, sometimes seconded to master students from curricula where the modelling method and Service had been presented and explored, confirm that many end-users are really considering the Service beyond curiosity. Many have a real strategy for testing and subsequently for supporting operational and policy developments, in particular regarding the consolidation of the hydrometric network, the alignment of water quantity and quality monitoring, and gap filling in data series and/or in catchment-wide approaches for water resources management, and for the assessment of hydrological driving forces of specific ecosystems. A more ambitious assessment of the Service-using is under consideration, which will develop on and beyond indicators displayed in Fig. 2. Yet, it is already obvious that the end-using is much more intense than it was and would have remained if the methodological development had been kept limited to the scientific team and to publications in scientific journals.

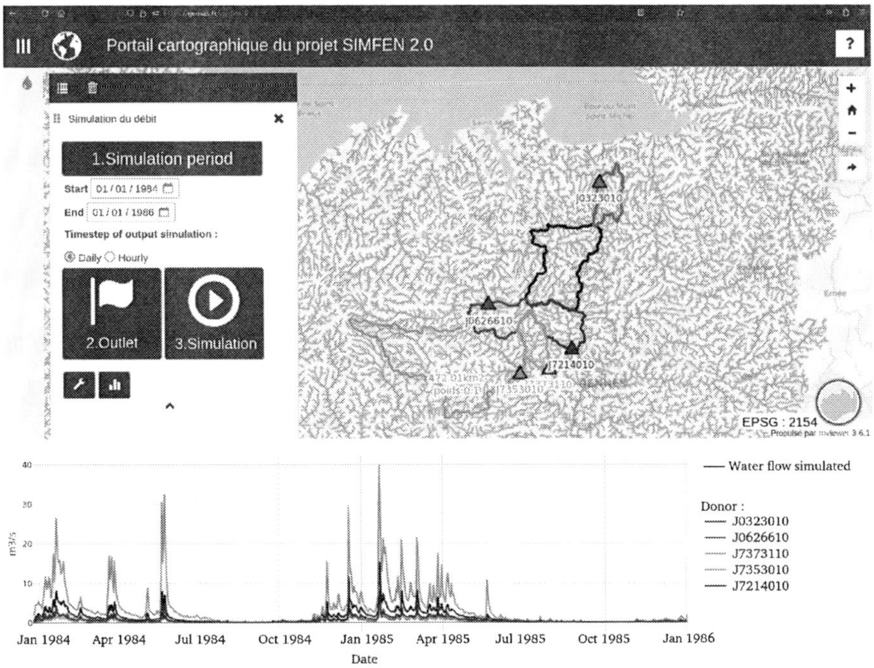

Figure 1. Screenshot of the SIMFEN Web Service interface with simulated discharge for a target catchment (black in the time series and on the map) obtained by transposing observed discharges from five surrounding gauging stations (donor catchments, in colour: J0323010, J0626610, J7373110, J7353010, J7214010).

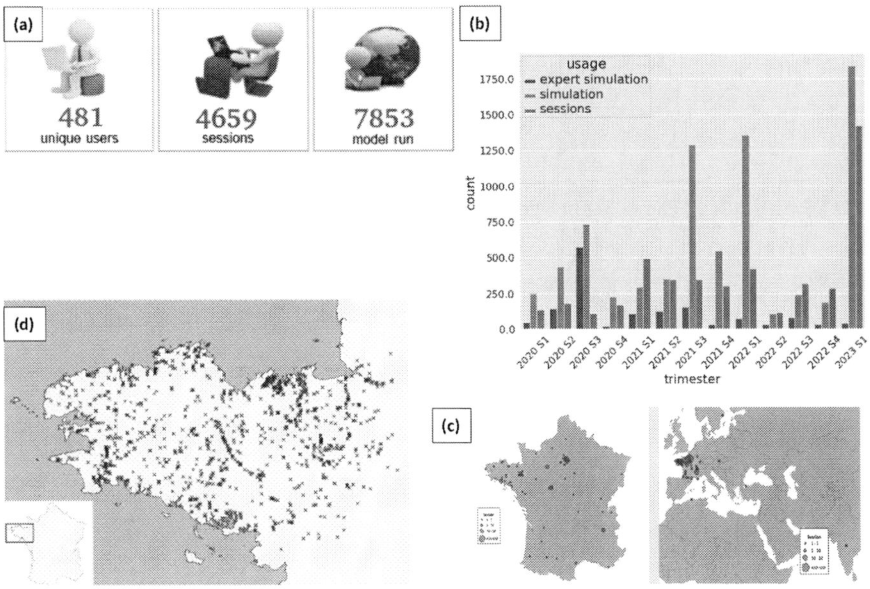

Figure 2. End-using indicators of the SIMFEN Web Service API over 1 January 2020–9 February 2023 period. Overall **(a)** and sequential **(b)** usage; Mapping of users **(c)** and catchment outlets **(d)**.

Code availability. The code has been made available as a R package by de Lavenne et al. (2021) at https://doi.org/10.57745/ZYNPKQ.

Data availability. Data used in this article are freely accessible through the SIMFEN Web Service (https://geosas.fr/simfen, SIMFEN, 2023).

Author contributions. CC designed the objectives and funding. AdL developed the modelling package. TL and HS developed the Web Service and User interface. JL organized the relationships with partners. All iterated on the Web Service design. CC drafted the article. TL, HS, JL and AdL contributed to the article refinement.

Competing interests. At least one of the (co-)authors is a guest member of the editorial board of *Proceedings of IAHS* for the special issue "IAHS2022 – Hydrological sciences in the Anthropocene: Past and future of open, inclusive, innovative, and society-interfacing approaches". The peer-review process was guided by an independent editor, and the authors also have no other competing interests to declare.

Disclaimer. Publisher's note: Copernicus Publications remains neutral with regard to jurisdictional claims in published maps and institutional affiliations.

Special issue statement. This article is part of the special issue "IAHS2022 – Hydrological sciences in the Anthropocene: Past and future of open, inclusive, innovative, and society-interfacing approaches". It is a result of the XIth Scientific Assembly of the International Association of Hydrological Sciences (IAHS 2022), Montpellier, France, 29 May–3 June 2022.

Financial support. This research has been supported by the Conseil Régional de Bretagne and Agence de l'Eau Loire Bretagne.

Review statement. This paper was edited by Salvatore Grimaldi and reviewed by two anonymous referees.

References

Bera, R., Squividant, H., Le Henaff, G., Pichelin, P., Ruiz, L., Launay, J., Vanhouteghem, J., Aurousseau, P., and Cudennec, C.: GéoSAS: A modular and interoperable Open Source Spatial Data Infrastructure for research, Proc. IAHS, 368, 9–14, https://doi.org/10.5194/piahs-368-9-2015, 2015.

Blöschl, G., Sivapalan, M., Wagener, T., Viglione, A., and Savenije, H.: Runoff Prediction in Ungauged Basins: Synthesis across Processes, Places and Scales, Cambridge University Press, Cambridge, https://doi.org/10.1017/CBO9781139235761, 2013.

Boudhraâ, H., Cudennec, C., Slimani, M., and Andrieu H.: Hydrograph transposition between basins through a geomorphology-based deconvolution-reconvolution approach, IAHS Publ., 333, 76–83, 2009.

Boudhraâ, H., Cudennec, C., Andrieu, H., and Slimani, M.: Net rainfall estimation by the inversion of a geomorphology-based transfer function and discharge deconvolution, Hydrolog. Sci. J., 63, 285–301, https://doi.org/10.1080/02626667.2018.1425801, 2018.

Ceola, S., Montanari, A., Krueger, T., Dyer, F., Kreibich, H., Westerberg, I., Carr, G., Cudennec, C., Elshorbagy, A., Savenije, H., van der Zaag, P., Rosbjerg, D., Aksoy, H., Viola, F., Petrucci, G., MacLeod, K., Croke, B., Ganora, D., Hermans, L., Polo, M. J., Xu, Z., Borga, M., Helmschrot, J., Toth, E., Ranzi, R., Castellarin, A., Hurford, A., Brilly, M., Viglione, A., Blöschl, G., Sivapalan, M., Domeneghetti, A., Marinelli, A., and Di Baldassarre, G.: Adaptation of water resources systems to changing society and environment – A statement by the International Association of Hydrological Sciences, Hydrolog. Sci. J., 61, 2803–2817, https://doi.org/10.1080/02626667.2016.1230674, 2016.

Cudennec, C.: Description mathématique de l'organisation du réseau hydrographique et modélisation hydrologique, PhD, ENSA Rennes, 2000.

Cudennec, C.: On width function-based unit hydrographs deduced from separately random self-similar river networks and rainfall variability, Hydrolog. Sci. J., 52, 230–237, https://doi.org/10.1623/hysj.52.1.230, 2007.

Cudennec, C. and Hubert, P.: Multi-objective role of HSJ in processing and disseminating hydrological knowledge, Hydrolog. Sci. J., 53, 485–487, https://doi.org/10.1623/hysj.53.2.485, 2008.

Cudennec, C., Lins, H., Uhlenbrook, S., and Arheimer, B.: Towards FAIR and SQUARE hydrological data, Hydrolog. Sci. J., 65, 681–682, https://doi.org/10.1080/02626667.2020.1739397, 2020.

Cudennec, C., Lins, H., Uhlenbrook, S., Amani, A., and Arheimer, B.: Operational, epistemic and ethical value chaining of hydrological data to knowledge and services: a watershed moment, Hydrolog. Sci. J., 67, 2363–2368, https://doi.org/10.1080/02626667.2022.2150380, 2022a.

Cudennec, C., Sud, M., and Boulton, G.: Governing Open Science, Hydrolog. Sci. J., 67, 2359–2362, https://doi.org/10.1080/02626667.2022.2086462, 2022b.

Dallery, D., Squividant, H., de Lavenne, A., Launay, J., and Cudennec, C.: An end-user-friendly hydrological Web Service for hydrograph Prediction in Ungauged Basins, Hydrolog. Sci. J., 67, 2420–2428, https://doi.org/10.1080/02626667.2020.1797045, 2022.

de Lavenne, A. and Cudennec, C.: Assessment of freshwater discharge into a coastal bay through multi-basin ensemble hydrological modelling, Sci. Total Environ., 669, 812–820, https://doi.org/10.1016/j.scitotenv.2019.02.387, 2019.

de Lavenne, A., Boudhraâ, H., and Cudennec, C.: Streamflow prediction in ungauged basins through geomorphology-based hydrograph transposition, Hydrol. Res., 46, 291–302, https://doi.org/10.2166/nh.2013.099, 2015.

de Lavenne, A., Skøien, J. O., Cudennec, C., Curie, F., and Moatar, F.: Transferring measured discharge time series: Large-scale comparison of Top-kriging to geomorphology-based inverse modeling, Water Resour. Res., 52, 5555–5576, https://doi.org/10.1002/2016WR018716, 2016.

de Lavenne, A., Loree, T., Squividant, S., and Cudennec, C.: transfR: Transfer of Hydrograph from Gauged to Ungauged Catchments, R package version 0.1.0, Recherche Data Gouv [code], https://doi.org/10.57745/ZYNPKQ, 2021.

de Lavenne, A., Loree, T., Squividant, H., and Cudennec, C.: The transfR toolbox for transferring observed streamflow series to ungauged basins based on their hydrogeomorphology, Environ. Model. Softw., 159, 105562, https://doi.org/10.1016/j.envsoft.2022.105562, 2023.

Di Baldassarre, G., Sivapalan, M., Rusca, M., Cudennec, C., Garcia, M., Kreibich, H., Konar, M., Mondino, E., Mård, J., Pande, S., Sanderson, M. R., Tian, F., Viglione, A., Wei, J., Wei, Y., Yu, D. J., Srinivasan, V., and Blöschl, G.: Socio-hydrology: Scientific challenges in addressing the sustainable development goals, Water Resour. Res., 55, 6327–6355, https://doi.org/10.1029/2018WR023901, 2019.

Dixon, H., Sandström, S., Cudennec, C., Lins, H., Abrate, T., Bérod, D., Chernov, I., Ravalitera, N., Sighomnou, D., and Teichert, F.: Intergovernmental cooperation for hydrometry – what, why, how?, Hydrolog. Sci. J., 67, 2552–2566, https://doi.org/10.1080/02626667.2020.1764569, 2022.

Ecrepont, S. and Cudennec, C.: Geomorphometry-based modelling of discharge series in ungauged basins – Robustness regarding DEM sources, Proc. IAHS, 383, 129–134, https://doi.org/10.5194/piahs-383-129-2020, 2020.

Ecrepont, S., Cudennec, C., Anctil, F., and Jaffrézic, A.: PUB in Québec: A robust geomorphology-based deconvolution-reconvolution framework for the spatial transposition of hydrographs, J. Hydrol., 570, 378–392, https://doi.org/10.1016/j.jhydrol.2018.12.052, 2019.

Hall, C. A., Saia, S. M., Popp, A. L., Dogulu, N., Schymanski, S. J., Drost, N., van Emmerik, T., and Hut, R.: A hydrologist's guide to open science, Hydrol. Earth Syst. Sci., 26, 647–664, https://doi.org/10.5194/hess-26-647-2022, 2022.

Heal, K. V., Bartosova, A., Hipsey, M. R., Chen, X., Buytaert, W., Li, H.-Y., McGrane, S. J., Gupta, A. B., and Cudennec, C.: Water quality: the missing dimension of water in the water-energy-food nexus?, Hydrolog. Sci. J., 66, 745–758, https://doi.org/10.1080/02626667.2020.1859114, 2021.

Hrachowitz, M., Savenije, H. H. G., Blöschl, G., McDonnell, J. J., Sivapalan, M., Pomeroy, J. W., Arheimer, B., Blume, T., Clark, M. P., Ehret, U., Fenicia, F., Freer, J. E., Gelfan, A., Gupta, H. V., Hughes, D. A., Hut, R. W., Montanari, A., Pande, S., Tetzlaff, D., Troch, P. A., Uhlenbrook, S., Wagener, T., Winsemius, H. C., Woods, R. A., Zehe, E., and Cudennec, C.: A decade of Predictions in Ungauged Basins (PUB) – a review, Hydrolog. Sci. J., 58, 1198–1255, https://doi.org/10.1080/02626667.2013.803183, 2013.

Koutsoyiannis, D., Blöschl, G., Bardossy, A., Cudennec, C., Hughes, D., Montanari, A., Neuweiler, I., and Savenije H.: Joint editorial – Fostering innovation and improving impact assessment for journal publications in hydrology, Water Resour. Res., 52, 2399–2402, https://doi.org/10.1002/2016WR018895, 2016.

McMillan, H. K.: A review of hydrologic signatures and their applications, WIREs Water, 8, e1499, https://doi.org/10.1002/wat2.1499, 2021.

Nardi, F., Cudennec, C., Abrate, T., Allouch, C., Annis, A., Assumpção, T. H., Aubert, A. H., Bérod, D., Braccini, A. M., Buytaert, W., Dasgupta, A., Hannah, D. M., Mazzoleni, M., Polo, M. J., Sæbø, Ø, Seibert, J., Tauro, F., Teichert, F., Teutonico, R., Uhlenbrook, S., Wahrmann Vargas, C., and Grimaldi, S.: Citizens AND HYdrology (CANDHY): conceptualizing a transdisciplinary framework for citizen science addressing hydrological challenges, Hydrolog. Sci. J., 67, 2534–2551, https://doi.org/10.1080/02626667.2020.1849707, 2022.

Neville, J. A., Guz, J., Rosko, H. M., and Owens, M. C.: Water quality inequality: a non-targeted hotspot analysis for ambient water quality injustices, Hydrolog. Sci. J., 67, 1011–1025, https://doi.org/10.1080/02626667.2022.2052073, 2022.

Pecora, S. and Lins, H. F.: E-monitoring the nature of water, Hydrolog. Sci. J., 65, 683–698, https://doi.org/10.1080/02626667.2020.1724296, 2020.

SIMFEN: Portrail cartographique du projet SIMFEN 2.0, SIMFEN [data set], https://geosas.fr/simfen, last access: 10 March 2023.

Squividant, H., Bera, R., Aurousseau, P., and Cudennec, C.: Online watershed boundary delineation: sharing models through Spatial Data Infrastructures, Proc. IAHS, 368, 144–149, https://doi.org/10.5194/piahs-368-144-2015, 2015.

UNESCO: Recommendation on Open Science, https://unesdoc.unesco.org/ark:/48223/pf0000379949.locale=en (last access: 10 March 2023), 2021.

WMO: Valuing weather and climate: economic assessment of meteorological and hydrological services, Geneva, Switzerland: World Meteorological Organization, 2015.

Young, G., Demuth, S., Mishra, A., and Cudennec, C.: Hydrological sciences and water security: An overview, Proc. IAHS, 366, 1–9, https://doi.org/10.5194/piahs-366-1-2015, 2015.

Proc. IAHS, 385, 91–96, 2024
https://doi.org/10.5194/piahs-385-91-2024

Open Access PIAHS

IAHS2022 – Past and future of open, inclusive, innovative, and society-interfacing approaches

Can we use indicator-based farm sustainability assessment tools for the WEFE Nexus?

Mohammad Merheb[1], **Christophe Cudennec**[1], **and Fernando Nardi**[2]

[1]Institut Agro, INRAE, SAS, 35000 Rennes, France
[2]Water Resources Research and Documentation Centre (WARREDOC), University for Foreigners of Perugia, Perugia, Italy

Correspondence: Mohammad Merheb (mohammad.merheb@agrocampus-ouest.fr)

Received: 21 June 2022 – Revised: 14 December 2022 – Accepted: 30 January 2023 – Published: 18 April 2024

Abstract. Fair and safe allocation of natural resources for the Euro-Mediterranean area, especially for semi-arid regions, strongly relies on the adoption of WEFE (Water Energy Food Ecosystem) Nexus strategies. Transitioning to WEFE Nexus requires novel quantifiable assessment for interlinked analysis of the four WEFE sectors. Several indicator-based tools exist for agricultural sustainability at the farm scale. This contribution investigates on the application of such tools in relation for WEFE Nexus approaches. The IDEA method was selected for extending its applicability as a novel WEFE Nexus indicator toolkit and the following challenges are identified: (1) some WEFE aspects need to be reinforced in order to expand the scope beyond the actual agro-ecological focus; (2) the application at the farm scale needs to be articulated with larger scales where the WEFE Nexus displays emerging consistencies; (3) Nexus interactions, trade-offs and synergies could be further accounted for. These three challenges help identify how the IDEA indicator-based tool could be adapted to assimilate the WEFE Nexus approach, and so to allow applications in new agro-hydrological contexts.

Keywords. UPH 22; SDG; indicator; Mediterranean; IDEA; WEFE Nexus

1 Introduction

In some contexts, in particular Mediterranean and semi-arid ones (Cudennec et al., 2007; Merheb et al., 2016), sustainability issues are in strong need of avoiding compartmentalization that actually impacts fair allocation of water, energy, food and environment resources among competing social and economic sectors. Sustainable and safe resource management thus requires the adoption of WEFE (Water Energy Food Ecosystem) Nexus strategies. Hoff (2011) underlines that "*A nexus approach can support a transition to sustainability*". This was further emphasized in the FAO reports which state "*the Water-Energy-Food Nexus is framed within the broader debate on sustainable development*" (FAO, 2014a, b). More recently, Liu et al. (2017) emphasized the critical role of the WEF Nexus in "*developing strategies for sustainable development*". This has been further advocated by Cudennec et al. (2018) and Heal et al. (2021) with additional perspectives on the epistemological and the water quality dimensions. The WEFE Nexus expands the WEF Nexus with the accounting of the multiple concurring Ecosystem dimensions. It elaborates on the complexity of interactions between different elements of the natural and human systems, and the necessity of considering such complexities for the well-being of present and future generations. The Ecosystem dimension, within the WEFE Nexus paradigm, may be further extended to consider the interlinkages between the bio-physical components of the WEF Nexus with a multiple conceptualization of the eco-social setting. As a result, the second E (Ecosystem) may be considered as a multi-scalar realization of the diverse and intertwined facets of the human-nature interface. In particular, this work envisages the triple conceptualization of the E – referring to the Environmental, Economic and Engagement (Society) facets – promoted within the Nexus-Ness project (https://prima-nexus-ness.org/nexus-ness/concept/, last access: June 2022), an innovation action co-funded by the Horizon 2020 and the Partnership on Research and Innovation in the Mediterranean Area (PRIMA) programs. Indeed, the WEFE Nexus is an integrated approach that focuses on the co-management of Water, Energy, Food, and Ecosystem resources and amenities co-demonstrating with citizens and stakeholders multiple benefits of Nexus strategies "*with a view to improving integrated solutions in the field that improve achievement of SDGs*" (Carmona Moreno et al., 2019).

The interlinkages (in particular trade-offs and synergies) between the various elements of the WEFE Nexus are difficult to grasp at a given scale, and so level of organization, and even more between the different scales/levels which can be concerned and nested in a particular scenario. Furthermore, beyond the diagnostic assessment of a peculiar condition, appreciating its sustainability, exploring future alternatives, innovating policies (Bazzana et al., 2023) and identifying transformation paths require a quantitative compromise between conceptualization and its feasibility towards operational implementation.

Assessing agricultural sustainability has a long history in contexts where the WEFE Nexus is less critical. Indicator-based agricultural sustainability assessment tools have been developed and already tested. This contribution seeks to verify if indicator-based farm sustainability assessment tools could be adapted to support adoption of WEFE Nexus strategies. We explore a particular method, IDEA, which is well consolidated for sustainability assessment and agro-ecological transition monitoring at the farm level.

In the following, we will present the IDEA method and discuss the challenges but also the prospects for its application for geographical, bio-physical and socio-economic scales of interest of WEFE Nexus assessment strategies.

2 The IDEA method

The IDEA method (in French "Indicateurs de Durabilité des Exploitations Agricoles", i.e. Indicators of Farms Sustainability) has been designed and circulating for more than 20 years. The first prototype was published in 1999. The method was then updated three times: IDEA1 in 2000, IDEA2 in 2002, and IDEA3 in 2008 (https://methode-idea.org/, last access: June 2022). Here we are referring to version 3 (Zahm et al., 2008) and to related applications and assessments (Binder et al., 2010; Marchand et al., 2014; De Olde et al., 2016). It includes the three dimensions (normative, systemic and procedural) of an operational sustainability assessment tool as defined by Binder et al. (2010). IDEA proved to be a robust and relatively easy-to-use method. The concepts and objectives are clear (De Olde et al., 2016), themselves rooted in the key concepts of agriculture sustainability (Zahm et al., 2008): environmental reproducibility, livability, and viability. We acknowledge that these three concepts respectively correspond to the environmental, social, and economic pillars of sustainability. In the IDEA terms, these are represented through the agro-ecological, socio-territorial and economic scales. The main concepts of the method are transcribed into 16 objectives, characterized with 41 indicators, themselves clustered into 10 components. The agro-ecological scale/level includes three components: diversity, organization of space and farming practices, assessed with 19 indicators. The socio-territorial scale/level includes three components: quality of the products and land, organization of space, and ethics and human development, assessed with 16 indicators. Finally, the economic scale/level is sub-divided into four components: economic viability, independence, transferability, and efficiency, assessed using six indicators. Table 1 summarizes the scales/levels, components and indicators of IDEA. A scale has a total score of 100. This score is calculated by summing the individual score of indicators in each component. It should be noted that a maximum score is set for each indicator and each component. Furthermore, indicators do not have the same score ranges, hence have different expert-based pre-defined weights. The overall farm sustainability score is equal to that of the scale with the lowest score. Finally, this method is based on predefined goals and set of indicators, and on a clear target group (farmers). This may reduce the stakeholders' participation in the goal settings and the elaboration of evaluation criteria. However, it allows more flexibility in comparing the system in time and space. Moreover, IDEA is flexible in its application to farms. Hundreds of French farms and various types of farms in Tunisia (M'Hamdi et al., 2009), Algeria (Bekhouche-Guendouz, 2011), Morocco (Baccar et al., 2019), Mexico (Salas-Reyes et al., 2015; Fadul-Pacheco et al., 2013), Senegal (Faye et al., 2020) and Lebanon (Srour et al., 2009) have been evaluated with the IDEA3 method.

3 Challenges and prospects for the WEFE Nexus assessment with the IDEA method

Three main challenges impact the assessing of the WEFE Nexus by means of the IDEA method, which are summarized in Table 1 as commentaries on the IDEA components and indicators. Some suggestions are also made in order to adapt IDEA for the WEFE assessment as hereafter briefly reported.

1. Some WEFE aspects need to be emphasized in order to widen the scope of IDEA beyond the agro-ecological focus. Indeed, if one looks more closely at the agro-ecological scale, IDEA emphasizes on the ecosystem in term of biodiversity, organization of space, pollution while the other Nexus aspects are under-represented. Indeed, it divides the total score of 100 unequally between the diversity and organization of space components on the one hand, and the farming practices component on the other hand. The first two relate more to the ecosystem aspect of the Nexus. Even some indicators in the farming practices can also be classified under the ecosystem aspect such as "soil protection". Hence,

Table 1. The scales/levels, components and indicators of the IDEA method (Zahm et al., 2008) along with the authors' commentaries.

Scale	Component	Indicators	Commentaries re. the WEFE Nexus
Agro-ecological	Diversity	Diversity of annual or temporary crops	These indicators relate to the ecosystem dimension of the WEFE Nexus. A single aggregated bio-diversity indicator might be considered.
		Diversity of perennial crops	
		Diversity of associated vegetation	
		Animal diversity	
		Enhancement and conservation of genetic heritage	
	Organization of space	Cropping pattern	These indicators relate to the food production aspect of the Nexus and would need to be reframed into the specific perspective.
		Dimension of fields	
		Stocking rate	
		Fodder area management	
		Organic matter management	These indicators relate to the ecosystem aspect.
		Ecological buffer zone	
		Measures to protect the natural heritage	This indicator relates to the cultural dimension.
	Farming practices	Fertilization	These indicators relate to the environmental and water quality aspects.
		Effluent processing	
		Pesticides and veterinary products	
		Animal well being	This indicator is related to food production and to ethics
		Soil resources protection	This indicator relates to the environment and to food production.
		Water resources protection	This is the only indicator that explicitly mentions water.
		Energy dependence	This is the only energy-related indicator.
Socio-Territorial	Quality of the products and lands	Quality of food stuffs produced	This indicator relates to the food production dimension.
		Enhancement of building and landscape heritage	This indicator relates to the cultural dimension.
		Processing of non-organic waste	This indicator relates to the environment dimension.
		Accessibility of space	These indicators are promising not only regarding the social dimension but also to articulate the farm scale/level with upper levels in the Nexus perspective.
		Social involvement	
	Organization of space	Short trade	
		Services, multi-activities	
		Contribution to employment	
		Collective work	
		Probable farm sustainability	
	Ethics and human development	Training	
		Labor intensity	
		Isolation	
		Reception hygiene and safety	
		Contribution to world food balance	This indicator proves how a WEFE indicator can be articulated between scales/levels.
Economic	Economic viability	Available income per worker compared with the national legal minimum wage	These indicators at the farm scale/level are entry points to articulate with the upper scales/levels.
		Economic specialization rate	
	Independence	Financial autonomy	
		Reliance on direct subsidies from CAP and indirect economic impact of milk and sugar quotas	
	Transferability	Total assets minus lands value by non-salaried worker unit	
	Efficiency	Operating expenses as a proportion of total production value	

this scale gives more values to the ecosystem over other aspects of the Nexus. Indeed, the energy aspect, and the water aspect are only explicitly represented in two indicators with a total score that cannot exceed 12 points out of 100. Therefore, there is an imbalance in the scoring of the indicators between different aspects of the WEFE Nexus.

The socio-territorial scale also includes indicators that can be well put under the WEFE components especially food quality. The indicator "processing of non-organic waste" can also be put under the ecosystem component. Yet, it appears that IDEA3 emphasizes more on the ecosystem aspect of the Nexus.

2. The application scale is the second challenge. The IDEA assesses sustainability at the farm scale/level while the Nexus is at play at various scales/levels. Therefore, the application on the farm scale needs to be articulated with larger scales where the Nexus displays emerging consistencies. Indeed, the Nexus includes interactions at various scales and there exist many externalities to a given Nexus system. These interactions are not explicitly accounted for in IDEA. However, the fact that IDEA is divided into agro-ecological, socio-territorial and economy scales, does permit to account for some interactions across scales, in particular the impact of the farm on the higher organizational scale. Indeed, while the agro-ecological scale "*analyses the propensity of the technical system to make efficient use of the environment at the lowest possible ecological cost*", the socio-territorial scale "*characterizes the integration of the farm within its landscape and in society*" (Zahm et al., 2008). Hence, the method does take into account certain interactions between the farm and its surroundings, however these are limited to very specific social, environmental and food quality aspects. Thus, if it is to be used for the WEFE Nexus assessment, the farm impact should be also assessed in terms of other aspects of the WEFE Nexus, including water and energy. Moreover, the impact of larger scales on the farm should also be accounted for. In addition, the economic scale of IDEA focuses on the farm economic viability, independence, inter-generational transferability and efficiency. While these are decisive aspects to focus on, they are limited to the farm itself and there is no assessment of the farm contribution to regional economy and vice versa. Another aspect that IDEA dismisses relates to the institutional and governance impacts. Furthermore, the method does not take into account the aggregated effect of multiple neighboring exploitations that share the same resources. The interactions of these exploitations affect the environment but also each individual farm. Moreover, the aggregated effects of multiple small scales can result in the emergence of new challenges at larger scales.

3. Third, an emerging challenge will appear if/when expanding the IDEA method to address the Nexus approach: it should be avoided to deal with different aspects of the Nexus on the one hand, and the sustainability domains on the other hand, in silos. Indeed, IDEA, like the majority of sustainability assessment tools, does not explicitly examine the synergies and trade-offs between the indicators within each scale or between scales whereas these are central in the Nexus approach.

4 Summary

The WEFE Nexus is certainly embedded in the question of sustainable development. It presents a most needed operational framework that can better account for and help understand and unravel the different interlinkages, synergies and trade-offs within but also between the natural and human systems in order to manage our resources in a sustainable manner; especially when dealing with vulnerable environments. However, for such a framework to become useful, one should be able to quantify its performance in terms of sustainability. Hence, the importance of the Nexus sustainability assessment. Yet, such an endeavor is not a simple task even if numerous tools exist for agriculture sustainability assessment. In this work, we highlighted some limitations but also prospects using the example of one particular tool: the IDEA. The main challenges relate to the under-representativeness of some aspects of the WEFE Nexus, the applicability scale, and the Nexus interactions, trade-offs and synergies (see Unsolved Problem in Hydrology (UPH) 22). Nevertheless, with some adaptation, such tools are a good avenue to explore sustainability assessment and transition design, with a focus on the Nexus as it is needed in some contexts. Their appeal relates to the fact that they cover all three dimensions of sustainability. Moreover, they are widely used with a large bibliographic record, which permits comparisons and discussions. Future research should certainly focus on enhancing these tools to better account for all WEFE aspects and to be able to capture in a more explicit way the intra-scale but also inter-scale synergies and trade-offs. The adaptation should certainly focus on adapting the list of indicators to better represent the different aspects of the WEFE. Moreover, a participatory approach in co-defining and co-weighing the indicators can certainly be beneficial when adapting such tools for different case studies. Finally, the insertion of new dynamic indicators can enhance the monitoring of the WEFE sustainability.

Data availability. No data sets were used in this article.

Author contributions. MM led the analysis and drafted the article. FN and CC initiated and consolidated the work.

Competing interests. At least one of the (co-)authors is a guest member of the editorial board of *Proceedings of IAHS* for the special issue "IAHS2022 – Hydrological sciences in the Anthropocene: Past and future of open, inclusive, innovative, and society-interfacing approaches". The peer-review process was guided by an independent editor, and the authors also have no other competing interests to declare.

Disclaimer. Publisher's note: Copernicus Publications remains neutral with regard to jurisdictional claims in published maps and institutional affiliations.

Special issue statement. This article is part of the special issue "IAHS2022 – Hydrological sciences in the Anthropocene: Past and future of open, inclusive, innovative, and society-interfacing approaches". It is a result of the XIth Scientific Assembly of the International Association of Hydrological Sciences (IAHS 2022), Montpellier, France, 29 May–3 June 2022.

Acknowledgements. Authors gratefully acknowledge the financial support of the NEXUS-NESS project that has received funding from the PRIMA Programme, an Art. 185 initiative supported and funded under Horizon 2020, the European Union's Framework Programme for Research and Innovation, with Grant Agreement no. 2042 (https://prima-nexus-ness.org/, last access: June 2022).

Financial support. This research has been supported by the PRIMA programme (grant no. 2042).

Review statement. This paper was edited by Salvatore Grimaldi and reviewed by two anonymous referees.

References

Baccar, M., Bouaziz, A., Dugué, P., Gafsi, M., and Le Gal, P.-Y.: The determining factors of farm sustainability in a context of growing agricultural intensification, Agroecol. Sustain. Food, 43, 386–408, https://doi.org/10.1080/21683565.2018.1489934, 2019.

Bazzana, D., Comincioli, N., El Khoury, C., Nardi, F., and Vergalli, S.: WEF Nexus Policy Review of Four Mediterranean Countries, Land 2023, 12, 473, https://doi.org/10.3390/land12020473, 2023.

Bekhouche-Guendouz, N.: Evaluation de la durabilité des exploitations bovines laitières des Bassins de la Mitidja et d'Annaba, PhD Thesis, Institut National Polytechnique de Lorraine, 2011.

Binder, C. R., Feola, G., and Steinberger, J. K.: Considering the normative, systemic and procedural dimensions in indicator-based sustainability assessments in agriculture, Environ. Impact Asses., 30, 71–81, https://doi.org/10.1016/j.eiar.2009.06.002, 2010.

Carmona-Moreno, C., Dondeynaz, C., and Biedler, M.: Position paper on water, energy, food and ecosystems (WEFE) nexus and sustainable development goals (SDGs), Publications Office of the European Union, Luxembourg, ISBN 978-92-76-00159-1, https://doi.org/10.2760/31812, JRC114177, 2019.

Cudennec, C., Leduc, C., and Koutsoyiannis, D.: Dryland hydrology in Mediterranean regions – a review, Hydrolog. Sci. J., 52, 1077–1087, https://doi.org/10.1623/hysj.52.6.1077, 2007.

Cudennec, C., Liu, J., Qi, J., Yang, H., Zheng, C., Gain, A. K., Lawford, R., de Strasser, L., and Yillia, P. T.: Epistemological dimensions of the water–energy–food nexus approach: reply to discussions of "Challenges in operationalizing the water–energy–food nexus", Hydrolog. Sci. J., 63, 1868–1871, https://doi.org/10.1080/02626667.2018.1545097, 2018.

De Olde, E. M., Oudshoorn, F. W., Sørensen, C. A. G., Bokkers, E. A. M., and de Boer, I. J. M.: Assessing sustainability at farm-level: Lessons learned from a comparison of tools in practice, Ecol. Indic., 66, 391–404, https://doi.org/10.1016/j.ecolind.2016.01.047, 2016.

Fadul-Pacheco, L., Wattiaux, M. A., Espinoza-Ortega, A., Sánchez-Vera, E., and Arriaga-Jordán, C. M.: Evaluation of Sustainability of Smallholder Dairy Production Systems in the Highlands of Mexico During the Rainy Season, Agroecol. Sustain. Food, 37, 882–901, https://doi.org/10.1080/21683565.2013.775990, 2013.

FAO (Ed.): Building a common vision for sustainable food and agriculture: principles and approaches, Food and Agriculture Organization of the United Nations, Rome, 50 pp., E-ISBN 978-92-5-108472-4, 2014a.

FAO: The Water-Energy-Food Nexus: a new approach in support of food security and sustainable agriculture, https://www.fao.org/policy-support/tools-and-publications/resources-details/en/c/421718/ (last access: in February 2023), 2014b.

Faye, A., Zucchini, E., Ngom, Y., Vignaroli, P., Tarchiani, V., and Dia, D.: Assessing the Sustainability of Horticultural Farms in Central Senegal: An Adaptation of the IDEA Method, J. Agric. Sustain., 13, https://doi.org/10.28924/ip/jas.1923, 2020.

Heal, K. V., Bartosova, A., Hipsey, M. R., Chen, X., Buytaert, W., Li, H.-Y., McGrane, S. J., Gupta, A. B., and Cudennec, C.: Water quality: the missing dimension of water in the water–energy–food nexus, Hydrolog. Sci. J., 66, 745–758, https://doi.org/10.1080/02626667.2020.1859114, 2021.

Hoff, H.: Understanding the Nexus, in: The Water, Energy and Food Security Nexus, Bonn, https://www.sei.org/publications/understanding-the-nexus/ (last access: 16 May 2022), 2011.

Liu, J., Yang, H., Cudennec, C., Gain, A. K., Hoff, H., Lawford, R., Qi, J., Strasser, L. de, Yillia, P. T., and Zheng, C.: Challenges in operationalizing the water–energy–food nexus, Hydrolog. Sci. J., 62, 1714–1720, https://doi.org/10.1080/02626667.2017.1353695, 2017.

Marchand, F., Debruyne, L., Triste, L., Gerrard, C., Padel, S., and Lauwers, L.: Key characteristics for tool choice in indicator-based sustainability assessment at farm level, Ecol. Soc., 19, 46, https://doi.org/10.5751/ES-06876-190346, 2014.

Merheb, M., Moussa, R., Abdallah, C., Colin, F., Perrin, C., and Baghdadi, N.: Hydrological response characteristics of Mediterranean catchments at different time scales: a meta-analysis, Hydrolog. Sci. J., 61, 2520–2539, https://doi.org/10.1080/02626667.2016.1140174, 2016.

M'Hamdi, N., Aloulou, R., Hedhly, M., and Ben Hamouda, M.: Evaluation de la durabilité des exploitations laitières tunisiennes par la méthode IDEA", BASE [En ligne], numéro 2, volume

13, 221–228, https://popups.uliege.be/1780-4507/index.php?id=3865 (last access: 16 May 2022), 2009.

Salas-Reyes, I. G., Arriaga-Jordán, C. M., Rebollar-Rebollar, S., García-Martínez, A., and Albarrán-Portillo, B.: Assessment of the sustainability of dual-purpose farms by the IDEA method in the subtropical area of central Mexico, Trop. Anim. Health Prod., 47, 1187–1194, https://doi.org/10.1007/s11250-015-0846-z, 2015.

Srour, G., Marie, M., and Abi Saab, S.: Evalu ation de la du rabilité des élevages de petits ru min an ts au Liban, in: Changes in sheep and goat farming systems at the beginning of the 21st century : research, tools, methods and initiatives in favour of a sustainable development, edited by: Pacheco, F. and Morand-Fehr, P., Zaragoza: CIHEAM/DRAP-Norte/FAO, 21–35 (Options Méditerranéennes: Série A. Séminaires Méditerranéens, n. 91, 2009.

Zahm, F., Viaux, P., Vilain, L., Girardin, P., and Mouchet, C.: Assessing farm sustainability with the IDEA method - from the concept of agriculture sustainability to case studies on farms, Sust. Dev., 16, 271–281, https://doi.org/10.1002/sd.380, 2008.

Proc. IAHS, 385, 97–102, 2024
https://doi.org/10.5194/piahs-385-97-2024

IAHS2022 – Variability and change across space, time, extremes, and interfaces

Assessment of climate change impact and comparison of downscaling approaches: a case study in a semi-arid river basin

Yali E. Woyessa

Department of Civil Engineering, Central University of Technology, Bloemfontein 9300, South Africa

Correspondence: Yali E. Woyessa (ywoyessa@cut.ac.za)

Published: 18 April 2024

Abstract. Rapid population increase, industrialization and pollution are putting a strain on available and diminishing freshwater resources. Recent climate projections suggest a drop of up to 10 % in precipitation in most of Southern Africa by 2050. The main aim of this paper is to assess the impact of climate change on water resources in a semi-arid river basin in South Africa using two downscaling approaches: statistical downscaling (SDE) and dynamic downscaling (CORDEX) approaches. Both SDE and CORDEX data were derived from the GCM simulations of the Coupled Model Inter-comparison Project Phase-5 (CMIP5) and across two greenhouse gas emission scenarios known as Representative Concentration Pathways (RCP) 4.5 and 8.5 with a spatial resolution 25 km × 25 km for SDE and 50 km × 50 km for CORDEX. Four GCM models were used for both approaches. SWAT hydrological model was run using these data for a period of 2020 to 2050. Varied results were obtained depending on the type of climate model used, but generally, the trends were similar in most cases. For SDE approach, the multi-model average showed a possible decrease in precipitation (by 14 %), a decrease in water yield (by 15 %) and an increase in potential evapotranspiration (by 10 %). For CORDEX data, the multi-model average showed a possible decrease in precipitation (up to −3 %), a decrease in water yield (up to −13 %) and an increase in potential evapotranspiration (ET) (up to +22 %). The latter is indicative of possible drought spells between rainy events. The SDE approach showed much more pronounced decrease of precipitation and water yield compared to the CORDEX approach. This difference could be attributed to the difference in spatial resolution of the two downscaling approaches. However, it is expected that the results of this study could assist in policy formulation to mitigate the negative impact of climate change in the region.

Keywords. Climate change; impact; hydrology; modelling; downscaling; South Africa

1 Introduction

Water security is reported to be one of the greatest challenges sub-Saharan African countries are facing. To ensure water security for all, freshwater must be available in acceptable quantity and quality as it is essential for health, livelihoods, ecosystems, and production (AMCOW, 2012). However, rapid population increase, industrialization and pollution are putting a strain on available and diminishing freshwater resources. The problem of water scarcity is likely to deepen according to some projected climate change scenarios. The Southern African region is projected to be generally drier by 2050 (DEA, 2013). Over the past five decades, research has shown that there is a tendency towards a rise in temperature in South Africa (MacKellar et al., 2014) where maximum and minimum temperatures have shown significant increases annually, and in almost all seasons. The frequency of annual high temperature extremes was found to have increased significantly whereas that of low temperatures have decreased significantly across the country (DEA, 2013). Past climate projection suggests that there could be a decrease of up to 10 % in precipitation in most of the Southern Africa by 2050 (Levina, 2006). Globally, it is estimated that almost 50 % the world population will be living in water stressed regions by the year 2025 (World Water Council, 2000). Furthermore, rapid population increase, industrialization and pollution are further straining the available and diminishing freshwater resources. Within this context, it is important to emphasize that development must be sustainable, adaptable and resilient to these global changes, such as climate change. The main aim of this paper is to assess the impact of climate change on water resources in a semi-arid river basin in South Africa using two climate downscaling approaches: statistical downscaling (SDE) and dynamic downscaling (CORDEX) approaches. The results of this study could assist in policy formulation to mitigate the negative impact of climate change in the region.

2 Methodology

This study is focused on a semi-arid river basin in the central region of South Africa (Free State province), known as the Modder River basin (Fig. 1). The Modder River basin has a total area of 17 366 km^2 and is located within the Upper Orange Water Management Area, Central South Africa (latitude 28°50" to 29°40" S and longitude 24°40" W to 27°00" E). The mean annual rainfall of the study area is 551 mm. The rainy season consists of January to April and September to December, leaving the months of May to August as dry winter season (Fig. 2). The highest maximum temperatures vary between 28 and 31 °C while the lowest minimum temperatures vary between −1 and 1 °C.

2.1 Climate modelling

A Global Climate Model (GCM) can provide reliable prediction information over large areas (around 1000 km × 1000 km) with vastly differing landscapes and greatly varying potential for floods, droughts, or other extreme climate events. However, Regional Climate Models (RCM) and Empirical Statistical Downscaling (SDE) can be applied over a limited area and can provide information on much smaller scales providing support for more detailed impact assessment and adaptation planning, which is vital in many vulnerable regions of the world (https://cordex.org/, last access: 15 January 2018).

Downscaled climate data at high resolution are freely available from two sources: the National Aeronautics and Space Administration (NASA) Earth Exchange Global Daily Downscaled Projections (NEX-GDDP) and Coordinated Regional Climate Downscaling Experiment (CORDEX). The NEX-GDDP dataset contains downscaled climate scenarios derived from the GCM simulations of the Coupled Model Inter-comparison Project Phase-5 (CMIP5) and across two greenhouse gas emission scenarios known as Representative Concentration Pathways (RCP) 4.5 and 8.5. The spatial resolution of the dataset is 0.25° × 0.25° (∼ 25 km × 25 km). These datasets provide a set of global, high resolution, bias-corrected climate change projections that can be used to evaluate climate change impacts on processes that are sensitive to finer-scale climate gradients and the effects of local topography on climate conditions on finer-scales. The Bias-Correction Spatial Disaggregation (BCSD) method used in generating the NEX-GDDP dataset is a statistical downscaling algorithm specifically developed to address the current limitations of global GCM outputs (Thrasher et al., 2012). Each of the climate projections includes daily precipitation and daily maximum and minimum temperatures for the periods from 1950 to 2005 (Retrospective Run) and from 2006 to 2099 (Prospective Run). However, only the period from 2006 to 2050 is considered in this study.

Recognizing the availability of limited studies on regional climate downscaling (RCD), the World Climate Research Program established a task force to develop a framework to evaluate and possibly improve RCD techniques for use in downscaling global climate projections. This task force, in consultation with the broader scientific community, initiated a framework called the Coordinated Regional Climate Downscaling Experiment (CORDEX). The downscaled climate data from CORDEX covers the period of 1951–2100, of which 1950–2005 is for Retrospective Run and 2020–2100 is for Prospective Run, and two greenhouse emission scenarios, namely RCP4.5 and RCP8.5. The spatial resolution for Africa is reported to be 50 km grid spacing.

For comparison of these two downscaling approaches, a sub-set of four GCMs (CNRM-CM5, IPSL-CM5A-LR, MIROC5, MPI-ESM-MR) was selected based on the correlation between historical measured rainfall data and simulated historical data (Retrospective Run) for the period of 1975–2005. Daily precipitation, and daily minimum and maximum temperatures for Prospective Run of 30 years (2020–2050) were considered for both datasets. In both cases, two greenhouse gas emission scenarios (RCP4.5 and RCP8.5) were considered, which are said to be the highest priority global model simulations within CIMP5.

2.2 Hydrological modelling

The Soil and Water Assessment Tool (SWAT) is a basin scale hydrological model that operates on a continuous daily time step and is designed to predict the impact of management on water, sediment and agricultural chemical yields in ungauged catchments. The model is physically based and capable of continuous simulation over long-time periods (Gassman et al., 2007). It integrates several parameters, such as weather, surface and groundwater hydrology, soil properties, plant growth, and land management practices to model processes within a watershed (Arnold et al., 1998). In SWAT, watersheds are divided into sub-basins based on outlet points along the stream network. Each sub-basin was then divided into hydrologic response units (HRUs), which are defined as areas with homogeneous soil type, land use, slope, and management practices. Water yields are calculated for each HRUs and summed up to determine the total sub-basin outputs, which in turn contributes to the total watershed yield out-

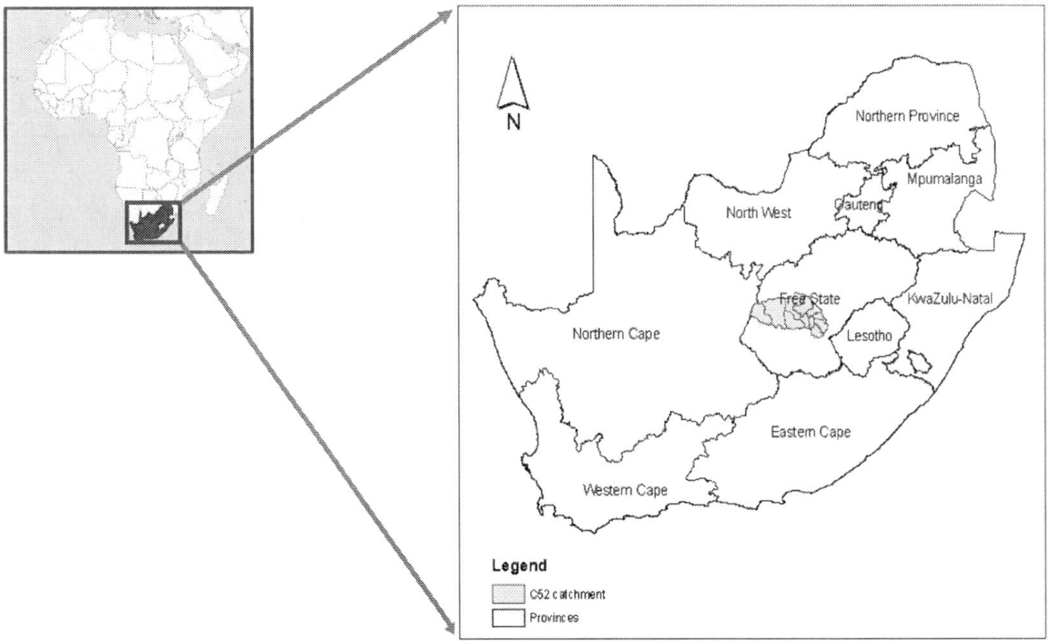

Figure 1. The study site (C52 Catchment) (© Google Maps).

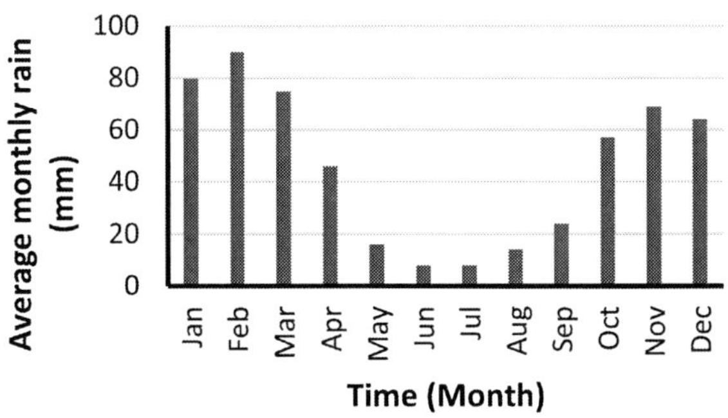

Figure 2. Long-term average monthly rainfall (1975–2005).

put. HRUs are not spatially defined within sub-basins but represented as percentages of total sub-basin area. Therefore, SWAT includes both spatially distributed parameterization at the sub-basin scale and lumped parameterization at the HRU scale (Gassman et al., 2007).

The SWAT model has been successfully applied to assess the impact of climate change and land management practices on water yields (Cousino et al., 2015; Ndlovu and Woyessa, 2020, 2021). For example, Stone et al. (2001, cited in Cousino et al., 2015) used SWAT to predict the impact of increase in atmospheric CO_2 concentration on water yield in a river basin. The authors reported that the overall water yield of the Missouri River basin will decrease by 10 %–20 % during spring and summer months but increase during the fall and winter months in response to doubling atmospheric CO_2 concentrations. Cousino et al. (2015) also used SWAT to model the effects of climate change on water, sediment, and nutrient yields from the Maumee River watershed. They reported that moderate climate change scenarios reduced annual flow (up to −24 %) and sediment (up to −26 %) yields, while a more extreme scenarios showed smaller flow reductions (up to −10 %) and an increase in sediment (up to +11 %).

In this study, the SWAT model was applied to simulate the impact of climate change on water balance components, such as runoff, water yield, evapotranspiration, and potential evapotranspiration. The model inputs are: Digital Elevation Model (DEM), land use and soil maps, daily precipitation and daily maximum and minimum temperatures (of future climate). Sensitivity and calibration analysis for parameters used in the model were carried out using SWAT statistical module during the previous study in the same river basin (Welderufael et al., 2013). Calibration was carried out on the most sensitive input parameters of the model, such as curve number, soil available water capacity, threshold depth of water in the shallow aquifer, etc. There were 13 sensitive parameters in total. In this calibration process, Nash and Sutcliffe efficiency (0.57) together with the coefficient of determination ($R^2 = 0.68$), agreement index (D-Index = 0.86), and residual mean square error (RMSE = 0.87) were used as measure of efficiency, all of which were found to be satisfactory (Welderufael et al., 2013).

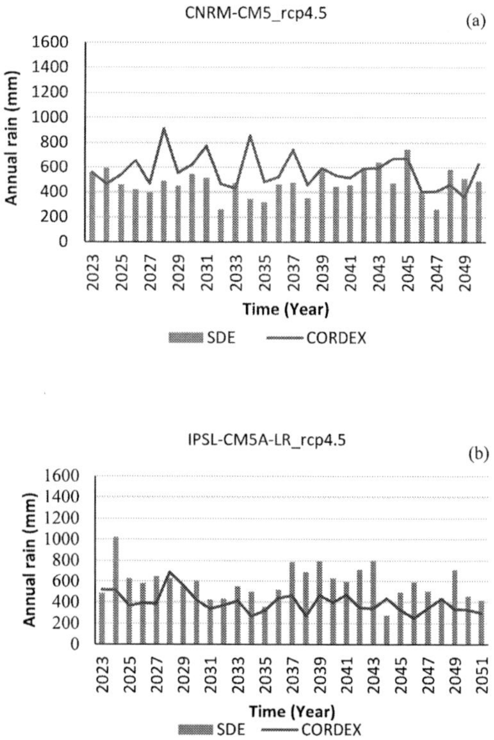

Figure 3. GCM projections of annual rainfall using the two downscaling approaches under RCP4.5 emission scenario.

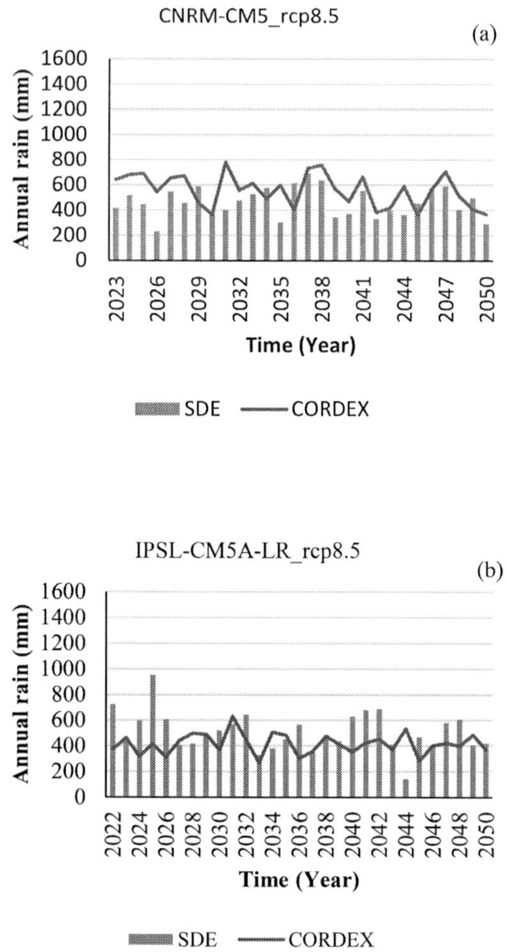

Figure 4. GCM projections of annual rainfall using the two downscaling approaches under RCP8.5 emission scenario.

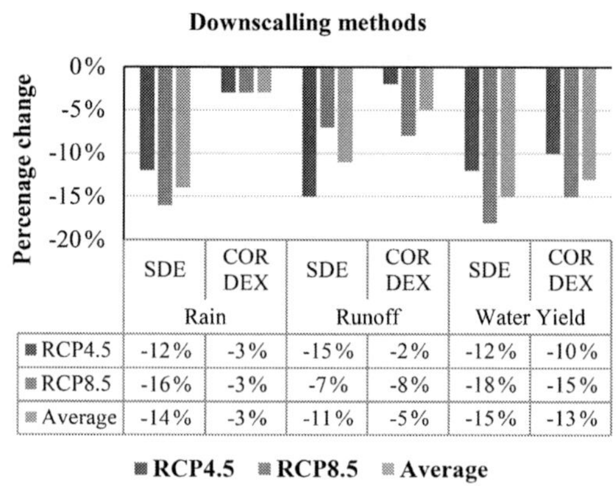

	Rain		Runoff		Water Yield	
	SDE	CORDEX	SDE	CORDEX	SDE	CORDEX
RCP4.5	-12%	-3%	-15%	-2%	-12%	-10%
RCP8.5	-16%	-3%	-7%	-8%	-18%	-15%
Average	-14%	-3%	-11%	-5%	-15%	-13%

Figure 5. Summary of climate change impact (percentage change) on water balance components according to the two downscaling approaches (SDE and CORDEX).

3 Results and discussion

The SWAT hydrological simulation was run using future climate data based on the four GCM models (under the two downscaling approaches: SDE and CORDEX), and two greenhouse gas emission scenarios (RCP4.5 and RCP8.5). Furthermore, analysis of the GCM projections of future rainfall data was done and compared with historical measured data in order to understand the trend by 2050. Analysis of GCM projections and hydrological simulations were conducted using the four GCMs that were selected and two emission scenarios (RCP4.5 and RCP8.5) for the period of 2020 to 2050. The first two years are considered by SWAT as a warm-up period and were not considered in the simulation. The figures presented hereunder are for two GCMs and two emissions scenarios using the two downscaling approaches with the main purpose of demonstrating the long-term climate using these two approaches and the two emission scenarios. The summary of the analysis of results based on multi-model averages are provided in Fig. 5. Figures 3 and 4 show the annual rainfall trends from 2022 to 2050 using both downscaling approaches and the two greenhouse gas emission scenarios. Closer inspection of these figures shows similar trends for both approaches with a slight overestimation using CNRM-CM5 under CORDEX approach with emission scenario of RCP4.5 and RCP 8.5 (Figs. 3a and 4a).

Figure 5 shows the summary of potential impact of climate change on future water resources availability. The SWAT model simulation was done using historical data as well as future climate data in order to analyze the expected changes in water balance components, namely rain, potential evapotranspiration (PET), surface runoff, and water yield. The PET has been analyzed using multi-model approach and was found that there will be an increase of 10 %–13 % using the SDE approach and 17 %–22 % using CORDEX.

The multi-model average shows that by 2050 there will be a reduction in rainfall, runoff and water yield by 12 %, 15 %, and 12 %, respectively for SDE under RCP4.5. However, these numbers decrease further to −16 % for rainfall and −18 % for water yield under RCP8.5. This result may not be surprising given the fact that the RCP8.5 is for high emission scenario. For CORDEX approach, slightly lower percentage change is observed but the trend is similar as one goes from RCP4.5 to RCP8.5.

4 Conclusions

Climate change impact assessment was undertaken in a semi-arid river basin in the central region of South Africa using two downscaling approaches, namely Statistical Downscaling (SDE) and Dynamic Downscaling from CORDEX under two greenhouse gas emission scenarios (RCP4.5 and RCP8.4) for a period of 2020 to 2050 using four GCMs. SWAT hydrological model was run using future climate data as input together with other biophysical data. The simulation results based on the two approaches revealed that there will be a reduction of all water balance components in general, and rainfall and water yield in particular. The multi-model average showed that there will be a reduction of rainfall by 14 % and 3 % using SDE and CORDEX, respectively. Similarly, the water yield is expected to decrease by 15 % and 13 % using SDE and CORDEX respectively. These expected decrease in rainfall coupled with increase in PET are going to be the two major drivers in exacerbating the existing problem of water scarcity in the region.

Data availability. The downscaled GCM (Global Circulation Model) data were obtained from from publicly available sources, such as Statistically Downscaled Data (SDE) from NASA Earth Exchange Global Daily Downscaled Projections (NEX-GDDP) (https://www.nccs.nasa.gov/services/data-collections/land-based-products/nex-gddp, NCCS, 2017), and Dynamically Downscaled Data from CORDEX (https://cordex.org/, CORDEX, 2017).

The hydrological model used (SWAT) is a freely available model from https://swat.tamu.edu/ (SWAT, 2017).

Competing interests. The author has declared that there are no competing interests.

Disclaimer. Publisher's note: Copernicus Publications remains neutral with regard to jurisdictional claims in published maps and institutional affiliations.

Special issue statement. This article is part of the special issue "IAHS2022 – Hydrological sciences in the Anthropocene: Variability and change across space, time, extremes, and interfaces". It is a result of the XIth Scientific Assembly of the International Association of Hydrological Sciences (IAHS 2022), Montpellier, France, 29 May–3 June 2022.

Review statement. This paper was edited by Christophe Cudennec.

References

African Ministers' Council on Water (AMCOW): Water Security and climate resilient development: Technical Background Document, African Ministers' Council on Water, https://www.gwp.org/globalassets/global/toolbox/references/technical-background-document-water-security-and-climate-resilient-development-wacdep-amcow-2012.pdf (last access: 10 September 2017), 2012.

Arnold, J. G., Srinivasan, R., Muttiah, R. S., and Williams, J. R.: Large-area hydrologic modelling and assessment: Part I. Model development, Journal of American Water Resources Association, 34, 73–89, 1998.

CORDEX (Coordinated Regional Downscaling Experiment): Regional Climate Change simulations for CORDEX domains, CORDEX [data set], https://cordex.org/data-access/regional-climate-change-simulations-for-cordex-domains/, last access: 4 October 2017.

Cousino, L. K., Becker, R. H., and Zmijewski, K. A.: Modelling the effects of climate change on water, sediment, and nutrient yields from the Maumee River watershed, Journal of Hydrology: Regional Studies, 4, 762–775, 2015.

DEA (Department of Environmental Affairs): Long-Term Adaptation Scenarios Flagship Research Programme (LTAS) for South Africa, Summary for Policy-Makers, Pretoria, South Africa, https://www.dws.gov.za/documents/Other/StrategicPlan/NWRS2-Final-email-version.pdf (last access: 20 July 2021), 2013.

Gassman, P. W., Reyes, M. R., Green, C. H., and Arnold, J. G.: The soil and water assessment tool: historical development, applications, and future research directions, Trans. ASABE, 50, 1211–1250, 2007.

Levina, E.: Domestic Policy Frameworks for Adaptation to Climate Change in the Water Sector Part II: Non-Annex I Countries. Lessons Learned from Mexico, India, Argentina and Zimbabwe, OECD, Paris, https://www.oecd.org/env/cc/37671630.pdf (last access: 15 October 2017), 2006.

MacKellar, N., New, M., and Jack, C.: Observed and modelled trends in rainfall and temperature for South Africa: 1960–2010, S. Afr. J. Sci., 110, 1–13, 2014.

NCCS (NASA Center for Climate Simulation): NASA Earth Exchange Global Daily Downscaled Projections (NEX-GDDP), NASA [data set], https://www.nccs.nasa.gov/services/data-collections/land-based-products/nex-gddp, last access: 25 September 2017.

Ndhlovu, G. Z. and Woyessa, Y. E.: Modelling impact of climate change on catchment water balance, Kabompo River in Zambezi River Basin, Journal of Hydrology: Regional Studies, 27, 100650, https://doi.org/10.1016/j.ejrh.2019.100650, 2020.

Ndhlovu, G. Z. and Woyessa, Y. E.: Evaluation of Streamflow under Climate Change in the Zambezi River Basin of Southern Africa, Water, 13, 3114, https://doi.org/10.3390/w13213114, 2021.

SWAT: Soil and Water Assessment Tool, https://swat.tamu.edu/, last access: 5 August 2017.

Thrasher, B., Maurer, E. P., McKellar, C., and Duffy, P. B.: Technical Note: Bias correcting climate model simulated daily temperature extremes with quantile mapping, Hydrol. Earth Syst. Sci., 16, 3309–3314, https://doi.org/10.5194/hess-16-3309-2012, 2012.

Welderufael, W. A., Woyessa, Y. E., and Edossa, D. C.: Impact of Rainwater harvesting on Water resources of the Modder River Basin, Central Region of South Africa, Journal of Agri Water Management, 116, 218–227, https://doi.org/10.1016/j.agwat.2012.07.012, 2013.

World Water Council: World Water Vision: Making Water Everybody's Business, 108 pp, ISBN 185383730X, Earthscan Publications Ltd, London, 2000.

Proc. IAHS, 385, 103–109, 2024
https://doi.org/10.5194/piahs-385-103-2024

Open Access PIAHS

Hydrological Characterization of Mediterranean Catchments

Antoine Allam[1,2], Roger Moussa[2], Wajdi Najem[1], and Claude Bocquillon[1]

[1]CREEN, ESIB, Univ. Saint Joseph, Beirut, Lebanon
[2]UMR LISAH, Univ. Montpellier, INRA, Montpellier, France

Correspondence: Antoine Allam (antoine.allam@net.usj.edu.lb)

Received: 29 May 2022 – Revised: 28 January 2023 – Accepted: 1 March 2023 – Published: 18 April 2024

Abstract. A climatic, physiographic, and hydrological homogeneity across the Mediterranean has been detected but not yet confirmed. Mediterranean climate is known for its precipitation seasonality and the alternation of humid winters and dry summers which conditions rivers flow regimes, landcover, agriculture and consequently any water resources management plan. Several physiographic traits could be also observed across the Mediterranean, like the elevated and exposed karstic features, and the cultivated and managed areas. Hydrologically, global river regimes were classified based on monthly average flows only, and Mediterranean regimes were identified under 3 of Haines' 15 global classes with a clear relation to Köppen's Mediterranean climate. Thus, we first studied the flow regimes of 55 Mediterranean catchments to verify if Mediterranean rivers fall into same regime class. Second, we characterized the Mediterranean hydrological response through different water balance functional models as advanced by Budyko, L'vovich and elaborated by Ponce & Shetty and Sivapalan. The water balance analysis highlighted the Mediterranean trend following the general climatic setting from the wet Northern region to the arid Southern region; it also showed hydrological homogeneity for mountainous karstic, and snow influenced catchments which yield the highest baseflows and runoff coefficients, especially if located in moderate climate.

Keywords. Mediterranean; functional modeling; hydrological homogeneity; UPH 6; PUB

1 Introduction

A climatic, physiographic, and hydrological homogeneity across the Mediterranean has been detected but not yet confirmed. Mediterranean climate is known for its precipitation seasonality and the alternation of humid winters and dry summers which conditions rivers flow regimes, landcover, agriculture and consequently any water resources management plan (Allam et al., 2020b). Several physiographic traits could be also observed across the Mediterranean, like the elevated and exposed karstic features, and the cultivated and managed areas that cover 30 % of the Mediterranean, with the symbolic olive tree that is considered one of its iconic bioindicators (Allam et al., 2020a). Hydrologically, global river regimes were classified based on monthly average flows only, and Mediterranean regimes were identified under 3 of the 15 global classes, finding a clear relation to Köppen's Mediterranean climate (Haines et al., 1988).

However, a specific Mediterranean hydrology is long to be defined yet and requires additional research to determine the relation between catchments climatic indices on one hand and the hydrological processes on the other. Since there is a great interest and benefit in sharing knowledge for sustainable water resources management between developed countries, mostly in the North, and underdeveloped countries in the South, a quest to understand the actual and future hydrological behaviour of Mediterranean catchments was launched through the analysis of hydrological similarity and variability. Thus, the objective is to carry out an extensive water balance analysis of Mediterranean catchments based on different water balance functional models to check the hydrological variability and similarity between all catchments and catchments within the same climatic classes, a similar work to the one carried out by Sivapalan on MOPEX dataset for the United States (Sivapalan et al., 2011). Thus, we characterised the Mediterranean hydrological response of 55 Mediterranean catchments through different water balance functional models as advanced by Budyko, L'vovich and elaborated by Ponce & Shetty and Sivapalan. The analysis of the functional models across the Mediterranean and through the climatic classes characterised furthermore the hydrological behaviour similarity and variability, and the competition between the water balance components of the Mediterranean catchments.

This paper is structured into 6 main sections; Sect. 1 Introduction, Sect. 2 presents the definition of the Mediterranean boundaries and a brief description of the hydrological datasets; Sect. 3 presents the methodology; Sect. 4 presents the hydrological characterisation results of the 55 selected Mediterranean catchments through water balance calculation based on the functional approach followed by a discussion in Sect. 5 and Conclusion in Sect. 6.

2 Study area and database

Mediterranean boundary, catchments, and hydrometric data

The Mediterranean refers to the sea and bordering region located in the middle of the European, African and Asiatic continents. With Köppen's classification the definition designated henceforth a moderate climate and extended geographically beyond the limits of the Mediterranean Sea (Köppen, 1936). However, the Mediterranean could be defined with several boundaries based on the field of practice; The hydrological boundary defined by the set of catchments draining towards the Mediterranean Sea was adopted for this study (Milano et al., 2013).

Since the geographic extent of the study is very wide, the delimitation of catchments was imported from international references. European catchments and some adjacent countries were imported from the Joint Research Centre (JRC) (De Jager and Vogt, 2010), catchments in the Middle East and Northern Africa were imported from HydroSHEDS database of World Wildlife Fund's project (Lehner and Grill, 2013).

The selection of catchments and their hydroclimatic data was very challenging as no single database gathers all the available stations in the region. The selected catchments cover a wide geographic domain spread across 15 countries in different physiographic settings. Despite the extensive work performed on data collection, some minor exceptions were still to be found. Some timeseries presented discontinuities and therefore only complete hydrological years were considered.

3 Methodology

3.1 Climatic classification

With multiple studies raising serious concerns of climate shifts and aridity expansion in the region, Allam has established a new high resolution classification for hydrology purposes based on Mediterranean specific climate indices (Allam et al., 2020b). The proposed approach includes the use of classic climatic indices and the definition of new climatic indices mainly precipitation seasonality index I_s or evapotranspiration threshold S_{PET} both in line with river flow regimes. The classification was set and validated by WorldClim-2 at 1 km high resolution gridded data for the 1970–2000 baseline period. Climatic classes coincided with a geographical distribution in the Mediterranean ranging from the most seasonal and dry class 1 in the south to the least seasonal and most humid class 5 in the North, showing up the climatic continuity from one place to another and enhancing the visibility of change trends. The K-means classification shown in Fig. 1 is distributed into five classes.

3.2 Functional modelling

The functional approach for annual water balance estimation took birth from the empirical model of L'vovich and the process-based approach of Horton; L'vovich advanced an empirical approach for the two-stage distribution of annual precipitation. The first stage separates the precipitation into surface runoff or quickflow S and wetting W with $P = S + W$ and the second stage partitioning separated the wetting W into Slow flow U and Vaporisation V with $W = U + V$ (L'vovich, 1979) while Horton introduced the ratio of the vaporisation V to the wetting W and named as Horton index, $H = V/W$ in a trial to search for catchments capacity to store the infiltrated water, depending on the soil type and the evaporation controlled by the vegetation type (Horton, 1933).

Ponce and Shetty have developed a simple conceptual model from the water balance equations to separate precipitation into different flows (Ponce and Shetty, 1995a, b). This model has been proposed as an alternative to empirical models, which have a limited applicability. Ponce et Shetty has adopted the same 2 partitioning stages as L'vovich and calibrated the initial abstraction coefficients surface runoff λ_s, baseflow λ_u, wetting potential W_p, and vaporisation potential V_p through the application of their equations on the same database and tented to classify the climatic regions according to the range of values they obtained. The mountainous conifer forests of North Africa close to the Mediterranean region has shown high W_p with low λ_s and a high V_p with high λ_u.

In the empirical approach, the annual water balance is estimated based on a systematic analysis of long-term rainfall and runoff series observed in different climatic regions

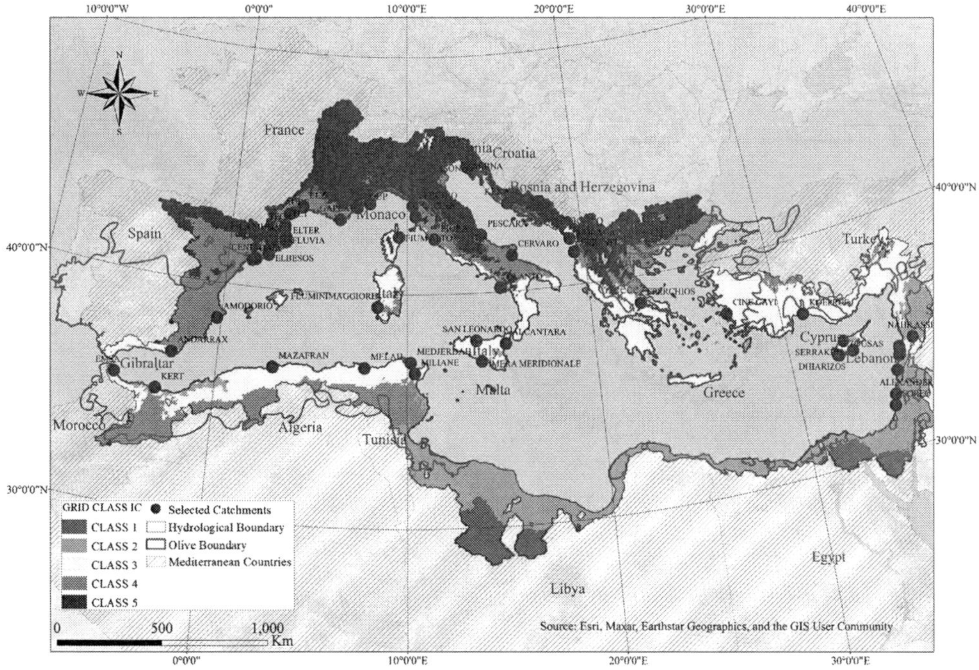

Figure 1. Geographical distribution of the Mediterranean climatic classes (Allam et al., 2020b).

and an empirical analysis of the effect of physiographic and climatic descriptors on the water balance (Sivapalan et al., 2011). The value of the empirical approaches lies in the simplicity of the relations which govern the annual assessment on the long term, and which appeared from the coevolution and the self-organization of climates, soils, topography and vegetations of the different regions.

4 Results

4.1 The water balance partitioning model of L'vovich

The water balance partitioning was carried out and reproduced accurately the empirical relations for most of the catchments. The charts have shown the well-marked interannual variability for all water balance components, with the same observation for the inter-catchment variability with different ranges between catchments. We set as an example Nahr Ibrahim catchment in Lebanon which is highly watered with a mean annual precipitation MAP of 1450 mm, yielding a mean annual runoff MAQ of 1052 mm; a surface flow S ranging between 100 and 550 mm, a baseflow U showing remarkable high values ranging between 200 and 1350 mm, and vaporisation V showing an upper limit, reached every year due to water availability in contrast with other catchments. The water balance partitioning charts are presented in Fig. 2.

4.2 The conceptual water balance model of Ponce and Shetty

The Ponce et Shetty models were fitted over L'vovich partitioning charts. Some catchments did not yield very good results, this inaccuracy returns to either interrupted flows by dams and lakes, catchments interconnection or simply poor quality of data. The sensitivity analysis conducted on the US MOPEX dataset, showed that Ponce and Shetty curves fitting was insensitive to extremes and that the distribution returned a linear fit (Sivapalan et al., 2011). Those extremes were observed in the Mediterranean wherever the flow regime was dominated by the groundwater runoff, case of Nahr el Assi in Lebanon; or wherever the precipitation was very low, case of Andarax in Spain, or Alcantara in Italy. The runoff and baseflow threshold of all the catchments according to Ponce and Shetty (1995a) are summarized in Table 1 and shown in red on the same charts for Ibrahim in Fig. 2.

The spatial distribution of Ponce and Shetty water balance parameters reflected a wide geographical variability across the Mediterranean without any clear pattern or regional clustering despite some adjacent catchments yielding neighbouring values. The French catchments yielded the lowest values for all parameters with both $\lambda_s W_p$ and $\lambda_u V_p$ below 100 mm except for Hérault river ($\lambda_s W_p = 122$ mm; $\lambda_u V_p = 1$ mm). In Cyprus, $\lambda_s W_p$ show close values ranging between 175 and 350 mm. These values are among the highest in the Mediterranean due to the high soil permeability on the island; this is confirmed with the high wetting potential W_p values rang-

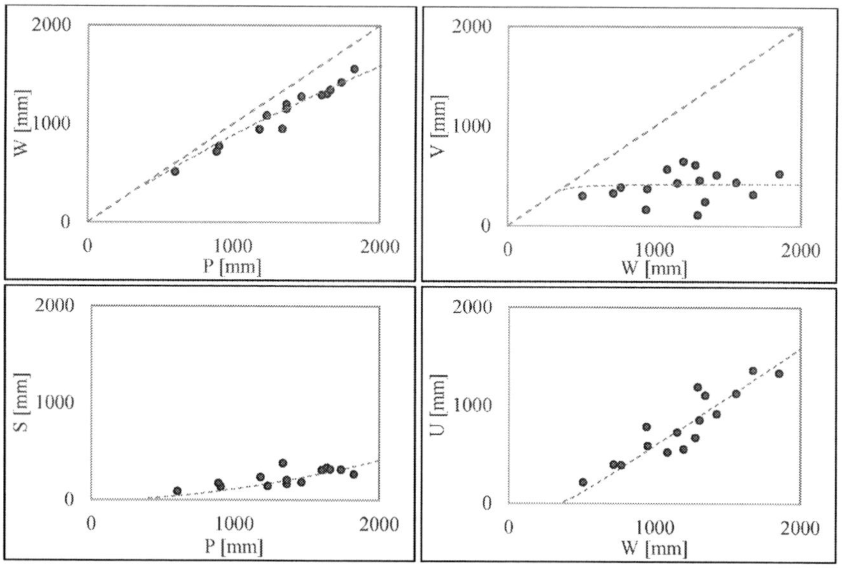

Figure 2. L'vovich water balance components proportional curves extracted for Ibrahim, Lebanon.

Table 1. Summary of the fitting curves parameters according to Ponce and Shetty (1995a).

Summary	λ_s	λ_u	W_p	V_p	$\lambda_s W_p$	$\lambda_u V_p$
Average	0.02	0.10	10 853	4213	88	118
Minimum	0.00	0.00	1897	182	3	0
Maximum	0.09	0.88	78 982	39 167	347	469
Standard Dev	0.02	0.18	14 854	7353	101	131
Abs. Confidence interval	0.01	0.05	3926	1943	27	35
Rel. Confidence interval	39 %	46 %	36 %	46 %	30 %	29 %

ing between 5000 and 9000 mm. The Lebanese coast have a wide variability of $\lambda_s W_p$ ranging from 7 to 270 mm between adjacent catchments as well as for W_p ranging from 5000 to 27 000 mm (except for Nahr el Assi in Lebanon); same observations could be described for $\lambda_u V_p$ and V_p variability. We noticed across the Mediterranean, some matching catchments which share remarkable water balance similarities despite their geographical, climatic, or landscape difference. These catchments could be considered twin catchments like Lez in Franc and Erzenit in Albania (CC4); Fium-alto in France and Fluminimaggiore in Italy (CC4); Mirna in Croatia and Alento in Italy (CC5). The water balance similarity between these catchments could be caused by an underlying physiographic similarity like snow or karst components which make these catchments interesting to examine closely.

The diversity of both microclimates and physiography between catchments stands behind the water balance variability as hydrological response on the same climatic forcing differs according to catchments' landform, landcover, soil types and other physiographic features, and despite some close catchments between Spain, France and Italy, no geographical pattern was detected.

4.3 The nondimensional water balance functional model of Sivapalan et al. (2011)

We verified the nondimensional formulation as suggested by Sivapalan et al. (2011) to check if any water balance pattern could be extracted from Mediterranean site-specific water balance results that express their common hydrological behaviour. The different non dimensional water balance components W^*, S^*, U^* and V^* were estimated from the values of W, S, U and V. A summary is shown here representing the annual average of all the catchments and coloured according to their climatic classes in Fig. 3 in a trial to discover any pattern behind, the average values of each climatic class were also added. The charts show the nature of competition between different water balance components of Mediterranean catchments as suggested by Sivapalan, where, at a first stage, quick flow S^* is competing with the storage W^* and at a second stage slow flow U^* is competing with vaporisation V^*. The climatic illustration of W^* and S^* vs $\tilde{P}$ shows that this

Table 2. Summary of the water balance components and nondimensional metrics averaged by climatic classes.

Climatic Class (55 Catchments)	P(mm)	Q(mm)	U(mm)	S(mm)	W(mm)	V(mm)	K_H	K_V	K_B	K_R
CC2 (3)	473	16	4	12	461	457	0.97	0.90	0.02	0.10
CC3 (17)	839	297	215	82	756	541	0.69	0.63	0.26	0.37
CC4 (17)	758	263	171	92	667	496	0.72	0.64	0.23	0.36
CC5 (18)	1047	522	365	158	885	521	0.57	0.50	0.35	0.50

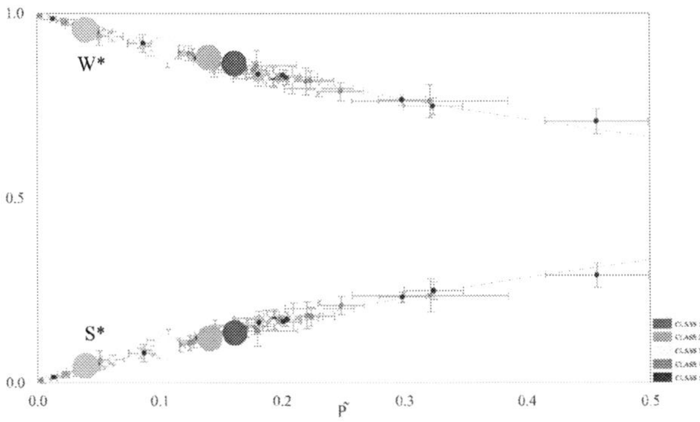

Figure 3. Inter-catchment variability of mean annual water balance nondimensional estimates of W^* and S^* versus annual climatic driver $\tilde{P}$ for all 55 catchments coloured according to climatic classification. Small points represent each catchment data and large points represent class averages and dashed lines are theoretical.

competition is gradually evolving from climatic CC2 in the South, where low precipitation and low quick flows dominate, to CC5 in the North where high precipitation and high quick flows dominate.

The maps in Fig. 4 show an interesting spatial distribution where K_H and K_V gradually decrease from South to North while K_B and K_R gradually decrease from North to South, nonetheless some exceptions were noticed where some Lebanese catchments showed an inverted distribution. The geographical distribution coincided with the Mediterranean climatic classification, therefore we searched for which climatic index is mostly affecting this distribution and found that the water balance metrics of all catchments are slightly correlated with the mean annual aridity index E_p/P ($R = 0.55$) which is also correlated to the non-dimensional aridity index φ with $R = 0.63$ ($r^2 = 0.4$), hence supporting Sivapalan finding on US MOPEX data where $r^2 = 0.46$. It is therefore possible to say, to certain extent, that annual water balance metrics K_H and K_V increase while K_B and K_R decrease when seasonality increases, which suggests that intra-annual variability is responsible for the inter-catchment variability. We also noticed that some CC3 and CC4 catchments yielded lower aridity φ than CC5 which is unusual. And looking to their physiography, we noticed that those were mountainous karstic catchments under snow influence, case of Nahr Ibrahim in North Lebanon which yielded the highest slow flow peaking at 0.71 and total runoff at 0.94, due to high snowmelt contribution and high karstification with a similar observation for Spercheios, Nahr el Kalb and other.

This finding completes the one advanced by Sivapalan et al. (2011) on baseflow metric variability stating that the wettest catchments have yielded the highest baseflow contribution and that soil types and topographic features govern the subsurface flow as part of the self-organisation of the vegetation with the climate as proposed by Horton (1933).

5 Discussion

L'vovich (1979) model showed a breakdown of the annual water balance advancing more info on the surface runoff S and groundwater runoff U which indicates the physiographic forcing on hydrology. It also permitted a precise interannual comparison with precious info on water resources. Ponce and Shetty (1995a) generalized L'vovich model from annual values to continuous curves which helps in predicting the available water resources throughout the years. The model equations are dependent from runoff threshold parameters, specific for each catchment which permits more precise inter catchment comparison, highlighting Mediterranean variability. In addition, these parameters also reflect the physiographic forcing on hydrology. Sivapalan et al. (2011) model went further with non dimensionalising Ponce and Shetty equations which permitted in addition to interannual and inter catchment comparison, an inter climatic and inter physiographic comparison, useful in our study to uncover the variability and homogeneity of Mediterranenan catchments through nondimensional metrics K_H, K_B, K_R and K_V. These metrics showed the special setting of Mediterranean catchments in correspondence to aridity index $\varphi = \tilde{V}/\tilde{P}$. Sivapalan model showed a space time symmetry where functional curves fit both to individual catchments for interannual analysis or to mean annual water balances for inter catchment analysis despite their variability.

The calculation of the dimensionless water balance metrics K_H, K_B, K_V, K_R endorsed the previous findings as aridity index $\varphi = \tilde{V}/\tilde{P}$ turned to be a primary determinant of the mean annual water balance. The charts showed that K_H and K_V were increasing with aridity, suggesting that in general, the vaporisation V (energy) takes over the wetting W and consequently the slow flow component U while moving towards arid catchments. In reverse K_B and K_R were decreas-

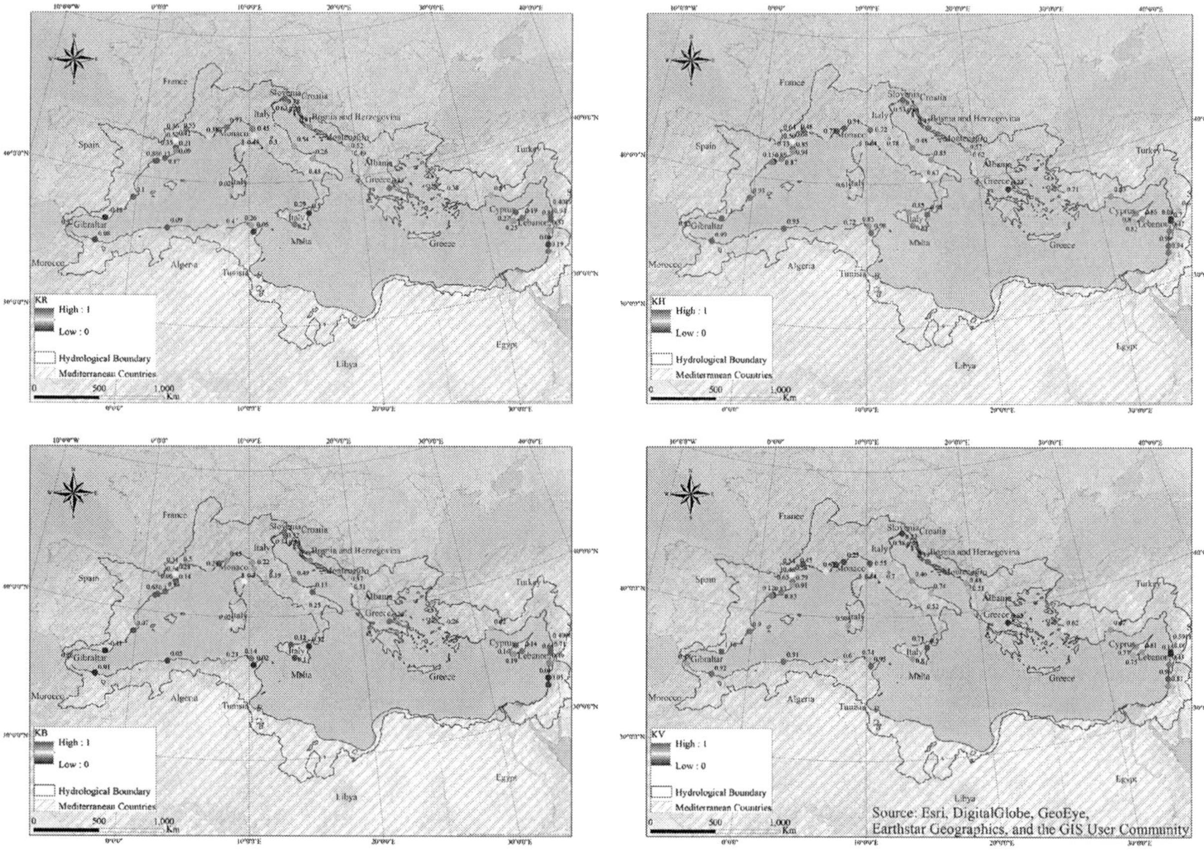

Figure 4. Spatial distribution of the water balance Runoff index K_{R}, Horton index K_{H}, Baseflow index K_{B}, and Vaporisation index K_{V}.

ing with aridity except for, once again, mountainous catchments which showed to be independent to climate variability. Therefore we state that "In the Mediterranean, the mountainous, snow and karst influenced catchments yield the highest slow flow and runoff especially if located in high seasonal region (Southern Mediterranean CC3 regions) where precipitation are concentrated in winter and runoff are concentrated in winter and spring seasons, when minimum vegetation cover the land and vaporisation (or evapotranspiration) is at lowest, part of the self-organisation of the vegetation with the climate as proposed by Horton (1933) but at the intra-annual scale."

6 Conclusions

The water balance models highlighted similar Mediterranean trends following the general Mediterranean climatic setting from the wet Northern region to the arid Southern region and showed hydrological homogeneity for specific physiographic classes like mountainous snow and karst influenced catchments. However, to characterise the Mediterranean hydrological variability and homogeneity, we will rely on baseflow and runoff coefficients K_{r} and K_u which are also useful to estimate catchments' resources.

These observations and findings shall be verified and completed with the Canonical Correlation Analysis to first discover which specific climatic and physiographic indices govern the Mediterranean hydrological behaviour, second predict the water balance coefficients values on all Mediterranean catchments and third check the climatic evolution impact on the hydrological behaviour of Mediterranean catchments in case of RCP scenarios.

Data availability. All the climatic data are freely available online. The WorldClim-2 data are available from the WorldClim website at https://www.worldclim.org/data/worldclim21.html (Fick and Hijmans, 2017). The ground weather station data are available from the Global Historical Climatology Network website at https://www.ncei.noaa.gov/products/land-based-station/global-historical-climatology-network-daily (NCEI, 2023). The ALADIN and CCLM simulations used in the current work can be downloaded from the Med-CORDEX database at https://www.medcordex.eu/ (CORDEX, 2023).

Streamflow time series data used in this study were collected from several national and regional hydrological agencies and services. Catalan Agency for Water – ACA, Hydrological Automatic Information System – SAIH (Spain); Banque HYDRO (France); Superior Institute for the Protection of Environmental Resources – ISPRA (Italy); Slovenian Republic Environment Agency – ARSO; Croatian Meteorological and Hydrological Service – DHMZ; Litani River Authority – LRA (Lebanon); Greek ministry of Environment and Energy – YPEKA; Système d'Informations Environnementales sur les Ressources en Eau et leur Modélisation – SIEREM (Algeria, Morocco, and Tunis); Global Runoff Discharge Centre – GRDC (Cyprus, Turkey, and Israel); Mediterranean Hydrological Cycle Observing System – MedHyCOS (Albania, Cyprus, and Montenegro) for three sub catchments.

Author contributions. All the authors (AA, RM, WN and CB) contributed to the conceptualization and methodology of the study as well as drafting, reviewing, and editing the article. AA developed the climatic classification and functional modeling. AA, RM, WN and CB carried out the validation and analysis of the results.

Competing interests. At least one of the (co-)authors is a guest member of the editorial board of *Proceedings of IAHS* for the special issue "IAHS2022 – Hydrological sciences in the Anthropocene: Variability and change across space, time, extremes, and interfaces". The peer-review process was guided by an independent editor, and the authors also have no other competing interests to declare.

Disclaimer. Publisher's note: Copernicus Publications remains neutral with regard to jurisdictional claims in published maps and institutional affiliations.

Special issue statement. This article is part of the special issue "IAHS2022 – Hydrological sciences in the Anthropocene: Variability and change across space, time, extremes, and interfaces". It is a result of the XIth Scientific Assembly of the International Association of Hydrological Sciences (IAHS 2022), Montpellier, France, 29 May–3 June 2022.

Acknowledgements. The authors would like to thank the editor Christophe Cudennec, first reviewer Mohammad Merheb and the anonymous second reviewer for their constructive comments. We would also like to thank the Med-CORDEX providers for making their Regional Climate Data available as well as all national and international hydrological agencies and services.

Review statement. This paper was edited by Christophe Cudennec and reviewed by Mohammad Merheb and one anonymous referee.

References

Allam, A., Moussa, R., Najem, W., and Bocquillon, C.: Specific climate classification for Mediterranean hydrology and future evolution under Med-CORDEX regional climate model scenarios, Hydrol. Earth Syst. Sci., 24, 4503–4521, https://doi.org/10.5194/hess-24-4503-2020, 2020a.

Allam, A., Moussa, R., Najem, W., and Bocquillon, C.: Hydrological Cycle, in: Water Resources in the Mediterranean Region, edited by: Zribi, M., Elsevier, 350, https://doi.org/10.1016/B978-0-12-818086-0.00001-7, 2020b.

CORDEX: https://www.medcordex.eu/, last access: 8 March 2023.

De Jager, A. L. and Vogt, J. V.: Development and demonstration of a structured hydrological feature coding system for Europe, Hydrol. Sci. J., 55, 661–675, https://doi.org/10.1080/02626667.2010.490786, 2010.

Fick, S. E. and Hijmans, R. J.: WorldClim 2: new 1-km spatial resolution climate surfaces for global land areas, Int. J. Climatol., 37, 4302–4315, 2017 (data available at: https://www.worldclim.org/data/worldclim21.html, last access: 8 March 2023.).

Haines, A., Finlayson, B., and McMahon, T.: A global classification of river regimes, Appl. Geogr., 8, 255–272, https://doi.org/10.1016/0143-6228(88)90035-5, 1988.

Horton, R. E.: The Rôle of infiltration in the hydrologic cycle, Eos, Transactions American Geophysical Union, 14, 446–460, https://doi.org/10.1029/TR014i001p00446, 1933.

Köppen, W.: Das geographische System der Klimate, Borntraeger, 1936.

Lehner, B. and Grill, G.: Global river hydrography and network routing: baseline data and new approaches to study the world's large river systems, Hydrol. Process., 27, 2171–2186, 2013.

L'vovich, M. I.: World water resources, present and future, GeoJournal, 3, 423–433, https://doi.org/10.1007/bf00455981, 1979.

Milano, M., Ruelland, D., Fernandez, S., Dezetter, A., Fabre, J., Servat, E., Fritsch, J.-M., Ardoin-Bardin, S., and Thivet, G.: Current state of Mediterranean water resources and future trends under climatic and anthropogenic changes, Hydrol. Sci. J., 58, 498–518, https://doi.org/10.1080/02626667.2013.774458, 2013.

NCEI: Ground weather station data, Global Historical Climatology Network [data set], https://www.ncei.noaa.gov/products/land-based-station/global-historical-climatology-network-daily, last access: 8 March 2023.

Ponce, V. M. and Shetty, A. V.: A conceptual model of catchment water balance: 1. Formulation and calibration, J. Hydrol., 173, 27–40, https://doi.org/10.1016/0022-1694(95)02739-C, 1995a.

Ponce, V. M. and Shetty, A. V.: A conceptual model of catchment water balance: 2. Application to runoff and baseflow modeling, J. Hydrol., 173, 41–50, https://doi.org/10.1016/0022-1694(95)02745-B, 1995b.

Sivapalan, M., Yaeger, M. A., Harman, C. J., Xu, X., and Troch, P. A.: Functional model of water balance variability at the catchment scale: 1. Evidence of hydrologic similarity and space-time symmetry, Water Resour. Res., 47, W02522, https://doi.org/10.1029/2010WR009568, 2011.

Proc. IAHS, 385, 111–115, 2024
https://doi.org/10.5194/piahs-385-111-2024

Open Access PIAHS

Conceptual principles of water resources management in irrigated agriculture

Mykhailo Romashchenko[1], **Vsevolod Bohaienko**[2], **Tetiana Matiash**[1], **Andrii Shatkovskyi**[1], **Serhii Kolomiets**[1], **and Iuliia Danylenko**[1]

[1]Institute of Water Problems and Land Reclamation of NAAN of Ukraine, Kyiv, Ukraine
[2]VM Glushkov Institute of Cybernetics of NAS of Ukraine, Kyiv, Ukraine

Correspondence: Vsevolod Bohaienko (sevab@ukr.net)

Received: 28 May 2022 – Revised: 17 February 2023 – Accepted: 1 March 2023 – Published: 18 April 2024

Abstract. Climate change causes a progressive deterioration of natural water supply increasing the need for irrigation and more rational use of water resources in it as the main mean for ensuring sustainable agriculture. Here we present the conceptual provisions intended to provide a capability to achieve maximum irrigation efficiency, which generalize our experience in irrigation water management in Ukraine. Main principles that should be integrated in an efficient decision support system include the use of a narrow high range of maintained moisture content in the root zone of soil; pressure as a criterion of moisture content level; differential, particularly fractional-order, models of moisture transfer for the prediction of irrigation schedules and rates; several methods of evapotranspiration assessment to improve the accuracy; remote sensing data to disseminate predictions to areas not covered by instrumental monitoring tools. The application of these principles support an increase in crop yields accompanied by irrigation water volumes reduction eliminating water losses for infiltration and the development of flooding and secondary salinization processes.

Keywords. Soil moisture; evapotranspiration; irrigation; decision support; water management

1 Introduction

Climate change, which complicates the conditions for agricultural production in most regions, is particularly significant for Ukraine where it is characterized by the fastest (over 0.6 °C in 10 years) increase of average annual air temperature in Europe (Krakovska et al., 2018).

The cultivation of crops in the Steppe and a large part of the Forest-steppe zone of Ukraine, which is 2/3 of the total cropped area, is carried out in the conditions of the significant deficit (from 150 to > 450–500 mm, Fig. 1) of climatic water balance (Romashchenko et al., 2020b) making urgent the expansion of irrigation.

The solution of this issue is provided by the "Strategy of Irrigation and Drainage in Ukraine for the period up to 2030" approved by the Cabinet of Ministers of Ukraine and the "Action Plan" for its implementation. Fulfilling their tasks in the aspect of increasing irrigation areas by 1.0–1.2 million ha will require the involvement of significant (> 4.0–4.5 billion m^3 considering the data given in Romashchenko et al., 2020b) additional volumes of water. Given the negative impact of climate change on the state of usable water resources, efficient water use is an important condition for increasing irrigation capacity.

The main components of solving the problem of efficient water use are the application of water-saving irrigation methods (varieties of drip irrigation and low-pressure sprinklers) and decision support systems (DSS) in irrigation management (Mainaa et al., 2014).

DSSs in irrigation are tools used in the management of irrigation processes aimed at regulating water supply to plants according to their needs, which contributes to the increase of water productivity. They are usually based on biomass accumulation and evapotranspiration models. The most common approach is grounded on the relationship between seasonal transpiration of crops under proper water supply and their biomass (de Wit, 1958).

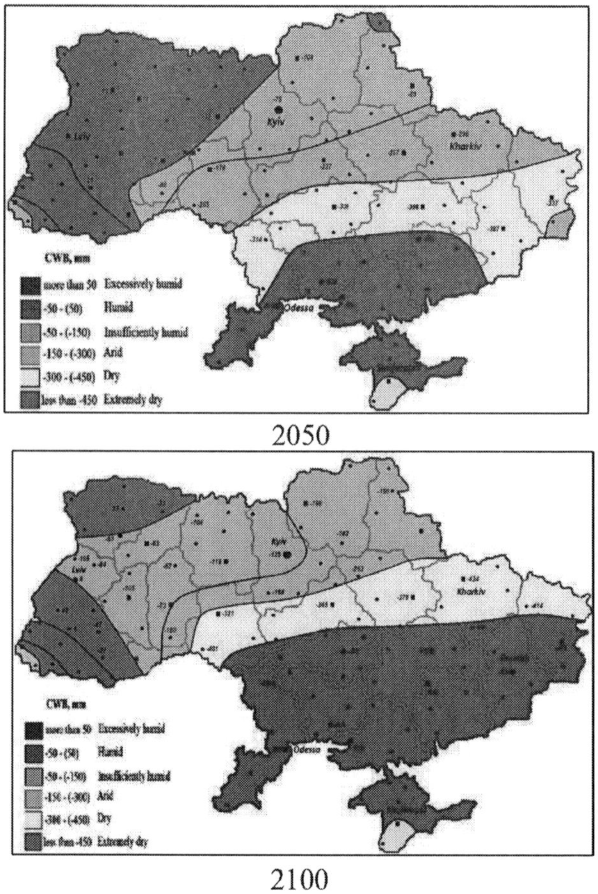

Figure 1. Forecast of moisture supply conditions in Ukraine according to annual climatic water balance. * source: Romashchenko et al. (2020b).

Necessary water supply regimes are provided using a variety of methods including an approach based on plant state assessment (Jones, 2004) and an approach based on the control of moisture supply to the root layer of soil (Campbell et al., 1982).

Methodological background of the decisions on the formation of irrigation regimes determines the level of efficiency and environmental friendliness of irrigation water usage and the ways to achieve it when managing irrigation using DSSs.

Thus, the aim of this study is to determine conceptual foundations of irrigation management using DSSs aimed at increasing the efficiency and environmental safety of irrigation. The use of these foundations should allow the fullest use of the potential (the maximum yield under close to optimal agricultural practices) of varieties and hybrids of crops with minimal irrigation water use per unit of yield. This should be accompanied by the minimization of infiltration water losses and, therefore, the risks of flooding and secondary salinization of irrigated lands. The scheme of core components of the considered methodology is depicted in Fig. 2.

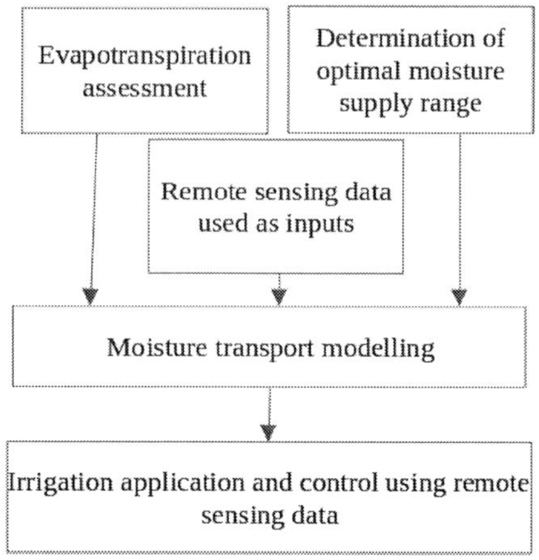

Figure 2. Core components of the methodology.

2 Principles of irrigation appointment

Irrigation should ensure the maintenance of moisture supply to the root layer of soil throughout the growing season in the optimal range for the growth and development of crops. Optimal one is such a level of moisture supply at which water consumption of irrigated crops is equal to or close to potentially possible under current conditions (Campbell et al., 1982).

According to the results of most studies, the upper limit of the optimal moisture supply range (OMSR) should be taken at the level of field capacity (FC), and the lower limit (pre-irrigation threshold) is a dynamic indicator that varies depending on the type of crop and its development stage (see, e.g., Wang et al., 2011). OMSR is an indicator that directly influences irrigation rate, hence the approach to its determination needs to be clarified.

The most common methods for determining FC by plots flooding in field conditions overestimate its value (Romashchenko et al., 2019) provoking a higher level of soil moisture during irrigation, causing the formation of easily mobile moisture, and creating preconditions for irrigation water losses on infiltrating below the root layer.

Therefore, in Romashchenko et al. (2019) a method for laboratory determination of FC on soil monoliths was proposed. The values of FC obtained using this technique are usually a few percent less than the values determined by plots flooding. So, for practical purposes, we propose to use as the upper limit of OMSR the values of FC determined by this method.

Regarding the lower limit of OMSR many experimental plant-related studies conducted for the specific crop and growing conditions propose the use of soil moisture values in the range from 60 % to 90 % FC (Muromtsev, 2011; Shatkovskyi et al., 2022). We propose to adhere a different approach (see, e.g., Shahnazari et al., 2007) that consists of maintaining moisture supply in the root layer in a narrow high range (usually above 80 % FC). According to its rationing this ensures the actual water consumption of crops to be equal to or close to potentially possible in current weather conditions.

The most accurate approaches to determining the lower limit of OMSR are related to the use of the water retention curves (WRC) and suction pressure as an indicator of moisture supply (Campbell et al., 1982). The use of WRC makes it possible to consider moisture content in combination with pore structure and to determine FC in terms of pressure. Further, the limits of OMSR are determined also in terms of pressure through a given range of per cents of FC. As an additional constraint on pre-irrigation threshold we propose it to be determined considering the reservation of soil moisture in the root layer of soil for at least one day of water consumption at the highest potentially possible level. This constraint is aimed at protecting plants from water stress in the conditions of rapid change of weather or potential failures in the application of watering.

Such an approach leads to the reduction of irrigation rates by 1.5–2.0 times making possible the implementation of environmentally friendly irrigation. It also contributes to obtaining yields close to the maximum possible at other unchanged elements of agricultural technology thus reducing the cost of irrigation water to form a unit of yield.

The presented conceptual principles were formed on the basis of the generalization of the results of long-term research of the authors performed in Ukraine and research data of other authors.

3 Principles of forecasting soil moisture dynamics

An important component of modern DSSs in irrigation management is the forecasting of soil moisture dynamics. Currently, the most used for these purposes is the Richards differential equation (Richards, 1931) that describes moisture transfer in soils with properties independent of scale. To increase modelling accuracy in porous media of fractal structure, it is proposed to use fractional-differential equations (Kavvas et al., 2017). Such models can, in particular, be considered as semi-empirical with the increased number of degrees of freedom while identifying their parameters in specific complex hydrogeological conditions (Romashchenko et al., 2021).

We also recommend to use equations that describe water transfer processes in terms of soil moisture potential and thus consider the vadose zone within the same hydrodynamic scheme with groundwater horizons.

Data from direct measurements of moisture level should be used to determine initial conditions, assess the quality of forecast, and calibrate the models. It is logical to use also the indicator of moisture potential and measure it using suction pressure sensors. The number of measurement points within a farm is determined according to the need to have at least one observation point for each "soil-crop" pair and perform measurements at two or more depths.

The use of moisture potential as an indicator of moisture supply level is substantiated by the fact that it is a direct indicator that characterizes the force with which soil moisture is retained in pores and does not depend on soil type. At moisture content level close to pre-irrigation threshold, its small changes may be accompanied by significant changes in pressure, which leads to higher accuracy in determining the state of moisture supply by pressure sensors (Degré et al., 2017).

4 Principles of evapotranspiration assessment

The accuracy of evapotranspiration estimates is one of the determining factors in modelling the state of moisture supply in the process of irrigation management. On the one hand, quantifying evapotranspiration intensity and more accurate modelling of moisture transfer requires integrated models that take into account soil and atmospheric physics along with plant physiology (Overgaard et al., 2006). Such models include the Penman-Monteith, Shuttleworth-Wallace, and Priestley-Taylor models and have numerous parameters that are often hard to assess. On the other hand, there is a large amount of easy to calculate empirical methods, the effectiveness of which is limited by specific agroclimatic conditions. The most used in Ukraine are, in particular, the methods of Dmytro Shtoyko and Mykola Ivanov (Romashchenko et al., 2020a).

The problem of determining the accuracy of evapotranspiration estimates remains relevant for each specific situation because under irrigation the maintained moisture ranges in the root layer of soil significantly influence the availability of water to plants, changing main factors influencing evapotranspiration. Thus, a specific empirical model or the parametrization of a phenomenological model that gives the best accuracy of pressure dynamics modelling should be determined in each particular case.

As currently no universal method exists for assessing evapotranspiration, in DSSs in irrigation we recommend (Romashchenko et al., 2020a) simultaneous usage of the models with the maximum number of parameters. It could be the Penman-Monteith model with mandatory determination of crop coefficients for each specific case (for Ukraine see Shatkovskyi et al., 2020) and empirical models developed for specific agroclimatic conditions. Such a combination makes it possible to more accurately determine evapotranspiration in the process of changes of crop development stages.

A critical input for the forecast calculations of evapotranspiration is weather forecast data. Our experience shows that with sufficient accuracy for the practice of irrigation management one can use the average weather forecast data from at least three of the most common weather information websites (Kovalchuk et al., 2018) and perform forecasting of irrigation schedules for up to 5 d.

5 Principles of remote sensing data usage

Another important aspect of modern methods for irrigation management is the intensive use of remote sensing data. The combination of hydrological models and remote sensing overcomes many of the shortcomings associated with low spatial resolution of these models and low temporal resolution of remote sensing data (Droogers and Bastiaanssen, 2002). This combination facilitates detailed spatial and temporal analysis while assessing irrigation efficiency, and can be used to identify areas where improved irrigation management is needed (Santos et al., 2008).

The ability of remote sensing to provide assessment of soil moisture content (Babaeian et al., 2019; Danylenko and Bohaienko, 2020) is highly important. When some fields have in-field monitoring tools and the other ones not, the usage of remote sensing allows irrigation management in fields not covered by monitoring tools using monitoring data from other fields to calibrate relevant models.

6 Conclusions

Progressive deterioration of natural moisture supply due to global warming requires the expansion of the use of irrigation as the main technological method for ensuring the sustainability of agriculture in the conditions of climate change.

In these conditions, the highest efficiency and environmental safety of irrigation is achieved when using for irrigation management the decision support systems that implement the above-described conceptual principles.

Based on the results of research carried out by the authors, the introduction of these principles in the practice of irrigation management in Ukraine ensures yields of irrigated crops at the level of 85 %–90 % of the potential of their varieties and hybrids (the maximum yield under close to optimal agricultural practices indicated in the passport of the variety (hybrid) in the State register of plant varieties suitable for distribution in Ukraine). It also reduces irrigation costs per unit of yield and allows conducting irrigation according to environmentally friendly standards minimizing or eliminating irrigation water losses on infiltration along with the development of flooding and secondary salinization processes.

Code availability. Supporting software codes are available from the authors upon reasonable request.

Data availability. The data are available from the authors upon reasonable request.

Author contributions. MR performed conceptualization; MR, VB, and SK designed methodology; VB, TM, AS, SK, and ID conducted an investigation; MR and VB wrote the manuscript; TM and ID reviewed and edited the manuscript.

Competing interests. The contact author has declared that none of the authors has any competing interests.

Disclaimer. Publisher's note: Copernicus Publications remains neutral with regard to jurisdictional claims in published maps and institutional affiliations.

Special issue statement. This article is part of the special issue "IAHS2022 – Hydrological sciences in the Anthropocene: Past and future of open, inclusive, innovative, and society-interfacing approaches". It is a result of the XIth Scientific Assembly of the International Association of Hydrological Sciences (IAHS 2022), Montpellier, France, 29 May–3 June 2022.

Review statement. This paper was edited by Christophe Cudennec and reviewed by two anonymous referees.

References

Babaeian, E., Sadeghi, M., Jones, S. B., Montzka, C., Vereecken, H., and Tuller, M.: Ground, proximal, and satellite remote sensing of soil moisture, Rev. Geophys., 57, 530–616, https://doi.org/10.1029/2018RG000618, 2019.

Campbell, G. S., Campbell, M. D., and Hillel, D.: Irrigation scheduling using soil moisture measurements: theory and practice, Advances in Irrigation, 1, 25–42, https://doi.org/10.1016/B978-0-12-024301-3.50008-3, 1982.

Danylenko, I. and Bohaienko, V.: Monitoring of soil moisture in the south of Ukraine using active and passive remote sensing data, Remote Sensing for Agriculture, Ecosystems, and Hydrology XXII, 11528, 1152807, https://doi.org/10.1117/12.2571049, 2020.

Degré, A., van der Ploeg, M. J., Caldwell, T., and Gooren, H. P.: Comparison of soil water potential sensors: a drying experiment, Vadose Zone J., 16, 1–8 https://doi.org/10.2136/vzj2016.08.0067, 2017.

de Wit, C. T.: Transpiration and crop yields, Versl. Landbouwk, Onderz. 64.6, Institute of Biological Chemistry Research on Field Crops and Herbage, Wageningen, The Netherlands, https://edepot.wur.nl/186445 (last access: 6 March 2023), 1958.

Droogers, P. and Bastiaanssen, W.: Irrigation performance using hydrological and remote sensing modeling, J. Irrig. Drain. E.-ASCE, 128, 11–18, https://doi.org/10.1061/(ASCE)0733-9437(2002)128:1(11), 2002.

Jones, H. G.: Irrigation scheduling: advantages and pitfalls of plant-based methods, J. Exp. Bot., 55, 2427–2436, https://doi.org/10.1093/jxb/erh213, 2004.

Kavvas, M. L., Ercan, A., and Polsinelli, J.: Governing equations of transient soil water flow and soil water flux in multi-dimensional fractional anisotropic media and fractional time, Hydrol. Earth Syst. Sci., 21, 1547–1557, https://doi.org/10.5194/hess-21-1547-2017, 2017.

Kovalchuk, V., Demchuk, O., Demchuk, D., and Voitovich, O.: Data mining for a model of irrigation control using weather web-services, in: International Conference on Computer Science, Engineering and Education, edited by: Hu, Z., Petoukhov, S., Dychka, I., and He, M., Advances in Intelligent Systems and Computing, Vol. 754, Springer, Cham, 133–143, https://doi.org/10.1007/978-3-319-91008-6_14, 2018.

Krakovska, S. V., Palamarchuk, L. V., Gnatiuk, N. V., and Shpytal, Ò. M.: Projections of air temperature and relative humidity in Ukraine regions to the middle of the 21st century based on regional climate model ensembles. Geoinformatika, 3, 62–77, http://www.geology.com.ua/en/7514-2/ (last access: 6 March 2023), 2018.

Mainaa, M. M., Amina, M. S. M., and Yazidb, M. A.: Web geographic information system decision support system for irrigation water management: a review, Acta Agr. Scand. B–S. P., 64, 283–293, https://doi.org/10.1080/09064710.2014.896935, 2014.

Muromtsev, N.: Plant moisture availability assessment, Dokuchaev Soil Bulletin, 67, 20–31, https://doi.org/10.19047/0136-1694-2011-67-20-31, 2011 (in Russian).

Overgaard, J., Rosbjerg, D., and Butts, M. B.: Land-surface modelling in hydrological perspective – a review, Biogeosciences, 3, 229–241, https://doi.org/10.5194/bg-3-229-2006, 2006.

Richards, L. A.: Capillary conduction of liquids through porous mediums, Physics, 1, 318–333, https://doi.org/10.1063/1.1745010, 1931.

Romashchenko, M. I., Kolomiets, S. S., and Bilobrova, A. S.: Laboratory diagnostic system for water-physical soil properties, Melioratisia i vodne hospodarstvo, 2, 199–208, https://doi.org/10.31073/mivg201902-193, 2019 (in Ukrainian).

Romashchenko, M. I., Bohaienko, V. O., Matiash, T. V., Kovalchuk, V. P., and Danylenko, I. I.: Influence of evapotranspiration assessment on the accuracy of moisture transport modeling under the conditions of sprinkling irrigation in the south of Ukraine, Arch. Agron. Soil Sci., 66, 1424–1435, https://doi.org/10.1080/03650340.2019.1674445, 2020a.

Romashchenko, M. I., Husiev, I. V., Shatkovskyi, A. P., Saidak, R. V., Iatsiuk, M. V., Shevchenko, A. M., and Matiash, T. V.: Impact of climate change on water resources and agricultural production, Melioratsiia i vodne hospodarstvo, 1, 5–22, https://doi.org/10.31073/mivg202001-235, 2020b (in Ukrainian).

Romashchenko, M. I., Bohaienko, V. O., Matiash, T. V., Kovalchuk, V. P., and Krucheniuk, A. V.: Numerical simulation of irrigation scheduling using fractional Richards equation, Irrigation Sci., 39, 385–396, https://doi.org/10.1007/s00271-021-00725-3, 2021.

Santos, C., Lorite, I. J., Tasumi, M., Allen, R. G., and Fereres, E.: Integrating satellite-based evapotranspiration with simulation models for irrigation management at the scheme level, Irrigation Sci., 26, 277–288, https://doi.org/10.1007/s00271-007-0093-9, 2008.

Shahnazari, A., Liu, F., Andersen, M. N., Jacobsen, S. E., and Jensen, C. R.: Effects of partial root-zone drying on yield, tuber size and water use efficiency in potato under field conditions, Field Crop. Res., 100, 117–124, https://doi.org/10.1016/j.fcr.2006.05.010, 2007.

Shatkovskyi, A., Romashchenko, M., Zhuravlov, O., Riabkov, S., Cherevychnyi, Y., and Hulenko, O.: Optimization of the parameters of drip irrigation regimes for crops in the steppe of Ukraine, Land Reclamation and Water Management, 2, 45–50, https://doi.org/10.31073/mivg202202-338, 2022.

Shatkovskyi, A. P., Romashchenko, M. I., Zhuravlov, O. V., Vasyuta, V. V., Melnychuk, F. S., Ovchatov, I. M., Yarosh, A. V., and Semenko, L. O.: Evaluation of the "Penman-Monteith" model for determination of soybeans' evapotranspiration in irrigated conditions of the Steppe of Ukraine, Modern Phytomorphology, 14, 111–113, https://doi.org/10.5281/zenodo.5078016, 2020.

Wang, F. X., Wu, X. X., Shock, C. C., Chu, L. Y., Gu, X. X., and Xue, X.: Effects of drip irrigation regimes on potato tuber yield and quality under plastic mulch in arid Northwestern China, Field Crop. Res., 122, 78–84, https://doi.org/10.1016/j.fcr.2011.02.009, 2011.

Proc. IAHS, 385, 117–120, 2024
https://doi.org/10.5194/piahs-385-117-2024

Open Access PIAHS

Mitigating Drought Financial Risk for Water Supply Sector through Index-Based Insurance Contracts

Gabriela C. Gesualdo, Marcos R. Benso, Fabrício A. R. Navarro, Luis M. Castillo, and Eduardo M. Mendiondo

São Carlos School of Engineering, University of Sao Paulo, Sao Carlos, Brazil

Correspondence: Gabriela C. Gesualdo (gabriela.gesualdo@usp.br)

Received: 13 June 2022 – Revised: 30 January 2023 – Accepted: 1 March 2023 – Published: 18 April 2024

Abstract. Drought management strategies have primarily focused on structural measures, which are insufficient to prevent water supply disruptions and economic losses. In this concept, adaptation entails anticipating the negative financial consequences of extreme weather events and taking appropriate measures to prevent and mitigate them. As a result, insurance is a valuable adaptation measure for compensating unexpected losses and preventing financial damage from becoming long-term economic damage. We simulated indexed insurance for the Cantareira Water Supply System (CWSS). The system is responsible for providing water to 7.2 million people in the Metropolitan Region of São Paulo (MRSP). Our methodology consists of three steps: (1) describing the indexed variable, (2) computing economic losses depending on event magnitude, and (3) evaluating risk premiums for low, medium, and high coverage levels. According to our findings, an annual fee (premium) of USD 0.43, 0.87, and 1.73 should be charged per person to obtain drought coverage for 3, 6, and 12 months (low, medium, and high levels), respectively. These values account for 1.7 %, 3.4 %, and 6.7 % of the annual costs paid by CWSS-supplied users, accordingly. The premium fee can be incorporated into water bills as a novel technique to pool the risk between supplied users and the utility, thereby protecting them against surcharge fluctuations.

Keywords. SDG 13; water management; adaptation measures; hydrological drought

1 Introduction

Droughts and floods caused direct losses of roughly USD 1.75 billion per year in Brazil between 1995 and 2014 (Agência Nacional de Águas e Saneamento Básico, 2017). These financial losses can be minimized and compensated through adaptive tools (Navarro et al., 2021), such as insurance against extreme events. In this context, insurance prevents short-term economic damage from becoming long-term economic damage (Seifert-Dähnn, 2018).

The traditional method of developing an insurance scheme entails calculating the insurance premium based on the expected disaster risk (Kunreuther and Michel-Kerjan, 2014). Similarly, the magnitude of the index variable, such as precipitation, streamflow, or reservoir level, triggers the indemnity in index insurance (index-based) (Mohor and Mendiondo, 2017). Despite several successful examples of index insurance application around the world (Le Den et al., 2017; Guillier, 2017; Hanger et al., 2018; Ruiz-Rivera and Lucatello, 2017) development in non-agricultural sectors remain limited.

Therefore, we present the development of index insurance for water utilities. We simulated index insurance for the Cantareira Water Supply System, responsible for providing water to 7.2 million people in the São Paulo Metropolitan Region (SPMR). The region is prone to droughts, the most recent of which occurred in 2014/2015, resulting in water restrictions and rationing in the SPMR. Since then, numerous studies have been conducted to better understand current and future hydrological behavior and to assess water security in the region (Gesualdo et al., 2019; J. S. Sone et al., 2022; Taffarello et al., 2016). Our focus is to identify highly correlated indexes and estimate the size of payouts in case of extreme hydrological conditions in the region. Our findings suggest the creation of a financial tool to reduce losses as an additional water resource management approach.

2 Methodology

We simulated index insurance to mitigate drought financial losses for the main contributor to the Cantareira water supply system (CWSS), which supplies the SPMR. The study area of the Jaguari basin is responsible for supplying 46 % of the Cantareira system, which provides water for approximately 7.2 million people in the SPMR (SABESP, 2021). Our methodology comprises (1) characterizing the index variable, (2) calculating economic losses by event magnitude, and (3) estimating risk premiums for low, medium, and high coverage levels.

The first step in creating an indexed insurance contract is determining an appropriate index. To be considered effective, it must be transparent, publicly available, difficult to manipulate, and strongly correlated with the actual financial losses covered by the insurance (Baum et al., 2018; Denaro et al., 2020). The index variable was chosen based on the highest correlation between the candidate's hydrometeorological variables and the reservoir level. The hydrometeorological variables of streamflow and reservoir information were derived from the SABESP, the water and waste management company of CWSS on a daily basis. The daily precipitation was from Brazilian Daily Weather Gridded Data (BR-DWGD) (Xavier et al., 2016), and ERA5 reanalysis replaces the ERA-Interim global climate data (Hersbach et al., 2020). Further information on the utility revenue data, such as Water tariff, Net revenue, Water consumption, and Population served, were obtained from the SABESP in a monthly resolution.

The insurance payment function translates the values of the index (i) into payments (p), which can take different forms. Payments are triggered once the index value is equaled to or exceeded, a strike value (k). This k value is defined as the threshold value of the index, therefore, it may vary according to the level of coverage offered by the insurance contract. For this work, we adopted a strike value of 10 % of the reservoir active volume, i.e., before reaching the so-called dead volume.

The financial risk was measured in terms of per capita per day residential economic losses, based onAubuchon and Morley (2013) equation, and linked to drought-related restrictions on water use. We consider that the contracts are offered by the insurer for a price, or premium (Pr), which can be estimated from the probability of payment distribution. The value of Pr is a function of the expected payouts ($E(p)$) plus the administrative fees (Af). To reduce the uncertainty of the calculations, we do not consider Af, thereby the values presented here represent the pure risk premium.

In risk contracts, the distribution of the probability of payment has a heavy tail, configuring high payments with a low probability of occurrence at the right end. To adjust this behavior and better indicate the price of risk to extremes, the Wang transformation can be used (Wang, 2002). This transformation adjusts the payout distribution by more heavily weighting low-probability events. Thus, the adjusted risk $F^*(p)$ can be determined as in Eq. (1):

$$F^*(p) = \varnothing\left(\varnothing^{-1}(F(p)) + \lambda\right) \quad (1)$$

where $\varnothing$ is the cumulative distribution function of payouts, and λ is the Sharpe ratio, representing market risk. We adopted λ as 0.25 based on weather derivative contracts (Wang, 2002). Thus, the adjusted premium (Pr adjusted) can be determined through the sum of expected payouts ($E(p)$) times the adjusted risk $F^*(p)$, given by Eq. (2):

$$\text{Pr ajusted} = \sum \left[E(p) \cdot F^*(p)\right] \quad (2)$$

To find the optimal premium value and cope with uncertainty, we need to generate a synthetic series of losses and assess its probability. We used the spatial stochastic model PRSim.wave (Phase Randomization Simulation using wavelets) proposed by Brunner and Gilleland (2020). The model was run 100 times for a 28-year sample of observed data (1985–2013) and obtained 2800 years of simulated data. In addition, we assumed three groups of annual coverage levels – high, medium, and low – representing respectively 12, 6, and 3 months of drought coverage during a 1-year contract. These groups represent the risk aversion of the policyholders, once it related the reservoir level for a given drought period (12, 6, and 3 months) to the economic losses. This exercise considers that the greater the risk aversion, the greater the Willingness to Pay, and consequently the premium value.

3 Results and discussion

In the same year, the association between reservoir level and hydrometeorological factors was minimal. However, because there is typically a large lag between the implementation of conservation measures and revenue losses, correlations between utility revenues and hydrometeorological factors are also assessed, with the index value lagging by 1 to 2 years.

The 2-year lag cumulative precipitation has a strong correlation (Spearman's correlation of 0.84) with the utility reservoir level and therefore the utility revenue, hence is used as the index variable in this study. The strike value (k) is given when the volume is 10 % before the dead volume, less than or equal to 385.7 hm^3, and this translates into an index of 2 years lag accumulated precipitation of 2424.3 mm

The observed statistical properties of the precipitation time series were satisfactorily reproduced by the stochastic simulation. We derived the losses distribution and average daily loss of USD 480 000.00 using the stochastic simulation (Data are in constant 2020 USD (United States dollar) – USD 1.00 = BRL 5.15). We computed the loss value for each

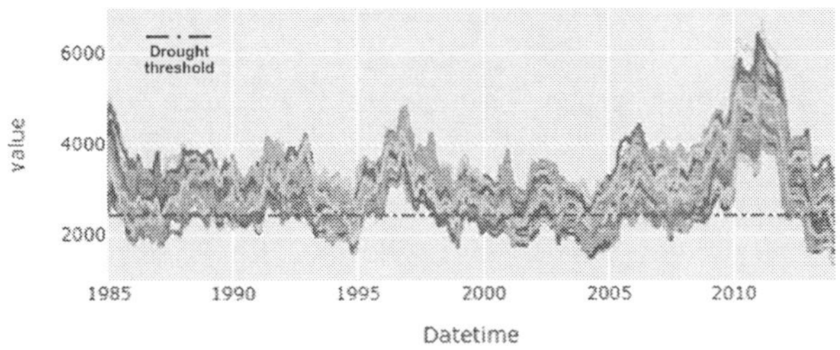

Figure 1. Synthetic series simulation for drought assessment.

day the reservoir is below the index (represented as a dashed line in Fig. 1).

Considering the three coverage levels, high, medium, and low, we find out an optimum premium value of USD 0.43, 0.87, and 1.73 for each water user and the respective coverage level. These values correspond to 1.7 %, 3.4 %, and 6.7 % of the annual fee paid in the annual water bills. The hydrologic risk management fee (premium fee) can be included in water bills as a new technique for pooling risk among users and utilities and protecting them from surcharge fluctuations.

The coverage level in this example will be assumed based on the water consumers' willingness to pay (WTP) and risk aversion. Water consumers that have previously encountered water restrictions and rationing are said to be WTP for preventative and adaptation efforts (J. Sone et al., 2022). As a result, we can predict that water consumers in SPMR are more willing to pay for index-based insurance because they have encountered numerous water restrictions over the last 20 years.

4 Conclusions

We simulate a drought index insurance for the Cantareira Water Supply system. Our results suggest that drought-related financial risks can be the management by water utilities and users to reduce the financial impact of droughts on water users via index-based insurance. As disruptions are difficult to remedy on a short-run basis, index insurance has the potential to significantly reduce the costs of managing financial risk and disruptions. Given the increased frequency and intensity of extreme weather events in the future, we acknowledge the critical necessity of financial mechanisms to minimize losses as an additional solution for water resource management. The findings revealed that utilities and users are exposed to a wide range of hydrologic and climate-related financial risks, particularly in recent years. A variety of factors could contribute to this, including climate change, which alters the frequency and intensity of droughts-related events, and the increasing population and demand in the region.

Code availability. The code used to process the data and to produce the figures can be requested from the first author.

Data availability. The daily streamflow time series and reservoir information used in this study are available via SABESP (https://mananciais.sabesp.com.br/Home, last access: 8 March 2023). Furthermore, the daily precipitation DWGD from Xavier et al. (2016, https://doi.org/10.1002/joc.4518) and ERA5 reanalysis from Hersbach et al. (2020, https://doi.org/10.1002/QJ.3803).

Author contributions. GCG: data curation, formal analysis, writing original draft preparation, and review & editing. MRB, FARN, and LMC: writing – review & editing. EMM: conceptualization, supervision, writing – review & editing.

Competing interests. The contact author has declared that none of the authors has any competing interests.

Disclaimer. Publisher's note: Copernicus Publications remains neutral with regard to jurisdictional claims in published maps and institutional affiliations.

Special issue statement. This article is part of the special issue "IAHS2022 – Hydrological sciences in the Anthropocene: Past and future of open, inclusive, innovative, and society-interfacing approaches". It is a result of the XIth Scientific Assembly of the International Association of Hydrological Sciences (IAHS 2022), Montpellier, France, 29 May–3 June 2022.

Acknowledgements. The authors would like to acknowledge the graduate program in Hydraulics and Sanitation Engineering of the University of Sao Paulo, São Carlos School of Engineering for the financial and organizational support. We appreciate the invaluable comments and careful reviews provided by the editors and anonymous reviewers.

Financial support. This research has been supported by the Coordination for the Improvement of Higher Education Personnel (CAPES) (finance Code 001), the National Institute of Science and Technology for Climate Change Phase 2 (INCT-II) under the Brazilian National Council for Scientific and Technological Development (CNPq) (grant no. 465501/2014-1), and the São Paulo Research Support Foundation (FAPESP) (grant no. 2014/50848-9).

Review statement. This paper was edited by Christophe Cudennec and reviewed by two anonymous referees.

References

Agência Nacional de Águas e Saneamento Básico (ANA): Conjuntura dos recursos hídricos no Brasil 2017: relatório pleno, ANA, Brasília, http://www.snirh.gov.br/portal/snirh/centrais-de-conteudos/conjuntura-dos-recursos-hidricos/relatorio-conjuntura-2017.pdf/view (last access: 31 May 2022), 2017 (in Portuguese).

Aubuchon, C. P. and Morley, K. M.: The economic value of water: Providing confidence and context to FEMA's methodology, J. Homel. Secur. Emerg., 10, 245–265, https://doi.org/10.1515/JHSEM-2012-0081, 2013.

Baum, R., Characklis, G. W., and Serre, M. L.: Effects of Geographic Diversification on Risk Pooling to Mitigate Drought-Related Financial Losses for Water Utilities, Water Resour. Res., 54, 2561–2579, https://doi.org/10.1002/2017WR021468, 2018.

Brunner, M. I. and Gilleland, E.: Stochastic simulation of streamflow and spatial extremes: a continuous, wavelet-based approach, Hydrol. Earth Syst. Sci., 24, 3967–3982, https://doi.org/10.5194/hess-24-3967-2020, 2020.

Denaro, S., Castelletti, A., Giuliani, M., and Characklis, G.: Insurance Portfolio Diversification Through Bundling for Competing Agents Exposed to Uncorrelated Drought and Flood Risks, Water Resour. Res., 56, e2019WR026443, https://doi.org/10.1029/2019WR026443, 2020.

Gesualdo, G. C., Oliveira, P. T., Rodrigues, D. B. B., and Gupta, H. V.: Assessing water security in the São Paulo metropolitan region under projected climate change, Hydrol. Earth Syst. Sci., 23, 4955–4968, https://doi.org/10.5194/hess-23-4955-2019, 2019.

Guillier, F.: French Insurance and Flood Risk: Assessing the Impact of Prevention Through the Rating of Action Programs for Flood Prevention, Int. J. Disast. Risk Sc., 8, 284–295, https://doi.org/10.1007/s13753-017-0140-y, 2017.

Hanger, S., Linnerooth-Bayer, J., Surminski, S., Nenciu-Posner, C., Lorant, A., Ionescu, R., and Patt, A.: Insurance, Public Assistance, and Household Flood Risk Reduction: A Comparative Study of Austria, England, and Romania, Risk Anal., 38, 680–693, https://doi.org/10.1111/risa.12881, 2018.

Hersbach, H., Bell, B., Berrisford, P., Hirahara, S., Horányi, A., Muñoz-Sabater, J., Nicolas, J., Peubey, C., Radu, R., Schepers, D., Simmons, A., Soci, C., Abdalla, S., Abellan, X., Balsamo, G., Bechtold, P., Biavati, G., Bidlot, J., Bonavita, M., De Chiara, G., Dahlgren, P., Dee, D., Diamantakis, M., Dragani, R., Flemming, J., Forbes, R., Fuentes, M., Geer, A., Haimberger, L., Healy, S., Hogan, R. J., Hólm, E., Janisková, M., Keeley, S., Laloyaux, P., Lopez, P., Lupu, C., Radnoti, G., de Rosnay, P., Rozum, I., Vamborg, F., Villaume, S., and Thépaut, J.-N.: The ERA5 global reanalysis, Q. J. Roy. Meteor. Soc., 146, 1999–2049, https://doi.org/10.1002/QJ.3803, 2020.

Kunreuther, H. and Michel-Kerjan, E.: Economics of Natural Catastrophe Risk Insurance, andbook of the Economics of Risk and Uncertainty, 1, 651–699, https://doi.org/10.1016/B978-0-444-53685-3.00011-8, 2014.

Le Den, X., Persson, M., Benoist, A., Hudson, P., de Ruiter, M., de Ruig, L., and Kuik, O.: Insurance of weather and climate-related disaster risk: inventory and analysis of mechanism to support damage prevention in the EU, https://doi.org/10.2834/40222, 2017.

Mohor, G. S. and Mendiondo, E. M.: Economic indicators of hydrologic drought insurance under water demand and climate change scenarios in a Brazilian context, Ecol. Econ., 140, 66–78, https://doi.org/10.1016/j.ecolecon.2017.04.014, 2017.

Navarro, F. A. R., Gesualdo, G. C., Ferreira, R. G., Rápalo, L. M. C., Benso, M. R., De Macedo, M. B., and Mendiondo, E. M.: A novel multistage risk management applied to water-related disaster using diversity of measures: A theoretical approach. Ecohydrology & Hydrobiology, 21, 443–453, https://doi.org/10.1016/j.ecohyd.2021.07.004, 2021.

Ruiz-Rivera, N. and Lucatello, S.: The interplay between climate change and disaster risk reduction policy: evidence from Mexico, Environ. Hazards-UK, 16, 193–209, https://doi.org/10.1080/17477891.2016.1211506, 2017.

SABESP: Relatório de Sustentabilidade, São Paulo, https://api.mziq.com/mzfilemanager/v2/d/9e47ee51-f833-4a23-af98-2bac9e54e0b3/f7132489-0141-cf70-79e6-cd9da896bef9?origin=1 (last access: 31 May 2022), 2021. SABESP: Portal dos Mananciais Sabesp – Home, https://mananciais.sabesp.com.br/Home, last access: 8 March 2023 (in Portuguese)

Seifert-Dähnn, I.: Insurance engagement in flood risk reduction – examples from household and business insurance in developed countries, Nat. Hazards Earth Syst. Sci., 18, 2409–2429, https://doi.org/10.5194/nhess-18-2409-2018, 2018.

Sone, J., Gesualdo, G., Schwamback, D., Wendland, E., and Brouwer, R.: Public perception of droughts and water shortages and metropolitan willingness to pay for water saving measures to improve water security, EGU General Assembly 2022, Vienna, Austria, 23–27 May 2022, EGU22-13392, https://doi.org/10.5194/egusphere-egu22-13392, 2022.

Sone, J. S., Araujo, T. F., Gesualdo, G. C., Ballarin, A. S., Carvalho, G. A., Oliveira, P. T. S., and Wendland, E. C.: Water Security in an Uncertain Future: Contrasting Realities from an Availability-Demand Perspective, Water Resour. Manag., 36, 2571–2587, https://doi.org/10.1007/S11269-022-03160-X, 2022.

Taffarello, D., Mohor, G. S., do Carmo Calijuri, M., and Mendiondo, E. M.: Field investigations of the 2013–14 drought through quali-quantitative freshwater monitoring at the headwaters of the Cantareira System, Brazil, Water Int., 41, 776–800, https://doi.org/10.1080/02508060.2016.1188352, 2016.

Wang, S. S.: A Universal Framework for Pricing Financial and Insurance Risks, ASTIN Bull., 32, 213–234, https://doi.org/10.2143/ast.32.2.1027, 2002.

Xavier, A. C., King, C. W., and Scanlon, B. R.: Daily gridded meteorological variables in Brazil (1980–2013), Int. J. Climatol., 36, 2644–2659, https://doi.org/10.1002/joc.4518, 2016.

Proc. IAHS, 385, 121–127, 2024
https://doi.org/10.5194/piahs-385-121-2024

Future shifting of annual extreme flows under climate change in the Volta River basin

Moctar Dembélé[1,2,7], Mathieu Vrac[3], Natalie Ceperley[7], Sander J. Zwart[1], Josh Larsen[4], Simon J. Dadson[5,6], Grégoire Mariéthoz[2], and Bettina Schaefli[2,7]

[1]Digital Innovation, International Water Management Institute (IWMI), CSIR Campus, No. 6 Agostino Neto Road, Accra, Ghana
[2]Institute of Earth Surface Dynamics (IDYST), Faculty of Geosciences and Environment, University of Lausanne, 1015 Lausanne, Switzerland
[3]Laboratoire des Sciences du Climat et de l'Environnement (LSCE-IPSL), CEA/CNRS/UVSQ, Université Paris-Saclay, Centre d'Etudes de Saclay, Orme des Merisiers, 91191 Gif-sur-Yvette, France
[4]School of Geography, Earth and Environmental Sciences, University of Birmingham, Birmingham, UK
[5]School of Geography and the Environment, University of Oxford, South Parks Road, Oxford OX1 3QY, UK
[6]UK Centre for Ecology and Hydrology, Wallingford, Oxfordshire OX10 8BB, UK
[7]Institute of Geography (GIUB) and Oeschger Centre for Climate Change Research (OCCR), University of Bern, 3012 Bern, Switzerland

Correspondence: Moctar Dembélé (moctar.dembele@cgiar.org)

Received: 29 May 2022 – Revised: 29 January 2023 – Accepted: 1 March 2023 – Published: 18 April 2024

Abstract. Global warming is projected to result in changes in streamflow in West Africa with implications for frequent droughts and floods. This study investigates projected shifting in the timing, seasonality and magnitude of mean annual minimum (MAM) and annual maximum flows (AMF) in the Volta River basin (VRB) under climate change, using the method of circular statistics. River flow is simulated with the mesoscale hydrologic model (mHM), forced with bias-corrected climate projection datasets consisting of 43 regional and global climate model combinations under three representative concentration pathways (RCPs). Projected changes indicate that AMF increases between + 1 % and +80 % across sub-basins, particularly in the near future (2021–2050), whereas MAM decreases between −19 % and −7 %, mainly from the late century (2071–2100), depending on RCPs. The date of occurrence of AMF is projected to change between −4 and +3 d, while MAM could shift between −4 and +14 d depending on scenarios over the 21st century. Annual high flows denote a strong seasonality with negligible future changes, whereas the seasonality of low flows has a higher variation, with a slight drop in the future.

Keywords. Hydroclimatic extremes; modelling; West Africa; PUB; UPH 1; UPH 9; SDG 6.4; SDG 13.1

1 Introduction

Human-induced climate warming has significantly altered the magnitude of mean and extreme river flows globally (Gudmundsson et al., 2021). River flow is a key indicator of water availability for people and the environment. Fluctuations in river water availability can undermine water supply and profoundly affect water and food security. Therefore, knowing trends in river flows is crucial for preserving livelihoods and safeguarding ecosystems. In West Africa, global warming has induced a significant increase in extreme rainfall, thereby affecting river flows (Chagnaud et al., 2022; Rameshwaran et al., 2021). Streamflow extremes in West Africa are non-stationary and associated with devastating floods and droughts that are becoming persistent and widespread (Elagib et al., 2021; Wilcox et al., 2018). As flood and drought are two extremes of the same hydrological cycle, it is important to integrate them in disaster reduction strategies and measures to better cope with current and future risks (Ward et al., 2020; Brunner et al., 2021; Kreibich et al., 2022). Understanding future dynamics in the occurrence of annual extreme river flows is critical to guide adaptation measures (Lane and Kay, 2021). However, studies usually focus on assessing changes in the magnitude of river flows while timing and seasonality are also important for water planning and management.

This study investigates the shifting of annual minimum and maximum flows timing, seasonality and magnitude using a large ensemble of regional and global climate models under multiple climate change scenarios in the Volta River basin (VRB), where continuous updates on the state of water resources is essential to cope with persisting water-related risks. Here, annual minimum and maximum flows are analysed instead of flow percentiles as done by Dembélé et al. (2022).

2 Study area

The Volta River Basin (VRB) covers approximately 415 600 km^2 shared among six countries (Benin, Burkina Faso, Côte d'Ivoire, Ghana, Mali, and Togo) in West Africa (Fig. 1). The land cover is dominated by savannah, with a maximum altitude of 940 m a.s.l. Rainfall is characterized by interannual and multidecadal variabilities, and ranges between 570 and 1200 mm yr^{-1} from north to south. Actual evaporation exceeds 80 % of the basin annual rainfall. The Black Volta (152 800 km^2), White Volta (113 400 km^2), Oti (74 500 km^2) and Lower Volta (74 900 km^2) constitute the four sub-basins of the drainage system. The Volta River flows over 1850 km and fills in the Lake Volta formed by the Akosombo dam before draining into the Atlantic Ocean (Dembélé, 2020).

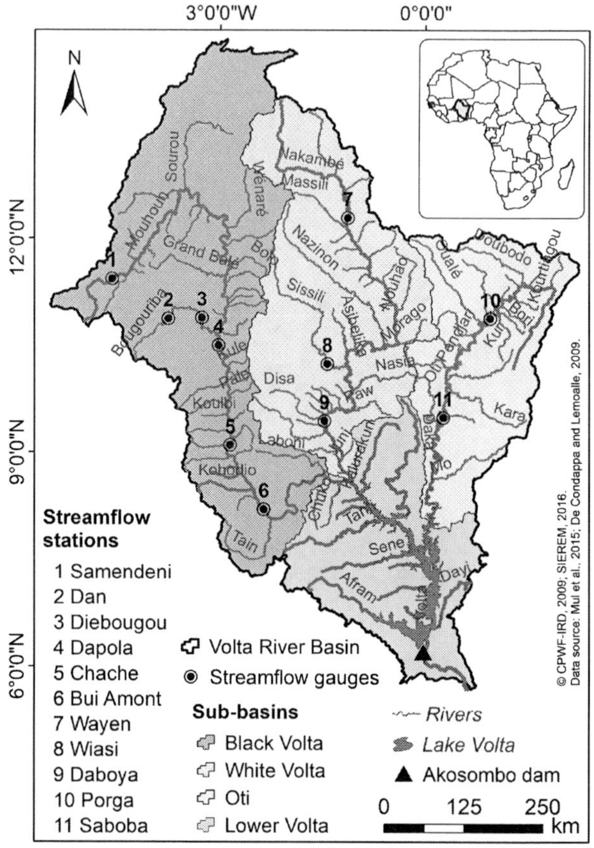

Figure 1. Hydrographic network of the Volta River basin. Adapted from Dembélé (2020).

3 Data and Methods

3.1 Climate projection data

Five Regional Climate Models (RCMs) from the Coordinated Regional-climate Downscaling Experiment (CORDEX) for Africa at a spatial resolution of 0.44° (∼ 50 km) are considered with twelve General Circulation Models (GCMs) from the fifth Coupled Model Intercomparison Project (CMIP5; Taylor et al., 2012) under three Representative Concentration Pathways (RCPs; Van Vuuren et al., 2011), resulting in 43 RCM-GCM combinations. The daily data include rainfall and air temperature (average, maximum and minimum). The historical runs consider 21 RCM-GCM combinations, whereas the future projections involve 18 RCM-GCM combinations for RCP8.5, 16 for RCP4.5 and 9 for RCP2.6 (Table 1). The climate projection datasets are evaluated with the best-performing satellite and reanalysis rainfall and temperature products in the VRB (Dembélé et al., 2020c), after a multivariate bias correction with the Rank Resampling for Distributions and Dependences (R2D2) method (Vrac and Thao, 2020).

3.2 Hydrological modelling

The fully distributed mesoscale Hydrologic Model (mHM) (Kumar et al., 2013; Samaniego et al., 2010) is used to simulate daily streamflow, using the bias corrected climate projection datasets as input. The model configuration and performance are given by Dembélé et al. (2020b). The periods 2021–2050, 2051–2080 and 2071–2100, representing the near-term future, the long-term future and the late-century, are considered to assess the impact of climate change on streamflow, relative to the historical period or baseline (1991–2020). The detailed methodology and the bias-correction results are given by Dembélé et al. (2022).

3.3 Timing of high and low flows

The mean annual minimum flow (MAM) of seven consecutive days is considered as low flow, while high flow is the

Table 1. CORDEX-Africa Regional Climate Models (RCMs) and General Circulation Models (GCMs) per Representative Concentration Pathways (RCP).

RCP	RCM	GCM
RCP8.5	CCLM4-8-17	MPI-ESM-LR, HadGEM2-ES, CNRM-CM5
	RACMO22T	EC-EARTH
	RCA4	IPSL-CM5A-MR, EC-EARTH, CanESM2, CNRM-CM5, CSIRO-Mk3-6-0, GFDL-ESM2M, NorESM1-M, MPI-ESM-LR, HadGEM2-ES, MIROC5
	REMO2009	MPI-ESM-LR, HadGEM2-ES, MIROC5, IPSL-CM5A-LR
RCP4.5	CCLM4-8-17	MPI-ESM-LR, HadGEM2-ES, CNRM-CM5
	CRCM5	CanESM2, MPI-ESM-LR
	RACMO22T	EC-EARTH
	RCA4	IPSL-CM5A-MR, CanESM2, CNRM-CM5, CSIRO-Mk3-6-0, GFDL-ESM2M, NorESM1-M, MPI-ESM-LR, HadGEM2-ES, MIROC5
	REMO2009	MPI-ESM-LR
RCP2.6	RCA4	NorESM1-M, MPI-ESM-LR, HadGEM2-ES, MIROC5
	REMO2009	MPI-ESM-LR, HadGEM2-ES, MIROC5, IPSL-CM5A-LR, GFDL-ESM2G

annual maximum flow (AMF) corresponding to the highest peak flow in a calendar year. High and low flows are estimated at Bui-Amont, Daboya and Saboba, which correspond to the outlets of the Black Volta, White Volta and Oti sub-basins (Fig. 1). The method of circular statistics (Mardia, 1972; Fisher, 1993) is used to estimate the timing of AMF and MAM based on the mean date of occurrence (D) and their seasonality based on the concentration of the dates of occurrence (R) for each of the 30-year historical and future periods (e.g., Blöschl et al., 2017; Hanus et al., 2021; Villarini, 2016). Julian dates are converted into angles that represent locations on the circumference of a circle to avoid concerns with calculating the mean when the day of occurrence comes around the start or end of a calendar year (Young et al., 2000). The angular value of the date of occurrence is obtained as follows:

$$\theta_i = D_i \cdot \frac{2\pi}{m_i}, \qquad 0 \leq \theta_i \leq 2\pi \tag{1}$$

Where D_i is the Julian date of occurrence of AMF or MAM varying between 1 and 365 (366 for leap years), m_i represents the number of days in a calendar year i, and θ_i is the angular date of occurrence in radians.

The mean date of occurrence $\overline{D}$ is calculated as:

$$\overline{D} = \begin{cases} \tan^{-1}\left(\frac{\overline{y}}{\overline{x}}\right) \cdot \frac{\overline{m}}{2\pi}, & \overline{x} > 0, \overline{y} \geq 0 \\ \tan^{-1}\left(\frac{\overline{y}}{\overline{x}}\right) \cdot \frac{\overline{m}}{2\pi} + \pi, & \overline{x} \leq 0 \\ \tan^{-1}\left(\frac{\overline{y}}{\overline{x}}\right) \cdot \frac{\overline{m}}{2\pi} + 2\pi, & \overline{x} > 0, \overline{y} < 0 \end{cases} \tag{2}$$

with

$$(\overline{x}, \overline{y}) = \left(\frac{1}{n}\sum_{i=1}^{n}\cos\theta_i, \frac{1}{n}\sum_{i=1}^{n}\sin\theta_i\right) \tag{3}$$

$$\overline{m} = \frac{1}{n}\sum_{i=1}^{n} m_i \tag{4}$$

where $\overline{m}$ is the average number of days in a year, n represents the total number of years and $\overline{x}$ and $\overline{y}$ correspond to the cosine and sine components of the mean date, respectively.

The mean resultant R gives the concentration of the dates of occurrence around the average date as follows:

$$R = \sqrt{\overline{x}^2 + \overline{y}^2}, \qquad 0 \leq R \leq 1 \tag{5}$$

A small value of R close to 0 suggests a high interannual variability of the date of occurrence, whereas R values near 1 denote a high seasonality in the date of occurrence.

4 Results

4.1 Changes in the magnitude of annual extreme flows

Considering all RCPs over the historical period, the average multi-model ensemble medians of MAM and AMF vary across sub-basins in the VRB as follows (Fig. 2): Black Volta (AMF $= 88.6 \pm 17\,m^3\,s^{-1}$, MAM $= 1.4 \pm 0.1\,m^3\,s^{-1}$), White Volta (AMF $= 79.3 \pm 3\,m^3\,s^{-1}$, MAM $= 1.5 \pm 0.1\,m^3\,s^{-1}$) and Oti (AMF $= 176.1 \pm 12\,m^3\,s^{-1}$, MAM $= 3.1 \pm 0.2\,m^3\,s^{-1}$). However, large differences are found in AMF across RCPs,

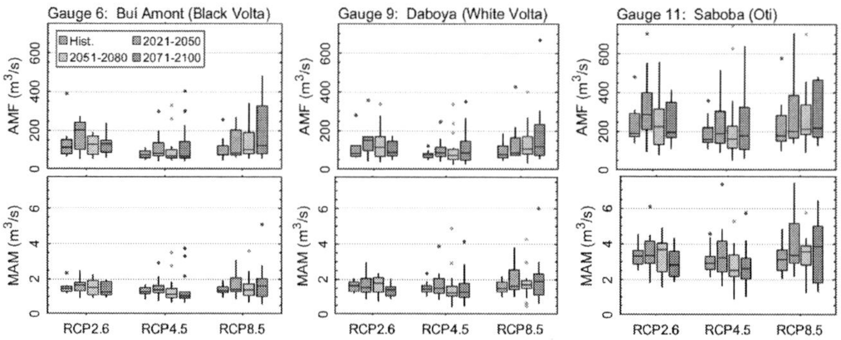

Figure 2. Mean annual minimum flows (MAM) and annual maximum flows (AMF) in the major sub-basins of the VRB (Black Volta, White Volta, Oti) over the historical (1991–2020) and future periods.

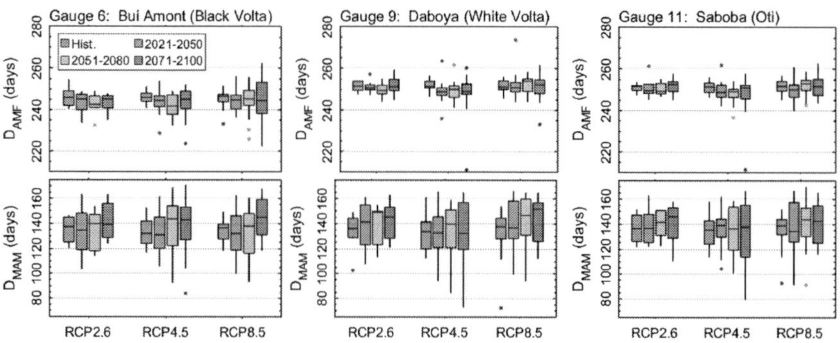

Figure 3. Mean Julian day of occurrence (D) of mean annual minimum flows (MAM) and annual maximum flows (AMF) in the major sub-basins of the VRB (Black Volta, White Volta, Oti) over the historical (1991–2020) and future periods.

with the highest values occurring under RCP2.6 and the lowest values under RCP4.5.

Future changes in annual high flows indicate an increase in median AMF in all the sub-basins under RCP2.6 and RCP8.5 over the 21st century, and also under RCP4.5 except over 2051–2080 in the White Volta and over 2051–2100 in the Black Volta. The increase in median AMF varies between +1 % in the Oti under RCP4.5 over 2051–2080 and +80 % in the White Volta under RCP2.6 over 2021–2050 with 56 % agreement on the direction of change among the RCM-GCMs. It is noteworthy, that AMF is projected to increase under all scenarios in the near future 2021–2050 in all sub-basins, with +11 % increase on average under RCP8.5, +15 % under RCP4.5 and +70 % under RCP2.6, with 76 %, 71 % and 59 % agreement on the direction of change of RCM-GCMs, respectively. An increase in flood occurrence is projected in all sub-basins under RCP2.6 and RCP8.5 over the 21st century, and mainly over 2021–2050 under RCP4.5.

For annual low flows, the projected decrease in median MAM varies between −19 % in the Black Volta under RCP4.5 over 2071–2100 and −7 % in the White Volta under RCP2.6 over 2021–2050. An overall median increase of MAM in all the sub-basins is projected under RCP8.5 by +16 % in the White Volta, +15 % in the Oti and +9 % in the Black Volta. Under RCP4.5, median MAM increases during 2021–2050 and decreases afterwards by −6 % on average in all sub-basins, whereas it increases until 2080 under RCP2.6 before decreasing by −15 % on average in the late 21st century. It is noteworthy that the median MAM increases under RCP2.6 by +8 % in the White Volta during 2051–2080. The agreement on the direction of change between the RCM-GCMs is on average 65 % for RCP8.5, 62 % for RCP4.5 and 59 % for RCP2.6. A higher likelihood for river droughts is projected from 2051 under RCP4.5 and from 2071 under RCP2.6 in all sub-basins.

4.2 Changes in the timing and seasonality of annual extreme flows

The median date of occurrence of AMF(D_{AMF}) varies on average between the Julian calendar days 246 and 252 (first dekad of September) over the historical period in all sub-basins, and it is projected to decrease by −2 d over the 21st century for all RCPs (Fig. 3). However, the highest reductions per sub-basin are −4 d in the Black Volta over 2051–2080 and −3 d in the White Volta and Oti over 2021–2050, all under RCP4.5. However, a rise by +3 d is projected in the White Volta over 2051–2080 under RCP8.5. Contrastingly,

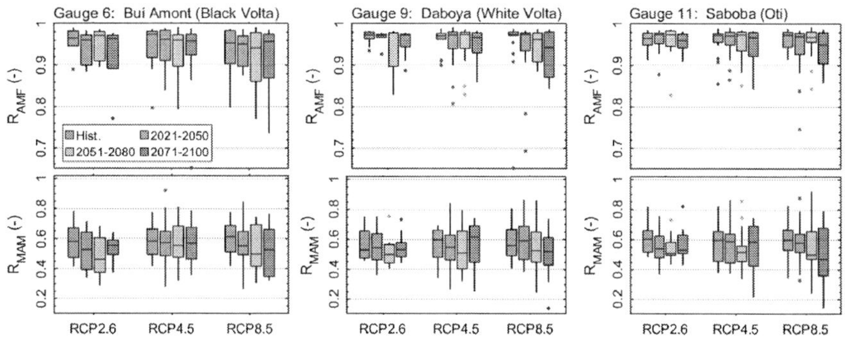

Figure 4. Concentration of the dates of occurrence (R) around the mean date for mean annual minimum flows (MAM) and annual maximum flows (AMF) in the major sub-basins of the VRB (Black Volta, White Volta, Oti) over the historical (1991–2020) and future periods.

the median D_{MAM} varies between 132 and 139 d (second dekad of May) over the historical period and increases on average by +6 d across sub-basins during the 21st century. However, notable rises in D_{MAM} are projected with +14 d under RCP8.5 over 2071–2100 in the White Volta, +11 d under RCP4.5 over 2051–2080 in the Black Volta and +10 d under RCP2.6 over 2071–2100 in the Oti, which might be the consequence of the forward shift of the rainy season (Dembélé et al., 2022). However, it is notable that D_{MAM} is projected to drop by −4 d over 2021–2050 under RCP8.5 in the Oti and the Black Volta.

The concentration of the date of occurrence (R) of D_{AMF} demonstrates a high seasonality in the occurrence of AMF ($R_{\mathrm{AMF}} = 0.97$) across sub-basins, and hardly changes over the 21st century (Fig. 4). The maximum change in R_{AMF} is −3 % over 2071–2100 in the White Volta under RCP8.5. The median R_{MAM} is 0.58 on average across sub-basins over the historical period and slightly decrease in the future, denoting a lower seasonality of MAM. A future reduction in R_{MAM} is observed in the Black Volta (−20 % to −2 %) and Oti (−22 % to −1 %) for all future climate scenarios, while changes are contrasted in the White Volta (−15 % to +6 %).

5 Limitations and future work

The large spread of AMF and MAM over 2071–2100 (Fig. 2), denoted by the moderate agreement on the direction of change of RCM-GCM combinations, can be explained by the limited capability of the models to project future trends, in addition to uncertainties associated with the hydrological model. Attention is solely given to climate-induced changes in river flow because climate was found to be the key driver in changes to global river flow (Gudmundsson et al., 2021). However, more local and regional changes in river flow might also result from other factors such as land and water management. Consequently, future studies in the VRB should also consider human interactions with the hydrological cycle (e.g. urbanization, agriculture, reservoirs) to better investigate the causes of changes in river flows (Yang et al., 2021; Yonaba et al., 2021). Furthermore, there is a need to improve hydrological modelling at large scale with earth observation data to better predict the spatial and temporal occurrence of extreme events (Dembélé et al., 2020a; Lindersson et al., 2020). Finally, uncertainties in hydrological modelling could be addressed by adopting multi-model approaches (Giuntoli et al., 2015; Vetter et al., 2017).

6 Conclusion

The analysis of projected shifting in timing, seasonality and magnitude of mean annual minimum flows and annual maximum flows in the Volta River basin under climate change reveals various changes over the 21st century. In general, projected annual high flows increase in all the sub-basins, particularly under RCP2.6 and RCP8.5, thereby leading to higher flood probability. Increase in river droughts are projected from 2051 under RCP4.5 and from 2071 under RCP2.6 in all sub-basins. High flows have a high seasonality with a mean date of occurrence that hardly changes in the future as opposed to low flows that show a future forward shift in time. These findings provide an important basis to inform research and climate change adaptation and mitigation strategies in the Volta River basin.

Code availability. The source code of the mHM model can be accessed at https://doi.org/10.5281/zenodo.1299584 (Samaniego et al., 2018).

Data availability. CORDEX data is available from the Earth System Grid Federation database at https://esgf-data.dkrz.de (ESGF, 2020). The database of the hydrological modelling is available at https://doi.org/10.5281/zenodo.3531873 (Dembélé, 2019).

Author contributions. MD designed the experiment, performed the analyses and drafted the paper. All co-authors contributed to the writing, review and editing.

Competing interests. The contact author has declared that none of the authors has any competing interests.

Disclaimer. Publisher's note: Copernicus Publications remains neutral with regard to jurisdictional claims in published maps and institutional affiliations.

Special issue statement. This article is part of the special issue "IAHS2022 – Hydrological sciences in the Anthropocene: Variability and change across space, time, extremes, and interfaces". It is a result of the XIth Scientific Assembly of the International Association of Hydrological Sciences (IAHS 2022), Montpellier, France, 29 May–3 June 2022.

Acknowledgements. We thank the streamflow data providers: Volta Basin Authority (VBA), Direction Générale des Ressources en Eau (DGRE) of Burkina Faso, Hydrological Services Department (HSD) of Ghana and the Direction Générale de l'Eau et de l'Assainissement (DGEA) of Togo. We are grateful to the developers of the global and regional climate models of CORDEX. We thank the developers of mHM at CHS/UFZ (Germany).

Financial support. Moctar Dembélé received funding by the Swiss Government Excellence Scholarship (grant no. 2016.0533/Burkina Faso/OP), the Swiss National Science Foundation (SNSF) Doc.Mobility fellowship (SNF, grant no. P1LAP2_178071), the Hydro-JULES visiting scientist fellowship (UKCEH; grant no. NERC NE/S017380/1) and the Sivapalan Young Scientists Travel Award (SYSTA) for participation to the 2022 IAHS Scientific Assembly. Bettina Schaefli was supported by a research grant of the SNSF (SNF, PP00P2_157611). Mathieu Vrac was supported by the COESION project funded by the French National program LEFE (Les Enveloppes Fluides et l'Environnement).

Review statement. This paper was edited by Christophe Cudennec and reviewed by Jean-Marie Kileshye-Onema and one anonymous referee.

References

Blöschl, G., Hall, J., Parajka, J., Perdigão, R. A., Merz, B., Arheimer, B., Aronica, G. T., Bilibashi, A., Bonacci, O., and Borga, M.: Changing climate shifts timing of European floods, Science, 357, 588–590, https://doi.org/10.1126/science.aan2506, 2017.

Brunner, M. I., Slater, L., Tallaksen, L. M., and Clark, M.: Challenges in modeling and predicting floods and droughts: A review, WIRES Water, 8, e1520, https://doi.org/10.1002/wat2.1520, 2021.

Chagnaud, G., Panthou, G., Vischel, T., and Lebel, T.: A synthetic view of rainfall intensification in the West African Sahel, Environ. Res. Lett., 17, 044005, https://doi.org/10.1088/1748-9326/ac4a9c, 2022.

Dembélé, M.: Database for the manuscript "Improving the predictive skill of a distributed hydrological model by calibration on spatial patterns with multiple satellite datasets", Zenodo [data set], https://doi.org/10.5281/zenodo.3531873, 2019.

Dembélé, M.: Spatially explicit hydrological modelling for water accounting under climate change in the Volta River Basin in West Africa, PhD, University of Lausanne, Lausanne, Switzerland, 271 pp., https://doi.org/10.13140/RG.2.2.15664.58885, 2020.

Dembélé, M., Ceperley, N., Zwart, S. J., Salvadore, E., Mariéthoz, G., and Schaefli, B.: Potential of Satellite and Reanalysis Evaporation Datasets for Hydrological Modelling under Various Model Calibration Strategies, Adv. Water Resour., 143, 103667, https://doi.org/10.1016/j.advwatres.2020.103667, 2020a.

Dembélé, M., Hrachowitz, M., Savenije, H. H. G., Mariéthoz, G., and Schaefli, B.: Improving the Predictive Skill of a Distributed Hydrological Model by Calibration on Spatial Patterns With Multiple Satellite Data Sets, Water Resour. Res., 56, e2019WR026085, https://doi.org/10.1029/2019wr026085, 2020b.

Dembélé, M., Schaefli, B., van de Giesen, N., and Mariéthoz, G.: Suitability of 17 gridded rainfall and temperature datasets for large-scale hydrological modelling in West Africa, Hydrol. Earth Syst. Sci., 24, 5379–5406, https://doi.org/10.5194/hess-24-5379-2020, 2020c.

Dembélé, M., Vrac, M., Ceperley, N., Zwart, S. J., Larsen, J., Dadson, S. J., Mariéthoz, G., and Schaefli, B.: Contrasting changes in hydrological processes of the Volta River basin under global warming, Hydrol. Earth Syst. Sci., 26, 1481–1506, https://doi.org/10.5194/hess-26-1481-2022, 2022.

Elagib, N. A., Al Zayed, I. S., Saad, S. A. G., Mahmood, M. I., Basheer, M., and Fink, A. H.: Debilitating floods in the Sahel are becoming frequent, J. Hydrol., 599, 126362, https://doi.org/10.1016/j.jhydrol.2021.126362, 2021.

ESGF: ESGF Node at DKRZ, ESGF [data set], https://esgf-data.dkrz.de, last access: 22 March 2020.

Fisher, N. I.: Statistical analysis of circular data, Cambridge University Press, 277 pp., https://doi.org/10.1017/CBO9780511564345, 1993.

Giuntoli, I., Vidal, J.-P., Prudhomme, C., and Hannah, D. M.: Future hydrological extremes: the uncertainty from multiple global climate and global hydrological models, Earth Syst. Dynam., 6, 267–285, https://doi.org/10.5194/esd-6-267-2015, 2015.

Gudmundsson, L., Boulange, J., Do, H. X., Gosling, S. N., Grillakis, M. G., Koutroulis, A. G., Leonard, M., Liu, J., Müller Schmied, H., and Papadimitriou, L.: Globally observed trends in mean and extreme river flow attributed to climate change, Science, 371, 1159–1162, https://doi.org/10.1126/science.aba3996, 2021.

Hanus, S., Hrachowitz, M., Zekollari, H., Schoups, G., Vizcaino, M., and Kaitna, R.: Future changes in annual, seasonal and monthly runoff signatures in contrasting Alpine catchments in Austria, Hydrol. Earth Syst. Sci., 25, 3429–3453, https://doi.org/10.5194/hess-25-3429-2021, 2021.

Kreibich, H., Van Loon, A. F., Schröter, K., Ward, P. J., Mazzoleni, M., Sairam, N., Abeshu, G. W., Agafonova, S., AghaKouchak, A., and Aksoy, H.: The challenge of unprecedented floods and droughts in risk management, Nature, 608, 80–86, https://doi.org/10.1038/s41586-022-04917-5, 2022.

Kumar, R., Samaniego, L., and Attinger, S.: Implications of distributed hydrologic model parameterization on water fluxes at multiple scales and locations, Water Resour. Res., 49, 360–379, https://doi.org/10.1029/2012wr012195, 2013.

Lane, R. A. and Kay, A. L.: Climate change impact on the magnitude and timing of hydrological extremes across Great Britain, Frontiers in Water, 3, 684982, https://doi.org/10.3389/frwa.2021.684982, 2021.

Lindersson, S., Brandimarte, L., Mård, J., and Di Baldassarre, G.: A review of freely accessible global datasets for the study of floods, droughts and their interactions with human societies, WIRES Water, 7, e1424, https://doi.org/10.1002/wat2.1424, 2020.

Mardia, K. V.: Statistics of directional data, Academic Press, London, 380 pp., ISBN 9781483218663, 1972.

Rameshwaran, P., Bell, V. A., Davies, H. N., and Kay, A. L.: How might climate change affect river flows across West Africa?, Climatic Change, 169, 21, https://doi.org/10.1007/s10584-021-03256-0, 2021.

Samaniego, L., Kumar, R., and Attinger, S.: Multiscale parameter regionalization of a grid-based hydrologic model at the mesoscale, Water Resour. Res., 46, W05523, https://doi.org/10.1029/2008wr007327, 2010.

Samaniego, L., Kumar, R., Thober, S., Rakovec, O., Schweppe, R., Schäfer, D., Schrön, M., Brenner, J., Demirel, C. M., Kaluza, M., Jing, M., Langenberg, B., and Attinger, S.: mesoscale Hydrologic Model (v5.9), Zenodo [code], https://doi.org/10.5281/zenodo.1299584, 2018.

Taylor, K. E., Stouffer, R. J., and Meehl, G. A.: An overview of CMIP5 and the experiment design, B. Am. Meteorol. Soc., 93, 485–498, https://doi.org/10.1175/BAMS-D-11-00094.1, 2012.

Van Vuuren, D. P., Edmonds, J., Kainuma, M., Riahi, K., Thomson, A., Hibbard, K., Hurtt, G. C., Kram, T., Krey, V., and Lamarque, J.-F.: The representative concentration pathways: an overview, Climatic change, 109, 5, https://doi.org/10.1007/s10584-011-0148-z, 2011.

Vetter, T., Reinhardt, J., Flörke, M., Van Griensven, A., Hattermann, F., Huang, S., Koch, H., Pechlivanidis, I. G., Plötner, S., and Seidou, O.: Evaluation of sources of uncertainty in projected hydrological changes under climate change in 12 large-scale river basins, Climatic Change, 141, 419–433, https://doi.org/10.1007/s10584-016-1794-y, 2017.

Villarini, G.: On the seasonality of flooding across the continental United States, Adv. Water Resour., 87, 80–91, https://doi.org/10.1016/j.advwatres.2015.11.009, 2016.

Vrac, M. and Thao, S.: R^2D^2 v2.0: accounting for temporal dependences in multivariate bias correction via analogue rank resampling, Geosci. Model Dev., 13, 5367–5387, https://doi.org/10.5194/gmd-13-5367-2020, 2020.

Ward, P. J., de Ruiter, M. C., Mård, J., Schröter, K., Van Loon, A., Veldkamp, T., von Uexkull, N., Wanders, N., AghaKouchak, A., and Arnbjerg-Nielsen, K.: The need to integrate flood and drought disaster risk reduction strategies, Water Security, 11, 100070, https://doi.org/10.1016/j.wasec.2020.100070, 2020.

Wilcox, C., Vischel, T., Panthou, G., Bodian, A., Blanchet, J., Descroix, L., Quantin, G., Cassé, C., Tanimoun, B., and Kone, S.: Trends in hydrological extremes in the Senegal and Niger Rivers, J. Hydrol., 566, 531–545, https://doi.org/10.1016/j.jhydrol.2018.07.063, 2018.

Yang, D., Yang, Y., and Xia, J.: Hydrological cycle and water resources in a changing world: A review, Geography and Sustainability, 2, 115–122, https://doi.org/10.1016/j.geosus.2021.05.003, 2021.

Yonaba, R., Biaou, A. C., Koïta, M., Tazen, F., Mounirou, L. A., Zouré, C. O., Queloz, P., Karambiri, H., and Yacouba, H.: A dynamic land use/land cover input helps in picturing the Sahelian paradox: Assessing variability and attribution of changes in surface runoff in a Sahelian watershed, Sci. Total Environ., 757, 143792, https://doi.org/10.1016/j.scitotenv.2020.143792, 2021.

Young, A. R., Round, C. E., and Gustard, A.: Spatial and temporal variations in the occurrence of low flow events in the UK, Hydrol. Earth Syst. Sci., 4, 35–45, https://doi.org/10.5194/hess-4-35-2000, 2000.

Proc. IAHS, 385, 129–134, 2024
https://doi.org/10.5194/piahs-385-129-2024

Open Access
PIAHS

UPH Problem 20 – reducing uncertainty in model prediction: a model invalidation approach based on a Turing-like test

Keith Beven[1], **Trevor Page**[1], **Paul Smith**[1,2], **Ann Kretzschmar**[1], **Barry Hankin**[1,3], **and Nick Chappell**[1]

[1]Lancaster Environment Centre, Lancaster University, Lancaster, UK
[2]Waternumbers, Lancaster, UK
[3]JBA Consulting, Warrington, UK

Correspondence: Keith Beven (k.beven@lancaster.ac.uk)

Received: 25 May 2022 – Revised: 28 January 2023 – Accepted: 1 March 2023 – Published: 18 April 2024

Abstract. This study proposes using a Turing-like test for model evaluations and invalidations based on evidence of epistemic uncertainties in event runoff coefficients. Applying the consequent "limits of acceptability" results in all the 100 000 model parameter sets being rejected. However, applying the limits, together with an allowance for timing errors, to time steps ranked by discharge, results in an ensemble of 2064 models that can be retained for predicting discharge peaks. These do not include any of the models with the highest (> 0.9) efficiencies. The analysis raises questions about the impact of epistemic errors on model simulations, and the need for both better observed data and better models.

Keywords. UPH20; epistemic uncertainty; model evaluation; limits of acceptability; event runoff coefficients

1 Introduction

We would like to use hydrological models that are fit for a particular purpose in making predictions. Traditionally, models have been calibrated against historical observed data and then some sample of those calibrated models used in prediction, with or without some additional uncertainty model. However, there has been little consideration of the aleatory and epistemic uncertainties of the pertinent observed hydrological data and processes and just how that might affect the way we assess "models as hypotheses" about how catchment systems work. Past work has revealed how the epistemic uncertainties associated with even the simplest water balance equation might have an impact on model calibration. In particular some events might introduce disinformation into the calibration process (e.g. Beven et al., 2011; Beven and Smith, 2015; Beven, 2016, 2019). In considering an alternative to this approach to model calibration, Beven and Lane (2019, 2022) have suggested taking a more Popperian approach to testing models as hypotheses in assessing when models should be considered as *not* fit for purpose. Model invalidation is, after all, a good thing in that it means we need to do better: that some improvements are required, either to the observed data, to the auxiliary relations, or to the model structures being used. The question is what is an appropriate methodology for such hypothesis testing when we *know* there are epistemic uncertainties associated with the observations? We consider this issue for the case of flood hydrograph simulation using Dynamic Topmodel, making use of a strategy of "limits of acceptability" for model simulations set prior to making model runs.

2 What should be the conditions for fitness-for-purpose?

Fitness-for-purpose is clearly a subjective concept, especially when faced with issues of epistemic uncertainty. Beven and Lane (2022) have suggested 8 principles for thinking about fitness-for-purpose as a qualitative Turing-like test for models as hypotheses about catchment responses as follows:

- to explore the definition of fitness with relevant stakeholders;

- to accept that models cannot be expected to perform better than allowed by the observed data used for simulations and evaluations;
- to ensure that models do not contradict secure evidence about the nature of the catchment response;
- to ensure that evaluations aim to get the right results for the right reasons;
- to allow for the possibility that all models might be rejected;
- to allow that the results of such tests will always be conditional;
- to allow that the evaluators might themselves need evaluating; and, finally,
- to ensure that there is a proper audit trail so that assumptions and decisions in evaluation processes can be reviewed and revisited by others.

Of these principles, one of the most interesting is the 2nd, which may be restated as models should not be expected to perform better than the observed data used for driving the simulations and evaluating the results. Beven (2019) suggested a way of using historical data sets in a somewhat different way to normal calibration procedures to produce limits of acceptability for models in a way that allowed for the epistemic uncertainties to be found in the rainfall and discharge observations. That methodology is based on the evaluation of event runoff coefficients, so is only practically applicable to catchments dominated by flashy responses with relatively low baseflow indices so that the total volumes of discharge expected from single events can be estimated using recession curve extrapolation as explained in Beven and Smith (2015) and Beven (2019).

Calculation of the event runoff coefficients over the historical record provides a distribution of estimates that will reflect the errors in the estimation of rainfall inputs to a catchment for different events (which can be large, particularly for convective events or in areas of high relief), and errors in the estimation of discharges for different events (which can be large, especially at low and high flows depending on the definition and stationarity of the rating curve, see e.g. Westerberg et al., 2011; McMillan et al., 2012; Coxon et al., 2015). This is reflected in the range of runoff coefficients that is found in an analysis of this type (Fig. 1; see also Beven, 2019; Beven et al., 2022). Note in particular that there are values greater than 1, sometimes significantly greater than 1, as well as some extremely low values even where the antecedent wetness of the catchment is quite high).

These extreme values reflect the potential for epistemic errors in observing the catchment water balance but will clearly create issues for the calibration of models that are constrained to maintaining a water balance in predicting the

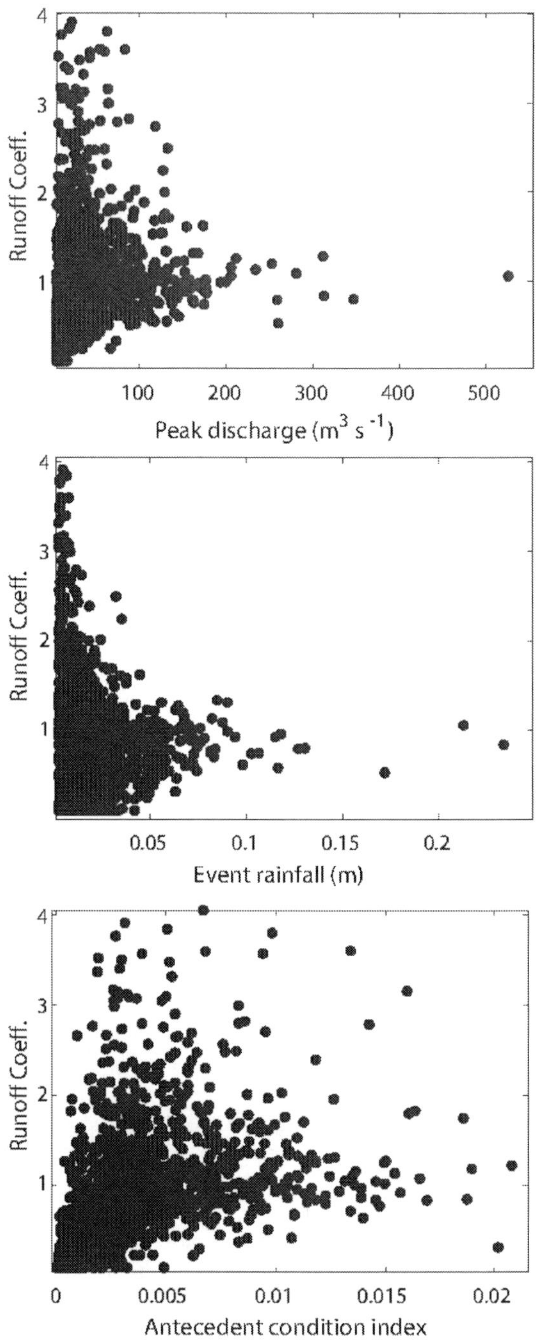

Figure 1. Event runoff coefficients for the River Kent at Sedgwick, Cumbria, UK, plotted against peak flow, event rainfall volume and antecedent flow index.

discharges. In effect, some events introduce disinformation into the traditional calibration process, which might or might not cancel out over a longer sequence of calibration data (e.g. Beven and Smith, 2015). There will be some element of error in the recession curve extrapolation to separate out the discharge associated with an event, but if care is taken not to separate events that have multiple peaks in quick succession, we would suggest that the extrapolation error will be small relative to the potential for disinformation in the observations.

3 Determination of Limits of Acceptability

As shown in Fig. 1, the range of historical event runoff coefficients can be expected to vary with the characteristics of the storm, here shown in terms of antecedent flow as a measure of the wetness of the catchment, the total event rainfall, and the peak discharge. In general, the range is greater for small storms and becomes more constrained for larger storms and peak discharges. Thus, in estimating whether model outputs for a particular event should be considered acceptable in terms of this past observed behaviour it is necessary to define some distribution of potential runoff coefficients given the event characteristics. In Beven (2019) and Beven et al. (2022) this has been done by taking a set of nearest neighbours for an event, weighted by the Mahalanobis distance for the rainfall volume, peak flow and antecedent flow values for the event of interest.

This provides a distribution of potential runoff coefficients for the event of interest, from which limits of acceptability for the model simulations can be defined. In doing so, the actual runoff coefficient for a calibration event can be used to scale the distribution from the nearest neighbour analysis since it is itself necessarily included in the nearest neighbours even if it is an extreme runoff coefficient event. The limits of acceptability can then be defined for that distribution (e.g. by taking the 95 % or 99 % range), treating the runoff coefficients (RCs) as multipliers of the recorded discharges for an evaluation event. This approach allows that models constrained to water balance should not be expected to predict runoff coefficients greater than 1.

Defining the nearest neighbours becomes more problematic however, for larger events where the sample size of similar events is necessarily smaller (Fig. 2). Beven et al. (2022) dealt with this by reducing the number of nearest neighbours considered as storm size increases, which has the effect of excluding some of the neighbours with more extreme RC values associated with smaller storms. This is in keeping with the concept of applying a "hydrologically sensible" Turing-like test in model evaluation.

4 The Application: Predicting the Hydrograph Impacts of Natural Flood Management in the Kent catchment

Here we consider again the application of the semi-distributed Dynamic Topmodel (Smith and Metcalfe, 2022) to the 210 km^2 River Kent catchment in Cumbria, England. This is an upland catchment not underlain by high baseflow primary aquifers, draining part of the English Lake District. It is therefore suitable for the runoff coefficient approach to defining limits of acceptability described above. The aim was to find acceptable parameter sets for Dynamic Topmodel that could later be used to predict the impacts of various Natural Flood Management measures in the catchment.

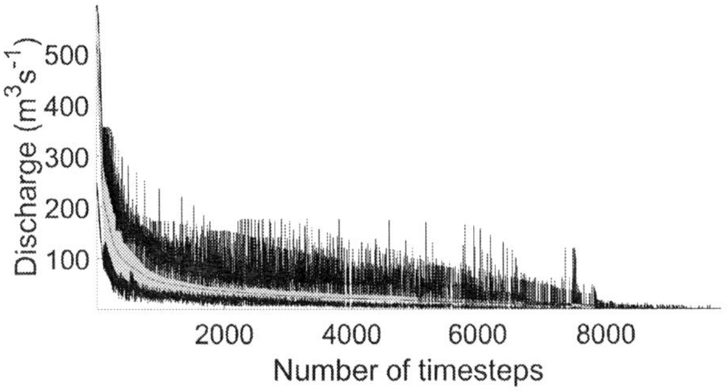

Figure 2. Upper and lower 95 % limits of acceptability (black lines) and range of acceptable model predictions (green shading) at time steps ranked by magnitude for the 2015 evaluation period.

The application has been described in more detail in Beven et al. (2022) where it was shown that applying the limits of acceptability defined in this way at every time step meant that every one of the 100 000 model runs (with randomly chosen parameter sets) were rejected. The model could therefore be invalidated on this basis. There were parameter sets giving high Nash-Sutcliffe or Kling-Gupta Efficiency value but these did not always give good predictions of the peaks that were of greatest interest in this application. Beven et al. (2022) went on to examine the peak predictions in more detail in ranked order, with the highest peak being Storm Desmond in 2015 that generated the largest peak on record in the catchment, with an estimated return period of 500 years. Desmond has a calculated runoff coefficient > 1.05 so that we should expect any model to underestimate the observed discharges. There were 3349 parameter sets that could be designated as acceptable within the runoff coefficient limits of acceptability for the top 26 peaks of the evaluation period together with an additional constraint of timing of ± 2 h (with a model time step of 15 min). This condition was introduced to allow for the fact that the runoff coefficient distributions allow only for volume errors, not timing errors in the recorded rainfall or discharge or discretisation errors of the data and model time steps. Application of these limits resulted in a strong constraint on the acceptable parameter values. Running those acceptable parameter sets for periods including the major flood events in 2005 and 2009 showed that they could provide a range of predictions that spanned the highest peaks in each of those evaluation periods. Here we expand on this approach to consider not only the peaks but all time steps ranked in order of magnitude.

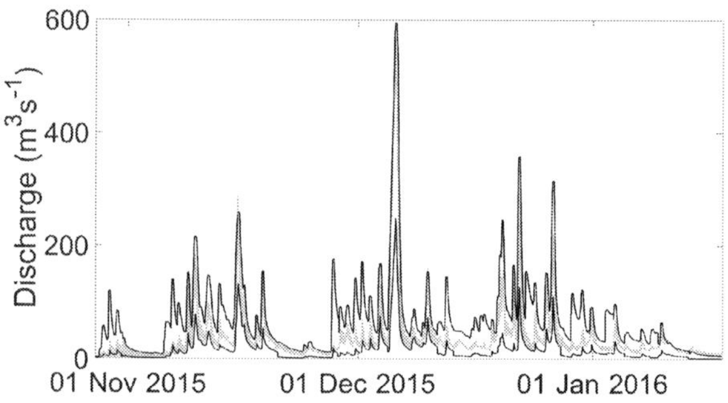

Figure 3. Upper and lower 95 % limits of acceptability (black lines) and range of acceptable model predictions (green shading) for part of the 2015 evaluation period.

5 Results

Figure 2 shows the Kent Catchment simulations for the time steps in the 2015 period ranked by observed discharge. The outer dark lines represent the limits of acceptability defined by the 2.5 and 97.5 quantiles of the nearest neighbour distributions of runoff coefficients for each event (the spikes in the limits are generally associated with the peaks of events, particularly those with low observed runoff coefficients. The green shading represents the range of the models considered acceptable in producing no more than 5 % of time steps outside the limits of acceptability at any given discharge threshold, again with an allowance of ±2 h for timing and discretisation errors. At a discharge threshold of $10\,m^3\,s^{-1}$, there are 2064 model parameter sets that survive these criteria. Figure 3 shows the runoff coefficient limits and outputs from the 2064 models for part of the 2015 period (including Storm Desmond at $527\,m^3\,s^{-1}$) in the more usual hydrograph form.

The result of applying this same ensemble of acceptable models to the 2005 and 2009 periods is shown in Figs. 4 and 5.

6 Discussion: Better Models or Better Data?

There have been a number of studies in the past that have tried to go beyond the type of statistical sum of squared errors criterion that have often been used in model calibration. Some have tried to use more formal statistical error models that more properly represent the heteroscedasticity and correlation in the model residuals, while others have argued that when the sources of uncertainty are more epistemic than aleatory then the application of statistical likelihoods or efficiency measures is not really appropriate (e.g. Beven et al., 2008). The type of model evaluation used here has some similarities with the fuzzy weighting method used in Pappenberger and Beven (2004) and elsewhere, but makes use of limits acceptability that are designed to reflect the epistemic uncertainties in the observed water balance for historical events.

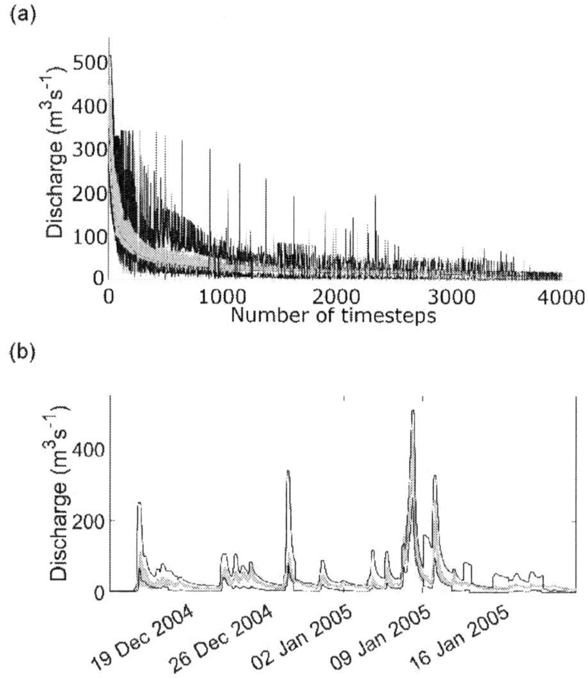

Figure 4. As for Figs. 2 and 3 but for 2005 simulations.

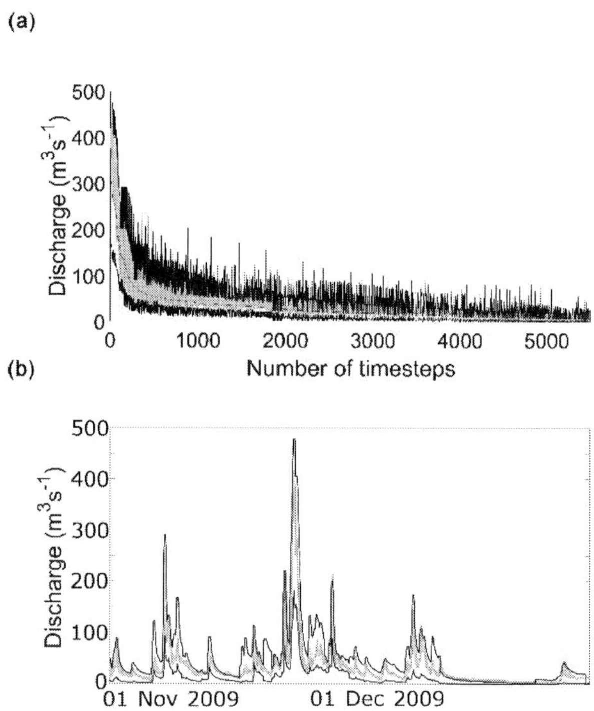

Figure 5. As for Figs. 2 and 3 but for 2009 simulations.

Strict application of the limits of acceptability defined in this way, without any allowance for timing errors, results in rejection of all the 100 000 parameter sets tried. This is despite the wide range of runoff coefficients estimated for many events and despite having model efficiency values for some parameter sets of over 0.9 (see Beven et al., 2022). We have here shown that some models can produce acceptable simulations over a large range of discharges, when some allowance in made for timing errors and the evaluations are made on time steps ranked by magnitude. This results in an ensemble of models that might still be fit-for-purpose for peak predictions, even if not for the full range of discharges. It is worth noting that this set of models does not include the highest efficiency values, as some of the highest efficiency simulations produce poorer peak predictions.

There is one aspect of this type of model evaluation that has not been considered explicitly. This is the effect of input errors for one event, having an impact on the simulations for subsequent events. An underestimate of the rainfall inputs, for example, will not only tend to produce an underestimate of the discharge peak for that event, but also result in drier antecedent conditions predicted for the next event. An overestimation of the inputs for a storm will have the opposite effect. The potential variability in peak predictions is implicit in the historical range of runoff coefficients, but not the effect on subsequent events (see also Beven and Smith, 2015).

In the final analysis, all the simulations tried could be rejected if the conditions are applied at all time steps (although it is worth noting that since the runoff coefficient distributions are applied as multipliers in setting the limits of acceptability, the relative errors in simulated values of lower discharges might be large, but the absolute errors will generally be small). This could imply that we need a better model, that can produce credible simulations over the complete range of discharges. However, from the analysis of the runoff coefficients shown in Fig. 1, it is also clear that *any* model is going to be subject to significant epistemic errors in the observations, particularly in a high-relief catchment like the Kent where raingauges do not exist at higher elevations. This alone is a good reason to use hydrological understanding in the Turing-like test approach to UPH20 being advocated here.

Code availability. The Dynamic Topmodel code used in this study is available at the R CRAN site https://cran.r-project.org/web/packages/dynatop/index.html (Smith and Metcalfe, 2022). The definition of the limits of acceptability is used as an example workflow in the Matlab CURE Uncertainty estimation toolbox available at https://www.lancaster.ac.uk/lec/sites/qnfm/credible/ (Page et al., 2022.

Data availability. The data used in the model runs is available on request from n.chappell@lancaster.ac.uk. It will eventually be stored as part of the Q-NFM Project archive on the NERC EIDC database.

Author contributions. KB conceived the use of runoff coefficient based limits of acceptability for model evaluation. PS, TP, AK and BH carried out the simulations and model evaluations within the NERC Q-NFM project led by NC. All the authors contributed to the writing and editing of the submitted paper.

Competing interests. The contact author has declared that none of the authors has any competing interests.

Special issue statement. This article is part of the special issue "IAHS2022 – Hydrological sciences in the Anthropocene: Variability and change across space, time, extremes, and interfaces". It is a result of the XIth Scientific Assembly of the International Association of Hydrological Sciences (IAHS 2022), Montpellier, France, 29 May–3 June 2022.

Acknowledgements. The Environment Agency is thanked for the release of rainfall, streamflow and raw rating data for the Kent catchment for the NERC Q-NFM project under Licence CL77737MG.

Financial support. This study has been carried out under the NERC funded Q-NFM project (grant no. NE/R004722/1).

Review statement. This paper was edited by Christophe Cudennec and reviewed by two anonymous referees.

References

Beven, K., Smith, P. J., and Wood, A.: On the colour and spin of epistemic error (and what we might do about it), Hydrol. Earth Syst. Sci., 15, 3123–3133, https://doi.org/10.5194/hess-15-3123-2011, 2011.

Beven, K. J.: EGU Leonardo Lecture: Facets of Hydrology – epistemic error, non-stationarity, likelihood, hypothesis testing, and communication, Hydrol. Sci. J., 61, 1652–1665, https://doi.org/10.1080/02626667.2015.1031761, 2016.

Beven, K. J.: Towards a methodology for testing models as hypotheses in the inexact sciences, Proceedings Royal Society A, 475, 2224, https://doi.org/10.1098/rspa.2018.0862, 2019.

Beven, K. J. and Lane, S.: Invalidation of models and fitness-for-purpose: a rejectionist approach, Chapter 5, in: Computer Simulation Validation – Fundamental Concepts, Methodological Frameworks, and Philosophical Perspectives, edited by: Beisbart, C. and Saam, N. J., Cham: Springer, 145–171, https://doi.org/10.1007/978-3-319-70766-2_6, 2019.

Beven, K. J. and Lane, S.: On (in)validating environmental models. 1. Principles for formulating a Turing-like Test for determining when a model is fit-for purpose, Hydrological. Process., 36, e14704, https://doi.org/10.1002/hyp.14704, 2022.

Beven, K. J. and Smith, P. J.: Concepts of Information Content and Likelihood in Parameter Calibration for Hydrological Simulation Models, ASCE J. Hydrol. Eng., 20, A4014010, https://doi.org/10.1061/(ASCE)HE.1943-5584.0000991, 2015.

Beven, K. J., Smith, P. J., and Freer, J.: So just why would a modeller choose to be incoherent?, J. Hydrol., 354, 15–32, 2008.

Beven, K. J., Page, T., Hankin, B., Smith, P., Kretzschmar, A., Mindham, D., and Chappell, N. A.: Deciding on fitness-for-purpose – of models and of natural flood management, Hydrol. Process., 36, e14752, https://doi.org/10.1002/hyp.14752, 2022.

Coxon, G., Freer, J., Westerberg, I. K., Wagener, T., Woods, R., and Smith, P. J.: A novel framework for discharge uncertainty quantification applied to 500 UK gauging stations, Water Resour. Res., 51, 5531–5546, 2015.

McMillan, H., Krueger, T., and Freer, J.: Benchmarking observational uncertainties for hydrology: rainfall, river discharge and water quality, Hydrol. Process., 26, 4078–4111, 2012.

Page, T., Smith, P., Beven, K., Pianosi, F., Sarrazin, F., Almeida, S., Holcombe, L, Freer, J., Chappell, N., and Wagener, T.: The CREDIBLE Uncertainty Estimation (CURE) Toolbox, Lancaster University [code], https://www.lancaster.ac.uk/lec/sites/qnfm/credible/ (last access: 10 March 2023), 2022.

Pappenberger, F. and Beven, K. J.: Functional Classification and Evaluation of Hydrographs based on Multicomponent Mapping (M^x), J. River Basin Management, 2, 89–100, 2004.

Smith, P. and Metcalfe, P.: dynatop: An Implementation of Dynamic TOPMODEL Hydrological Model in R, CRAN Repository [code], https://cran.r-project.org/web/packages/dynatop/index.html, last access: 10 March 2023.

Westerberg, I., Guerrero, J.-L., Seibert, J., Beven, K. J., and Halldin, S.: Stage-discharge uncertainty derived with a non-stationary rating curve in the Choluteca River, Honduras, Hydrol. Process., 25, 603–613, https://doi.org/10.1002/hyp.7848, 2011.

Proc. IAHS, 385, 135–140, 2024
https://doi.org/10.5194/piahs-385-135-2024

Open Access
PIAHS

Tracing and hydraulic modelling to assess the hydraulic performance of a constructed wetland

Bruno J. Lemaire[1,2], **Cédric Chaumont**[1], **Julien Tournebize**[1], **and Hocine Henine**[1]

[1]Université Paris-Saclay, INRAE, Hydrosystèmes Continentaux Anthropisés – Ressources, Risques, Restauration, 92761, Antony, France
[2]Université Paris-Saclay, AgroParisTech, 75005, Paris, France

Correspondence: Bruno J. Lemaire (bruno.lemaire@agroparistech.fr)

Received: 17 June 2022 – Revised: 7 March 2023 – Accepted: 10 March 2023 – Published: 18 April 2024

Abstract. Constructed wetlands are widely used to protect sensitive water resources from non-point source pollution from agriculture. Their potential to remove nitrate and pesticides increases with the water residence time and a uniform distribution of the inflow over the wetland area. Over the hydrological season, inflow variations greatly modify the theoretical residence time. The knowledge of the corresponding variations of the hydraulic performance constitutes a gap for the better management of treatment wetlands, especially for wetland with heterogeneous vegetation implementation. The aim of this work is to investigate how the hydraulic performance changes with the flow rate in a partly vegetated wetland. The study site, a 0.5 ha wetland, is located in an area of intensive cereal crop production in Northern France. The three-dimensional hydrodynamic model Delft3D-Flow was used to simulate flow through vegetation, forced by observed meteorological conditions. It was calibrated on continuous outflow concentration measurements and unmanned aerial vehicle (UAV) images during a 13 d tracing experiment with rhodamine WT. The simulated hydraulic performance indicators matched satisfactorily with observed values thanks to the detailed description of the vegetation. Simulations for the locally usual flow range and for a fixed water depth showed a limited increase of the hydraulic performance with the flow rate. This shows that conducting a tracing at low flow is sufficient to assess the average hydraulic performance of a wetland.

Keywords. SDG 6.6; Field Observation; Agricultural Drainage; Three-Dimensional Modelling

1 Introduction

A promising measure against non-point source pollution from agriculture, complementary to the reduction of fertilizers and pesticides applications, is the implementation of constructed wetlands intercepting polluted surface water (Tournebize et al., 2017). They are widely used to protect sensitive water resources (e.g. aquifers, lakes or rivers), especially when grass strips are bypassed by a subsurface drainage network. Constructed wetlands, intermittently fed during autumn and spring in a temperate climate, are composed of three interacting compartments: sediment layer, vegetation and water column. Pollutant mitigation increases with the time spent by water in contact with the sediment and with vegetation. The pollutant removal potential is mainly driven by the inflow rate and by its uniform distribution over the wetland area. When sparse or in patches, vegetation makes the flow path complex and can cause bypasses and dead zones in the wetland, and consequently affects its hydraulic performance. The scientific challenge is to link the yearly patterns of discharge, hydraulic performance and removal rate, in order to provide guidelines for the design of constructed wetlands.

Most literature about waste stabilization ponds and other constructed wetlands is based on laboratory experiments and two-dimensional hydraulic simulations at constant flow rate and homogeneous mass density (for a review: Ho et al., 2017). For surface free wetland, vertical flow stratification and mixing due to bathymetry, inflows, weather conditions and vegetation are often neglected, while they can have considerable consequences: Sweeney et al. (2003)'s three dimensional simulations showed that wind could change the residence time distribution from quasi plug flow into quasi fully mixed flow in a large waste stabilization pond, with high effects on the treatment efficiency; Adamsson et al. (2005) attributed the unexpectedly high residence time of a tracer at low flow in a laboratory tank to the deep storage of the cool inflow. Bulat et al. (2019) applied a three-dimensional model for the flow within rigid emergent and submerged vegetation to a large fluvial lake and showed the improvement of stratified flow description compared to the usual representation with a Manning-Strickler bottom roughness.

Holland et al. (2004) found experimentally that the flow rate had little impact on the hydraulic performance of a non-vegetated constructed wetland. The aim of this work is to investigate whether this is still true in a vegetated shallow wetland, using a three-dimensional model.

2 Material and methods

2.1 Study site and instrumentation

The study site is the wetland (0.53 ha) constructed in 2010 in Rampillon, 60 km South-East of Paris, France (Fig. 1, Tournebize et al., 2012). This wetland is a buffer zone between a 355 ha drained agricultural catchment and a direct recharge zone into a karstic aquifer through sinkholes (Mander et al., 2021). At full capacity, the depth of the compartments ranges from 0.5 (for the main compartment) to 1.2 and 1.5 m (for the inlet and outlet compartments). While the inlet is located above the water surface, the outlet collects water at mid-depth to avoid the clogging by floating vegetation. Baffles are intended to avoid short-circuits between inlet and outlet. Since the initial planting (*juncus*, *carex* and *typha*), the vegetation has developed naturally and, partially degraded by animals, became sparse and highly heterogeneous. At time experiment, vegetation cover was estimated to 39 % of total wetland area.

2.2 Tracing experiments

204 g of the conservative tracer Rhodamine WT were injected on 22 March 2016, between 07:10 and 10:15 local time. The outflow rate, recorded every 15 min during the tracing (Sigma 950 AV flow meter), varied from 3 to 10 $\mathrm{L\,s^{-1}}$ over the first nine days (average 6.6 $\mathrm{L\,s^{-1}}$), then peaked at 70 $\mathrm{L\,s^{-1}}$ after a rain event (Fig. 2). The theoretical residence time computed on these first nine days from the mean volume and flow rate was around 3.5 d. The tracer concentration was continuously monitored for 13 d with a fluorimeter (GGUN FL30), with a recovery rate of 94 %.

UAV images (11 images during the first day, and one each of the following 2 d) captured the flow paths within the wetland. The average water depth during the first nine days was only 0.1 m in the main compartment, so that the velocity averaged across the main flow direction was in the range of $10^{-3}\,\mathrm{m\,s^{-1}}$ (Reynolds number around 100, laminar flow). It increased to around $10^{-2}\,\mathrm{m\,s^{-1}}$ in preferential flow areas (turbulent flow).

2.3 Model configuration

The three-dimensional hydrodynamic model Delft3D-Flow (version 4.01.01.rc.03, Deltares, 2014) was used to simulate flow within the constructed wetland. This model has proven ability to simulate the hydrodynamics, the stratification and the mixing of shallow lakes (Soulignac et al., 2017). It describes the flow through emergent and submerged vegetation (Bulat et al., 2019).

The model assumes a hydrostatic pressure distribution and solves the Reynolds-averaged Navier-Stokes equations for an incompressible fluid with the Boussinesq approximation and a turbulence-closure model, here the k-ε model. The mesh was composed of about 2500 1.5 m $\times$ 1.5 m cells and 10 horizontal layers. Their thickness increases exponentially from top to bottom (6 % to 16.5 % of the total height) for the description of flow stratification.

In this model configuration, viscosity and diffusion are isotropic; they are the sums of a molecular term, kinematic viscosity and molecular diffusivity, and of the k-ε eddy viscosity and diffusivity respectively. Contrary to other studies with more turbulent flow (e.g., Soulignac et al., 2017), turbulence within the large cells was neglected (background horizontal eddy viscosity and diffusivity). Rhodamine WT was simulated as a passive tracer.

Due to unreliable measurements of the inflow rate, it was computed from the outflow rate and water depth measurements. Both inflow and outflow rates were forced in single cells at the inlet and outlet to ensure numerical stability; the temperature measured in the inlet pipe was used as input in the model.

Surface energy balance is represented by Delft3D "Ocean model". It was forced by daily SAFRAN reanalyses (air humidity and temperature, solar radiation) (Raimonet et al., 2017) and by hourly wind speed and direction from the nearest weather station of Melun, 32 km away.

Delft3D rigid vegetation module represents flow in vegetation (Bulat et al., 2019): the drag caused by vegetation is proportional to the stem density and diameter and creates a rather uniform current speed profile along the stems as opposed to the usual logarithmic profile. We placed patches of submerged (20 cm high) and emergent vegetation according to UAV images and in situ observation, we fixed their stem diameter to 5 mm and kept their stem densities as calibration parameters (Fig. 1a).

The roughness of the bottom of the wetland was represented by a Manning coefficient computed after Arcement and Schneider (1989), with a value of $0.3\,\mathrm{s\,m^{-1/3}}$ under submerged vegetation, else of $0.01\,\mathrm{s\,m^{-1/3}}$ (settled fine sediment). The free slip condition was adopted for banks and baffles, the latter represented as thin dams. Due to animals

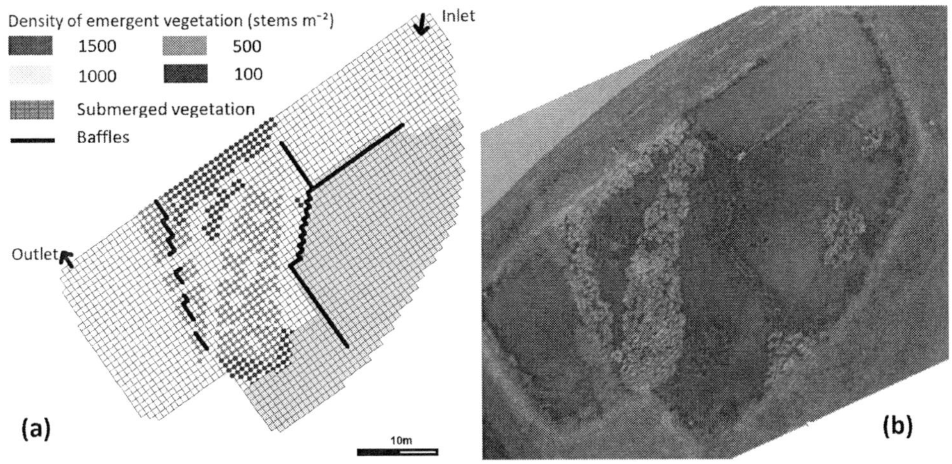

Figure 1. **(a)** Mesh and representation of the baffles and of vegetation inside the wetland (The pink arrow represents the longitudinal cross-section used in Fig. 4), **(b)** UAV image of the wetland, 48 h after the tracer injection.

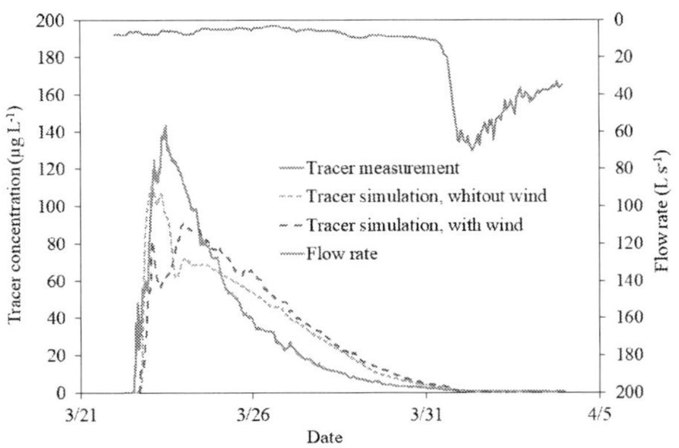

Figure 2. Residence time distribution measured during the winter tracing (2016) and simulated by the model and flow rate entering the wetland.

(*nutria, Myocastor coypus*) digging breaches in the baffles, short-circuits are clearly visible on UAV images (not shown). These breaches were represented in the model (Fig. 1a).

Water was initially set at rest and at the daily air temperature (7.5 °C); the warming up period was 8 h. The time step was 6 s to fulfill the Courant-Friedrichs-Lewy condition.

2.4 Hydraulic performance indicators

This subsection presents the extraction of hydraulic performance indicators from the residence time distribution $C(t)$, where C is the tracer concentration and t the time elapsed since the beginning of injection. Dimensionless flow-weighted time φ and concentration C' make it possible to compute the theoretical and the effective mean residence time in varying outflow rate Q and retention volume V (Holland et al., 2004):

$$\varphi = \int_0^t \frac{Q(t')}{V(t')} \mathrm{d}t' \tag{1}$$

$$C'(\varphi) = \frac{C(\varphi)\,V(\varphi)}{M_{\text{out}}} \tag{2}$$

where M_{out} is the total mass recovered over the tracer monitoring time T.

The first indicator is the dimensionless mean residence time φ_{m}, also called effective volume ratio:

$$\varphi_{\text{m}} = \int_0^T \varphi C'(\varphi)\mathrm{d}\varphi \tag{3}$$

The second one, a short-circuit index, is the dimensionless time of the 1st decile of the tracer mass, $\varphi_{10\,\%}$. The third one, a mixing indicator, is the dimensionless delay between the 1st and last deciles of the tracer mass, $\varphi_{90\,\%} - \varphi_{10\,\%}$. Teixeira and do Nascimento Siqueira (2008) experimentally showed that these were more reliable than most other indicators.

2.5 Model calibration and wetland hydraulic performance

During the calibration, only the vegetation stem density was triggered, in steps of 100 stems m^{-2}. In a first phase, we reproduced qualitatively the flow pattern visible in the UAV images and the overall shape of the residence time distribution, especially the timing of the peaks (Fig. 2). Then, a second phase consisted in fitting the hydraulic indicators, with and without wind input. The water level evolution was checked at the end of modelling.

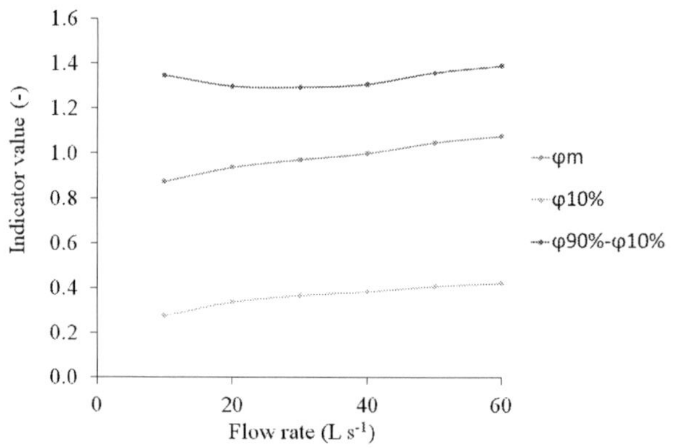

Figure 3. Evolution of simulated indicators (mean residence time, short-circuit and mixing indicators) with the flow rate.

The calibrated model was then used to test whether hydraulic performance changes with the flow rate within its annual range in the study area, excluding high flows. Tracing simulations were run for different constant flow rates from 10 to 60 $L\,s^{-1}$, with the average water depth and the weather conditions observed during the tracing, but without wind.

3 Results

3.1 Performance of the calibrated model

Model calibration made it possible to fit the simulated residence time distribution with the observations (Fig. 2). The recovery rate of the tracer is satisfactory (97 % in the simulations). While results were poor with a Manning-Strickler description of vegetation (data not shown), the simulated residence time distribution and hydraulic indicators matched observations after calibrating the vegetation density. Suppressing the wind input improved the shape of the residence time distribution (Fig. 2). The dimensionless mean residence time φ_m got closer to its observed value of 0.84, changing from 0.96 to 0.88. These high values of φ_m are close to the ideal value of 1 for the plug flow and the fully stirred tank. In the rest of the article, no wind is applied in the simulations.

As for short-circuits, the tracer began to leave the zone slightly later than observed but the 1st decile of the tracer mass occurred slightly earlier (simulated $\varphi_{10\,\%}$ is 0.21 vs. 0.24); the main peak occurred earlier but the tail of the distribution arrived later, so that the mixing indicator values were acceptably close (simulated $\varphi_{90\,\%} - \varphi_{10\,\%}$ is 1.50 vs. 1.28). The second simulated peak indicated either a partial storage and later release of the tracer, or a second and longer flow path.

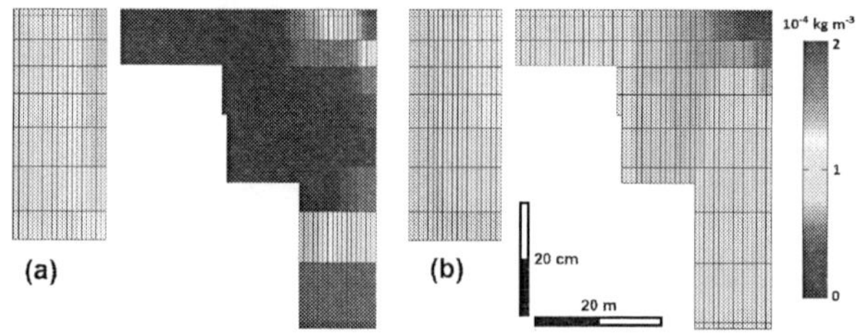

Figure 4. Tracer concentration on the longitudinal cross-section of Fig. 1a, for a tracing simulation with a constant flow rate, 17 h **(a)** and 41 h **(b)** after injection.

3.2 Evolution of hydraulic performance with the flow rate scenarios

Whereas the input flow rate increased by a factor 6, the hydraulic performance indicators changed moderately (Fig. 3). The short-circuit index exhibited the sharpest increase (by 53 %). The dimensionless effective residence time (effective volume ratio) increases slightly (by 23 %) and became larger than 1, for flow rates greater than 40 $L\,s^{-1}$, indicating that the tracer spent more time in the wetland than it would be in a perfectly stirred system. Finally, the mixing indicator remained stable (less than 10 % variation), with a minimum around 30 $L\,s^{-1}$.

4 Discussion and conclusions

4.1 Quality of the wetland hydrodynamic model

The agreement of three-dimensional simulation with observations was satisfactory, in line with results reported for a laboratory tank (Adamsson et al., 2005). The detailed representation of vegetation contributes to the modelling agreement with observations (Bulat et al., 2019). While vegetation was homogeneous at the construction, its evolution over time can create preferential flow paths which degrade the hydraulic efficiency, here the short-circuit index. The seasonal evolution of vegetation should also be taken into account in a simulation over several months.

4.2 Effect of the flow rate on the hydraulic performance

Results showed that short-circuit index and the dimensionless mean residence time increase slightly with the flow rate whereas the mixing index remains stable, which is consistent with the experimental results of Holland et al. (2004). When, the flow rate is greater than 40 $L\,s^{-1}$, the dimensionless mean residence time exceeds 1, which is not possible when using 2D simulation models (Persson et al., 1999).

To explain this behavior, we can observe in the longitudinal cross-section (see Fig. 1a) that the tracer can remain partly stored deeply in the 1st compartment of the wetland, even almost two days after the injection (Fig. 4b). Since the inflow comes from agricultural subsurface drains through soil profile, its temperature is quasi constant, whereas water temperature follows the dynamics of air temperature and sunlight. A cold inflow during tracer injection plunged to the depth where wetland water had the same temperature. Later, a warmer inflow remained close to the surface (Peeters and Kipfer, 2009). In such a stratified flow, the mean residence time can be higher than the theoretical residence time.

The results of this study show that, in wetlands with a controlled water level, a tracing at low flow provides reliable information about the average hydraulic performance, especially the mean residence time and mixing which are of major interest for treatment. This is all the more interesting since environmental tracings require quasi steady flow conditions and UAV observations in good weather conditions. Periods of several days of both relatively stable high flow and good weather are rare.

Holland et al. (2004) showed that water depth has a greater influence on hydraulic performance than the flow rate. Thus, if the management allows regulation of the water height, its variations have to be considered in an annual (or multiannual) assessment.

Code and data availability. Tracing data are on a repository https://doi.org/10.57745/CLNZYQ (Chaumont, 2022).

Author contributions. CC and JT designed the tracing and CC carried it out and analysed it, HH and BJL supervised the simulation work of interns Agathe Bertrand and Yue Zhang, BJL prepared the manuscript with contributions from all co-authors.

Competing interests. The contact author has declared that none of the authors has any competing interests.

Disclaimer. Publisher's note: Copernicus Publications remains neutral with regard to jurisdictional claims in published maps and institutional affiliations.

Special issue statement. This article is part of the special issue "IAHS2022 – Hydrological sciences in the Anthropocene: Variability and change across space, time, extremes, and interfaces". It is a result of the XIth Scientific Assembly of the International Association of Hydrological Sciences (IAHS 2022), Montpellier, France, 29 May–3 June 2022.

Acknowledgements. We thank interns Agathe Bertrand and Yue Zhang, who carried out the modelling part of this study, for the quality of their work.

Financial support. This work was financially supported by the French National Research Agency ANR, as a part of the PESTIPOND project (grant no. ANR-18-CE32-0007), and by the Ile-de-France Environmental Research Federation (FIRE (grant no. FR-3020)).

Review statement. This paper was edited by Christophe Cudennec and reviewed by two anonymous referees.

References

Adamsson, Å., Bergdahl, L., and Lyngfelt, S.: Measurement and three-dimensional simulation of flow in a rectangular detention tank, Urban Water J., 2, 277–287, https://doi.org/10.1080/15730620500386545, 2005.

Arcement, G. J. and Schneider, V. R.: Guide for selecting Manning's roughness coefficients for natural channels and flood plains, US geological survey, 38 p., https://doi.org/10.3133/wsp2339, 1989.

Bulat, M., Biron, P. M., Lacey, J. R. W., Botrel, M., Hudon, C., and Maranger, R.: A three-dimensional numerical model investigation of the impact of submerged macrophytes on flow dynamics in a large fluvial lake, Freshwater Biol., 64, 1627–1642, https://doi.org/10.1111/fwb.13359, 2019.

Chaumont, C.: Tracing Experiment Rampillon RWT 2016 [data set], Data INRAE, https://doi.org/10.57745/CLNZYQ, 2022.

Deltares: Delft3D-Flow [code] User Manual, Deltares, 725, https://content.oss.deltares.nl/delft3d4/Delft3D-FLOW_User_Manual.pdf (last access: 1 September 2022), 2014.

Ho, L. T., Van Echelpoel, W., and Goethals, P. L. M.: Design of waste stabilization pond systems: A review, Water Res., 123, 236–248, https://doi.org/10.1016/j.watres.2017.06.071, 2017.

Holland, J. F., Martin, J. F., Granata, T., Bouchard, V., Quigley, M., and Brown, L.: Effects of wetland depth and flow rate on residence time distribution characteristics, Ecol. Eng., 23, 189–203, https://doi.org/10.1016/j.ecoleng.2004.09.003, 2004.

Mander, Ü., Tournebize, J., Espenberg, M., Chaumont, C., Torga, R., Garnier, J., Muhel, M., Maddison, M., Lebrun, J. D., Uher, E., Remm, K., Pärn, J., and Soosaar, K.: High denitrification potential but low nitrous oxide emission in a constructed wetland treating nitrate-polluted agricultural run-off, Sci. Total Environ., 779, 146614, https://doi.org/10.1016/j.scitotenv.2021.146614, 2021.

Peeters, F. and Kipfer, R.: Currents in stratified water bodies 1: Density-driven flows, in: Encyclopedia of inland waters, edited by: Likens, G. E., Academic Press, London Boston, 530–538, https://doi.org/10.1016/B978-012370626-3.00080-6, 2009.

Persson, J., Somes, N. L. G., and Wong, T. H. F.: Hydraulics efficiency of constructed wetlands and ponds, Water Sci. Technol., 40, 291–300, https://doi.org/10.1016/S0273-1223(99)00448-5, 1999.

Raimonet, M., Oudin, L., Thieu, V., Silvestre, M., Vautard, R., Rabouille, C., and Le Moigne, P.: Evaluation of Gridded Meteorological Datasets for Hydrological Modeling, J. Hydrometeorol., 18, 3027–3041, https://doi.org/10.1175/JHM-D-17-0018.1, 2017.

Soulignac, F., Vinçon-Leite, B., Lemaire, B. J., Scarati Martins, J. R., Bonhomme, C., Dubois, P., Mezemate, Y., Tchiguirinskaia, I., Schertzer, D., and Tassin, B.: Performance Assessment of a 3D

Hydrodynamic Model Using High Temporal Resolution Measurements in a Shallow Urban Lake, Environ. Model. Assess., 22, 309–322, https://doi.org/10.1007/s10666-017-9548-4, 2017.

Sweeney, D., Cromar, N., Nixon, J., Ta, T., and Fallowfield, H.: The spatial significance of water quality indicators in waste stabilization ponds - Limitations of residence time distribution analysis in predicting treatment efficiency, Water Sci. Technol., 48, 211–8, https://doi.org/10.2166/wst.2003.0123, 2003.

Teixeira, E. C. and do Nascimento Siqueira, R.: Performance Assessment of Hydraulic Efficiency Indexes, J. Environmen. Eng., 134, 851–859, https://doi.org/10.1061/(ASCE)0733-9372(2008)134:10(851), 2008.

Tournebize, J., Gramaglia, C., Birmant, F., Bouarfa, S., Chaumont, C., and Vincent, B.: Co-design of constructed wetlands to mitigate pesticide pollution in a drained catch-Basin: A solution to improve groundwater quality, Irrig. Drain., 61(SUPPL.1), 75–86, https://doi.org/10.1002/ird.1655, 2012.

Tournebize, J., Chaumont, C., and Mander, Ü.: Implications for constructed wetlands to mitigate nitrate and pesticide pollution in agricultural drained watersheds, Ecol. Eng., 103, 415–425, https://doi.org/10.1016/j.ecoleng.2016.02.014, 2017.

Proc. IAHS, 385, 141–146, 2024
https://doi.org/10.5194/piahs-385-141-2024

IAHS2022 – Variability and change across space, time, extremes, and interfaces

Use of the HEC RAS model for the analysis of exceptional floods in the Ouémé basin

Ernest Amoussou[1,2], Félix Toundé Amoussou[2], Aymar Yaovi Bossa[3], Domiho Japhet Kodja[2], Henri Sourou Totin Vodounon[1,2], Constant Houndénou[2], Valérie Borrell Estupina[4], Jean-Emmanuel Paturel[5], Gil Mahé[5], Christophe Cudennec[6], and Michel Boko[2]

[1]Department of Geography and Regional Planning (DGAT), University of Parakou, Parakou, Benin
[2]Laboratoire Pierre Pagney: Climate, Water, Ecosystem and Development (LACEEDE), University of Abomey-Calavi, Abomey-Calavi, Benin
[3]National Water Institute (INE), University of Abomey-Calavi, Abomey-Calavi, Benin
[4]G-EAU, Univ. Montpellier, IRD, INRAE, Montpellier, France
[5]Montpellier HydroSciences Laboratory, University of Montpellier 2, Montpellier, France
[6]The Agro Institute Agrocampus Ouest, Rennes, France

Correspondence: Ernest Amoussou (ajernest@yahoo.fr) and Félix Toundé Amoussou (felixamoussou96@gmail.com)

Received: 10 June 2022 – Revised: 16 November 2022 – Accepted: 14 March 202 – Published: 18 April 2024

Abstract. The Ouémé River basin extends over almost half of Benin's territory, entirely located in a humid tropical climate. This river system includes a deltaic zone (delta of the Ouémé) known for its high agricultural potential and thus subject to a socio-economic development agenda. The Ouémé delta is facing recurrent floods that maintain rural agricultural population into a retrograding crisis with significant damages such as losses of properties. The objective of this study is to improve decision-making in the Ouémé basin through the simulation of exceptional floods using the HEC-RAS model.

The HEC RAS model is a conceptual model, which works through mathematical and physical formulas to implement environmental phenomena for forecasting, understanding and analysis purposes. The model inputs used are basin GIS data, hydro-meteorological data, characteristics of existing hydraulic structures, etc. The targeted outputs include 1D/2D/3D view plans with support of satellite images, tables, graphs and curves. It is worth mentioning that the model provides outputs compatible with other tools, such as civil engineering (Civil 3D, Revit, Infraworks, etc.) and GIS, that help to expand the valorization fields.

The implementation of the model in the Ouémé basin has made it possible to note: (i) that the recurring effect of losses and damages is justified by the settlement of the population on the river banks; (ii) that there is an important agricultural production in areas of high flood risk; (iii) that depending on the occurrence of the phenomenon, the flooded extent and the height of submersion remains variable, and more important for extreme flooding; (iv) about 12.07 % occurrence of river flood against 13.24 % for flash flood at a return period of 30 years. Moreover, it is very relevant to note that most of flood waters converge to the western part of the basin (an area with a low risk of flooding, stretched over 63.68 km^2) and to the eastern part around the Damè-Wogon depression (an area at high risk of flooding, stretched over 10.49 km^2).

Keywords. Flood risk; flood damage estimation; Ouémé River Basin; HEC-RAS model; exceptional flood

1 Introduction

Water, at the heart of any sustainable development process, is also a source of disasters for most human communities and ecosystems. These disasters are increasing day by day, an effect of global warming that the planet is currently experiencing. These hydroclimatic consequences result in exceptional floods and extreme droughts (IPCC, 2007; Marengo and Espinoza, 2015). A natural hazard for human society and its ecosystem that is difficult to predict but can be anticipated to mitigate economic, health and other losses (FAO, 2008; Shen et al., 2007). Benin, a West African country is not spared, like the most watered basins in the golf of guinea (Degan et al., 2019; Kodja, 2018; Mahé, 2006). However, Bonou district, is very heavily watered during the monsoon period (Bani, 2006), experiences periodic exceptional flooding situations which plunge a population with a low standard of living into a very unpleasant state. Moreover, with this potential for sustainable agricultural development and poverty reduction according to the Millenium Municipality Project (PCM, 2015), it is urgent to this end, a flood management plan to better take advantage of these nationally recognized assets. It is in this preventive clause that this research entitled, use of the HEC RAS model for the analysis of exceptional floods in the Ouémé basin in the Bonou district is organized. It will make it possible to show the causes of the losses and damages; and propose long-term preventive measures to minimize the damage faced with the occurrence of exceptional floods in the locality.

2 Methodology

2.1 Modeling exceptional floods

An exceptional flood is a phenomenon deemed to be rare and of extreme intensity but whose occurrence is more or less unpredictable on a temporal scale (FAO, 2008). It is often accompanied by such extraordinary damage when vulnerable to a high intensity downpour (Amoussou et al., 2014). However, with modeling using physics, mathematics and computer tools, it is possible to understand the causes and effects of a phenomenon. Thus, the modeling of exceptional floods on the district used statistical hydrology tools to determine the exceptional events characterizing the district; because hypothetically it is obvious that the extreme values are at the origin of floods. The surface hydraulic modeling tool HEC RAS is used to model the exceptional floods for different return periods, because it is likely that the extent of the flooded areas is a function of the return period. The susceptibility of losses and damages is assessed through mapping since land use is a factor reacting to remarkable losses and damages during floods.

2.2 Characterization of extreme events in Bonou district

Bonou district (Fig. 1) is located in the department of Ouémé in the south-east of the Republic of Benin between 6°47′40″ and 6°57′38″ N of the Northern Latitude and between 2°23′70″ and 2°35′01″ E of East Longitude.

The average annual rainfall is 1121 mm. It is distributed on:

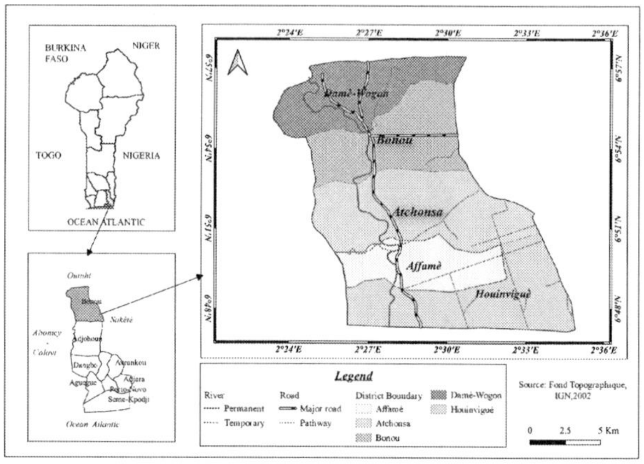

Figure 1. Administrative map of the study area.

- two (02) rainy seasons from April to July and from September to August when rain floods occur especially from June to July.
- Two (02) dry seasons from December to March and from July to August when a river flood is observed by the overflow of the Ouémé River during the month of September with the contribution of rainfall from the North of the country (Totin et al., 2016).

Indeed, the characterization of the exceptional events of the district will take into account the runoff aspect (river flood) and the rainfall aspect (rain flood). The determination of the peak flows of the Ouémé River and the estimation of maximum rainfall are made following the statistical laws of GEV (Generalized Extreme Values) including the Gumbel law (Gradex method) being the best suited in our region (Lawin et al., 2011). Base on Gumbel equation

$$F(\mathrm{x}) = e^{\left(-e\left(-\frac{(x-a)}{b}\right)\right)}, \tag{1}$$

and the Hazen formula

$$F(x_{Ri}) = \frac{(R_i - 0.5)}{N}, \tag{2}$$

the maximum daily rainfall and flow are estimated for return period of $T = 2$, 10 and 30 years (Table 1). It shows that there is a 50 % chance each year of recording flows of around 778.67 $\mathrm{m^3\,s^{-1}}$ against once in thirty years (approximately 3.33 %) for a flow of 1274.81 $\mathrm{m^3\,s^{-1}}$ (respectively 114.74 mm of rainfall).

2.3 Modeling of extreme floods by HEC RAS at Bonou

The modeling of extreme floods by HEC RAS results in a spatialized hydraulic implementation of the values obtained by characterizing the extreme events observables in the district. The output will map, analyze and assess flood risk.

Table 1. Results of the estimation of flow and precipitation values for the return period of 2, 10 and 30 years.

T (i years)	2	10	30
F (i)	0.5	0.9	0.9667
U (i)	0.367	2.250	3.384
Q_{dmax} (i)	778.65 $m^3 s^{-1}$	1088.38 $m^3 s^{-1}$	1274.81 $m^3 s^{-1}$
P_{dmax} (i)	68.72 mm	97.28 mm	114.47 mm

Table 2. Land cover and manning coefficient.

Land Cover	Manning coefficient
Tree cover areas	0.150
Built up areas	0.028
Cropland	0.030
Shrubs cover areas	0.035

(Source: U.S. Army Corps of Engineers, HEC5, Hydraulic Reference Manuel, CPD-69, 3–16 p., 2016).

Beyond the Digital Elevation Model (DEM, resolution = 30 m × 30 m) of the district, the estimated P_{dmax} and Q_{dmax} values, the boundary conditions (precipitation = P_{dmax}, upstream flow = Q_{dmax}, downstream = normal depth), the Manning coefficient (Table 2) according to land use, the hydraulic model is computed.

However, a study focused on the flood hydrograph of the catastrophic year 2010 (GFDRR, 2011), like one of the recent possible considerations. Indeed, the daily flows of 2010 in Bonou shows an annual maximum fluctuating around 1026 $m^3 s^{-1}$ which remains well above the estimated biennial flow (778.65 $m^3 s^{-1}$), closer to the ten-year flow (1088.38 $m^3 s^{-1}$) and distant from the one in their thirties.

From this hydrograph, it has been shown that the biennial flood lasts about 68 d and about 20 d for the ten-year and thirty-year ones (Fig. 2). These daily intervals are considered for the river simulation and a distributed hyetograph of the P_{dmax} value on 1 h for the rainfall.

3 Results

3.1 Modeling

The results show us the following:

The compilation of observations from the results can be summarized as follows (Tables 3, 4):

Ultimately, it is very remarkable that the waters are concentrated much further west of the river and upstream to the east in the Damè-Wogon depression.

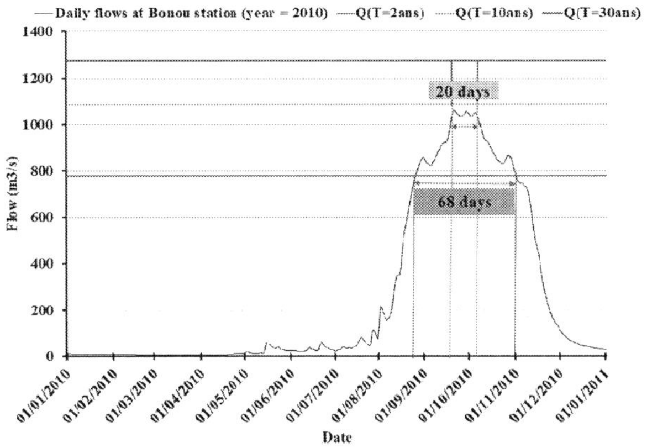

Figure 2. Stationarity of daily flows at the Bonou station in 2010 (Source: DG-Eau, 2012).

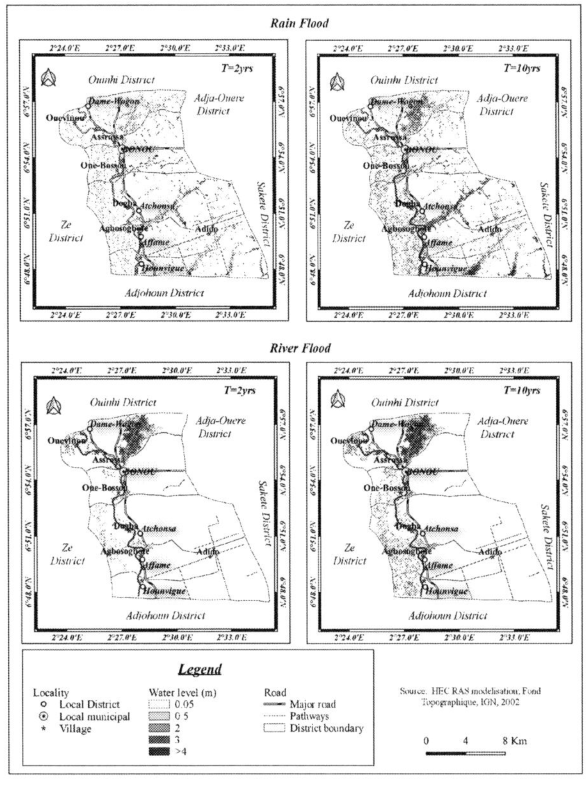

Figure 3. River and rain flooding map (T = 2, 10 years).

3.2 Susceptibility of losses and damages in the district

Following the land cover map (ESA Sentinel 2A 2016), from which the spatial distribution map of agglomerations and crops in the district are extracted. From the latter, we were able to highlight the relativity of the damage linked to extreme floods. This leads to major conclusions such as:

Table 3. Statistics on modeling results for River flood (Fig. S3 – River flood).

Type of flood	River flood		
Return period	2 years	10 years	30 years
Flooded area (km^2)	10, 49	14, 17 (5, 67 %)	30, 18 (12, 07 %)
Remarks	– overflow of water on both sides of the Ouémé river but very concentrated towards the west – very substantial overflow of the river towards the Damè-Wogon depression (9.7 km^2) to the east of the river. – agglomeration and culture will be affected		

Table 4. Statistics on modeling results for Rain flood (Fig. S3 – Rain flood).

Type of flood	Rain flood		
Return period	2 years	10 years	30 years
Flooded area (km^2)	26, 39 (10, 55 %)	42, 23 (16, 89 %)	98, 10 (39, 24 %)
Remarks	– four (04) meters of water depth in the shallows – affected infrastructure such as the main paved road linking Akpro-Misserete to Dasso – the agglomerations and property of the localities of Ahouanzoumè, Damè-Wogon, Gnanhoui (Damè-Wogon district); Atchabita, Bonou center, Oué-Bossou (district of Bonou); Agomahan, Dogba (district of Atchonsa); Affamè center, Affamè Sota (district of Affamè); Abèokuta, Agbosognota, Hounviguè (Hounviguè district) will be flooded.		

- settlement of the population along the banks of the Ouémé River,
- a preponderant domination of agricultural activities in areas of recurrent flooding (western part of the river) and depression (Damè-Wogon).

3.3 Delimitation of areas at risk of flooding

The compilation of the various results of the modeling of extreme floods and the susceptibility of losses and damages in the Bonou district makes it possible to generate a map (Fig. 4) of the flood zone at different levels of flood risk.

Indeed, based on the water level and land uses, we distinguish mainly two zones; an unwatered area (104.32 km^2) and an area at risk of flooding (145.68 km^2). However, we can divide into three risks flooding levels (river and rain) in the district, like:

- *high risk* (*20.13 km^2*): Ahouanzoumè, Asrossa and Gnanhoui-zoumè,
- *medium risk* (*63.68 km^2*): Atchabita, Affamè centre, Abèokuta, Avlankamè, Agbosota, Dogba, Ouvienou and Oué-bossou.
- *low risk* (*61.87 km^2*): rest of the localities. Most crops such as: maize, tomato, cassava, pepper, etc. are grown in high and medium risk areas; a cause of losses recorded during floods.

4 Proposal of territorial development solutions

Following these different results, some solutions have been proposed in a territorial management context. Preventive measures not vigorously studied, but serving as a front were mentioned.

Base on those results, for human security and agricultural production, we suggest:

4.1 a. better territorial occupation

- leave the middle riverbed, a natural water habitat,
- practice perennial or non-vulnerable crops to low propagation wave floods in high-risk areas: palm groves, teak, mahogany, etc. allowing the flood hydrograph to be rolled downstream,
- Cultivate food crops in low-risk and unwatered areas.

4.2 b. artificial lake creation (Fig. 5)

A preliminary study has shown us that an artificial lake will retain the excess water capable of flooding the population and

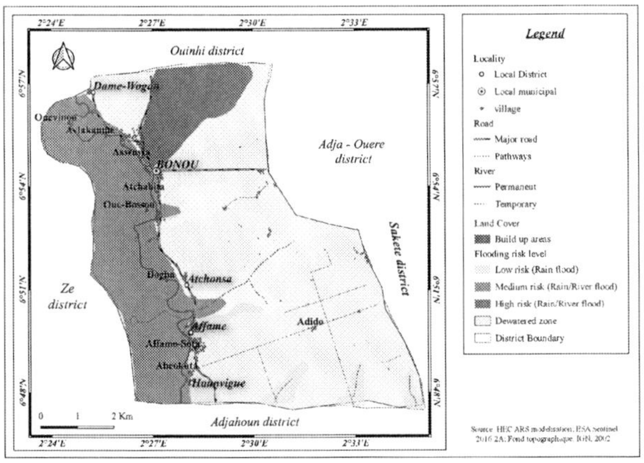

Figure 4. Map of areas at risk of flooding in Bonou district (Source: HEC RAS, modelisation, 2020; ESA, Sentinel 2A, 2016).

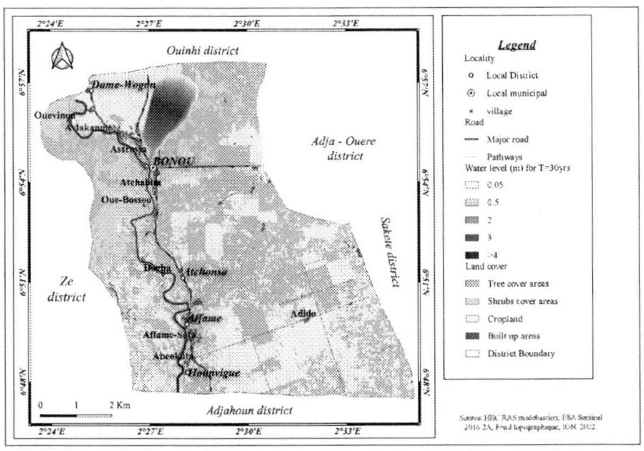

Figure 5. Results of an exceptional flood of $T = 30$ years of 1 d at constant flow (Source: HEC RAS, modelisation, 2020; ESA, Sentinel 2A, 2016).

crops. Well placed in the depression of Damé-wogon, it is located upstream of the cultures of the district. A simulation study confirms the effectiveness of its role of protection and rolling of river flood waves. This is a clearing of a side of 5 m in direct communication with the middle bed of the river.

We notice;

- A probable optimal protection of crops and agglomerations for floods of
- $T <= 10$ years
- a probable extension of the duration of the cropping seasons during the recession by exploitation of the stored water.
- a limitation of the storable volume in the cuvette for $T >= 10$ years.

4.3 c. Modification of the section of the watercourse

The activity of dredging the section of the river will widen and increase the draft. Therefore, increase the volume of water to be evacuated from the district.

4.4 d. Installation of channel networks

To limit the damage linked to rain floods in the fields, networks of meshed channels and directly connected to the middle bed of the river can be envisaged. They will make it possible to evacuate the water from the fields to the dredging river or vice versa.

5 Conclusions

This work on the modeling of exceptional floods in Bonou district is essentially based on the various tools and data from the cartography, on hydroclimatic data and on the use of a hydraulic model. The methodology is based on the use of various sources of information, statistical analyses, cartography and hydraulic modelling. It obviously made it possible to understand that the extreme values of flow and rainfall are at the origin of the floods. But specifically, it is the persistence of fluctuations in these values over time that amplifies this cause.

From this study, it is noted that the extent of the areas flooded during a rainfall flooding exceeds the river flooding. The extent of these zones decreases with the decrease of the return period. This confirms the variation of floodplains according to the type of flood and according to the duration of occurrence.

The occupation of floodplains, the banks of the river and the cultivation of crops in the plains are the main causes that intensify the losses and damage observed during floods. Hence, the occupation of the land is a factor at the base of the losses and damages in the district. Moreover, the enhancement of this work requires the establishment of guidelines to manage floods. Among so many recommendations, we suggest for the safety of goods and services: the transformation into an artificial lake/temporary reservoir area, the plain of Dame-wogon; to encourage the activity of extracting sand from the bed of the Ouémé River and the establishment of channel network systems.

However, the best way to stay safe from river flooding is to settle in the eastern part of the river, far from the bank.

Data availability. No data sets were used in this article.

Author contributions. Conducted this research: EA. Conceptualization: EA, FTA, AYB, DJK and HSTV. Methodology and modelling and data processing: EA, FTA, AYB, DJK, HSTV and CH. Writing-Original Draft Preparation: EA, FTA, AYB, DJK, HSTV, CH, J-EP and GM. Writing-Review and Editing: VBE, J-EP, GM,

CC and MB. All authors have read and agreed to the published version of the manuscript.

Competing interests. At least one of the (co-)authors is a guest member of the editorial board of *Proceedings of IAHS* for the special issue "IAHS2022 – Hydrological sciences in the Anthropocene: Variability and change across space, time, extremes, and interfaces". The peer-review process was guided by an independent editor, and the authors also have no other competing interests to declare.

Disclaimer. Publisher's note: Copernicus Publications remains neutral with regard to jurisdictional claims in published maps and institutional affiliations.

Special issue statement. This article is part of the special issue "IAHS2022 – Hydrological sciences in the Anthropocene: Variability and change across space, time, extremes, and interfaces". It is a result of the XIth Scientific Assembly of the International Association of Hydrological Sciences (IAHS 2022), Montpellier, France, 29 May–3 June 2022.

Review statement. This paper was edited by Salvatore Grimaldi and reviewed by two anonymous referees.

References

Amoussou, E., Camberlin, P., Totin Vodounon, S. H., Tramblay, Y., Houndenou, C., Mahé, G., and Boko, M.: Evolution des précipitations extrêmes dans le bassin-versant du Mono (Bénin – Togo) en contexte de la variabilité et de changement climatique, in: Climat: Système & Interactions 27éme Colloque de l'Association Internationale de Climatologie, Dijon, France, 331–337, ISBN 978-2-907696-20-3, 2014.

Bani, G.: Monographie de la commune de Bonou, Afrique conseil, 25 p., https://docplayer.fr/79623514-Monographie (last access: March 2019), 2006.

Degan, B. A., Gbaguidi, V., Karambiri, H., Alamou, E., and Afouda, A. A.: Impact de la variabilité climatique sur l'évolution des crues du fleuve Ouémé au Bénin de 1952 à 2011, EPAC/UAC, Benin, 204 p., https://biblionumeric.epac-uac.org:9443/jspui/handle/123456789/2216 (last access: March 2019), 2019.

DG-Eau: Carte hydrogéologique du Bénin: Echelle 1: 500000, Rapport de recherche sur la carte hydrogéologique du Bénin, Ministère des Mines, de l'Energie et de l'Eau, République de Bénin, 2012.

FAO: Climate change and Food security: a Framework for Action, Rome, conference de Bali, 78 p., https://www.fao.org/3/au035e/au035e.pdf (last access: March 2019), 2008.

Fatma, C. S. E.: Délimitation des zones inondables par l'utilisation combiné du logiciel HEC-RAS et HEC GEORAS cas du centre "AIN Jemaa", Master Sciences et Techniques, 97 p., http://www.secheresse.info/spip.php?article72316 (last access: July 2019), 2013.

GFDRR: Rapport d'Evaluation des Besoins Post Catastrophiques liées à l'inondation, Bénin, 75 p., https://www.gfdrr.org/sites/default/files/publication/pda-2011-benin-fr.pdf (last access: March 2019), 2010.

HEC-RAS River Analysis System: 2D Modeling user's manual, US Army Corps of Engineers, Institute for Water Ressources Hydrologic Engineering Center, CPD-68A, 171 p., https://www.hec.usace.army.mil/software/hec-ras/documentation/HEC-RAS5.0UsersManual.pdf (last access: 25 August 2019), 2016.

IPCC: Changements climatiques, rapport de synthèse, 103 p., https://www.ipcc.ch/site/assets/uploads/2018/02/ar4_syr_fr.pdf (last access: April 2019), 2007.

Kodja, J. D.: Indicateurs des événements hydroclimatiques extrêmes dans le bassin-versant de l'Ouémé à l'exutoire de Bonou en Afrique de l'Ouest, Thèse de Doctorat, Université de Montpellier, 214 p., https://tel.archives-ouvertes.fr/tel-01869842 (last access: July 2019), 2018.

Lawin, A. E., Afouda, A., and Lebel, T.: Analyse de la variabilité du régime pluviométrique dans la région agricole d'Ina au Bénin, European Journal of Scientific Rechearch, ISSN 1450-216 X Vol.50, No.3, 425–439, https://www.scirp.org/(S(lz5mqp453edsnp55rrgjct55))/reference/referencespapers.aspx?referenceid=2107220 (last access: Febuary 2020), 2011.

Mahé, G.: Variabilité pluie-Débit en Afrique de l'Ouest et Centrale au 20éme siècle: changements hydroclimatiques, occupation du sol et modélisation hydrologique, université des Sciences et Techniques Montpellier, Ecole Doctorale 148 Terre, Eau, Espace, 161 p., https://horizon.documentation.ird.fr/exl-doc/pleins_textes/divers11-03/010043107.pdf (last access: January 2021), 2006.

Marengo, A. J. and Espinoza, J. C.: Extreme seasonal droughts and floods in Amazonia: causes, trends and impacts, Int. J. Climatol., 36, 1033–1050, https://doi.org/10.1002/joc.4420, 2015.

Pomerleau, J.: Changements climatiques et sécurité en Afrique, centre universitaire de formation en environnement, Université de Sherbrooke, 99 p., https://savoirs.usherbrooke.ca/bitstream/handle/11143/7402/cufe_Pomerleau_essai97.pdf?sequence=1&isAllowed=y (last access: March 2018), 2009.

Projet Commune du Millénaire de Bonou pour un Développement Durable (PCM-Bonou), 92 p., https://bj.chm-cbd.net/sites/bj/files/2022-09/Projet (last access: 14 May 2019), 2015.

Shen, C., Wang, WC., Hao, Z., and Gong, W.: Exceptional drought events over eastern China during the last five centuries, Clim. Change, 85, 453–471, https://doi.org/10.1007/s10584-007-9283-y, 2007.

Totin, V. S., Amoussou, E., Odoulami, L., Boko, M., and Blivi, B.: Seuils pluviométriques des niveaux de risques d'inondation dans le bassin de l'Ouémé au Bénin, XXIXème colloque de l'Association de Climatologie, Lausanne – Besançon, 369–374, http://www.climato.be/aic/colloques/LausanneBesancon (last access: 18 April 2019), 2016.

U.S. Army Corps of Engineers: HEC5, Hydraulic Reference Manuel, CPD-69, 3–16 p., https://www.hec.usace.army.mil/software/hec-ras/documentation/HEC-RAS5.0ReferenceManual.pdf (last access: 25 August 2019), 2016.

Proc. IAHS, 385, 147–153, 2024
https://doi.org/10.5194/piahs-385-147-2024

Open Access PIAHS

Rainfall data augmentation in Northern Italy through daily extremes and the Hershfield factor

Paola Mazzoglio, Ilaria Butera, and Pierluigi Claps

Department of Environment, Land and Infrastructure Engineering, Politecnico di Torino, Torino, 10129, Italy

Correspondence: Paola Mazzoglio (paola.mazzoglio@polito.it)

Received: 28 May 2022 – Revised: 26 January 2023 – Accepted: 29 March 2023 – Published: 18 April 2024

Abstract. The majority of rainfall measurements in the world is at the daily scale, i.e. related to a specific calendar day and measured over fixed 24 h. On these data, daily annual maximum rainfall depths (F-maxima) series are easily obtained. On the other hand, 24 h annual maximum rainfall depths (S-maxima), which refer to a period starting at any instant, are more useful indicators. S-maxima values cannot be less than the F-maxima, and are generally higher. The ratio between these extremes, called Hershfield factor (H), has been studied to move from F-maxima to S-maxima, allowing to take advantage of the relevant amount of information included in historical records of daily extremes. For instance, before 1980, in the Italian Hydrological Yearbooks only a subset ($< 50\,\%$) of the rain gauges was equipped with a recording device, from which annual maxima over 1, 3, 6, 12 and 24 consecutive hours can be derived. In this study we investigate the possibility of using F-maxima to complement the S-maxima records related to the Po river basin and the Liguria region (North of Italy). As a first step we retrieved from official databases all the daily rainfall measurements, available over this area, from early 1900 until today and we quality-controlled the measurements. We then computed the annual H for all the stations and all the years where both the F- and S-maxima were available, to obtain data that can be analyzed in their temporal and spatial variability. The spatial distribution of the Hershfield factor shows values similar to the ones suggested in the literature and is related to the geographic position of the stations, allowing the possibility to identify some distinct areas with positive or negative anomalies. The obtained map of the H factor, with interpolated local anomalies, allows to reconstruct the missing S-maxima in stations with only F-maxima, and improve the knowledge of the spatial variability of sub-daily rainfall extremes.

Keywords. UPH 17; field observation; Italy; Po river basin; rainfall; extremes

1 Introduction

For decades, several hydrological variables as the rainfall depths were measured on a calendar day basis, usually at 09:00 a.m. local time (Morbidelli et al., 2021). In Italy, those daily measurements are published in the Hydrological Yearbooks, a collection of data usually available only in printed format. The availability of shorter duration records was limited by the requirement of continuous monitoring rainfall gauges. After the replacement of the mechanical rain gauges with the automatic ones in the last decades of the 20th century, the reference time for the calendar day rainfall measure was shifted to 00:00 a.m. However, the daily measurement approach was never ended, even if it is widely known that such discretization in time leads to bias when dealing with extreme events (Papalexiou et al., 2016). Considering the vast heritage of measurements and its simplicity, the majority of annual maximum rainfall depths in the world are at the daily scale, i.e. expressed over fixed 24 h (F-maxima).

It is clear, however, that sliding 24 h maxima obtained from continuous measurements (S-maxima) are more useful values. They refer to 24 h periods starting at any instant and cannot be less than the F-maxima. An analysis conducted over the United States proved that the S-maxima statistics like the median, the mean and the standard deviation are higher than those of F-maxima and have a greater dispersion; the difference between the F- and S-maxima statistics increases as the time interval increases and appears to become constant for time scales larger than 36 h (Papalexiou et al., 2016).

The ratio between F- and S-maxima, named Hershfield factor (H), has been suggested to correct the incongruences induced by the different time discretization, allowing to take advantage of the relevant amount of information included in historical records of F-maxima (Hershfield, 1961; Harihara Ayyar and Tripathi, 1973; van Montfort, 1990; Dwyer and Reed, 1995; van Montfort, 1997; Papalexiou et al., 2016; Llabrés-Brustenga et al., 2020).

H acts as a multiplier that aims at correcting the underestimation caused by the daily time sampling. Hershfield itself suggested to use a value equal to 1.13 after an analysis carried out using data covering the United States (Hershfield, 1961). In a recent analysis in the United States, Papalexiou et al. (2016) obtained a value equal to 1.14. An application over India suggested a value equal to 1.15 (Harihara Ayyar and Tripathi, 1973). Over China, van Montfort (1997) obtained values in the range 1.04–1.17. Dwyer and Reed (1995) suggested a mean value of 1.16, while proposing differences according to different regimes: a correction factor of 1.15 for rainfall regimes with more frequent short-duration concentrated events, and 1.17 for sites with rainfall regimes with more frequent longer events, such as those produced by large frontal systems. Llabrés-Brustenga et al. (2020) obtained a mean value of 1.125 over Catalonia (Spain) and highlighted considerable variations within the year: summer is the season in which lower H (1.093) was obtained, justified by the fact that rainfall is often raised by the influence of the diurnal cycle and the S-maxima are not usually split by daily measuring in the morning; in spring, a higher H (1.161) was obtained, possibly determined by the fact that rainfall events are more often split when F-maxima measurements are taken at 08:00 a.m.

All these studies (except Papalexiou et al., 2016 and Llabrés-Brustenga et al., 2020) are based on old records. In our study, we investigated the possibility of using ≈ 1000 stations with at least 10 years of both F- and S-maxima for the same years, covering the period from early 1900 until today, to enrich the availability of S-maxima rainfall measurements, by selecting the data-rich Po river basin and the Liguria region (North of Italy) as a case study. The objective is to obtain mean values of H for the stations where both F- and S-maxima data are available and make these value available, through spatial interpolation, to F-maxima stations where S-maxima data are unavailable.

2 Data management and quality-control

2.1 Area of interest

In this study we analysed an area of about 90 000 km^2 extent located in the North of Italy, that coincides with the Po river basin and the Liguria region. The different administrative regions included in this area and the elevation data are represented in Fig. 1.

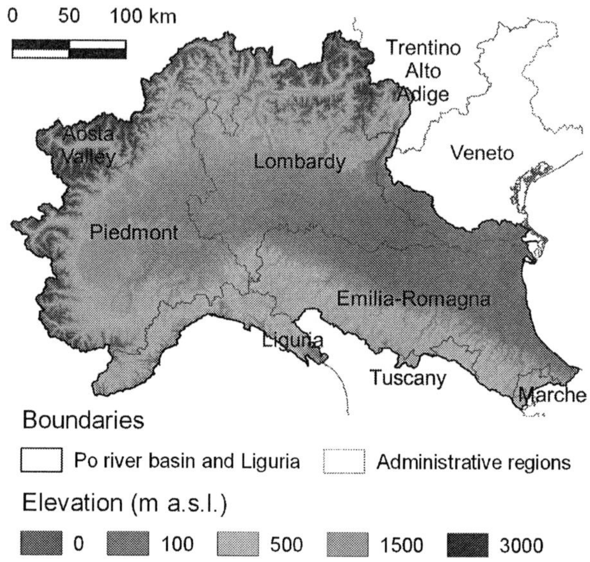

Figure 1. Elevation data with borders and names of the administrative regions. Source: Shuttle Radar Topography Mission (Farr et al., 2007).

2.2 Validated S- and F-maxima datasets

Annual maximum rainfall depths related to a 24 h time interval (i.e., the S-maxima) are derived from the Improved Italian – Rainfall Extreme Dataset (I^2-RED; Mazzoglio et al., 2020), a collection of rainfall extremes recorded since 1916 by more than 5200 rain gauges installed all over Italy. In this analysis we used only the time series of the 1713 rain gauges installed in the study area.

A set of validated F-maxima (i.e., quality-controlled daily annual maxima published in a dedicated table of annual maxima) is available in digital format from: ARPA Piemonte for the years 1988–2018 (ARPA Piemonte, 2022), the Functional Center of the Autonomous Aosta Valley Region for the years 1994–2016 (Centro Funzionale Regione Autonoma Valle d'Aosta, 2022), the VAPI Piemonte project for the years 1913–1986 (Villani, 2003), Meteotrentino for the years 1921–2009 (Meteotrentino, 2022) and ARPA Veneto for the years 1984–2021 (ARPA Veneto, 2022).

2.3 Continuous time series datasets

The spatio-temporal coverage of the validated datasets (Fig. 2) is very limited with respect to the spatio-temporal coverage of the printed version of the Hydrological Yearbooks. The digitization of the Yearbooks tables containing validated F-maxima would require a massive effort, out of the scope of this work. To fill the spatio-temporal gaps caused

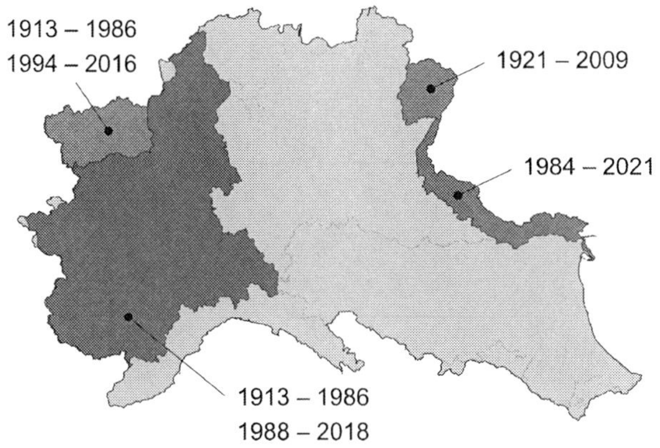

Figure 2. Spatio-temporal availability of the validated F-maxima. Grey colour indicates areas where a database of daily annual maxima in digital format is not available.

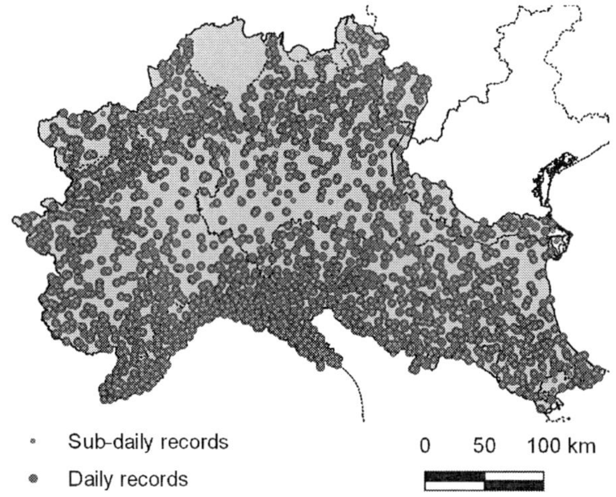

Figure 3. Spatial distribution of the sub-daily and daily records. Source: I^2-RED data (Mazzoglio et al., 2020).

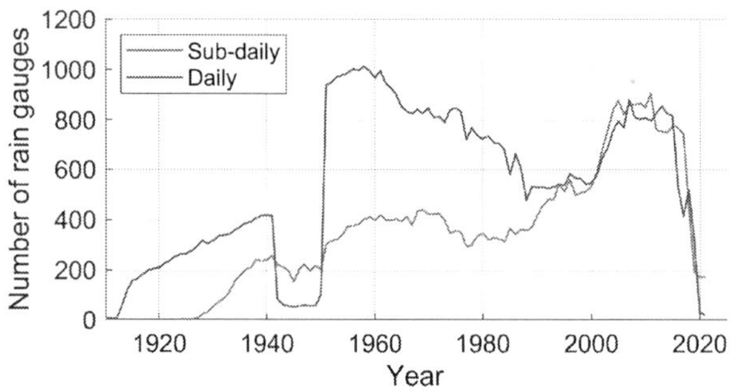

Figure 4. Time history of the number of active rain gauges.

by missing validated data, we resorted to available historical continuous daily rainfall time series, available in digital format, that are partially uncontrolled. The main data source that we used is the SCIA database (ISPRA, 2022). It has been integrated with few daily time series, coming from: ARPA Piemonte for 1988 onwards (ARPA Piemonte, 2022), the "Banca dati meteoclimatica" of Liguria Region (Regione Liguria and ARPAL, 2022), the "Atlante climatico della Liguria" for the years 1960–2010 (Agrillo and Bonati, 2013) and the Dext3r webapp (ARPAE-SIMC, 2022).

In this study we did not consider the influence of the change of the time of the daily measurements, because precise and extensive information on the years in which the hydrological agencies started to record the daily measurement at 00:00 a.m. is not available. However, it should be noted that Llabrés-Brustenga et al. (2020) pointed out that very high H is obtained only in the case of F-maxima taken with starting time in the afternoon. They also argue that F-maxima is usually more precise if sampled at 00:00 a.m., even though 08:00 a.m. can be considered as a good sampling time.

All the continuous daily time series of the datasets mentioned before were merged into a unique dataset. The SCIA database was found lacking some of the most relevant daily maxima recorded in recent years in the Liguria region: a manual gap-filling procedure was thus performed after having identified the time series that contains only 1 missing daily data (i.e., the actual F-maxima). In this case, the missing measurement (that was probably removed by too strict quality-control procedure applied while building the SCIA dataset) was retrieved from the printed version of the Hydrological Yearbooks or from ARPA Liguria website. Afterwards, to verify the correctness of the continuous daily data, high precipitation values (greater than 500 mm) were manually controlled, to verify if they were realistic. In a few cases, these high values were removed, because caused by a mistake with decimals in the table.

After this merging and control operations, only continuous time series with at least 330 out of 365 daily measurements were used to compute the annual F-maxima. These were integrated with the already validated F-maxima.

2.4 Final datasets

The data merging procedure allowed us to reach a total of about 1000 stations with at least 10 years of both F- and S-maxima for the same years. Nevertheless, the total number of F-stations is significantly higher: in this region, 2149 stations with F-maxima are available, allowing to have a data density of 1 rain gauge every 42 km^2 (the data density of the sub-daily time series is of 1 rain gauge every 53 km^2). In Fig. 3 it is possible to observe the spatial distribution of the F- and S-maxima. From the figure one can realize that the data density is not uniform: even including the F-maxima, several locations in the lowlands or at high elevations are still poorly sampled.

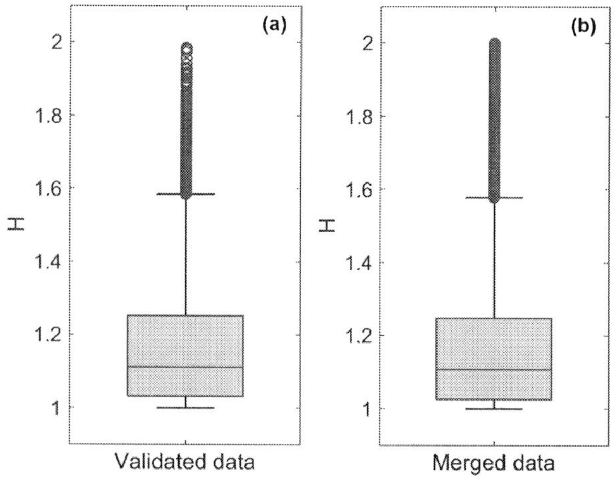

Figure 5. Box plot of the yearly Hershfield factors computed using the validated data **(a)** and from the merged dataset **(b)**. Each box plot displays the following information: the median, the lower and upper quartiles (blue box), any outlier (values that are more than 1.5 times the interquartile range, depicted as a blue circle), and the minimum and maximum values that are not outliers (black whiskers).

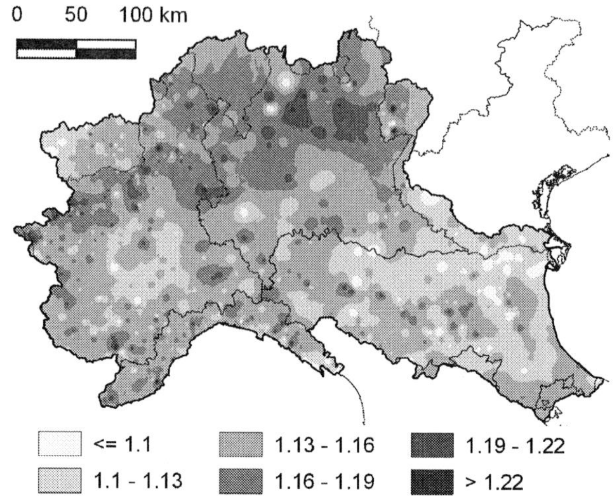

Figure 6. Spatial distribution of the Hershfield factor obtained by applying and Ordinary Kriging with spherical variogram.

In Fig. 4 it is possible to see the time history of the available stations. From this figure it clearly emerges that, for instance, before 1980 only a subset (< 50 %) of the rain gauges was equipped with a continuous recording device.

3 Evaluation of the Hershfield factor

For all the stations and all the years where both the F-maxima and S-maxima are available, the Hershfield factor was computed according to its definition:

$$H = \frac{h_{24}}{h_g}, \qquad (1)$$

where h_{24} is the 24 h annual maximum and h_g is the daily maximum in a given year.

As a general rule, $1 \leq H \leq 2$ (van Montfort, 1997). It can be equal to 1 if the S-maxima occurred inside the same calendar day, while it can be equal to 2 if the extreme event is equally divided into two days.

In a first step, we analysed H using only the validated data, whatever the record length, to obtain reliable results that would be compared to those obtained using the entire dataset.

Even in validated data we found cases in which F-maxima were higher than S-maxima. This inconsistency is not unusual, and can be attributed to errors in the digitization, due to the manual daily recording. These occurrences, and the others in which we found $H > 2$, were removed. Figure 5a shows the box plot of the individual H values obtained year by year: the overall median H is 1.112, while the whiskers ends are, respectively, 1 and 1.585. The overall mean value obtained for H is 1.170, somewhat higher than those in the literature.

Having removed only the undeniable anomalies ($H < 1$ or $H > 2$) does not exclude the possibility that other anomalies may have occurred. We then used the limits of the box plot to filter out possible outliers in the process of extending the computation to the entire dataset. Therefore, when computing the H values for all the stations using the entire dataset, made of validated and not validated data, we used series with at least 10 years of data and we used the whiskers ends of the validated data (1 and 1.585) as thresholds to remove the outliers. After this computation, the median H resulted of 1.144 and the mean resulted equal to 1.146, which are in line with the values reported in the literature. By increasing the minimum record length to 20 years we obtained an average of H equal to 1.142, and a median equal to 1.141, suggesting that even a time series of only 10 years could provide reliable estimates.

Examinig the box plots in Fig. 5 one can recognize that the variability of the H computed from the entire dataset turns out to be similar to the one of the validated data, despite a higher number of outliers. This evidence confirms the suitability of using also non-validated data to guarantee an higher spatio-temporal coverage of F-maxima.

After the computation and validation of the H values in the so-reached $\sim$ 1000 stations, we proceeded to a spatial interpolation at 1 km resolution of the mean H. This procedure was realized using the Ordinary Kriging (Fig. 6). By analyzing the spatial distribution of the interpolated Hershfield factor within the study area, we found variability patterns similar to those available in the literature, i.e. the mean (at station) H values turn out to be non uniform over the space. They seem influenced by the geographic position of the stations, and the patterns entail the possibility to identify some dis-

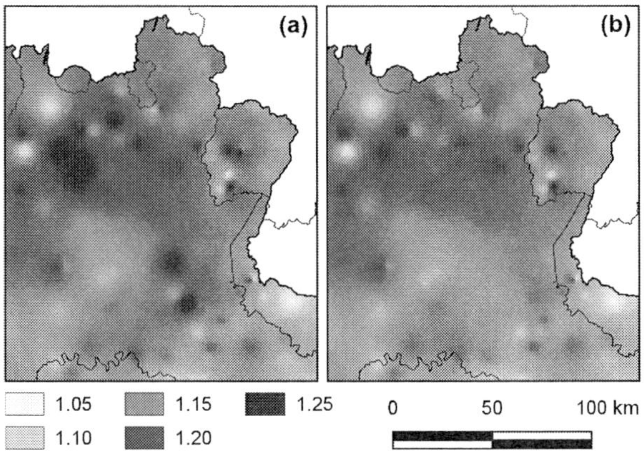

Figure 7. Spatial distribution of the Hershfield factor before **(a)** and after **(b)** the site-specific data-cleaning operation. Focus over the Eastern part of Lombardy region.

tinct areas with a positive or negative anomaly with respect to the spatial average. More specifically, higher values emerge in the Northern part of the examined region, mainly in Piedmont and Lombardy. The Aosta Valley, despite its location in the northern parts, shows different anomalies. This difference could probably be addressed to different prevailing rainfall mechanisms occurring in the Aosta Valley, possibly due to the presence of high peaks at the borders between Piedmont and the Aosta Valley, which act as a barrier. An influence of the geographic position was also pointed out over Catalonia (Spain) by Llabrés-Brustenga et al. (2020), and over China, where van Montfort (1997) analyzed monthly H values in relation to the location of the station, the autocorrelation within its daily measurements and the fraction of wet days.

The exam of the spatial distribution of H has suggested the necessity to perform an additional data control, as some suspicious isolated anomalies clearly emerged (Fig. 7a). This happened especially in areas where official and quality-controlled annual maxima were not available, so that F-maxima were computed from continuous daily records. For these isolated "peaks" of H, the F-maxima time series of the rain gauges have been checked and the incorrect F-maxima data have been replaced using the values reported in the printed version of the Hydrological Yearbooks, as depicted in Fig. 7b. Note that Fig. 6 already report the fully corrected version of the map.

4 Discussion

To increase the spatial detail of the extremes of precipitation over the North of Italy, a consistent database including daily precipitation (F-maxima) can be used. The analysis of the ratio between 24 h and daily extremes (Hershfield factor) throughout the investigated area can allow to estimate annual S-maxima (i.e. over 24 h) in the stations that lack this information. In some areas (mountain regions and the lowlands between Piedmont and Lombardy) the increase in detail is noteworthy, and can provide important improvement over the spatial analysis of 24 h extremes.

Before proceeding with the use of these reconstructed data, some care must be exerted about the reliability of these newly-computed maxima. In these terms, it is worth mentioning that Koutsoyiannis and Iliopoulou (2022) pointed out that there is no theoretical basis behind the "inflation" through the Hershfield factor, even though it can represent an acceptable estimate from an engineering point of view. The authors pointed out that all realizations are stochastically equivalent and there is no theoretical basis for their correction. By correcting the series, its stochastic properties can be distorted.

In terms of series properties, indeed, Papalexiou et al. (2016) had previously noted that the ratios between the mean values of F- and S-maxima deviate from the standard deviation ratios: this implies that the application of the H factor cannot appropriately preserve the standard deviation of the F-maxima. These authors have then introduced an approach that simultaneously correct both the mean and the standard deviation errors.

Despite the caveats, however, we cannot ignore that the possibility to include data recorded in decades at the beginning of the past century (see Fig. 4) can produce relevant benefit in the spatial detail of the process. This suggests us to proceed with the application of the H factor to include the F-maxima in the existing 24 h records.

On the other hand, when looking for trends in the data, to protect the reconstructed data from the distortion mentioned by Koutsoyiannis and Iliopoulou (2022) robust methods can be used, as Sen's method, the non-parametric Mann-Kendall test and the Spearman rank correlation test that all exhibit negligible sensitivity to data distortion (Morbidelli et al., 2021).

In this work and in all the other ones cited in this paper the H values were considered as constant over time. Nevertheless, future works could be directed to investigate if the variations observed all over Italy in the 24 h extremes (Mazzoglio et al., 2022) can be detected also in the H values.

5 Conclusions

In this study we performed a complex data merging operation to recover all the F-maxima recorded starting from the 20th century by 2149 stations. This dataset was then used to evaluate the Hershfield factor, that is a coefficient computed as the ratio between the S-maxima and the F-maxima that can be relevant for the reconstruction of missing sub-daily annual maxima. The interpolation realized, reported with a map in Fig. 6, allows to reconstruct the missing S-maxima in stations with only F-maxima, and improves the knowledge of the spatial variability of sub-daily rainfall extremes.

In view of the extension of this data-recovering procedure to the entire Italy, further investigations are to be directed to the correction of distortion effects noted in the statistical properties of the reconstructed series. In that sense, it is important to consider the trade-off between the increase of uncertainty on the 24 h series properties and the increase in the spatial detail of S-maxima in data-scarce regions.

Data availability. The daily time series can be downloaded from the original repositories. Full credits are reported in Sect. 2.

The 24 h rainfall data were obtained from the I^2-RED database. Although Italian law requires an open-source policy for all public data, this right has not yet been implemented by all the Italian agencies involved in the management of the rain gauge network. The agreements we signed with some of these agencies, aimed at monitoring the correct use of the data, restricted their use to the aims of the authors' project. As a result of these legal restrictions, a complete version of I^2-RED can only be provided to two groups of people: members of the authors' research group (who are already fully authorized to use the data) and people who can prove they have received clearance from the regional authorities. The entire quality-controlled database is available on Zenodo (https://doi.org/10.5281/zenodo.4269509, Mazzoglio et al., 2020), albeit with restricted access. The data can be used by third parties, for an indefinite timeframe, upon having completed an agreement with the authors and with the regional agencies involved in the data collection. The raw data availability depends on the region: a complete description of how to access these data is reported in Mazzoglio et al. (2020).

Author contributions. PC, IB and PM designed the experiments and PM carried them out. PM performed the data merging operation and developed the code. PM prepared the manuscript with contributions from all co-authors.

Competing interests. The contact author has declared that none of the authors has any competing interests.

Disclaimer. Publisher's note: Copernicus Publications remains neutral with regard to jurisdictional claims in published maps and institutional affiliations.

Special issue statement. This article is part of the special issue "IAHS2022 – Hydrological sciences in the Anthropocene: Variability and change across space, time, extremes, and interfaces". It is a result of the XIth Scientific Assembly of the International Association of Hydrological Sciences (IAHS 2022), Montpellier, France, 29 May–3 June 2022.

Acknowledgements. The authors acknowledge the regional agencies involved in the management of the rain gauge networks that provided the rainfall measurements included in I^2-RED. Full credits are reported in Mazzoglio et al. (2020).

Review statement. This paper was edited by Christophe Cudennec and reviewed by two anonymous referees.

References

Agrillo, G. and Bonati, V.: Atlante climatico della Liguria, ARPAL – Centro Funzionale Meteoidrologico di Protezione Civile, 127 pp., https://www.arpal.liguria.it/contenuti_statici//clima/atlante/Atlante_climatico_della_Liguria.pdf (last access: 3 April 2023), 2013.

ARPAE-SIMC: Dext3r, ARPAE-SIMC [data set], https://simc.arpae.it/dext3r/, last access: 25 April 2022.

ARPA Piemonte: Banca dati meteorologica, ARPA Piemonte [data set], http://www.arpa.piemonte.it/rischinaturali/accesso-ai-dati/annali_meteoidrologici/annali-meteo-idro/banca-dati-meteorologica.html (last access: 3 April 2023), 2022.

ARPA Veneto: Dati delle precipitazioni di massima intensità, ARPA Veneto [data set], https://www.arpa.veneto.it/bollettini/storico/precmax/ (last access: 3 April 2023), 2022.

Centro Funzionale Regione Autonoma Valle d'Aosta: Dati osservati del Centro Funzionale RAVDA, Centro Funzionale Regione Autonoma Valle d'Aosta [data set], https://presidi2.regione.vda.it/str_dataview_download (last access: 3 April 2023), 2022.

Dwyer, I. J. and Reed, D. W.: Correcting mean annual maxima for data discretization, Preprints 6th Int. Meet. on Statistical Climatology, Galway, Ireland, 447–450, http://imsc.pacificclimate.org/proceedings/6IMSC.pdf (last access: 3 April 2023), 1995.

Farr, T. G., Rosen, P. A., Caro, E., Crippen, R., Duren, R., Hensley, S., Kobrick, M., Paller, M., Rodriguez, E., Roth, L., Seal, D., Shaffer, S., Shimada, J., Umland, J., Werner, M., Oskin, M., Burbank, D., and Alsdorf, D.: The Shuttle Radar Topography Mission, Rev. Geophys., 45, RG2004, https://doi.org/10.1029/2005RG000183, 2007.

Harihara Ayyar, P. S. and Tripathi, N.: Relationship of the clock-hour to 60-min and the observational day to 1440-min rainfall, Ind. J. Meteorol. Geophys, 24, 279–282, 1973.

Hershfield, D. M.: Technical Paper No. 40: Rainfall frequency atlas of the United States, Department of Commerce, Weather Bureau, Washington, DC, 1961.

ISPRA: SCIA Sistema nazionale per l'elaborazione e diffusione di dati climatici, ISPRA [data set], http://www.scia.isprambiente.it/wwwrootscia/Home_new.html# (last access: 3 April 2023), 2022.

Koutsoyiannis, D. and Iliopoulou, T.: Ombrian curves advanced to stochastic modeling of rainfall intensity, in: Rainfall, edited by: Morbidelli, R., Elsevier, 261–284, https://doi.org/10.1016/B978-0-12-822544-8.00003-2, 2022.

Llabrés-Brustenga, A., Rius, A., Rodríguez-Solà, R., and Casas-Castillo, M. C.: Influence of regional and seasonal rainfall patterns on the ratio between fixed and unrestricted measured intervals of rainfall amounts, Theor. Appl. Climatol., 140, 389–399 https://doi.org/10.1007/s00704-020-03091-w, 2020.

Mazzoglio, P., Butera, I., and Claps, P.: I^2-RED: a massive update and quality control of the Italian annual extreme rainfall dataset, Water, 12, 3308, https://doi.org/10.3390/w12123308, 2020.

Mazzoglio, P., Ganora, D., and Claps, P.: Long-term spatial and temporal rainfall trends over Italy, Environmental Sciences Proceedings, 21, 28, https://doi.org/10.3390/environsciproc2022021028, 2022.

Meteotrentino: Dati storici, Meteotrentino [data set], https://www.meteotrentino.it/#!/content?menuItemDesktop=111 (last access: 3 April 2023), 2022.

Morbidelli, R., Saltalippi, C., Dari, J., and Flammini, A.: A review on rainfall data resolution and its role in the hydrological practice, Water, 13, 1012, https://doi.org/10.3390/w13081012, 2021.

Papalexiou, S. M., Dialynas, Y. G., and Grimaldi, S.: Hershfield factor revisited: Correcting annual maximum precipitation, J. Hydrol., 542, 884–895, https://doi.org/10.1016/j.jhydrol.2016.09.058, 2016.

Regione Liguria and ARPAL: Banca dati meteoclimatica, Regione Liguria [data set], https://www.regione.liguria.it/servizi/item/14279-banca-dati-meteoclimatica.html (last access: 25 April 2022), 2022.

van Montfort, M. A. J.: Sliding maxima, J. Hydrol., 118, 77–85, https://doi.org/10.1016/0022-1694(90)90251-R, 1990.

van Montfort, M. A. J.: Concomitants of the Hershfield factor, J. Hydrol., 194, 357–365, https://doi.org/10.1016/S0022-1694(96)03212-X, 1997.

Villani, P. (Ed.): Rapporto sulla Valutazione delle piene in Piemonte, in: Relazione delle attività del CUGRI fino al 2001, 89–118, Ed. Del Paguro, ISBN 88-87248-35-4, 2003.

Proc. IAHS, 385, 155–162, 2024
https://doi.org/10.5194/piahs-385-155-2024

Modelling human impacts on surface and subsurface hydrological drought

Tesfaye Belay Senbeta, Krzysztof Kochanek, Emilia Karamuz, and Jaroslaw Jan Napiorkowski

Department of Hydrology and Hydrodynamics, Institute of Geophysics,
Polish Academy of Sciences, Warsaw, 01-452, Poland

Correspondence: Tesfaye Belay Senbeta (tsenbeta@igf.edu.pl)

Received: 30 May 2022 – Revised: 20 February 2023 – Accepted: 29 March 2023 – Published: 18 April 2024

Abstract. The scientists tend to assume that climatic factors are the only drivers in the development of drought. However, this classical assumption is no longer valid in the Anthropocene, as most catchments have been significantly modified by human factors. Recent theoretical developments have shown that there are mutual interactions and feedback mechanisms – drought shapes humans and vice versa. Therefore, this study aims to assess the role of human impacts on hydrological drought (HD) characteristics by considering runoff and baseflow separately. For this purpose, the two rainfall-runoff models: SWAT (Soil and Water Assessment Tool) and HBV (*Hydrologiska Byrans Vattenbalansavdelning*) models are compared. The results show significant differences in the magnitude of change between the two models, suggesting that the ensemble method of multiple models need to be considered for modelling human impact. However, the results confirm that despite significant differences in the magnitude of change, both hydrological models indicate similar impacts, i.e. positive and negative.

Keywords. Human factors; Panta Rhei; observation-modelling framework; UPH9; SWAT; HBV; hydrological drought

1 Introduction

Catchment responses to precipitation inputs depend on geophysical characteristics, climatic conditions, and human factors (Kubiak-Wójcicka and Bąk, 2018). These factors affecting hydrological processes are mainly divided into two categories: natural factors (climatic factors, watershed characteristics) and human factors (Harrigan et al., 2014; Jiang et al., 2015). Watershed characteristics include watershed size, terrain slope, natural vegetation, geology, orientation, and prevailing soil moisture conditions, while human factors include development of irrigation, water transfer, water abstraction and land cover dynamics (B. Liu et al., 2016). Local climate variability and its impact on droughts have increased and are likely to increase further due to the release of greenhouse gases that have caused global warming (Kang and Sridhar, 2017). In northern Europe, human-modified climate change is likely to be responsible for recent drought (Gudmundsson and Seneviratne, 2016). Frequent drought is also a serious problem for the environment and socio-economic situation in Poland (Bąk and Kubiak-Wójcicka, 2017), with severe droughts occurring frequently in the central part of the River Vistula basin (Karamuz et al., 2021; Bąk and Kubiak-Wójcicka, 2017). Studies have also identified the impact of human factors on HD as either exacerbating (Wanders and Wada, 2015) or ameliorating (Jiao et al., 2020). Human factors impact the development of HD, by accumulating, attenuating, delaying, prolonging, etc. (Y. Liu et al., 2016). Which of these factors – climate or human – determines the characteristics of HD depends on the spatial and temporal scales considered for studies. This means that climate is mainly the dominant factor at the global level and on a longer-term scale (Y. Liu et al., 2016), while human factors are mainly dominant at the local level and on a short-term scale (Zou et al., 2017). Therefore, site-specific water management measures are recommended to address the different HD problems in different river reaches (Qiu et al., 2016).

A drought is generally when available water (i.e. precipitation, soil moisture, surface water, and groundwater) is significantly scarce for an extended period. The overall impact on the components of the hydrological cycle can lead to a long-term socio-economic drought, which in turn leads to water scarcity, i.e. an imbalance between water demand and available water supply (van Loon et al., 2016a). The study of drought has a long tradition of assuming that climatic factors are the only driving factors for its development (Nasiri et al., 2022). However, this classical assumption is no longer valid in the Anthropocene, in which the Earth's systems have been significantly modified by human impacts (van Loon et al., 2016b). This led to a redefinition of drought by van Loon et al. (2016b) that takes human responses into account. The existence of interactions between drought impacts and human responses suggests a reciprocal influence – drought impacts shape human societies and vice versa (Di Baldassarre et al., 2017). Moreover, the IAHS has proclaimed the period 2013–2022 as the Scientific Decade, which will focus on research into the interactions and feedbacks between humans and the environment under the title "Panta Rhei – everything flows" (Montanari et al., 2013). Furthermore, Blöschl et al. (2019) identified the occurrence of droughts and their changes as an "Unsolved Problem in Hydrology" (UPH9), which states: "How do droughts occur, do they change and if so, why?".

Although the studies on droughts are not new, the dynamics of drought impacts and human responses, as well as the modelling of this feedback, have not been sufficiently explored (Di Baldassarre et al., 2017). For instance, Wu et al. (2019) and Rangecroft et al. (2019) have used total runoff to determine HD, while other studies consider HD by looking separately at the components of the hydrological process, particularly baseflow drought (e.g. Bazrkar and Chu, 2020). Our study, therefore, focuses on assessing the impact of human factors on HD characteristics considering (sub)surface runoff separately. Two rainfall-runoff (RR) models: SWAT (Arnold et al., 1998) and HBV (Bergström and Forsman, 1973) were applied for reconstructing (simulating) the "proxy" natural runoff during the human-modified period, which was compared with the observed runoff (human-modified runoff) by applying the observation – modelling framework proposed by van Loon and van Lanen (2013). Naturalized and observed runoff and the corresponding filtered baseflow were used as indicators of HD.

The aim of this study is therefore to model the impact of human factors on the characteristics of HD by comparing simulation-based and observed-based using standardized HD indices. We also considered two RR models of different complexity, the lumped version of the HBV model and the semi-distributed SWAT model, to determine the role of using different RR models of different complicities in identifying HD characteristics.

2 Data and methods

2.1 Study area

The Kamienna catchment is located in Central Poland, with a catchment area of about 2000 km^2, and river length of about 156 km. It is a tributary of the Vistula River basin and lies at geographical coordinates between 20.60–21.70° E and 50.79–51.22° N (Fig. 1). The topography of the catchment ranges from about 130 m to just over 600 m (Senbeta and Romanowicz, 2021), with a mean elevation of 260 m above mean sea level. The catchment has experienced the greatest change in land cover from agriculture to forests and natural areas, as well as a significant change in the hydrological regime (Senbeta and Romanowicz, 2021).

2.2 Data

Two rainfall-runoff (RR) models were used in our study: the SWAT and the HBV models with different complexity and input data requirements. Climate variables such as precipitation and temperature for the period 1968–2018 were used for both models, while additional spatial GIS data (topography, land use, and soil map) were used for the SWAT model; see Senbeta and Romanowicz (2021) for more details. Potential evapotranspiration (E_o) was calculated externally with the temperature-based method for the HBV model. Precipitation and E_o at each station were then aggregated to catchment areal-average value using the Thiessen polygon and used as input to the HBV model. River runoff at the Czekarzewice gaging station was also used as an additional input for model calibration, validation, and further HD analysis. The periods 1971–1976 and 1977–1982 were used for model calibration and validation for both models, while the entire period, 1971–2018, was used for HD analysis.

2.3 Human impacts modelling approach

The study applied "observational-modelling" comparative approach (van Loon and van Lanen, 2013) that compares observed river runoff (i.e. human-modified runoff) with simulated (naturalized) runoff using the RR models to determine the human impact on the hydrological processes. In this approach, the observed runoff is considered as an indicator of the combined impacts of human and climatic factors, while the simulated runoff is an indicator of only the impact of climate variability. This means that the difference between the observed and simulated runoff is an indicator of human influence on river runoff. The selected RR models – the SWAT and HBV models – calibrated and validated in the reference (natural conditions) period (1971–1982) were applied to model natural runoff during the human period (1983–2018). This significant change in the river runoff regime in the early 1980s in the Kamienna catchment was noted in an earlier study (Senbeta and Romanowicz, 2021). Details on the setup, calibration, and validation of the SWAT can also be found in

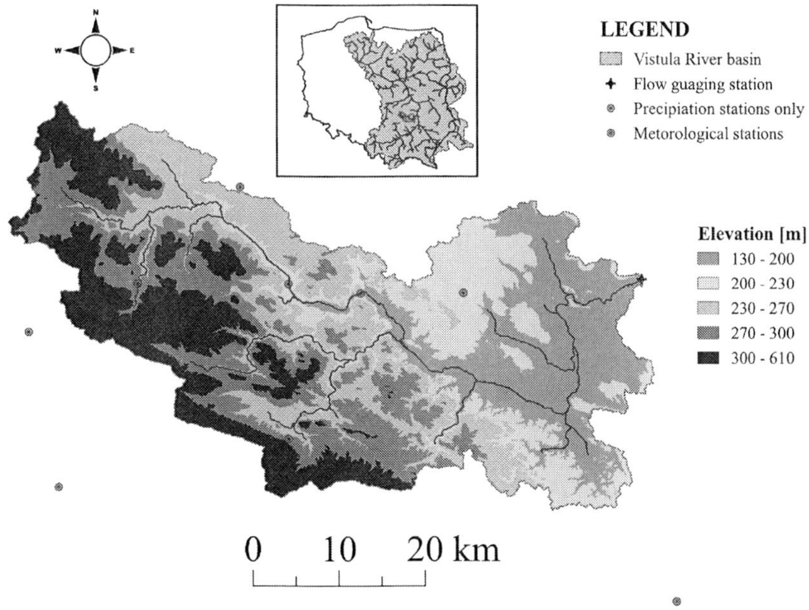

Figure 1. Geographical location of the study area and distribution of the hydroclimatic measuring stations.

Senbeta and Romanowicz (2021), while details on the HBV modelling processes are described in Sect. 2.4.

The study also aims to consider baseflow as a proxy indicator for groundwater drought, but no observed baseflow data were available throughout the study period. Therefore, the baseflow data were filtered out from the observed and RRs simulated runoff using the Wittenberg algorithm (Wittenberg, 1999). This baseflow filtering method was proposed based on a non-linear reservoir algorithm and the calculation is done backwards, starting with the last value of the baseflow recession as follows:

$$Q_t = Q_o\left[1 + \frac{(1-b)\,Q_o^{1-b}}{ab}\Delta t\right]^{\frac{1}{b-1}} \quad (1)$$

$$Q_{t-\Delta t} = \left[Q_t^{b-1} + \frac{(1-b)}{ab}\right]^{\frac{1}{b-1}} \quad (2)$$

where Q_o is an initial discharge [$\mathrm{mm\,d^{-1}}$], Q_t is a baseflow recession at time t, and a is a parameter with the dimension [$\mathrm{mm}^{1-b}\,\mathrm{d}^b$]; b is a dimensionless exponent [0.5], Δt is a time step [day].

2.4 HBV model

The HBV model (Bergström and Forsman, 1973), originally developed by the Swedish Meteorological and Hydrological Institute is a rainfall-runoff model that simulates river runoff at daily time scales. The conceptual model consists of three reservoirs: the soil box, the upper layer, and the lower layer, representing respectively the soil moisture, the fast response of the subsurface, and the slow response of the groundwater. The model uses the degree-day method to simulate snowmelt. The model is lumped when a single input is used for the entire catchment, while it is considered semi-distributed when a catchment is divided into sub-catchments based on elevation and vegetation zones. In the study, the lumped version of HBV-Light was used. For more details on this version of the model, Seibert and Vis (2012), is recommended.

2.5 Hydrological drought identification method

The parametric standardized approach was applied to determine the Streamflow Drought Index (SDI; Nalbantis and Tsakiris, 2009) and the Standardized Baseflow Index (SBFI; Bazrkar and Chu, 2020) as indicators of surface and subsurface HD, respectively. In this approach, both drought indices (DI) are derived based on the Log-normal distribution function. The HD characteristics were derived from DI, based on the run theory with the threshold level (TL) of -1. This means that a DI of less than -1 indicates drought, while a value above -1 indicates either mild (normal) drought or wet conditions. Furthermore, drought can be classified into three states: moderate ($-1.5 < \mathrm{DI} \leq -1$), Severe ($-2.0 < \mathrm{DI} \leq -1.5$) and extreme ($\mathrm{DI} \leq -2.0$) (Nalbantis and Tsakiris, 2009). The number of months for a particular drought event is referred to as the duration of drought (DD), the sum of the DI values for a single drought event indicates the severity of the drought (S), and the ratio of S to DD indicates the intensity of the drought (I) (Mckee et al., 1993).

Table 1. Change in hydrological drought characteristics based on streamflow drought index (SDI).

Flow	DD [months]	$\lvert S\rvert$ [–]	ND [–]	$\lvert I\rvert$ [per month]	Remarks
SDI_O	108	140.1	17	1.3	
SDI_N	73	117.4	10	1.61	SWAT
SDI_N	92	127.5	14	1.39	HBV
Human impacts on hydrological drought characteristics [%]					
HI[a]	47.9	19.3	70.0	−19.3	SWAT
HI[b]	17.4	9.9	21.4	−6.5	HBV

Note. DD = drought duration, S = drought severity, ND = number of drought events, I = drought intensity, SDI_N = Naturalized-based streamflow drought index, SDI_O = Observation-based streamflow drought index, [a] Human impact based on the SWAT model simulation, [b] Human impact based on the HBV model simulation

Table 2. Frequency of hydrological drought based on observed and naturalized runoff.

Conditions	Hydrological drought frequency (%)				Remarks
	Q_O	Naturalized runoff			
		SWAT	HBV	Average*	
Wet	48.0	46.2	48.5	45.7	N
Normal	32.9	40.9	35.2	42.1	N
Subtotal	80.9	87.1	83.7	87.8	
Moderate	15.8	8.8	10.8	7.8	D
Severe	3.4	0.4	5.1	2.7	D
Extreme	0.0	3.7	0.4	1.8	D
Subtotal	19.1	12.9	16.3	12.2	
Total	100.0	100.0	100.0	100.0	

Note. Q_O is observed runoff, N indicates no or normal drought conditions, D indicates hydrological drought. * Based on an average of the SWAT and HBV models naturalized runoff.

The change in the characteristics of HD is quantified by comparing observation-based and simulation-based results as follows:

$$\Delta HD_C = \frac{|HDI_O| - |HDI_N|}{|HDI_N|} \times 100\,\% \qquad (3)$$

where ΔHD_C refers to change in HD characteristics [%], HDI_O refers to observation-based HD index (SDI/SBFI), and HDI_N refers to simulation-based (naturalized) HD index (SDI/SBFI).

3 Results

3.1 Comparison of the RR models in HD analysis

Changes in HD characteristics based on SWAT and HBV-based runoff are compared with observation-based runoff to model human impacts, as shown in Table 1.

The duration of drought (DD), the severity of the drought (S), the number of drought events (ND), and the intensity of the drought (I) changed by 47.9 %, 19.3 %, 70.0 %, and −19.3 % based on the SWAT model and by 17.4 %, 9.9 %, 21.4 %, −6.5 % based on the HBV model, respectively, indicating that there are significant differences in human impacts on HD characteristic between the two RR models. Even though the magnitude of human impacts differs significantly between the two models, both models show the same direction of human impacts, i.e. either positive or negative changes for each of the considered HD characteristic.

3.2 Hydrological drought analysis

The HD indices were calculated based on the observed and naturalized runoff and the respective filtered baseflow to determine the change in the characteristics of HD using the TL value of −1. The results show that 17 HD events in different drought categories (i.e., moderate, severe, or extreme) occurred during the human-influenced period, which is more than the 12 HD events for naturalized conditions. Due to the significantly different results found between the SWAT and HBV models (Sect. 3.1), the study used the arithmetic mean of the two RR models for further analysis of human impacts on HD. The HD events of 1973–1974, 1984–1985, 1989, 1990, 1992, 2006, 2007, and 2008 were sustained streamflow droughts of more than 5 consecutive months based on observations, while the HD events of 1973–1974, 1984–1985, 1989–1990, and 2012–2013 were streamflow droughts of more than 5 months under naturalized conditions. Moreover, the results show variations in the magnitude of drought severity between natural and human-modified conditions for both surface and subsurface HD (Fig. 2). The early 1980s and 1990s were the most severely affected decades with successive HD events of longer DD (15 and 17 months) and I (−1.4 and −1.47 intense per month) based on observed runoff (not shown). Unlike the other decades considered in this study, the results for the 2010s decade show a greater number of HD events for the naturalized condition than for the observation-based analysis.

During the perturbed period (1983–2018), HD characteristics such as ND events, mean DD, and S increased by 83.1 %, 18.5 %, and 5.6 % for observation-based runoff and 25.3 %, 39.3 %, and 41.6 % for observation-based baseflow as compared to the baseline period (1971–1982) (not shown).

The result also shows that the frequency of HD increases from 12.2 % to 19.1 % due to the influence of human interventions (Table 2), indicating a 6.9 % increase in HD frequency.

4 Discussion

In this study, two RR models (SWAT and HBV) were applied for modelling the impacts of humans on the characteristics of HD, using the "observational-modelling" comparison ap-

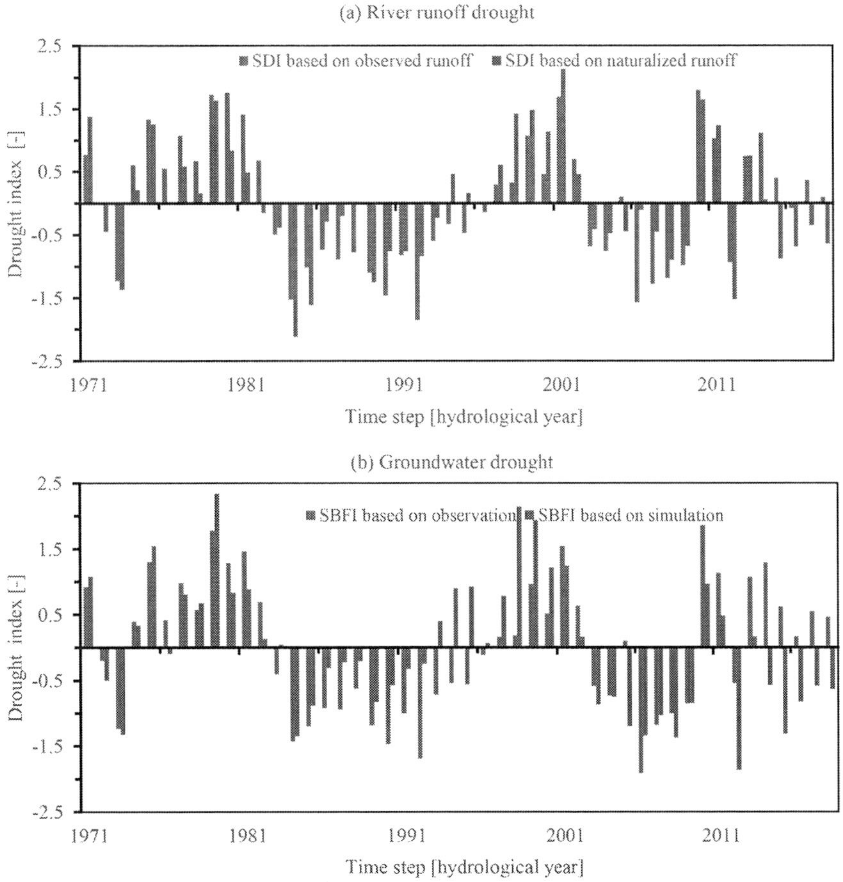

Figure 2. Hydrological drought analysis **(a)** Streamflow Drought Index (SDI); **(b)** the Standardized Baseflow Index (SBFI).

proach. A comparison of the RR models revealed that naturalized runoff based on SWAT and HBV produced significantly different results in quantifying changes in HD characteristics (Table 1). Obviously, the differences in results were generally expected, but contrary to expectations, too large differences, e.g., up to 30 % and 50 % attribution of human impacts on surface HD characteristics for DD and ND events shown between the two models, respectively. This implies that the percentage change of HD characteristics in this approach depends on the RR models applied, their performance during calibration and validation (van Loon and van Lanen, 2013), and the ability of the model to simulate runoff during the dry season. However, the study confirms that a similar direction of impact for all HD characteristics is achieved, i.e. positive changes (worsening of drought) or negative (weakening of drought). The negative value of drought intensity (I) could be due to the buffering impact that human interventions (such as the release of water reservoirs) have during the DD. It is also worth noting that, human-modified conditions reduced the maximum duration of drought as well as the number of drought events in the 2010s. This is related to the fact that the Wiory reservoir, the largest reservoir in the study area, was put into operation a few years later. Reservoirs increases in baseflow index (BFI) in long-term, which is a positive effect of water storage (Marcinkowski and Grygoruk, 2017; Raczyński, 2020). However, the applied approach suffers from the limitation that observed and simulated runoff were compared, which contains different uncertainties, and that only two RR models are used. Therefore, reconstruction (simulation) of observed runoff during the human-modified period (Nasiri et al., 2022) and multi-model ensemble methods and/or integrated surface-subsurface hydrological models are recommended for the future to overcome the respective limitations.

5 Conclusions

This study has evidently shown that modelling human impacts on HD with two RR models – the SWAT and the HBV model – of different complexity led to significant differences, although both models indicate similar impacts, i.e. positive or negative. We, therefore, conclude that modelling human impacts on HD based on comparing observations and simulations by applying multi-models need be considered in the future to draw reliable conclusions from the results. Future research on human influence modelling could therefore extend this approach by considering an ensemble of multiple models. In addition, further research is needed to confirm this new finding by applying the method to different case studies.

Code availability. The SWAT and HBV models were utilized in this research. The source code for both models is open-source and can be downloaded from the following websites: https://swat.tamu.edu/software/arcswat/ (TAMU, 2021) and https://www.geo.uzh.ch/en/units/h2k/Services/HBV-Model/HBV-Download.html (UZH, 2021). The version of SWAT used in this research was ArcSWAT 2012.10_7.24 and the version of HBV was the HBV-light.

Data availability. The hydroclimatic variables used in this study were obtained from the Institute of Meteorology and Water Management (IMGW-PIB). The topographic and soil data used in this study were prepared by Piniewski et al. (2017), and the land use data was obtained from Corine data (https://land.copernicus.eu/pan-european/corine-land-cover, Copernicus, 2020).

Author contributions. Conceptualization: TBS, KK, EK and JJN; methodology: TBS, KK, EK and JJN; data curation: TBS and EK; writing – original draft preparation: TBS; writing – review and editing: TBS, KK, EK and JJN; supervision: KK and EK; funding acquisition: JJN. All authors have read and agreed to the published version of the manuscript.

Competing interests. The contact author has declared that none of the authors has any competing interests.

Disclaimer. Publisher's note: Copernicus Publications remains neutral with regard to jurisdictional claims in published maps and institutional affiliations.

Special issue statement. This article is part of the special issue "IAHS2022 – Hydrological sciences in the Anthropocene: Variability and change across space, time, extremes, and interfaces". It is a result of the XIth Scientific Assembly of the International Association of Hydrological Sciences (IAHS 2022), Montpellier, France, 29 May–3 June 2022.

Acknowledgements. This work is part of the HUMDROUGHT project conducted at the Institute of Geophysics of the Polish Academy of Sciences. The project website can be found at https://humdrought.igf.edu.pl/ (last access: 4 April 2023). We would like to thank the Institute of Meteorology and Water Management (IMGW-PIB) for providing hydroclimatic data. We also thank the two anonymous reviewers and the Guest Editor of *Proceedings of the International Association of Hydrological Sciences*, Christophe Cudennec, for all valuable comments and suggestions that helped us to improve the quality of the manuscript.

Financial support. This research has been supported by the National Science Center (grant no. 2018/30/Q/ST10/00654).

Review statement. This paper was edited by Christophe Cudennec and reviewed by two anonymous referees.

References

Arnold, J. G., Srinivasan, R., Muttiah, R. S., and Williams, J. R.: Large area hydrologic modeling and assessment part I: model development, Journal of the Americal Water Resources Association, 34, 73–89, 1998.

Bąk, B. and Kubiak-Wójcicka, K.: Impact of meteorological drought on hydrological drought in Toruń (central Poland) in the period of 1971–2015, Journal of Water and Land Development, 32, 3–12, https://doi.org/10.1515/jwld-2017-0001, 2017.

Di Baldassarre, G., Martinez, F., Kalantari, Z., and Viglione, A.: Drought and flood in the Anthropocene: feedback mechanisms in reservoir operation, Earth Syst. Dynam., 8, 225–233, https://doi.org/10.5194/esd-8-225-2017, 2017.

Bazrkar, M. H. and Chu, X.: New Standardized Base Flow Index for Identification of Hydrologic Drought in the Red River of the North Basin, Nat. Hazards Rev., 21, 05020011, https://doi.org/10.1061/(asce)nh.1527-6996.0000414, 2020.

Bergström, S. and Forsman, A.: Development of a conceptual deterministic rainfall-runoff model, Hydrol. Res., 4, 147–170, https://doi.org/10.2166/nh.1973.0012, 1973.

Blöschl, G., Bierkens, M. F. P., Chambel, A., Cudennec, C., Destouni, G., Fiori, A., Kirchner, J. W., McDonnell, J. J., Savenije, H. H. G., Sivapalan, M., Stumpp, C., Toth, E., Volpi, E., Carr, G., Lupton, C., Salinas, J., Széles, B., Viglione, A., Aksoy, H., Allen, S. T., Amin, A., Andréassian, V., Arheimer, B., Aryal, S. K., Baker, V., Bardsley, E., Barendrecht, M.H., Bartosova, A., Batelaan, O., Berghuijs, W. R., Beven, K., Blume, T., Bogaard, T., Borges de Amorim, P., Böttcher, M. E., Boulet, G., Breinl, K., Brilly, M., Brocca, L., Buytaert, W., Castellarin, A., Castelletti, A., Chen, X., Chen, Y., Chen, Y., Chifflard, P., Claps, P., Clark, M. P., Collins, A. L., Croke, B., Dathe, A., David, P. C., de Barros, F. P. J., de Rooij, G., Di Baldassarre, G., Driscoll, J. M., Duethmann, D., Dwivedi, R., Eris, E., Farmer, W. H., Feiccabrino, J., Ferguson, G., Ferrari, E., Ferraris, S., Fersch, B., Finger, D., Foglia, L., Fowler, K., Gartsman, B., Gascoin, S., Gaume, E., Gelfan, A., Geris, J., Gharari, S., Gleeson, T., Glendell, M., Gonzalez Bevacqua, A., González-Dugo, M. P., Grimaldi, S., Gupta, A. B., Guse, B., Han, D., Hannah, D., Harpold, A., Haun, S., Heal, K., Helfricht, K., Herrnegger, M., Hipsey, M., Hlaváčiková, H., Hohmann, C., Holko, L., Hopkinson, C., Hrachowitz, M., Illangasekare, T. H., Inam, A., Innocente, C., Istanbulluoglu, E., Jarihani, B., Kalantari, Z., Kalvans, A., Khanal, S., Khatami, S., Kiesel, J., Kirkby, M., Knoben, W., Kochanek, K., Kohnová, S., Kolechkina, A., Krause, S., Kreamer, D., Kreibich, H., Kunstmann, H., Lange, H., Liberato, M. L. R., Lindquist, E., Link, T., Liu, J., Loucks, D. P., Luce, C., Mahé, G., Makarieva, O., Malard, J., Mashtayeva, S., Maskey, S., Mas-Pla, J., Mavrova-Guirguinova, M., Mazzoleni, M., Mernild, S., Misstear, B. D., Montanari, A., Müller-Thomy, H., Nabizadeh, A., Nardi, F., Neale, C., Nesterova, N., Nurtaev, B., Odongo, V. O., Panda, S., Pande, S., Pang, Z., Papacharalampous, G., Perrin, C., Pfister, L., Pimentel, R., Polo, M. J., Post, D., Prieto Sierra, C., Ramos, M. H., Renner, M., Reynolds, J. E., Ridolfi, E., Rigon, R., Riva, M., Robertson, D. E., Rosso, R., Roy, T., Sá, J. H. M., Salvadori, G., Sandells, M., Schaefli, B., Schumann, A., Scolobig, A., Seibert, J., Servat, E., Shafiei, M., Sharma, A., Sidibe, M., Sidle, R. C., Skaugen, T., Smith, H., Spiessl, S. M., Stein, L., Steinsland, I., Strasser, U., Su, B., Szolgay, J., Tarboton, D., Tauro, F., Thirel, G., Tian, F., Tong, R., Tussupova, K., Tyralis, H., Uijlenhoet, R., van Beek, R., van der Ent, R. J., van der Ploeg, M., Van Loon, A. F., van Meerveld, I., van Nooijen, R., van Oel, P. R., Vidal, J. P., von Freyberg, J., Vorogushyn, S., Wachniew, P., Wade, A. J., Ward, P., Westerberg, I. K., White, C., Wood, E. F., Woods, R., Xu, Z., Yilmaz, K. K., and Zhang, Y.: Twenty-three unsolved problems in hydrology (UPH) – a community perspective, Hydrolog. Sci. J., 64, 1141–1158, https://doi.org/10.1080/02626667.2019.1620507, 2019.

Copernicus: CORINE Land Cover, Copernicus [data set], https://land.copernicus.eu/pan-european/corine-land-cover, last access: 4 November 2020.

Gudmundsson, L. and Seneviratne, S. I.: Anthropogenic climate change affects meteorological drought risk in Europe, Environ. Res. Lett., 11, 044005, https://doi.org/10.1088/1748-9326/11/4/044005, 2016.

Harrigan, S., Murphy, C., Hall, J., Wilby, R. L., and Sweeney, J.: Attribution of detected changes in streamflow using multiple working hypotheses, Hydrol. Earth Syst. Sci., 18, 1935–1952, https://doi.org/10.5194/hess-18-1935-2014, 2014.

Jiang, C., Xiong, L., Wang, D., Liu, P., Guo, S., and Xu, C. Y.: Separating the impacts of climate change and human activities on runoff using the Budyko-type equations with time-varying parameters, J. Hydrol. (Amst), 522, 326–338, https://doi.org/10.1016/j.jhydrol.2014.12.060, 2015.

Jiao, D., Wang, D., and Lv, H.: Effects of human activities on hydrological drought patterns in the Yangtze River Basin, China, Nat. Hazards, 104, 1111–1124, https://doi.org/10.1007/s11069-020-04206-2, 2020.

Kang, H. and Sridhar, V.: Combined statistical and spatially distributed hydrological model for evaluating future drought indices in Virginia, J. Hydrol. Reg. Stud., 12, 253–272, https://doi.org/10.1016/j.ejrh.2017.06.003, 2017.

Karamuz, E., Bogdanowicz, E., Senbeta, T. B., Napiórkowski, J. J., and Romanowicz, R. J.: Is It a Drought or Only a Fluctuation in Precipitation Patterns? – Drought Reconnaissance in Poland, Water (Basel), 13, 807, https://doi.org/10.3390/w13060807, 2021.

Kubiak-Wójcicka, K. and Bąk, B.: Monitoring of meteorological and hydrological droughts in the Vistula basin (Poland), Environ. Monit. Assess., 190, 87–100, https://doi.org/10.1007/s10661-018-7058-8, 2018.

Liu, B., Zhou, X., Li, W., Lu, C., and Shu, L.: Spatiotemporal Characteristics of Groundwater Drought in Jiangsu Province, China, Water (Basel), 8, 480, https://doi.org/10.3390/w8110480, 2016.

Liu, Y., Ren, L., Zhu, Y., Yang, X., Yuan, F., Jiang, S., and Ma, M.: Evolution of Hydrological Drought in Human Disturbed Areas: A Case Study in the Laohahe Catchment, Northern China, Adv. Meteorol., 2016, 5102568, https://doi.org/10.1155/2016/5102568, 2016.

Marcinkowski, P. and Grygoruk, M.: Long-term downstream effects of a dam on a lowland river flow regime: Case study of the upper narew, Water (Switzerland), 9, 783, https://doi.org/10.3390/w9100783, 2017.

Mckee, T. B., Doesken, N. J., and Kleist, J.: The relationship of drought frequency and duration to time scales, in: Proceedings of the Eighth Conference on Applied Climatology, 17–22 January 1992, Anaheim, California, American Meteorological Society, 179–184, https://www.droughtmanagement.info/literature/AMS_Relationship_Drought_Frequency_Duration_Time_Scales_1993.pdf (last access: 4 April 2023), 1993.

Montanari, A., Young, G., Savenije, H. H. G., Hughes, D., Wagener, T., Ren, L. L., Koutsoyiannis, D., Cudennec, C., Toth, E., Grimaldi, S., Blöschl, G., Sivapalan, M., Beven, K., Gupta, H., Hipsey, M., Schaefli, B., Arheimer, B., Boegh, E., Schymanski, S. J., Di Baldassarre, G., Yu, B., Hubert, P., Huang, Y., Schumann, A., Post, D. A., Srinivasan, V., Harman, C., Thompson, S., Rogger, M., Viglione, A., McMillan, H., Characklis, G., Pang, Z., and Belyaev, V.: "Panta Rhei-Everything Flows": Change in hydrology and society-The IAHS Scientific Decade 2013–2022, Hydrolog. Sci. J., 58, 1256–1275, https://doi.org/10.1080/02626667.2013.809088, 2013.

Nalbantis, I. and Tsakiris, G.: Assessment of hydrological drought revisited, Water Resour. Manag., 23, 881–897, https://doi.org/10.1007/s11269-008-9305-1, 2009.

Nasiri, N., Asghari, K., and Besalatpour, A. A.: Quantitative analysis of the human intervention impacts on hydrological drought in the Zayande-Rud River Basin, Iran, J. Water Clim. Change, 13, 3473–3495, https://doi.org/10.2166/wcc.2022.188, 2022.

Piniewski, M., Szcześniak, M., Kardel, I., Berezowski, T., Okruszko, T., Srinivasan, R., Vikhamar Schuler, D., and Kundzewicz, Z. W.: Hydrological modelling of the Vistula and Odra river basins using SWAT, Hydrolog. Sci. J., 62, 1266–1289, https://doi.org/10.1080/02626667.2017.1321842, 2017.

Qiu, L., Peng, D., Xu, Z., and Liu, W.: Identification of the impacts of climate changes and human activities on runoff in the upper and middle reaches of the Heihe River basin, China, J. Water Clim. Change, 7, 251–262, https://doi.org/10.2166/wcc.2015.115, 2016.

Raczyński, K.: Influence of a Multipurpose Retention Reservoir on Extreme River Flows, a Case Study of the Nielisz Reservoir on the Wieprz River (Eastern Poland), Water Resour., 47, 29–40, 2020.

Rangecroft, S., Van Loon, A. F., Maureira, H., Verbist, K., and Hannah, D. M.: An observation-based method to quantify the human influence on hydrological drought: upstream–downstream comparison, Hydrolog. Sci. J., 64, 276–287, https://doi.org/10.1080/02626667.2019.1581365, 2019.

Seibert, J. and Vis, M. J. P.: Teaching hydrological modeling with a user-friendly catchment-runoff-model software package, Hydrol. Earth Syst. Sci., 16, 3315–3325, https://doi.org/10.5194/hess-16-3315-2012, 2012.

Senbeta, T. B. and Romanowicz, R. J.: The role of climate change and human interventions in affecting watershed runoff responses, Hydrol. Process., 35, e14448, https://doi.org/10.1002/hyp.14448, 2021.

TAMU (Texas A&M University): ArcSWAT: ArcGIS interface for SWAT 2012, TAMU [code], https://swat.tamu.edu/software/arcswat/, last access: 23 September 2021.

UZH (University of Zurich): HBV-light, UZH [code], https://www.geo.uzh.ch/en/units/h2k/Services/HBV-Model/HBV-Download.html, last access: 24 October 2021.

van Loon, A. F. and van Lanen, H. A. J.: Making the distinction between water scarcity and drought using an observation-modeling framework, Water Resour. Res., 49, 1483–1502, https://doi.org/10.1002/wrcr.20147, 2013.

Van Loon, A. F., Stahl, K., Di Baldassarre, G., Clark, J., Rangecroft, S., Wanders, N., Gleeson, T., Van Dijk, A. I. J. M., Tallaksen, L. M., Hannaford, J., Uijlenhoet, R., Teuling, A. J., Hannah, D. M., Sheffield, J., Svoboda, M., Verbeiren, B., Wagener, T., and Van Lanen, H. A. J.: Drought in a human-modified world: reframing drought definitions, understanding, and analysis approaches, Hydrol. Earth Syst. Sci., 20, 3631–3650, https://doi.org/10.5194/hess-20-3631-2016, 2016a.

van Loon, A. F., Gleeson, T., Clark, J., van Dijk, A. I. J. M., Stahl, K., Hannaford, J., Di Baldassarre, G., Teuling, A. J., Tallaksen, L. M., Uijlenhoet, R., Hannah, D. M., Sheffield, J., Svoboda, M., Verbeiren, B., Wagener, T., Rangecroft, S., Wanders, N., and Van Lanen, H. A. J.: Drought in the Anthropocene, Nat. Geosci., 9, 89–91, https://doi.org/10.1038/ngeo2646, 2016b.

Wanders, N. and Wada, Y.: Human and climate impacts on the 21st century hydrological drought, J. Hydrol. (Amst), 526, 208–220, https://doi.org/10.1016/j.jhydrol.2014.10.047, 2015.

Wittenberg, H.: Baseflow recession and recharge as nonlinear storage processes, Hydrol. Process., 13, 715–726, https://doi.org/10.1002/(SICI)1099-1085(19990415)13:5<715::AID-HYP775>3.0.CO;2-N, 1999.

Wu, J., Chen, X., Yu, Z., Yao, H., Li, W., and Zhang, D.: Assessing the impact of human regulations on hydrological drought development and recovery based on a "simulated-observed" comparison of the SWAT model, J. Hydrol. (Amst), 577, 123990, https://doi.org/10.1016/j.jhydrol.2019.123990, 2019.

Zou, L., Xia, J., and She, D.: Analysis of Impacts of Climate Change and Human Activities on Hydrological Drought: a Case Study in the Wei River Basin, China, Water Resour. Manag., 32, 1421–1438, https://doi.org/10.1007/s11269-017-1877-1, 2017.

Proc. IAHS, 385, 163–173, 2024
https://doi.org/10.5194/piahs-385-163-2024

Open Access
PIAHS

IAHS2022 – Variability and change across space, time, extremes, and interfaces

Developing Non-Stationary Frequency Relationships for Greater Pamba River basin, Kerala India incorporating dominant climatic precursors

Arathy Nair, Adarsh Sankaran, Meera Geetha Mohan, and Sreedevi Vijayalakshmi

TKM College of Engineering, Kollam, Kerala, PIN-691005, APJ Abdul Kalam Technological University, Kerala, India

Correspondence: Adarsh Sankaran (adarsh_lce@yahoo.co.in)

Received: 15 June 2022 – Revised: 19 February 2023 – Accepted: 29 March 2023 – Published: 18 April 2024

Abstract. Global climate changes significantly contribute to increased frequency of hydrologic extremes. This significantly underestimates the hydrologic design parameters, bringing of hydro systems to increased failure risk. In order to address this concern, the current practice of development of hydrologic frequency tools need to be updated accounting for non-stationarity. This study first considered a diverse set of statistical tests to examine the trend, change points, non-stationarity and randomness of streamflow, rainfall and temperature time series of scales ranging from daily to annual. The annual maxima time series indicated non stationarity against the stationary behaviour of daily series of hydro-meteorological datasets of the basin. Subsequently, this study developed the Temperature Duration Frequency (TDF), Rainfall Intensity Duration Frequency (IDF) and Flood Frequency (FF) curves of Greater Pamba river basin in Kerala India, the part of which was most severely affected by the near century return period flood event of 2018. The analysis was performed for a multitude of combinations of variations in distribution parameters with time and climatic drivers as physical covariates in the extreme value formulations. The study proposed a novel wavelet coherence (WC) based driver selection of most dominant combination of climatic precursors in developing FF and IDF relations of three locations of Kalloopara, Malakkara and Thumpamon and TDF curve of Kuttanad region in the basin, considering data of 1985–2015 period. The proposed WC framework considers bi-multi-and partial effects of climatic oscillations (COs) like El Niño Southern Oscillation (ENSO), Indian Ocean Dipole (IOD), Pacific Decadal Oscillation (PDO) and North Atlantic Oscillation (NAO) in identifying potential drivers. The different WC formulations captured in-phase relationships of streamflows and rainfall with COs at intra-annual, annual and inter annual scales up to 4 years. The methods showed that addition of climatic precursors improved the NS estimates of flood and rainfall quantiles by more accurately capturing the magnitudes of extreme streamflows and rainfalls of 2018, 2021 than the time covariate formulations. However, the role of COs on extreme temperature is not found to be influential in developing TDF relationships, which needs further investigation.

Keywords. Streamflow; non-stationarity; wavelet; frequency; climatic oscillation

1 Introduction

Unpredictable climate variations result in the occurrence of hydrological extremes like floods, droughts etc prompting the exploration of spatiotemporal variability of future extreme streamflow (SF), rainfall (RF) and temperature (Temp) events. To evaluate the consequences on such extreme events due to the uncertainty of projected climate changes, the study on SF, RF and Temp frequency analysis (FA) is indispensable. The hydrologic design parameters are gravely understated by these extremes, increasing the danger of failure for the hydro systems. Taking into account non-stationarity in the distribution parameters will allow the existing method of developing hydrologic frequency tools to be modified in order to meet this issue. The introduction of covariates such

Published by Copernicus Publications on behalf of the International Association of Hydrological Sciences.

as climatic oscillations (COs) and time into the parameters of distribution, so as to deal with the non-stationarity in the hydro-meteorological variable data set is considered to be a widely used technique (Katz et al., 2002; Khaliq et al., 2006; Adlouni et al., 2007). Applications for a wide variety of hydro-climatic extremes, including RF (Ouarda et al., 2019), floods (Villarini et al., 2009), and air temperatures (Ouarda and Charron, 2018), has already been established using this approach.

The wavelet analysis methods are gaining much acceptance while dealing with the analysis of highly irregular, complex, and intermittent non-stationary (NS) time series. For capturing the dominating periodicity and multiscale relationship between two time series, the continuous wavelet transform and bivariate wavelet coherence (BWC) analysis are frequently utilised (Torrence and Compo, 1998; Grinsted et al., 2004). Hu and Si (2016) proposed the multiple wavelet coherence (MWC) approach as an extension of BWC. Mihanovic et al. (2009) introduced the concept of partial wavelet coherence (PWC), thereby evaluating the standalone dependence of two variables via a statistical way, after eliminating the effect of other likely governing variables. The concept of wavelet coherence is applied in this study for estimating the teleconnections between different COs and hydro-meteorological variables resulting in the identification of prevalent climatic precursors which can be employed as covariates in the distribution parameters for the development of NS models for non-stationary frequency analysis (NSFA).

India is a country with an extensive river network and SF is considered as the major source of water supply for irrigation and industrial purposes. It is crucial to understand the RF and SF dynamics along with Temp variations to quantify the influence of global climatic drivers at basin scale for improved decision-making. It is well understood that a diverse set of Climatic drivers are influencing the hydro-climatic process of India (Adarsh and Janga Reddy, 2016). Kerala meteorological subdivision of southern India is becoming climatically sensitive in the recent past and the signatures of the same are evident in the form of extreme floods (Anandalekshmi et al., 2019). This study aims to conduct the FA of SF, RF, Temp extremes of Greater Pamba river basin (GPRB), Kerala, India, which is one of the most affected regions in the extreme rare Kerala flood event of 2018, with covariate dependent distribution functions to characterize these extreme events in a NS framework. This study aims (i) to propose a wavelet-based framework for identification of dominant covariate in NS frequency formulations; (ii) to develop NS Intensity-Duration-Frequency (IDF) curves, Flood Frequency (FF) curves and Temperature-Duration-Frequency (TDF) for GPRB, incorporating the dominant climatic precursors obtained from proposed WC framework.

2 Study Area and Data used

The Greater Pamba river basin consists of rivers Pampa, Manimala and Achencovil rivers flowing through Kerala encompassing the districts of Pathanamthitta, Alappuzha and Kottayam. The three basins of Achenkovil (1484 km^2), Pamba (2235 km^2) and Manimala (847 km^2) together form the GPRB with total drainage area of 4566 km^2 and it is classified as a medium-sized basin. The distinct characteristics of three different basins comprised within the GPRB highlight the need for conducting NSFA for the same hydro-meteorological variable recorded in multiple stations. Three SF stations, seven RF stations and one Temp station are considered for the study. The daily SF data for SF1-Kalloopara, SF2-Malakkara and SF3-Thumpamon stations were obtained from India Water Resource Information System (IWRIS) (https://indiawris.gov.in/wris, last access: 3 February 2021). The daily RF and Temp data were gathered from India Meteorological Department (IMD) at a high-resolution of $0.25° \times 0.25°$ (Pai et al., 2014) and $1° \times 1°$ (Srivastava et al., 2009) gridded data set respectively. All the datasets were collected for a span of 31 years (1985–2015). In case of NSFA, the daily SF data was converted to annual maximum SF for the development of FF curves. While the daily RF data was used to extract annual maximum series for 10 durations (1, 3, 6, 12, 18, 24, 36, 48, 60, 72 h) to develop IDF curves and the daily maximum Temp data was extracted to annual maximum Temp over 6 durations (1, 2, 4, 6, 8, 10 d) for developing TDF curves. The locations of the hydro-meteorological stations and grid points are indicated in Fig. 1. Four large-scale COs namely, El Niño Southern Oscillation (ENSO), North Atlantic Oscillation (NAO), Pacific Decadal Oscillation (PDO) and Indian Ocean Dipole (IOD) are considered as potential drivers for teleconnection analysis (Krishnamurthy and Goswami, 2000; Krishnan and Sugi, 2003; Mokhov et al., 2012; Li and Chen, 2014; Azad and Rajeevan, 2016; Rathinasamy et al., 2019; Yeditha et al., 2022). The climatic indices are taken from https://psl.noaa.gov/gcos_wgsp/Timeseries/ (last access: 3 February 2021) for the period 1985 to 2015.

3 Methodology

3.1 Statistical analysis

A number of statistical tests are used to analyse the temporal variance in the SF, RF, and Temp time series. The study of descriptive statistical analysis includes looking at concepts like trend, stationarity, change point, and randomness. Trend in the timeseries is estimated using MK test (Mann, 1945; Kendall, 1975) and Sen's slope test (Sen, 1968). The trend analysis shows general trend in a time series, but the stationarity test shows how the mean and variance vary with time. The Augmented Dickey-Fuller (ADF) (Dickey and Fuller, 1979) and KPSS test (Kwiatkowski et al., 1992) are used

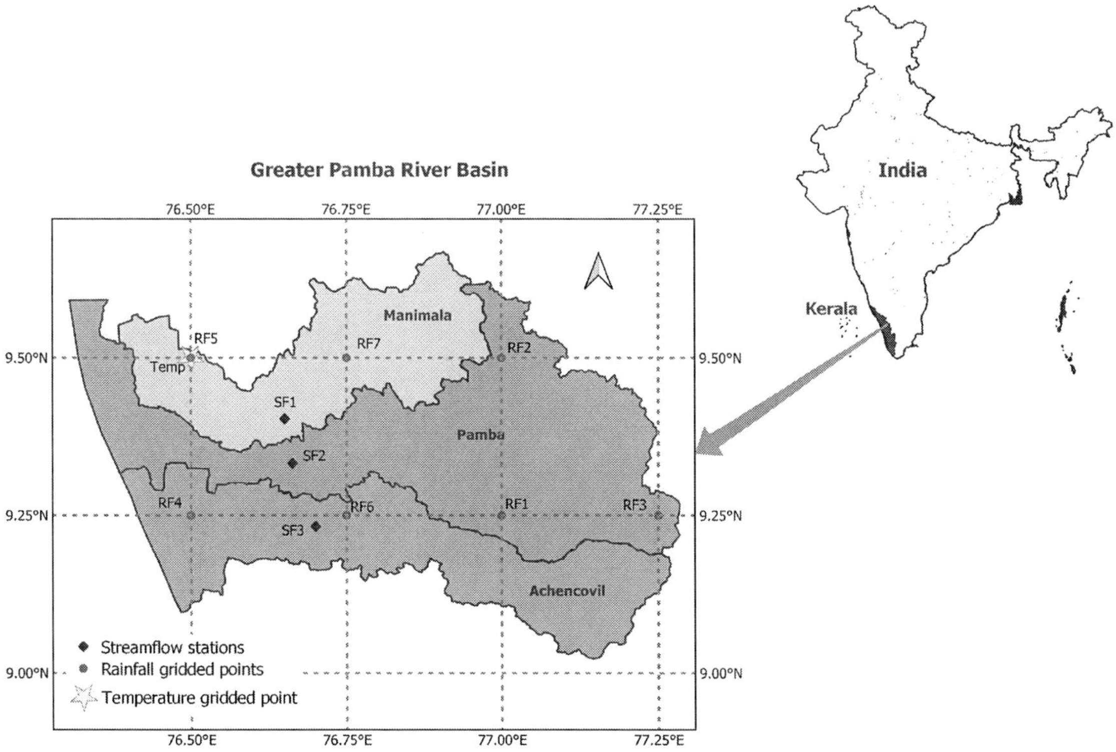

Figure 1. Location of hydro-meteorological stations and grid points in GPRB, Kerala, India.

for testing stationarity of different time series, at 5 % significance. The significant homogeneous and non-homogeneous nature of the times series was tested by Pettit's change point test (Pettit, 1979), Standard Normal Homogeneity test (Alexandersson and Moberg, 1997) and Buishand range test (Buishand et al., 2008). Randomness of the chosen variable time series of each station were tested using Runs test.

3.2 Wavelet Coherence analysis

A wavelet is defined as a small wave which has good ability in representing a signal locally in both time and frequency domains. The fundamental principle of wavelet transforms (WT) is to decompose the time series into its wavelets which are scaled and translated version of the mother wavelet (Torrence and Compo, 1998; Grinsted et al., 2004). The key benefit of using WT is that it reveals the hidden non-stationarity information in the original variable. The continuous wavelet transform (CWT) has been widely used for hydro-climatic teleconnection studies, while discrete wavelet transform (DWT) is popular for hydrological forecasting (Sang, 2013; Nourani et al., 2014). While CWT work on all the scales to analyse time series but the coherence will be significant only at certain time scale and their strength also differ with time scale. Considering the benefits of WT, methods like cross-wavelet analysis (XWT) and wavelet coherence (WC) have also become effective instruments for examining potential connections between two signals (Grinsted et al., 2004). XWT identifies only the regions with a high common power in the two-time series and WC measures how coherent XWT is in the time-frequency domain.

BWC quantifies the relationship between two variables, i.e., one hydro-meteorological variable and one CO. MWC is important to assess the simultaneous influence of more than one teleconnection at various timescales. Apart from this, PWC is accomplished to determine the standalone relationship between a CO and hydro-meteorological variable. For instance, two COs, ENSO and IOD, which are both highly connected, may have an impact on streamflow (SF2). In this case, WC analysis of the link between SF2 and IOD automatically takes ENSO's impact into account. This could lead to a misinterpretation of the true connection between SF2 and IOD. Hence solitary impact of each CO on hydro-meteorological variable must therefore be investigated using PWC.

In the case of teleconnection investigations, it is desirable to examine the relationship between two time series (hydro-meteorological variable and CO) in the time-frequency domain. However, the use of all scales results in redundant data, which significantly raises the cost of calculation and lowers the model's performance. In this study, application is for identifying the dominant driver using wavelet analy-

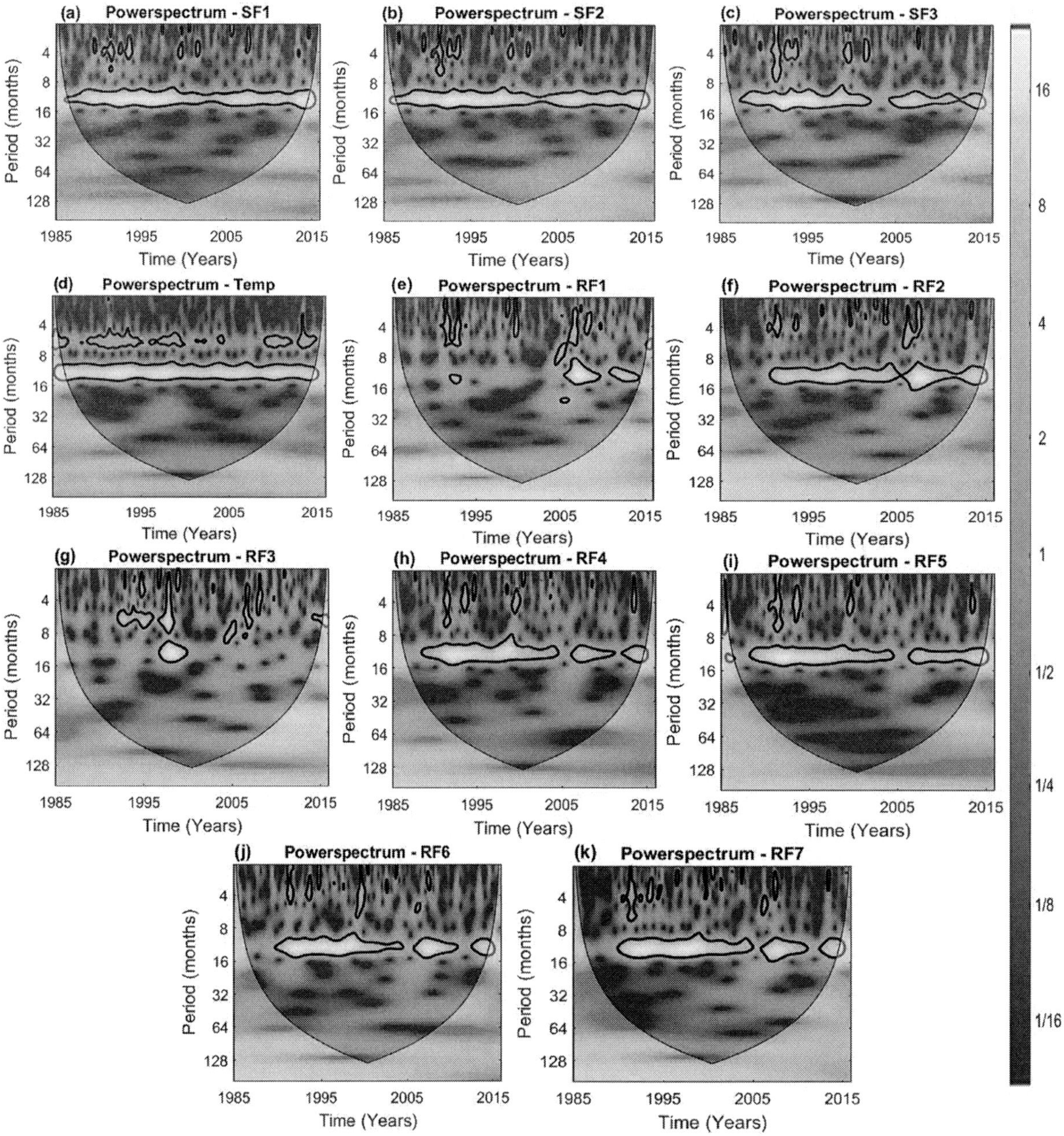

Figure 2. Time–frequency spectra of hydro-meteorological variables.

sis, for which, quantifying the strength is essential. Therefore, the relative dominance of single or multiple COs on the hydro-meteorological variable is measured using the Average Wavelet Coherence (AWC) and the Percentage of Significant Coherence (PoSC) (Song et al., 2020; Sreedevi et al., 2022). By averaging the WC generated across all dominant scales using the obtained coherence values, the AWC may be determined. The PoSC can be calculated by dividing the total number of power values created in the WC computation by the number of significant power values. Greater dominance is indicated by higher PoSC and AWC scores. Even though the coherence between COs and hydro-meteorological variables may grow with the addition of more COs, an increase in the PoSC value of at least 5 % must be seen in order to draw the conclusion that the new addition has any practical impact.

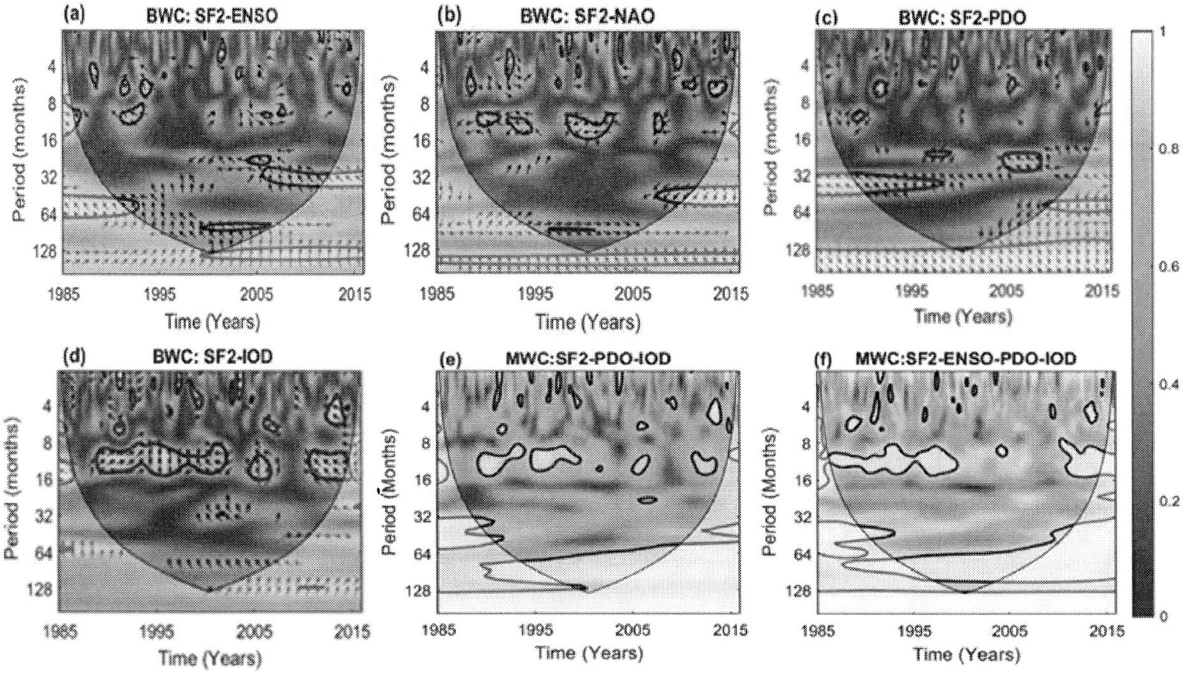

Figure 3. BWC and MWC spectrum of COs with SF2-Malakkara station.

3.3 Development of S and NS models

The Cumulative Distribution Function (CDF) of Generalized Extreme Value (GEV) distribution used for developing FF, IDF and TDF relationship is

$$F(x, \mu, \sigma, \xi) = e^{-Q(x)} \tag{1}$$

Where,

$$Q(x) = \begin{cases} \left[1 + \xi\left(\frac{x-\mu}{\sigma}\right)\right]^{-1/\xi}, \xi \neq 0 \\ \exp\left[-\frac{(x-\mu)}{\sigma}\right], \xi = 0 \end{cases}$$

Where $\mu =$ location parameter, $\sigma =$ scale parameter and $\xi =$ shape parameter.

In the NS case, the distribution parameters namely location and scale parameters are made dependent on covariates which can represent a CO or time. The shape parameter of the GEV is kept as a constant. The COs required for developing NS models were chosen based on the dominant COs obtained from the WC analysis. Table 1 indicates the best NS models used in this study. The method of Maximum Likelihood estimation (MLE) (Khaliq et al., 2006) is used for parameter estimation and the best fit model is identified with the help of Akaike Information Criterion (AIC) (Akaike, 1974). The model resulting in smallest AIC value is considered to be the best performing model.

The NS return levels for various return periods, RPs (T in years) are computed using Eq. (2). In the case of NS models, 95 percentiles of the covariate dependent parameters from the past observations are taken as a representative value for the location and scale parameters.

Table 1. Development of Stationary and Non-Stationary models.

Model name	Model Parameters
SGEV	μ, σ, ξ
NSGEV1	$\mu_t = \mu_0 + \mu_1$ Time, σ, ξ
NSGEV2	$\mu_t = \mu_0 + \mu_1$ ENSO, σ, ξ
NSGEV3	$\mu_t = \mu_0 + \mu_1$ NAO, σ, ξ
NSGEV4	$\mu_t = \mu_0 + \mu_1$ PDO, σ, ξ
NSGEV5	$\mu_t = \mu_0 + \mu_1$ IOD, σ, ξ
NSGEV6	$\mu_t = \mu_0 + \mu_1$ ENSO $+ \mu_2$ NAO, σ, ξ
NSGEV7	$\mu_t = \mu_0 + \mu_1$ ENSO $+ \mu_2$ PDO, σ, ξ
NSGEV8	$\mu_t = \mu_0 + \mu_1$ ENSO $+ \mu_2$ IOD, σ, ξ
NSGEV9	$\mu_t = \mu_0 + \mu_1$ NAO $+ \mu_2$ PDO, σ, ξ
NSGEV10	$\mu_t = \mu_0 + \mu_1$ NAO $+ \mu_2$ IOD, σ, ξ
NSGEV11	$\mu_t = \mu_0 + \mu_1$ PDO $+ \mu_2$ IOD, σ, ξ

$$I = \begin{cases} \hat{\mu} + \frac{\hat{\sigma}}{\hat{\xi}}\left[\left(-\log\left(1 - \frac{1}{T}\right)\right)^{-\hat{\xi}} - 1\right], \hat{\xi} \neq 0 \\ \hat{\mu} + \hat{\sigma}\left[-\log\left(-\log\left(1 - \frac{1}{T}\right)\right)\right], \hat{\xi} = 0 \end{cases} \tag{2}$$

4 Results and Discussions

4.1 Statistical analysis

The mean, standard deviation, skewness, kurtosis, minimum and maximum values of the chosen SF, RF and Temp stations belonging to GPRB was computed. All the datasets are positively skewed to the right with high skewness except for Temp data set which is slightly skewed. In terms of kurtosis, all the data sets have a leptokurtic distribution ($k > 3$; sharply peaked with heavy tails) while Temp data set have a platykurtic distribution ($k \sim 0$; flat peak and has more dispersed values with lighter tails). Sen's slope test was used to evaluate the size of the change while the MK test was used to determine the significance of the trend. All the RF grid points and SF1 station was found to have an increasing trend, while Temp and SF2, SF3 stations followed a decreasing trend. According to ADF test, SF1, RF4 and RF5 were found to be stationary while others were non-stationary whereas in the case of KPSS test, all the stations except for RF6 and RF7 were found to be stationary. The results of the change point analysis pointed to a considerable sudden change at each location. All the data sets were proved to be non-random in nature of occurrence with the help of Runs test. The overall results of the statistical analysis pave way to the presence of trend, non-stationarity, inhomogeneity and non-randomness in the SF, RF and Temp data sets.

4.2 Wavelet Coherence analysis

The WC analysis observed annual periodicity (8–16 months) for most of the time series for all stations, except for RF1 and RF3. Lower periodicities (2, 4 and 6 months) are observed mainly for Temp station, but there is localized presence in various time domain for the same periodicities for SF and RF stations as well. A pictorial representation of the same is given in Fig. 2. The results of BWC analysis of SF2 (Malakkara) station with the COs are presented in Fig. 3 along with two samples of MWC as well.

Due to space constraints, the BWC, PWC and MWC results obtained for one sample station; SF2 (Malakkara) station is only discussed in detail. BWC analysis for SF2-ENSO and SF2-NAO doesn't show any significant periodicity, however some localized contours were found around the periodicity of 2–48 months. For SF2-IOD across the time periods of 1990–2003, 2005–2007, and 2011–2014, the coherence correlations predominately range between periodicities of 8–16 months. In the case of SF2-PDO inter-decadal ranges were observed for the period 1988–1998 with a periodicity of 32–64 months. Higher AWC and PoSC values were obtained for SF2-IOD (AWC = 0.478 and PoSC = 25.03). The combined influence of PDO and IOD produced the highest coherence values for SF2 station for MWC-two factor analysis (AWC = 0.75, PoSC = 41.71). Among the three-factor scenarios, the ENSO, PDO and IOD combination produced the highest coherence score. On considering all the COs, the coherence value was at its maximum (AWC = 0.93, PoSC = 48.03).

Table 2. Comparison of dominant COs by WC analysis and best NS models.

Hydro-meteorological variables	Dominant COs by WC analysis	Best NS models
SF1	PDO	NSGEV4
SF2	IOD	NSGEV5
SF3	IOD	NSGEV5
Temp	ENSO	NSGEV8
RF1	PDO	NSGEV7
RF2	PDO	NSGEV4
RF3	IOD	NSGEV5
RF4	IOD	NSGEV5
RF5	IOD	NSGEV5
RF6	IOD	NSGEV5
RF7	IOD	NSGEV5

All of the PWC's AWC and PoSC values were lower than the BWC readings. The maximum reduction in PoSC value resulted in ENSO (from 23.36 %) after the removal of NAO (to 5.44 %) and PDO (to 7.32 %) from it. The removal of IOD had very less effect on ENSO. The influence of PDO on SF2 station is highly influenced by ENSO, NAO and IOD. PWC results of IOD and NAO showed less reduction in PoSC values which also indicates that IOD and NAO are not being strongly modulated by other COs. Similarly bi-multi-and partial effects of COs on all the other hydro-meteorological variables were evaluated. IOD, PDO, ENSO and its combinations were found to be more influential for all the hydro-meteorological variables in general.

4.3 Stationary and Non-Stationary models

The S and NS analysis was carried out for all the hydro-meteorological variables by developing GEV models as mentioned in Sect. 3.3 using all possible combinations of Time and CO as covariate and the best fit model was identified with the help of AIC values (Table 2). The FF, IDF and TDF curves developed for corresponding stations based on the best fit NS model is given in Figs. 4 and 5. It can be seen that all the best fit NS models were based on CO formulations. Additionally, it should be highlighted that the CO, which showed a strong teleconnection with the hydro-meteorological variable in relation to AWC and PoSC values, emerged as the best covariate that addresses the non-stationarities in terms of climate-induced NS model for FF, IDF and TDF analyses. For instance, the best NS model obtained for SF2-Malakkara station is location parameter linearly varying with IOD as covariate and it can be clearly seen that the most dominant CO for SF2 station is IOD as per AWC and PoSC values of WC analysis as discussed

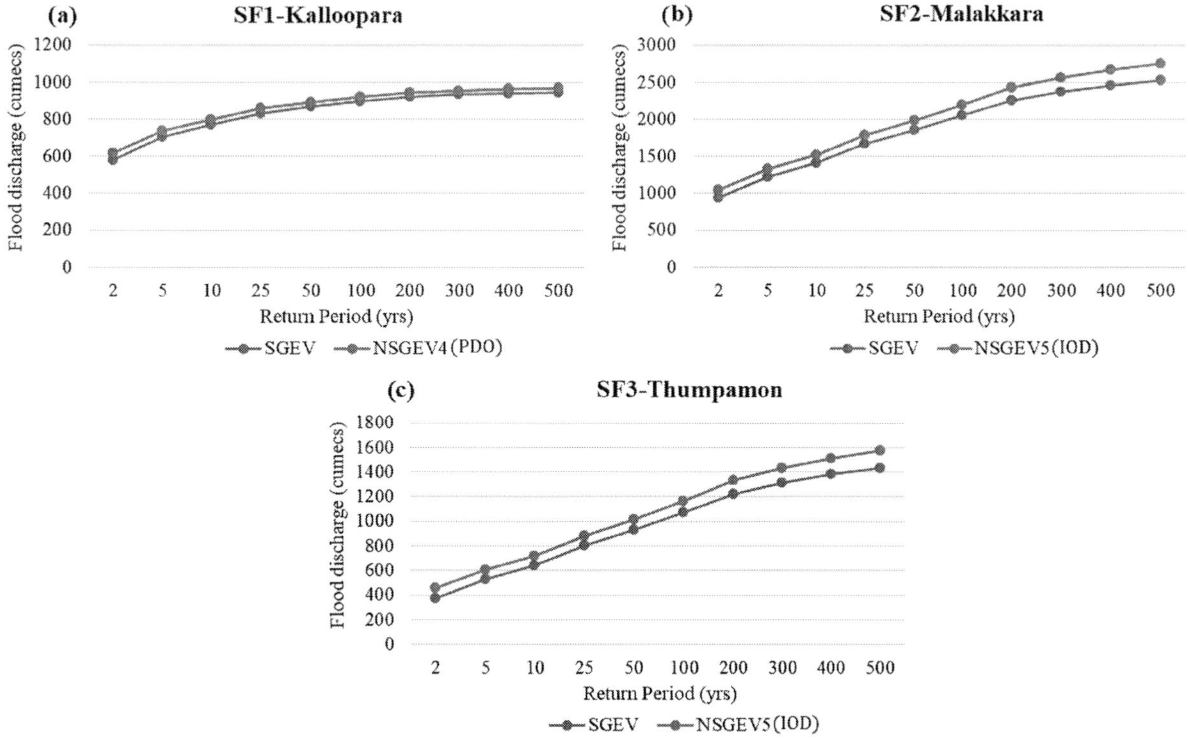

Figure 4. S and NS FF curves of SF stations.

in Sect. 4.2. The same was applicable for all the hydrometeorological stations. The findings, which are summarised in Table 2, support the conclusion that the dominating COs derived from the WC analysis turned out to be the best covariate to describe NS behaviour of the respective hydrometeorological stations. As a result of this study, it can be inferred that using a wavelet-based framework to identify dominant covariates will significantly help in determining the best covariate to describe each station's non-stationary behaviour. Also, teleconnection investigations over various frequency-time scales using BWC, MWC and PWC helped in the identification of key temporal features or events in a signal, such as discontinuities, rapid leaps and shifts, and patterns of dominant variability over the chosen time period.

The physical connections between the covariates and the local hydrology of GPRB was also examined to validate our findings. For instance, Maximum flood discharge of the basin (1988 cumecs) was recorded in the year 1994, a year with a predominately positive IOD event (Kumar et al., 2019). Also, the peak annual maximum temperature was observed in the year 1997, where the combined effect of ENSO and IOD was significantly predominant (Kumar et al., 2019). The results obtained using the best-fitting NS model, NSGEV8, for the Temp grid point, where the location parameter varies linearly with ENSO and IOD, validates the same. Hence, it can be concluded that the annual maximum value of the hydrometeorological variable peaked in the year where their respective best fit CO (Tables 1 and 2) effect was predominant. Additionally, the SF, RF and Temp return levels for various RPs were predicted using S and NS model with time and COs as covariates. A sample of 2018 and 2021 floods predicted for SF2 station using all the three models is shown in Fig. 6. According to the Central Water Commission's (CWC) "Report on Kerala Flood and Solutions" from 2018, the return period of the 2018 Kerala flood was unquestionably greater than 100 years, leading to the conclusion that the addition of climatic precursors improved the prediction of NS estimates of the SF and RF quantiles when compared to S and time covariate formulations (Fig. 6), but that this was not the case for the temperature return level, which requires further research.

This study proposed a novel framework in developing NS models based on the identification of best covariates using WC analysis, based on the assumption that the non-stationarity of the signal is well explained by the CO. It is worth noting that the model complexity increases while trying to incorporate non-stationarities related to climate change and variability. Hence, it is quite evident that model uncertainty will also increase as the complexity increases. Uncertainty analysis of NS models is an area which requires further exploration and future works are needed to access the uncertainty at covariate and parameter scale for developing reliable NS models.

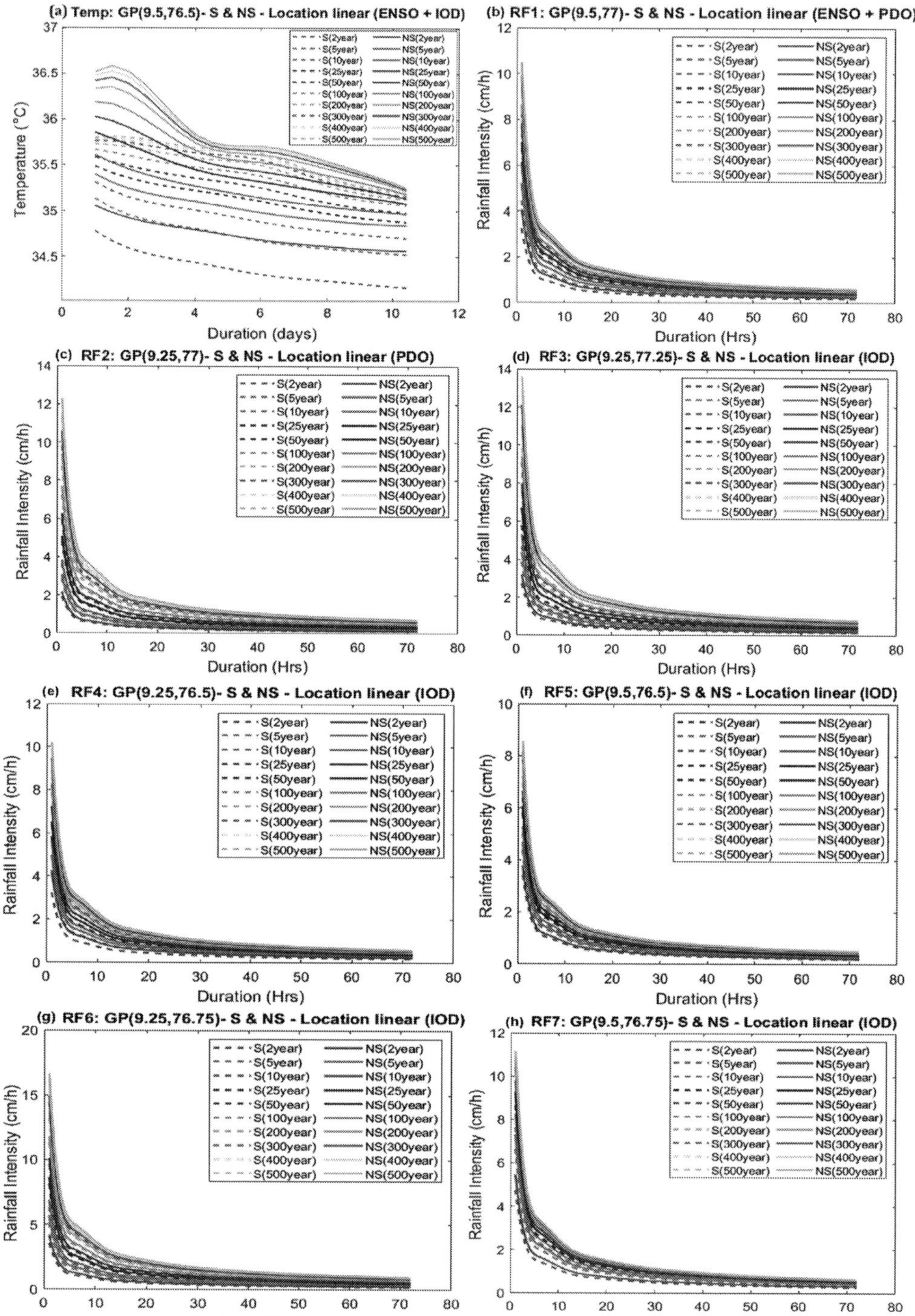

Figure 5. S and NS TDF and IDF curves of Temp and RF stations.

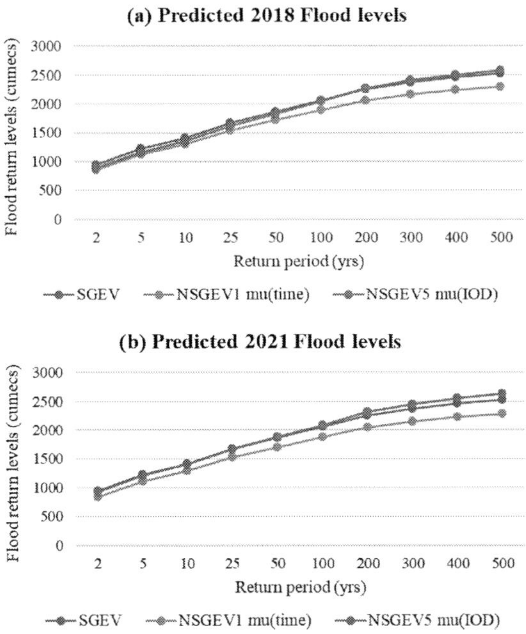

Figure 6. 2018 and 2021 Predicted flood levels of SF2 (Malakkara) station.

5 Conclusion

The hydro-meteorological variables namely, three SF, one Temp and seven RF stations belonging to GPRB were found to have significant trend, non-stationarity, inhomogeneity and non-randomness in the dataset. The main focus of the study is to develop FF, IDF and TDF relationships in a NS framework by incorporating COs and Time as covariates in the distribution parameters. Out of the four chosen COs, the most dominant climatic precursors were identified using WC analysis. A multitude of combinations of COs with the hydro-meteorological variables were analysed via CWT, BWC, MWC and PWC. According to AWC and PoSC values of wavelet coherence analysis, it is discovered that IOD has a major impact on the basin scale hydrology. The creation of the NS model for FA was greatly aided by the covariate selection made using the WC based driver selection of the most dominating combination of climatic precursors. It was found that NS models where location parameter is varying in correspondence with IDO, PDO, ENSO and its combinations turned out to be the best fit models for the respective stations. In-phase correlations of SF and RF with COs at intra-annual, annual, and inter-annual periods up to 4 years were captured by several WC formulations. The NS flood and RF return level estimates were better with the addition of climatic antecedents than with the time covariate formulations. The same was verified by precisely simulating the severe SF and RF magnitudes of 2018 and 2021 utilising CO as a covariate in NS models.

Code and data availability. All the data used for this work are freely available and the resource agency which provides the same are mentioned in Sect. 2 (Gridded rainfall data – Pai et al., 2014; Gridded temperature data – Srivastava et al., 2009; Streamflow data – https://indiawris.gov.in/wris, last access: 3 February 2021, India-WRIS, 2021; Climatic indices data – https://psl.noaa.gov/gcos_wgsp/Timeseries/, last access: 3 February 2021, GCOS, 2020). The software codes used in the study are developed by the authors. The current work is a part of the funded project sanctioned by Science Engineering and Research Board, Department of Science and Technology (DST-SERB) India. The project work is still under progress; hence the codes cannot be made available in public. However, under genuine and reasonable requests, the codes will be made available by the corresponding author.

Author contributions. AS, AN and MGM conceptualized the idea. AN, MGM and SV done the formal analysis and plotting, AN and MGM drafted the manuscript, AS supervised and revised the manuscript. All authors reviewed the manuscript.

Competing interests. The contact author has declared that none of the authors has any competing interests.

Disclaimer. Publisher's note: Copernicus Publications remains neutral with regard to jurisdictional claims in published maps and institutional affiliations.

Special issue statement. This article is part of the special issue "IAHS2022 – Hydrological sciences in the Anthropocene: Variability and change across space, time, extremes, and interfaces". It is a result of the XIth Scientific Assembly of the International Association of Hydrological Sciences (IAHS 2022), Montpellier, France, 29 May–3 June 2022.

Acknowledgements. The authors acknowledge DST-SERB, India for sanctioning the project entitled "Developing Non-Stationary Frequency Relationships of Hydro-climatic Extremes of Kerala Meteorological Subdivision under Changing Climate Scenario", (File Number: CRG/2021/003688) for which the present work will function as a component. We would like to thank India-WRIS for giving access to daily streamflow data, IMD for providing gridded daily maximum temperature and rainfall data and also GCOS for supplying climate indices data for the years 1985–2015. Authors thank IAHS for supporting the corresponding author with SYSTA to present the work in the 11th IAHS Scientific Assembly held at Montpellier France during 29 May–3 June 2022.

Review statement. This paper was edited by Christophe Cudennec and reviewed by two anonymous referees.

References

Adarsh, S. and Janga Reddy, M.: Analysing the hydroclimatic teleconnections of summer monsoon rainfall in Kerala, India using Multivariate Empirical Mode Decomposition and time dependent intrinsic Correlation, IEEE Geosci. Remote Sens. Lett. 13, 1221–1225, 2016.

Adlouni, S. E., Ouarda, T. B. M. J., Zhang, X., Roy, R., and Bobée, B.: Generalized maximum likelihood estimators for the nonstationary generalized extreme value model, Water Resour. Res., 43, W03410, https://doi.org/10.1029/2005WR004545, 2007.

Akaike, H.: A new look at the statistical model identification, IEEE T. Automatic Control, 19, 716–23, 1974.

Alexandersson, H. and Moberg, A.: Homogenization of Swedish Temperature data. Part I: homogeneity test for linear trends, Int. J. Climatol., 17, 25–34, 1997.

Anandalekshmi, A., Panicker, S. T., Adarsh, S., Siddik, M. A., Aloysius, S., and Mehjabin, M.: Modeling the concurrent impact of extreme rainfall and reservoir storage on Kerala Floods 2018: A Copula approach, Model. Earth Syst. Environ., 5, 1283–1296, 2019.

Azad, S. and Rajeevan, M.: Possible shift in the ENSO-Indian monsoon rainfall relationship under future global warming, Sci. Rep.-UK, 6, 20145, https://doi.org/10.1038/srep20145, 2016.

Buishand, T. A., de Haan, L., and Zhou, C.: On spatial extremes: with application to a rainfall problem, Ann. Appl. Stat., 2, 624–642, 2008.

Dickey, D. A. and Fuller, W. A.: Distribution of the estimators for autoregressive time series with a unit root, J. Am. Stat. Assoc., 74, 427–431, 1979.

GCOS: Download Climate Timeseries, GCOS [data set], https://psl.noaa.gov/gcos_wgsp/Timeseries/ (last access: 3 February 2021), 2020.

Grinsted, A., Moore, J. C., and Jevrejeva, S.: Application of the cross wavelet transform and wavelet coherence to geophysical time series, Nonlin. Processes Geophys., 11, 561–566, https://doi.org/10.5194/npg-11-561-2004, 2004.

Hu, W. and Si, B. C.: Technical note: Multiple wavelet coherence for untangling scale-specific and localized multivariate relationships in geosciences, Hydrol. Earth Syst. Sci., 20, 3183–3191, https://doi.org/10.5194/hess-20-3183-2016, 2016.

India-WRIS: India Water Resources Information System, India-WRIS [data set], https://indiawris.gov.in/wris, last access: 3 February 2021.

Katz, R. W., Parlang, M. B., and Naveau, P.: Statistics of extremes in hydrology, Adv. Water Resour., 25, 1287–1304, 2002.

Kendall, M. G.: Rank Correlation Methods, 4th edn., Charles Griffin, London, 1975.

Khaliq, M. N., Ouarda, T. B. M. J., Ondo, J. C., Gachon, P., and Bobée, B.: Frequency analysis of a sequence of dependent and/or non-stationary hydro-meteorological observations: a review, J. Hydrol., 329, 534–552, 2006.

Krishnamurthy, V. and Goswami, B. N.: Indian Mon soon–ENSO Relationship on Interdecadal Timescale, J. Climate, 13, 579–595, 2000.

Krishnan, R. and Sugi, M.: Pacific decadal oscillation and variability of the Indian summer monsoon rainfall, Clim. Dynam., 21, 233–242, 2003.

Kumar, P., Kaur, S., Weller, E., and Min, S. K.: Influence of natural climate variability on the extreme ocean surface wave heights over the Indian Ocean, J. Geophys. Res.-Oceans, 124, 6176–6199, https://doi.org/10.1029/2019JC015391, 2019.

Kwiatkowski, D., Phillips, P. C. B., Schmidt, P., and Shin, Y.: Testing the null hypothesis of stationarity against the alternative of a unit root how sure are we that economic time series have a unit root?, J. Econ., 54, 159–178, 1992.

Li, Q. and Chen, J.: Teleconnection between ENSO and climate in South China, Stoch. Environ. Res. Risk. Assess., 28, 927–941, 2014.

Mann, H. B.: Non-parametric tests against trend, Econometrica, 13, 163–171, 1945.

Mihanovic, H., Orlic, M., and Pasaric, Z.: Diurnal thermocline oscillations driven by tidal flow around an island in the Middle Adriatic, J. Marine Syst., 78, S157–S168, 2009.

Mokhov, II., Smirnov, D. A., Nakonechny, P. I., Kozlenko, S. S., and Kurths, J.: Relationship between El-Niño/Southern oscillation and the Indian monsoon, Izv. Atmos. Ocean Phys., 48, 47–56, https://doi.org/10.1134/S0001433812010082, 2012.

Nourani, V., Hosseini Baghanam, A., Adamowski, J., and Kisi, O.: Applications of hybrid wavelet – Artificial Intelligence models in hydrology: A review, J. Hydrol., 514, 358–377, 2014.

Ouarda, T. B. M. J. and Charron, C.: Nonstationary Temperature-duration-frequency curves, Sci. Rep., 8, 15493, https://doi.org/10.1038/s41598-018-33974-y., 2018.

Ouarda, T. B. M. J., Yousef, L. A., and Charron, C.: Non-stationary intensity–duration-frequency curves integrating information concerning teleconnections and climate change, Int. J. Climatol., 39, 2306–2323, 2019.

Pai, D. S., Latha, S., Rajeevan, M., Sreejith, O. P., Satbhai, N. S., and Mukhopadhyay, B.: Development of a new high spatial resolution (0.25° × 0.25°) Long period (1901–2010) daily gridded rainfall data set over India and its comparison with existing data sets over the region, MAUSAM, 65, 1–18, https://doi.org/10.54302/mausam.v65i1.851, 2014.

Pettitt, A. N.: A non-parametric approach to the change-point problem, Appl. Stat., 28, 126–135, 1979.

Rathinasamy, M., Agarwal, A., Sivakumar, B., Marwan, N., and Kurths, J.: Wavelet analysis of precipitation extremes over India and teleconnections to climate indices, Stoch. Environ. Res. Risk Assess., 33, 2053–2069, https://doi.org/10.1007/s00477-019-01738-3, 2019.

Sang, Y. F.: A review on the applications of wavelet transform in hydrology time series analysis, Atmos. Res., 122, 8–15, 2013.

Sen, P. K.: Estimates of the regression coefficient based on Kendall's tau, J. Am. Stat. Assoc., 63, 1379–1389, 1968.

Song, X., Zhang, C., Zhang, J., Zou, X., Mo, Y., and Tian, Y.: Potential linkages of precipitation extremes in Beijing-Tianjin-Hebei region, China, with large-scale climate patterns using wavelet-based approaches, Theor. Appl. Climatol., 141, 1251–1269, 2020.

Sreedevi, V., Adarsh, S., and Nourani, V.: Multiscale Coherence Analysis of Reference Evapotranspiration of North Western Iran Using Wavelet Transform, J. Wat Clim. Change, 13, 505–521, 2022.

Srivastava, A. K., Rajeevan, M., and Kshirsagar, S. R.: Development of High Resolution Daily Gridded Temperature Data Set

(1969–2005) for the Indian Region, Atmos. Sci. Lett., 10, 249–254, https://doi.org/10.1002/asl.232, 2009.

Torrence, G. and Compo, G. P.: A practical guide to wavelet analysis, B. Am. Meteorol. Soc., 79, 61–78, 1998.

Villarini, G., Smith, J. A., Serinaldi, F., Bales, J., Bates, P. D., Krajewski, W. F.: Flood frequency analysis for nonstationary annual peak records in an urban drainage basin, Adv. Water Resour., 32, 1255–1266, 2009.

Yeditha, P. K., Pant, T., Rathinasamy, M., and Agarwal, A.: Multiscale investigation on streamflow temporal variability and its connection to global climate indices for unregulated rivers in India, J. Wat. Clim. Change, 13, 735–757, 2022.

Proc. IAHS, 385, 175–180, 2024
https://doi.org/10.5194/piahs-385-175-2024

Open Access
PIAHS

Flood monitoring system in the Dakar agglomeration (Senegal)

Laurent Pascal Diémé[1], **Christophe Bouvier**[2], **Ansoumana Bodian**[1], **and Alpha Sidibé**[3]

[1]Laboratoire Leïdi "Dynamique des Territoires et Développement", Université Gaston Berger (UGB), Saint Louis, Sénégal
[2]IRD UMR 5151 CNRS-IRD-UM, HSM, Montpellier, France
[3]DPGI "Direction de la Prévention et de la Gestion des Inondations au Sénégal", MEA, Dakar, Sénégal

Correspondence: Laurent Pascal Diémé (dieme.laurent-pascal-malang@ugb.edu.sn)

Received: 26 May 2022 – Revised: 29 January 2023 – Accepted: 29 March 2023 – Published: 18 April 2024

Abstract. In recent decades, African cities have been confronted with series of floods linked to rapid urbanisation, intensification of heavy rains and the failure of the storm drainage system. Developing tools to characterise floods and reduce their impact is essential to facilitate decision support in a complex and vulnerable context. This study, conducted in the urban periphery of the Dakar metropolitan area, aims to propose a fine resolution (5 m) model of flows and overflows of rainwater drainage network in a pilot area. Two methodological steps are combined to achieve this objective: (1) the construction of the urban drainage topology to reconstitute the water drainage directions, taking into account the buildings, artificial channels and retention basins, using algorithms developed for this purpose, (2) simulations of the flows in real time or in project mode, using a parsimonious spatial model adapted to the local context, coupling a hydrological model (SCS-LR) on the scale of small basins with a hydraulic model (kinematic wave) for the propagation through the hydrological network. The former ensures the speed of the calculation, and the latter provides precise information on the behaviour of the network during a rainfall event. The overflow points of the network are detected by the difference between the maximum flow and the capacity of the network to evacuate floods. This modelling provides sufficiently informative simulations to guide the deployment of emergency services in the field, in real time, or to evaluate the efficiency of infrastructures in project mode, in a context of limited data. The model also provides boundary conditions for applying more complex hydraulic models to determine the impact of overtopping on limited areas.

Keywords. UPH 21; modelling; floods; drainage topology; Dakar; Athys

1 Introduction

Floods are recurrent in most urban areas in West Africa such as Nouakchott, Ouagadougou, Accra, Douala, Lagos, or Dakar, with impacts on populations, public health, economy and living environment (Satterthwaite, 2017). These cities have a high level of vulnerability and appear less resilient to floods due to their limited coping capacities (Bruckman et al., 2019). The trend towards increased intense precipitations, attributed to climate change (Panthou et al., 2018; Chagnaud et al., 2022), is expected to increase the recurrence of urban flooding. This brings with it challenges in terms of socio-economic development and flood risk management in these cities where urban sprawl is rapid and sometimes unregulated (Jaglin et al., 2018; Williams et al., 2019). It is expected that the impacts of future urban flooding can be limited not only by risk reduction measures but also by the implementation of forecasting systems that are essential to facilitate decision support in this complex and vulnerable context. One of the challenges is to understand and anticipate the risks of stormwater overflow (Renard and Riquier, 2008). Several approaches exist in the literature to identify the areas likely to be flooded by runoff. They range from methods strictly based on the analysis of topographic information to locate areas of water concentration (Pons et al., 2010), to methods introducing hydrological (Landemaine, 2016), 1D hydraulic (Pons et al., 2014) or 2D (Pons et al., 2021) modelling. However, these methods are relatively limited and do not always take into account the effects of urbanisation or the various urban infrastructures that modify the directions of surface runoff.

The present study, carried out in the urban periphery of the Dakar conurbation, aims firstly to construct a topology that integrates urban objects modifying the drainage directions and secondly to propose a calculation method that makes it possible to locate the overflows of the stormwater drainage network on the scale of the agglomeration, with calculation times compatible with real-time forecasting.

2 Urban context of Dakar

Dakar, the capital of Senegal, hosts 50 % of the country's urban population over only 0.3 % of the national territory (ANSD, 2013). Its urbanisation took place rapidly, in a few decades, largely sustained by the flows of rural exodus following the drought of the 1970s. The largest concentrations of population (Fig. 1) are found in the suburbs and the new extension areas of the city (Lessault and Imbert, 2013). Vulnerability is further exacerbated by the poor coverage of adequate rainwater drainage systems. Thus, since 2005, Dakar has been subject to floods that have strongly affected the population and their living environment, the infrastructures, the economy and the environment. There is a growing interest in characterising and reducing the impacts of floods in a context of intense rainfall and rapid urbanisation.

3 Data

The study area covers the communes of Pikine-Guédiawaye (Fig. 2), where the data were most detailed. These data come from the Dakar urban database set up at the DPGI (Direction de la Prévention et de la Gestion des Inondations au Sénégal) and the DTGC (Direction des Travaux Géographique et de la Cartographie). These are: (i) the Digital Terrain Model (DTM) of 10 m resolution from which the water flow paths are determined, (ii) the location of buildings, which constitute obstacles to the surface water flow, (iii) the buried and surface drainage channels and pipes which can modify or replace natural pathways, (iv) the rainwater retention basins.

4 Methodology

The methodology for processing the DTM and modifying it to integrate the drainage directions induced by both the location of buildings, drains and retention basins has been described in detail by Diémé et al. (2022). The general principles are recalled in the following.

4.1 DTM processing

The DTM is the basis for the method of constructing the drainage topology. From this DTM, the natural drainage direction in one of the 8 directions is determined for each grid cell over the test area (Jenson and Domingue, 1988). All the drainage directions are then forced to integrate successively the obstacles to water flow, the stormwater collectors and the retention basins.

4.1.1 Include water flow obstacles in the DTM

The principle for forcing flows through obstacles is to raise the elevations corresponding to these obstacles (here the urban blocks) by 20 m in the DTM. This will have the effect of preventing flows from crossing these obstacles. This classic operation can be carried out with GIS tools (ArcGIS, Qgis) or in the Vicair module of the ATHYS modelling platform (http://www.athys-soft.org/, last access: 22 May 2022).

4.2 Integration of the stormwater drainage

The drainage model forced by the building is used as a basis for integrating the modifications according to the location of the stormwater drainage. The algorithm developed for this purpose in the Vicair module considers the rasterised plan of the drains, identifies the two upstream and downstream ends of each section and then orients the drainage directions along the section from the higher end to the lower end. The elevations are read from either the DTM or an attribute table entered to take into account buried collectors. The final result is a drainage model forced by both the urban blocks and the pluvial drains.

4.3 The integration of retention basins

The retention basins are located in the previously defined drainage model using the rasterised plan of their location. The algorithm developed in Vicair determines the outlet of each basin by searching for a channel mesh whose direction of flow drains the basin; all the basin meshes are then oriented towards this outlet. The method thus allows the construction of a drainage model forced by urban blocks, drainage channels and retention basins (Fig. 3).

5 Flow modelling

The study area was discretised into elementary catchments with an urban area of approximately 10 ha, an algorithm in the Vicair module. The flow network is defined as the meshes that drain more than one hectare.

The hydrological model SCS-LR (Soil Conservation Service – Lag and Route) first calculates the flood hydrographs at the outlet of each catchment, taking into account the type

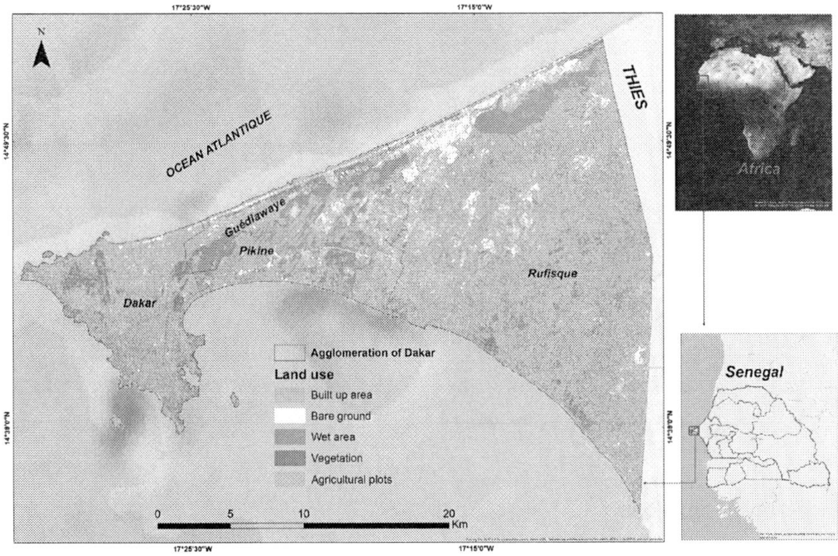

Figure 1. Location of the Dakar agglomeration.

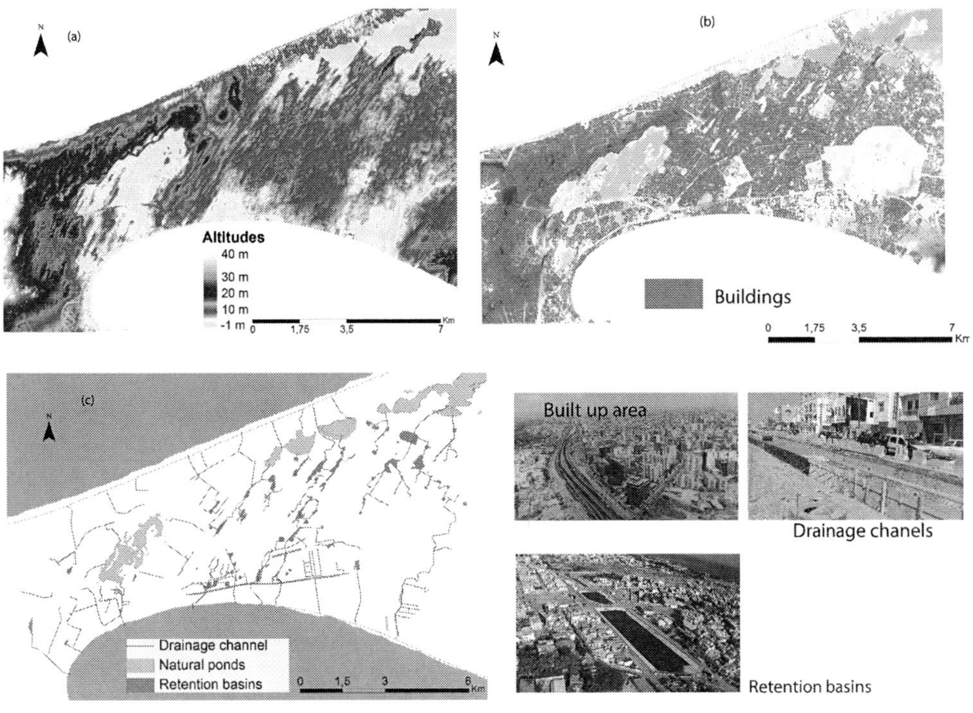

Figure 2. Geographical data modifying flow directions: **(a)** DTM, **(b)** buildings, **(c)** channels and retention basins.

of land use. These hydrographs are then injected and propagated in the network by the KW (Kinematic Wave) hydraulic model.

The parameters required for the SCS production model are S (maximum capacity of soil water) and Ia (losses before runoff). These parameters were calibrated based on rainfall-runoff observations made by Bassel (1996) in the Fann-Mermoz experimental basin. According to these data, the decennial runoff coefficient in the basin is equal to the building coefficient (0.2), which led to consider $S = 120$ mm and

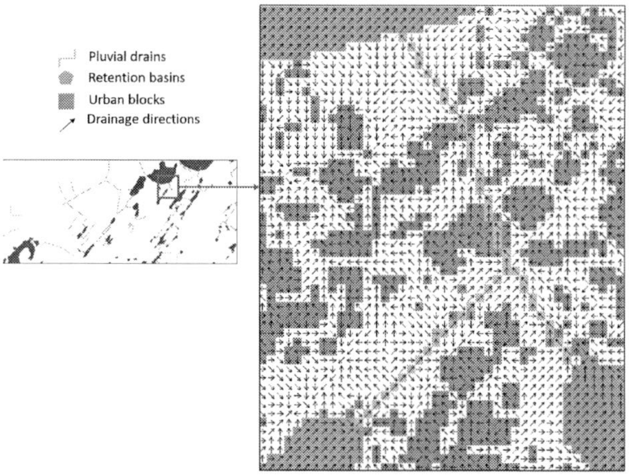

Figure 3. Forced drainage directions according to urban blocks (grey), drains (yellow) and retention basins (blue).

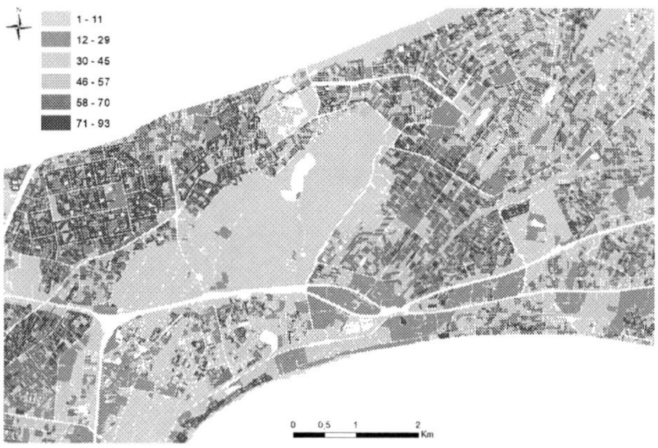

Figure 4. Density of urbanisation at the level of each urban block (%).

Ia = 0.2 mm. The parameterization and spatialization of the production on the study area was done using the GIS layers of the buildings to estimate the density of the urbanization at the scale of each urban block (Fig. 4) and to classify the production units according to the types of density: not urbanized, not very dense urbanization, medium dense and very dense. For very dense urbanisation class (75 % to 93 %) we find $S = 5$ mm, Ia = 0.2 mm which corresponds to a runoff coefficient of 0.92.

The LR transfer model applied for the elementary basins has two parameters Vo which is the transfer velocity (in $\mathrm{m\,s^{-1}}$) on each mesh and Ko the proportionality coefficient between translation and diffusion. Based on the study by Bassel (1996), the parameter Vo was set to $1\,\mathrm{m\,s^{-1}}$ and Ko was left at 0.7 by default.

The flow through the collectors is taken into account by the KW hydraulic model by considering the hydraulic properties of the collectors (roughness, width and depth) and the slope of the land. The slopes were smoothed by moving average over 50 meshes downstream of the mesh considered. The roughness Strickler coefficient was set at $50\,\mathrm{m^{1/3}\,s^{-1}}$ for all the collectors (297). The dimensions of the drains are available in the attribute table of the drains layer and 6 classes (Table 1) of drains were finally considered.

Table 1. Constitution of the width and depth classes of storm drains.

Classes	Strickler	Width	Depth
1	50	0.5	0.5
2	50	1	1
3	50	2	1
4	50	3	1
5	50	4	1
6	50	5	1

The functioning of the reservoirs is regulated by a volume-height-discharge rate law, specific to each reservoir. The absence of information on their storage capacity affects the results of the simulations and could therefore overestimate or underestimate their true operation.

The outputs of the simulations are the hydrographs at the outlet of the elementary basins and the water heights in the reservoirs (Fig. 5), and the mapping of the overflows (Fig. 6).

The overflow points are identified by the difference between the flows simulated by the model and the capacity of the network to evacuate the simulated flows (Fig. 6). This capacity was estimated by applying the Kinematic Wave model at full load. The example in Fig. 6 corresponds to a 100-year design rainfall applied as input to the model. The design rainfall model was built according Desbordes and Raous (1980), combining a total rainfall duration of 4 h and an intense rainfall phase of 1 h, for return periods of 2 to 100 years. The maximum intensities were deduced from the IDF curves produced for Senegal by Sane et al. (2018).

6 Discussion

The method provides maps of the severity of overflows from drainage structures for different rainfall intensities and taking into account local urban conditions. The model has the advantage of covering an entire agglomeration with a fine resolution (5 m), short calculation times (typically 5 min) compatible with real-time forecasting applications of possible overflow risks (monitoring of alert coasts). It is also applicable in project mode to assess the impact of future urban developments and to develop flood risk prevention policies. However, there are some limitations in this work. For the construction of the drainage topology, not all obstacle objects have been taken into account (gullies, walls, road interchanges). The use of a lidar, which is emerging in African cities, would allow to refine the construction of the drainage topology. The implementation of the hydrolog-

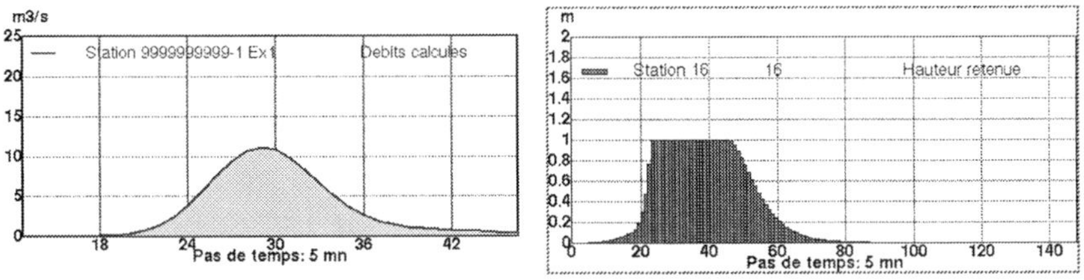

Figure 5. Elementary hydrographs and water levels in storage basins.

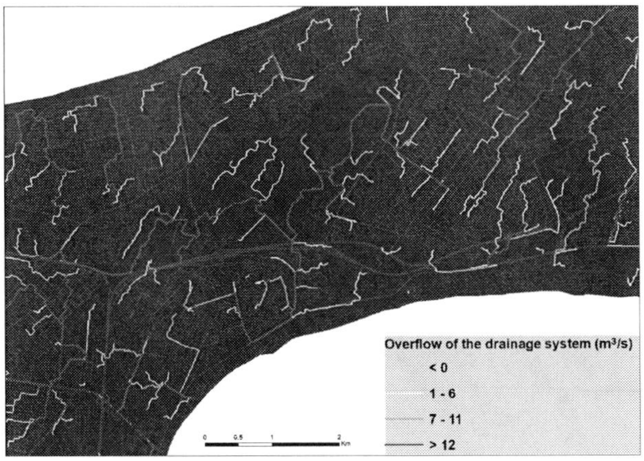

Figure 6. Identification of the overflow points of the drains according to different flow rates ($m^3\ s^{-1}$).

ical production-transfer (SCS-LR) and hydraulic propagation (KW) models applied to this topology can be improved through the equipment of experimental basins to better fix the values of the different parameters to be calibrated, insofar as there are uncertainties on the historical data exploited for this study. Also, the simulations were done without a perfect knowledge of the dimensions (width, depth, ...) of all storm drains. This raises the question of the constitution of a quality urban spatial database to help improve the results of the simulations. The validation of the model remains a perspective of this work through feedback (comparing the simulations with recent flood situations).

7 Conclusions

The method is based on 2 steps, consisting of reconstructing the drainage directions on natural and urban object modified terrain, and then applying hydrological and hydraulic modelling. The method can be applied either for real time forecast or for design mode). It covers the whole agglomeration at a fine spatial scale (5 m), with fast calculation. Its main purpose is to produce simulations of flows and network overflow points that are sufficiently informative in real time or in project mode, to guide the deployment of emergency services in the field, in real time or to diagnose the operation of the network in project mode, or to initiate actions at strategic locations. The representation of the overflow points is associated with a one-dimensional 1D model, and provides the boundary conditions for applying 2D hydraulic models in order to determine locally the impact of the overflow of the structures on limited areas.

The method seems flexible and adaptable to different contexts, and facilitated by the development of open-source global platforms and the possibility of accessing spatial data.

Code availability. The codes are implemented in the Athys modelling platform which is a free and open source software.

Data availability. The data used in this article is available from the first author, Laurent Pascal Diémé on the basis of a reasonable request.

Author contributions. LPD has conceived and wrote the paper. CB has implemented the codes on Athys. AS has provided the datas and AB has contributed to the data analysis.

Competing interests. The contact author has declared that none of the authors has any competing interests.

Disclaimer. Publisher's note: Copernicus Publications remains neutral with regard to jurisdictional claims in published maps and institutional affiliations.

Special issue statement. This article is part of the special issue "IAHS2022 – Hydrological sciences in the Anthropocene: Variability and change across space, time, extremes, and interfaces". It is a result of the XIth Scientific Assembly of the International Association of Hydrological Sciences (IAHS 2022), Montpellier, France, 29 May–3 June 2022.

Acknowledgements. The authors are thanking all the data providers.

Review statement. This paper was edited by Christophe Cudennec and reviewed by Abdoulaye Faty and one anonymous referee.

References

ANSD: Recensement Général de la Population et de l'Habitat, de l'Agriculture et de l'Élevage (RGPHAE), final report, 418 pp., 2013.

Bassel, M.: Eaux et environnement à Dakar – Pluies, ruissellement, pollution et évacuation des eaux, Contribution à l'étude des problèmes d'environnement liés aux eaux dans la région de Dakar, PhD Thesis, Department of Geography, University Cheikh Anta Diop of Dakar, 1996.

Bruckmann, L., Amanejieu, A., Moffo, M. O. Z., and Ozer, P.: Analyse géohistorique de l'évolution spatio-temporelle du risque d'inondation et de sa gestion dans la zone urbaine de Douala (Cameroun), Physio-Géo, 13, 91–113, https://doi.org/10.4000/physio-geo.8038, 2019.

Chagnaud, G., Panthou, G., Vischel, T., and Lebel, T. A.: synthetic view of rainfall intensification in the West African Sahel, Environ. Res. Lett., 17, 044005, https://doi.org/10.1088/1748-9326/ac4a9c, 2022.

Desbordes M. and Raous P.: Fondement de l'élaboration d'une pluie de projet urbaine: méthode d'analyse et application à la série de Montpellier – Bel-Air, La Météorologie, 6, 317–336, 1980.

Diémé, L. P., Bouvier, C., Bodian, A., and Sidibé, A.: Construction de la topologie de drainage à fine résolution spatiale en milieu urbain: exemple de l'agglomération de Dakar, LHB, 108, 2061313, https://doi.org/10.1080/27678490.2022.2061313, 2022.

Jaglin, S., Didier, S., and Dubresson, A.: Métropolisations en Afrique subsaharienne: au menu ou à la carte?, Métropoles, Hors-série, https://doi.org/10.4000/metropoles.6065, 2018.

Jenson, S. K. and Domingue, J. O.: Extracting topographic structure from digital elevation data for geographic information system analysis, Photogramm. Eng. Rem. S., 54, 1593–1600, 1988.

Landemaine, V.: Érosion des sols et transferts sédimentaires sur les bassins versants de l'Ouest du Bassin de Paris: Analyse, quantification et modélisation à l'échelle pluriannuelle, PhD thesis, University of Rouen-Normandie, 237 pp., 2016.

Lessault, D. and Imbert, C.: Mobilité résidentielle et dynamique récente du peuplement urbain à Dakar (Sénégal), Cybergeo: European Journal of Geography, Espace, Société Territoire, 662, 1–18, https://doi.org/10.4000/cybergeo.26146, 2013.

Panthou, G., Lebel, T., Vischel, T., Quantin, G., Sane, Y., Ba, A., and Diopkane, M.: Rainfall intensification in tropical semi-arid regions: The Sahelian case, Environ. Res. Lett., 13, 064013, https://doi.org/10.1088/1748-9326/aac334, 2018.

Pons, F., Delgado, J. L., Guero, P., and Berthier, E.: ExZeco: A GIS and DEM based method for pre-determination of flood risk related to direct runoff and flash floods, in: 9th International Conference on Hydroinformatics, Tianjin, China, Vol. 25, 2010.

Pons, F., Laroche, C., Fourmigue, P., and Alquier, M.: Cartographie des surfaces inondables extrêmes pour la directive inondation: Cas de la Nartuby, LHB, 2, 34–41, https://doi.org/10.1051/lhb/2014014, 2014.

Pons, F., Alquier, M., Paya, E., Moulin, C., Panier, N., and Chollet, A. E.: Premiers tests de la méthode Cartino2D sur le territoire de Toulon Provence Méditerranée, LHB, 107, 1–13, https://doi.org/10.1080/00186368.2021.1912968, 2021.

Renard, F. and Riquier, J.: Analyse territorialisée du risque de débordements de réseau d'assainissement liés aux eaux pluviales: application au Grand Lyon, Norois – Environnement, aménagement, société, 208, 45–60, https://doi.org/10.4000/norois.2181, 2008.

Sane, Y., Panthou, G., Bodian, A., Vischel, T., Lebel, T., Dacosta, H., Quantin, G., Wilcox, C., Ndiaye, O., Diongue-Niang, A., and Diop Kane, M.: Intensity–duration–frequency (IDF) rainfall curves in Senegal, Nat. Hazards Earth Syst. Sci., 18, 1849–1866, https://doi.org/10.5194/nhess-18-1849-2018, 2018.

Satterthwaite, D.: The impact of urban development on risk in sub-Saharan Africa's cities with a focus on small and intermediate urban centres, Int. J. Disast. Risk Red., 26, 16–23, https://doi.org/10.1016/j.ijdrr.2017.09.025, 2017.

Williams, D. S., Manez Costa, M., Sutherland, C., Celliers, L., and Scheffran, J.: Vulnerability of informal settlements in the context of rapid urbanization and climate change, Environ. Urban., 31, 157–176, https://doi.org/10.1177/0956247818819694, 2019.

Proc. IAHS, 385, 181–187, 2024
https://doi.org/10.5194/piahs-385-181-2024

Open Access PIAHS

Pros and cons of various efficiency criteria for hydrological model performance evaluation

Charles Onyutha

Department of Civil and Environmental Engineering, Kyambogo University,
P.O. Box 1, Kyambogo, Kampala, Uganda

Correspondence: Charles Onyutha (conyutha@kyu.ac.ug)

Received: 25 May 2022 – Revised: 28 January 2023 – Accepted: 30 March 2023 – Published: 18 April 2024

Abstract. Confidence in hydrological predictions is linked to the model's performance in reproducing available observations. However, judgment of a model's quality is challenged by the differences which exist among the available efficiency criteria or objective functions. In this study, model outputs based on several objective functions were compared and found to differ with respect to various circumstances of variability, number of outliers, and model bias. Computational difficulty or speed of a model during calibration was shown to depend on the choice of the efficiency criterion. One source of uncertainty in hydrological modelling is the selection of a particular calibration method. However, this study showed that the choice of an objective function is another sub-source of calibration-related uncertainty. Thus, tackling the issue of uncertainties on model results should comprise combination of modelled series obtained based on (i) various objective functions separately applied to calibrate a model, (ii) different calibration methods, and (iii) several hydrological models. The pros and cons of many new and old efficiency criteria which can be found explored in this study highlight the need for modellers to understand the impact of various calibration-related sub-sources of uncertainties on model outputs.

Keywords. UPH 20; Modelling; Efficiency criteria; Model evaluation; Model calibration uncertainty

1 Introduction

Model quality can be judged in terms of "goodness-of-fit" (GOF) or how well a model fits through observations or measured data points (Onyutha, 2022). Mathematical measures of model quality can be regarded as efficiency criteria (Beven, 2012). There are several efficiency criteria such as coefficient of determination (R^2) (also called R-squared), revised R-squared (RRS) (Onyutha, 2022), index of agreement (IOA) (Willmot, 1981), Nash Sutcliffe efficiency (NSE) (Nash and Sutcliffe, 1970), Kling Gupta efficiency (KGE) (Gupta et al., 2009), hydrological model skill score or Onyutha Efficiency (OE) (Onyutha, 2022), Liu mean efficiency (LME) (Liu, 2020), Taylor skill score (TSS) (Taylor, 2001), Root mean squared error (RMSE), and Mean absolute error (MAE). Differences exist among these various efficiency criteria and/or objective functions. Furthermore, each efficiency criterion has its advantages and disadvantages. Thus, outputs of a model calibrated using various objective functions can also differ.

Question number 20 of the unsolved problems in hydrology (UPH 20) (Blöschl et al., 2019) deals with the need to reduce model uncertainty. As highlighted in UPH20, uncertainty in model results can be due to model structure, parameter, or inputs. However, other sources of model uncertainty also exist as explored in this study.

It is worth noting that many old and new efficiency criteria can be found in literature. However, no any paper (which explored the pros and cons of the various old and new efficiency criteria) could be found in literature by the time of conducting this study. Therefore, this study was aimed at filling this knowledge gap to give modellers wide-ranging information which could influence decision regarding the choice of the best performing model in making hydrological prediction.

2 Materials and methods

2.1 Data and models

To evaluate efficiency criteria, several series were required. One way to obtain the required series for analysis was to make use of outputs from hydrological models. Here, conceptual models were preferred to physical models. This was because conceptual models (i) have few parameters, (ii) require few inputs, and (iii) are easy to calibrate compared with physical models. In this line, two lumped conceptual hydrological models including Hydrological Model focusing on Sub-flows' Variation (HMSV) (Onyutha, 2019) and Nedbør-Afstrømnings-Model (NAM) (Nielsen and Hansen, 1973) were applied to model river flow of the Jardine River catchment (with area 2500 km^2) in Australia. Each of the selected models require catchment-wide averaged rainfall and potential evapotranspiration as inputs.

2.2 Application of the selected hydrological models

Each selected model was calibrated using the Generalized Likelihood Uncertainty Estimation (Beven and Binley, 1992) framework. To do so, one of the metrics (R^2, RRS, IOA, NSE, KGE, OE, LME, TSS, RMSE, and MAE) was selected as the objective function for model optimization by running HMSV or NAM using 10 000 sets of parameters randomized within the stipulated parameter space. The optimal set of parameters was that which yielded the best value of the objective function or the most promising fit between the observed and modelled series. The procedure was repeated to ensure each of the efficiency criteria was used to generate a set of model outputs. Finally, model outputs based on the various efficiency criteria were compared.

To compute the various efficiency criteria, consider $\bar{X}$ and $\bar{Y}$ as the mean of the observed (X) and modelled (Y) series, respectively. Take r as the coefficient of correlation between X and Y. Let r_{a} denote the maximum attainable r value (and it was taken to be 0.9975 in this study). Other terms to note include sample size (n), standard deviation of X (s_x), variance of $X (s_x^2)$, standard deviation of $Y (s_y)$, variance of $Y (s_y^2)$, normalized s_y $(\hat{s}_y)$, distance covariance of X (d_{xx}), distance covariance of Y (d_{yy}), and distance covariance of X and Y (d_{xy}). We can compute the selected efficiency criteria using

$$R^2 = \sum\left(X-\bar{X}\right)\left(Y-\bar{Y}\right) / \sqrt{\sum\left(X-\bar{X}\right)^2 \sum\left(Y-\bar{Y}\right)^2} \quad (1)$$

$$\mathrm{RRS} = |r| \times \frac{\min(s_x, s_y)}{\max(s_x, s_y)} \times \frac{\min(s_x^2, s_{yc}^2)}{\max(s_x^2, s_{yc}^2)} \quad (2)$$

where $s_{yc}^2 = (n-1)^{-1} \times \sum\left(Y-\bar{X}\right)^2$

$$\mathrm{IOA} = 1 - \frac{\sum(X-Y)^2}{\sum\left(\left|\left(X-\bar{X}\right)\right| + \left|\left(Y-\bar{Y}\right)\right|\right)^2} \quad (3)$$

$$\mathrm{NSE} = 1 - \frac{\sum(X-Y)^2}{\sum\left(X-\bar{X}\right)^2} \quad (4)$$

$$\mathrm{KGE} = 1 - \sqrt{(r-1)^2 + \left(\frac{s_y}{s_x} - 1\right)^2 + \left(\frac{\bar{Y}}{\bar{X}} - 1\right)^2} \quad (5)$$

$$\mathrm{OE} = r_d \times \frac{\min(d_{xx}, d_{yy})}{\max(d_{xx}, d_{yy})} \times \frac{\min(s_x^2, s_{yc}^2)}{\max(s_x^2, s_{yc}^2)} \quad (6)$$

such that $r_d = d_{xy} \times \left(\sqrt{d_{xx} \times d_{yy}}\right)^{-1}$

$$\mathrm{LME} = 1 - \sqrt{\left(\frac{s_y}{s_x} \times r - 1\right)^2 + \left(\frac{\bar{Y}}{\bar{X}} - 1\right)^2} \quad (7)$$

$$\mathrm{TSS} = \frac{4(1+Q_d)}{\left(\hat{s}_x + \frac{1}{\hat{s}_y}\right)^2 (1+r_{\mathrm{a}})} \quad (8)$$

while $Q_d = \sum\left(X-\bar{X}\right)\left(Y-\bar{Y}\right) / \left(n \times s_x s_y\right)$

$$\mathrm{RMSE} = \sqrt{n^{-1} \times \sum(X-Y)^2} \quad (9)$$

$$\mathrm{MAE} = n^{-1} \times \sum(|X-Y|) \quad (10)$$

Note that in Eqs. (2) and (6), when $s_x = s_{yc} = 0$, RRS = OE = 0. Results of the models were analysed while exploring the advantages and disadvantages of each of the considered efficiency criteria.

2.3 Simulation experiments

To demonstrate the pros and cons of the various efficiency criteria or objective functions, several experiments were conducted using synthetic series. This was to investigate the influence of various factors on model outputs obtained using each objective function. These factors included variability, bias, and presence of outliers. Various efficiency criteria were compared with respect to the computation time under varying sample sizes. In the experiment involving outliers, one series (A) of sample size n was generated. Here, the standard deviation and mean of A were purposely made to be approximately 0.25 and 0.5, respectively. Another series (B) was then obtained as a copy of A. One data point of A was made as an outlier and the resultant series was termed A_1. An outlier was obtained by making a selected data point to be at least three times the 75th percentile in the series. The various efficiency criteria were applied to A_1 and B. The outliers' extent was computed as the ratio of the number of outliers to n in percentage. While keeping B unchanged, the procedure was repeated with the number of outliers in A increased to two, three, four, ..., G, to obtain A_2, A_3, A_4, ..., A_G, respectively. Here, G is an integer which makes the outliers' extent to be approximately 5 % of n.

For experiments involving biases, the starting point was series A. Using small increments in terms of $\Delta = 0.05$, several other series (B's) were obtained using $B_{i,j} = (\Delta \times j + A_j)$ while considering $(1 \leq j \leq F)$ and $(1 \leq i \leq n)$ where F is a whole number such that the percentage bias is equal to or approximately 10 %. Efficiency criteria were applied to A and each of the B's. Like for biases, the experiments for variability were based on series A as the starting point. Several other series (B's) were generated in terms of $\gamma = 0.001$ using $B_{i,j} = (\gamma \times j \times A_i)$ where $(1 \leq j \leq W)$ and $(1 \leq i \leq n)$. Here, W is a whole number which makes the standard deviation of B_j equal to 1. Relative coefficient of variation (RCV) was computed as the ratio of the standard deviation of B_j to the mean of A. Values of efficiency criteria applied to A and each of the B's were compared. When $\gamma \times j = 1$, it meant, $B_j = A$ and here, RCV ≈ 0.5 while all the efficiency criteria were expected to be at their values which indicate an ideal model performance.

3 Results and discussion

3.1 Comparison of various GOF metrics

Figure 1 shows result for comparison of the GOF values based on various criteria. The computation time (in seconds) for OE was slightly larger than those for other metrics (Fig. 1a). This is because OE comprises distance correlation, a term which has substantial run time of old algorithms. The fastest algorithm (known by the time of writing this paper) for computing distance correlation is the one provided by Chaudhuri and Hu (2019). Computation times for the other efficiency criteria are comparable. For series of n up to 1×10^6, it took less than one second to obtain a value of each efficiency criterion.

The values of R^2 remained constant regardless of the variation in the RCV (Fig. 1a). As the RCV increased from zero to 0.5, values of the other efficiency criteria (NSE, KGE, OE, RRS, TSS, IOA) increased. On the other hand, as the RCV increased from 0.5 to one, the values of NSE, KGE, OE, RRS, TSS, and IOA decreased. MAE and RMSE decreased linearly to zero as RCV increased from zero to 0.5. However, both MAE and RMSE increased linearly as RCV increased from 0.5 to one. The best value of each GOF metric was obtained when the RCV was approximately 0.5 (or when the two series being compared were identical).

Generally, increase in the number of outliers leads to poor performance of the model (Fig. 1c–d). Except for TSS and R^2, GOF metrics which occur over the range 0–1 reduced as the extent of outliers increased (Fig. 1c). However, an increase in the number of outliers makes the values of NSE, KGE, and LME less negative while MAE and RMSE increase in magnitudes (Fig. 1d). As bias increases, the magnitudes of all the selected GOF metrics (except for TSS and R^2) generally decrease (Fig. 1e) or increase (Fig. 1f) thereby showing reduction in model quality. TSS and R^2 do not quantify bias (Fig. 1e).

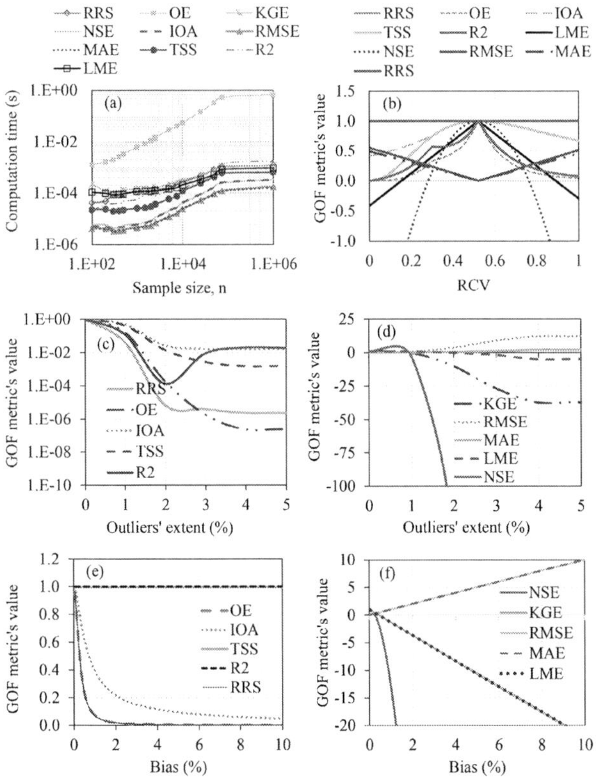

Figure 1. Comparing GOF metrics based on **(a)** computational load, **(b)** RCV, **(c–d)** outliers' extent, and **(e–f)** model bias.

3.2 Application of HMSV and NAM

Figure 2 shows impact of the choice of an efficiency criterion on the modelled output. Modelled series based on various efficiency criteria resonated with observations to various extents (Fig. 2a–b). Considering HMSV (Fig. 2a), observed flow was more over-estimated by the modelled results based on R^2 than those obtained using other objective functions. However, modelled outputs based on RMSE exhibited the largest under-estimation compared with results from other objective functions. While considering NAM (Fig. 2b), results based on calibration using almost all selected objective functions resulted into under-estimation of the maximum observed flow. The largest and smallest under-estimation was from results based on R^2 and TSS, respectively.

The water balance closure was noted to depend on the selected efficiency criterion. The dissimilarity in the extent of spread of the right tails of the cumulative flow generated by HMSV and that of NAM was due to the difference in the structures of the models (Fig. 2c–d). One key area in which

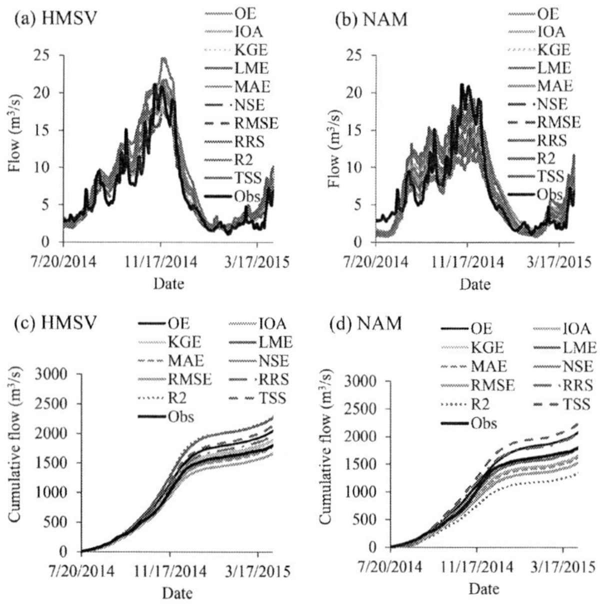

Figure 2. Observed (Obs) flow versus outputs of **(a, c)** HMSV and **(b, d)** NAM.

these hydrological models tended to differ was in reproducing observed peak flows.

3.3 Discussion

Known sources of uncertainty in hydrological models include model structure, input data errors, parameters, and calibration methods. However, this study, based on results from Figs. 1 and 2, showed that the choice of an objective function is another sub-source of calibration-related uncertainty. In other words, the choice of an efficiency criterion influences judgment of a model's quality (Onyutha, 2016). Expressly, the choice of an objective function influences the outputs of a model. Thus, it is important for a modeller to know their pros and cons of the various GOF metrics or objective functions.

3.3.1 Coefficient of determination (R^2)

The metric R^2 (Eq. 1) varies from 0 to 1. Advantages of R^2 are that: it is popular, can be computed fast, and R^2 does not have the interpretability limitations. Interpretation of R^2 is straightforward in terms of the percentage of the variability explained by the model. Furthermore, R^2 is suitable for variables which are linearly related when there is no bias. There are several disadvantages of R^2. It does not quantify bias (Onyutha, 2022) and can be sensitive to outliers in the sample. It provides invalid results when the data has measurement errors (Cheng et al., 2014). There are several common misunderstandings on the use of R^2; thus, the use and interpretation of R^2 is confusing (Alexander et al., 2015). The metric R^2 can be low and high for an accurate and imperfect model, respectively (Onyutha, 2022). The most commonly used R^2 formula or the version based on correlation is unsuitable for analysis of relationships which are not linear. It yields the same value when we regress X on Y and *vice versa* (Onyutha, 2022) thereby invalidating the terming of R^2 as coefficient of determination. Finally, R^2 lacks justification for its use as a descriptive statistic (Cameron, 1993).

3.3.2 Revised R-squared (RRS)

The metric RRS (Eq. 2) (Onyutha, 2022) varies from 0 to 1 and it has a number of advantages. Its computation is fast. We get different values in the two cases when regressing X on Y and *vice versa* (unlike R^2). Furthermore, RRS does not have interpretability limitations (Onyutha, 2022). The metric RRS indicates the amount of the total variance in observations explained by the model. A modeller can use RRS to evaluate model performance in terms of bias, correlation and variability (Onyutha, 2022). The formula for RRS does not comprise direct squaring of the error term, an aspect responsible for making other metrics (such as NSE) sensitive to large model errors. The main disadvantage of RRS is that it assumes linear relationship between X on Y.

3.3.3 Index of agreement (IOA)

The metric IOA (Eq. 3) (Willmot, 1981) ranges from 0 to 1 and the advantages of IOA are that it is widely applied and can be computed fast. The limits of the IOA do not have the interpretability limitations. Furthermore, IOA can detect additive differences in the means and variances of X and Y (Moriasi et al., 2015). The disadvantages of IOA are that it can yield high values even for poorly fit model (Krause et al., 2005; Onyutha, 2022). It is sensitive to extreme values in the sample. Furthermore, it lacks physical meaning since it does not have unit. In other words, the IOA values between zero and one are difficult to interpret.

3.3.4 Nash Sutcliffe efficiency (NSE)

The metric NSE (Eq. 4) (Nash and Sutcliffe, 1970) yields values which range from $-\infty$ to 1 and the advantages of NSE are that it is simple and popular. Its computation can be fast. Values of NSE in the two cases when we regress X on Y and *vice versa* are different (unlike R^2). The metric NSE has several disadvantages. It has interpretability limitation compared to other metrics which vary from zero to one (Onyutha, 2022). Furthermore, NSE lacks physical meaning since it does not have a unit. Values between the limits $-\infty$ and 1 are difficult to interpret. The metric NSE can be sensitive to outliers. The sampling uncertainty in NSE estimator is substantial (Clark et al., 2021). The reliance of NSE on mean of observations leads to exaggerated model efficiency when analysing highly seasonal river flow (Gupta et al., 2009). The metric NSE also does not indicate bias (Jackson et al., 2019) and it is not suitable for single-event simulations since it can be inadequate in quantifying differences in the time and magnitude of peak flows (Jackson et al., 2019).

3.3.5 Kling Gupta efficiency (KGE)

The metric KGE (Eq. 5) (Gupta et al., 2009) varies from $-\infty$ to 1 and it has a number of advantages. It is becoming popular. It considers correlation, variability, and bias measures. It takes little time to be computed. The value of KGE in the two cases when we regress X on Y and *vice versa* are different (unlike R^2). On the other hand, there are many disadvantages of KGE. It varies from $-\infty$ to 1 and this engenders interpretability limitations compared to other metrics which vary from zero to one (Onyutha, 2022). A model's improvement starts from KGE equal to -0.41 even if the KGE values are still negative (Knoben et al., 2019). It can be sensitive to outliers. The sampling uncertainty of KGE is substantial (Clark et al., 2021). The metric KGE assumes linear relationships among variables and it can be inadequate in evaluating model performance when this assumption is violated. Furthermore, KGE assumes that the data is normally distributed and has no outliers; thus, non-normal distribution and presence of outliers all affect the metric. While optimizing KGE, the means of the simulation or values which under-estimate $\bar{X}$, especially in the high flows will tend to be favourably selected (Liu, 2020). The metric KGE lacks physical meaning since it does not have unit. Values between the limits $-\infty$ to 1 are difficult to interpret (not as it is clear for the values of R^2).

3.3.6 Hydrological model skill score

The metric OE (Eq. 6) (Onyutha, 2022) varies from 0 to 1 and its advantages are that it does not assume linear relationship among variables. The values of OE in the two cases when we regress X on Y and *vice versa* are different (unlike R^2). The metric OE allows model performance evaluation in terms of bias, correlation and variability (Onyutha, 2022). It does not involve direct squaring of the error term, an aspect responsible for making other metrics such as NSE sensitive to large model residuals. Like other metrics, OE has a few disadvantages. It is slightly computationally slower than other metrics such as NSE. As mentioned before, the component of OE which slightly increases the computational time is the distance correlation.

3.3.7 Liu mean efficiency (LME)

The metric LME (Eq. 7) (Liu, 2020) varies from $-\infty$ to 1. Its advantages are that it can be computed fast. It considers correlation, variability, and bias measures. The metric LME has a number of disadvantages. It assumes linear relationship. Furthermore, it is characterized by underdetermined solutions mainly approaching the excessive flow variation (Lee and Choi, 2022). Thus, the maximum potential LME can be characterized by an infinite number solutions (Lee and Choi, 2022). The variation of LME from $-\infty$ to 1 brings about the interpretability limitations compared to other metrics which vary from 0 to 1. Like a few other metrics, LME lacks physical meaning since it does not have a unit. Values other than one are difficult to interpret. The sampling uncertainty of LME could be substantial just like that of NSE or KGE (following the methodology of Clark et al., 2021).

3.3.8 Taylor skill score (TSS)

The metric TSS (Eq. 8) (Taylor, 2001) varies from 0 to 1. The advantages of TSS are that it is popular and can be computed fast. Limits of TSS are straightforward to interpret. However, TSS has are a number of disadvantages. It does not quantify bias. It can yield high values even for poorly fit model (Onyutha, 2022). It can also be affected by the presence of outliers in the data. The formula for TSS has a term which requires case-specific calibration to determine the maximum possible correlation attainable. Furthermore, the metric TSS assumes linear relationship. TSS also lacks physical meaning since it does not have a unit. Values between zero and one are difficult to interpret.

3.3.9 Root mean squared error (RMSE)

The metric RMSE (Eq. 9) varies from 0 to $+\infty$ and it has several advantages. It has the same unit as the variable being modelled; thus, easy to interpret given the physical meaning. It is popular, simple and can be computed fast. The main disadvantages of RMSE are that its values remain the same in the two cases when we regress X and Y and *vice versa* (like that of R^2). The metric RMSE can be affected by the presence of outliers in the data.

3.3.10 Mean absolute error (MAE)

The metric MAE (Eq. 10) varies from 0 to $+\infty$. Its advantages are that it is simple and can be computed fast. It has the same unit as the variable being modelled; thus, easy to interpret given its physical meaning. There are a few disadvantages of MAE. The two cases of regressing X on Y and *vice versa* lead to the same value of the metric MAE. Furthermore, MAE can be dominated by even one outlier or large error.

4 Conclusions

The pros and cons of the various efficiency criteria or objective functions presented in this study highlight the need to understand the impact of each efficiency criterion on model performance assessment (or the effect of the choice of an objective function on model calibration results) before making hydrological predictions. Several sources of uncertainty in hydrological models are known including model structure, input data errors, parameters, and calibration methods. However, this study showed that the choice of an objective function for calibrating a model is also another sub-source of uncertainty related to model calibration. Therefore, a modeller should make use of several efficiency criteria to judge the quality of a model. Finally, the modeller should obtain an ensemble of modelled series using results based on various (i) objective functions used individually to calibrate a model (ii) calibration methods, and (iii) hydrological models.

Code availability. MATLAB codes to compute OE and RRS can found in a freely downloadable supplementary material from Onyutha (2022). To obtain HMSV, please contact the author (conyutha@kyu.ac.ug).

Data availability. For more information about the used data, please contact the author (conyutha@kyu.ac.ug).

Competing interests. The author has declared that there are no competing interests.

Disclaimer. Publisher's note: Copernicus Publications remains neutral with regard to jurisdictional claims in published maps and institutional affiliations.

Special issue statement. This article is part of the special issue "IAHS2022 – Hydrological sciences in the Anthropocene: Variability and change across space, time, extremes, and interfaces". It is a result of the XIth Scientific Assembly of the International Association of Hydrological Sciences (IAHS 2022), Montpellier, France, 29 May–3 June 2022.

Acknowledgements. The author acknowledges that modelling datasets from Onyutha (2022) were used in this study.

Financial support. This research has been supported by the International Association of Hydrological Sciences (IAHS) through an IAHS SYSTA award granted to the author to participate in the IAHS 2022 Scientific Assembly in Montpellier, France, where this paper was presented.

Review statement. This paper was edited by Christophe Cudennec and reviewed by two anonymous referees.

References

Alexander, D. L. J., Tropsha, A., and Winkler, D. A.: Beware of R2: simple, unambiguous assessment of the prediction accuracy of QSAR and QSPR models, J. Chem. Info. Mod., 55, 1316–1322, 2015.

Beven, J. K.: Rainfall-Runoff Modelling – The Primer, 2nd Edition, Wiley-Blackwell, 488 pp., ISBN 978-0-470-71459-1, 2012.

Beven, K. J. and Binley, A. M.: The future role of distributed models: model calibration and predictive uncertainty, Hydrol. Process., 6, 279–298, 1992.

Blöschl, G., Bierkens, F. P., Chambel, A., et al.: Twenty-three unsolved problems in hydrology (UPH) – a community perspective, Hydrolog. Sci. J., 64, 1141–1158, https://doi.org/10.1080/02626667.2019.1620507, 2019.

Cameron, S.: Why is the R squared adjusted reported?, J. Quant. Econ., 9, 183–186, 1993.

Chaudhuri, A. and Hu, W.: A fast algorithm for computing distance correlation, Comput. Stat. Data Anal., 135, 15–24, 2019.

Cheng, C.-L., Shalabh, and Garg, G.: Coefficient of determination for multiple measurement error models, J. Multivar. Anal., 126, 137–152, 2014.

Clark, M. P., Vogel, R. M., Lamontagne, J. R., Mizukami, N., Knoben, W. J. M., Tang, G., Gharari, S., Freer, J. E., Whitfield, P. H., Shook, K. R., and Papalexiou S. M.: The abuse of popular performance metrics in hydrologic modelling, Water Resour. Res., 57, e2020WR029001, https://doi.org/10.1029/2020WR029001, 2021.

Gupta, H. V., Kling, H., Yilmaz, K. K., and Martinez, G. F.: Decomposition of the mean squared error and nse performance criteria: implications for improving hydrological modeling, J. Hydrol., 377, 80–91, https://doi.org/10.1016/j.jhydrol.2009.08.003, 2009.

Jackson, E. K., Roberts, W., Nelsen, B., Williams, G. P., Nelson, E. J., and Ames, D. P.: Introductory overview: error metrics for hydrologic modelling – a review of common practices and an open source library to facilitate use and adoption, Environ. Modell. Softw., 119, 32–48, 2019.

Knoben, W. J. M., Freer, J. E., and Woods, R. A.: Technical note: Inherent benchmark or not? Comparing Nash–Sutcliffe and Kling–Gupta efficiency scores, Hydrol. Earth Syst. Sci., 23, 4323–4331, https://doi.org/10.5194/hess-23-4323-2019, 2019.

Krause, P., Boyle, D. P., and Bäse, F.: Comparison of different efficiency criteria for hydrological model assessment, Adv. Geosci., 5, 89–97, https://doi.org/10.5194/adgeo-5-89-2005, 2005.

Lee, J. S. and Choi, H. I.: A rebalanced performance criterion for hydrological model calibration, J. Hydrol., 606, 127372, https://doi.org/10.1016/j.jhydrol.2021.127372, 2022.

Liu, D.: A rational performance criterion for hydrological model, J. Hydrol., 590, 125488, https://doi.org/10.1016/j.jhydrol.2020.125488, 2020.

Moriasi, D. N., Gitau, M. W., Pai, N., and Daggupati, P.: Hydrologic and water quality models: performance measures and evaluation criteria, T. ASABE, 58, 1763–1785, 2015.

Nash, J. E. and Sutcliffe, J. V.: River flow forecasting through conceptual models part I – a discussion of principles, J. Hydrol., 10, 282–290, 1970.

Nielsen, S. A. and Hansen, E.: Numerical simulation of the rainfall-runoff process on a daily basis, Nordic Hydrol., 4, 171–190, 1973.

Onyutha, C.: Influence of hydrological model selection on simulation of moderate and extreme flow events: a case study of the Blue Nile basin, Adv. Meteorol., 2016, 1–28, https://doi.org/10.1155/2016/7148326, 2016.

Onyutha, C.: Hydrological model supported by a step-wise calibration against sub-flows and validation of extreme flow events, Water, 11, 244, https://doi.org/10.3390/w11020244, 2019.

Onyutha, C.: A hydrological model skill score and revised R-squared, Hydrol. Res., 53, 51–64, 2022.

Taylor, K. E.: Summarizing multiple aspects of model performance in a single diagram, J. Geophys. Res., 106, 7183–7192, 2001.

Willmott, C. J.: On the validation of models, Phys. Geogr., 2, 184–194, https://doi.org/10.1080/02723646.1981.10642213, 1981.

Proc. IAHS, 385, 189–196, 2024
https://doi.org/10.5194/piahs-385-189-2024

Open Access PIAHS

IAHS2022 – Variability and change across space, time, extremes, and interfaces

Semi-automated morphological characterization using South Rivers Toolbox

Santiago Yépez[1], Francisca Salas[2], Andrea Nardini[3], Noelia Valenzuela[1], Víctor Osores[4], José Vargas[2], Rolando Rodríguez[1], and Hervé Piégay[5]

[1]Departamento Manejo de Bosques y Medio Ambiente, Facultad de Ciencias Forestales, Universidad de Concepción, Concepción, 4070386, Chile
[2]Facultad de Ingeniería, Departamento de Ingeniería Civil, Universidad de Concepción, Concepción, 4070386, Chile
[3]Fundación CREACUA, Riohacha, La Guajira, 440001, Colombia
[4]Departamento de Matemática, Física y Estadística, Facultad de Ciencias Básicas, Universidad Católica del Maule, Talca, 3466706, Chile
[5]UMR 5600 CNRS EVS, ISIG platform, University of Lyon, Site ENS de Lyon, 15 Parvis René Descartes, 69362 Lyon, France

Correspondence: Santiago Yépez (syepez@udec.cl)

Received: 30 May 2022 – Revised: 22 February 2023 – Accepted: 29 March 2023 – Published: 18 April 2024

Abstract. Automatic morphological characterization of river systems is important because it provides valuable information on river behavior, helps quantify fluvial changes, improves model accuracy, and supports the management and restoration of river systems. In recent years, scientific interest in the development and use of automated tools for the geomorphological characterization of rivers has increased. The objective of this study was to characterize in a semi-automated way the River Styles of the Duqueco River in south-central Chile. To achieve this objective, an experimental complement within QGIS called South Rivers Toolbox (SRT) was developed. This toolbox allows users to characterize and classify River Styles based on hydromorphological metrics extracted at segment scale, implementing a semi-automatic approach that no longer relies on subjective expert judgment. As a result of the analysis, 43 Rivers Styles have been classified and characterized using the SRT. The upper zone is characterized by being mostly confined, single-channel, with a bed dominated by boulders and blocks, this section of the river is highly affected by hydropower generation plants; the middle zone is confined and single-channel, with a change in the granulometry more dominated by boulders. The lower zone is more diverse in styles, unconfined, with a high presence of geomorphic units that give way to multichannel styles and transition zones. Using the SRT program it is possible to retrospectively analyze the evolution of the river, identifying sections sensitive to geomorphological modification, which is very useful to synoptically evaluate the flood risk, given that, traditionally, for this type of studies only hydraulic modeling is implemented at a river section scale, often with total lack of knowledge of how the river functions and evolves.

Keywords. New monitoring; Modelling; River Styles; Fluvial Morphodynamics; Duqueco River; Chile

1 Introduction

Studies have shown that there has been an increase in the frequency and intensity of heavy precipitation events in many parts of the world in recent decades, and that this trend is likely to continue as temperatures rise further. This can lead to more frequent and severe flooding, as well as other impacts on the environment and human infrastructure (Duarte et al., 2006; Eissa and Zaqui, 2011; IPCC, 2014). In Chile, the factors that contribute to flooding can vary depending on the region and the specific climatic conditions. For example, in the northern regions of Chile, floods can occur when heavy rainfalls overcome the riverbeds capacity, leading to flash floods. In central Chile, floods can be caused by heavy rainfall, but can also be linked to the El Niño Southern Oscillation (ENSO), which can lead to changes in precipitation patterns and more extreme weather events (Rojas, 2015; Urrutia de Hazbun and Lanza Lazcano, 1993). It is important to have a comprehensive approach to mitigating the effects of extreme events, such as floods. The DOH (Chilean Directorate of Hydraulic Works) in Chile has been moving towards such a direction. Developing riverbed management plans can help to identify areas that are particularly vulnerable to flooding and prioritize measures such as the construction of works to reduce the impact of floods. However, a tool that supports the assessment of fluvial morphodynamics is essential for this management problem in order to identify measures compatible with the character and behavior of the river, not only for what concerns flood risk, but also for the correct design and proper management of reservoirs and dams. The original objective of our study was to characterize geomorphologically the Duqueco River (Biobío Region, Chile), to understand its character and behavior by adopting the River Styles Framework (Brierley and Fryirs, 2005). For this purpose, the SRT program was designed as an experimental complement within the free software QGIS interface. This toolbox was specifically designed in this study to characterize and classify River Styles based on hydromorphological attributes measured at the segment scale, implementing a semi-automated approach that reduces expert judgment according to the framework developed by Nardini et al. (2020a). This approach allows the character, behavior and evolution of rivers to be assessed throughout the basin, with the objective of sustainable management.

2 Study Area

The study area is located along the Duqueco River, located in the upper Biobío River basin in the Biobío Region, between 37.30 and 37.36° S latitude and 71.29 and 72.30° W longitude (Fig. 1).

The river has a pluvio-nival hydrological regime, and its flows and chemical water quality are monitored by the "Villucura" and "Cerrillos" stations. According to Chile's water balance (DGA, 1987), the area's isotherms vary between 4–15 °C and the isohyets between 1200–4000 mm yr^{-1}, with a strong orographic gradient in the mountainous zone.

Four hydroelectric power plants with no reservoirs operate in the basin, two conventional (including an intake barrage, lateral canals, an external power plant and a restitution canal), "Mampil" and "Peuchén", and two run-of-river, "Los Padres" and "Lleurequeo". Two dams and intakes, "Las Juntas" and "El Trintre", and the "Rucacura" counter-dam are located on the mainstream, whose purpose is to re-establish the river's natural regime.

3 Materials and Methods

This chapter describes the methodology implemented in this study, firstly detailing the available data, then describing the processing of satellite images and DEMs (Digital Elevation Models), which allows the extraction of the data necessary to implement the methodology, based on the disaggregation-aggregation framework of Alber and Piegay (2011), followed by Stage 1 of the River Styles framework for the classification of fluvial styles (Brierley and Fryirs, 2005). Simultaneously, the use of GIS tools (Geographic Information Systems), the application of tools from the South Rivers Toolbox and the associated templates programmed in Excel are highlighted.

3.1 Data available

This study used a 10 m TanDEM-X radar interferometric digital elevation model from the DRL (Deutsches Zentrum für Luft- und Raumfahrt – German Aerospace Center) in Germany, satellite images from the Planet Scope nanosatellite constellation with a pixel resolution of 3 m (Planet Team, 2021), and channel grain size data from 4 gauging stations (located between the fluviometric stations). In addition, a photographic record of the channel for the field inspection and validation phase was used.

3.2 Satellite imagery processing

The processing of satellite images allows the generation of Unitary Geographic Objects (UGO) relevant to the analysis, with greater efficiency than a manual delimitation, these objects are: the active channel of the Duqueco River in high waters, bars, islands, and fluvial elements of the floodplain. In ENVI-IDL 5.6 software, a multispectral mosaic of the basin in high water (30 June to 5 July 2019) and low water condition (29 to 31 January 2019) is constructed, and the Normalized Difference Water Index (NDWI) (McFeeters, 1996) is calculated, which distinguishes areas with water, from those with dry soil and vegetation, influenced by the spectral separation of these materials. Finally, images at different dates with high and low water conditions are processed and combined to better identify islands and bars of the active channel.

3.3 Disaggregation and aggregation framework for the fluvial corridor

The framework developed by Alber and Piégay (2011), consists of the delimitation and characterization of fluvial metrics from raw data at regional scale e.g., digital elevation models, aerial photos, ancient maps, among other inputs. It consists of four main steps which are explained below:

a. *Unit Geographic Objects (UGO)*: these are linear and polygonal spatial units that delimit biophysical components. In this work, the following units are extracted:

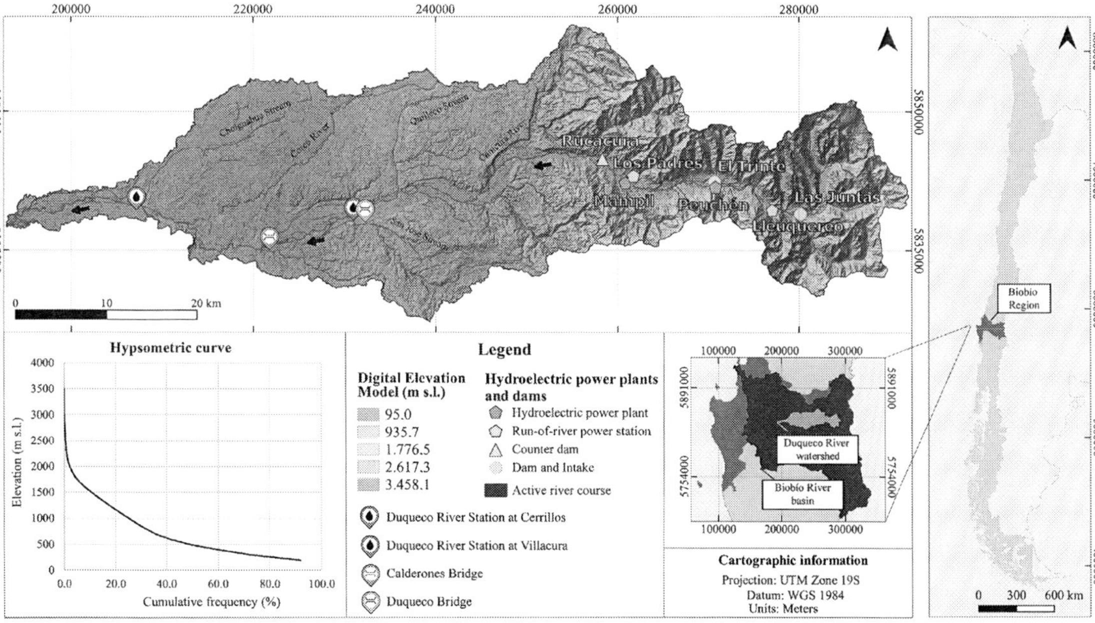

Figure 1. Study area in the Duqueco River watershed. Spatial location of the main hydroelectric power plants and dams. Source: created by the authors.

active channels and their envelope; geomorphological units of the bed (bars and islands) and floodplain (paleochannels, cut meanders, among others); floodplain; and the centerlines of the active channel and floodplain (or Valley Bottom – VB) (Gilbert et al., 2016).

b. *Disaggregated Geographic Objects (DGO):* these are the UGOs discretized longitudinally. In our case, segmentation is performed every 50 m, on the axis of the centerline of the active channel envelope, established based on a 1 : 1 ratio with the average width of the active channel.

c. *Local scale metrics:* attributes of the fluvial corridor and its floodplain are calculated systematically in each DGO. These may include: floodplain and active channel envelope width, number of active channels, sinuosity, drainage area, elevation, slope, planform, degree of confinement and cause, bed material and specific stream power.

   In the case of planform, the algorithm developed by Nardini and Brierley (2021) is applied, which automates the identification of planform by evaluating the degree of similarity of each segment to an archetype, these are a modified version of those proposed by Kleinhans and Van den Berg (2011).

d. *Aggregate Geographic Objects (AGO):* These are segments of the riverbed that present a certain degree of homogeneity with respect to an attribute. These homogeneous segments are determined statistically by using the "Hubert test" tool in the SRT toolbox using continuous data (Hubert, 2000). River reaches, which are the scale of analysis adopted in the following, are identified by a geographical operation, by intercepting the planform AGOs with the bed material AGOs, as suggested in Nardini et al. (2020c).

3.4 River Styles Framework: Stage 1

The first stage of the River Styles framework corresponds to understanding the character and behaviour of the river through the definition and characterization of the River Styles. For this, the attributes of the procedural tree in Fig. 2, are evaluated in each river reach; this tree is a modified version of the original one and it is explicitly conceived to be suitable for automation of the process (Brierley and Fryirs, 2005; Nardini et al., 2020a); it includes the following attributes:

- *Confinement:* the percentage of contact of either bank between the active channel and the flood- plain, which is used to classify the section of the channel as: confined, partially confined or laterally unconfined.
- *Cause of confinement:* it discerns whether the cause of river confinement is due to an alluvial form (terraces, fans, etc.), to the valley (hillslope) or to human intervention (river defences, levees, roads).
- *Planform:* the watercourse is classified according to the planform archetypes defined by Nardini and Brier-

ley (2021), the process is carried out through an algorithm that gathers local metrics of the watercourse, and then gives a score according to the degree of similarity with each type of planform.

- *Bed material texture:* in our application it is based on granulometry data obtained on the banks of the Duqueco River in four sectors distributed downstream of the fluviometric station "Duqueco River in Villucura" and photographs of the riverbed: a class is assigned to each section, according to the classification of Wentworth (1922).
- *Geomorphic Units (UG):* the presence of bars and islands in each DGO is evaluated with a binary attribute 1/0, for presence/absence, and then an overall judgment of presence is given over each river reach.

The set of attributes described above is subjected to an unsupervised classification (clustering), by means of the "Attribute based clustering" tool of QGIS, which generates groups with the same combination of attributes, called River Styles.

3.5 Semi-automated process using South Rivers Toolbox

The number of attributes that have been described in the previous sections, and which must be determined along the channel, require a great deal of time and effort. Thus, if they were established by manual measurements alone, geomorphological characterization using the River Styles framework would be impractical for large basins.

In 2020, within the GeoMagda project – financed by The Nature Conservancy NGO – different algorithms and morphometric analysis programs used worldwide at that time in the extraction of fluvial metrics were compared. Some improvements were also proposed by some of the authors of this paper and their own developments were programmed in Python (Nardini et al., 2020a, b, c; Nardini and Brierley, 2021).

The readjustments made it possible to generate an innovative integrated toolbox, composed of a sequence of tools oriented to the application of the River Styles Framework in a semi-automated way. In this context, later on the South Rivers Toolbox has been developed, allowing a considerable reduction in work time, ensuring a holistic and objective view in the calculation of fluvial parameters, as well as the possibility of replicating the analysis in any other river. It is currently open source and available as an experimental add-on for QGIS (Osores and Yépez, 2023), in the repository: https://github.com/vosores/SouthRiversToolbox (last access: 11 October 2022).

4 Results and Discussion

It is possible to visualize the pattern of local metrics of the Duqueco River (slope, elevation, drainage area, width of the floodplain and the active channel, confinement, specific stream power, among others) along the channel and how the variables are interacting.

In Fig. 3a, it is possible to visualize how the drainage area increases in a staggered manner, due to the entry of the main tributaries. As for the slope, it is steep in the first 36 km, with values reaching $0.15\,\mathrm{m\,m^{-1}}$, and then progressively decreases until the confluence with the Biobío River. Based on the above, the specific streampower of the channel (Fig. 3c), that depends on the slope and discharge of the channel, reaches its highest values in the middle-upper zone of the channel, where the combination of these variables is at its maximum.

Figure 3b shows the width of the floodplain and the confinement, with a very narrow floodplain at the head of the channel, practically confined along its entire length up to kilometer 42, from where the confinement is lost, showing a greater extension of the floodplain, and more freedom of movement of the channel.

As a result of the unsupervised classification that groups the attributes of the channel, there are 43 main River Styles and 57 complementary styles, by classifying on a second level with the floodplain geomorphic units, which are shown in Fig. 4. From the analysis of their lengths, 56.9 % of the fluvial styles are characterized by a single-channel planform, 37.4 % by a transitional type and only 5.6 % by a multi-channel type. As a visual aid, all the single-channel and transitional fluvial styles have been represented in green and blue tones, and the multichannel ones in red, so that the predominance of the first two over the small multichannel sections can be graphically appreciated.

RS_11 and RS_12 (both single channel), which predominate in the upper zone, are completely confined by the valley. The predominant fluvial styles are of transitional type, as is the case of RS_23 and RS_28, all of them with a "wandering" type planform, characterized by the predominance of the single channel type, with high frequency of lateral bars, a dominant arm and absence of abrupt widening in the channel envelope. Finally, in the few areas with multi-channel fluvial styles, R_40 and R_41 predominate, characterized by an "anabranching" type planform, which is characterized by separated channels, with one dominant arm, high island density and little or no presence of bars. Overall, the river can be classified into 3 main zones:

- *Duqueco River in the upper zone:* it is characterized by the presence of RS_11 and RS_12, very similar, with a river totally confined, with a single channel planform and constrained sinuosity; there is not significant presence of any type of geomorphic unit and the riverbed is characterized by boulders and blocks, differing only

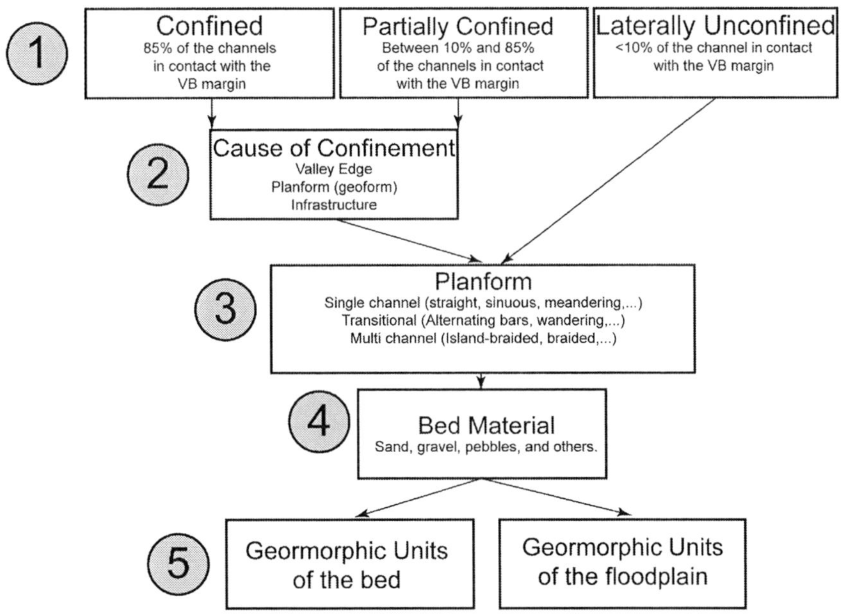

Figure 2. Modified procedural tree of the River Styles framework (Modified from Nardini et al., 2020a). Source: created by the authors.

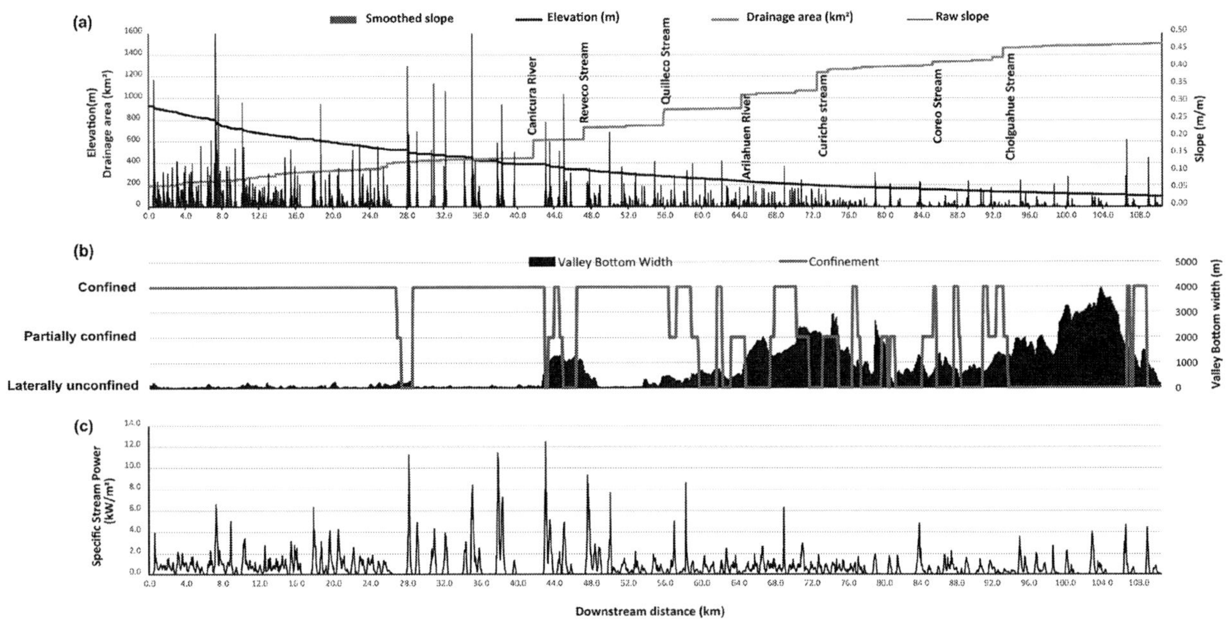

Figure 3. Spatial evolution of local metrics along the Duqueco River. Source: created by the authors.

by the cause of the confinement. The uppermost zone of the channel (RS_11) is limited by hillsides of the Andes Mountain range, downstream the hillsides are farther apart, and it is observed that RS_12 is confined by terraces. The specific stream power is not valid in this zone as it is strongly affected by the dams. Even so, the high slope value and low presence of geomorphic units identify it as a sediment transport zone. In the area downstream of the counter-embayment, the highest values of specific stream power are found, which makes it a zone of high erosion and sediment transport, but due to the confinement of the area, the river has little or no space to adjust, so it is not very sensitive to changes after an extreme event.

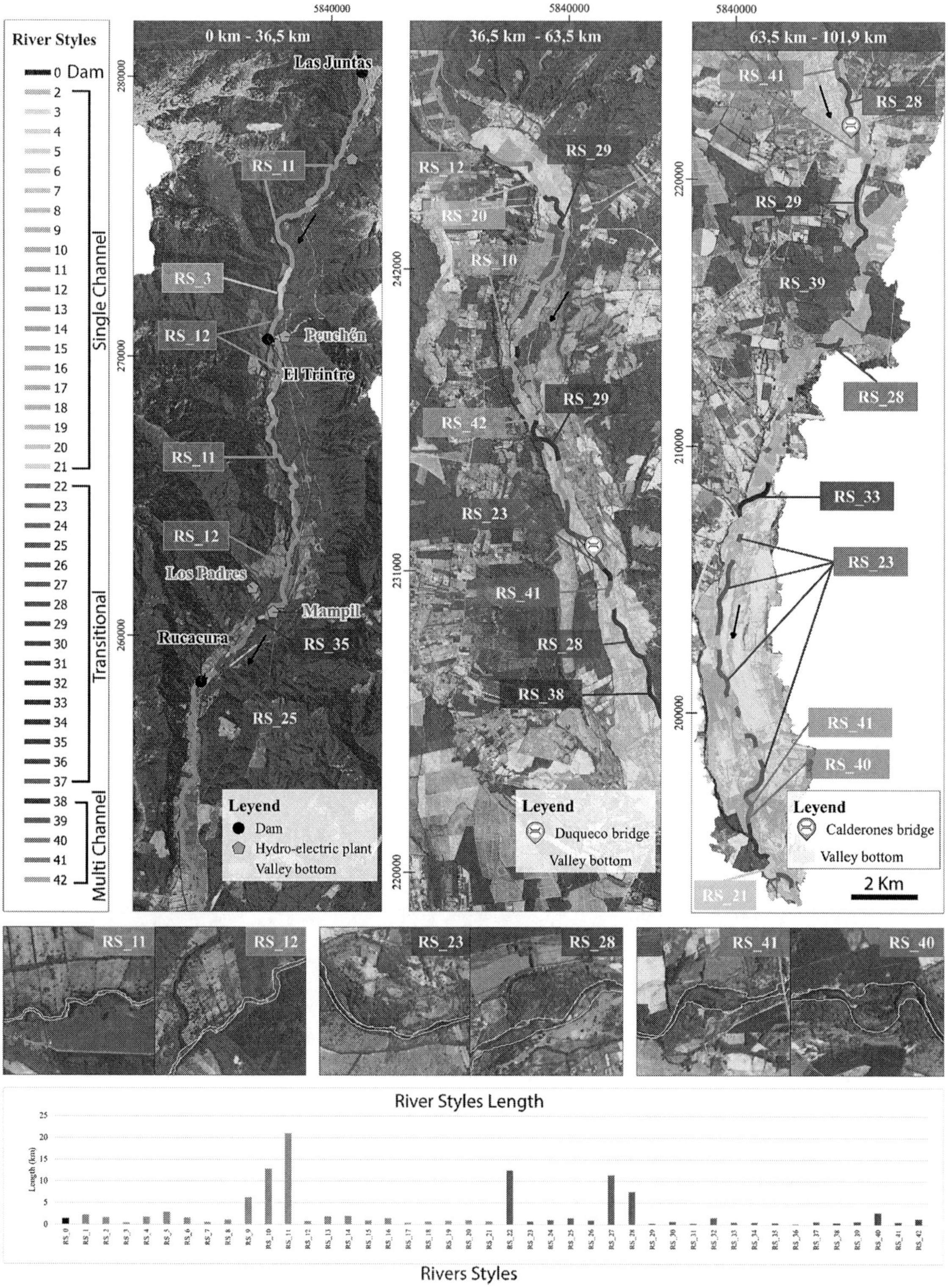

Figure 4. Distribution of River Styles in the Duqueco River in 2019. Source: created by the authors.

- *Duqueco River in the middle zone:* area between the first widening of the floodplain up to the Duqueco bridge, characterized by the presence of the RS_10, RS_23, RS_28 and RS_29 fluvial styles. From this zone on, the riverbed is only boulders, still mostly confined by terraces, but the first unconfined or partially confined segments appear (e.g., RS_23, RS_28 and RS_29). The prevailing planform for RS-10 is of the confined sinuous single-channel type, and transitional for the other styles, due to the presence of bars, which is accentuated in the area near the Duqueco Bridge. In terms of behaviour, the specific stream power reaches its highest values at the beginning of the middle zone. Like the high zone, these points are confined, so erosion does not affect the shape of the river (lateral migrations). There is only one point with high specific stream power located near the Duqueco bridge, which is not confined (RS_23), so it is free to adjust its shape in the long term or during extreme events.
- *Duqueco River in the lower zone:* this is the most extensive zone from the Calderones Bridge to the confluence with the Biobío River. It is characterized by the presence of R_23, R_28 and R_29 as transitional channels, with the occasional appearance of multi-channels such as RS_41. From this zone onwards, there is a marked widening of the floodplain, it is mostly unconfined, and multichannel segments increase, although single-channel type still prevails. The variety of fluvial styles is much wider, due to the higher frequency of UGs that give way to transitional or multichannel segments. In terms of behaviour, it is more prone to adjustments to the river's planform, being largely unconfined and possessing freedom of movement. On the other hand, the specific stream power has a low value, so it is a sedimentation zone (hence the large formation of UGs).

5 Conclusions

The tool we called South Rivers Toolbox, that supports the application of the aggregation and disaggregation framework, allowed us to obtain the hydromorphological attributes of the Duqueco River at different spatial scales, in a semi-automated way, with objective criteria; it ensures a higher degree of objectivity (making it replicable for other users) when compared to an analysis based solely on expert judgment and it makes it possible to develop this type of analysis for large basins.

In particular, the easiest hydromorphological attributes to extract are channel width, number of channels, sinuosity, and local confinement, all of which are automated by SRT tools. Their good computational performance depends only on the delimitation of Unitary Geographic Objects. On the other hand, the attributes of elevation, slope, drainage area and specific stream power, imply the "burning" of the DEM at an earlier stage, what generates a degree of inaccuracy in the measurements, which however, for practical purposes, is negligible. In the case of the specific stream power, it is called "coarse", because it lacks the bankfull flow data along the river, and an estimate is made based on the empirical relationship between the discharge and the drainage area. It is concluded that, despite the lack of precision in its estimation, it provides an indication on erosive zones, which is fundamental to understand the behaviour of the river. Regarding the Planform, the programmed Excel spreadsheet assertively predicts its shape in most of the channel sections, with small manual corrections and expert judgment, with a performance like the work developed in the Baker River (Chile) by Nardini and Brierly (2021). This allows us to conclude that, despite requiring a significant amount of input data, the preparation is relatively easy to conduct, and the result is reliable, objective and, above all, replicable. Continuing with the hydromorphological attributes, but this time calculated on the homogeneous segments of the river, the confinement is obtained in a fully automated way with the SRT tools; the identification of its "cause" is an innovative process that has not been implemented in any other algorithm so far, so it has an added value.

By implementing the first stage of the River Styles framework in the Duqueco River, 43 main River Styles and 57 complementary styles were determined. The Duqueco River is characterized by a high rate of sediment transport in the upper and upper-middle reaches of the basin, up to the Duqueco Bridge area, while in the lower part of the basin, specific streampower decreases, sediment transport decreases and deposition zones progressively increase. The confinement restricts the lateral movement of the river in the upper and upper-middle zone, while freeing it in the lower zone. It is concluded that the area near the Calderones Bridge (Fig. 1) is one of the most critical since it has a high specific stream power and is not confined. This coincides with the existing evidence of the flood that occurred in July 2006, with the loss of connectivity of the bridge, and the flooding of the riparian population of the Llano Blanco sector (downstream of the Calderones Bridge).

In summary, we can state that this modified approach to River Styles classification, supported by SRT, as proposed by previous papers and this paper, provides a conceptual and working framework for understanding the complexity of the fluvial continuum, in addition to its usefulness as an important support tool for other important issues such as flood risk assessment.

Data availability. The South Rivers Toolbox is freely available on the website: https://doi.org/10.5281/zenodo.7810989 (Osores and Yépez, 2023). This study's produced datasets are no longer available with the updating of the model. The complete and updated datasets are available from the corresponding author on request.

Author contributions. SY, FS, JV and RR planned the campaign; SY, FS, VO, AN, JV and RR performed the measurements; SY, FS, NV, AN and HP analyzed the data; SY and FS wrote the manuscript draft; SY, FS, VO, AN, HP and NV reviewed and edited the manuscript.

Competing interests. The contact author has declared that none of the authors has any competing interests.

Disclaimer. Publisher's note: Copernicus Publications remains neutral with regard to jurisdictional claims in published maps and institutional affiliations.

Special issue statement. This article is part of the special issue "IAHS2022 – Hydrological sciences in the Anthropocene: Variability and change across space, time, extremes, and interfaces". It is a result of the XIth Scientific Assembly of the International Association of Hydrological Sciences (IAHS 2022), Montpellier, France, 29 May–3 June 2022.

Acknowledgements. We would like to thank Planet for providing the RapidEYE and Planet Scope (nanosatellite) images used in this study free of charge. Also, to the TanDEM-X DEM GEOL08450 project for providing the high spatial resolution terrain elevation model.

Financial support. This research has been supported by the Chilean economic development agency (CORFO) (grant no. 19BP-117424).

Review statement. This paper was edited by Christophe Cudennec and reviewed by two anonymous referees.

References

Alber, A. and Piégay, H.: Spatial disaggregation and aggregation procedures for characterizing fluvial features at the network-scale: Application to the Rhône basin (France), Geomorphology, 125, 343–360, https://doi.org/10.1016/j.geomorph.2010.09.009, 2011.

Brierley, G. and Fryirs, K.: Geomorphology and river management: applications of the river styles framework, John Wiley and Sons, ISBN 978-1-405-11516-2, 2005.

Dirección General de Aguas (DGA): Balance hídrico de Chile, Santiago, Chile, 1987.

Duarte, C. M., Alonso, S., Benito, G., Dachs, J., Montes, C., Pardo Buendía, M., Ríos, A. F., Simó, R., and Valladares, F.: Cambio Global, Impacto de la actividad humana sobre el sistema Tierra, CSIC, Consejo superior de investigaciones científicas, ISBN 978-84-00-08452-3, 2006.

Eissa, A. E. and Zaki, M. M.: The impact of global climatic changes on the aquatic environment, Procedia Environ. Sci., 4, 251–259, https://doi.org/10.1016/j.proenv.2011.03.030, 2011.

Gilbert, J. T., Macfarlane, W. W., and Wheaton, J. M.: The Valley Bottom Extraction Tool (VBET): A GIS tool for delineating valley bottoms across entire drainage networks, Comput. Geosci., 97, 1–14, https://doi.org/10.1016/j.cageo.2016.07.014, 2016.

Hubert, P.: The segmentation procedure as a tool for discrete modeling of hydrometeorological regimes, Stoch. Env. Res. Risk A., 14, 297–304, https://doi.org/10.1007/PL00013450, 2000.

IPCC: Cambio climático 2014: Informe de síntesis. Contribución de los Grupos de trabajo I, II y III al Quinto Informe de Evaluación del Grupo Intergubernamental de Expertos sobre el Cambio Climático, Equipo principal de redacción, edited by: Pachauri, R. K. and Meyer, L. A., IPCC, Ginebra, Suiza, 157 pp., ISBN 978-92-9169-343-6, 2014.

Kleinhans, M. G. and van den Berg, J. H.: River channel and bar patterns explained and predicted by an empirical and a physics-based method, Earth Surf. Proc. Land., 36, 721–738, https://doi.org/10.1002/esp.2090, 2011.

McFeeters, S. K.: The use of the Normalized Difference Water Index (NDWI) in the delineation of open water features, Int. J. Remote Sens., 17, 1425–1432, https://doi.org/10.1080/01431169608948714, 1996.

Nardini, A. and Brierley, G.: Automatic river planform identification by a logical-heuristic algorithm, Geomorphology, 375, 107558, https://doi.org/10.1016/j.geomorph.2020.107558, 2021.

Nardini, A., Yepez, S., Zuniga, L., Gualtieri, C., and Bejarano, M. D.: A Computer Aided Approach for River Styles–Inspired Characterization of Large Basins: The Magdalena River (Colombia), Water, 12, 1147, https://doi.org/10.3390/w12041147, 2020a.

Nardini, A., Yepez, S., and Bejarano, M. D.: A Computer Aided Approach for River Styles — Inspired Characterization of Large Basins: A Structured Procedure and Support Tools, Geosciences, 10, 231, https://doi.org/10.3390/w12041147, 2020b.

Nardini, A., Yépez, S., Mazzorana, B., Ulloa, H., Bejarano, M. D., and Laraque, A.: A Systematic, Automated Approach for River Segmentation Tested on the Magdalena River (Colombia) and the Baker River (Chile), Water, 12, 2827, https://doi.org/10.3390/w12102827, 2020c.

Osores, V. and Yépez, S.: South Rivers Toolbox, Zenodo [software], https://doi.org/10.5281/zenodo.7810989, 2023.

Planet Team: Planet Imagery Product Specifications, March 2021, [software], https://www.planet.com/products/, last access: 22 March 2021.

Rojas, O.: Cambios Ambientales y Dinámica de Inundaciones Fluviales en una Cuenca Costera del Centro Sur de Chile (Tesis Doctoral), Universidad de Concepción, Concepción, http://repositorio.udec.cl/jspui/handle/11594/1773 (last access: 22 March 2021), 2015.

Urrutia de Hazbún, R. and Lanza Lazcano, C.: Catástrofes en Chile 1541–1992, in: Catástrofes en Chile 1541–1992, 439–439, https://bibliotecadigital.ciren.cl/handle/20.500.13082/18899 (last access: 22 March 2021), 1993.

Wentworth, C. K.: A scale of grade and class terms for clastic sediments, J. Geol., 30, 377–392, https://doi.org/10.1086/622910, 1922.

Proc. IAHS, 385, 197–201, 2024
https://doi.org/10.5194/piahs-385-197-2024

The importance of retention times in Natural Flood Management interventions

Elizabeth Follett[1], **Keith Beven**[2], **Barry Hankin**[2,3], **David Mindham**[2], **and Nick Chappell**[2]

[1]Royal Academy of Engineering Research Fellow, Department of Civil Engineering & Industrial Design, School of Engineering, University of Liverpool, Liverpool, UK
[2]Lancaster Environment Centre, Lancaster University, Lancaster, UK
[3]JBA Consulting, Warrington, UK

Correspondence: Keith Beven (k.beven@lancaster.ac.uk)

Received: 25 May 2022 – Revised: 28 January 2023 – Accepted: 30 March 2023 – Published: 18 April 2024

Abstract. The starting point for this study is the simulation study of Metcalfe et al. (2018) which suggested that retention times of the order of 10 h are required for natural flood management storage features to have a maximum effect on large flow peaks. A analysis of the celerity characteristics for some log jams at Tebay Gill, Cumbria (upland UK), suggests that the impacts of storage in slowing the celerities is only of the order of minutes. An analysis of storage-discharge dynamics based on observations at 4 jams reveals that the dynamics of storages can be represented with time constants of between 3 and 213 min, still well short of those required to maximise the effects for larger flood flows. That is not to say that there will not be reduction in flood peaks for smaller events, only that for large events the effects will be limited. A spreadsheet tool for retention times has been developed to help in the design of new schemes that is freely available.

Keywords. Flood risk; flood damage estimates; instream wood jams; natural flood management

1 Introduction

One of the strategies for natural flood management in the UK, that has been widely applied over the last few years, is the use of instream barriers to retain water during flood events with the aim of reducing peak magnitude, flooding of houses and businesses and consequently reduce damages. This intervention aims to increase water retention in upstream catchments via enhanced in-channel or floodplain storage, aiming to reduce burden on downstream flood infrastructure (Dadson et al., 2017; Lane, 2017). In a modelling study, Metcalfe et al. (2018) demonstrated how reduction in flood peak magnitude could depend on design of the structures used. If barriers filled and drained too quickly, the effect on the hydrograph would be minimal, even for large amounts of potential storage. The retention time in the storage was too short. If, however, barriers filled but did not drain between events, then a sequence of events before a major flood peak could mean that potential storage would have filled and that the effect on the coming flood peak would be minimal. In this case, the retention time in the storage was too long. The modelling results of Metcalfe et al. (2018) suggested that a mean retention time of 1 h was too short, and that a mean retention time of 100 h was too long. However, a retention time on the order of 10 h (600 min) could produce a significant reduction in a major flood peak that produced major flood damages (Storm Desmond, December 2015). More recent work has also reinforced the requirement for large amounts of storage to be made available, with design incorporating suitable retention times for peak flow reductions to be significant (Beven et al., 2022).

There is an issue, however, in that few, if any, of the instream log jams (or runoff attenuation features, RAFs) that have been installed to date in the UK have taken the concept of retention time into account (e.g. Figs. 1, 2 from the sequence of log jams at Tebay in Cumbria are a case in point). In this paper we provide a spreadsheet tool that might help improve the design of such in-stream RAFs of different designs (notably diverting flood-flows into bunded floodplain

Figure 1. Jam 16 of 77 installed at Tebay Gill, Cumbria.

Figure 2. The Tebay jams 19-17 in action, 11 January 2020 (photo courtesy of Tim Winder).

stores, rather than solely holding water in the channel) to produce retention times of the right order of magnitude. The RAFs within this study primarily hold water in the channel, typical of those throughout the upland UK, and are located at 54°24′49.55″ N, 2°34′45.53″ W (jam 16) within a channel incised into a diamicton surficial geology, over a Silurian sedimentary geology comprising of sandstone, siltstone and mudstone. The spreadsheet allows for a single RAF with simplified geometry that is treated as a brush or log jam spanning the channel. Other work is concerned with the representation of multiple RAFs in series down a headwater channel reach (Follett and Hankin, 2022).

2 The Retention Time Spreadsheet

This spreadsheet (Follett and Beven, 2023) calculates the upstream depth, backwater volume, and residence times in

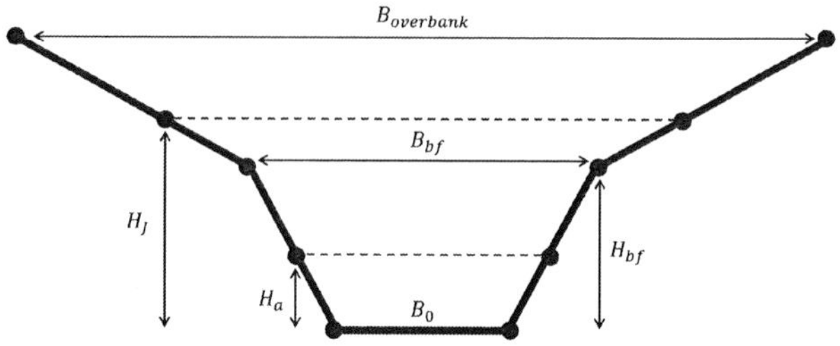

Figure 3. Schematic diagram of channel cross section showing channel bottom width B_0, bankfull width B_{bf}, overbank width 1 m above the channel bankfull depth, vertical distance of lower gap above the channel bed H_a, and vertical distance between the channel bed and top edge of the jam H_J.

the upstream backwater region behind a log jam from user-input channel and jam parameters for steady flow states. A compound trapezoidal channel is used based on specification of the channel bottom width, bankfull width, and estimated width of the channel 1 m above bankfull depth (Fig. 3).

Uniform flow is found from an input Chézy coefficient and slope, assuming a wide channel with low influence of side-wall resistance. The jam model accommodates jams that span the vertical distance of the channel and jams with a lower underflow gap (Follett et al., 2020, 2021). Flow over the jam once the storage has reached the top of the jam is assumed to follow a sharp-crested weir relationship (Munson et al., 2013; Hankin et al., 2020). Jam construction is related to relative magnitude of the jam-generated backwater rise. The backwater rise is given by the ratio of uniform flow depth to the flow depth generated upstream of a jam that spans the vertical channel extent, from which the dimensionless jam structural parameter, C_A, is found.

3 Theory and Methods

3.1 Unobstructed open channel flow

The effect of a jam-generated backwater was investigated in a trapezoidal open channel of slope S, bankfull width B_{bf}, bankfull depth H_{bf}, and coefficient of gravitational acceleration g. Uniform flow depth h_0 resulted from a balance between the weight of water on bed slope S balanced by bed friction, represented by a dimensionless Chézy coefficient Cz (Julien, 2010):

$$Q = \bar{B}\left(h_0^3 g S C z^2\right)^{1/2} \tag{1}$$

The channel followed a compound trapezoidal shape with bottom channel width B_0 and overbank width 1 m above the channel bankfull depth B_{overbank}. The channel width $B(z)$ was then,

$$\begin{cases} B(z) = \frac{2z}{\tan\theta_1} + B_0, & 0 < z \leq H_{\mathrm{bf}} \\ B(z) = \frac{2(z-H_{\mathrm{bf}})}{\tan\theta_2} + B_{\mathrm{bf}}, & H_{\mathrm{bf}} < z \end{cases} \tag{2}$$

for the in-channel and overbank sections.

3.2 Jam-generated upstream backwater rise

Flow through a jam generated momentum loss proportional to the number, size, and packing density of the logs comprising the jam (Follett et al., 2020, 2021), elevating the water depth upstream of the jam relative to unobstructed open channel flow. Flow through a jam with a lower gap was a combination of unit discharge $q = Q/B$ passing through the jam and lower gap sections:

$$q = \left[\frac{2g(h_\mathrm{J}-a)^3}{3\sqrt{3}C_A}\right]^{1/2} + \left[\frac{C_{p0}}{\left(1+C_b\frac{a}{h_\mathrm{J}}\right)}ga^2h_\mathrm{J}\right]^{1/2} \tag{3}$$

with vertical distance of the lower gap a, dimensionless jam structural parameter C_A related to the jam length L_J, drag coefficient for a rigid circular cylinder $C_\mathrm{d} \cong 1$, frontal area density a_f, and solid volume fraction ϕ ($C_A = L_\mathrm{J}C_\mathrm{d}a_\mathrm{f}/(1-\phi)^3$; Follett et al., 2020), gate discharge coefficient $C_{p0} = 2/3$ and $C_b = \frac{C_{p0}C_\mathrm{f}}{S} - 1$ (Malcherek, 2018; Follett et al., 2021). The dimensionless friction coefficient C_f may be related to the Chézy coefficient, $C_\mathrm{f} = 1/C_z^2$ (Julien, 2010). C_A increases with the amount of solid wood present in the jam, and higher C_A results in higher jam-generated backwater rise for a given discharge. Equation (3) corresponds to the upstream depth for a channel-spanning jam (Follett et al., 2020) when the lower gap vertical distance $a = 0$,

$$q = \left[\frac{2g(h_\mathrm{J})^3}{3\sqrt{3}C_A}\right]^{1/2}. \tag{4}$$

C_A was related to the relative magnitude of jam-generated backwater rise by comparing the upstream backwater depth generated by a channel-spanning jam (Eq. 4) and unobstructed uniform flow (Eq. 1), for the same unit discharge,

$$C_A = \frac{2}{3\sqrt{3}} \cdot \frac{C_{\mathrm{f}0}}{S(h_0/h_\mathrm{J})^3} \tag{5}$$

with the relationship between C_A and relative ratio h_0/h_J independent of unit discharge (Follett and Hankin, 2022). Discharge passing over the jam top edge H_J was represented by a sharp-crested weir (Munson et al., 2013; Hankin et al., 2020),

$$q = \frac{2}{3}\sqrt{2g}(h_\mathrm{J}-H_\mathrm{J})^{3/2} + Q(h_\mathrm{J} = H_\mathrm{J}) \tag{6}$$

with discharge when upstream water depth is equal to the jam top edge $Q(h_\mathrm{J} = H_\mathrm{J})$ given by Eq. (3). The upstream backwater generated by the jam was assumed to occupy a simplified triangular shape, with backwater length $L_\mathrm{bw} = (h_\mathrm{J}-h_0)/S$ and backwater volume

$$V_\mathrm{bw} = \int_{h_0}^{h_\mathrm{J}} \frac{z-h_0}{S}B(z)\mathrm{d}z \tag{7}$$

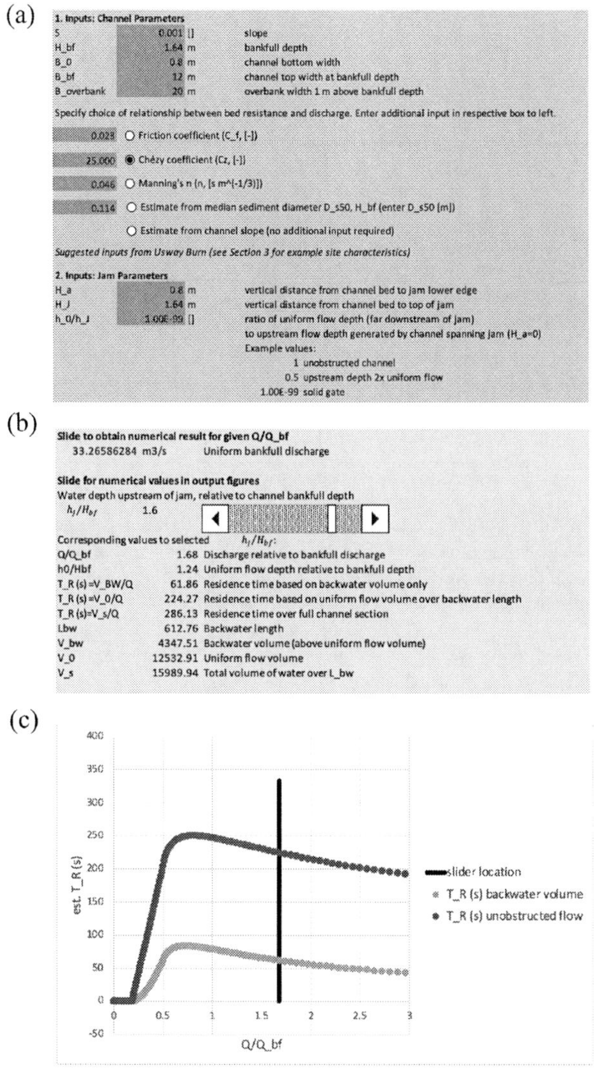

Figure 4. **(a)** input data block, **(b)** output data block, and **(c)** part of graphical output block of spreadsheet tool, representing jam of dimensions similar to Fig. 1 acting as a solid gate.

above the channel uniform flow volume $V_0 = L_\mathrm{bw}\bar{B}h_0$. The total volume of water in the segment was then $V_\mathrm{s} = V_0 + V_\mathrm{bw}$. The storage retention time was estimated from V_s/Q.

4 Applying the spreadsheet

If we take a jam of dimensions similar to that of Fig. 1 at Tebay Gill we can use the spreadsheet to calculate the expected wave retention times. The data are input in a block as shown in Fig. 4a with the logs assumed to act as a solid gate.

Outputs are given both as numeric values for a given discharge or as graphical outputs over a range of discharge values (Fig. 5). Figure 4b, c shows the outputs for calculated

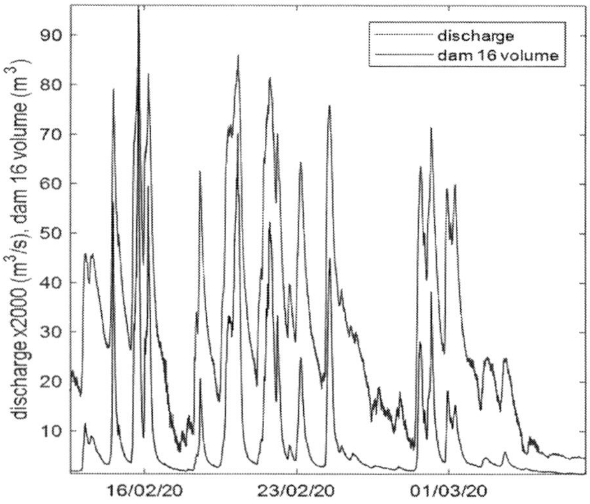

Figure 5. Part of the time series of upstream flume discharge and storage volume at Tebay Jam 16 used in the DBM model analysis (including an approximately 1-in-1 year daily discharge event on 21 February 2020: derived from 4 years of record).

retention times for storage behind the jam based on celerities over the backwater profile and using average backwater depth in comparison with the case of uniform flow without the jam in place. Other plots of the storage and water depths for different flow rates are also provided. The calculations for Jam 16 at Tebay Gill shown in Fig. 4 suggest that the peak flood wave retention time is on the order of minutes, peaking at about half the bankfull flow, not the hours suggested as most effective in the Metcalfe et al. (2018) study. The celerities are always slower than for uniform flow at the same discharge but give response times much less than the 1 h that Metcalfe et al. (2018) suggested would have relatively small effects on larger flood peaks.

5 Dynamics of the Storage Volumes

This analysis suggests that the dynamics of the storage volumes generated by RAFs are more important than the slowing of the celerities in determining the overall impacts of the jam. This, in effect, was also the basis of the Metcalfe et al. (2018) simulations, which treated each of many thousands of storage elements in the catchment as a linear store with an assumed time constant. For 4 of the jams at Tebay, including Jam 16, we can determine the time constants for the storage volume by fitting a Data-based Mechanistic (DBM) model (e.g. Young and Beven, 1994; Magliano et al., 2019) to the time series of discharges measured at an upstream flume and the volumes behind each jam, estimated from pressure transducer recordings of water level and channel geometry.

Figure 5 shows part of the time series available for a succession of events in February 2020. It is worth noting here that, as shown in Fig. 2, the storage volumes behind the jams

Table 1. Results of first order DBM analysis of discharge/storage relationship for the jams with level recorders at Tebay.

Jam No.	DBM time constant (mins)	DBM time delay (mins)	R^2 fit
8	3.2	0	0.921
10	20.7	0	0.948
16	213.3	0	0.899
17	20.4	0	0.978

are limited by the jam spacing, which will have an impact on the response (see Follett and Hankin, 2022, for an analysis of sequences of jams). Table 1 shows the results for each of the 4 jams with recorded water level. These time constants are longer than the estimated flood wave response based on the celerities, but are still less than the suggested 10 h required to maximise the impact on large floods suggested by Metcalfe et al. (2018). Jam 16, with the largest length upstream to the next jam (19 m) and critically the only jam with significant floodplain engagement, has the longest storage retention time at 3.5 h.

These results give a basis for estimating the volumes of storage that might be required to obtain the magnitude of retention times behind jams to achieve the maximum impact for large flood events such as Storm Desmond. The differences are striking. For example, if we take the Metcalfe et al. (2018) suggestion of 10 h (600 min) as of the right order, then between 2.8 and 187 times more storage would be required at each site.

6 Conclusions

A spreadsheet analysis of the effects of natural flood management flow retention features, allowing for underflow, seepage and weir overtopping under steady flow conditions shows that the celerities at higher flows are not sufficiently slowed to have a major impact on large or small flood events. This leads to the conclusion that any effects will be dominated by the dynamics of the storage behind the retention features. For the jams with level recorders, this is assessed in terms of the time constants calibrated for a first order DBM model. Comparison with the suggested retention time of approximately 10 h from the simulations by Metcalfe et al. (2018) that, for maximum impact at higher flood flows, retention times should be neither too long nor too short, reveals that the storages should be much larger. That is not to say that there will not be reduction in discharge peaks for frequently occurring rain-events, only that for events that produce flooding the effects will be limited.

Code availability. The spreadsheet is freely available under hosting from the JBA Trust at https://www.jbatrust.

org/how-we-help/publications-resources/rivers-and-coasts/nfm-leaky-barrier-retention-times (last access: 6 April 2023) and is archived in the Zenodo repository at https://doi.org/10.5281/zenodo.7806786 (Follett and Beven, 2023).

Data availability. No additional data are required to run the spreadsheet. Data for the Tebay dams are available on request from Nick Chappell (n.chappell@lancaster.ac.uk).

Author contributions. KB conceived the idea for a software tool to calculate retention times for in-stream NFM measures. Most of the programming was done by EF. NC, DM and BH contributed to both the field and data analysis that were used to check the calculations. All authors contributed to the writing and editing of the paper.

Competing interests. The contact author has declared that none of the authors has any competing interests.

Disclaimer. Publisher's note: Copernicus Publications remains neutral with regard to jurisdictional claims in published maps and institutional affiliations.

Special issue statement. This article is part of the special issue "IAHS2022 – Hydrological sciences in the Anthropocene: Variability and change across space, time, extremes, and interfaces". It is a result of the XIth Scientific Assembly of the International Association of Hydrological Sciences (IAHS 2022), Montpellier, France, 29 May–3 June 2022.

Financial support. This study has been carried out under the NERC funded Q-NFM project (grant no. NE/R004722/1). The first author has received funding from the Royal Academy of Engineering's Research Fellowships programme and the European Regional Development Fund through the Welsh Government Sêr Cymru programme (grant no. 80762-CU-241).

Review statement. This paper was edited by Christophe Cudennec and reviewed by two anonymous referees.

References

Beven, K. J., Page, T., Hankin, B., Smith, P., Kretzschmar, A., Mindham, D., and Chappell, N. A.: Deciding on fitness-for-purpose – of models and of natural flood management, Hydrol. Process., 36, e14752, https://doi.org/10.1002/hyp.14752, 2022.

Dadson, S. J., Hall, J. W., Murgatroyd, A., Acreman, M., Bates, P., Beven, K., Heathwaite, L., Holden, J., Holman, I. P., Lane, S. N., O'Connell, E., Penning-Rowsell, E., Reynard, N., Sear, D., Thorne, C., and Wilby, R.: A restatement of the natural science evidence concerning catchment-based "natural" flood management in the UK, P. R. Soc. A, 473, 20160706, https://doi.org/10.1098/rspa.2016.0706, 2017.

Follett, E. and Beven, K.: Design tool: Leaky barrier retention times for Natural Flood Management interventions, Zenodo [code], https://doi.org/10.5281/zenodo.7806786, last access: 6 April 2023.

Follett, E. and Hankin, H.: Investigation of effect of logjam series for varying channel and barrier physical properties using a sparse input data 1D network model, Environ. Modell. Softw., 158, 105543, https://doi.org/10.1016/j.envsoft.2022.105543, 2022.

Follett, E., Schalko, I., and Nepf, H.: Momentum and energy predict the backwater rise generated by a large wood jam, Geophys. Res. Lett., 47, e2020GL089346, https://doi.org/10.1029/2020GL089346, 2020.

Follett, E., Schalko, I., and Nepf, H.: Logjams with a lower gap: Backwater rise and flow distribution beneath and through logjam predicted by two-box momentum balance, Geophys. Res. Lett., 48, e2021GL094279, https://doi.org/10.1029/2021GL094279, 2021.

Hankin, B., Hewitt, I., Sander, G., Danieli, F., Formetta, G., Kamilova, A., Kretzschmar, A., Kiradjiev, K., Wong, C., Pegler, S., and Lamb, R.: A risk-based network analysis of distributed in-stream leaky barriers for flood risk management, Nat. Hazards Earth Syst. Sci., 20, 2567–2584, https://doi.org/10.5194/nhess-20-2567-2020, 2020.

Julien, P. Y.: Erosion and sedimentation, 2nd Edn., Cambridge University Press, Cambridge, https://doi.org/10.1017/CBO9780511806049, 2010.

Lane, S. N.: Natural flood management, WIREs Water, 4, e1211, https://doi.org/10.1002/wat2.1211, 2017.

Magliano, P. N., Mindham, D., Tych, W., Murray, F., Nosetto, M. N., Jobbagy, E. G., Rufino, M. C., and Chappell, N. A.: Hydrological functioning of cattle ranching impoundments in the Dry Chaco rangelands of Argentina, Hydrol. Res., 50, 1596–1608, https://doi.org/10.2166/nh.2019.149, 2019.

Malcherek, A.: Eine neue, auf der impulsbilanz basierende theorie zur hydraulik des schützes – Ein diskussionsbeitrag, Wasser-Wirtschaft, 5, 40–69, https://doi.org/10.1007/s35147-018-0215-8, 2018.

Metcalfe, P., Beven, K., Hankin, B., and Lamb, R.: A new method, with application, for analysis of the impacts on flood risk of widely distributed enhanced hillslope storage, Hydrol. Earth Syst. Sci., 22, 2589–2605, https://doi.org/10.5194/hess-22-2589-2018, 2018.

Munson, B. R., Okiishi, T. H., Huebsch, W. W., and Rothmayer, A. P.: Fluid Mechanics, 7th Edn., Wiley and Sons, Singapore, ISBN-13 9781118116135, 2013.

Young, P. C. and Beven, K. J.: Data-based mechanistic modelling and the rainfall-flow non-linearity, Environmetrics, 5, 335–363, https://doi.org/10.1002/env.3170050311, 1994.

Proc. IAHS, 385, 203–209, 2024
https://doi.org/10.5194/piahs-385-203-2024

Open Access
PIAHS

Assessment of Hydrological Changes in Godavari River Basin Under the Impacts of El-Niño

Chandni Thakur[1], **Kasiapillai Sudalaimuthu Kasiviswanathan**[1], **Claudia Teutschbein**[2], **Bankaru-Swamy Soundharajan**[3], **M M Diwan Mohaideen**[1], **and Venkatesh Budamala**[1]

[1]Department of WRD&M, Indian Institute of Technology Roorkee, Roorkee, 247667, India
[2]Department of Earth Sciences, Uppsala University, Uppsala, 75236, Sweden
[3]Department of Civil Engineering, Amrita Vishwa Vidyapeetham, Coimbatore, 641112, India

Correspondence: Chandni Thakur (chandni@wr.iitr.ac.in)

Received: 25 May 2022 – Revised: 27 January 2023 – Accepted: 30 March 2023 – Published: 18 April 2024

Abstract. El Niño–Southern Oscillation (ENSO) is the most prominent driver of the inter-annual variability of Indian summer monsoon rainfall (ISMR). This study focuses on understanding the hydrological variations in Godavari River Basin (GRB) due to the weakening of ISMR during El Niño years (1980–2008), using the variable infiltration capacity (VIC) model. The entire basin was discretized into 1325 uniform grids of resolution $0.15^\circ \times 0.15^\circ$ (about 16.65 km), and hydrological parameters of the basin were analysed at each grid level for various El Niño events. Based on the Oceanic Niño Index (ONI), obtained from National Oceanic and Atmospheric Administration (NOAA), El Niño events occurred in the past were subclassified into weak (2004 and 2006), moderate (1986, 1994 and 2002), strong (1987 and 1991) and very strong (1982, 1987) events. For this study, VIC model was run for the period 1980–2008 and a composite of El Niño and normal years (1981, 1985, 1989, 1990, 1992, 1993, 1996, 2001 and 2003) was prepared to assess the impacts of El Niño events on the hydrology of GRB. Our results showed a negative correlation of precipitation, abstractions and soil moisture with the increasing magnitude of El Niño events. The quantum of precipitation was reduced during El Niño years compared to normal years, which showed the basin's exposure to more frequent droughts during El Niño events.

Keywords. El Niño impacts (SDG 13.1); ISMR variability; hydrological changes (UPH 01); VIC

1 Introduction

Indian summer monsoon rainfall (ISMR), also known as Southwest monsoon falls between the months of June and September and accounts for 70 %–90 % of the annual rainfall in India (Varikoden and Preethi, 2013). The inter-annual variability of ISMR has a profound impact on Indian agriculture and economy, being the principal source of rainfall for the Indian subcontinent. El Niño–Southern Oscillation (ENSO), a recurring inter-annual phenomenon caused by sea surface temperature (SST) anomalies over the equatorial Pacific Ocean, is identified as the most prominent climate driver of ISMR (Ihara et al., 2007; Rajeevan and Pai, 2007; Varikoden and Preethi, 2013; Schulte et al., 2020; Lenka et al., 2022). Scientific progress and ongoing research on ISMR have revealed the manifestation of droughts across India due to the occurrence of El Niño events (Varikoden et al., 2015; Duhan et al., 2018; Singh and Shukla, 2020; Dixit and Jayakumar, 2021; Rajbanshi and Das, 2021). Kumar et al. (2006) analysed a 132-year historical rainfall record and observed that El Niño events accompanied the majority of the meteorological droughts over India. In addition, numerous researchers reported a reduction in rainfall during El Niño years resulting in the weakening of ISMR with few exceptions (Geethalakshmi et al., 2009; Varikoden et al., 2015). The most recent El-Niño event during 2015–2016 ranked among the strongest three events in the past 65 years. With an ONI of 2.6, it was categorized as "very strong" and had enormous effects on rainfall and water resources in South Asia. During the 2015 monsoon season, negative rainfall anomalies were responsible for deficits of monsoon rainfall over the southeast Indian ocean (Kakatkar et al., 2018). In addition, the observed deficit in ISMR was 16 % in 2015, and the year turned out to be a drought year with adverse effects on water resources and agricultural productivity (Kulkarni et al., 2016). Hence, it is essential to investigate rainfall reductions to understand potential changes to the hydrological system during El-Niño events. We believe that such an investigation is crucial for a sustainable development of South Asian regions given the importance of reservoirs and dams for a continuous year-round water supply, irrigation and hydropower generation.

Specifically, for a country like India with rapid economic development, a crucial assessment of El Niño on hydrological processes can assist in monitoring water resources for long-term sustainability. Therefore, the present study considered the Godavari River basin (GRB), India's second-largest river basin and significant breadbaskets. In support of this study, Dixit et al. (2021) performed a frequency analysis of drought events in GRB for future scenarios and demonstrated the basin's exposure to more frequent droughts in the future (2053–2099). This calls for an urgent need to understand the impacts of El Niño events on the GRB hydrological phenomena. The inferences of the present investigation can be utilized to foster new policies in GRB to cope with future El Niño events in balancing the competing demands of the water and food sectors. In this study, the VIC model was employed to assess the catchment response of El Niño events. The modelling framework was tested using the data collected from the Godavari River Basin (GRB), located in the southern part of India.

2 Data and methods

2.1 Description of the study area

The GRB lies between the longitudes 73°21′–83°09′ E and latitudes 16°07′–22°50′ N, as shown in Fig. 1. GRB is the largest peninsular river, and the catchment accounts for 9.5 % of India's geographical area (i.e., 312 812 km^2). The catchment area extends over the states of Maharashtra, Madhya Pradesh, Andhra Pradesh, Chhattisgarh, Orrisa, and Karnataka. The basin receives 80 % of its annual rainfall from ISMR, with an average annual rainfall of about 1097 mm (1971–2005). However, GRB has a tropical climate with temperatures varying between 26 and 44 °C (1969–2004) (Mishra et al., 2020). Also, GRB has more than 15 major water infrastructures built across the river with a combined storage capacity of more than 21 BCM to serve irrigation, water supply, and hydropower production. Moreover, the installed hydropower generation capacity is more than 2.7 GW, and the total irrigated area is around 6000 km^2. Similar to many large river basins which experience the impacts of extreme climate events, GRB is also vulnerable to extreme calamities. For instance, the downstream regions of GRB has been experiencing the floods while severe droughts were reported in the significant portion of the upstream of the basin.

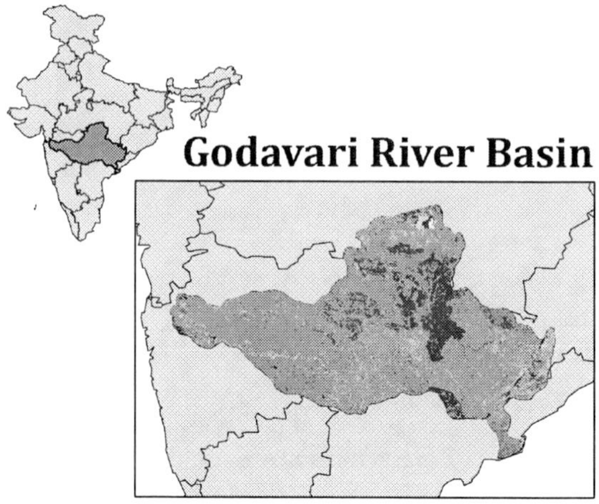

Figure 1. Map showing location of GRB in India.

2.2 Datasets

A wide range of inputs, such as meteorological forcing and physiographic features (i.e., elevation, soil, land cover and vegetation information), are necessary to setup and run the VIC model. Meteorological forcing variables such as rainfall, minimum temperature (T_{min}), and maximum temperature (T_{max}) were obtained from India Meteorological Department (IMD). The gridded rainfall datasets with a spatial resolution of 0.25° × 0.25° (Pai et al., 2014), daily minimum and maximum air temperature datasets of 1° × 1° resolution (Srivastava et al., 2009) were used as the observed forcing to simulate VIC model.

Digital Elevation Model (DEM) of 30 m spatial resolution developed by Shuttle Radar Topographic Mission (SRTM) was used to delineate and extract terrain-related information of GRB. The Decadal Land Use Land Cover (LULC) map of 100 m spatial resolution for 2005 was obtained from NASA's ORNL DAAC portal. The soil information was extracted from a 30 arcsec Harmonized World Soil Database (HWSD), obtained from the Food and Agriculture Organization (FAO). HWSD raster comprising of soil attributes up to a depth of one meter, bifurcating the soil layer into topsoil (0–30 cm) and subsoil (30–100 cm) was used. The observed daily streamflow data for GRB was obtained from 1980 to 2008 from the Central Water Commission (CWC), Government of India, to validate the VIC model simulations. ENSO conditions were defined based on the mean monthly ONI values from National Oceanic and Atmospheric Administration (NOAA) (https://origin.cpc.ncep.noaa.gov/products/analysis_monitoring/ensostuff/ONI_v5.php, last access: 15 April 2022).

2.3 Variable Infiltration Capacity Model (VIC)

VIC model is a macroscale and physically-based semi-distributed model which simulates water and energy fluxes independently at each grid cell (Liang et al., 1994). Prominent features of the model include (a) simulation of each grid cell independently, (b) explicit representation of sub-grid vegetation and elevation heterogeneity, and (c) consideration of non-linear baseflow. The water balance at each time step in the VIC model is based on the continuity equation (Liang et al., 1994) as described below:

$$\partial S/\partial t = P - E - R \quad (1)$$

where $\partial S/\partial t$ is the change in water storage, P is precipitation, E is evapotranspiration and R is runoff. The units of all the above-mentioned variables are mm within chosen timestep (daily scale was adopted in this paper).

VIC curve accounts for the spatial heterogeneity of runoff generation in the model, whereas baseflow generation is based on the Arno Model conceptualization (Franchini and Pacciani, 1991). The Penman-Monteith equation estimates the evapotranspiration (ET) for each vegetation type defined in the model. The simulated runoff and baseflow fluxes from each grid cell are utilized by a separate routing model based on the linearized Saint–Venant equation to get the total discharge at the selected gauge stations (Lohmann et al., 1996, 1998).

2.4 Modelling framework

The flowchart (Fig. 2) represents the methodology adopted in this study. The VIC model was employed in the study for hydrological modelling of GRB. The total area of the basin is 312 812 km^2 and discretized into uniform grids of resolution $0.15° \times 0.15°$ (about 277.22 km^2, a total of 1325 grids) and all input parameters were established at each grid level. Genetic algorithm (GA) optimization was employed for finding the optimal parameter set of VIC model (Whitley, 1994). Parameterization at each grid level was derived from thematic layers such as LULC map, FAO-HWSD global soil map, and DEM to obtain model parameters in the prescribed format (ASCII file format). Grid-wise flux outputs computed from VIC-3L were fed to a standalone routing model to simulate daily streamflow at the outlet (Polavaram). Different performance metrics like the coefficient of determination (R^2) and Nash–Sutcliffe Efficiency (NSE), with a threshold limit of above 0.75, were adopted to analyse the model behaviour for calibration and validation. After calibration of the VIC model, grid-wise flux outputs, namely surface runoff, baseflow, soil moisture, and evapotranspiration, were analysed for various El Niño episodes. The present study applied the ONI, which represents the running 3-month mean of sea surface temperature (SST) anomalies for the Niño 3.4 region to identify El Niño years. El Niño events were defined as five consecutive overlapping 3-month periods at or above the +0.5 anomaly, and the events were further subclassified into weak (with a 0.5 to 0.9 SST anomaly), moderate (1.0 to 1.4), strong (1.5 to 1.9) and very strong (≥ 2.0) events (Table 1). Finally, the percentage change in water balance components of GRB, namely precipitation, soil moisture and abstractions, were compared with normal years for various classes of El Niño events.

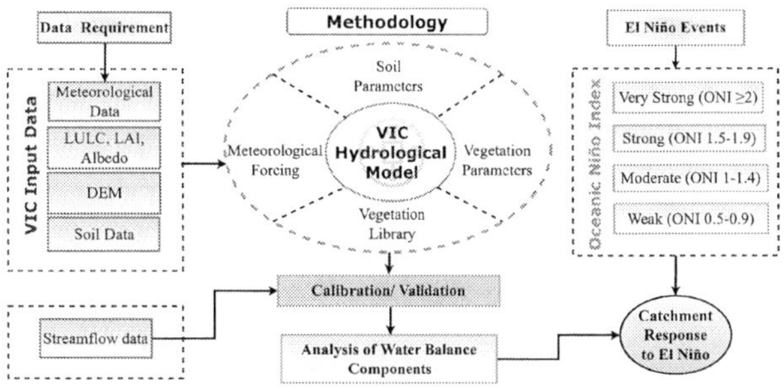

Figure 2. Flowchart describing the methodology.

Table 1. El-Niño classification based on ONI and corresponding years (1980–2008).

El-Niño Events	Years
Weak	2004, 2006
Moderate	1986, 1994, 2002
Strong	1987, 1991
Very Strong	1982, 1997
Normal (without El-Niño Effect)	1981, 1985, 1989, 1990, 1992, 1993, 1996, 2001, 2003

3 Results and discussion

The model parameters were calibrated using GA and the results are not presented here for brevity. After that, simulations were performed and compared with observed streamflow. The observed daily streamflow data collected for the period 1980–2008 was used to develop the model. The first 21 years of data were used for the calibration (1980–2000) and the remaining 8 years (2001–2008) of data were used for validation of the model. The R^2 and NSE for model calibration resulted 0.85 and 0.78 respectively. In the model validation, the values were 0.83 and 0.76 respectively. By looking at the similar values of NSE and R^2, it is clear that the overall model generalization is very good, which is also clear from the scatter plot (Fig. 3). A composite of weak, moderate, strong, and very strong El-Niño and normal years was prepared from calibrated VIC model simulations for various hydrological components at each grid level. Percentage of change (POC) in precipitation, abstractions and soil moisture for El Niño years was computed and compared to normal years which is explained in subsequent sections.

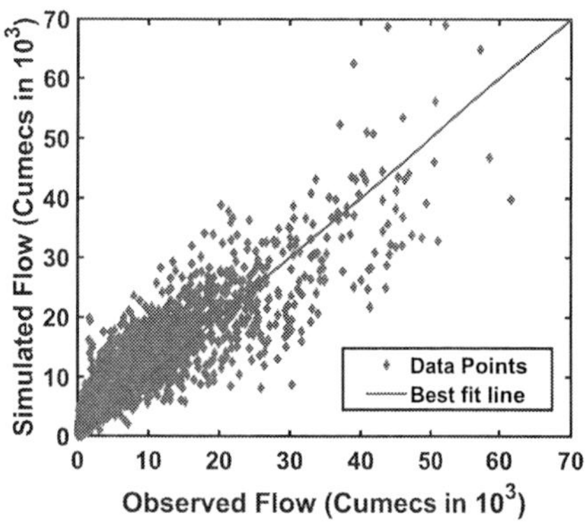

Figure 3. Comparison of calibrated model with observed data through scatter plot.

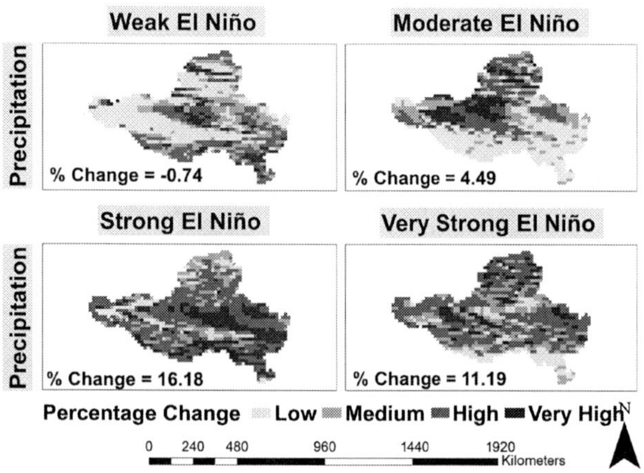

Figure 4. Percentage change of precipitation for various El Niño events comparing with normal years.

3.1 Precipitation: El-Niño years vs. Normal years

An inverse relationship between the average precipitation and El Niño intensity from moderate to very strong was noticed for GRB during 1980–2008 (Fig. 4). The positive correlation in average precipitation for weak El-Niño years could be due to spatial variations in impacts of the events in different basin regions, resulting in an overall increase of precipitation in the basin. POC in precipitation was observed with the index of −0.74 for weak El-Niño years, and a significant increase in POC was noticed with increasing ONI. This implies that the amount of precipitation received during El Niño years was lower than normal years, and the total precipitation would have been even less with the increasing magnitude of El Niño events. These findings very well match with previous studies (Ihara et al., 2007; Rajeevan and Pai, 2007).

Spatial variation of precipitation over the basin revealed that the upstream region of GRB was less influenced during weak El Niño episodes than the middle and downstream regions. In contrast, the POC in precipitation was noticeably higher for upstream and middle Godavari regions for moderate El Niño episodes. Furthermore, the impact of El Niño events was becoming more prominent when there was an increasing magnitude of the event. However, POC was observed lower for strong El Niño than the very strong El Niño episodes. It may be noted that the study did not quantify the relative impacts of tropical climate patterns such as the Indian Ocean Dipole (IOD), and the tropical Atlantic variability (ATL) on the ISMR. Therefore, a possible reason for lower POC in the case of a strong ENSO than the very strong ENSO for precipitation may be due to the co-occurrence of the El Niño and IOD events in the Indo-Pacific Ocean sector or over India. Ashok et al. (2004) showed that IOD significantly reduces the impact of ENSO on the ISMR whenever IOD and ENSO events co-occur. Cherchi and Navarra (2013) also showed large anomalies in the characteristics of ISMR due to the occurrence of ENSO and IOD events.

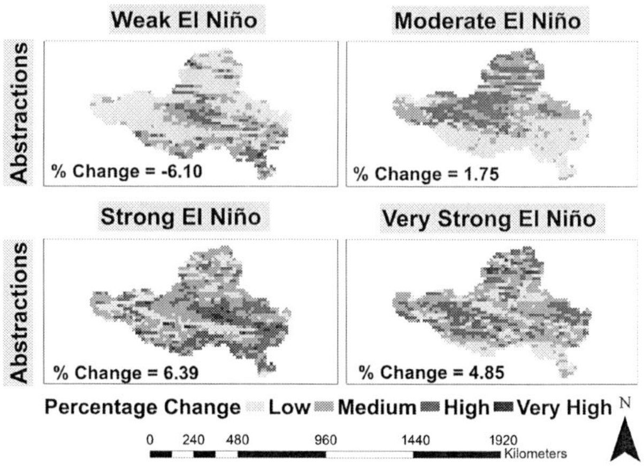

Figure 5. Percentage change of abstractions for various El Niño events comparing with normal years.

3.2 Abstractions: El-Niño years vs. Normal years

An apparent increase in POC for abstractions was observed with the increasing intensity of El Niño from moderate to very strong (Fig. 5). In contrast, POC in abstractions was observed to be −6.10 for weak El-Niño years due to increased precipitation indicating more water available for losses. A significant increase in POC for abstractions was noticed with increasing ONI from moderate to very strong El Niño. This can be attributed to the fact that there was less soil moisture availability during the El Niño years due to the reduction in precipitation, which resulted in decreased evapotranspiration and other losses.

3.3 Soil Moisture: El-Niño years vs. Normal years

POC in soil moisture was highest for very strong El Niño, which means soil moisture availability is reduced during El

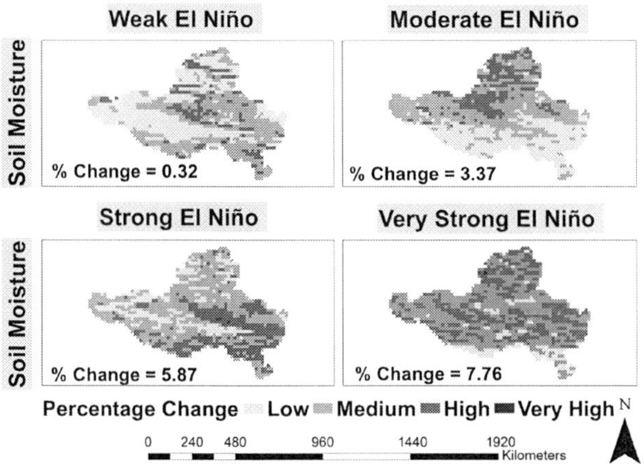

Figure 6. Percentage change of soil moisture for various El Niño events comparing with normal years.

Table 2. Summary statistics of the hydrological components of GRB for El-Niño and Normal years from 1980 to 2008.

Years	Statistics	Rainfall (mm)	Soil moisture (mm)	Flow ($m^3 s^{-1}$)
Normal	Mean	1131.65	309.25	2845.06
	Median	1140.94	306.82	2718.26
	Highest	1539.65	359.67	5797.76
	Lowest	913.35	279.84	1792.03
	SD	175.12	22.06	1224.57
	CV	15.47	7.13	43.04
El Niño	Mean	1050.82	295.55	2422.13
	Median	955.10	287.58	1810.71
	Highest	1351.75	330.53	4558.78
	Lowest	895.67	279.96	1150.83
	SD	168.98	18.25	1229.79
	CV	16.08	6.18	50.77

Niño years (Fig. 6). This can be explained relating the low quantum of precipitation for the El-Niño years, which resulted in the reduction of soil water recharge, less percolation, and hence less soil moisture availability in soil layers. A similar finding in the spatial variability of soil moisture was observed in various regions of GRB. Overall, it was found that El Niño-induced impacts were high for intense El Niño episodes throughout the basin.

3.4 Summary statistics of hydrological components: El-Niño years vs. Normal years

The analysis of summary statistics computed from the model simulations corresponding to the normal as well as ENSO years (Table 2) revealed that the average annual precipitation for El Niño years (1050.82 mm) is lesser than normal years (1131.65 mm). Similar to our findings, Bothale and Katpatal (2016) showed a precipitation deficit during El Niño years over the Pranhita catchment (a sub-catchment of GRB). Our analysis also highlighted a significant reduction in annual average streamflow from 2845.06 $m^3 s^{-1}$ in normal years to 2422.13 $m^3 s^{-1}$ during El-Niño years (Table 2). Additionally, we compared the rainfall intensity obtained from 1 and 3 d rainfall depths for El Niño and normal years from 1980–2008. For normal years, the mean and standard deviation of maximum daily rainfall intensity was found to be 121.56 and 86.10 mm d^{-1}, respectively, while 87.21 and 26.47 mm d^{-1} were reported for El-Niño years. Similarly, the mean and standard deviation of maximum 3 d rainfall intensity was found to be 61.87 and 51.72 mm d^{-1} (normal years), respectively, while 46.59 and 18.82 mm d^{-1} were reported for El-Niño years. In addition, the number of rainy days were computed for El Niño and normal years from 1980–2008. The mean and standard deviation of rainy days was found to be 81 and 4.83 d for normal years, respectively, while 70 and 13.01 d were observed during El-Niño years. The results showed the reduction in rainfall intensity and the associated number of rainy days for El-Niño years than normal years. Furthermore, the analysis revealed that rainfall deficits during El Niño years could be the major reason for the reduction in soil water recharge, less percolation, and hence less soil moisture availability in soil layers. Consequently, insufficient soil moisture during the crop season has a high risk of leading to agricultural drought and crop failure (Sai et al., 2016). Numerous researchers have shown the occurrence of droughts in GRB due to El Niño events, particularly in the Marathwada region of Maharashtra (Dixit and Jayakumar, 2021; Chavadekar and Kashid, 2021). Additionally, Kulkarni et al. (2016) demonstrated a dramatic 52 % decline in pulse (Tur, Moong and Urad Dal) output in Maharashtra during the drought of 2015 caused by the El-Niño event. Hence, El Niño events have adverse impacts on the water resources and agricultural productivity in GRB.

4 Conclusions

This study primarily focuses on advancing the current understanding of the impacts of the El Niño events on GRB. El Niño events significantly impacted the various hydrological components of the basin such as precipitation, abstractions, and soil moisture. Analysis of hydrological components for a composite of weak, moderate, strong, and very strong El-Niño against the normal years showed a negative correlation of average precipitation, abstractions, and soil moisture with increasing magnitude of El Niño events. In particular, El Niño years received less precipitation and the maximum percentage decrease was observed to be 16.18 % for strong El Niño events. The results also revealed a reduction in rainfall intensity (1 and 3 d rainfall intensity) and the associated rainy days, resulting in an annual average precipitation deficit from 1131.65 to 1050.82 mm for the period 1980–2008. The reduced quantum of precipitation and consequently insufficient soil moisture during the El Niño years indicated adverse impacts on the water resources and agricultural productivity of the basin. Hence, it can be concluded that the El-Niño impact assessment is essential when analysing the changes in the hydrological response of the basin. Also, El Niño-induced consequences in hydrological components of GRB can help water resources managers to minimize potential future impacts and facilitate adaptation to future El Niño events.

Data availability. Most of the datasets in this study are available online. The specific link for ONI values is provided in Sect. 2.2. All datasets including the observed discharge data for Godavari River Basin are available from the corresponding author upon reasonable request.

Author contributions. ChT, KSK, BSS and ClT conceptualized the modeling framework presented in the paper. ChT and KSK wrote the initial draft of the paper and interpreted the findings. While MMDM assisted in setting up the VIC model, VB contributed in preparing the figures and analyzed the validity of results. BSS and ClT provided valuable inputs and critically reviewed the paper.

Competing interests. The contact author has declared that none of the authors has any competing interests.

Disclaimer. Publisher's note: Copernicus Publications remains neutral with regard to jurisdictional claims in published maps and institutional affiliations.

Special issue statement. This article is part of the special issue "IAHS2022 – Hydrological sciences in the Anthropocene: Variability and change across space, time, extremes, and interfaces". It is a result of the XIth Scientific Assembly of the International Association of Hydrological Sciences (IAHS 2022), Montpellier, France, 29 May–3 June 2022.

Acknowledgements. The authors are thankful to Swedish Research Council (SRC) for providing financial support.

Financial support. This research has been supported by the Swedish Research Council (SRC) during the study period (grant no. 2018-05848).

Review statement. This paper was edited by Christophe Cudennec and reviewed by Rajendra Prasad and one anonymous referee.

References

Ashok, K., Guan, Z., Saji, N. H., and Yamagata, T.: Individual and combined influences of ENSO and the Indian Ocean Dipole on the Indian summer monsoon, J. Climate, 17, 3141–3155, https://doi.org/10.1175/15200442(2004)017<3141:IACIOE>2.0.CO;2, 2004.

Bothale, R. V. and Katpatal, Y. B.: Trends and Anomalies in Extreme Climate Indices and Influence of El Niño and La Niña over Pranhita Catchment in Godavari Basin, India, J. Hydrol. Eng., 21, 05015023, https://doi.org/10.1061/(asce)he.1943-5584.0001283, 2016.

Chavadekar, A. U. and Kashid, S. S.: Meteorological drought prediction of marathwada subdivision based on hydro-climatic inputs using genetic programming, ISH Journal of Hydraulic Engineering, 27, 229–241, https://doi.org/10.1080/09715010.2019.1620647, 2021.

Cherchi, A. and Navarra, A.: Influence of ENSO and of the Indian Ocean Dipole on the Indian summer monsoon variability, Clim. Dynam., 41, 81–103, https://doi.org/10.1007/s00382-012-1602-y, 2013.

Dixit, S. and Jayakumar, K. V.: A study on copula – based bivariate and trivariate drought assessment in Godavari River basin and the teleconnection of drought with large – scale climate indices, Theor. Appl. Climatol., 1335–1353, https://doi.org/10.1007/s00704-021-03792-w, 2021.

Dixit, S., Tayyaba, S., and Jayakumar, K. V.: Spatio-temporal variation and future risk assessment of projected drought events in the Godavari River basin using regional climate models, J. Water Clim. Change, 12, 3240–3263, https://doi.org/10.2166/wcc.2021.093, 2021.

Duhan, D., Pandey, A., and Srivastava, P.: Rainfall variability and its association with El Niño Oscillation in Tons River Basin, India, Meteorol. Atmos. Phys., 130, 405–425, https://doi.org/10.1007/s00703-017-0525-x, 2018.

Franchini, M. and Pacciani, M.: Comparative analysis of several conceptual rainfall-runoff models, J. Hydrol., 122, 161–219, https://doi.org/10.1016/0022-1694(91)90178-K, 1991.

Geethalakshmi, V., Yatagai, A., Palanisamy, K., and Umetsu, C.: Impact of ENSO and the Indian Ocean Dipole on the north-east monsoon rainfall of Tamil Nadu State in India, Hydrol. Process., 647, 633–647, https://doi.org/10.1002/hyp.7191, 2009.

Ihara, C., Kushnir, Y., Cane, M. A., and Peña, V. H. D. L.: Indian summer monsoon rainfall and its link with ENSO and Indian Ocean climate indices, Int. J. Climatol., 187, 179–187, https://doi.org/10.1002/joc, 2007.

Kakatkar, R., Gnanaseelan, C., Chowdary, J. S., Parekh, A., and Deepa, J. S.: Indian summer monsoon rainfall variability during 2014 and 2015 and associated Indo-Pacific upper ocean temperature patterns, Theor. Appl. Climatol., 1235–1247, https://doi.org/10.1007/s00704-017-2046-4, 2018.

Kulkarni, A., Gadgil, S., and Patwardhan, S.: Monsoon variability, the 2015 Marathwada drought and rainfed agriculture, Curr. Sci. India, 111, 1182–1193, https://doi.org/10.18520/cs/v111/i7/1182-1193, 2016.

Kumar, K. K., Rajagopalan, B., Hoerling, M., and Cane, M.: Unraveling the Mystery of Indian Monsoon Failure During El Niño, Science, 314, 115–119, https://doi.org/10.1126/science.1131152, 2006.

Lenka, S., Devi, R., Joseph, C. M., and Gouda, K. C.: Effect of large – scale oceanic and atmospheric processes on the Indian summer monsoon, Theor. Appl. Climatol., 1561–1576, https://doi.org/10.1007/s00704-021-03896-3, 2022.

Liang, X., Lettenmaier, D. P., Wood, E. F., and Burges, S. J.: A simple hydrologically based model of land surface water and energy fluxes for general circulation models, J. Geophys. Res., 99, 14415–14428, https://doi.org/10.1029/94jd00483, 1994.

Lohmann, D., Nolte-holube, R., and Raschke, E.: A large-scale horizontal routing model to be coupled to land surf ace parametrization schemes, Tellus A, 48, 708–721, https://doi.org/10.3402/tellusa.v48i5.12200, 1996.

Lohmann, D., Raschke, E., Nijssen, B., and Lettenmaier, D. P.: Regional scale hydrology : I . Formulation of the VIC-2L model coupled to a routing model, Hydrolog. Sci. J., 43, 131–141, https://doi.org/10.1080/02626669809492107, 1998.

Mishra, V., Shah, H., López, M. R. R., Lobanova, A., and Krysanova, V.: Does comprehensive evaluation of hydrological models influence projected changes of mean and high flows in the Godavari River basin?, Climatic Change, 163, 1187–1205, https://doi.org/10.1007/s10584-020-02847-7, 2020.

Pai, D. S., Sridhar, L., Rajeevan, M., Sreejith, O. P., Satbhai, N. S., and Mukhopadhyay, B.: Development of a new high spatial resolution ($0.25° \times 0.25°$) long period (1901–2010) daily gridded rainfall data set over India and its comparison with existing data sets over the region, MAUSAM, 65, 1–18, https://doi.org/10.54302/mausam.v65i1.851, 2014.

Rajbanshi, J. and Das, S.: The variability and teleconnections of meteorological drought in the Indian summer monsoon season: Implications for staple crop production, J. Hydrol., 603, 126845, https://doi.org/10.1016/j.jhydrol.2021.126845, 2021.

Rajeevan, M. and Pai, D. S.: On the El Niño-Indian monsoon predictive relationships, Geophys. Res. Lett., 34, 1–4, https://doi.org/10.1029/2006GL028916, 2007.

Sai, M. S., Murthi, C. S., Chandrasekar, K., Jeyaseelan, A. T., Diwakar, P. G., and Dadhwal, V. K.: Agricultural drought: Assessment & monitoring, MAUSAM, 67, 131–142, https://doi.org/10.54302/mausam.v67i1.1155, 2016.

Schulte, J., Policielli, F., and Zaitchik, B.: A skewed perspective of the Indian rainfall–El Niño–Southern Oscillation (ENSO) relationship, Hydrol. Earth Syst. Sci., 24, 5473–5489, https://doi.org/10.5194/hess-24-5473-20200, 2020.

Singh, R. M. and Shukla, P.: Drought Characterization Using Drought Indices and El Niño Effects, Natl. Acad. Sci. Lett., 43, 339–342, https://doi.org/10.1007/s40009-019-00870-6, 2020.

Srivastava, A. K., Rajeevan, M., and Kshirsagar, S. R.: Development of a high resolution daily gridded temperature data set (1969–2005) for the Indian region, Atmos. Sci. Let., 10, 249–254, https://doi.org/10.1002/asl.232, 2009.

Varikoden, H. and Preethi, B.: Wet and dry years of Indian summer monsoon and its relation with Indo-Pacific sea surface temperatures, Int. J. Climatol., 1761–1771, https://doi.org/10.1002/joc.3547, 2013.

Varikoden, H., Revadekar, J. V., Choudhary, Y., and Preethi, B.: Droughts of Indian summer monsoon associated with El Niño and Non-El Niño years, Int. J. Climatol., 1925, 1916–1925, https://doi.org/10.1002/joc.4097, 2015.

Whitley, D.: A genetic algorithm tutorial, Stat. Comput., 4, 65–85, https://doi.org/10.1007/BF00175354, 1994.

Proc. IAHS, 385, 211–218, 2024
https://doi.org/10.5194/piahs-385-211-2024

Evaluating the skill of the mesoscale Hydrologic Model (mHM) for discharge simulation in sparsely-gauged basins in Nigeria

Kingsley Nnaemeka Ogbu[1,2,4], Oldrich Rakovec[2], Luis Samaniego[2,5], Gloria Chinwendu Okafor[3], Bernhard Tischbein[1], and Hadush Meresa[1]

[1]Center for Development Research, University of Bonn, Bonn, Germany
[2]Department of Computational Hydrosystems, UFZ-Helmholtz Centre for Environmental Research, Leipzig, Germany
[3]Department of Meteorology and Climate Change, Nigeria Maritime University, Okerenko, Nigeria
[4]Department of Agricultural and Bioresources Engineering, Nnamdi Azikiwe University, Awka, Nigeria
[5]University of Potsdam, Institute of Environmental Science and Geography, Am Neuen Palais 10, 14469 Potsdam, Germany

Correspondence: Kingsley Nnaemeka Ogbu (kn.ogbu@unizik.edu.ng)

Received: 29 May 2022 – Revised: 4 February 2023 – Accepted: 29 March 2023 – Published: 18 April 2024

Abstract. Predictive hydrologic modelling to understand and support agricultural water resources management and food security policies in Nigeria is a demanding task due to the paucity of hydro-meteorological measurements. This study assessed the skill of using different remotely sensed rainfall products in a multi-calibration framework for evaluating the performance of the mesoscale hydrologic Model (mHM) across four different data-scarce basins in Nigeria. Grid-based rainfall estimates obtained from several sources were used to drive the mHM in different basins in Nigeria. Model calibration was first performed using only discharge records, and also by using a combination of discharge and actual evapotranspiration, forced with different rainfall products. The mHM forced with CHIRPS produced reasonable Kling-Gupta efficiency KGE) results ($0.5 > \mathrm{KGE} < 0.85$) under both calibration frameworks. However, constraining model parameters under a multi-calibration arrangement showed no significant discharge simulation improvement in this study. Results show the utility of the mHM for discharge simulation in data-sparse basins in Nigeria.

Keywords. UPH 20; MPR; discharge; CHIRPS

1 Introduction

Modelling of hydrologic systems is crucial to understanding how climatic variables drive hydrologic responses, which are highly sensitive to land use/cover changes and population dynamics. Quantifying the components of the water balance model is strategic to policymakers and water resources managers for developing key water resources management projects (Nesru et al., 2020). In Nigeria, in-situ measurement of hydrologic variables is largely constrained by financial instability, weak institutions and political instability resulting in a steady decline and uneven distributions of existing hydro-meteorological networks (Poméon et al., 2018; Adeoti, 2020). Precipitation acts as a key forcing in the hydrologic system and affects the spatial and temporal variability of other hydrologic processes. Gridded rainfall products provide continuous and spatially-homogenous estimates and have become an alternative, especially in data-scarce regions (Ayehu et al., 2018). However, their ability to reproduce observed hydrologic processes is a precondition for use in water resources modelling (Dembélé et al., 2020).

The availability of remotely-sensed datasets (elevation, soil, land use/cover, climatic variables) led to the development of complex hydrologic models but these efforts did not impact model results (Poméon et al., 2018). This is because the traditional method of model calibration which involves determining the best model parameter set could reproduce observed hydrologic variables but misrepresent important watershed processes (Rajib et al., 2018). The importance of understanding hydrologic processes and improving their mathematical representation is well highlighted during the International Association of Hydrologic Sciences (IAHS) scientific decade (2003–2012) (Hrachowitz et al., 2013). In other to overcome the problems of over-parameterization and equifinality, the Mesoscale Parameter Regionalization (MPR) proposed by Samaniego et al. (2010) presents a technique which links model parameters at a coarser scale with their counterparts at a finer resolution using pedotransfer functions, whereby only the global parameters that define these relationships are obtained through calibration. In comparison with other regionalization methods (e.g. standard regionalization), MPR showed superiority in preserving the spatial variability of state variables and overall performance of model hydrologic processes simulation (Samaniego et al., 2010; Kumar et al., 2010). The MPR technique reduces the number of free mHM calibration parameters and seeks to address Question 20 (reducing model uncertainty) of the Unsolved Problems in Hydrology (UPH 20) (Blöschl et al., 2019). In this study, the key research question is: what is the performance of the mHM-MPR technique in reproducing the temporal variation of the streamflow process under a paucity of input-data conditions? This is the first attempt in applying the mHM-MPR technique for hydrologic modelling in data-scarce basins within Nigeria. This study assessed the suitability of using gridded-rainfall products in a multi-calibration framework for evaluating the performance of mHM for river discharge simulations across four (4) different data-scarce basins in Nigeria.

2 Methodology

2.1 Study Area

The study area consists of four (4) river basins located in the northern part of Nigeria (Lat 4.00–14.00° N, Lon 2.00–15.00° E), which were selected based on the availability of discharge data: Jamaare (13 929.711 km^2), Hadejia (16 820.336 km^2), Kaduna (64 848.594 km^2) and Oroo (4500.174 km^2) as shown in Fig. 1. Rainfall in this region is unimodal and is impacted by the movement of the Inter-Tropical Discontinuity (ITD). Jamaare, Hadejia and Kaduna basins receive an annual mean rainfall of about 434–969 mm while the Oroo basin is characterized by an annual mean of 897–1535 mm (Gbode et al., 2019; Ogbu et al., 2020).

2.2 The Mesoscale Hydrologic Model (mHM)

The mHM is a grid-based, conceptual and fully distributed hydrologic model which simulates various hydrologic processes (evapotranspiration, infiltration, surface and subsurface runoff, etc.) that are formulated based on the HBV model (Kumar et al., 2013; Samaniego et al., 2010). Spatial-temporal simulations of hydrologic processes in the mHM are processed at the grid/cell scale. Three levels of gridded information are required for mHM set-up; Level-0 (basin characteristics), Level-1 (dominant hydrological processes) and Level-2 (meteorological datasets), to account for sub-grid variability (Kumar et al., 2013). The MPR proposed by a study (Samaniego et al., 2010) is the main feature in the mHM model and serves to bridge the gap between observations and the basin scale (Rakovec et al., 2019). The MPR involves a two-step parameterization procedure (see Fig. 2); (1) model parameters at Level-0 are regionalized by linking them with their corresponding basin characteristics through linear or non-linear transfer functions; (2) In this stage, effective parameters are obtained by linking the regionalized parameters with their corresponding one at Level-0 through an upscaling operator. The major goal of the MPR is to derive global model parameters that are spatially seamless, scale-independent and transferable across locations (Rakovec et al., 2019).

2.3 Data

In this study, mHM was set up using grid-based meteorological, soil, land use and morphological datasets. Rainfall estimates were obtained from the Climate Hazards Group InfraRed Precipitation with Station data (CHIRPS), Climate Prediction Center (CPC), European Center for Medium-Range Weather Forecast (ECMWF) Reanalysis 5th Generation (ERA5), Global Precipitation Climatological Center (GPCC) and Multi-Source Weighted Ensemble Precipitation (MSWEP) and used as model forcing. Potential evapotranspiration was computed with the Hargreaves method using daily temperature data obtained from the ERA5 product. These climatic products were selected based on their performances in previous studies (Poméon et al., 2017; Dembélé et al., 2020; Ogbu et al., 2020, 2022; Hounguè et al., 2021) within the West African region. Soil attributes for six different soil layers were extracted from the Harmonized World Soil Database, version 1.2 while Land use and cover information were obtained from Globecover product. Slope, aspect, flow accumulation and flow direction were derived from a 90 m resolution digital elevation model obtained from the Shuttle Radar Topographic Mission database. mHM calibration and validation were performed using discharge and AET information obtained from the German-developed Global Runoff Data Center (GRDC) and Global Land Evaporation Amsterdam Model (GLEAM) databases, respectively.

2.4 Model set-up

mHM Version 5.11 (Samaniego et al., 2021) was set up in four (4) data-scarce basins while varying rainfall inputs for each setup. In the first case, calibration and validation were

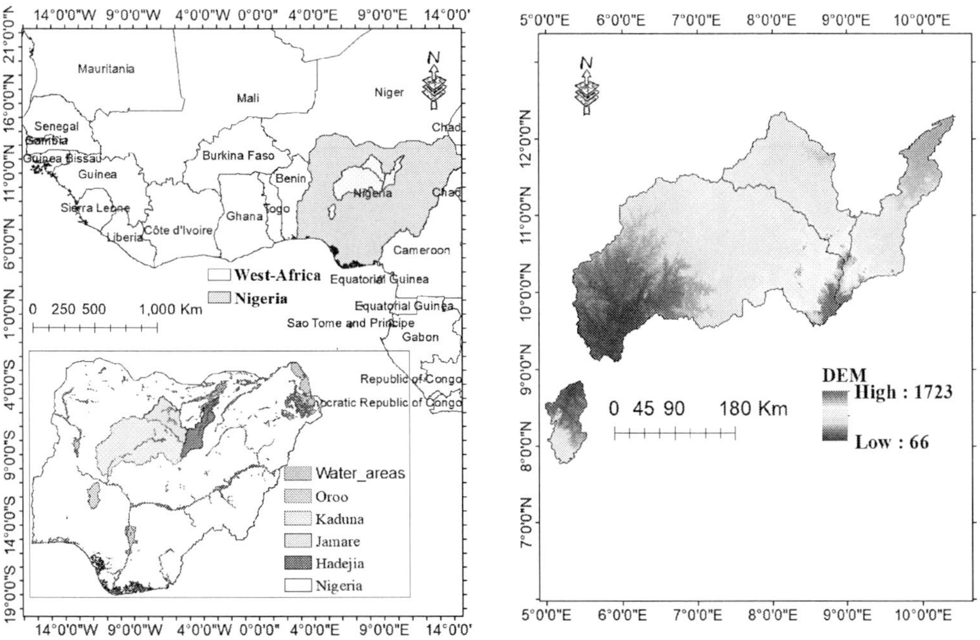

Figure 1. Study locations – Jamaare, Hadejia, Kaduna and Oroo River Basins.

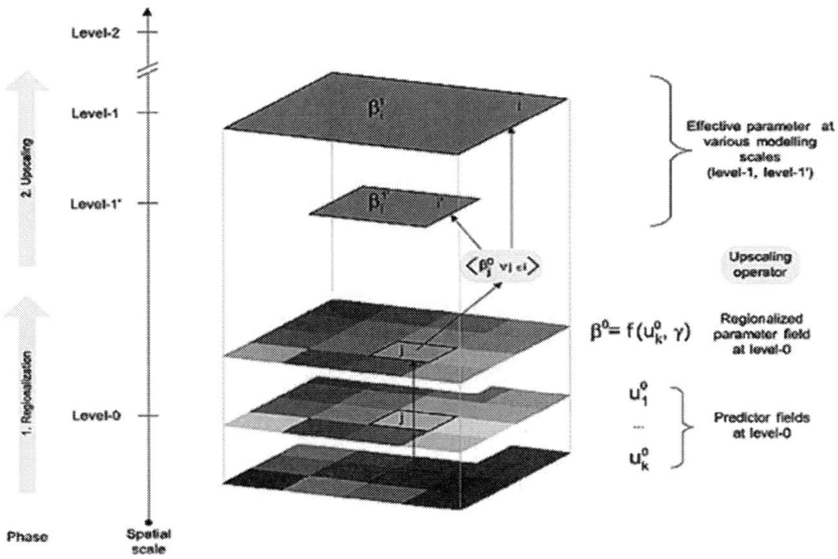

Figure 2. Schematic diagram of MPR (Samaniego et al., 2010).

performed using only Qobs while in the second case, they were performed using Qobs and AET. Simulation and calibration periods vary for all basins due to the paucity of data and large gaps in existing discharge time series. Discharge (Q) optimization was performed using the Dynamically Dimensioned Search (DDS) (Tolson and Shoemaker, 2007) algorithm (4000 iterations), based on the Kling-Gupta Efficiency (KGE) (Kling et al., 2012; Gupta et al., 2009) metric at a daily time step. Best model parameter sets are obtained with the DDS algorithm by using about 10 %–20 % of the number of iterations required by the Shuffle Complex Evolution optimization method (Rakovec et al., 2019). The KGE is an improved version of Nash Sutcliffe Efficiency (NSE) and constitutes correlation (r), variability and mean bias as

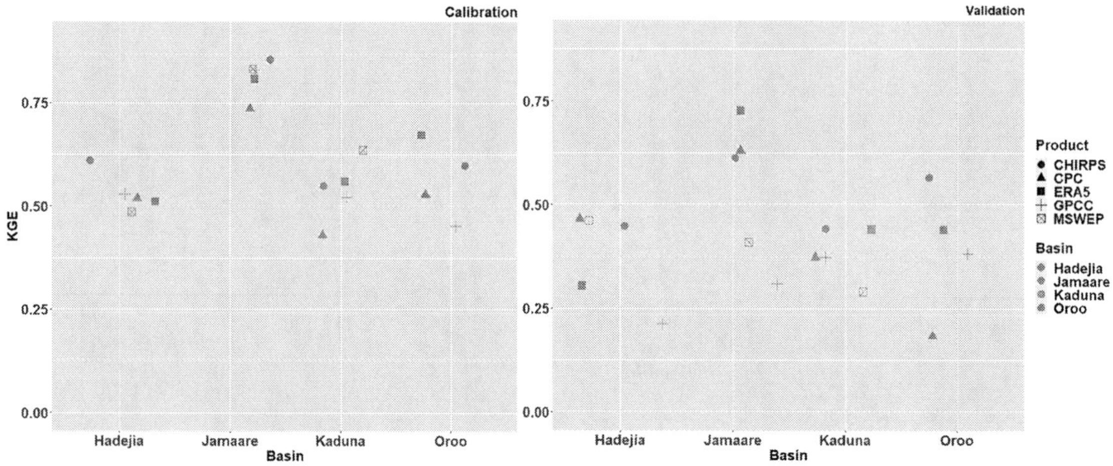

Figure 3. Qobs calibration and validation for gridded products in each basin.

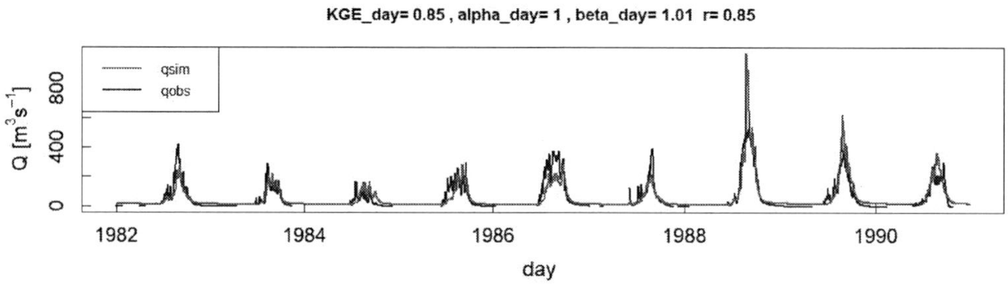

Figure 4. Daily hydrograph simulation when calibration is performed using only Qobs for Jamaare.

shown in Eq. (1).

$$\mathrm{KGE} = 1 - \sqrt{(r-1)^2 + (\alpha - 1)^2 + (\beta - 1)^2} \quad (1)$$

where, r = linear correlation, α = measure of flow variability error, β = bias.

The multivariable calibration, using Q and domain average AET (Eq. 2) was used to complement Qobs to assess if a more realistic result will be achieved.

$$\mathrm{SO}_{30} = [1 - \mathrm{KGE}(Q)] \times \mathrm{RMSE}\left(\overline{\text{basin ETA}}\right) \quad (2)$$

where, SO_{30} = mHM objective function Number 30, KGE = Kling-Gupta Efficiency, RMSE $\left(\overline{\text{basin ETA}}\right)$ = root mean square error of the basin average actual ET simulation.

3 Results and Discussion

3.1 Results

3.1.1 Model Performance for Discharge

Generally, model results varied across rainfall datasets for all domains during calibration and validation periods for Qobs simulation (Fig. 3). Daily discharge simulations showed reasonable results (KGE > 0.5) in all domains except for MSWEP (in Hadejia), GPCC (in Oroo) and CPC (in Kaduna) during model calibration. On the other hand, validation results showed KGE > 0.5 in Jamaare Basin (for ERA5, CHIRPS and CPC) and Oroo Basin (for CHIRPS).

An example of daily mHM simulated hydrographs for Jamaare River Basin during model calibration (Fig. 4) and validation (Fig. 5) showed an acceptable fit with the Qobs time series. However, the KGE value decreased from 0.85 (during model calibration) to 0.61 (during model validation) (see Figs. 4 and 5).

3.1.2 Model Performance for Qobs and AET Model Calibration Scheme

In the Qobs/AET model calibration setup, daily streamflow simulations exhibited the same trend if not slightly worse as in the Qobs model calibration scheme (Fig. 6). This result is similar to a previous mHM study (Poméon et al., 2018) in West Africa, with the Qobs model calibration scheme showing improved discharge predictions. Unsatisfactory KGE results (< 0.5) were obtained in Oroo Basin (driven by GPCC) and Kaduna Basin (forced with GPCC) during model calibra-

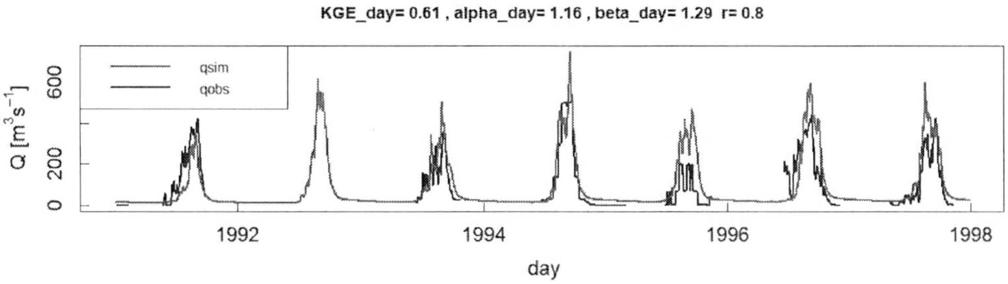

Figure 5. Daily hydrograph simulation when validation is performed using only Qobs for Jamaare.

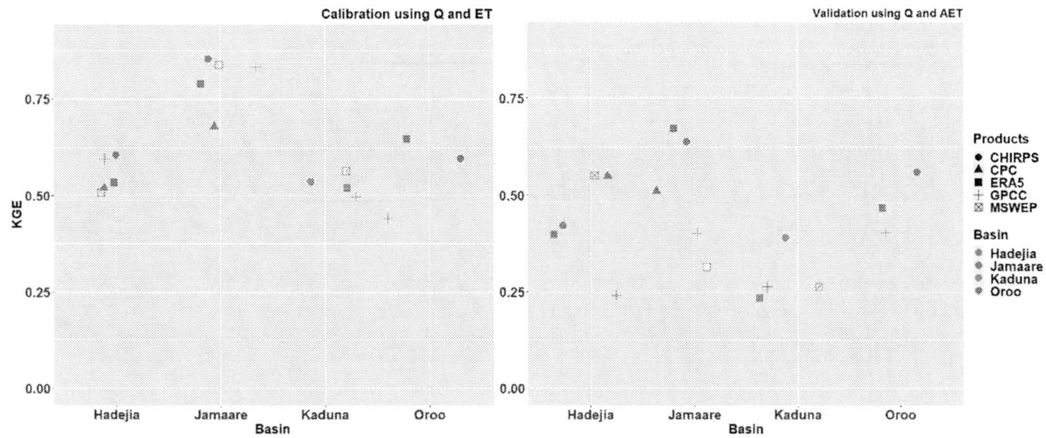

Figure 6. Calibration and validation results for gridded products using Qobs/AET calibration setup.

tion. For the validation period, discharge simulations showed acceptable results ($SO_{30} > 0.5$) in Oroo (driven by CHIRPS), Hadejia (CPC and MSWEP) and Jamaare (ERA5, CHIRPS, 0.51068). We showed example hydrographs for the Jamaare River Basin produced using Qobs/AET model calibration setup (Fig. 7) and for model validation (Fig. 8). For this example, SO_{30} value decreased from 0.85 (calibration) –0.64 (validation) while correlation followed the same trend from 0.85–0.78.

3.2 Discussions

The flow simulation performances driven by the different gridded rainfall products under the different optimization frameworks vary across the different domains, modelled in this study. Overall, simulated streamflow exhibited acceptable KGE values (KGE > 0.50) during calibration periods than during model validation for most of the rainfall products. These poor performances could be attributed to gaps in discharge observations, which exist more within validation periods. Small KGE improvements achieved when mHM was calibrated with both Qobs and AET were only evident in Hadejia and Jamaare basins and do not reflect mHM robustness for this study. This poor performance in discharge simulation displayed when model parameters are constrained under a multivariable calibration scheme was also reported in another study (Poméon et al., 2018) conducted within the West Africa region. Visual inspection of simulated and observed hydrographs showed that the CHIRPS outperformed other rainfall products in mimicking observed discharge trends as shown in Figs. 4, 5, 7 and 8. The performances of the CHIRPS dataset in most of the domains are in line with its ability in reproducing observed rainfall at a point-to-pixel scale in the West African region as shown in several studies (Ogbu et al., 2020; Poméon et al., 2017; Dembélé and Zwart, 2016). In this study, satisfactory KGE scores obtained in most domains while using remotely sensed rainfall datasets, especially during model calibration could be attributed to the implementation of the MPR technique within the mHM structure, which ensured that the number of free calibration parameters was reduced while preserving its spatial variability.

4 Conclusions

In this study, the discharge simulation skill of the mHM was evaluated in four data-limited basins located in Nigeria, under multivariable optimization setups. The MPR technique,

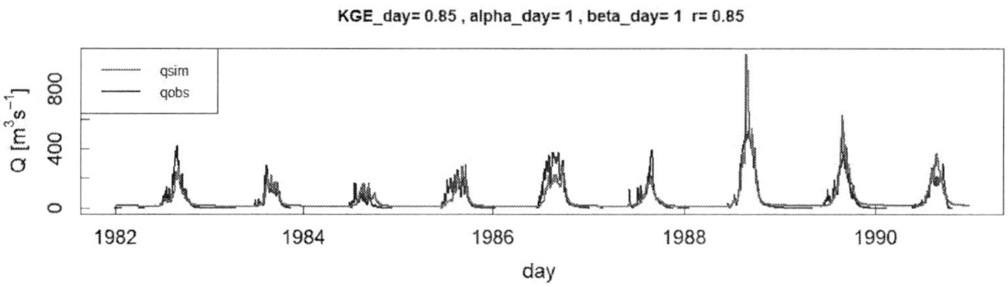

Figure 7. Optimized daily hydrograph simulation using Qobs/AET mHM model setup for Jamaare.

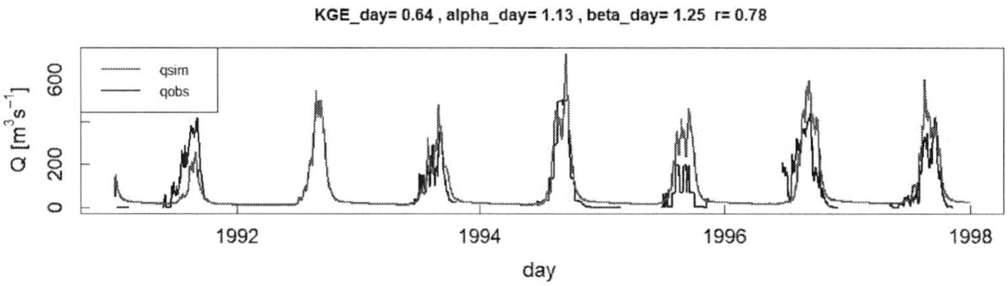

Figure 8. Daily hydrograph validation using Qobs/AET mHM model setup for Jamaare.

which integrates the spatial heterogeneity of a domain's physiographic characteristics and overcomes the problem of model over-parameterization, is well suited for application in this data-scarce region. Notwithstanding the large gaps in Qobs, the reanalysis product (ERA5) and the satellite rainfall dataset (CHIRPS) were consistent in satisfactory discharge simulations in most of the domains. However, significant improvements in discharge simulations were not observed when mHM was calibrated using Qobs and AET. Currently, the mHM lacks a reservoir component and several dams, which exist within the study domains, were not included during model setups. Furthermore, mHM accepts only three land use and land cover classes (impervious, pervious, and forest) which do not represent different existing classes. This study presents the utility of different gridded-rainfall datasets for discharge simulation in data-scarce regions in Nigeria. However, investments in hydro-climatic instrumentations should be given top priority by the government at all levels in Nigeria as remote-sensing rainfall products can only complement in-situ records.

Code and data availability. The mHM code, version 5.10 (accessed on 1 December 2020) can be obtained from Zenodo (https://doi.org/10.5281/zenodo.1069202, Samaniego et al., 2021).

Author contributions. Conceptualization: KNO, LS, BT. Formal analysis: KNO, OR. Methodology: KNO, OR, HM. Supervision: LS, BT. Visualization: KNO, OR, GCO. Writing: KNO, OR, LS, BT, GCO, HM.

Competing interests. The contact author has declared that none of the authors has any competing interests.

Disclaimer. Publisher's note: Copernicus Publications remains neutral with regard to jurisdictional claims in published maps and institutional affiliations.

Special issue statement. This article is part of the special issue "IAHS2022 – Hydrological sciences in the Anthropocene: Past and future of open, inclusive, innovative, and society-interfacing approaches". It is a result of the XIth Scientific Assembly of the International Association of Hydrological Sciences (IAHS 2022), Montpellier, France, 29 May–3 June 2022.

Acknowledgements. The first author acknowledges the providers of all freely available gridded climate dataset used in this study and the mHM team at the Department of Computational Hydrosystems, Helmholtz Centre for Environmental Research (UFZ), Germany for the computing resources used in this research.

Financial support. This work is a part of the doctoral research of the first author and was funded by Nnamdi Azikiwe University,

Awka, Nigeria and the Centre for Development Research (ZEF), University of Bonn, Germany.

Review statement. This paper was edited by Christophe Cudennec and reviewed by two anonymous referees.

References

Adeoti, O.: Constraints on data collection implementation at the river basin level in Nigeria, J. Hydrol. Reg. Stud., 32, 100738, https://doi.org/10.1016/j.ejrh.2020.100738, 2020.

Ayehu, G. T., Tadesse, T., Gessesse, B., and Dinku, T.: Validation of new satellite rainfall products over the Upper Blue Nile Basin, Ethiopia, Atmos. Meas. Tech., 11, 1921–1936, https://doi.org/10.5194/amt-11-1921-2018, 2018.

Blöschl, G., Bierkens, M. F. P., Chambel, A., Cudennec, C., Destouni, G., Fiori, A., Kirchner, J. W., McDonnell, J. J., Savenije, H. H. G., Sivapalan, M., Stumpp, C., Toth, E., Volpi, E., Carr, G., Lupton, C., Salinas, J., Széles, B., Viglione, A., Aksoy, H., Allen, S. T., Amin, A., Andréassian, V., Arheimer, B., Aryal, S. K., Baker, V., Bardsley, E., Barendrecht, M. H., Bartosova, A., Batelaan, O., Berghuijs, W. R., Beven, K., Blume, T., Bogaard, T., Borges de Amorim, P., Böttcher, M. E., Boulet, G., Breinl, K., Brilly, M., Brocca, L., Buytaert, W., Castellarin, A., Castelletti, A., Chen, X., Chen, Y., Chen, Y., Chifflard, P., Claps, P., Clark, M. P., Collins, A. L., Croke, B., Dathe, A., David, P. C., de Barros, F. P. J., de Rooij, G., Di Baldassarre, G., Driscoll, J. M., Duethmann, D., Dwivedi, R., Eris, E., Farmer, W. H., Feiccabrino, J., Ferguson, G., Ferrari, E., Ferraris, S., Fersch, B., Finger, D., Foglia, L., Fowler, K., Gartsman, B., Gascoin, S., Gaume, E., Gelfan, A., Geris, J., Gharari, S., Gleeson, T., Glendell, M., Gonzalez Bevacqua, A., González-Dugo, M. P., Grimaldi, S., Gupta, A. B., Guse, B., Han, D., Hannah, D., Harpold, A., Haun, S., Heal, K., Helfricht, K., Herrnegger, M., Hipsey, M., Hlaváčiková, H., Hohmann, C., Holko, L., Hopkinson, C., Hrachowitz, M., Illangasekare, T. H., Inam, A., Innocente, C., Istanbulluoglu, E., Jarihani, B., Kalantari, Z., Kalvans, A., Khanal, S., Khatami, S., Kiesel, J., Kirkby, M., Knoben, W., Kochanek, K., Kohnová, S., Kolechkina, A., Krause, S., Kreamer, D., Kreibich, H., Kunstmann, H., Lange, H., Liberato, M. L. R., Lindquist, E., Link, T., Liu, J., Loucks, D. P., Luce, C., Mahé, G., Makarieva, O., Malard, J., Mashtayeva, S., Maskey, S., Mas-Pla, J., Mavrova-Guirguinova, M., Mazzoleni, M., Mernild, S., Misstear, B. D., Montanari, A., Müller-Thomy, H., Nabizadeh, A., Nardi, F., Neale, C., Nesterova, N., Nurtaev, B., Odongo, V. O., Panda, S., Pande, S., Pang, Z., Papacharalampous, G., Perrin, C., Pfister, L., Pimentel, R., Polo, M. J., Post, D., Prieto Sierra, C., Ramos, M. H., Renner, M., Reynolds, J. E., Ridolfi, E., Rigon, R., Riva, M., Robertson, D. E., Rosso, R., Roy, T., Sá, J. H. M., Salvadori, G., Sandells, M., Schaefli, B., Schumann, A., Scolobig, A., Seibert, J., Servat, E., Shafiei, M., Sharma, A., Sidibe, M., Sidle, R. C., Skaugen, T., Smith, H., Spiessl, S. M., Stein, L., Steinsland, I., Strasser, U., Su, B., Szolgay, J., Tarboton, D., Tauro, F., Thirel, G., Tian, F., Tong, R., Tussupova, K., Tyralis, H., Uijlenhoet, R., van Beek, R., van der Ent, R. J., van der Ploeg, M., Van Loon, A. F., van Meerveld, I., van Nooijen, R., van Oel, P. R., Vidal, J. P., von Freyberg, J., Vorogushyn, S., Wachniew, P., Wade, A. J., Ward, P., Westerberg, I. K., White, C., Wood, E. F., Woods, R., Xu, Z., Yilmaz, K. K., and Zhang, Y.: Twenty-three unsolved problems in hydrology (UPH) – a community perspective, Hydrol. Sci. J., 64, 1141–1158, https://doi.org/10.1080/02626667.2019.1620507, 2019.

Dembélé, M. and Zwart, S. J.: Evaluation and comparison of satellite-based rainfall products in Burkina Faso, West Africa, Int. J. Remote Sens., 37, 3995–4014, https://doi.org/10.1080/01431161.2016.1207258, 2016.

Dembélé, M., Schaefli, B., van de Giesen, N., and Mariéthoz, G.: Suitability of 17 gridded rainfall and temperature datasets for large-scale hydrological modelling in West Africa, Hydrol. Earth Syst. Sci., 24, 5379–5406, https://doi.org/10.5194/hess-24-5379-2020, 2020.

Gbode, I. E., Adeyeri, O. E., Menang, K. P., Intsiful, J. D. K., Ajayi, V. O., Omotosho, J. A., and Akinsanola, A. A.: Observed changes in climate extremes in Nigeria, Meteorol. Appl., 26, 642–654, https://doi.org/10.1002/met.1791, 2019.

Gupta, H. V., Kling, H., Yilmaz, K. K., and Martinez, G. F.: Decomposition of the mean squared error and NSE performance criteria: Implications for improving hydrological modelling, J. Hydrol., 377, 80–91, https://doi.org/10.1016/j.jhydrol.2009.08.003, 2009.

Houngnè, N. R., Ogbu, K. N., Almoradie, A. D. S., and Evers, M.: Evaluation of the performance of remotely sensed rainfall datasets for flood simulation in the transboundary Mono River catchment, Togo and Benin, J. Hydrol. Reg. Stud., 36, 100875, https://doi.org/10.1016/j.ejrh.2021.100875, 2021.

Hrachowitz, M., Savenije, H. H. G., Blöschl, G., McDonnell, J. J., Sivapalan, M., Pomeroy, J. W., Arheimer, B., Blume, T., Clark, M. P., Ehret, U., Fenicia, F., Freer, J. E., Gelfan, A., Gupta, H. V., Hughes, D. A., Hut, R. W., Montanari, A., Pande, S., Tetzlaff, D., Troch, P. A., Uhlenbrook, S., Wagener, T., Winsemius, H. C., Woods, R. A., Zehe, E., and Cudennec, C.: A decade of Predictions in Ungauged Basins (PUB)–a review, Hydrolog. Sci. J., 58, 1198–1255, https://doi.org/10.1080/02626667.2013.803183, 2013.

Kling, H., Fuchs, M., and Paulin, M.: Runoff conditions in the upper Danube basin under an ensemble of climate change scenarios, J. Hydrol., 424–425, 264–277, https://doi.org/10.1016/J.JHYDROL.2012.01.011, 2012.

Kumar, R., Samaniego, L., and Attinger, S.: The effects of spatial discretization and model parameterization on the prediction of extreme runoff characteristics, J. Hydrol., 392, 54–69, https://doi.org/10.1016/j.jhydrol.2010.07.047, 2010.

Kumar, R., Samaniego, L., and Attinger, S.: Implications of distributed hydrologic model parameterization on water fluxes at multiple scales and locations, Water Resour. Res., 49, 360–379, https://doi.org/10.1029/2012WR012195, 2013.

Nesru, M., Shetty, A., and Nagaraj, M. K.: Multi-variable calibration of hydrological model in the upper Omo-Gibe basin, Ethiopia, Acta Geophys., 68, 537–551, https://doi.org/10.1007/s11600-020-00417-0, 2020.

Ogbu, K. N., Houngnè, N. R., Gbode, I. E., and Tischbein, B.: Performance evaluation of satellite-based rainfall products over Nigeria, Climate, 8, 103, https://doi.org/10.3390/cli8100103, 2020.

Ogbu, K. N., Rakovec, O., Shrestha, P. K., Samaniego, L., Tischbein, B., and Meresa, H.: Testing the mHM-MPR Reliability for Parameter Transferability across Locations in North–Central Nigeria, Hydrology, 9, 1–23, https://doi.org/10.3390/hydrology9090158, 2022.

Poméon, T., Jackisch, D., and Diekkrüger, B.: Evaluating the performance of remotely sensed and reanalysed precipitation data over West Africa using HBV light, J. Hydrol., 547, 222–235, https://doi.org/10.1016/j.jhydrol.2017.01.055, 2017.

Poméon, T., Diekkrüger, B., and Kumar, R.: Computationally efficient multivariate calibration and validation of a grid-based hydrologic model in sparsely gauged West African river basins, Water, 10, 1418, https://doi.org/10.3390/w10101418, 2018.

Rajib, A., Evenson, G. R., Golden, H. E., and Lane, C. R.: Hydrologic model predictability improves with spatially explicit calibration using remotely sensed evapotranspiration and biophysical parameters, J. Hydrol., 567, 668–683, https://doi.org/10.1016/j.jhydrol.2018.10.024, 2018.

Rakovec, O., Mizukami, N., Kumar, R., Newman, A. J., Thober, S., Wood, A. W., Clark, M. P., and Samaniego, L.: Diagnostic Evaluation of Large-Domain Hydrologic Models Calibrated Across the Contiguous United States, J. Geophys. Res.-Atmos., 124, 13991–14007, https://doi.org/10.1029/2019JD030767, 2019.

Samaniego, L., Kumar, R., and Attinger, S.: Multiscale parameter regionalization of a grid-based hydrologic model at the mesoscale, Water Resour. Res., 46, 1–25, https://doi.org/10.1029/2008WR007327, 2010.

Samaniego, L., Brenner, J., Craven, J., Cuntz, M., Dalmasso, G., Demirel, C. M., Döring, N., Jing, M., Kaluza, M., Kumar, R., Langenberg, B., Mai, J., Müller, S., Musuuza, J., Prykhodko, V., Rakovec, O., Schäfer, D., Schneider, C., Schrön, M., Schüler, L., Schweppe, R., Shrestha, P. K., Spieler, D., Stisen, S., Thober, S., Zink, M., and Attinger, S.: mesoscale Hydrologic Model – mHM v5.11.2, Zenodo, https://doi.org/10.5281/zenodo.1069202, 2021.

Tolson, B. A. and Shoemaker, C. A.: Dynamically dimensioned search algorithm for computationally efficient watershed model calibration, Water Resour. Res., 43, 1–16, https://doi.org/10.1029/2005WR004723, 2007.

Proc. IAHS, 385, 219–224, 2024
https://doi.org/10.5194/piahs-385-219-2024

Data rescue and valorization of old climatological data – application to rainfall data in Burkina Faso and Ivory Coast using NUNIEAU software

Jean-Emmanuel Paturel[1], **Bérenger Kouacou**[2], **Franck Lohou**[2], **Frédéric Pons**[3], **Kouakou Dje**[2], **Naky Coulibaly**[2], **Harouna Karambiri**[4], **Valérie Borrell**[5], **Andrew Ogilvie**[5], and **Eric Servat**[6]

[1]HSM, Univ. Montpellier, CNRS, IMT Mine Alès, IRD, Montpellier, France
[2]DMN, SODEXAM, Abidjan, Côte d'Ivoire
[3]CEREMA, Aix-en-Provence, France
[4]2iE, Ouagadougou, Burkina Faso
[5]G-EAU, Univ. Montpellier, INRAE, IRD, Montpellier, France
[6]ICIREWARD, Montpellier, France

Correspondence: Jean-Emmanuel Paturel (jean-emmanuel.paturel@ird.fr)

Received: 29 May 2022 – Revised: 15 March 2023 – Accepted: 29 March 2023 – Published: 18 April 2024

Abstract. In 2011, the XVI World Meteorological Congress Resolution 16 gives priority to the rescue and digitization of climate archives and to promote global and regional initiatives to collaborate on DARE (data rescue) and the exchange of scientific knowledge and related technological advances. The Congress urged Members to make every effort to prevent the deterioration of climate-relevant data and to make these data available to support climate change analyses and relevant climate services. In the framework of operational and scientific research actions around the updating of hydrological standards in West Africa (Actualisation des Normes hydrologiques en Afrique de l'Ouest – ActNAO Project) and in response to the WMO call, we used the NUNIEAU software (NUmérisation de NIveaux d'EAU en mer ou en rivières – free software from CEREMA) which allows the digitization of different types of paper documents by automatic recognition. This software has been used on rainfall pluviograms in Burkina Faso and Côte d'Ivoire. The use of this tool allows rainfall analysis on intraday time steps, a time step that is very useful for the knowledge and management of flash-flood events in small basins, especially urban basins. Illustrated by a few scientific results, access to these data, which remains very limited in West Africa, opens up fields of analysis that are not only disciplinary but also, and above all, inter or transdisciplinary.

Keywords. Field observation; data rescue; pluviogram digitization; Burkina Faso; Côte d'Ivoire; NUNIEAU software

1 Introduction

In West Africa, urban flooding has been in the medias for several years due to poorly controlled and planned urbanization and a non-stationary climate on a human scale marked by a succession of rainy periods and dry periods without any cycle.

Knowledge of short-duration rainfall is useful information if one is interested in runoff and potential overflow in small watersheds (Puech and Chabi-Gonni, 1984). For many years, rain gauges have been installed throughout the sub-region in synoptic stations to measure rainfall at sub-day time steps: yesterday in paper form (pluviograms), today, more generally in digital form, even if there are still many recording rain gauges in paper form.

2 Data rescue

2.1 DARE Project

Data Rescue (DARE) belongs to the long-standing strategic priorities of WMO. The Eighth Commission for Climatology in 1982, urged donor countries to support the African Dare I project (https://insitu.copernicus.eu/library/reports/copy_of_20211217_EEA_Report_DataRescue_v8_sent.pdf, last access: 12 April 2023). In 2011, the World Meteorological Organization reiterated its request to WMO Members to give priority to the rescue and digitization of climate archives to make these data available (WMO, 2011).

The project has continued year after year and has allowed Côte d'Ivoire to do a lot of work archiving its climate data in 2018. For various reasons, the work was limited to scans of the Monthly Climatological Tables (MCT). It is important to remember that these scans do not make the data usable, it is only a photograph of documents. This would require finding tools to digitize these documents. On the other hand, nothing has been done for the pluviograms. The Direction Nationale de la Météorologie from Côte d'Ivoire would archive nearly 300 000 pluviograms.

2.2 Processing and valorisation of rainfall data

However, these pluviograms are a mine of information for scientific knowledge that allow, inter alia, the knowledge of rainfall dynamics, the analysis of climate change, the production of relevant climatological services. Yet, these pluviograms remain most of the time in cupboards and are only too seldom examined – when they have not been lost or deteriorated by humidity, heat, dust, mould… floods!

Around 1980, the Comité Interafricain d'Etudes Hydrauliques (CIEH) decided to carry out this analysis for extreme events in a large number of its member countries, i.e. about ten African countries. A document was produced as a result of this work which presents the height-duration-frequency (HDF) curves for West and Central Africa for rainfall from 5 min to 24 h (Puech and Chabi-Gonni, 1984). Since then, no large-scale updates have been made and the rainfall records have continued to be stored in the cupboards.

Some countries, such as Senegal and Côte d'Ivoire, have updated this process as part of their research. In Côte d'Ivoire, Soro (2011) compiled a number of rainfall records up to 2001. This process was done manually and only a selection of pluviograms could be considered due to the difficulty of the task.

To our knowledge, there are only a few software programs that try to simplify this task by automating it as much as possible. The NUNIEAU software (Numérisation de Niveaux d'EAU en mer ou en rivières) allows this semi-automation. We used it for this study.

2.3 NUNIEAU software

The NUNIEAU software was developed by the Hydraulic Service of Cerema Méditerranée under Windows environment and coded under MatLab (https://www.cerema.fr/fr/actualites/logiciel-numerisation-enregistrements-graphiques-niveaux-eau, last access: 12 April 2023). CEREMA has register and deposit the software code with APP (Agence pour la Protection des Programmes – European organism specialized in the protection of software, mobile applications, databases, websites and strategic data – https://www.app.asso.fr/, last access: 12 April 2023). The executable code is available for free.

Its principle is simple: on a previously scanned image, the software locates the image in a marker and recovers the signal recorded on the paper, made up of pixels whose color is included in a predefined range: a paper data is thus transformed into a digital data then usable.

The quality of the data generated from the NUNIEAU software is dependent of the quality of the rainfall data; however this latter aspect is not a subject of this article.

3 Examples of application of NUNIEAU

Two examples of application of NUNIEAU immediately come to mind for hydrologists: the fine analysis of rainfall events to better understand their dynamics but also, and above all, the elaboration of Height-Duration-Frequency (HDF)/Intensity-Duration-Frequency (IDF) curves – and more generally the elaboration of project synthetic rainfall. These various curves are essential elements in the design of many hydraulic infrastructures. Thus, we propose:

- to compare the HDF curves obtained with NUNIEAU on the most recent possible extreme event data with existing historical curves;
- to show some fine analyses of the intraday rainfall at the Abidjan and Ouagadougou stations.

The objective is to illustrate what access to these data can open up as fields of analysis in Hydrology because the results presented do not always respect the usual scientific rules such as working on identical sample sizes.

3.1 NUNIEAU data

Within the framework of the ActNAO project (Actualisation des Normes hydrologiques en Afrique de l'Ouest), work with NUNIEAU has begun on data from the rainfall stations of Ouagadougou in Burkina Faso (events qualified as extreme) and Abidjan-Aéroport in Côte d'Ivoire (all events recorded). The pluviograms were made available to us by the Direction de la Météorologie Nationale.

For Ouagadougou, a selection of rainfall records was made by Traoré (2012). With NUNIEAU, 541 extreme rainfall events (in terms of accumulation and intensity) were analysed at a time step of 5 min over the period 1977–2011, i.e. about 30 years with gaps.

For Abidjan, all rainfall records were processed by NUNIEAU at a time step of 5 min over the period 1988–1999, i.e. about 10 years with some gaps. It should be noted that since all rainfall records are processed, this allows for continuity of rainfall (or non-rainfall) measurements, thus facilitating work on rainy events as well as on drought events.

3.2 Historical data

The historical data for the HDF curves for Ouagadougou and Abidjan are taken from the CIEH document. The quality of the reproduction of the document (even the original) is not very good and we had to redo some processing following the methodology described in the source document in order to validate some of the "decoded" figures. For Ouagadougou, the period covered is 29 years between 1954 and 1982; for Abidjan, the period covered is 28 years between 1951 and 1978.

The document by Soro (2011) allowed a first update of the IDF curves of the Abidjan station with the consideration of rainfall patterns between 1957 and 2001.

3.3 Working methodologies

3.3.1 HDF curves

In order to make the results as comparable as possible, it is necessary to apply the same methodology for the different calculations made. We will therefore briefly present the methodology explained by the CIEH (Puech and Chabi-Gonni, 1984):

- "we only recorded storm events that could provide information";
- the sample includes all values above a threshold. The threshold is: for "n" years of observations, we retain "m" values with $m = 3n$;
- the adjustment of the sampled rainfall is done using an exponential law and the method of moments to calibrate its parameters;
- the HDF curves were approximated by fitting a "Montana" type law: $H = at^{1-b}$ (H in mm, t in mn) with a 2-step adjustment: $t < 1$ h and $t > 2$ h.

Scripts were coded in R to be able to repeat these different processing steps.

Soro (2011) used a different methodology that must have impacts on the results obtained: also values above a threshold but with a differently defined threshold, applications of different rainfall fitting laws according to rainfall durations (Gumbel, GEV and Log-Normal), one fitting of a "Montana" type law to the HDF curves for rainfall durations less than 2 h.

Table 1. Montana coefficients (a and b) for two fits of HDF curves for Ouagadougou (CIEH: 1954–1982; NUNIEAU: 1977–2011).

Return period	a CIEH	b CIEH	a NUNIEAU	b NUNIEAU
< 1 h				
1 year	5.6	0.5	4.7	0.5
10 years	9.4	0.5	7.0	0.4
> 2 h				
1 year	31.0	0.9	32.8	0.9
10 years	49.0	0.9	68.1	0.9

3.3.2 Intraday analysis of rainfall events

This depends on the quantity and quality of the data. But access to intraday data opens up many fields of analysis. For example, it can be used to:

- statistical analyses of time series of variables deduced from rainfall: cumulative total, number of rainy days, number of dry days...
- event analyses: fine characterization of the event such as duration, start time, end time, time of the position of the rainfall maximum within the event, its position in relation to the total duration of the event, determination of the rainiest time slots...

We will present here some results that result from an event-driven analysis. Scripts have been coded in R.

4 Results

4.1 Ouagadougou HDF curves

Table 1 compares the Montana coefficients of the CIEH work on years prior to 1982 with those obtained by NUNIEAU from data between 1977 and 2011. The b coefficients do not vary much, unlike the a coefficients: for very similar b coefficients, for rainfall of less than 1 h duration, the CIEH a coefficients are higher than those of the new adjustment, whereas for rainfall of more than 2 h duration, the CIEH a coefficients are lower than those of the new adjustment.

Figure 1 shows what these differences between Montana's parameters give in terms of height values for different event durations and different return periods. The two adjustments are finally quite close with however for the period of return 1 year and for events higher than 2 h heights less strong now.

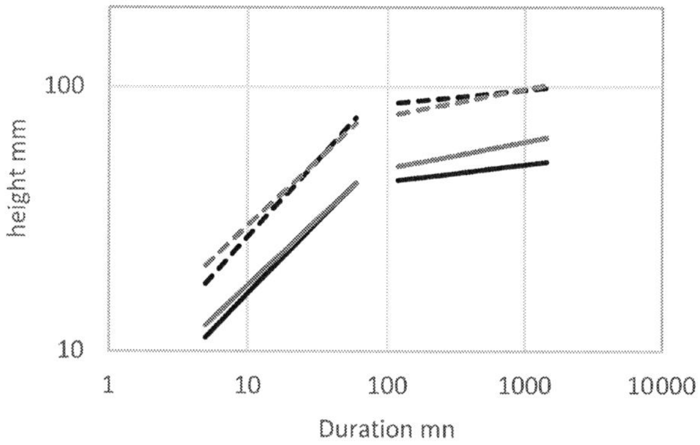

Figure 1. Height-Duration-Frequency curves for two fits of HDF curves for Ouagadougou (CIEH grey: 1954–1982; NUNIEAU black: 1977–2011; solid line: 1 year, dotted line: 10 years) – Attention, it is a log-log scale!

Table 2. 1 year return period rainfall quantiles adjusted by a Montana law – Soro (1957–2001), CIEH (before 1982) and NUNIEAU (1988–1999).

Duration	Soro_1year (45 years)	CIEH_1year (28 years)	NUNIEAU_1year (10 years)
5	16.3	13.1	12.3
10	23.2	21.1	19.1
15	28.5	27.8	24.8
30	40.6	44.9	38.6
45	49.9	59.3	50.0
60	57.8	72.3	60.1
90	71.1		
120	82.4	104.3	68.4
240		116.4	75.7
720		138.7	88.7
1440		154.9	98.1

4.2 Intraday analysis Ouagadougou

Figure 2 shows some of the characteristics of the 541 rainfall events analysed.

The median duration of the extreme rainfall events is 175 min, i.e. slightly less than 3 h. Extreme events generally began in the late afternoon, which may be related to the maximum convection at the origin of much of the rainfall over Ouagadougou. However, we note that at other times of the night or day, there are extreme events that begin which may correspond to events of a rather stratiform nature. The shape of the extreme rainfall is rather triangular, with a very large shower body in the first part of the rainfall, which decreases as the extreme event develops.

4.3 Abidjan HDF curves

Table 2 compares the quantiles estimated by the Montana laws calculated by Soro (2011), by the CIEH (Puech and

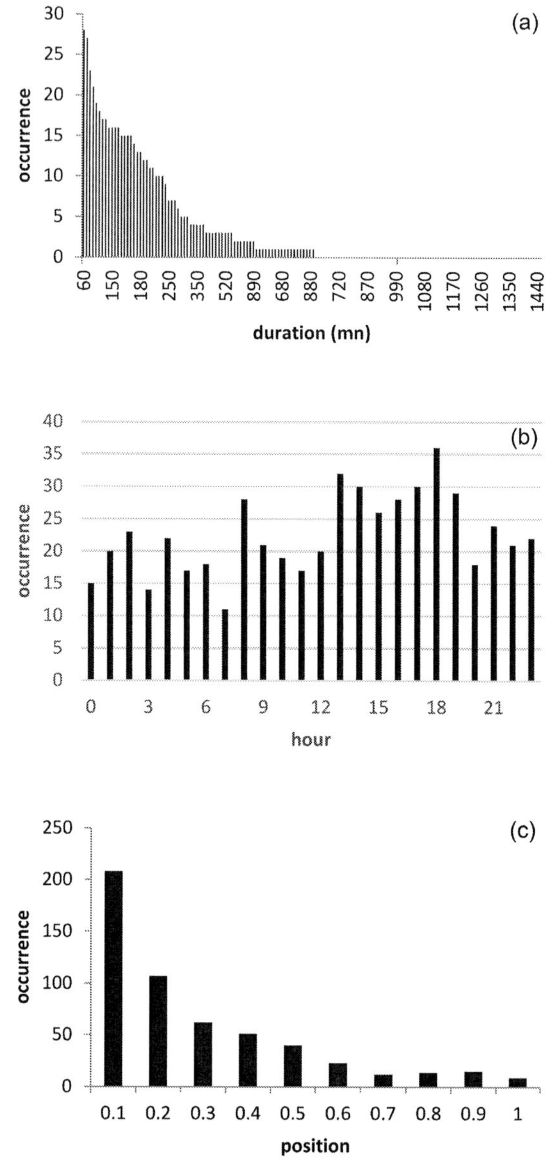

Figure 2. Some characteristics of the 541 rainfall events recorded over the city of Ouagadougou between 1977 and 2011: **(a)** occurrence of the duration of the events; **(b)** occurrence of the starting time of the events; **(c)** occurrence of the position of the maximum rainfall over 5 min within a rainfall event of normalized duration at 1 unit time.

Chabi-Gonni, 1984) and by NUNIEAU. Because of the very short period (10 years) of the data processing that could be done using NUNIEAU on the Abidjan station, we limited ourselves to the quantiles of return period 1 year.

More or less important differences appear. Soro (2011) thus highlights stronger rainfall quantiles on short event durations (< 30 min) while on longer events, the quantiles were lower. However, these discrepancies may also be related to several factors: the definition of the samples is not the same, the adjustment laws differ from one author to another (Soro, 2011, and CIEH), the durations of the reference period are very different (from 10 to 45 years).

5 Conclusion

The few scientific results presented must be viewed with great caution because a good number of "scientific rules" have not been followed: for example, the duration of the samples is not the same, the adjustment laws are different. The objective of this work was not to present innovative scientific results but rather to show what access to these data can open up in terms of disciplinary analysis but also, and above all, inter or transdisciplinary analysis: some of these results could allow for climatological services, others for agronomic services, others for civil security.

Access to sub-daily data opens up fields of analysis that are not always explored in Africa because of the difficulty of obtaining such data. In particular, the analysis of extreme events at these time steps will allow to better fight against the flooding episodes that African cities are increasingly facing. The vulnerability of these cities is more or less identified (poor governance, lack of infrastructure, lack of human and financial resources, non-systemic approaches to the problems . . .) and that of the pluviometric hazard/risk is still too badly perceived. The use of tools such as NUNIEAU should help to overcome this misperception.

The use of NUNIEAU required scanning paper documents (thus saving them for many years in a digital format) and transforming the paper data into usable digital data. Even if the arduous task of scanning is simplified, it is still long. For Côte d'Ivoire, two years ago, DNM estimate that it would take 2.5 years and 10 full-time staff to process their 300 000 pluviograms. But every month, the technicians improve and the working time is drastically reduced. In every way, there is a price to pay to make real progress in the knowledge and the fight against extreme events: a drop in the bucket compared to the costs of their impacts!

Code availability. CEREMA has register and deposit the software code of NUNIEAU with APP (Agence pour la Protection des Programmes – European organism specialized in the protection of software, mobile applications, databases, websites and strategic data). The number of the deposit is IDDN.FR.001.510017.000.S.P.2005.000.30200. The executable code is available for free with access to the community https://www.expertises-territoires.fr/jcms/pl1_69057/fr/nunieau-logiciel-de-numerisation-des-enregistrements-graphiques-de-niveaux-d-eau (last access: 13 April 2023; Expertises Territoires, 2023). The executable code is available for free.

Data availability. The data are not publicly accessible. They are the property of the Agence Nationale de la Météorologie of Burkina Faso and of the Direction de la Météorologie Nationale of Côte d'Ivoire.

Author contributions. JEP is the project administrator under the supervision of FP. Funds were acquired by ES, VB, AO and JEP. Resources in Côte d'Ivoire and Burkina Faso were provided by KD and HK respectively. NC, BK and FL were responsible for data preparation and software processing.

Competing interests. The contact author has declared that none of the authors has any competing interests.

Special issue statement. This article is part of the special issue "IAHS2022 – Hydrological sciences in the Anthropocene: Past and future of open, inclusive, innovative, and society-interfacing approaches". It is a result of the XIth Scientific Assembly of the International Association of Hydrological Sciences (IAHS 2022), Montpellier, France, 29 May–3 June 2022.

Acknowledgements. We thank the Agence Nationale de la Météorologie of Burkina Faso for allowing us in 2012 to scan rainfall pluviograms as part of a 2iE M2 internship.

We thank the Direction de la Météorologie Nationale of Côte d'Ivoire for supporting our work every day since 2020.

Financial support. The project ActNAO (Actualisation des Normes Hydrologiques en Afrique de l'Ouest) is funded by the Institute of Research for Development, the University of Montpellier and the UNESCO ICIREWARD Center and is a response to the WMO call.

Review statement. This paper was edited by Christophe Cudennec and reviewed by two anonymous referees.

References

Expertises Territoires: nunieau – logiciel de numerisation des enregistrements graphiques de niveaux d'eau, https://www.expertises-territoires.fr/jcms/pl1_69057/fr/nunieau-logiciel-de-numerisation-des-enregistrements-graphiques-de-niveaux-d-eau, last access: 13 April 2023.

Puech, C. and Chabi-Gonni, D.: Courbes Hauteur de Pluie-Durée-Fréquence – Afrique de l'Ouest et Centrale pour des pluies de 5 minutes à 24 heures, CIEH, Ouagadougou, 1984.

Soro, G. E.: Modélisation statistique des pluies extrêmes en Côte d'Ivoire, Thèse unique, Université Abobo-Adjamé, Abidjan, 2011.

Traoré, M.: Changement climatique : nouvelles courbes Intensité-Durée-Fréquence au Burkina Faso (Ouagadougou), Master Ingénierie, 2iE, Ouagadougou, 2012.

WMO: Seizième Congrès météorologique mondial, Genève, 1 May–3 June 2011, Rapport final abrégé et résolutions, OMM-No. 1077, https://library.wmo.int/?lvl=notice_display&id=7378#.Y8EcYnbP201 (last access: 12 April 2023), 2011.

Proc. IAHS, 385, 225–229, 2024
https://doi.org/10.5194/piahs-385-225-2024

Open Access PIAHS

Effects of the latest drought on the alluvial aquifer of a semiarid region in northeastern Brazil

Rafaela da S. Alves[1], **Samiramisthaís S. Linhares**[1,2], **Gilberto Möbus**[1], **Hela Gasmi**[1,3], **Eduardo S. P. R. Martins**[1,3], **Renan V. Rocha**[1], **and Alyson B. S. Estácio**[1,3]

[1]Research Institute for Meteorology and Water resources (FUNCEME), Fortaleza, 60.115-221, Brazil
[2]University of Bahia, Post-Graduate Program in Geology, Salvador, 40170-110, Brazil
[3]Department of Hydraulic and Environmental Engineering, Federal University of Ceará, Fortaleza, 60020-181, Brazil

Correspondence: Rafaela da S. Alves (rafaela.alves@funceme.br)

Received: 30 May 2022 – Revised: 16 February 2023 – Accepted: 29 March 2023 – Published: 18 April 2024

Abstract. Climate change has modified global and local water cycle patterns inducing more intense and severe drought events. In the Ceará state (Northeast Brazil), a semiarid region very vulnerable to droughts, groundwater acts as a strategic reserve to ensure water security for small rural communities. Well drilling jointly with the building of dams and cisterns are the main policy performed by the public authority to cope with the droughts. This study performed a preliminary qualitative analysis to investigate the effects of dam building in the piezometric level of an alluvial aquifer in the Ceará state during the most recent drought (2012–2016). The micro-basins of the Algodão and Forquilha creek were selected as study area, in which the alluvial aquifer has been monitored by Funceme since 2010. The alluvial aquifer was extensively exploited for local agriculture before the drought (2012–2016) due to public policies of well drilling (2001–2002). During the most recent drought, the construction of new dams, coupled with reduced rainfall, compromised the recharge of the alluvial aquifer. We evaluate these effects through the groundwater level. The results revealed a continuous drawdown of groundwater in the monitored wells. Since 2017 the piezometric level started to rise, however, the graphs reveal a different behavior of the piezometric levels monitored after the latest drought, with sudden changes between the periods of recharge and drawdown of the alluvial aquifer. Field diagnostics with local stakeholders revealed how the latest drought changed the dynamics of the activities developed by the rural communities. Before the drought the aquifer supplied the local agriculture. The drought has affected the recharge of the alluvial aquifer and compromised local agriculture. Local stakeholders have adapted, and agriculture has been replaced by livestock. The heterogeneity of this environment and of its water users requires efficient management policies adapted for the local physical and socioeconomic conditions to avoid overexploitation and maintain the recharge of the alluvial aquifer. These policies at the local scale can be the identification of local actors who are willing to develop their own management of surface and groundwater resources.

Keywords. Alluvial aquifer; drought; local scale; public policies

1 Introduction

Climate change has modified global and local water cycle patterns inducing more intense and severe drought events (IPCC, 2021). This scenario is intensified in semiarid regions, where rainfall distribution is irregular and frequent drought events occur (Nunes Carvalho et al., 2022), leaving these regions very vulnerable. In the Ceará state about 75 % of its territory corresponds to crystalline. And in the semiarid region the precipitation is less than 500 mm (Vasconcelos Junior et al., 2021). In this scenario, groundwater acts as a strategic reserve to ensure water security for small rural communities, especially alluvial aquifer. The alluvial aquifer is characterized by its porosity, good potential for water storage, follow drainages, good quality water, and generally low cost access. This resource is important to ensure water supply, especially in periods of multi-year drought (Burte et al., 2009).

To face the recurrent droughts, the Ceará's state public authority increased the building of small dams and cisterns and also was noticeable an increase in the groundwater demand for domestic supply.

The objective of this work was to perform a qualitative analysis of the effects of the latest drought with a focus in the dam building and reduction of aquifer recharge, especially for the period of 2012 to 2016, on the piezometric level of the alluvial aquifer monitored by Funceme in the Forquilha and Algodão Watershed (Fig. 1).

Research area context

The micro-basins of the Forquilha and Algodão creek were selected as study area. We called the Experimental and Representative basin of Funceme. The area is in the semiarid region of the Ceará state. The geology is represented by a crystalline basement and alluvial sediments follow the main creek. We called the Experimental and Representative basin of Funceme. The area is in the semiarid region of the Ceará state. The geology is represented by a crystalline basement and alluvial sediments follow the main creek.

The network gauges and the alluvial aquifer have been monitored by Funceme since 2010 (Fig. 1). Nowadays, we are monitoring 43 points (Fig. 1a): 20 wells in the alluvial aquifer, 13 wells in crystalline aquifer and 10 reservoirs. On these points we get water level, pH, temperature, and electric conductivity measures.

Several hydrological studies were already developed in the region, notably the research related to the alluvial aquifer developed by Burte (2008).

Figure 2 shows the photo monitoring in one reservoir. We take pictures every 15 d to follow water surface variation. In this case it is also important to note that infiltration through the dam wall occurs in this reservoir, supplying the downstream alluvial aquifer.

Public studies (governmental reports, thesis) till 2020 reveal 117 wells in Forquilha and Algodão Watershed (Fig. 1b), demonstrating that this alluvial aquifer (Fig. 1) has a great importance for multiple uses (Fig. 4), e.g. local agricultural development and the water supply for domestic activities (Burte, 2008). The alluvial aquifer increased the resilience of these communities. However, a preliminary field diagnostic observed (Fig. 3) conflicts between upstream and downstream in the basin as the aquifer needs the river streamflow to be recharged, but the dams are preventing this. These conflicts have already been reported by Burte (2008).

2 Methodology

We collected data of the number of wells with state and federal institutions. In the physical context, we worked on groundwater level data, from wells which Funceme monitors since 2010. The piezometric level of alluvial aquifer monitored allowed us to visualize the variation of groundwater level.

In the social context, we made preliminary field diagnostics with local stakeholders to understand how the decrease in aquifer recharge has changed local productive activities. The field diagnostics was performed through some interviews and workshops.

3 Results

As a result of the droughts, the main policy performed by the public authority to cope with the droughts was building of small dams and cisterns. In addition, new wells have been constructed as well as reactivated to ensure water supply. As the alluvial aquifer is recharged directly by the river flow, and the water flow is dammed, groundwater recharge is compromised.

Figure 1b shows that there are approximately 117 wells drilled in the area, 70 wells drilled in alluvial aquifer and 47 drilled in crystalline aquifer. The infrastructure projects encouraged the use of the alluvial aquifer, as this water is easily accessible and low cost, and boosted irrigated agriculture. The increase in the of number families and the development of local economics were the effects reported by stakeholders (Fig. 3).

The survey of groundwater uses of the alluvial aquifer indicated that 62.86 % of the wells are used for irrigation. Secondly, 29.52 % of the wells are used for multiple uses, 6.67 % for domestic supply and and less than 1 % for industrial supply.

The precipitation reduction and the dam construction have significantly compromised the recharge of the alluvial aquifer. The stakeholders interviewed in this territory report that irrigated agriculture has decreased, and was replaced with livestock. Between the years 2001 and 2002, 59 wells were drilled in the alluvial aquifer due to public projects. On the other hand, between the years 2016 and 2018, the state drilled 23 wells in the crystalline aquifer to face the drought. The state continues to drill wells but has invested in wells that catch water from the crystalline aquifer.

When we analyze the data of the piezometric level of the alluvial aquifer monitored by Funceme, we observe that the recharge and drawdown periods are different before the drought (2012) and after the drought in 2017. Therefore, the groundwater level configuration has changed. We can see sharper peaks which show a match with rainy periods in Ceará State (February to May).

From June 2014 until February 2017, the groundwater level has been drawdown. After that, the piezometric level

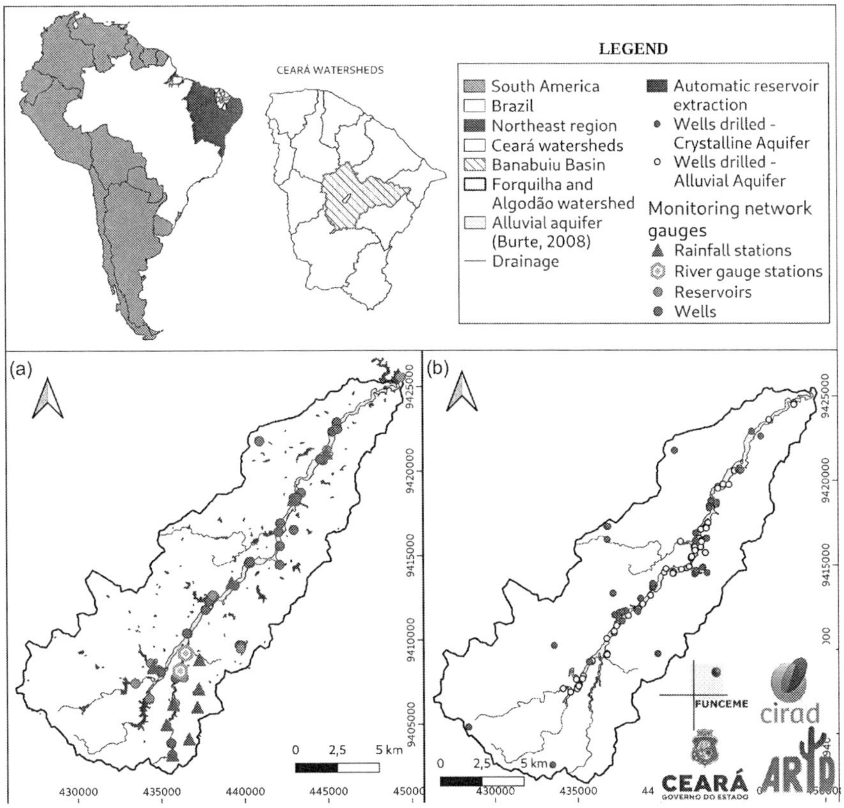

Figure 1. Monitoring network gauges **(a)** and wells drilled in alluvial and crystalline aquifer **(b)** in Forquilha and Algodão watershed.

Figure 2. Monitoring of a reservoir by photos.

Figure 3. Field diagnostics with local stakeholders.

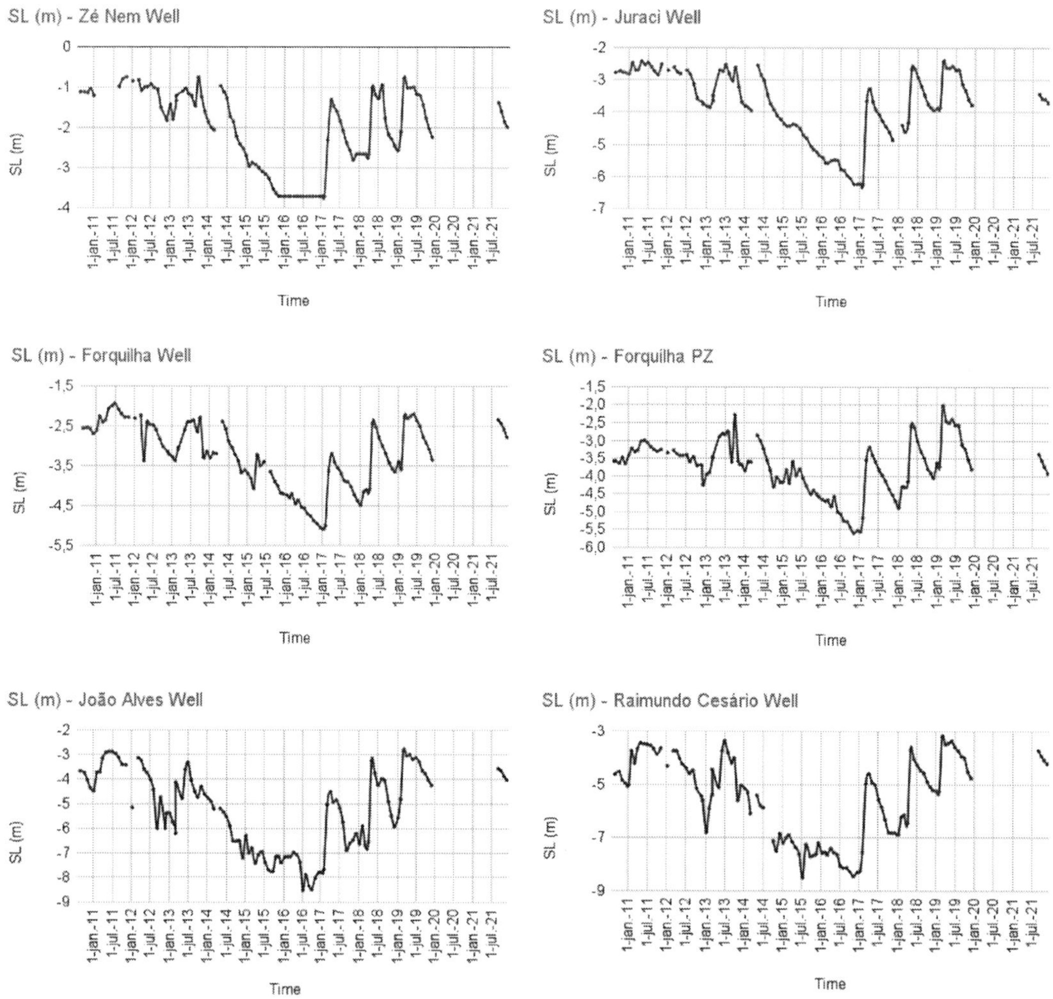

Figure 4. Groundwater level of the wells monitored in Forquilha and Algodão watershed. SL: Static Level.

has started to rise. However, the monitoring reveals a different pattern of the piezometric levels after the latest drought. With sudden changes between the periods of recharge and drawdown of the aquifer (Fig. 4).

The justification for the lack of data between February 2020 and July 2021 is the pandemic of COVID-19.

4 Conclusions

We can conclude that the precipitation reduction and the dams affect the recharge of the alluvial aquifer. This behavior has been observed through the groundwater level.

Specifically on the alluvial aquifer, continuous monitoring on a daily or sub-daily scale is needed, as they are smaller, heterogeneous, and shallow and can be quickly recharged. The heterogeneity of this environment and its water users requires efficient management policies suitable to physical and socioeconomic local conditions. These policies can be to develop a system of territorial intelligence, seeking to engage actors willing to manage local surface and underground water resources. Integration among actors in order to promote local governance and increase the resilience of diffuse communities, through Citizen Science.

Data availability. Data can be made available upon request directly to the authors.

Author contributions. RdSA, SSL, HG and RVR wrote the original draft manuscript; RdSA and RVR reviewed and edited the manuscript; RdSA, SSL, GM and HG performed the measurements; RdSA, SSL, GM, HG, ESPRM, RVR and ABSE built the methodology and the conceptualization.

Competing interests. The contact author has declared that none of the authors has any competing interests.

Disclaimer. Publisher's note: Copernicus Publications remains neutral with regard to jurisdictional claims in published maps and institutional affiliations.

Special issue statement. This article is part of the special issue "IAHS2022 – Hydrological sciences in the Anthropocene: Past and future of open, inclusive, innovative, and society-interfacing approaches". It is a result of the XIth Scientific Assembly of the International Association of Hydrological Sciences (IAHS 2022), Montpellier, France, 29 May–3 June 2022.

Acknowledgements. The authors would like to thank Institute Research Institute for Meteorology and Water resources (FUNCEME), ARID as a researchers' network and French Agricultural Research Centre for International Development (CIRAD).

Review statement. This paper was edited by Christophe Cudennec and reviewed by two anonymous referees.

References

Burte, J. D. P.: Les petits aquifers alluviaux dans les zones cristallines semi-arides: fonctionnements et strategies de gestion, Etude de cas dans le Nordeste brésilien, Thesis, Fortaleza-Ceará, https://repositorio.ufc.br/handle/riufc/17088 (last access: 21 April 2023), 2008.

Burte, J. D. P., Kosuth, P., Möbus, G., and Martins, E. S. P. R.: MNT for mapping and assessing water potential of small aluvial aquifers on regional scale. [Uso de MNT para mapeamento e avaliação do potencial hídrico dos pequenos aquíferos aluviais em escala regional] (in portuguese), XVIII Simpósio Brasileiro de Recursos Hídricos, https://doi.org/10.13140/2.1.3371.7763, 2009.

IPCC: Climate Change 2021: The Physical Science Basis. Contribution of Working Group I to the Sixth Assessment Report of the Intergovernmental Panel on Climate Change, edited by: Masson-Delmotte, V., Zhai, P., Pirani, A., Connors, S. L., Péan, C., Berger, S., Caud, N., Chen, Y., Goldfarb, L., Gomis, M. I., Huang, M., Leitzell, K., Lonnoy, E., Matthews, J. B. R., Maycock, T. K., Waterfield, T., Yelekçi, O., Yu, R., and Zhou, B., Cambridge University Press, Cambridge, United Kingdom and New York, NY, USA, ISBN 978-92-9169-158-6, https://www.ipcc.ch/report/ar6/wg1/ (last access: 21 April 2023), 2021.

Nunes Carvalho, T. M., Lima Neto, I. E., and Souza Filho, F. d.: Uncovering the influence of hydrological and climate variables in chlorophyll-A concentration in tropical reservoirs with machine learning, Environ. Sci. Pollut. Res., 29, 74967–74982, 2022.

Vasconcelos Junior, F. C., Jones, C., Gandu, A. W., and Martins, E. S. P. R.: Impacts of the Madden-Julian Oscillation on the intensity and spatial extent of heavy precipitation events in northern Northeast Brazil, Int. J. Climatol., 41, 3628–3639, https://doi.org/10.1002/joc.7039, 2021.

Proc. IAHS, 385, 231–237, 2024
https://doi.org/10.5194/piahs-385-231-2024

Open Access
PIAHS

IAHS2022 – Variability and change across space, time, extremes, and interfaces

Isotope study of monthly rainfall and its response in the Santos Formation phreatic aquifer, Mesa de Los Santos, Santander (Colombia)

Maria Alejandra Cetina[1], **Jean-Denis Taupin**[2], **Sully Gómez**[1], **and Nicolas Patris**[2]

[1]Grupo de investigación en Recursos Hídricos-GPH, Escuela de Ingeniería Civil, Universidad Industrial de Santander, Bucaramanga, Colombia
[2]Hydrosciences, Univ. Montpellier, IRD, CNRS, 34090 Montpellier, France

Correspondence: Maria Alejandra Cetina (mariacetina26@gmail.com)

Received: 29 May 2022 – Revised: 6 April 2023 – Accepted: 21 April 2023 – Published: 18 April 2024

Abstract. Mesa de Los Santos is an elevated plateau located in the Eastern Cordillera of Colombia (Andean region) in the Department of Santander (altimetric variation 300 to 1800 m a.s.l.). The region is marked by a shortage of potable surface water. The isotopic study of the rain and its relation to Los Santos Formation phreatic and fractured aquifer was carried out. Four rain stations were installed on top of the plateau (1275 to 1684 m a.s.l.) to collect monthly rainwater samples for stable isotopes. The rainwater stations recorded 1469 to 764 mm from North to South, and the seasonal patterns of stable isotopes were similar in all stations. The preliminary Local Meteoric Water Line is $\delta^2H = 8.22 \times \delta^{18}O + 13.9$, slightly above the GMWL with an intercept of +13.9 ‰, possibly indicating continental vapor recycling. The rain-weighted annual isotope means showed more depleted values in the southmost rain station ($\delta^{18}O = -8.99$ ‰, 1275 m a.s.l.). From the phreatic and fractured aquifer, 35 groundwater points were monitored bimonthly for stable isotopes ($n = 134$) and three trends were differentiated in the isotopic content value of $\delta^{18}O$. Large temporal variability characterized 10 groundwater points (amplitudes greater than 1 ‰ in $\delta^{18}O$ for each point) consistently with the seasonal behaviour of the rain, indicating flows with rapid transit on a monthly scale. 21 groundwater points showed a low temporal variability with year-round differences under 1 ‰ and presenting a spatial distribution of $\delta^{18}O$ with enriched values towards the north (between −6.71 ‰ and −6.00 ‰) and depleted values towards the south (−8.97 ‰ to −8.14 ‰), which tends to be also consistent with the rainwater isotope distribution observed. The stability of year-round groundwater isotopic values is a sign of efficient mixing of groundwater and a slower transit. Finally, 4 groundwater points presented signs of evaporation, showing *d*-excess values between −6.4 ‰ and +4.3 ‰, in connection with surface water bodies that undergo evaporation before infiltration.

Keywords. Rainwater; sedimentary groundwater; spatial and temporal isotope variability; Colombian Andes

1 Introduction

Deuterium and Oxygen-18, stable isotopes of water, are broadly distributed throughout the terrestrial system and are commonly used as conservative tracers (Clark and Fritz, 1997). The isotopic fractionation of Deuterium and Oxygen-18 in rainfall, resulting from factors such as temperature, mass or continentality effect, latitude and altitude, can be used to determine the connection between groundwater and its recharge zone, the prevailing climatic conditions during infiltration, the dynamics of the water cycle, etc. (Blavoux and Letolle, 1995; Gonfiantini et al., 2001; Binet et al., 2017; Saranya et al., 2020).

Mesa de Los Santos is an elevated plateau located in the Eastern Cordillera (Andean region, Fig. 1a) of Colombia in the Department of Santander (Fig. 1a). The study zone covers an area of 430 km^2, has a population of approximately 18 850 inhabitants CNPV, 2018; CNA, 2014), and features a semi-arid temperate to cold-dry climate, with a temperature difference of 10 °C between the bottom (28 °C) and top areas (18 °C) (Ingeoexploraciones-CAS, 2016). The rainfall regime is bimodal, with two low-rain seasons (December–February and June to August) and two rainy seasons (April–May and September to November), the mean annual rainfall is 1145 mm yr^{-1} (calculated over a period of 20 years, from 1995 to 2014), highlighting a critical situation in the southwestern zone with a decrease to 863 mm yr^{-1}, and the actual evapotranspiration is 815 mm yr^{-1} (Becerra and Parra, 2016).

Historically, this region has been suffering from shortage of potable surface water (mainly in the southwestern) (FUNDEMESA, 2009). The hydrographic network is made up of the tributaries of Sogamoso (Honda, Los Angelinos, Del Potrero, Santera, Chivatera and El Zanjón Aguas Gordas streams) and Chicamocha (Chinavega and La Mojarra streams) rivers (Alcaldia Los Santos, 2003; Fig. 1b). The region is geographically isolated resulting from its elevated plateau landform nature, which restricts the surface water resources to only the contribution of rainfall that occurs in the area. During the dry seasons, (from 1 to 4 months without rain), the supply of surface water is limited, resulting in shortages. Considering this setting, groundwater is one of the main resources of water supply for the population.

Geologically, in the Mesa de Los Santos emerge Paleozoic metamorphic rocks of the Silgará Formation (schists), Jurassic igneous rocks Granito de Pescadero unit (granites), Jurassic sedimentary rocks Jordán (quartzarenite, siltstones and sandstone) and Girón (conglomeratic sandstones) formations and Lower Cretaceous sedimentary rocks Los Santos (quartzarenite, siltstones and conglomerates), Rosa Blanca (limestone), Paja (shale) and Tablazo (Sandy limestone) formations (Ingeominas-UIS, 2007, Fig. 1b). Metamorphic, igneous and Jurassic sedimentary rocks emerge on the moderately to steeply sloping hillsides and escarpments of the plateau boundaries. The Lower Cretaceous sedimentary sequence forms the top of the plateau where most of the people live. Los Santos Formation is the geological unit with the greatest area exposure at the top of the Mesa and is divided into three members: the Lower Member (conglomerates and quartz sandstones) with 56 to 152 m thickness, the Middle Member (siltstones, claystones and sandstones) with a thickness of 22 to 41 m, and the Upper Member (quartzarenites well cemented), 33 to 54 m thick (Ingeominas-UIS, 2007). The Upper Member of the Los Santos Formation presents the greatest hydrogeological potential of the study area and configures a phreatic and fractured aquifer (Ingeominas-UIS, 2007; Díaz et al., 2009). The aim of the present study is analyzing the isotopic composition of rainfall in the Mesa de Los Santos area and the isotope response of groundwater flows in the phreatic and fractured aquifer of the Los Santos Formation Upper Member sandstones, to contribute to the knowledge of groundwater circulation and provide tools for water resource management.

2 Materials and methods

The present study is based mainly on rain and groundwater isotopic data acquired during 2020–2021 with complementary chemical data and Tritium to refine the interpretation of flow circulation. On the plateau, 4 rainwater stations (Rs) were installed on the plateau top (Fig. 1b) to collect monthly rainwater samples for stable isotopes ($n = 45$) from September 2020 to August 2021. From the phreatic aquifer, 35 groundwater points (22 wells, 8 boreholes, and 5 springs) were monitored to measure water level, electrical conductivity, and stable isotopes ($n = 134$, Fig. 1b, Table 2) in the months of December 2020, February, April, June, and August 2021. The tritium units (TU) were measured ($n = 20$) in December 2020 and the major ion concentrations were analyzed ($n = 30$) in October 2021. The isotopic analyses were carried out at the LAMA Stable Isotope Laboratory of HydroSciences Montpellier in France, using an ISOPRIME isotope ratio mass spectrometer (IRMS) with an analytical error of ± 0.8 ‰ for δ^2H and ± 0.05 ‰ for δ^{18}O; values are expressed in ‰ vs. VSMOW. Tritium was measured at the LHA-Avignon (analytical error 0.2–0.6 TU). The major ions were determined at the Industrial Consulting Chemical Laboratory of the Santander Industrial University (UIS).

3 Results and discussion

3.1 Rainwater isotopic composition

The monthly isotopic composition of rainwater (δ^{18}O and δ^2H), the monthly precipitation amounts and the rain-weighted annual average isotopic values in each station are shown in Table 1. The seasonal behaviour of δ^{18}O and δ^2H is similar in the four stations and is rather consistent with the precipitation amount effect which is predominant in the tropical zone (Clark and Fritz, 1997), with a moderate correlation of R^2: Rs1 = 0.41; Rs2 = 0.36; Rs3 = 0.14; and Rs4 = 0.38. The preliminary Local Meteoric Water Line (LMWL) calculated for the 12 months of isotope rainfall monitoring is $\delta^2\text{H} = 8.22 \times \delta^{18}\text{O} + 13.9$ (Fig. 2), which presents a similar slope than the Global Meteoric Water Line (GMWL, $\delta^2\text{H} = 8 \times \delta^{18}\text{O} + 10$) (Craig, 1961), the LMWL determined in Santander Massif (Gómez et al., 2015, $\delta^2\text{H} = 8.15 \times \delta^{18}\text{O} + 12.81$), and the most of the LMWL determined in the Central Andes (Valdivielso et al., 2020). The LMWL show a greater deuterium excess compared to the GMWL (Craig, 1961; Rozanski et al., 1993), possibly indicating continental vapor recycling process (e.g., Gallaire et al., 2010). Gómez et al. (2015) determined an altitudinal δ^{18}O gradient of -0.16 ‰ per 100 m in the Santander Massif. In our case no correlation was found with elevation, probably due to the low elevation range (ΔAltitude = 408 m). However, the values shown an isotopic composition more depleted in lowest one in elevation toward the south (Rs4).

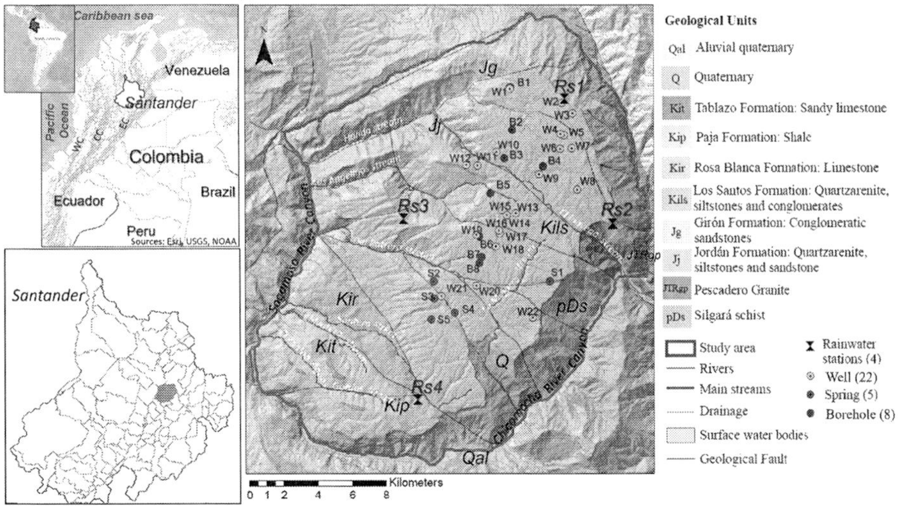

Figure 1. (a) Study area location (highlighted in orange), EC: Eastern Cordillera, CC: Central Cordillera, WC: Western Cordillera. **(b)** geological map (Ingeominas-UIS,2007), rainwater stations, and groundwaters point location.

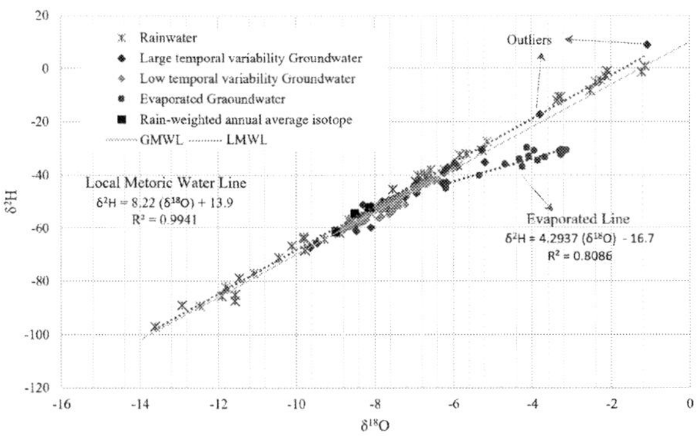

Figure 2. Global meteoric water line (GMWL), local meteoric water line (LMWL), rain, groundwater, and rain-weighted annual average isotopic values.

The upper limit of the range of d-excess values in each rain station (Rs1: +10.9 to +15.8; Rs2: +5.0 to +15.0; Rs3: +9.5 to +16.0; Rs4: +7.4 to +12.0) showed values probably affected by continental vapor recycling process. A few monthly samples (d excess under +8 ‰) showed signs of evaporation, concerning months with lower rain amounts. In almost every month (except April 2021) a marked difference is observed in the deuterium excess values of the southern station (Rs4) which are generally lower than the other stations, which could indicate two moisture mass trajectories present in the area, possibly one from Santander Massif originated from the north and the second from canyons of the Chicamocha and Sogamoso rivers. The previous hypothesis could also explain the differences in the fractionation process of the stable isotopes ($\delta^{18}O$) between north and south zone.

3.2 Groundwater isotopic composition

The Upper Member of the Los Santos Formation is proposed as a phreatic and fractured aquifer. Despite the lithology composed of quartzarenites-type sedimentary rocks, the high content of siliceous cement reduces the primary porosity and allows to define a circulation through fractures (Díaz et al., 2009). Further, the top of the plateau presents an undulating geomorphology, which may control the directions of the groundwater and configuring disconnected reservoirs. Although the systems are complex the isotopic composition allowed to define preliminary groundwater behaviors. Three behaviors in the isotopic composition of $\delta^{18}O$ (V-SMOW) can be identified: Large temporal variability, Low temporal variability, and evaporation marks (Table 2, Fig. 2). These behaviours can be observed throughout the entire study area (with no discernible spatial relationship).

For 10 groundwater points: 7 wells (W1, W2, W4, W7, W11, W19 and W21), 1 borehole (B1), and 2 springs (S2, S5) (Table 2), a relatively large temporal isotope variability prevails showing amplitudes greater than 1 ‰ in $\delta^{18}O$. The wells and the borehole show a shallow water level less than 10 m deep (water table averages between 1.4 and 5.1 m). The $\delta^{18}O$ signal amplitude over the year varies between 1.67 ‰ (W7) and 8.39 ‰ (W21), this last point also showing the most enriched non-evaporated $\delta^{18}O$ values (−1.07 ‰ and −3.80 ‰). The high temporal variability observed is consistent with the isotopic monthly behaviour of the rain. The TU values show a high activity in 3 samples (W11 = 2.5 TU, W19 = 2.1 TU, and W21 = 1.9 TU) supporting a rapid transit process with young water (Table 2). This implies a rapid infiltration possibly occurring through the fractures network, and water levels close to the surface possibly limiting the

Table 1. Isotopic compositions ($\delta^{18}O$ and $\delta^{2}H$) of rainwater stations.

	d-exc Rs1				Rs2 (1678 m a.s.l.)				Rs3 (1433 m a.s.l.)				d-exc Rs4			
	P (mm)	$\delta^{18}O \pm$ 0.05 ‰	$\delta^{2}H \pm$ 0.8 ‰	d exc ‰	P (mm)	$\delta^{18}O \pm$ 0.05 ‰	$\delta^{2}H \pm$ 0.8 ‰	d exc ‰	P (mm)	$\delta^{18}O \pm$ 0.05 ‰	$\delta^{2}H \pm$ 0.8 ‰	d exc ‰	P (mm)	$\delta^{18}O \pm$ 0.05 ‰	$\delta^{2}H \pm$ 0.8 ‰	d exc ‰
Sep	59	−6.71	−41.0	12.74	54	−6.89	−40.3	14.90	18	−5.83	−32.4	14.25	20	−5.88	−36.5	10.57
Oct	212	−7.53	−45.4	14.81	160	−6.74	−39.6	14.28	166	−6.57	−38.1	14.44	100	−6.21	−38.2	11.48
Nov	263	−12.92	−89.0	14.33	177	−11.82	−82.3	12.23	78	−11.46	−78.9	12.79	78	−11.77	−82.9	11.30
Dec	70	−3.27	−10.7	15.40	69	−2.14	−3.0	14.19	57	−3.33	−10.6	15.99	0	–	–	
Jan	35	−3.36	−12.2	14.66	MD	–	–	–	38	−2.27	−4.5	13.60	24	−1.14	0.9	9.95
Feb	99	−2.10	−1.0	15.75	182	−1.22	−1.4	8.35	124	−2.39	−5.1	14.00	60	−2.53	−8.3	11.98
Mar	116	−5.14	−27.4	13.66	294	−8.11	−54.3	10.61	43	−5.66	−32.1	13.17	33	−5.26	−30.8	11.28
Apr	134	−7.95	−52.7	10.90	58	−8.89	−62.0	9.17	109	−6.95	−42.6	13.05	79	−9.29	−64.2	10.05
May	104	−10.45	−71.1	12.48	MD	–	–	–	145	−11.91	−85.8	9.50	202	−12.48	−89.5	10.39
Jun	131	−11.08	−77.0	11.69	45	−11.57	−87.6	5.00	143	−13.62	−97.1	11.87	41	−11.57	−85.1	7.44
Jul	85	−8.66	−56.6	12.65	119	−8.07	−52.7	11.93	41	−8.94	−61.0	10.55	66	−8.23	−55.4	10.47
Aug	160	−10.11	−66.8	14.00	115	−9.81	−63.5	14.97	70	−9.80	−64.1	14.24	60	−9.78	−68.5	9.70
Total	1469	−8.51*	−54.5*		–	–	–	–	1030	−8.13*	−52.2*		764	−8.99*	−61.4*	

Rs = Rainwater station, P = Precipitation, MD = Missing data, * Rain-weighted annual average isotopes values, m a.s.l. = meters above sea level.

mixing in the unsaturated zone. The electrical conductivity data show three types of groundwater mineralization: low (17 to 86 µS cm^{-1}), medium (99 to 344 µS cm^{-1}) and high (640 to 1629 µS cm^{-1}). The first and second modes are consistent with the lithology (quartzarenites) that implies low to moderate mineralization, while the third seems linked possibly to a contamination with anomalies in Cl^-, SO_4^{2-}, Na^+, and K^+.

Twenty-one groundwater points show a low temporal variability (differences less than 1 ‰ over the year): 12 wells (W3, W5, W6, W10, W12, W13, W14, W15, W16, W17, W18 and W22), 6 boreholes (B2, B3, B4, B6, B7 and B8), and 3 springs (S1, S3, S4). The wells water levels average between 2.2 and 7.1 m, with extremes values of 11.7 and 0.46 m. There is no information about the static level of all boreholes (because they have sealed borehead, except B4), but the depth of the submersible pumps evidence samples taken between 21 and 59 m from the surface. The amplitudes of $\delta^{18}O$ time series vary between 0.06 (B8) and 0.71 ‰ (W3), indicating an efficient mixing of the groundwater. Spatially, this set of values present a distribution of $\delta^{18}O$ with enriched values towards the north (between −6.71 ‰ and −6.00 ‰) and depleted values towards the south (−8.97 ‰ to −8.14 ‰), which tends to be also consistent with the rain isotope distribution also marked with depleted values towards the south. The TU values show activities less than 2.0 TU, with two samples close to 1.0 TU (W16 = 1.0 TU, and B4 = 0.8 TU) confirming an older water component with respect to the first group (Table 2) implying a slower transit, with deeper water levels and longer residence times. The electrical conductivity shows low (13 to 89 µS cm^{-1}) to medium (94 to 339 µS cm^{-1}) groundwater mineralization depending on lithology with no evidence of contamination, except for the point W5 where the moderate electrical conductivity (281 µS cm^{-1}) is related to positive anomalies for Cl^-, and Na^+ and K^+, implying a local contamination problem related to a very shallow phreatic level (less than 1 m).

The last 4 groundwater points present signs of evaporation (Fig. 2): 3 wells (W8, W9 and W20) and 1 borehole (B5) which show a shallow water level, with $\delta^{18}O$ values between −6.2 ‰ and −3.1 ‰, d-excess values between −6.4 ‰ and +4.3 ‰. These points (Fig. 2) are situated on a local evaporation line: $\delta^2H = 4.29 \times \delta^{18}O\text{-}16.7$. The 3 wells are close (distance less than 5 m) to small surface water bodies and present a continuous pumping favouring the drainage of surface water which is submitted to evaporation before infiltration. The borehole is located inside an upper sub-basin, near a small water body, and also subjected to frequent pumping.

4 Conclusions

The isotopic composition of rainwater in the Mesa de Los Santos area shows a geographic distribution with more depleted values towards the south of the area, suggesting possibly two different moisture pathways to the plateau (the San-

Table 2. Groundwater characteristics and isotopic compositions ($\delta^{18}O$ and $\delta^{2}H$).

point	Groundwater Type	Altitude (m a.s.l.)	Well/ borehole depth (m)	Submersible pumps depth (m)	Groundwater depth (m)	EC (μ S cm^{-1})	$\delta^{18}O$ ± 0.05 ‰	$\delta^{2}H$ ± 0.8 ‰	*d* exc ‰	UT UT	Error ±
Large temporal variability											
W1	Well	1609	7	–	1.21 to 1.4	181 to 238	−8.03 to −6.14	−37.2 to −52.3	11.9 to 12.0		
W2	Well	1666	7	–	0.6 to 4.02	138 to 247	−8.17 to −5.96	−53.2 to −35.6	11.6 to 12.1		
W4	Well	1693	8	–	0.9	302 to 640	−8.42 to −5.99	−55.2 to −36.9	10.7 to 12.1		
W7	Well	1703	10	–	1.56 to 2.56	781 to 1629	−7.50 to −5.90	−49.0 to −36.8	10.4 to 11.0		
W11	Well	1619	9	–	3.26 to 6.44	17 to 83	−8.54 to −6.17	−59.4 to −42.9	6.4 to 9.5	2.5	0.3
W19	Well	1676	11.5	–	5.04 to 7.67	142 to 313	−9.64 to −6.64	−67.4 to −42.1	9.8 to 11.0	2.1	0.3
W21	Well	1669	4.8	–	1.39 to 3.56	99 to 706	−9.46 to −1.07	−65.6 to 8.9	10.1 to 17.5	1.9	0.3
B1	Borehole	1609	27	18	8.06 to 8.26	50 to 116	−8.29 to −6.84	−51.8 to −42.3	12.4 to 15.1	0.9	0.3
S2	Spring	1657	–	–	–	17 to 45	−8.50 to −5.27	−58.9 to −30.6	9.0 to 11.5		
S5	Spring	1642	–	–	–	29 to 86	−8.47 to −5.19	−61.2 to −35.3	5.0 to 6.6		
Low temporal variability											
W3	Well	1689	8.8	–	0.84 to 1.42	27 to 53	−6.71 to −6.00	−43.0 to −37.7	10.3 to 11.0		
W5	Well	1686	10	–	0.46 to 0.87	102 to 315	−6.96 to −6.73	−44.3 to −42.6	11.2 to 11.4		
W6	Well	1703	8.3	–	3.76 to 8.06	43 to 108	−7.14 to −6.51	−46.9 to −41.7	9.6 to 10.4		
W10	Well	1628	12	–	4.26 to 8.43	48 to 132	−7.39 to −6.86	−47.8 to −44.5	10.1 to 11.7	1.7	0.3
W12	Well	1624	7	–	3.3 to 4.75	21 to 87	−7.96 to −7.53	−51.9 to −48.9	11.1 to 12.1	1.4	0.4
W13	Well	1695	4.95	–	1.5 to 4.12	78 to 162	−7.19 to 6.59	−43.6 to −42.4	9.8 to 11.2	1.9	0.2
W14	Well	1691	22	20	4.98 to 11.7	21 to 63	−7.64 to 7.47	−50.1 to −49.2	10.5 to 11.7		
W15	Well	1687	20	18	1.78 to 7.14	35 to 88	−7.76 to −7.42	−51.0 to −49.1	10.3 to 11.4		
W16	Well	1686	9	–	7*	14 to 152	−8.57 to −8.25	−58.3 to −55.9	9.5 to 10.3	1.0	0.3
W17	Well	1688	11	–	7.23 to 9.96	158 to 339	−8.65 to −8.36	−58.9 to −58.0	8.9 to 10.3		
W18	Well	1679	15	14.5	13.45*	13 to 63	−7.57 to −7.20	−53.3 to −51.1	6.6 to 7.5	1.1	0.3
W22	Well	1734	6	–	0.76 to 3.51	16 to 30	−8.14 to −7.77	−54.5 to −50.6	10.2 to 10.6	1.6	0.3
B2	Borehole	1658	22	21	–	270 to 594	6.49 to −6.37	−42.3 to −41.6	9.2 to 10.2	1.6	0.4
B3	Borehole	1629	55	50	–	112 to 221	−7.42 to −7.36	−50.3 to 50.2	8.6 to 9.1		
B4	Borehole	1678	63	59	37.62 to 39.39	89 to 189	−7.83 to −7.72	−53.4 to −52.3	9.2to 10.0	0.8	0.2
B6	Borehole	1676	48	40	–	77 to 171	−8.31 to −8.23	−57.3 to −57.2	8.5 to 9.2		
B7	Borehole	1678	29	24	–	19 to 37	−7.89 to −7.54	−56.2 to −53.9	6.2 to 6.9	1.2	0.3
B8	Borehole	1676	33	30	–	40 to 74	−8.63 to −8.57	−59.5 to −59.4	9.2 to 9.6		
S1	Spring	1713	–	–	–	37 to 94	−8.21 to −7.74	−55.6 to −52.7	9.3 to 10.2		
S3	Spring	1650	–	–	–	20 to 43	−8.77 to −8.14	−59.9 to −54.8	10 to 12.4	1.7	0.6
S4	Spring	1617	–	–	–	15 to 29	−8.97 to −8.93	−62.4 to −60.9	9.2 to 9.3		
Evaporated signs											
W8	Well	1704	8	–	4.87 to 6.47	134 to 344	−6.18 to −3.23	−45.1 to 31.8	−6.4 to 4.3		
W9	Well	1636	5.5	–	1.27 to 2.47	105 to 209	−3.67 to −3.12	−33.4 to −30.6	−5.7 to −4.0	1.8	0.3
W20	Well	1680	4.3	–	0.73 to 2.5	108 to 414	−4.24 to −3.25	−37.0 to −30.5	−4.5 to −3.1		
B5	Borehole	1629	53	50	–	14 to 33	−5.34 to −3.95	−40.2 to −29.7	−0.4 to 3.3	1.9	0.3

EC = electrical conductivity, m a.s.l. = meters above sea level, * Dinamic level (afected by pumping).

tander Massif and the canyons Sogamoso and Chicamocha Rivers). The isotopic signatures of rainwater and groundwater allow us to recognize the rapid response of the underground system in the phreatic Upper Member of the Los Santos Formation due to the fractures network controlling the percolation and circulation. The groundwater points with large temporal isotopic variability suggest a flow with rapid transit occurring at the monthly scale with a water table near the surface. The groundwater points with variability stable isotopic signature consistent with the rain isotopic distribution, indicate a good mix of water within the aquifer, with slower flow transit and groundwater circulations deeper than the previous behaviour. Finally, the isotopic behaviour of a number of points evidenced the mixing of evaporated surface waters into the ground, promoted by active pumping.

Code availability. The map used in this paper was created using Esri's ArcMap software, which was licensed through Universidad Industrial de Santander. The software was used to create a geospatial representation of the data collected for the study. The underlying code of the ArcMap software is not publicly accessible.

Data availability. Tables 1 and 2 presented in this study, display a subset of the data analyzed. The complete dataset is currently unavailable as it forms part of the ongoing doctoral thesis study of the primary author.

Author contributions. MC, SG and JT planned the campaign; MC performed the measurements and sampling; JT and NP determination of isotopes; MC, SG and JT analyzed the data; MC wrote the manuscript draft; JT, SG and NP reviewed and edited the manuscript.

Competing interests. The contact author has declared that none of the authors has any competing interests.

Disclaimer. Publisher's note: Copernicus Publications remains neutral with regard to jurisdictional claims in published maps and institutional affiliations.

Special issue statement. This article is part of the special issue "IAHS2022 – Hydrological sciences in the Anthropocene: Variability and change across space, time, extremes, and interfaces". It is a result of the XIth Scientific Assembly of the International Association of Hydrological Sciences (IAHS 2022), Montpellier, France, 29 May–3 June 2022.

Review statement. This paper was edited by Christophe Cudennec and reviewed by two anonymous referees.

References

Alcaldia Los Santos: Esquema de Ordenamiento Territorial del Municipio de Los Santos-EOT (Territorial Planning Scheme of the Municipality of Los Santos), 2003.

Becerra, N. J. and Parra, C. G.: Balance Hídrico para estimar recarga potencial en la Mesa de Los Santos y dirección de flujo de Agua Subterránea (Water Balance to estimate potential recharge in Mesa de Los Santos and groundwater flow direction), Undergrade Thesis, Universidad Industrial de Santander, Bucaramanga, Colombia, http://tangara.uis.edu.co/biblioweb/ (last access: 15 May 2022), 2016.

Binet, S., Joigneaux, E., Pauwels, H., Albéric, P., Fléhoc, C., and Bruand, A.: Water exchange, mixing and transient storage between a saturated karstic conduit and the surrounding aquifer: Groundwater flow modeling and inputs from stable water isotopes, J. Hydrology, 544, 278–289, 2017.

Blavoux, B. and Letolle, R.: Apports des techniques isotopiques á la connaissance des eaux souterraines (Contribution of isotopic tools to the knowledge of groundwaters), Geochronique, 54, 12–15, 1995.

Clark, I. D. and Fritz, P.: Environmental isotopes in hydrogeology. Submitted. New York, Boca Ratón, Lewis Publishers, 328 pp., 1997.

CNPV: Censo Nacional de Población y Vivienda (National Population and Housing Census), https://terridata.dnp.gov.co/index-app.html#/perfiles/68418/2 (last access: 20 March 2022), 2018.

CNA: Censo Nacional Agropecuario (National Agricultural Census), https://datosabiertos.esri.co/datasets/esri-colombia::censo-nacional-agropecuario-hogares (last access: 20 March 2022), 2014.

Craig, H.: Isotopic variations in meteoric waters, Science, 133, 1702–1703, 1961.

Díaz, E. J., Contreras, N. M., Pinto, J. E., Velandia, F., Morales, C. J., and Hincapie, G.: Preliminary hydrogeological evaluation from geological formations of Mesa de Los Santos, Santander, Boletín de Geología, 31, 61–70, 2009.

FUNDEMESA-Fundación para el Desarrollo de la Mesa de los Santos: Mesa de Los Santos, Memorias del paraíso (Mesa de Los Santos, Memories of paradise), Bucaramanga, 96 p., 2009.

Gallaire, R., Taupin, J. D., and Coudrain, A.: Recent developments in the cryo-climatic parameters and isotopic content of precipitation in the Bolivian Andes: La Paz and the Zongo Glacier, Hydr. Sci. J., 55, 467–483, 2010.

Gómez, S., Taupin, J. D., and Rueda, A.: Hydrodynamic, geochemical and isotopic study of formations aquifers in the region of Bucaramanga (Colombia), Rev. Peru. Geo-Atmos. RPGA, 61, 44–61, 2015.

Gonfiantini, R., Roche, M. A., Olivry, J. C., Fontes, J. C., and Zuppi, G. M.: The altitude effect on the isotopic composition of tropical rains, Chem. Geol., 181, 147–167, 2001.

Ingeoexploraciones-CAS: Elaboración y formulación del plan de manejo de aguas subterráneas en el sector de La Mesa, Municipio de Los Santos. Fase II (Preparation and formulation of the groundwater management plan in the La Mesa sector, Municipality of Los Santos. Phase II), Corporación Autónoma Regional de Santander, 263 p., 2016.

Ingeominas-UIS: Memoria explicativa – Investigación geológica e hidrogeológica en la Mesa de Los Santos y sector nordeste de Curití (Explanatory report – Geological and hydrogeological investigation in the Mesa de Los Santos and northeast sector of Curití), Instituto de Investigaciones e Información Geocientífica Minero-Ambiental y Nuclear (Ingeominas, now Servicio Geológico Colombiano) and Universidad Industrial de Santander, Bucaramanga, 145 p., https://miig.sgc.gov.co/Paginas/advanced.aspx (last access: 15 May 2022), 2007.

Rozanski, Z., Araguás-Araguás, L., and Gonfiantini, R.: Isotopic patterns in precipitation, Geophys. Monogr., 78, 165–169, 1993.

Saranya, P., Krishnakumar, A., Kumar, S., and Krishnan, K. A. Isotopic study on the effect of reservoirs and drought on water cycle dynamics in the tropical Periyar basin draining the slopes of Western Ghats, J. Hydrol., 581, 124421, https://doi.org/10.1016/j.jhydrol.2019.124421, 2020.

Valdivielso, S., Vázquez-Suñé, E., and Custodio, E. Origin and variability of oxygen and hydrogen isotopic composition of precipitation in the Central Andes: A review, J. Hydrol., 587, 124899, https://doi.org/10.1016/j.jhydrol.2020.124899, 2020.

Proc. IAHS, 385, 239–246, 2024
https://doi.org/10.5194/piahs-385-239-2024

Evaluating the spatial and temporal variation in baseflow across headwater streams in the Jonkershoek valley, South Africa

Retang Anna Mokua[1], **Julia Glenday**[2], **and Dominic Mazvimavi**[1]

[1]Department of Earth Sciences, Environmental and Water Sciences, University of the Western Cape, Bellville, 7535, South Africa
[2]South African Environmental Observation Network (SAEON), Fynbos Node, Newlands, 7708, South Africa

Correspondence: Retang Anna Mokua (mokuar@gmail.com)

Received: 28 May 2022 – Revised: 30 March 2023 – Accepted: 21 April 2023 – Published: 18 April 2024

Abstract. An understanding of processes and factors accounting for spatiotemporal variations in baseflow is necessary for sustainable water resources management, particularly in semi-arid regions. This study assessed the spatial and temporal variations in baseflow across three mountain streams within the Jonkershoek valley, in the Table Mountain Group (TMG) geological region of South Africa. A two-parameter recursive digital filtering method (RDF) was combined with conductivity mass balance (CMB) using electrical conductivity to separate baseflow from daily streamflows during the 1946–2019 period. The results indicated that annual baseflow contributions varied spatially. Baseflows were estimated to contribute 38 % to 86 % of the annual streamflows across the three sub-catchments. This notable variation could be explained by differences in topography, geology and fracturing, and vegetation cover. All three sub-catchments had higher proportional contributions from baseflow in the dry summer compared to the wet winter, while interannual variations were relatively minor.

Keywords. UPH 5; baseflow; field observation; recursive digital filter; conductivity mass balance; Jonkershoek

1 Introduction

An understanding of processes and factors accounting for spatiotemporal variations in baseflow is necessary to develop appropriate water quantity and quality management strategies (Santhi et al., 2008). This is particularly the case in semi-arid regions, where water resources are under severe stress. The southwestern part of the Western Cape Province (WC) of South Africa experiences a Mediterranean climate typified by long dry summers. Consequently, many of the rivers are non- perennial. However, in some areas the deeply fractured Table Mountain Group (TMG) quartzitic sandstone aquifer contributes baseflow to surface water resources. The fractured nature of this geology also favours the formation of interflow within the shallow weathered zone where there is a high density of fractures (Xu et al., 2003; Roets et al., 2008). Baseflow from the TMG maintains streamflow in the mountainous catchments during both the dry and wet seasons (Saayman et al., 2003). These rivers are critical for water supply throughout the Western Cape Province. However, processes and factors accounting for the variability of baseflows in rivers in the TMG geology region are not adequately understood.

Streamflow hydrographs can be separated into quickflow and baseflow. The separation of baseflow can provide insights into streamflow generation mechanisms and by extension chemical transport and water quality effects of processes within the catchment (Cartwright et al., 2014; Bosch et al., 2017).

Previous studies investigating baseflows in the TMG geological region have mainly focused on a regional scale across meso-scale catchments (Le Maitre and Colvin, 2008; Smakhtin, 2001; Jia, 2007). For example, Jia (2007) investigated baseflow characteristics in quaternary catchments with areas ranging from 37 to 3000 km^2 within the TMG region. The estimated baseflow index (BFI) ranged from 4 % to 30 %.

The present study has taken a step further by investigating baseflow characteristics at a headwater sub-catchment scale ($< 10\,\mathrm{km}^2$), exploring variability across the Jonkershoek Valley. The study aims to characterise baseflow across three neighbouring headwater streams to determine spatial and temporal variations that may exist at this scale in the TMG region. The main objectives are, to quantify baseflow and estimate the baseflow index (BFI), to assess if the conductivity mass balance (CMB) method using electrical conductivity (EC) can improve baseflow estimation in the TMG catchments, and to investigate the spatial and temporal variations in baseflow characteristics across the sub-catchments.

2 Study area

The study was carried out in three neighbouring sub- catchments of the Jonkershoek Valley with perennial streams that are tributaries of the main Jonkershoek River (Fig. 1). The Jonkershoek catchment has a total area of $\sim 146\,\mathrm{km}^2$. It is located near the town of Stellenbosch (33°57′ S, 18°15′ E), in the Western Cape Province. A summary of the sub-catchment characteristics is given in Table 1.

The variation in morphology, land cover, rainfall, and streamflow, coupled with the long-term hydrometric records, made ideal study sites (Slingsby et al., 2021). The Bosboukloof sub-catchment consists mostly of afforested slopes, with *Pinus radiata* covering nearly 58 % of the area since 1940. Tierkloof had pine plantations covering up to 36 % from 1956 to 2014, but has been dominated by native fynbos shrubland since fires in 2015. Langrivier has been kept as a fynbos-dominated control site (Scott et al., 2000). The planting of pines was found to reduce the streamflow yield in the afforested sub-catchments (Scott et al., 2000). In all three sub-catchments, the upper portion is occupied by precipitous cliffs of highly fractured, TMG quartzitic sandstone of the Peninsula formation. These have little or no vegetation or soil (Hewlett and Bosch, 1984). This TMG unit overlies the basement bedrock of Malmesbury Shale with major granite intrusions (Cape Granite). Much of this basement is covered by talus and scree on the lower slopes and valley floor. The TMG outcrop covers about 40 % of the Langrivier catchment compared to 26 % in Tierkloof and 12 % in Bosboukloof (Table 1). According to the conceptual models (Roets et al., 2008; Saayman et al., 2003), the mountainous catchments across the TMG are dominated by sub-surface flow pathways. The streamflow appears to be mainly sourced from springs and seeps near the base of the TMG outcrop. Interflow was found to be a significant streamflow contributor across the sub-catchments, occurring as preferential flow through fractured rock, the scree-talus, and at the rock-soil (Mokua et al., 2020).

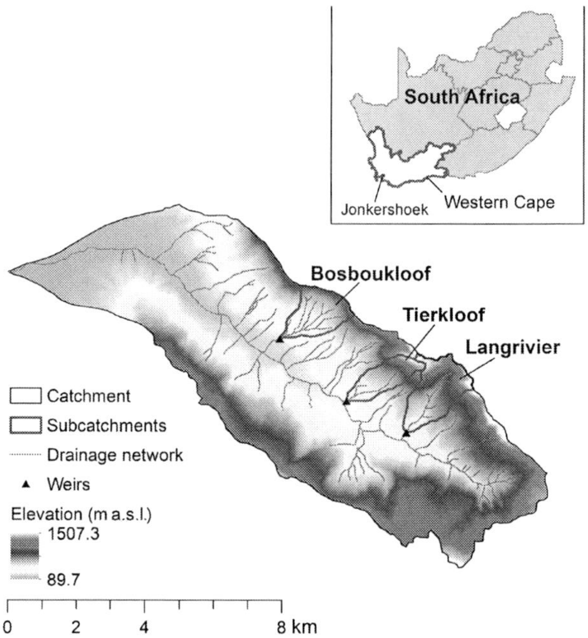

Figure 1. Jonkershoek Valley catchment including the location of the study sites and stream guaging stations.

3 Materials and Methods

3.1 Data acquisition and measurement campaign

Streamflow and rainfall data from 1946 to 2019 were obtained from the South African Observation Network (SAEON). The instrumentation and datasets are described in Slingsby et al. (2021). Each sub-catchment has a V-notch gauging weir with continuous logging. The specific conductance (electrical conductivity referenced to 25 °C, $\mu\mathrm{S\,cm}^{-1}$) of streamflow at each weir was measured monthly from 2018 to early 2020 using a multi-parameter probe (YSI professional Plus, YSI Inc) resulting in 27 readings per stream. EC of rainfall was measured from cumulative rainfall samples in each sub-catchment.

3.2 Baseflow separation

3.2.1 Recursive Digital Filtering (RDF)

In this study, a two-parameter RDF (Eckhardt, 2005) was used for continuous hydrograph separation. The filter equation was applied as follows.

$$Q_{\mathrm{B}j} = \frac{(1-\mathrm{BFI}_{\max})aQ_{\mathrm{B}j-1} + (1-a)\mathrm{BFI}_{\max}Q_{\mathrm{T}j}}{1-a\mathrm{BFI}_{\max}} \quad (1)$$

where $Q_{\mathrm{B}j}$ ($\mathrm{m}^3\,\mathrm{s}^{-1}$) is the baseflow on day j, $Q_{\mathrm{T}j}$ is the total discharge ($\mathrm{m}^3\,\mathrm{s}^{-1}$) on day j, $\mathrm{BFI}_{\max}$ and a are the filter parameters. The $\mathrm{BFI}_{\max}$ is the maximum value of the base-

Table 1. Summary of the study sites including mean Annual precipitation (MAP) and Mean Annual Runoff (MAR) for the hydrological years 1946–2019 (SAEON, 2019).

	Bosboukloof	Tierkloof	Langrivier
Area (km^2)	2.09	1.45	2.46
Altitude (m a.m.s.l.)	274–1026	280–1530	366–1460
Average slope (°)	26	49	40
MAR ($mm\,yr^{-1}$)	409	881	1461
MAP ($mm\,yr^{-1}$)*	1104	1244	1539
Vegetation	*Pinus radiata* (58 %) & fynbos	(2015–2021) fynbos (1954–2014) *Pinus radiata*	fynbos
TMG outcrop (% of area)	12	26	41

* Note catchment average rainfall estimated by Thiessen polygon method

flow index (i.e the ratio of baseflow to streamflow) and the a is the recession constant.

Appropriate filter parameters values vary depending on the hydrological and hydrogeological characteristics of the catchment (Kouanda et al., 2018). As such, the BFI_{max} and a were determined for each sub-catchment separately. Initial BFI_{max} was determined using the backwards filtering method using the daily streamflow data (Collischonn and Fan, 2013) implemented in U.S.G.S Groundwater Toolbox (Barlow et al., 2017). The a was estimated using three methods namely, the correlation method (Tallaksen, 1995), mean value (Saraiva Okello et al., 2018) and master recession curve (Eckhardt, 2005). The average a value of the three methods was applied. The RDF separation was applied to the daily streamflow time series for the period 1946 to 2019 using U.S. Geological Survey Groundwater Toolbox 1.3.1 (Barlow et al., 2017).

3.2.2 Conductivity Mass Balance (CMB) and RDF calibration

Considering the high sensitivity of the filtering parameters, Eckhardt (2005) proposed that the RDF methods should be validated using tracer-based hydrograph separations. The integration of recursive digital filter (RDF) with the conductivity mass balance method (CMB) has been shown to improve baseflow estimation in many headwater catchments (Kouanda et al., 2018; Saraiva Okello et al., 2018; Lott and Stewart, 2016). The RDF method only requires streamflow data, while the CMB method relies on hydrochemical measurements. The CMB method was used to calibrate the BFI_{max} parameter. The CMB method was applied using the following equation (Lott and Stewart, 2016):

$$Q_{Bj} = Q_{Tj}\left[\frac{C_{Tj} - C_S}{C_B - C_S}\right] \qquad (2)$$

where Q_T is the measured streamflow discharge ($m^3\,s^{-1}$) on day j, C_{Tj} is the measured specific conductance of the streamflow at a given timestep given in micro Siemens per centimeter ($\mu S\,cm^{-1}$), C_B and C_S are the estimated specific conductance ($\mu S\,cm^{-1}$) of baseflow and quickflow components, respectively. The C_B values were assumed to be the maximum EC value measured for each stream, which occurred during the dry season. Due to the relatively low EC values during the high flow season, especially in Langrivier (Mokua et al., 2020), the quickflow component was represented by the EC value of the winter rainfall ($20\,\mu S\,cm^{-1}$) for all sub-catchments, assuming that quickflow had EC approximately equal to that of rainfall. This assumes little to no EC change on the quickflow pathway(s), i.e., that it is direct flow.

The RDF predictions were compared to the CMB baseflow values for the sample days. Three goodness of fit methods, percent bias (PBIAS), Root-Mean-Square Error (RMSE) and coefficient of determination were used the evaluate the performance of the RDF. BFI_{max} parameter values were adjusted to improve performance.

4 Results and discussion

4.1 RDF vs. CMB baseflow estimates for sampled days

The uncalibrated RDF tended to estimate greater BFI values than the CMB. This was most notable for Langrivier (Table 2). The calibrated BFI_{max} values improved the RDF output in that most of the estimated BFIs showed better agreement with the CMB method results (Table 2). However, while the correlations between the CMB and RDF outputs were improved by adjusting the BFI_{max} parameter, the R^2 values remained low. The PBIAS results show that the calibrated RDF generally estimated more baseflow than the CMB on sampled days for Bosboukloof and Tierkloof, while for Langrivier there was a shift from overestimation to slight underestimation. Given these different biases, the magnitudes of the BFI differences between the three catchments, as predicted by the RDF, are rough estimates. Nevertheless, the differences between the catchments' mean annual BFIs were large in relation to these biases, as described below.

4.2 RDF parameter values

The filter parameters exhibited differences among the sub-catchments, particularly the calibrated BFI_{max}. The recession constant (a) showed minor spatial variation ranging from 0.967 to 0.974. These values were consistent with those of Jia (2007) for meso-scale catchments across the TMG outcrop areas: 0.945 to 0.985. The adjusted long-term BFI_{max} value was higher for Bosboukloof, at 0.86 compared to Tierkloof at 0.68, and Langrivier at 0.54. The different values imply the influence of catchment characteristics such as drainage patterns, vegetation, topography and hydrogeological characteristics (e.g., bedrock fractures and, degree of weathering) (Price, 2011; Kouanda et al., 2018).

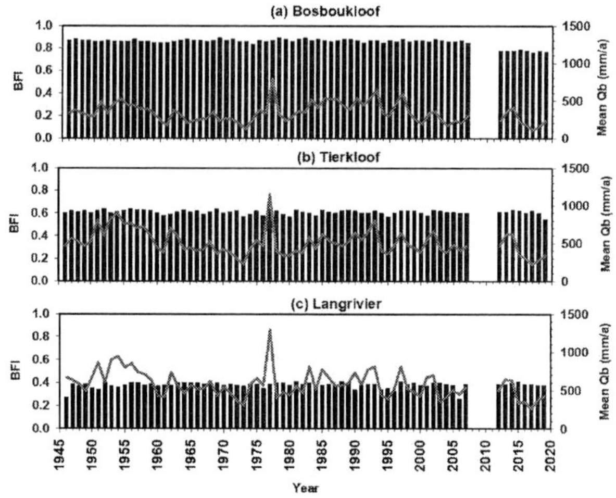

Figure 2. Temporal variations in annual BFI estimates and baseflow (Q_b) for the period 1946–2019 at **(a)** Bosboukloof, **(b)** Tierkloof and **(c)** Langrivier.

4.3 Spatial and temporal variation of baseflow

4.3.1 Annual scale

Annual BFI values were calculated for the period 1946–2019. The mean annual BFI and baseflow volume estimated for the study sites are shown in Fig. 2. The BFI estimates and magnitudes of baseflow showed marked spatial variation. The mean annual BFI was highest in Bosboukloof at 0.86 (range: 0.75–0.89), lower in Tierkloof at 0.63 (range: 0.55–0.64), and lowest in Langrivier at 0.38 (range: 0.25–0.41). Langrivier had the highest inter-annual BFI variability. The BFI values are in agreement with those of previous studies in the small catchments of the southwestern Cape region of South Africa (Brendenkamp et al., 1995). However, they were higher than those of the quaternary catchments (average scale: $400\,km^2$) of the TMG outcrop area (Jia, 2007) which ranged from 0.30 to 0.48 and the value of 0.30 estimated for the fractured metasedimentary rock in general (LeMaitre and Colvin, 2008). The variability across the three sub-catchments could be explained by factors such as topography, spatial irregularities in geological structures (e.g. fractures) and vegetation cover (Bloomfield et al., 2009; Price, 2011). The sub-catchments with more scree-talus material cover and pine forest, Bosboukloof and Tierkloof, had higher baseflow proportions, than Langrivier, which has a larger area of steep rocky cliffs. The scree-talus is likely to have high infiltration rates and act as a shallow aquifer providing an additional source of baseflow (Harris and Diamond, 2013). The low BFI in Langrivier could potentially be explained by the steep topography suggesting that rainfall is likely to be transmitted faster to the stream via both surface flow and preferential pathways through the fractured rock (Hewlett and Bosch, 1984), increasing proportion of quickflow.

Figure 2 showed that Bosboukloof generates the lowest mean annual baseflow of $342\,mm\,yr^{-1}$ compared to $511\,mm\,yr^{-1}$ for Tierkloof and $578\,mm\,yr^{-1}$ for Langrivier. This was expected considering the low mean annual runoff (MAR) of Bosboukloof (Table 1). The higher baseflow volumes in Tierkloof and Langrivier could be attributed to their larger TMG outcrops and the higher rainfall amount in the upper steep areas of these catchments. For small catchments ($<200\,km^2$) in the United States, Santhi et al. (2008) found that landscape characteristics such as relief and gradient are highly correlated with BFI ($R^2 = 0.79$) and baseflow volume ($R^2 = 0.93$), respectively. High baseflow was also found to be correlated with the amount of rainfall. Considering land-cover, a positive relationship between BFI and pine forest area was reported by Pramono et al. (2017) in the Kedungbulus sub-watershed (Indonesia), who found that a catchment with 75 % pine forest cover had a higher BFI (0.75) than a comparable catchment with 43 % pine cover (BFI of 0.55). By promoting interception and infiltration, treed cover may reduce the proportion of rainfall that becomes quickflow during storm events.

However, when considering baseseflow volume, Price (2011) found a positive correlation between higher forest cover and lower baseflow amounts, which was attributed to high evapotranspiration and infiltration rates of the forest. This could be the case in the Jonkershoek where the afforested Bosboukloof and Tierkloof had low baseflow volumes compared to the non-afforested Langrivier. The infiltrating water is likely to recharge the high draining scree-cover and shallow weathered bedrock, and the pines tap into this groundwater, reducing baseflow. Annual BFI

Table 2. Summary of the goodness of fit indicators results between the uncalibrated or calibrated recursive digital filter and CMB method daily BFI values, BFI ($n = 27$). Analysis period (2018–2019, 2020).

Site		BFI_{max}	RDF mean BFI	CMB mean BFI	p-value	RMSE	R^2	PBIAS (%)
Bosboukloof	Uncalib	0.86	0.89	0.82	0.164	0.16	0.01	3.2
	Calib	0.85	0.83	0.82	0.263	0.19	0.06	1.2
Tierkloof	Uncalib	0.68	0.81	0.76	0.042	0.22	0.01	8.8
	Calib	0.65	0.75	0.76	0.974	0.27	0.11	9.2
Langrivier	Uncalib	0.54	0.67	0.51	0.040	0.25	0.101	9.6
	Calibr	0.40	0.50	0.51	0.971	0.24	0.174	−0.3

Note: Uncalib = Uncalibrated, Calib = Calibrated

values showed only minor temporal variations, with Bosboukloof and Tierkloof showing the least variability (Fig. 2). Baseflow is consistently a major component of streamflow in these sub-catchments. Langrivier on the other hand had a higher proportion of quickflow. These results could be viewed as a contradiction to previous research, as peak flows during storm events in this catchment have been shown to be dominated by "old water" (e.g. Midgley and Scott, 1994; Mokua et al., 2020). These findings indicate that the quickflow may include rapid sub-surface flows from the shallow weathered zones that push older stored water into the stream. Therefore, the assumption that rain chemistry represents quick flow may not be a good fit for this site, hence the poor correlation between CMB and RDF (Table 2).

4.3.2 Monthly and seasonal scale

BFI values for monthly and seasonal scales were also calculated. The mean monthly BFIs and baseflow discharge for the three sub-catchments are presented in Fig. 3. Similar to the annual BFI estimates, these values exhibit large spatial variation and seasonality. All three sub-catchments had higher proportional contributions from baseflow in the dry summer (October–March) compared to wet winter (April–September), as expected given seasonality of rainfall. Baseflow peaks were detected during the wet winter months of July and August but over 50 % of the dry season flow was attributed to baseflow (BFI > 0.5). For the Bosboukloof, baseflows contribute 89 % of summer flows and 82 % in winter, Tierkloof 73 % and 60 %, while these were 50 % and 34 % for Langrivier. According to Kelly et al. (2019), the high BFI indicates permeable catchment conditions, whereby the catchment water is stored during the wet season and is discharged to the streams during the dry season. Also, the drier antecedent conditions could have resulted in lower quickflow generation during episodic summer events, which would increase BFI. In Bosboukloof and Tierkloof the pines have been found to have impacted the dry season flow to a greater extent compared to the annual and wet season runoff, potentially due to high evapotranspiration rates (Scott et al., 2000).

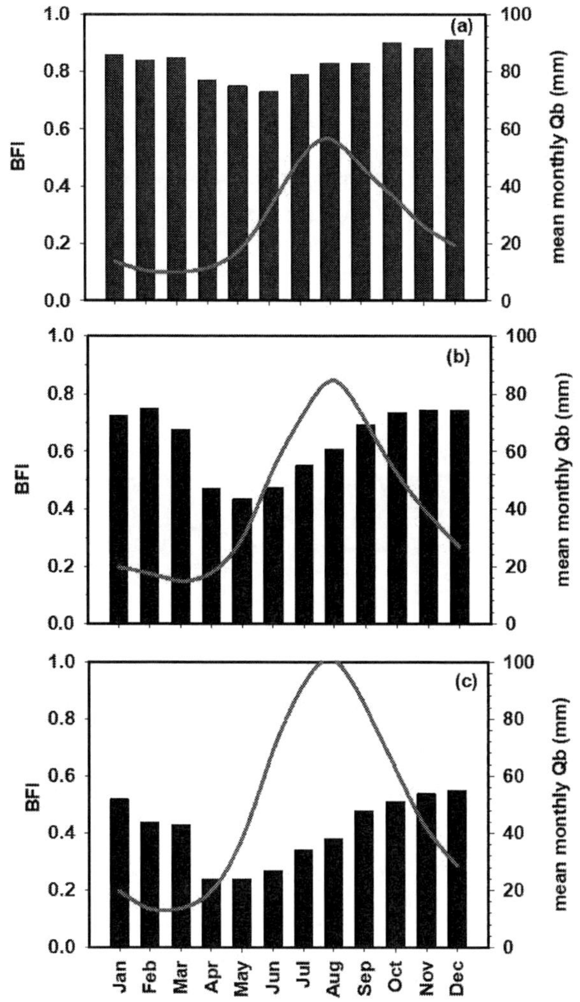

Figure 3. Mean monthly BFIs and baseflow (Q_b) for the 1946–2019 period at the studied sub-catchments at **(a)** Bosboukloof, **(b)** Tierkloof and **(c)** Langrivier.

The larger baseflow volume during the wet season months (July–September) indicates wetter antecedent conditions in the scree-soil cover below the TMG and likely lower evapotranspiration (ET) from both pines and fynbos. As a result, more quick flow was generated by the steep upper catchment areas during the wet season, hence lower BFI. This is consistent with other research that suggest catchment wetness and/or water storage capacity is an important control on baseflow (Yao et al., 2021).

The high BFI observed during the dry season across the sub-catchments indicates the importance of baseflow in maintaining dry season flows.

5 Conclusions

This study applied the RDF method, calibrated based on CMB separation, to separate baseflow for three headwater streams in the Jonkershoek catchment for the period 1946–2019. Calibration of the BFI_{max} filter parameter did improve the fit of the RDF baseflow predictions to the CMB outputs for the sampled days. However, while the two methods produced similar average baseflow estimates, the correlation remained low. The dominant contribution of "old water" even in storm responses that has been observed in these streams (Midgley and Scott, 1994) may mean that the rainfall EC values used in the CMB do not adequately represent what is considered the "quickflow" response.

RDF outputs indicated spatial variations in annual and monthly BFIs across the sub-catchments. This is likely to be a function of several factors such as the heterogeneity in bedrock geology, topography and vegetation cover across the sub-catchments. The results suggest that more than half the annual and summer streamflow in Bosboukloof and Tierkloof is derived from baseflow. Although, Langrivier's flow was dominated by quickflow, baseflow accounted for half of the summer flow on average. This highlights the importance of baseflow in sustaining the streamflow during the dry season in the Jonkershoek catchment. The seasonal variability shown in the monthly BFI and baseflow amounts can be attributed to low rainfall and high ET rates during summer. The results demonstrate that a high degree of variability can exist in the contribution of baseflow between neighbouring headwater streams, likely linked to catchment geomorphic characteristics and land cover. This study provides a useful reference for understanding baseflow characteristics in headwater catchments of the TMG and can contribute to informed water resources management.

Data availability. The raw data is publicly accessible from the South African Environmental Observation Network (SAEON), https://catalogue.saeon.ac.za/ (de Buys, 2019a, b, c, d, e, f, g, h) and https://catalogue.saeon.ac.za/records (SAEON, 2019).

Author contributions. RAM conducted the research, and investigation process, prepared manuscript, carried out the fieldwork, hydrochemical analysis and interpretation of data. JG was involved in the fieldwork, conceptualization and reviewing of the manuscript. Prof DM assisted with the reviewing of the manuscript and formatting.

Competing interests. The contact author has declared that none of the authors has any competing interests.

Disclaimer. Publisher's note: Copernicus Publications remains neutral with regard to jurisdictional claims in published maps and institutional affiliations.

Special issue statement. This article is part of the special issue "IAHS2022 – Hydrological sciences in the Anthropocene: Variability and change across space, time, extremes, and interfaces". It is a result of the XIth Scientific Assembly of the International Association of Hydrological Sciences (IAHS 2022), Montpellier, France, 29 May–3 June 2022.

Acknowledgements. We acknowledge the National Research Foundation for the financial support and South African Environmental Observation Network, Fynbos Node (SAEON) for making this study possible. We kindly thank Abri de Buys for assisting with field data collection and processing.

Financial support. This research has been supported by the National Research Foundation Professional Development Programme (PDP) PhD grant (no. MND19053044376) under the South African Environmental Observation Network,Fnyos Node (SAEON).

Review statement. This paper was edited by Christophe Cudennec and reviewed by Jean-Marie Kileshye-Onema and one anonymous referee.

References

Barlow, P. M., Cunningham, W. L., Zhai, T., and Gray, M.: U.S. Geological Survey Groundwater Toolbox version 1.3.1, a graphical and mapping interface for analysis of hydrologic data: U.S. Geological Survey Software Release [software], https://doi.org/10.5066/F7R78C9G, 2017.

Bloomfield, J. P., Allen, D. J., and Griffiths, K. J.: Examining Geological Controls on Baseflow Index (BFI) Using Regression Analysis: An Illustration from the Thames Basin, UK, J. Hydrol., 373, 164–176, https://doi.org/10.1016/j.jhydrol.2009.04.025, 2009.

Bosch, D. D., Jeff, G. A., Allen, P. G., Lim, K., and Youn Shik, Y.: Regional Studies Temporal variations in baseflow for the Little

River experimental watershed in South Georgia, J. Hydrol., 10, 110–121, 2017.

Brendenkamp, D. B., Botha, L. J., Van Tonder, G. J., and Van Rensburg, H. J.: Manual on quantitative estimation of groundwater recharge and aquifer storativity: Based on Practical Hydrological methods, Water Research Commission, TT 73/95, 1995.

Cartwright, I., Gilfedder, B., and Hofmann, H.: Contrasts between estimates of baseflow help discern multiple sources of water contributing to rivers, Hydrol. Earth Syst. Sci., 18, 15–30, https://doi.org/10.5194/hess-18-15-2014, 2014.

Collischonn, W. and Fan, F. M.: Defining Parameters for Eckhardt's Digital Baseflow Filter, Hydrol. Process., 27, 2614–2622, https://doi.org/10.1002/hyp.9391, 2013.

de Buys, A.: Langrivier rain gauge 8B (366 m), event rainfall, 06/2011–09/2019, SAEON (South Africa Environmental Observation Network) Observations Database [data set], https://doi.org/10.15493/SAEON.FYNBOS.10000002, 2019a.

de Buys, A.: Langrivier rain gauge 14B (472 m), event rainfall, 06/2011–06/2019, SAEON (South Africa https://doi.org/10.15493/SAEON.FYNBOS.10000003, 2019b.

de Buys, A.: Bosboukloof rain gauge 11B (400 m), event rainfall, 06/2011–09/2019, SAEON (South Africa Environmental Observation Network) Observations Database [data set], https://doi.org/10.15493/SAEON.FYNBOS.10000004, 2019c.

de Buys, A.: Langrivier 700 m fog station, event rainfall, 05/2014–07/2019, SAEON (South Africa Environmental Observation Network) Observations Database [data set], https://doi.org/10.15493/SAEON.FYNBOS.10000005, 2019d.

de Buys, A.: Langrivier gauging weir hourly stage and estimated flow, 08/2011–09/2019, SAEON (South Africa Environmental Observation Network) Observations Database [data set], https://doi.org/10.15493/SAEON.FYNBOS.10000006, 2019e.

de Buys, A.: Bosboukloof gauging weir hourly stage and estimated flow, 08/2011–09/2019, SAEON (South Africa Environmental Observation Network) Observations Database [data set], https://doi.org/10.15493/SAEON.FYNBOS.10000007, 2019f.

de Buys, A.: Dwarsberg weather station (1214 m), event rainfall, 03/2013–09/2019, SAEON (South Africa Environmental Observation Network) Observations Database [data set], https://doi.org/10.15493/SAEON.FYNBOS.10000008, 2019g.

de Buys, A.: Dwarsberg manual gauge, monthly rainfall, 1945–1991, SAEON (South Africa Environmental Observation Network) Observations Database [data set], https://doi.org/10.15493/SAEON.FYNBOS.10000009, 2019h.

Eckhardt, K.: How to Construct Recursive Digital Filters for Baseflow Separation, Hydrol. Process., 19, 507–515, https://doi.org/10.1002/hyp.5675, 2005.

Harris, C. and Diamond, R. E.: The Use of Isotope Hydrology to Characterize and Assess Water Resources in Southern Africa, in: The use of isotope hydrology to characterize and assess water resources in South(ern) Africa, edited by: Abiye, T., Water Research Commission Report No. TT570/13, 38–52, 2013.

Hewlett, J. D. and Bosch, J. M.: The Dependence of Storm Flows on Rainfall Intensity and Vegetal Cover in South Africa, J. Hydrol., 75, 365–381, 1984.

Jia, H.: Groundwater Resource Evaluation in Table Mountain Group Aquifer Systems, Ph.D. thesis, Department of Earth Science, University of the Western Cape, Bellville, 205 pp., 2007.

Kelly, L., Kalin, R. M., Bertram, D., Kanjaye, M., Nkhata, M., and Sibande, H.: Quantification of Temporal Variations in Base Flow Index Using Sporadic River Data: Application to the Bua Catchment, Malawi, Water, 11, 2–17, 2019.

Kouanda, B., Coulibaly, P., Niang, D., Fowe, T., Karambiri, H., and Paturel, J. E.: Analysis of the Performance of Base Flow Separation Methods Using Chemistry and Statistics in Sudano-Sahelian Watershed, Burkina Faso, Hydrology: Current Research, 9, 300, https://doi.org/10.4172/2157-7587.1000300, 2018.

Le Maitre, D. C. and Colvin, C. A.: Assessment of the Contribution of Groundwater Discharges to Rivers Using Monthly Flow Statistics and Flow Seasonality, Water SA., 34, 549–564, https://doi.org/10.4314/wsa.v34i5.180652, 2008.

Lott, D. A. and Stewart, M. T.: Base Flow Separation: A Comparison of Analytical and Mass Balance Methods, J. Hydrol., 535, 525–33, https://doi.org/10.1016/j.jhydrol.2016.01.063, 2016.

Midgley, J. J. and Scott, D. F.: The Use of Stable Isotopes of Water (D and 18O) in Hydrological Studies in the Jonkershoek Valley, Water SA., 20, 151–154, 1994.

Mokua, R. A., Glenday, J., Nel, J., and Butler, M.: Combined use of stable isotopes and hydrochemical characteristics to determine streamflow sources in the Jonkershoek Catchment, South Africa, Isotopes Environ. Health Stud., 56, 1–22, https://doi.org/10.1080/10256016.2020.1760861, 2020.

Pramono, I. B., Budiastuti, M. T., Gunawan, S., and Wiryanto, T.: Base Flow from Various Area of Pine Forest at Kedungbulus Sub Watershed, Kebumen District, Central Java, Indonesia, Int. J. Dev. Sustain., 6, 99–114, 2017.

Price, K.: Effects of Watershed Topography, Soils, Land Use, and Climate on Baseflow Hydrology in Humid Regions: A Reviw, Prog. Phys. Geo, 35, 465–492, 2011.

Price, K., Jackson, C. R., Parker, A. J., Reitan, T., Dowd, J., and Cyterski, M.: Effects of Watershed Land Use and Geomorphology on Stream Low Flows during Severe Drought Conditions in the Southern Blue Ridge Mountains, Georgia and North Carolina, United States, Water Resour Res., 47, 2516–2534, 2011.

Roets, W., Xu, Y., Raitt, L., and Brendonck, L.: Groundwater Discharges to Aquatic Ecosystems Associated with the Table Mountain Group (TMG) Aquifer: A Conceptual Model, Water SA., 34, 77–88, 2008.

Saayman, I. C., Scott, D. F., Prinsloo, F. W., Moses, G., Weaver, J. M. C., and Talma, S.: Evaluation of the Isotopes in the identification of the streamflow generation in the TMG Catchments, Water Research Commission Report, No. 1234/1/03, 2003.

SAEON (South African Environmental Observation Network and South African Forestry Research Institute and Council for Scientific and Industrial Research): SAEON DATA PORTAL [data set], https://catalogue.saeon.ac.za/records (last access: January 2022), 2019.

Santhi, C., Allen, P. M., Muttiah, R. S., Arnold, J. G., and Tuppad, P.: Regional Estimation of Base Flow for the Conterminous United States by Hydrologic Landscape Regions, J. Hydrol., 351, 139–153, 2008.

Saraiva Okello, A. M. L., Uhlenbrook, S., Jewitt, G. P. W., Masih, I., Riddell, E., and van der Zaag, P.: Hydrograph Separation Using Tracers and Digital Filters to Quantify Runoff Components in a Semi-Arid Mesoscale Catchment, Hydrol. Process., 32, 1334–1350, https://doi.org/10.1002/hyp.11491, 2018.

Scott, D. F., Prinsloo, F. W., Moses, G., Mehlomakulu, M., and Simmers, A. D. A.: A Re-Analysis of the South African Catchment Afforestation Experimental Data, Water Research Commission, 810, 1–138, 2000.

Slingsby, J. A., Buys, A., Adrian, D. A., Simmers, A., Prinsloo, E., Forsyth, G. G., Glenday, J., and Allsopp, N.: Jonkershoek: Africa's Oldest Catchment Experiment – 80 Years and Counting, Hydrol. Process., 35, 1–7, 2021.

Smakhtin, V. U.: Estimating Continuous Monthly Baseflow Time Series and Their Possible Applications in the Context of the Ecological Reserve, Water SA., 27, 213–217, 2001.

Tallaksen, L. M.: A Review of Baseflow Recession Analysis, J. Hydrol., 165, 349–370, https://doi.org/10.1016/0022-1694(94)02540-R, 1995.

Xu, Y., Wu, Y., and Beekman, H. E.: The role of interflow in estimating recharge in mountainous catchments, in: Groundwater Recharge Estimation in Southern Africa, edited by: Xu, Y. and Beekman, H. E., UNESCO IHP Series No. 64, 135–145, 2003.

Yao, L., Sankarasubramanian, A., and Wang, D.: Climatic and Landscape Controls on Long-Term Baseflow, Water Resour. Res., 57, 1–20, https://doi.org/10.1029/2020WR029284, 2021.

Proc. IAHS, 385, 247–252, 2024
https://doi.org/10.5194/piahs-385-247-2024

Open Access PIAHS

Flood impact assessment in urban settings with porous buildings – insights from a fine-scale hydraulic-economic model

David Nortes Martinez[1], Frédéric Grelot[1], Cécile Choley[2], and Pascal Finaud-Guyot[3,4]

[1]G-EAU, Univ. Montpellier, AgroParisTech, BRGM, CIRAD, INRAE, Institut Agro, IRD, Montpellier, France
[2]ENGEES, Icube, Strasbourg, France
[3]HSM, Univ. Montpellier, CNRS, IRD, Montpellier, France
[4]Inria, Team Lemon, Montpellier, France

Correspondence: David Nortes Martinez (david.nortes-martinez@inrae.fr) and Frédéric Grelot (frederic.grelot@inrae.fr)

Received: 14 June 2022 – Revised: 17 April 2023 – Accepted: 21 April 2023 – Published: 18 April 2024

Abstract. Classical hydraulic approaches of urban floods consider buildings as obstructions to water flow, without considering the flow exchanges between streets and buildings. Since the hydraulic behavior within the building is not considered, there exists a gap on the available information regarding the fine characterization of material damage and human exposure within buildings. Because of this gap, damage estimation usually assumes that the floodwater depths inside and outside the buildings are the same. However, as flood damage functions are very elastic with respect to floodwater depth, especially in the lower values, relatively small differences in water depth can lead to large differences in the assessment of material damage. Not considering street-building flow exchanges might be introducing a bias in the estimation of property damage.

In this paper, we propose to analyse how fine-scale hydraulic and economic modelling approaches considering street-building flows can influence the characterization of material damage at a larger scale (district). To do so, we couple a hydraulic model with an economic model to simulate water depths and to estimate flood damage under: (i) a classical approach using non-porous buildings and (ii) an alternative approach explicitly considering street-building flow exchanges. The results obtained show rather notable differences in floodwater depth and economic damage at both building and district level.

Keywords. UPH 21; SDG 11.B; SDG 11.5; modelling; flood; flood damage assessment; building; opening; vulnerability

1 Introduction

Estimating the impacts of flooding in urban areas is of primary importance when qualifying the exposure of territories. One of the difficulties encountered is that the processes at stake involve fine scales (building scale).

Classical approaches in flood modeling in the urban environment integrate buildings as obstructions to the water flow (Schubert and Sanders, 2012). However, these approaches fail to simulate water flows inside buildings (Bellos and Tsakiris, 2015). By disregarding street-building water flow exchanges, classical approaches are not able to explain the potential role of buildings for water storage and the consequent potential effect on the reduction of the peak discharge (Gems et al., 2016). Furthermore, secondary connections between streets through buildings and fine simulation of floodwater depth within the building are out of the scope of classical modeling practices in urban flood modeling (Sturm et al., 2018). Despite those drawbacks, experiments considering water intrusion within buildings are seldom in the literature (Bellos and Tsakiris, 2015; Choley et al., 2021).

The absence of such fine simulations leads to the need for assumptions when estimating urban flood damage. For instance, considering buffers around building edges to estimate the floodwater depth inside from the depth outside.

Such methods may lead to overestimations of flood damage because of the underlying hypotheses (e.g. independently of the building and severity of the event, the water fills the building and the same depth is reached everywhere). Hence, we argue that finer flood damage estimations of urban floods and the design of better emergency responses would benefit from more accurate simulations of the hydraulic behavior within buildings.

In our work, we propose a coupled hydraulic-damage model for the simulation of flood events and the estimation of material damage in porous urban environments. In this paper, we use this model to determine whether to consider openings in the hydraulic modeling influences the damage estimation at a district level.

2 Method

Our method couples a hydraulic model with a damage model. The hydraulic model represents both the street water flows using classical 2D shallow water equations and the mass conservation in each building with a reservoir model assuming no velocity in average within the building. Street shallow water equations are discretized using a classical finite volume Godunov scheme based on 2D unstructured mesh. Street and building models are computed using an Euler explicit approach such as preserving the stability and the water depth positivity. Those models are implemented in the sw2d software (https://sw2d.inria.fr/, last access: 28 April 2023).

An opening is used to connect a cell and a building. It should be noted that each building can be connected to different cells of the mesh and conversely, a cell can be connected to several buildings. The opening discharge depends on the water level in the street and in the building allowing to represent a nil discharge if the same water level is reached on both side of an opening (see Choley et al., 2021, for more details).

To assess flood damage, we use two tools from the *floodam* tool set: *floodam.building* and *floodam.spatial*. The *floodam* tool set is a collection of libraries written in R language, developed within the French working group "GT AMC" which aims at developing methodologies for the economic appraisal of flood management policies (with the support of the French Ministry in charge of the Environment).

The *floodam.building* tool (Grelot and Richert, 2019; Grelot et al., 2022) produces flood damage functions for built assets, such as buildings or dwellings. To do so, the *floodam.building* tool counts on a library of elementary damage functions for each component of a building, for all possible combinations of flood duration and floodwater depth, established from expert knowledge. Thus, departing from an inventory of elementary components of a given building or dwelling – e.g. floors, ceilings, coatings, walls, etc. –, the *floodam.building* tool enables us to estimate the building's damage function as a combination of the damage functions of elementary components. In addition, *floodam.building* also implements the average national french damage functions (see Rouchon et al., 2018).

The *floodam.spatial* tool (Grelot and Richert, 2021) is intended for flood damage estimation from spatial data. The tool enables us to transform spatialized flood parameters, such as floodwater depth associated to a spatial point or polygon, into monetary estimates of the damage caused based on the assets involved.

The workflow to couple all these tools is displayed in Fig. 1: on the one hand, departing from information on the built area and on the event, the hydraulic model estimates the floodwater depth in the simulation perimeter considering either nonporous (i.e. no openings) and porous (i.e. with openings) buildings. The estimation of the floodwater depth in the interior of each building in the simulation perimeter depends on whether buildings are considered porous or nonporous. In the former case, the hydraulic model is able to calculate the floodwater depth inside each building, whereas in the latter case, the floodwater depth in the interior of the building should be estimated indirectly. To do so, we follow the following procedure: (i) identification of the building; (ii) identification of the grid cells adjacent to the building; (iii) calculation/estimation of the potential gap between the building's floor level and the street level; and (iv) estimation of the floodwater depth inside the buildings as the difference between the exterior floodwater level and the building's floor level.

The damage function issued from *floodam.building*, the estimates of floodwater depth within each building calculated from the hydraulic model and the duration of the flood are fed to *floodam.spatial*. Using *floodam.spatial*, we estimate the damage endured by each building in the simulation perimeter.

3 Experimental design

Our experimental design uses the 1988 flood event in Nîmes (southeastern France) as reference. This major event was originated by heavy rainfall upstream the city (420 mm in 8 h). It registered waters depths up to 3 m in the streets of the city. Around 45 000 people were impacted and the total damage rose to more than 600 millions of euros (Mignot et al., 2006; Fabre, 1989, 1990).

The simulation perimeter (Fig. 2a and b) corresponds to the *Richelieu* district, delimited by the streets Vincent Faïta, Sully and Pierre Semard. This perimeter was severely affected during the 1988 flood event. The street layout is rather simple within the perimeter, with mostly 90° intersections. The average slope, oriented north to south, is higher than 1 % (Mignot et al., 2006; Choley et al., 2021).

The boundary conditions used in the modeling are as follows: Upstream, the discharge is injected either at the upstream end of streets Sully and Vincent Faïta or at the connection between the Vincent Faïta street and the ones coming

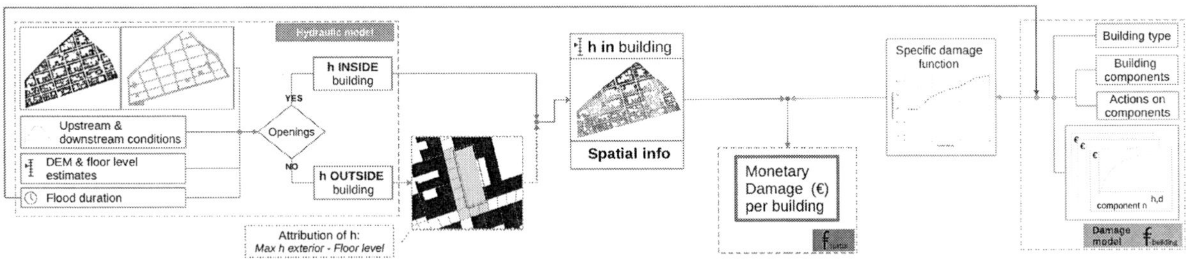

Figure 1. Coupled model flowchart.

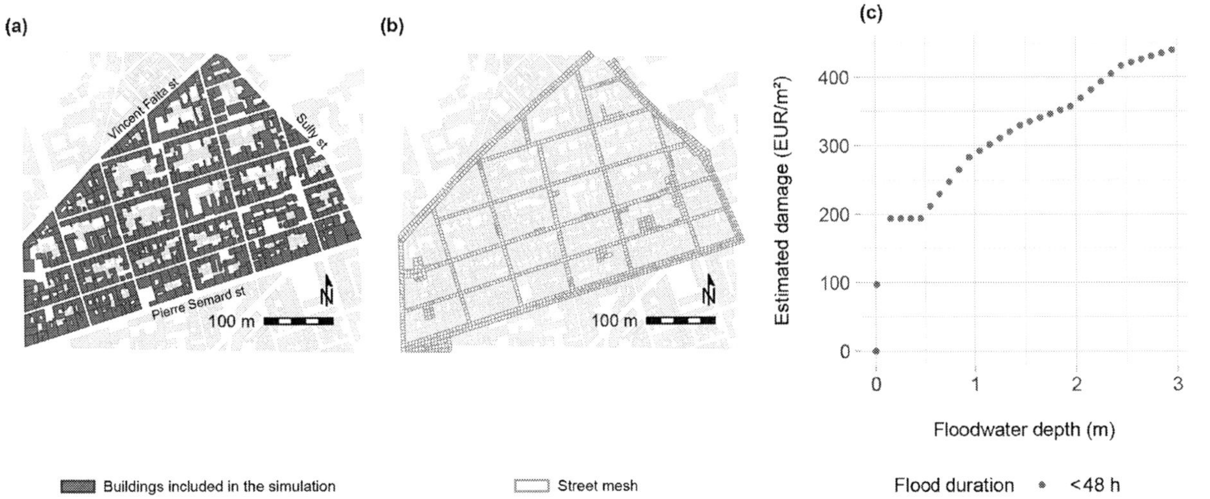

Figure 2. Experimental design: (a) buildings considered in the simulation perimeter; (b) street mesh in the simulation perimeter; (c) average national French damage function for short flood events (less than 48 h).

from the North. At the peak of the flood, the total injected discharge is 176 $m^3 s^{-1}$ with up to 92 $m^3 s^{-1}$ in the Sully street and 47 $m^3 s^{-1}$ in the Faïta street. At the downstream end, the free surface elevations imposed at the limits corresponding to an outlet street vary in the range [0.46, 1.77] m (Mignot, 2005; Choley et al., 2021).

The perimeter includes 438 buildings; courtyards are not taken into account. The floor level of buildings can be higher, lower or at the same level of the street. The simulation lasts 55 000 s, and a uniform Strickler friction coefficient $K = 40\,m^{1/3}\,s^{-1}$ is applied (Mignot, 2005).

We simulate two scenarios: (i) the *baseline* scenario, which integrates nonporous buildings as obstacles, not considering street-building flow exchanges; and (ii) the *indoor* scenario, which integrates porous buildings as reservoirs, considering street-building flow exchanges through openings.

There are two openings per facade adjacent to a street. The width of the openings is variable and related to the width of the facade. One of the openings is always located at the same level as the building floor, while the other is located at a fixed height above the street level. Thus, since the floor level of each building is located at a variable elevation from the street, the second opening of the pair is located at a variable height from the floor level of each building. Openings are considered fully open, i.e. obstacles to the street-building water flow, such as doors or windows, are not considered. The total number of the openings in a building depends on the number of facades adjacent to streets: a building with one facade adjacent to a street will have 2 openings, a building with 2 facades, 4 openings, etc.

To determine the economic damage to each building we use the average national French damage function for short flood events (less than 48 h), as displayed in Fig. 2c. It showcases a direct relationship between the monetary value of the flood damage and the floodwater depth: as the floodwater depth increases, the monetary value of damage increases. Notwithstanding, the elasticity damage-depth is variable: in the intervals (0, 0.15] and (2, 2.5] m the sensitivity of the monetary damage to changes in floodwater depth is more accentuated than in the intervals (0.15, 2] and (2.5, 3] m. Furthermore, in the interval (0, 0.15] m monetary damage is extremely sensitive to changes in floodwater depth.

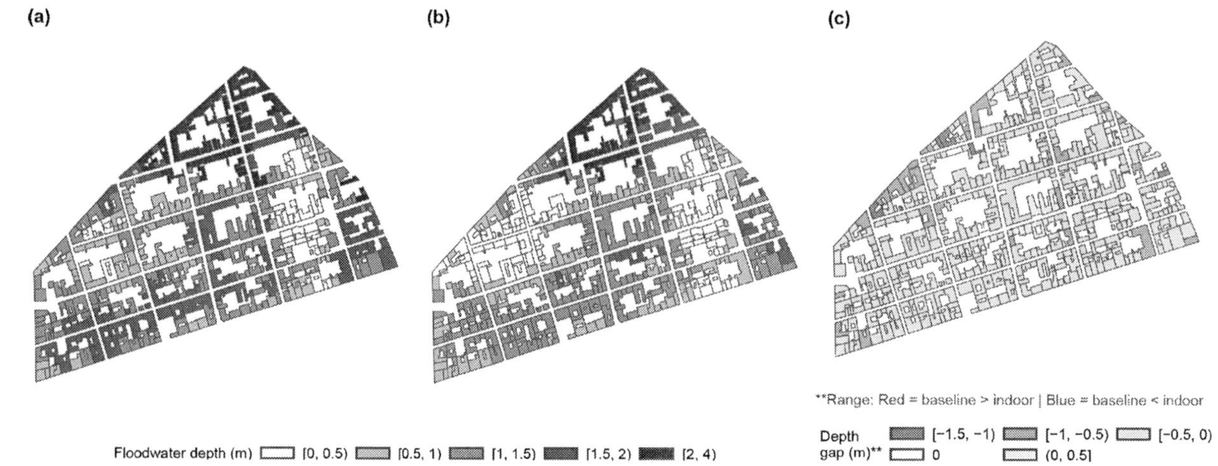

Figure 3. Geographical distribution of estimates of floodwater depth associated to each building in the simulation perimeter: **(a)** *baseline* scenario; **(b)** *indoor* scenario; and **(c)** differences in floodwater depth between *baseline* and *indoor* scenarios.

Table 1. Estimated monetary damage per scenario in the simulation perimeter.

	Baseline	Indoor
Monetary damage (Millions of EUR)	15.3	13.7
Average damage per m^2 (EUR m^{-2})	309.0	277.0

4 Results

4.1 Floodwater depth

Figure 3a and b display the maximum floodwater depth per building in the *baseline* and the *indoor* scenarios, respectively.

In the *baseline* scenario, 80 % of floodwater depths concentrate in the range [0, 1.75] m, with the most extreme simulated floodwater depths reaching 3.5 m. In the *indoor* scenario, the 80 % of the observations concentrates in the range [0, 1.5] m and the most extreme floodwater depth reaches 3.1 m. Comparing both scenarios (Fig. 3c), we observe that the *baseline* scenario overestimates the floodwater depth in 95 % of the buildings in comparison with the *indoor* scenario. In terms of magnitude, the depth gap between the two scenarios concentrates in the interval [−0.5, 0.5] m for 92 % of the buildings, while the remaining 8 % fall in the interval [−1.5, −0.5) m.

4.2 Economic damage

Table 1 shows the monetary estimation of flood damage for the ensemble of the simulation perimeter, calculated according our two scenarios, *baseline* and *indoor*. The estimated flood damage in the *baseline* and in the *indoor* scenarios are 15.3 and 13.7 millions of euros, respectively. In other words, not considering street-building exchanges leads us to overestimate the damage caused by the flood event by 10 %.

Figure 4 shows a comparison of monetary damage per squared meter estimated with the *baseline* and *indoor* scenarios. Although both scenarios present the same range of monetary damage (between EUR 0 and 439 per m^2), 90 % of the buildings display differences between the *indoor* and the *baseline* scenarios. Indeed, the *baseline* scenario overestimates the monetary damage in 86 % of the buildings, underestimates the monetary damage in 4 % of the buildings and remains the same as the *indoor* scenario in the remaining 10 %. The interval EUR [−50, 50] per m^2 accounts for 84 % of the buildings. In terms of spatial distribution (Fig. 4a), the differences in economic damage follow, although not systematically, a pattern similar to the differences in floodwater depth from Fig. 3c.

5 Discussion and conclusion

In our simulation perimeter, almost all buildings show differences in floodwater depth between the *baseline* and the *indoor* scenarios (only one building remains unchanged). In the case of estimated monetary damage, 90 % of the buildings differ between the scenarios. Thus, at individual building level, differences between scenarios are notable. Looking at the ensemble, not considering street-building exchanges leads to an overestimation of damage at district level, although the magnitude of the overestimation relative to the magnitude of the total damage is rather small.

These results are nonetheless coherent with the type of *indoor* scenario simulated. When considered, openings are set as fully open, thus water does not found any obstacle to fill the building. In these conditions, minimal differences between the *baseline* scenario and the *indoor* scenario are to be expected.

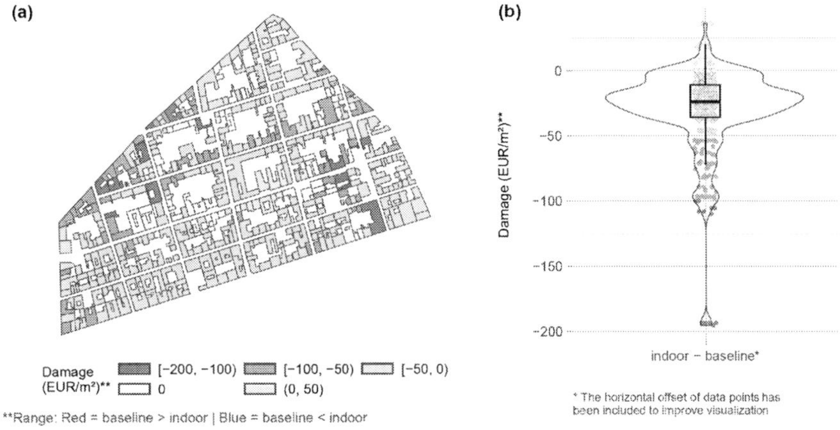

Figure 4. Differences in the estimates of damage per squared meter in the simulation perimeter between the *baseline* scenario and the *indoor* scenario: **(a)** geographical distribution; and **(b)** density-boxplot.

More in-depth studies conducted at building level, considering both the building layout and the state of the openings, show that significant differences in the damage level exist when openings are close though not waterproof (Nortes Martinez et al., 2022). Consequently, the ongoing work focuses, on the one hand, on adapting the model to consider alternative opening states, and, on the other hand, on carrying out complementary analyses at higher levels of resolution, considering the influence of the hydraulic behavior inside the buildings not only in the estimation of monetary damages but also in terms of danger.

Code availability. The hydraulic model as well as the models *floodam.building* and *floodam.spatial* are available upon request.

Data availability. The French National damage functions are available at https://www.ecologie.gouv.fr/levaluation-economique-des-projets-gestion-des-risques-naturels (Ministère de la Transition écologique et de la Cohésion des territoires, 2023). The data generated by the simulation models are part of an ongoing research project and are therefore not yet publicly available. Please contact the authors to inquire about availability.

Author contributions. CC and PFG developed the hydraulic models and calculate the physical flood parameters. FG developed the damage model. DNM coupled both models and performed the floodwater depth and damage analysis. DNM and FG wrote the first draft of the article. All authors discussed and edited the paper.

Competing interests. The contact author has declared that none of the authors has any competing interests.

Disclaimer. Publisher's note: Copernicus Publications remains neutral with regard to jurisdictional claims in published maps and institutional affiliations.

Special issue statement. This article is part of the special issue "IAHS2022 – Hydrological sciences in the Anthropocene: Variability and change across space, time, extremes, and interfaces". It is a result of the XIth Scientific Assembly of the International Association of Hydrological Sciences (IAHS 2022), Montpellier, France, 29 May–3 June 2022.

Acknowledgements. The authors acknowledge the financial support offered by the French Ministry of the Environment and the French National Research Agency (ANR). David Nortes Martínez received a postdoctoral fellowship from the French National Research Institute for Agriculture, Food and the Environment (INRAE). Cécile Choley received a doctoral fellowship from the French National School for Water and Environmental Engineering (ENGEES).

Financial support. This research has been supported by the French National Research Agency (ANR) for the project DEUFI (DEtailing Urban Flood Impact) (grant no. ANR-18-CE01-0020) and the French Ministry of the Environment (INRAE-DGPR agreement no. 2103343215).

Review statement. This paper was edited by Christophe Cudennec and reviewed by two anonymous referees.

References

Bellos, V. and Tsakiris, G.: Comparing Various Methods of Building Representation for 2D Flood Modelling In

Built-Up Areas, Water Resour. Manage., 29, 379–397, https://doi.org/10.1007/s11269-014-0702-3, 2015.

Choley, C., Finaud-Guyot, P., Garambois, P.-A., and Mosé, R.: An Effective Urban Flood Model Accounting for Street-Building Exchanges, in: Simhydro 2021, 6th International Conference Models for complex and global water issues – Practices and expectations, 16–18 June 2021, Sophia Antipolis, France, https://hal.archives-ouvertes.fr/hal-03520964 (last access: 28 April 2023), 2021.

Fabre, G.: Les inondations catastrophiques de Nîmes et de sa région du lundi 3 octobre 1988/The catastrophic floods of Nîmes and its area on 3rd October 1988, Géocarrefour, 64, 224–230, https://doi.org/10.3406/geoca.1989.5696, 1989.

Fabre, G.: La catastrophe hydrologique éclair de Nîmes (3 octobre 1988) (The flash flood disaster at Nîmes on 3 octobre 1988), Bulletin de l'Association de Géographes Français, 67, 113–122, https://doi.org/10.3406/bagf.1990.1522, 1990.

Gems, B., Mazzorana, B., Hofer, T., Sturm, M., Gabl, R., and Aufleger, M.: 3-D hydrodynamic modelling of flood impacts on a building and indoor flooding processes, Nat. Hazards Earth Syst. Sci., 16, 1351–1368, https://doi.org/10.5194/nhess-16-1351-2016, 2016.

Grelot, F. and Richert, C.: Modelling Flood Damage functions of buildings. Manual for floodam v1.0.0, Research report, IRSTEA, https://hal.inrae.fr/hal-02609309 (last access: 28 April 2023), 2019.

Grelot, F. and Richert, C.: floodam.spatial: Estimation of flood damage in a given territory, http://www.floodam.org/ (last access: 28 April 2023), r package version 0.9.0 [code], 2021.

Grelot, F., Gontrand, F., Richert, C., Boisgontier, H., and Nortes Martinez, D.: floodam.building: Creation of flood damage functions for buildings, http://www.floodam.org/ (last access: 28 April 2023), r package version 1.0.0.2 [code], 2022.

Mignot, E.: Etude expérimentale et numérique de l'inondation d'une zone urbanisée : cas des écoulements dans les carrefours en croix, Theses, Thèse en Mécanique, Ecole centrale de Lyon, https://hal.inrae.fr/tel-02586882 (last access: 28 April 2023), 2005.

Mignot, E., Paquier, A., and Haider, S.: Modeling floods in a dense urban area using 2D shallow water equations, J. Hydrol., 327, 186–199, https://doi.org/10.1016/j.jhydrol.2005.11.026, 2006.

Ministère de la Transition écologique et de la Cohésion des territoires: L'évaluation économique des projets de gestion des risques naturels, Gouvernement de France [data set], https://www.ecologie.gouv.fr/levaluation-economique-des-projets-gestion-des-risques-naturels (last access: 28 April 2023), 14 March 2023.

Nortes Martinez, D., Grelot, F., Choley, C., and Finaud-Guyot, P.: Flood impact assessment in urban context: Coupling hydraulic and economic models for a fine scale damage assessment, EGU General Assembly 2022, Vienna, Austria, 23–27 May 2022, EGU22-3984, https://doi.org/10.5194/egusphere-egu22-3984, 2022.

Rouchon, D., Christin, N., Peinturier, C., and Nicklaus, D.: Analyse multicritères des projets de prévention des inondations, Guide méthodologique 2018, Théma – balises, Paris, France, 2018.

Schubert, J. E. and Sanders, B. F.: Building treatments for urban flood inundation models and implications for predictive skill and modeling efficiency, Adv. Water Resour., 41, 49–64, https://doi.org/10.1016/j.advwatres.2012.02.012, 2012.

Sturm, M., Gems, B., Keller, F., Mazzorana, B., Fuchs, S., Papathoma-Köhle, M., and Aufleger, M.: Experimental analyses of impact forces on buildings exposed to fluvial hazards, J. Hydrol., 565, 1–13, https://doi.org/10.1016/j.jhydrol.2018.07.070, 2018.

Proc. IAHS, 385, 253–258, 2024
https://doi.org/10.5194/piahs-385-253-2024

Open Access
PIAHS

IAHS2022 – Past and future of open, inclusive, innovative, and society-interfacing approaches

Water scarcity conflicts in a community managed irrigation system in Northern Tunisia: Supporting dialogue and negotiation

Intissar Ferchichi[1], **Abdelaziz Zairi**[1], **and Serge Marlet**[2]

[1]Institut National de Recherche en Génie Rural, Eaux et Forêts (INRGREF), Ariana, 2080, Tunisia
[2]Centre de Coopération Internationale en Recherche Agronomique pour le Développement (CIRAD), Montpellier, 34398, France

Correspondence: Intissar Ferchichi (ferchichiintissar@hotmail.fr)

Received: 29 May 2022 – Revised: 19 April 2023 – Accepted: 21 April 2023 – Published: 18 April 2024

Abstract. Conflict over water resources emerges from complex interactions among biophysical, social, and economic processes operating at multiple scales. The concomitant use of surface and groundwater managed for multiple uses is a classic coping strategy to manage the water scarcity yet a relevant example of such conflict. Managing these resources is thus a dynamic decision-making process involving actors with different perceptions of the situation, who adapt different strategies to satisfy their objectives and interests. This paper presents the results of an effort to address these conflicts through reinforcing the adaptive capacity of different actors to cope with water scarcity in a multi-scale irrigation system. The study site focuses on the irrigated area of Zaouiet Jedidi in Northeast Tunisia. This region is characterized by an intensive irrigated agriculture, mostly through an excessive pumping of groundwater resources. Individual strategies to face the water scarcity are dominating leading to a rapid deterioration of the groundwater resources. Irrigation through individual and informal wells is an example of such strategies. As the demand for agricultural and urban water has concurrently continued to develop, the groundwater has continued to drop. Despite the State's efforts to increase the water supply to the region through the transfer of surface water from the dams of northwestern Tunisia, the overexploitation of groundwater resources exceeded 200 %. A participatory process was implemented to initiate a process of interaction and coordination between the different stakeholders involved in the management of the irrigated area. We analyzed to what extent and by what mechanisms, a participatory problem-solving process can facilitate the emergence of collection action to face the water scarcity.

Keywords. Water scarcity; UPH 22; participatory modelling; capacity development; SDG16

1 Introduction

There is a grown concern about the lack of effective options to manage water scarcity. These options are usually beyond the social or institutional capacity or carry too high an economic cost to be affordable in developing countries (Frederick, 1997). Ohlsson and Turton (1999) argued that we need to view the water scarcity in relation to present water practices and management rules. This shifted the focus on the need to find the appropriate societal tools for dealing with the consequences of natural water scarcity. Much emphasis is placed today on the necessity to understand stakeholders' perception as a pre-request to identify the social bottlenecks which stand in the way of new kinds of adaptive measures to water scarcity. The complexity of water resource management issues stems from the irreducible plurality of standpoints. It involves stakeholders who frame issues in various and sometimes conflicting ways, they employ different strategies to satisfy their goals and interests. Differences in perceptions reflect differences in the underlying value system and may often lead to communication barriers and policies controversies (Elsawah et al., 2015).

Many studies emphasized the need to establish participatory learning environments where individuals can meet, interact, learn, and take collective decisions (Keen et al., 2005). According to Van den Hove (2006), problem solving processes, needed to tackle the sustainable development issues, are in fact dynamic processes of capacity building. These processes should allow the integration of different sources of information, values, and logics of various actors. It is deemed that participatory approaches have the potential to meet these problem-solving requirements. Several studies have demonstrated the effectiveness of simulation tools and role-playing games (RPG) to facilitate dialogue, negotiation and collective decision making among multiple stakeholders (Etienne, 2009). This study presents the results of an effort to reinforce the adaptive capacity of different actors to cope with water scarcity in the irrigated area of Zaouiet Jdidi located at north Tunisia. A participatory process was implemented to initiate a process of interaction and coordination between local stakeholders.

2 Materials and methods

2.1 Context and case study

The Zaouiet Jdidi irrigation scheme is located in Cap Bon region (Nabeul governorate), northern Tunisia, and covers 1042 ha (Fig. 1). The whole area is actually cropped by citrus orchards. This system is irrigated from a collective irrigation network since 1969 to save citrus orchards because of previous overexploitation of the Grombalia aquifer from individual wells. Since 1985, irrigation has been supplied from the Medjerda River by the Canal of Medjerda-Cap Bon (CMCB). The irrigation water was pumped by the Belly pumping station into the buffer reservoir of Sidi Toumi, then distributed through pipelines in the command area. The irrigation scheme is divided into 30 hydraulic sectors. Until 1998, allocation and distribution of irrigation water were managed by the Commissariat Reìgional au Deìveloppement Agricole (CRDA). In 1998, irrigation management was transferred to the water users' association (WUA). The CRDA retained a control function on the implementation of the management contract but was not involved in the routine operation of the system that was under the responsibility of the WUA. Currently, This system operated under a situation of water scarcity that resulted from (i) priority given to drinking water supplies which continued to increase and (ii) development of irrigated areas due to the continuous deterioration of the groundwater's quality (Ferchichi et al., 2017).

2.2 Participatory intervention process

The purpose of this participatory process is to support the negotiation and discussion between different actors and to facilitate the emergence of collective actions to face the problem of water scarcity. The approach is based on five stages (Fig. 2). A role-playing game was designed based on (i) description of the system (infrastructure, farms' typology, regulations, and governing rules) and (ii) participatory modelling of farmers' representation of the water scarcity in the irrigated area of Zaouiet Jedidi. The role-playing game was used in eight participatory workshops organized with different actors (farmers, WUA, CRDA), from January to March 2015. The first workshop was organized first with the WUA, followed by six workshops organized with the WUA and farmers from different sectors. The final workshop was organized with water managers from the CRDA. A restitution workshop was organized in May 2015 with WUA technical stuff, representatives of farmers from each sector and managers from the CRDA. Proposed solutions during previous workshops by each actor were discussed. Participants proposed to test innovative planning of the water distribution between users of the collective irrigation network. Two workshops with farmers and WUA were organized for this planning. During the implementation of this process, observations, structured and semi-structured interviews were conducted to assess impacts of such intervention of actors' perception and strategies

3 Results

3.1 Initial actors' perception of the origins of the water scarcity

During interviews farmers referred to the problem of water scarcity as a situation where: (i) water is rarely available when it is most needed; (ii) interruptions of the water supply are frequent and unpredicted; (iii) low pressure and (iv) insufficient quantities and extended irrigation intervals. While the managers, represented by the WUA and the administration, defined the water scarcity as a situation where they can no longer satisfy the farmers water demand.

The role-playing game represented a sector covered with citrus orchards and irrigated both from collective network of the WUA and individual wells. The role session comprised a total of six game turns which represent the three-peak period of water consumption (June, July, and August), each game turn represents 15 d.

3.2 Participatory workshops with local stakeholders using role playing games

Participants proposed solutions to improve the irrigation water management in their system. These proposals focus on three components:

1. Improvement of the management of the collective network mainly through testing the reorganization of water turn between farmers, keeping the irrigation network in charge during peak period hours and making sure that water turn schedules are respected by farmers.

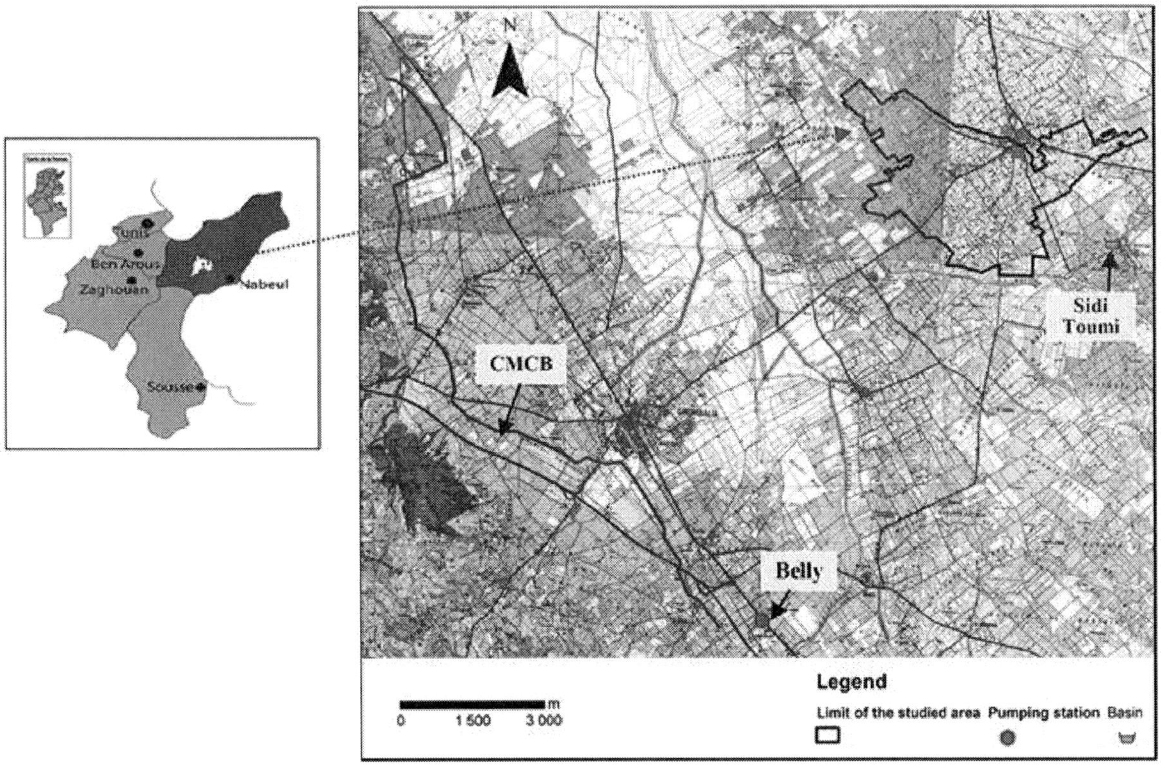

Figure 1. The irrigated area of Zaouiet Jdidi, Cap Bon (Ferchichi et al., 2017).

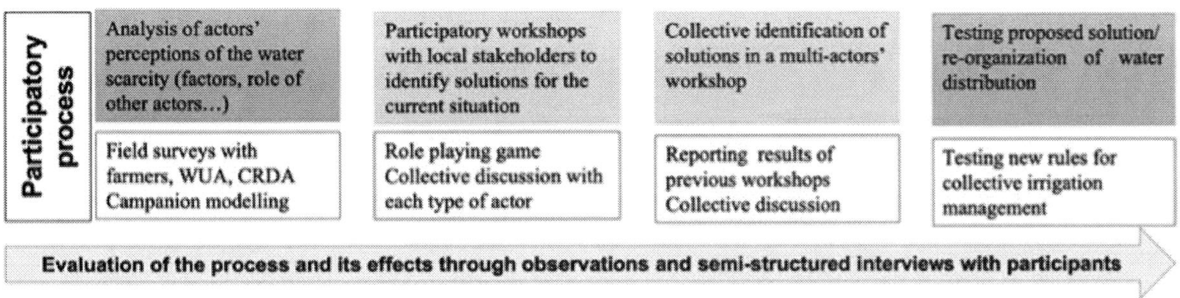

Figure 2. Presentation of the participatory process implemented in the irrigated area of Zaouiet Jedidi.

2. Improvement of groundwater management through reinforcing the conjunctive use of surface and groundwater.

3. Reinforcing communication between managers and farmers through regular meetings between farmers and the WUA.

During the multi-actors' workshop, the relevance and feasibility of the proposed actions were classified by participants. Reorganizing irrigation turns between farmers and improving the communication between farmers and managers were identified as the most feasible and relevant actions. Participants propose testing new rules to schedule irrigation turns between farmers in a pilot sector. Farmers from sector 15 volunteered to test this reform.

3.3 Testing new collective rules for irrigation schedules

A technical support for the reform of the irrigation schedule in the pilot sector was provided through two workshops organized with farmers and the WUA. The water needs of each farm were estimated and the total needs were compared to the current capacity of the network. It turns out that only 42 % of the needs of this sector can be met from the network.

Figure 3. Photos of farmers and WUA discussing new irrigation management rules.

The access to groundwater resources and the irrigation technique were considered in this process. Each farmer indicated his desired duration and frequency of irrigation (Fig. 3).

The main decisions made by the participants are to:

1. Organize irrigation schedules according to the water needs of each farmer rather than the application of the WUA's "first come first served" rule.
2. Revise the modalities of access to the collective network: only two farmers can irrigate at the same time in the same sector and only one farmer can irrigate during peak periods of electricity consumption.
3. Farmers must sign a commitment to ensure that they will respect these rules otherwise sanctions can be applied by the WUA.

4 Discussion

The results of the role-playing game with farmers showed that the construction of a model based on the stakeholders' perception of the water scarcity can be a strong support to trigger group discussion and negotiation and to explore possible solutions to cope with the problem of water scarcity. Despite the existence of a large set of sophisticated model-based tools developed for tackling water management problems, many studies argued that they did not consider the requirement of the user group and failed to support decision making processes (Bekius et al., 2022). By combining a simple simulation model and a role-playing game, tensions on the water resources and difficulties in such context allowed participants to shift from the virtual world to the real word and to propose solutions that can be applied in their system. Prell et al. (2007) argued that decisions outputs of models that have been developed through collaboration with stakeholders are less likely to be contested than those based on the outputs of "black box" models. Legitimacy building of this process with farmers was achieved through the participative modelling because it fulfilled an important condition defined by Suchman (1995): "to provide legitimacy an account must mesh both with larger belief systems and with the experienced reality of the audience's daily life".

The participative process not only succeeded in changing the social representation of farmers and eliciting individual learning, but interactions between participants also facilitated the collective learning. As a result, participants become aware of the interdependences between different actors. They shifted from the upstream management level controlled by the administration, considered initially as the only level to define solutions, to the downstream level managed by the WUA and farmers, to propose achievable solutions. In this study, we considered that group discussion and interactions reinforced the collective action through two processes: (1) consensus- oriented process and (2) compromise-oriented process. For the first process participants were able to share a common vision of the problem, they reached a consensus and proposed collective actions to face the water scarcity. For the second process, conflicts of interests exist due to the plurality of participants' standpoints. Compromises were established while defining some rules. As an example, farmers without access to the groundwater pressured the others to use this alternative resource and decreasing their water demand from the collective irrigation network. Van den Hove (2003) criticized the expanding discourse associated with participatory approaches, based on reaching consensus, and where the unavoidable conflicts tend to be occulted. The author recommended to consider participatory approaches as a combination of consensus-oriented processes in the pursuit of a common interest and compromise-oriented processes aiming at the adjustment of particular interests.

5 Conclusion

The main contribution of this study is the development of a learning approach in order to change the perception of local actors of the current situation of water scarcity and to reflect on the construction of collective actions for the improvement of the quality of irrigation service. Through this participatory process, it is estimated that the participants have acquired negotiation techniques allowing them to face the problems of water management in their system. The assessment of learning evidence and its effects on participants' perceptions leads us to conclude the necessity to understand the social context, in which the problem-solving process is anchored, and to analyze the underlying interests and values beyond apparent perceptions and discourses.

Code availability. The source code used for the participatory modelling can be obtained by writing to the corresponding authors.

Data availability. The literature data base can be obtained by writing to the corresponding authors. The spatial data used to map the irrigated area are obtained from data collected from the public administration (Ministry of Agriculture, CRDA Nabeul). The description of the roleplaying game (design method, rules, tests, etc.) is available in Ferchichi (2017).

Author contributions. The three authors developed the methodology and organised the different workshops. IF prepared the datasets, performed the analysis of the workshops results and wrote the manuscript. AZ and SM reviewed the manuscript. All authors provided critical feedback and helped shape the research, analysis and manuscript.

Competing interests. The contact author has declared that none of the authors has any competing interests.

Disclaimer. Publisher's note: Copernicus Publications remains neutral with regard to jurisdictional claims in published maps and institutional affiliations.

Special issue statement. This article is part of the special issue "IAHS2022 – Hydrological sciences in the Anthropocene: Past and future of open, inclusive, innovative, and society-interfacing approaches". It is a result of the XIth Scientific Assembly of the International Association of Hydrological Sciences (IAHS 2022), Montpellier, France, 29 May–3 June 2022.

Acknowledgements. This work was carried out in the framework of PAP-AGIR which aimed to support water users' associations in Tunisia. The project was managed by CIRAD and INRGREF under the supervision of the Tunisian Ministry of Agriculture and with funding from AFD. The authors would like to thank the farmers and the GDA of Zaouiet Jdidi for their support which was crucial for this study.

Financial support. This research was funded by the project PAP-AGIR which was supported by the French Development Agency AFD (agreement no. CTN 1101 02 W of 23 April 2009). The publication of this study was funded by the project Massire which is supported by the International Fund for Agricultural Development FIDA (grant no. 2000002013).

Review statement. This paper was edited by Christophe Cudennec and reviewed by Mohammad Merheb and one anonymous referee.

References

Bekius, F., Meijer, S., and Thomassen, H.: A Real Case Application of Game Theoretical Concepts in a Complex Decision-Making Process: Case Study ERTMS, Group Decision Negotiation, 31, 153–185, https://doi.org/10.1007/s10726-021-09762-x, 2022.

Elsawah, S., Guillaume, J. H. A., Filatova, T., Rook, J., and Jakeman, A. J.: A methodology for eliciting, representing, and analysing stakeholder knowledge for decision making on complex socio-ecological systems: From cognitive maps to agent-based models, J. Environ. Manage., 151, 500–516, https://doi.org/10.1016/j.jenvman.2014.11.028, 2015.

Etienne, M. (Ed.): La modélisation d'accompagnement: Une démarche participative en appui au développement durable, Éditions Quae, Paris, France, https://doi.org/10.35690/978-2-7592-0621-6, 2009.

Ferchichi, I.: Evaluation des services de l'eau agricole et renforcement des interactions des acteurs de la gestion de l'eau-Cas d'un système à multi niveaux de gestion (périmètre public irrigué de Zaouiet Jedidi, Cap Bon-Tunisie), Doctoral thesis, Institut National Agronomique en Tunisie, 2017.

Ferchichi, I., Marlet, S., and Zairi, A.: How farmers deal with water scarcity in community-managed irrigation systems: A case study in northern Tunisia, Irrig. Drain., 66, 556–566, https://doi.org/10.1002/ird.2135, 2017.

Frederick, K.: Adapting to climate impacts on the supply and demand for water, Clim. Change, 37, 141–156, https://doi.org/10.1023/A:1005320504436, 1997.

Keen, M., Brown, V. A., and Dyball, R (Eds.).: Social learning in environ- mental management: Towards a sustainable future, 1st edn., Routledge, https://doi.org/10.4324/9781849772570, 2005.

Ohlsson, L. and Turton, A. R.: The turning of a screw: Social resource scarcity as a bottleneck in adaptation to water scarcity, Occasional Paper Series, School of Oriental and African Studies Water Study Group, University of London, 1999.

Prell, C., Hubacek, K., Reed, M., Quinn, C., Jin, N., Holden, J., Burt, T., Kirby, M., and Sendzimir, J.: If you have a hammer everything looks like a nail: traditional versus participatory model building, Interdisciplinary Science Reviews, 32, 263–282, 2007.

Suchman, M.: Managing Legitimacy: Strategic and Institutional Approaches, The Academy of Management Review, 20, 571, https://doi.org/10.2307/258788, 1995.

Van den Hove, S.: Participatory approaches for environmental governance: Theoretical justifications and practical effects, Stakeholder Involvement Tools: Criteria for Choice and Evaluation Ed. NEA, Report NEA/RWM/FSC, 10, 2003.

Van den Hove, S.: Between consensus and compromise: acknowledging the negotiation dimension in participatory approaches, Land use policy, 23, 10–17, 2006.

Proc. IAHS, 385, 259–265, 2024
https://doi.org/10.5194/piahs-385-259-2024

Open Access PIAHS

Rainfall risk over the city of Abidjan (Côte d'Ivoire): first contribution of the joint analysis of daily rainfall from a historical record and a recent network of rain gauges

Christian Yao[1,3], Modeste Kacou[1], Ehouman Serge Koffi[2], Amidou Dao[2], Clément Dutremble[3], Maurice Guilliod[3], Bamory Kamagaté[2], Jean-Louis Perrin[3], Christian Salles[3], Luc Neppel[3], Jean-Emmanuel Paturel[3], Eric Pascal Zahiri[1], and Luc Séguis[3]

[1]LASMES, Félix Houphouët Boigny University, Abidjan, Côte d'Ivoire
[2]Laboratoire Géosciences et Environnement, Nangui Abrogoua University, Abidjan, Côte d'Ivoire
[3]HydroSciences Montpellier, University of Montpellier, CNRS, IRD, Montpellier, France

Correspondence: Christian Yao (christian.yao@ird.fr)

Received: 28 May 2022 – Revised: 11 April 2023 – Accepted: 21 April 2023 – Published: 18 April 2024

Abstract. Every year, rains cause material damage and human losses, in Abidjan (Côte d'Ivoire). The objective of this study is to contribute to the characterization of the rain hazard in the District of Abidjan. The available data are made up of daily rainfall from a historical station "Abidjan airport" (1961–2014) and an academic network of rain gauges (21) progressively implemented in Abidjan since 2015. A descriptive analysis (date of occurrence, rainfall depth, mean wet days intensity and number of rainy days) on the Highest Cumulative Rainfall Periods (HCRP: 60 d) is conducted on the long-term station. The periods of highest risk of flooding during the long and short rainy seasons are characterized. The Experimental variograms of extreme rainfalls derived from the current network, allow to evaluate their extensions according to the rainy season.

Keywords. SDG11; Field observation; Rainfall risk; Abidjan; Côte d'Ivoire

1 Introduction

Globally, climate change has increased the occurrence of natural disasters, primarily floods, resulting in loss of life and property (Alfieri et al., 2017). In African metropolises, stormwater drainage infrastructure rarely matches the rapid urbanisation associated with high population growth rates in cities (Amoako and Inkoom, 2018). Imperviousness and poor drainage systems combine to increase flood risk (Duan et al., 2016). In terms of climate change, increased rainfall is also likely to lead to increased flooding. In the Sahel, the contribution of extreme rainfall days to annual totals has increased in recent decades (Panthou et al., 2018). Further South, on the edge of Gulf of Guinea and in particularly in Côte d'Ivoire, the rainfall estimation product CHIRPS, Didi et al. (2020) show the absence of a significant increase in rainfall intensity in agreement with the results of the study by Goula et al. (2012) based on the Côte d'Ivoire rainfall network.

In order to provide more relevant information for policy makers in terms of local flood risk, this study focuses on rainfall in the city of Abidjan. Every year, about ten people lose their lives due to flooding (OCHA, 2013). While previous studies have characterised rainfall over fixed periods (annual or seasonal), this study deals with the Highest Cumulative Rainfall Periods (HCRP 60 d) likely to generate floods, using data from the Abidjan airport long-term station and an academic network of rain gauges progressively installed since 2015 (Fig. 1). Specifically, the temporal stationarity, rainfall indices qualifying the risk of flooding, the distribution of heavy rainfall and their spatial variations will be analysed on the HCRPs.

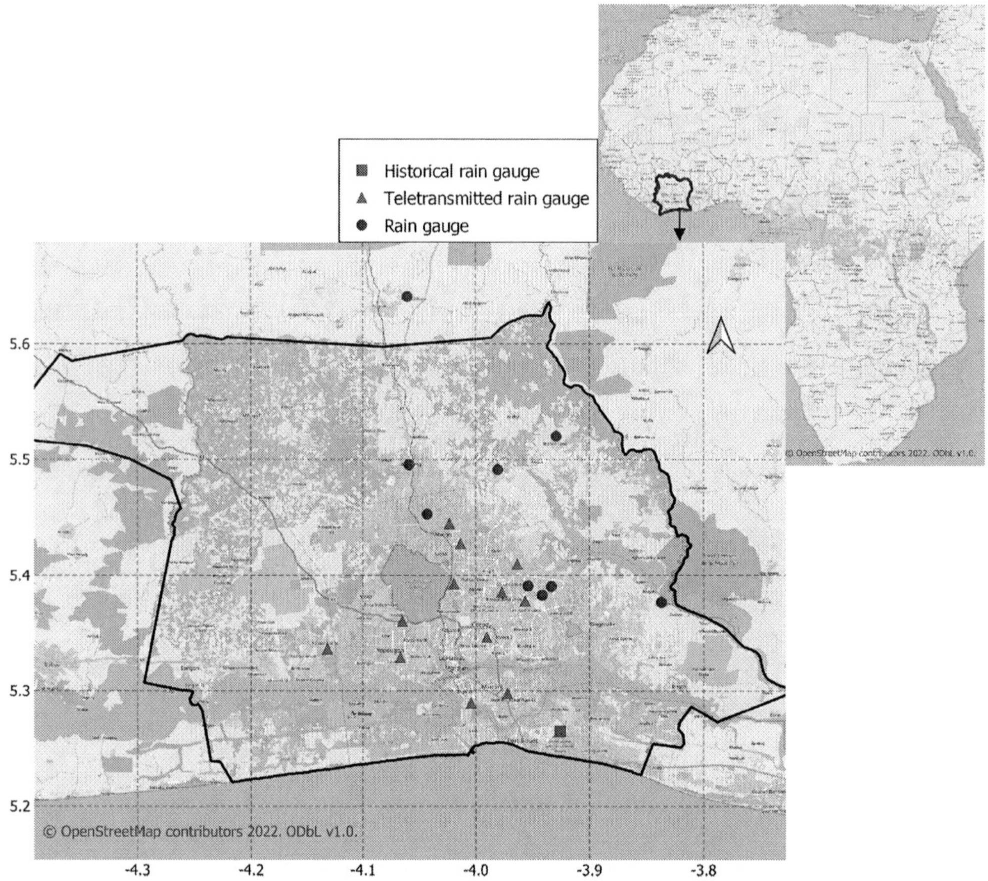

Figure 1. University network of rain gauges deployed in Abidjan since 2015 and historical rain gauge of the National Meteorology. © OpenStreetMap contributors 2022. Distributed under the Open Data Commons Open Database License (ODbL) v1.0.

2 Materials and methods

2.1 Presentation of the study area

The climate in Abidjan is sub-equatorial, hot and humid. It has a Long Rainy Season (LRS, generally from May to July) and a Short Rainy Season (SRS, generally from September to November). The average annual rainfall is 1819 mm from 1961 to 2015 (Météorologie Nationale).

2.2 Daily rainfall data

2.2.1 The academic rain gauge network

21 Precis Mécanique® tipping bucket rain gauges are distributed mainly in the urbanised area of Abidjan district with a higher density in the most flood-prone area of the Cocody commune. The criticised daily rainfall (Zahiri et al., 2021) shows a gap rate of about 12 % between 2015 and 2021. They were filled, for each station, by linear regression with neighbouring stations without gaps (this work is not described in this study).

2.2.2 Abidjan Airport Station

The Airport station is a National Meteorological station with a daily rainfall record from 1961 to 2015 with gaps for the years 2002, 2003 and 2004. This historical series constitutes the reference climatological data for this study.

2.3 Methods

2.3.1 Calculation of the Highest Cumulative Rainfall Periods (HCRP)

The HCRPs are calculated for the Abidjan airport series, taking the highest values of 60 d moving sums, before and after August. They are determined separately for the Long and Short Rainy Seasons (LRS and SRS). The significance of the trends is assessed using the Mann Kendal test. The Kolmogorov-Smirnov test is used to see to what extent the daily rainfall distributions during the LRS, SRS and the rest of the year are significantly different or not.

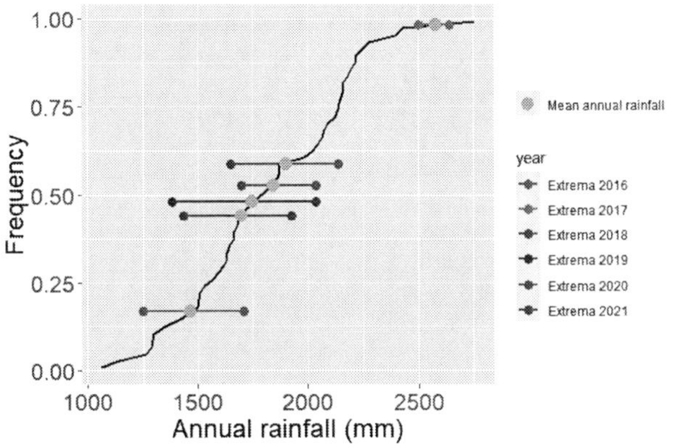

Figure 2. Interannual and spatial variability of annual rainfall of the academic network (median and annual extremes) plotted on the annual rainfall cumulative distribution function curve of the historical station Abidjan Airport.

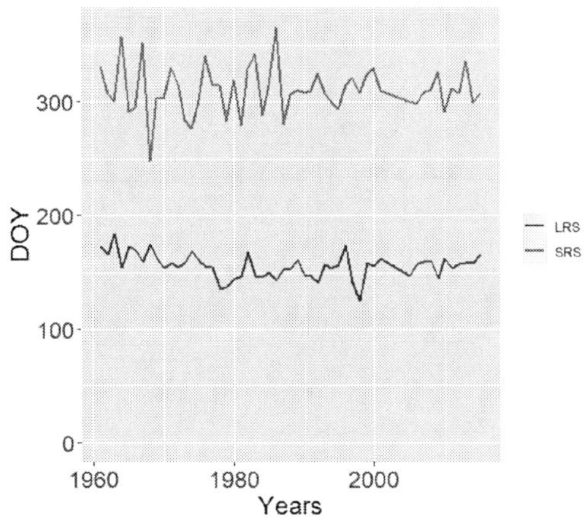

Figure 3. DOY of HCRP during LRS and SRS on the long-term daily rainfall series of Abidjan Airport.

2.3.2 Calculation of climate change detection indices

ETCCDI (Expert Team on Climate Change Detection and Indices, Zhang et al., 2011) indices such as total rainfall (PRCPTOT), number of rainy days (RR1), average daily rainfall (SDII) and daily rainfall not exceeded 90 % of the time (R90) were calculated in the HCRP at the historical station and at each station in the academic network on rainy days (≥ 1 mm) (Nkrumah et al., 2019). These indices describe a set of parameters that can characterise the occurrence or non-occurrence of flooding.

2.3.3 Spatial analysis

The variogram is used to study the spatial variability of observations in a data set. It is calculated as follows:

$$\gamma(h) = \frac{1}{2N(h)} \sum_{i=1}^{N(h)} [z(x_i) - z(x_i + h)]^2 \quad (1)$$

where $\gamma(h)$ is the variogram; $N(h)$ is the number of pairs $(z(x_i), z(x_i + h))$ separated by the distance h; x_i is the position in space of the value z and z the quantity studied (R90 in the HCRP).

These variograms are constructed using the rainy days during which at least one station in the network recorded a cumulative rainfall greater than or equal to 47 mm. The 47 mm threshold represents the 90th quantile (recurrence 10 times a year) of daily rainfall not exceeded from 1961 to 2015 at the Abidjan airport station. This value is close to the 50 mm used by Goula et al. (2012) to characterise extreme rainfall in Côte d'Ivoire.

3 Results

3.1 Temporal and spatial variability of annual rainfall

The annuals rainfalls (average and extremes) of the network stations are plotted on the cumulative frequency distribution of the annual rainfall series at the historical station (Fig. 2). The 6 years (2016–2021) of observations in the network already allow us to approach the extremes of the long-term chronicle. There is significant variability from one year to the next. 2016 was a dry year with an inter-station average of 1508 mm exceeded in about one year out of five. 2017 was very wet (2458 mm), exceeded only once in the Abidjan airport series. The other years show average inter-station rainfall close to the inter-annual average for Abidjan airport.

3.2 Dates of occurrence of HCRPs in the historical series

Figures 3 and 4 show respectively the central dates of occurrence (in DOY) of the HCRP from 1961 to 2015 and the associated boxplots. In both figures, there is a higher variability of the HCRP during the short rainy season than during the long rainy season.

The Mann Kendal test shows that there is no trend in the dates of occurrence of HCRPs during the long and short rainy seasons (with p-values of 0.16 and 0.85 respectively).

3.3 Rainfall indices during HCRPs

In absolute values, the HCRP indices of the long and short rainy seasons (Fig. 5) are strongly differentiated: PRCPTOT is more than twice as strong during the long rainy season (954 mm versus 366 mm). This is also the same ratio for SDII, the average daily rainfall (28 mm d^{-1} versus 16 mm d^{-1}). R90 is equal to 68 mm (interannual average of

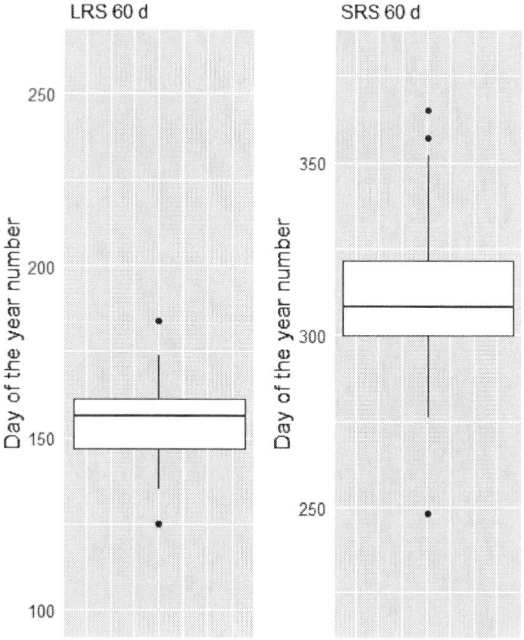

Figure 4. HCRP boxplots during LRS and SRS.

Table 1. Trends (slope and p-value) for cumulative rainfall (PRCPTOT), number of rainy days (RR1), mean rainfall (SDII) and daily rainfall of quantile 0.90 (R90) during the HCRPs of the 2 rainy seasons.

	HCRP LRS		HCRP SRS	
	Slope	P-value	Slope	P-value
PRCPTOT	−0.22	0.02**	0.04	–
RR1	−0.26	0.007**	0.08	–
SDII	−0.13	0.15*	−0.02	–
R90	−0.11	–	0.005	–

** Significant (p-value ≤ 5 %). * Weakly significant. – not significant (p-value > 20 %).

annual R90) during the long rainy season for about 30 mm during the short rainy season.

At the scale of the whole series, the deficit signal of the years 1983–1995 is particularly visible on PRCPTOT, SDII and R90 of the HCRP of LRS. This deficit period is not present on the same indices describing the HCRP of SRS.

In terms of trends from 1961 to 2015, the following points can be noted for each season:

- No significant trend for the SRS indices,

- Significant decreases in PRCPTOT and RR1 accompanied by a less significant decrease in SDII for the LRS. There is also no trend for R90.

Table 2. Statistics of extreme rainfalls.

		% of rainy days in		
daily rainfall (mm)	Return period	HCRP LRS	HCRP SRS	Rest of the year
≥ 74	4.5 per year	77	12	11
≥ 100	2 per year	78	15	7
≥ 150	1 every 2 years	88	8	4
≥ 191	1 every 10 years	100	0	0

3.4 Distribution of daily rainfall

The non-parametric Kolmogorov Smirnov test carried out in pairs between the 3 distributions (LRS, SRS and rest of the year) gave p-values in the order of 10^{-10} well below the significance level of 10^{-2} (Fig. 6). The 3 distributions are significantly different. It is during the HCRP of the LRS that the daily rainfall is the highest.

Over the year, intense, extreme and very extreme daily rainfall occurs from 77 % to 100 % during the HCRPs of LRS (Table 2).

3.5 Spatial variability

3.5.1 Variographic analysis of heavy rainfall

Figure 7a and b show the variograms of rainfall greater than 47 mm at least one network station during the HCRPs 2019–2021 of the LRS and SRS.

Both variograms show low nugget effects with 0.1 for the LRS variogram and 0.125 for the SRS variogram. This reflects a weak discontinuity in the small-scale data. There is no spatial dependence beyond 22.5 km during the LRS and 15 km during the SRS. The asymptotic sill reached beyond 36 km (equivalent to the extent of the network) during the LRS shows less spatial variability of rainfall than during the SRS where the sill is reached at about 17 km. These results show that rainfall is more widespread during the LRS than the SRS over the Abidjan district.

3.5.2 Spatial analysis of rainfall totals

The annual rainfall over the academic network in 2019, 2020 and 2021 does not show a directional gradient over the district (result not shown). The same is true for the cumulative extreme rainfall (above 70 mm) per year (see Fig. 8).

4 Discussion and conclusion

The results show great differences between the HCRPs, which are the real rainy heart of the rainy seasons, and this has consequences for the risk of flooding. Soils are wetter and more prone to runoff during the HCRP of the LRS than

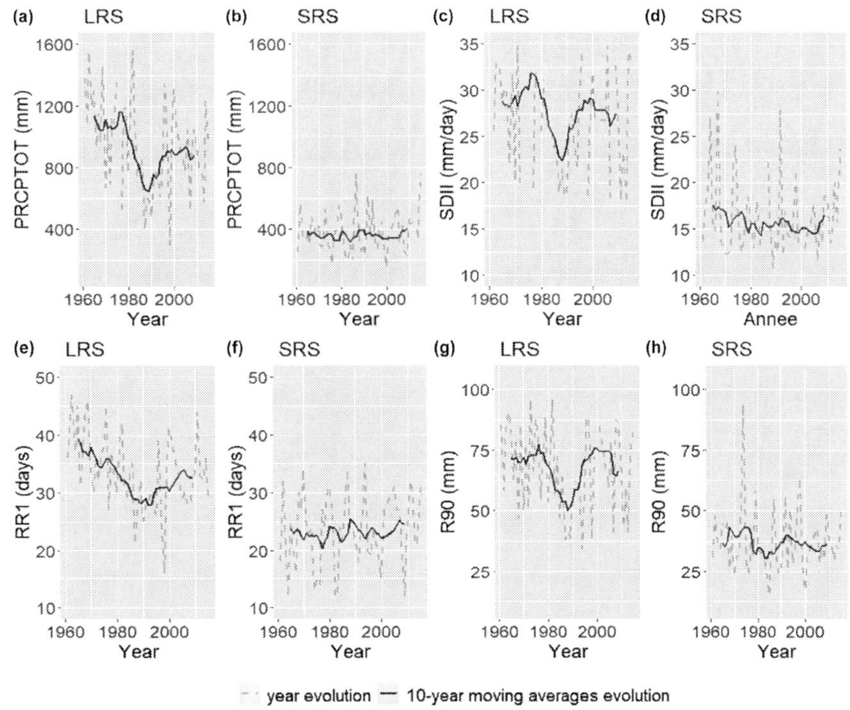

Figure 5. Temporal evolution of rainfall indices of the 10-year moving averages of total rainfall (PRCPTOT) **(a, b)**; mean intensity of wet days (SDII) **(c, d)**; number of wet days (RR1) **(e, f)**, and 90th percentile (R90) **(g, h)** during the Highest Cumulative Rainfall Periods of Long Rain Season (LRS) and Short Rain Season (SRS).

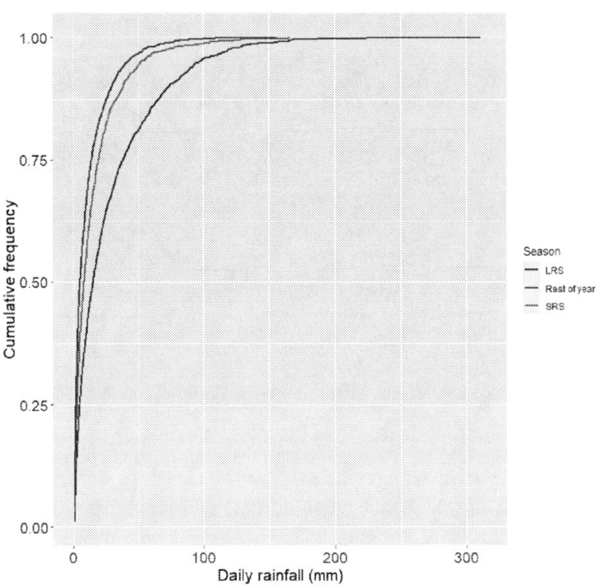

Figure 6. Distribution of daily rainfall during the HCRP 60 of both rainy seasons (LRS and SRS) and during the rest of the year (dry period).

during the HCRP of the SRS. The risk of flooding is also more limited in time during the LRS than during the SRS.

Another factor aggravating the risk of flooding lies in the distribution of daily rainfall. Rainfall is heavier during the HCRP of the LRS, especially the extremes. The tendency of the LRS HCRP to cause flooding is confirmed by the register of accidents and disasters kept by the military fire brigade group of Côte d'Ivoire. Over the period 2014 to 2021, 100 % of the floods recorded occur during the main rainy season between 28 April and 12 July – the bounds of the 50 % occurrence interval of the LRS's HCRP.

The variograms of heavy rainfall over the last 3 years of network operation indicate a greater spatial extent of rainfall events during the HCRPs of the LRS. Flooding can occur in a larger number of catchments at the same time, making Civil Protection responses more challenging.

The increase in urban flooding observed in Abidjan over the long term (since 1961) is not attributable to climate change through an intensification of rainfall but rather to urbanisation and the organisation of drainage networks.

The strong differences observed between the HCRPs of the LRS and the SRS are certainly to be found in the origin of the two rainy seasons. According to Bichet and Diedhiou (2018), the two rainy seasons clearly visible in Abidjan correspond to two different phases of the West African Monsoon. The first phase from April to June is oceanic (linked to

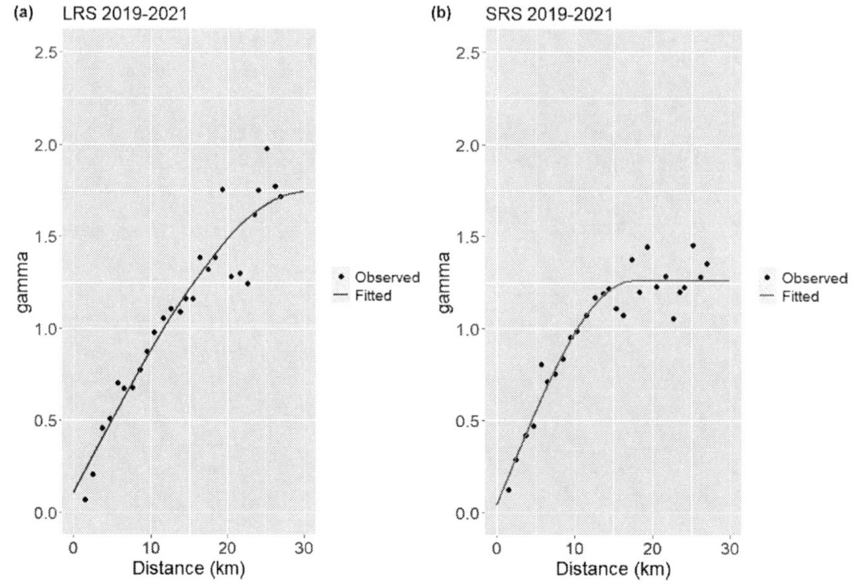

Figure 7. Variograms of LRS and SRS over all years of operation of the academic network.

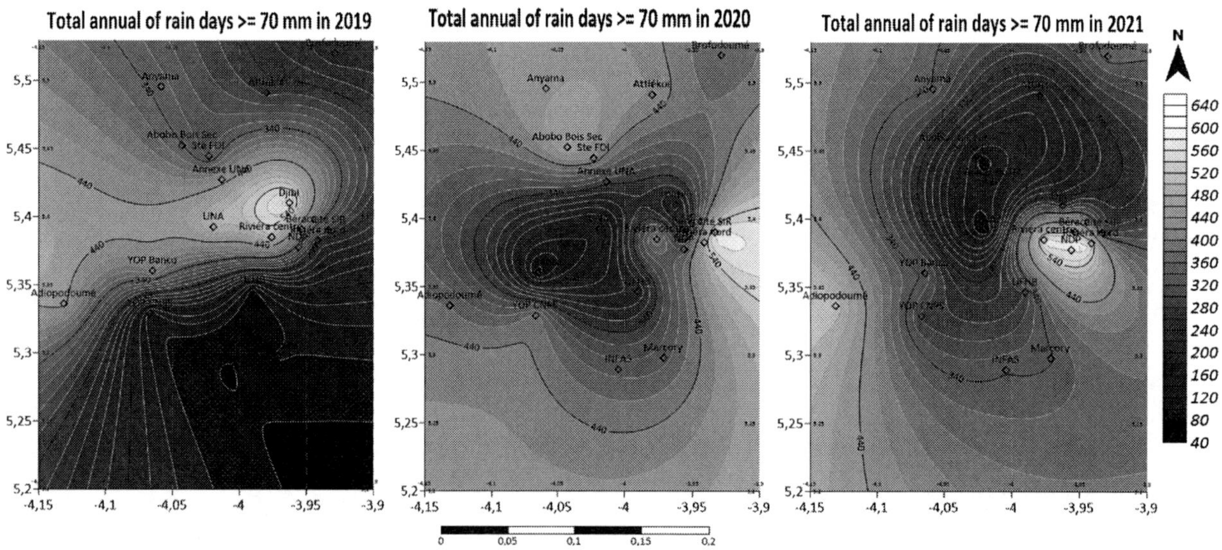

Figure 8. Spatial distribution of annual cumulative of daily rainfall greater than or equal to 70 mm over the Abidjan district.

the surface temperature of the ocean in the Gulf of Guinea) and the second phase which is continental appears from July to November.

This study has shown the key role of the long rainy season and in particular the highest cumulative rainfall periods over 60 d in the risk of flooding. It is also shown that extreme rainfall is less severe during the short rainy season. The ETCCDI parameters calculated on the HCRPs show that the increase in flooding in Abidjan District is not due to an intensification of daily rainfall. Soil sealing linked to urbanisation and the weaknesses of drainage infrastructures certainly contribute to increase flooding, but other rainfall factors should be explored, such as the number of rainy periods (for *n* consecutive days) in the HCRPs, and above all, the analysis of rainfall intensities within the events should be carried out, provided that a historical chronicle with high temporal resolution is available.

The next step will be to determine the characteristics of rainfall events at the sub-daily scale in order to update the intensity-duration-frequency curves and the design storm, which is a prerequisite for a better estimation of floods.

Data availability. Daily data from the rain gauge network are freely available at https://doi.org/10.23708/AN9SKZ (Zahiri et al., 2021). The historical data of the airport station was provided by the national meteorological agency SODEXAM.

Author contributions. LS, MK and EPZ defined the frame of the paper. CY carried out the study. Rainfall from the raingauge academic network were collected and criticized by ESK, AD, CD, MG, JLP and LS. CY prepared the manuscript with contributions from all co-authors.

Competing interests. The contact author has declared that none of the authors has any competing interests.

Disclaimer. Publisher's note: Copernicus Publications remains neutral with regard to jurisdictional claims in published maps and institutional affiliations.

Special issue statement. This article is part of the special issue "IAHS2022 – Hydrological sciences in the Anthropocene: Variability and change across space, time, extremes, and interfaces". It is a result of the XIth Scientific Assembly of the International Association of Hydrological Sciences (IAHS 2022), Montpellier, France, 29 May–3 June 2022.

Acknowledgements. We acknowledge EVIdENCE Project and SODEXAM for providing dataset used in the framework of this research. Christian Yao benefits from an IRD doctoral grant.

Financial support. This research has been supported by the "Partenariat rénové pour la Recherche au Service du Développement de la Côte d'Ivoire" (grant no. D18.64/JMH/kf; 6 mars 2018, project EVIdENCE).

Review statement. This paper was edited by Christophe Cudennec and reviewed by two anonymous referees.

References

Alfieri, L., Bisselink, B., Dottori, F., Naumann, G., de Roo, A., Salamon, P., Wyser, K., and Feyen, L.: Global projections of river flood risk in a warmer world: RIVER FLOOD RISK IN A WARMER WORLD, Earth's Future, 5, 171–182, https://doi.org/10.1002/2016EF000485, 2017.

Amoako, C. and Inkoom, D. K. B.: The production of flood vulnerability in Accra, Ghana: Re-thinking flooding and informal urbanisation, Urban Stud., 55, 2903–2922, https://doi.org/10.1177/0042098016686526, 2018.

Bichet, A. and Diedhiou, A.: Less frequent and more intense rainfall along the coast of the Gulf of Guinea in West and Central Africa (1981–2014), Clim. Res., 76, 191–201, https://doi.org/10.3354/cr01537, 2018.

Didi, S. R. M., Ly, M. , Kouadio, K., Bichet, A., Diedhiou, A., Coulibaly, H. S. J., Kouadio, K. C. A., Coulibaly, T. J. H., Obahoundje, S., and Savané, I.: Using the CHIRPS Dataset to Investigate Historical Changes in Precipitation Extremes in West Africa, Climate, 8, 84, https://doi.org/10.3390/cli8070084, 2020.

Duan, W., He, B., Nover, D., Fan, J., Yang, G., Chen, W., Meng, H., and Liu, C.: Floods and associated socioeconomic damages in China over the last century, Nat. Hazards, 82, 401–413, https://doi.org/10.1007/s11069-016-2207-2, 2016.

Goula, B. T. A., Soro, E. G., Kouassi, W., and Srohourou, B.: Trends and breaks in extreme daily rainfall in Côte d'Ivoire (West Africa), Hydrolog. Sci. J., 57, 1067–1080, https://doi.org/10.1080/02626667.2012.692880, 2012.

Nkrumah, F., Vischel, T., Panthou, G., Klutse, N. A. B., Adukpo, D. C., and Diedhiou, A.: Recent Trends in the Daily Rainfall Regime in Southern West Africa, Atmosphere, 10, 741, https://doi.org/10.3390/atmos10120741, 2019.

OCHA: Côte d'Ivoire: Flood and cholera risk zones, http://www.unocha.org/cotedivoire (last access: 18 April 2022), 2013.

Panthou, G., Lebel, T., Vischel, T., Quantin, G., Sane, Y., Ba, A., Ndiaye, O., Diongue-Niang, A., and Diopkane, M.: Rainfall intensification in tropical semi-arid regions: the Sahelian case, Environ. Res. Lett. 13, 064013, https://doi.org/10.1088/1748-9326/aac334, 2018.

Zahiri, É., Séguis, L., Kacou, M., Kamagaté, B., Dao, A., Perrin, J.-L., Koffi, S. E., and Guilliod, M.: Daily rainfall in Abidjan district (Côte d'Ivoire) (2015–), IRD/DataSuds [data set], V7, https://doi.org/10.23708/AN9SKZ, 2021.

Zhang, X., Alexander, L., Hegerl, G. C., Jones, P., Tank, A. K., Peterson, T. C., Trewin, B., and Zwiers, F. W.: Indices for monitoring changes in extremes based on daily temperature and precipitation data, WIREs Clim. Change, 2, 851–870, https://doi.org/10.1002/wcc.147, 2011.

Proc. IAHS, 385, 267–273, 2024
https://doi.org/10.5194/piahs-385-267-2024

Seasonal precipitation forecasting with large scale climate predictors: a hybrid ensemble empirical mode decomposition-NARX scheme

Rim Ouachani[1,2], **Zoubeida Bargaoui**[2], **and Taha Ouarda**[3]

[1]Higher Institute of Transport and Logistics of Sousse, Sousse University, Sousse, 4023, Tunisia
[2]Ecole Nationale d'Ingénieurs de Tunis, Université de Tunis El Manar, Tunis, 1002, Tunisia
[3]INRS-ETE, National Institute of Scientific Research, Quebec City, QC G1K 9A9, Canada

Correspondence: Rim Ouachani (rim.ouachani@istls.u-sousse.tn)

Received: 15 June 2022 – Revised: 1 April 2023 – Accepted: 21 April 2023 – Published: 18 April 2024

Abstract. Much of northern Tunisia regularly experiences extremes of drought and flooding, with high rainfall variability. The development of reliable and accurate seasonal rainfall forecasts can provide valuable information to help mitigate some of the outcome of floods and enhance water management and monitoring, particularly for agriculture. Whether climate indices oscillations contain some information to be useful for hydrological forecasting is worth investigating. Ensemble monthly rainfall forecasts are carried out using a hybrid neural network model. The hybrid model called EEMD-NARX based on a non-linear autoregressive network with exogenous inputs (NARX) coupled to Ensemble Empirical Mode Decomposition (EEMD) method is developed in this work. First, the EEMD is performed to extract significant information from modes of variability (IMF) associated to climate indices and precipitation. Each IMF of selected indices as well as precipitation IMFs are then used as inputs to the NARX forecasting model to forecast each IMF of precipitation. To make forecasts operational, we reconstruct precipitation by summing of all forecasted IMFs to make comparison with observed precipitation in the Medjerda river basin located in north Tunisia. Results show that IMFs of MEI and SOI indices can be distinguished from a white noise at the 95 % level. It is also found that an oscillatory forcing coming from the Atlantic influences the precipitation in the Mediterranean basin. The results indicate that exogenous inputs like climatic indices improve the accuracy of forecasts in some in some precipitation stations. The correlation coefficient between observed and forecasted monthly precipitation is ranging from 0.7 to 0.8. EEMD allows extracting significant components from exogenous inputs like climate indices that help reducing predictive uncertainty as well as improving forecasts of a NARX model at longer lead-times.

Keywords. Rainfall forecasting; flood monitoring; UPH17; climate indicators; data-preprocessing; NARX

1 Introduction

Precipitation forecasting is useful for water resources and hydraulic structures management. Extreme events have devastating consequences that disfigure the nature and cause thousands of casualties. For new monitoring, good management and forecasting of floods, it is essential to go through a precipitation forecasting. Seasonal forecasting is useful for the agricultural sector which is among the key sectors of the Tunisian economy. The precipitation process is difficult to understand and to model because of the complexity of the phenomena and the atmospheric processes that generate it. Then a reliable forecast of the precipitation remains a challenge. Several methods could be used for precipitation forecasting from statistical to empirical models through the artificial intelligence models. Artificial neural networks (ANN) are robust tools for modeling and forecasting many of the nonlinear hydrological processes but their performance remains linked to the complexity of the network. Improving the performance of a neural network model can be obtained with a preprocessing of the input and output data. Among the signal preprocessing methods we found the Empirical Mode Decomposition (EMD) (Huang et al., 1998). It is one of the methods in the frequency domain that can process non-linear and non-stationary data. An application of this approach applied to environmental data analysis: rainfall, temperature, wind and streamflow analysis is presented in Rao and Hsu (2008). Kisi et al. (2014) reported the use of ANNs in forecasting hydrological variables as well as combined methods with ANNs. In their work, the EMD-ANN model was compared with the single ANN model. The optimal test results were obtained for the three-input to EMD-ANN model. Liang et al. (2021) proposed three hybrid models that couple varied pre-processing methods, which are empirical mode decomposition (EMD), ensemble empirical mode decomposition (EEMD), and empirical wavelet transform (EWT), with the nonlinear autoregressive networks with exogenous inputs (NARX) were applied to forecast tidal level.

In spite of the generalization ability of ANNs and due to the nonlinear and non-stationary nature of the rainfall time series, it is necessary the search for analysis alternatives that improve the accuracy of predictions. Basha et al. (2015) developed a stochastic model that reproduces non-stationary oscillation (NSO) processes by employing ensemble empirical mode decomposition (EEMD) and non-parametric techniques to predict the evolution of temperature, precipitation and soil moisture. Ouachani et al. (2013) studied the effect of climate variability on precipitation in the Medjerda basin and found that indices related to ENSO as well as Mediterranean Oscillation have potential power in forecasting.

Could exogenous inputs such as climate indices, defined as difference between sea surface temperature or sea level pressure between two different localizations in the sea, add some additional information to an ANN based EEMD model is the question that we will try to respond.

The remaining part of the paper is organized as follows. The proposed methodology is detailed in Sect. 2, where a brief description of Empirical Mode Decomposition (EMD), EEMD, backpropagation scheme and NARX are also presented. Section 3 presents the hydro-climatic data and Sect. 4 the obtained results. Finally, the paper is concluded in Sect. 5.

2 Methodology

This paper proposes a hybrid model for long term rainfall forecasting, adopting the Ensemble Empirical Mode Decomposition (EEMD), and as a forecasting tool, the Nonlinear AutoRegressive with eXogenous (NARX) input network. The two methods are described below.

2.1 The Ensemble Empirical Mode Decomposition (EEMD) method

As discussed in Blöschl et al. (2019), in UPH18, we need to extract information from available data in order to inform the building process of hydrological forecasting models. EMD is a nonparametric method that aims to decompose a signal into a set of meaningful components called intrinsic mode functions (IMFs). The complete mathematical description of the empirical mode decomposition is beyond the scope of this article, but can be found in Huang et al. (1998). A brief description of the algorithm can be made. The EMD extracts a series of IMFs that must respect two criteria: (1) for each IMF, the number of local extrema and the number of zero-crossings are equal or differ at most by one; (2) at any time and for every IMF, the mean value of the envelope defined by the local minima and the local maxima is zero. The criterion (1) forces an IMF to evolve as a series of periodic fluctuations and prevents the superposition of multiple oscillations. The criterion (2) imposes a null trend to the IMFs and is necessary to obtain IMFs with periodic zero-crossings as imposed by the criterion (1). The process to extract an IMF is called sifting. The sifting process gives the following final decomposition:

$$X = \sum\nolimits_{j=1}^{q} C_j + r. \tag{1}$$

Where C_j with $j = 1, 2, \ldots, q$ are the IMFs and r is the residue trend.

To verify whether IMFs contain exact signal or simply noise, Wu and Huang (2004) established a statistical significance test (at any given statistical confidence level) based on the relationship between the energy and the mean period of each component. Wu and Huang (2009) proposed the ensemble method by sifting an ensemble of white noise-added signal and treats the mean as the final true result. The EEMD requires two parameters: the variance of the added noise and the number of samples. To calculate the average of each IMF, their total number must be known. The approximate dyadic properties of EMD suggest that the total number of IMFs should be close to $\log 2(N)$, where N is the number of observations (Wu and Huang, 2009). The stopping criterion is generally a fixed number of iterations (e.g. 10).

2.2 Artificial Neural Network

The ANNs have been widely used in the scientific field of time series prediction due to their inherent nonlinearity and high robustness in noise. Typically, the challenge task of time series prediction can be expressed as finding the appropriate function F so as to acquire an estimate $\hat{y}(t+D)$ of the time series y at time $t+D$ $(D = 1, 2\ldots)$ given the past values of y up to time t, plus the values of exogenous input x:

$$\hat{y}(t+D) = F\left(y(t), \ldots, y(t-d_y), x(t), \ldots, x(t-d_x)\right) \tag{2}$$

where $y(t)$ and $x(t)$ represent the values of y and x in time t respectively. The variables d_y and d_x are the *lag time* parameters of model and in case of $D = 1$ we have the one step ahead prediction of time series y.

In this paper, we apply the backpropagation NN learning algorithm, which includes four main steps as: feed forward computation, backpropagation to the output layer, backpropagation to the hidden layer and weight updates. This forecasting engine has good abilities for dealing with nonlinear systems, such as forecasting problem of precipitation. The structure of the NN is a three layered back propagation NN with one input layer, one hidden layer and one output layer.

For each hidden neuron j (in the hidden layer), the input xj and output yj are defined as:

$$\hat{x}_i = \sum_{i=1}^{n} w_{ij} y_i \tag{3}$$

$$\hat{y}_j = F\left(\hat{x}_j + h_j\right) \tag{4}$$

where w_{ij} is the weight between the ith neuron in the input layer and j^{th}neuron in the hidden layer; $F(.)$ is the activation function of the hidden neurons; $\hat{x}_i$ and $\hat{y}_j$ are the output of input neuron i and hidden neuron j; h_j is the bias of hidden neuron j. The initial number of neurons in the hidden layer is considered. In this work, the learning rule used to adjust the NAR weights is based on the Levenberg-Marquardt method, one of the BP algorithms. It is being more powerful and faster than the conventional gradient descent techniques.

2.3 Proposed EEMD-NARX model

The overall procedure for EMD-NARX is given below:

1. Decompose the concerned time series (X_t) into a finite number of intrinsic mode functions (IMFs).
2. Select significant IMF components using significance test.
3. Calibrate the NARX model using each IMF_i of the selected indices and precipitation as inputs to predict $\text{IMF}_i$of the precipitation. Validate the model using data from the validation period.
4. Predict the IMF components using the calibrated and validated NARX model with the best performance in the validation period.
5. Sum up the forecasted IMFs from each model.

The model is run 100 times using the bootstrap procedure that adds noise with known standard deviation equal to 0.2 to the output. The mean of the obtained ensembles represent the forecasting value. For all the 100 experiments, the performance measure of the mean is computed and thus a reliable estimate of the performance is obtained. For NARX, three layers are considered: one input layer, one hidden layer and one output layer. Several network architectures are tested by varying the number of nodes in the hidden layer from 5 to 20. The architecture with the best performance in the validation period is used to make forecasts.

Among the model efficiency criteria in the literature, the coefficient of determination (R^2) and the Nash–Sutcliffe coefficient (NASH) (Eq. 5) of efficiency are the most commonly employed performance evaluation criteria and are also found to be good evaluation criteria by experts.

$$\text{NASH} = 1 - \frac{\sum_{i=1}^{n}(y_i - \hat{y}_i)^2}{\sum_{i=1}^{n}(y_i - \overline{y})^2} \tag{5}$$

where n refers to the total number of observations; $\hat{y}_i$, y_i, $\overline{y}$ represent the predicted monthly precipitation, observed precipitation, and the mean observed precipitation data, respectively.

The Willmott index of agreement (IOA) (Eq. 6) is also used

$$\text{IAO} = 1 - \frac{\sum_{i=1}^{n}(\hat{y}_i - y_i)^2}{\sum_{i=1}^{n}\left(\left|\hat{y}_i - \overline{y}\right| + \left|y_i - \overline{y}\right|\right)^2} \tag{6}$$

IOA equal to 1 is being perfect score. It is sensitive to the difference between the mean of $\hat{y}_i$ and y_i as well as the difference between the standard deviation of $\hat{y}_i$, y_i.

3 Hydro-climatic data

The analysis of the monthly precipitation is carried out for six rainfall stations reported in Ouachani et al. (2013) associated with the upper part of the Medjerda river basin (Fig. 1), a trans-boundary river, located in northern Tunisia and which accounts for the Mediterranean water budget in the Blue Plan (Margat and Treyer, 2004). The water resources and agricultural potential of this region is crucial for the Tunisian economy. Therefore, the new monitoring, modelling and optimal management of these resources is of primary importance. Field observation of rainfall is provided by the National Water Resources Division of Tunisia. These rainfall stations were chosen for their long-term records (generally exceeding 50 years) and for their good data quality. Four climate indices; the Multivariate Enso Index (MEI), Southern Oscillation Index (SOI), North Atlantic Oscillation (NAO) and Mediterranean Oscillation Index (MOAC) are used in this work as suggested by Ouachani et al. (2013). The common period 1950 to 2011 between series is chosen. Before analysis, the precipitation and climate indices time series are standardized, by subtracting the mean and dividing by the standard deviation, to ensure a comparable scale.

4 Results

4.1 Decomposition of precipitation using EEMD

The data is decomposed into IMFs by using the Matab code of EEMD method provided by Wu and Huang (2009) with

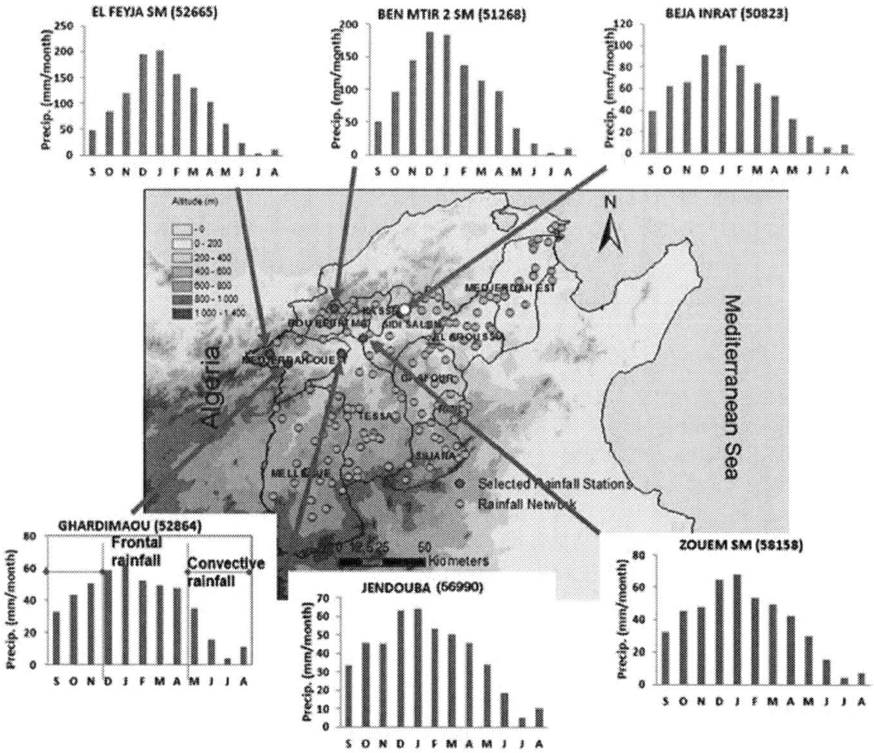

Figure 1. Geographic map of selected Rainfall stations in the Medjerda river basin with each seasonal precipitation.

100 ensembles and a noise level of 0.2. The monthly precipitation data recorded at Ghardimaou (52864) for the period 1950–2011 is shown in Fig. 2 as well as a total of 8 components and the residue. The first component (C1) which has a very high frequency is generally considered as noise. The last component (C8) represents the trend. C4 to C6 show a long-term oscillation with asymmetrical changes. While C7 represents the 40-year oscillation. The IMFs are subjected to the significance test and are plotted in Fig. 3a. The result of the significance test indicates that the three IMFs C2, C3 and C5 can be considered as real oscillatory components and are distinguishable from random noise. However, the other components are not statistically significant. The test of significance is performed also for climate indices. I can be shown for example from Fig. 3b and c that all IMFs related to MEI can be distinguished from a white noise at the 95 % level while only the second component IMF2 of SOI is not significant. All these components can be used as inputs to the forecasting model. Here we can focus on the UPH17 (Blöschl et al., 2019) and how data preprocessing methods can help extract significant components from traditional hydrological observations. NAO and MOAC components on medium time scales (IMF3, IMF5) are well correlated to precipitation components with 6 months delay time. It can be concluded that the negative phase of NAO as a tendency to generate precipitations in the winter season. As discussed in the previous section, the significant components as well as the other components are forecasted by a NARX model.

4.2 Forecasting model based EEMD

The data are divided by block into three parts for training, validation and test (forecast), respectively assigning 60 %, 20 %, 20 %. Two models are compared. The first one (EEMD-NARX) uses a set of inputs constructed from climate indices and precipitation with time delay ranging from 1 to 12 months. The second model (EEMD-ANN) has only precipitation as inputs. Table 1 summarizes the performance criteria estimates for the (EEMD-NARX), which is compared against an EEMD-ANN model for the original signal without exogenous inputs. From this we can observe good forecasts. All performance indices are in accordance. It can be observed good IAO coefficients exceeding 0.8. When the EEMD-ANN scheme is unable to give good forecasts, the EEMD-NARX outperforms a considerable difference. This is shown in the case of station Jendouba (56990) and Ghardimaou (52864) in gray color row. The IAO criteria grow from 0.80 to 0.85.

Regressions of forecasted precipitations using (EEMD-NARX) versus observed data precipitation shown in Fig. 4 reveal good forecasts of the monthly precipitation with correlation exceeding 0.7 for all stations. Results are compara-

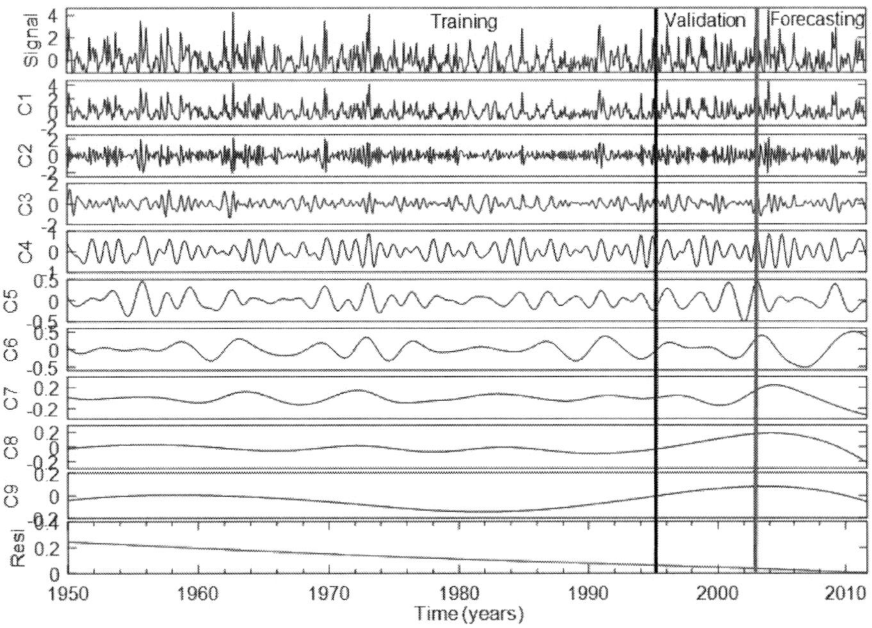

Figure 2. Monthly precipitation anomaly data recorded from 1950 to 2011 at station Ghardimaou (52864) in the Medjerda basin decomposed into IMFs (blue lines) and a residue (red line) using EEMD. Where the calibration period from 1950 to 1995. The period 1995–2003 is for the validation and the period 2003–2011 is used to validation.

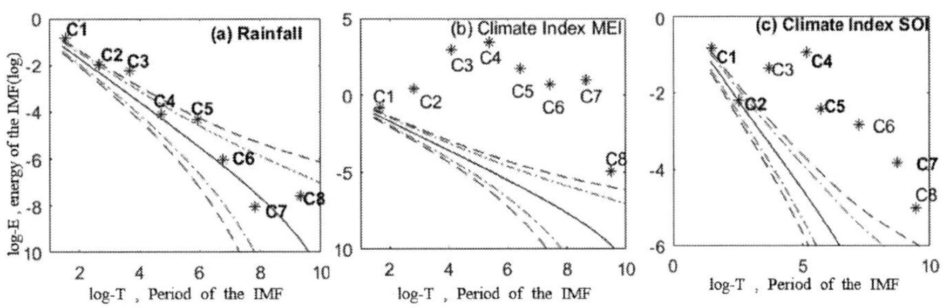

Figure 3. Significance test of the extracted IMFs for **(a)** precipitation station Ghardimaou (52864), **(b)** climate index MEI and **(c)** climate index SOI with 95 % (red line) and 99 % (blue line) confidence limits. Each point (∗) below the lines indicates that the hypothesis that the corresponding IMF of the observed series is not distinguishable from the corresponding IMF of a random noise series cannot be rejected with the confidence levels (95 % and 99 % respectively).

ble to those found by Kisi et al. (2014). However we notice the disability of the model to forecast the highest values of precipitation as the regression line is under the $y = x$ line. In addition the model overestimate low values as zero crossing value of the regression line is positive in all stations. All IFMs are well forecasted except IMF1 who can be distinguished from the others. The first IMF is the decomposition of the signal in the very high frequency and contains some noise even if it explains the precipitation signal as shown by the significance test. As reported by Kisi et al. (2014), because, the IMF1 component is characterized by higher mean frequencies and include noise component because EMD acts a filter bank for Gaussian noise, white noise and turbulence of time series. Therefore, forecasting of IMF1 component is difficult.

5 Conclusions

To forecast the future tendencies of nonlinear and nonstationary precipitation time series, a hybrid intelligent forecasting model is proposed, which is based on EEMD and ANN. EEMD allows extracting significant components to help reducing predictive uncertainty as well as improving forecasts of a neural network model. According to the obtained results, the EEMD-NARX scheme improves the forecasting results and offers a simple approach for the stable prediction of

Table 1. Performance evaluation of the precipitation forecasts by the EEMD-NARX and EEMD-ANN models.

STATION	Qobsmoy	SigmQobs	EEMD-NARX			EEMD-ANN		
			NODES	NASH	IAO	NODES	NASH	IAO
52864	43.67	42.85	14	0.52	0.85	5	0.26	0.80
51268	98.09	107.54	14	0.67	0.89	6	0.75	0.93
52665	86.09	86.85	15	0.35	0.82	7	0.45	0.86
50823	51.06	49.53	14	0.40	0.84	6	0.57	0.86
56990	42.98	40.11	13	0.49	0.84	5	0.42	0.79
58158	37.56	38.53	13	0.52	0.84	6	0.61	0.89

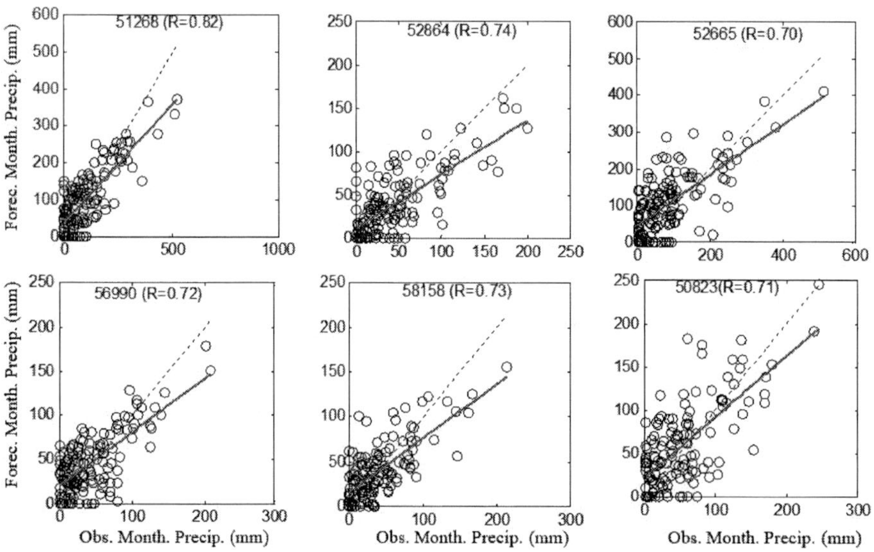

Figure 4. Regressions of forecasted precipitation using EEMD-NARX model versus observed precipitations (circles) in the period 2003 to 2011 with the correlation coefficient for each station used in the study. The red line is the linear regression while the black dot line is the first bisector.

non-stationary data when an EEMD-ANN is unable to give good forecasts. It can be concluded that exogenous inputs like climate indices can add some additional information to enhance monthly precipitation forecasts. As a future work, it would be interesting to explore the possibility of employing different aggregation methods. Because the forecast of IMF1 is difficult, it would be useful to perform a comparison with wavelet decomposition. The hybrid model has a good generalization performance and long-term forecasting ability.

Code availability. All codes are developed using MATLAB software. For the NARX method we used the Deep learning toolbox of MATLAB (https://de.mathworks.com/products/deep-learning.html, MathWorks, 2023.). The EEMD program is web-accessible. The program and its instructions can be downloaded from http://rcada.ncu.edu.tw/ (NCU, 2023).

Data availability. All Climate indices are freely accessed through the Web while the precipitation data cannot be accessed for free. It is provided by the National Water Resources Division of Tunisia (http://www.agridata.tn/, last access: 14 May 2023).

Author contributions. The paper was written by RO under the supervision of ZB and TO.

Competing interests. The contact author has declared that none of the authors has any competing interests.

Disclaimer. Publisher's note: Copernicus Publications remains neutral with regard to jurisdictional claims in published maps and institutional affiliations.

Special issue statement. This article is part of the special issue "IAHS2022 – Hydrological sciences in the Anthropocene: Variability and change across space, time, extremes, and interfaces". It is a result of the XIth Scientific Assembly of the International Association of Hydrological Sciences (IAHS 2022), Montpellier, France, 29 May–3 June 2022.

Acknowledgements. The authors would like to acknowledge the IAHS and the SYSTA for the grant given to Rim Ouachani to attend the IAHS 2022 scientific assembly.

Financial support. This research has been supported by the Tunisian Ministry of Higher Education and the Higher Institute of Scientific Research of Canada. The attempt of the IAHS2022 conference was supported by the International Association of Hydrological Sciences.

Review statement. This paper was edited by Christophe Cudennec and reviewed by two anonymous referees.

References

Blöschl, G., Bierkens, M. F. P., Chambel, A., et al.: Twenty-three unsolved problems in hydrology (UPH) – a community perspective, Hydrolog. Sci. J., 64, 1141–1158 https://doi.org/10.1080/02626667.2019.1620507, 2009.

Huang, N. E., Shen, Z., Long, S. R., Wu, M. C., Shih, E. H., Zheng, Q., Tung, C. C., and Liu, H. H: The empirical mode decomposition method and the Hilbert spectrum for nonstationary time series analysis, P. Roy. Soc. Lond. A Mat., 454, 903–995, 1998.

Kisi, O. Basari, C. and Latifoğlu, L.: Investigation of Empirical Mode Decomposition in Forecasting of Hydrological Time Series. Water Resour. Manag., 28, 4045–4057, https://doi.org/10.1007/s11269-014-0726-8, 2014.

Liang, B. X., Hu, J. P., Liu, C., and Hong, B.: Data pre-processing and artificial neural networks for tidal level prediction at the Pearl River Estuary, J. Hydroinform., 23, 368–382, 2021.

Margat, J. and Treyer, S.: L'eau Des Méditerranéens, Editions L'Harmattan, Paris, France, Vol. 158, ISBN 2296600255, 2004.

MathWorks: Deep Learning Toolbox – MATLAB, MathWorks, https://de.mathworks.com/products/deep-learning.html, last access: 14 May 2023.

National Water Resources Division of Tunisia: Open Data Portal, http://www.agridata.tn/, last access: 14 May 2023.

NCU (National Central University): EEMD program, NCU, Taiwan, http://rcada.ncu.edu.tw/, last access: 14 May 2023.

Ouachani, R., Bargaoui, Z., and Ouarda, T.: Power of teleconnection patterns on precipitation and streamflow variability of upper Medjerda Basin, Int. J. Climatol., 33, 58–76, https://doi.org/10.1002/joc.3407, 2013.

Rao, A. R. and Hsu, E.-C.: Hilbert-Huang transform analysis of hydrological and environmental time series, in: Water Science and Technology Library, Vol. 60, Springer, Netherlands, ISBN: 1402064535, 2008.

Wu, Z. and Huang, N. E.: A Study of the Characteristics of White Noise Using the Empirical Mode Decomposition Method, P. Roy. Soc. Lond. A Mat., 460, 1597–1611, https://doi.org/10.1098/rspa.2003.1221, 2004.

Wu, Z. and Huang, N. E.: Ensemble empirical mode decomposition: A noise-assisted data analysis method, Advances in Adaptive Data Analysis, 1, 1–41, 2009.

Proc. IAHS, 385, 275–279, 2024
https://doi.org/10.5194/piahs-385-275-2024

Open Access
PIAHS

IAHS2022 – Variability and change across space, time, extremes, and interfaces

Impact of the storm Alex on water exchanges between the Roya River and its alluvial aquifer

Sandra Lanini[1,3], **Bernard Ladouche**[1,3], **Benoit Dewandel**[1,3], **Melissande Ibba**[2], **Vincent Bailly-Comte**[1,3], **and Marie Genevier**[2]

[1]BRGM, Univ. Montpellier, Montpellier, France
[2]BRGM, Marseille, France
[3]G-EAU, UMR 183, INRAE, CIRAD, IRD, AgroParisTech, Supagro, BRGM, Montpellier, France

Correspondence: Sandra Lanini (s.lanini@brgm.fr)

Received: 20 May 2022 – Revised: 30 March 2023 – Accepted: 21 April 2023 – Published: 18 April 2024

Abstract. The alluvial aquifer of the transnational Roya River watershed is an important water resource for drinking water supply. Through successive European projects, a monitoring network has been implemented over the alluvial plain to improve the understanding of the functioning of this aquifer. For instance, studies highlighted the predominant role of surface water in the recharge of the aquifer. Following the storm Alex and the resulting exceptional flood event in the Roya valley in October 2020, a general decrease of the piezometric levels was observed in the alluvial aquifer. Changes in the river morphology and in the granulometry of the hyporheic zone have impacted surface water – groundwater exchanges and reduced the aquifer recharge.

Keywords. UPH12; Roya River; alluvial aquifer; aquifer recharge; storm Alex; Field observation

1 Context

The alluvial aquifer of the transnational Roya River watershed is exploited for drinking water to supply both French and Italian coastal cities. In 2019, 36 Mm^3 water have been withdrawn from the alluvial aquifer to provide water to 350 000 inhabitants of the French and Italian riviera. Almost a third of this volume (14 Mm^3) comes from the Porra wellfield located in the upstream area of the Roya alluvial plain (Fig. 1).

The storm Alex, which passed through Europe in early October 2020, led to a high precipitation event in the Alpes-Maritimes county (South-East of France) on 3 October 2020. The meteorological stations recorded exceptional precipitation amounts reaching 663 mm in 24 h at the Mesches dam of Tende (Payrastre et al., 2022) and 284.1 mm at Breil-sur-Roya (Chochon et al., 2021). These precipitations caused flash floods in several valleys, notably those of the Vésubie, the Roya and the Tinée Rivers. The peak flow has been estimated to 1400 $m^3\,s^{-1}$ (with an uncertainty that can reach ±50 %) for the Roya River at Breil-sur-Roya (Payrastre et al., 2022). The human and material damages were considerable. The storm also branded the landscapes with riverbeds moving. A less visible consequence of this event is the decrease of the piezometric levels in the alluvial aquifer, which began in the following days and induced operating difficulties in the Porra wellfield (Genevier et al., 2020).

The Roya transnational valley and its alluvial aquifer have been extensively studied since 2006 within successive Interreg European research projects (EUROBASSIN, RISKNAT, EUR-EAU-PA, CONCERT-EAUX). The last one focused on surface water – groundwater exchanges (Lanini and Wuillemier, 2020). Relying on the data collected and analysed within these projects, the present paper aims at describing the changes in the hydrosystem functionning caused by the storm Alex.

2 Case study description

The Roya watershed covers 669 km^2 from the upper part in the French Alps to the Italian Mediterranean coast. The downstream part of the catchment linked to the Roya reach globally comprised between the Italian cities of Trucco

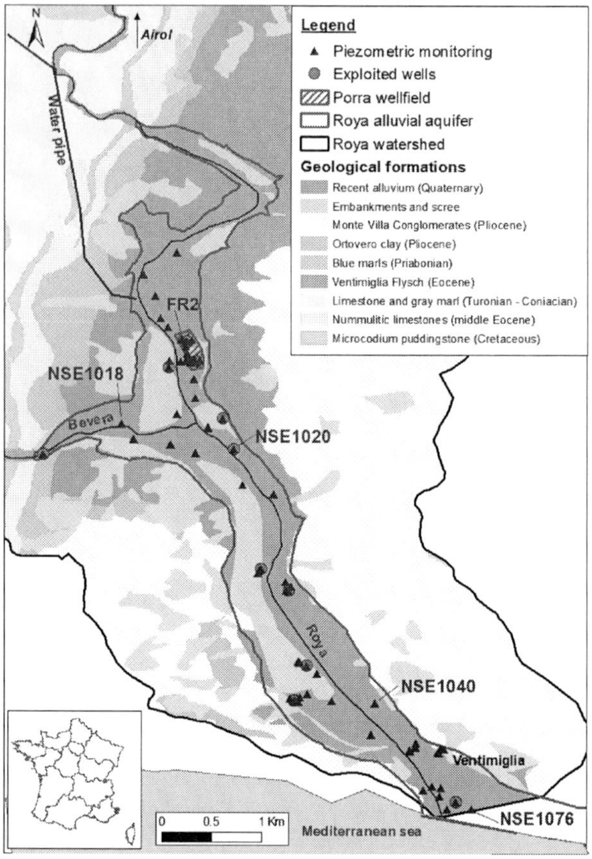

Figure 1. Geology of the Roya alluvial plain and location of monitoring points.

and Ventimiglia, includes a small elongate alluvial plain of 3.9 km^2 (Fig. 1). At the Porra wellfield, the recent alluvium (sand with clay lenses) forms a 60 m thick layer resting on an impermeable bedrock composed of flysch.

The climate is Mediterranean in the alluvial plain, with mean annual precipitation of 791 mm yr^{-1} and mean annual temperature of 14.5 °C (from SAFRAN reanalysis data on the 2001–2019 period, Vidal et al., 2010).

The Roya River mean annual discharge is around 12 $m^3\,s^{-1}$ at Airol, but around a quarter of it is diverted toward a powerplant and released in the Roya River just upstream the Porra wellfield (Fig. 1). Measurements of the flow along different profiles have shown the existence of flow losses in the Roya River between upstream and downstream. The Bevera River, which joins the Roya on the right bank downstream of Porra, also has significant losses.

3 Methods

The study of the alluvial aquifer is based on the monitoring of the groundwater level in a large number of piezometers (32) distributed on both sides of the Roya River (Fig. 1). In the Porra wellfield catchment, one observation well is equipped for monitoring temperature and electrical conductivity at seven different depths. Daily temperature data are also available from 2012 to 2018 in one of the pumping well, as well as punctual measures in a close piezometer and in the river. Two sampling campaigns of surface water and groundwater were conducted in low and high water context (respectively in October 2018 and May 2019) to perform chemical and isotopic analyses ($\delta^{18}O$, δ^2H, ^{87}Sr / ^{86}Sr, CFC / SF6) in order to better understand the exchange between surface and groundwater (Ladouche et al., 2020).

A specific study of the aquifer recharge was also conducted (Lanini et al., 2020). The Water Table Fluctuation method (Healy and Cook, 2002) was applied to estimate the recharge of the aquifer using the daily data of six piezometers of the monitoring network that are not influenced by groundwater pumping. In parallel, the effective precipitation infiltration on the alluvial plain catchment has been evaluated thanks to several water budget methods using the ESPERE application (Lanini and Caballero, 2021). Comparing the results provided by the two approaches, the part of the aquifer recharge that is "seen" by the aquifer but does not come directly from the effective precipitation infiltration is estimated.

4 Results: functioning of the alluvial aquifer

4.1 Before the storm Alex

In the observation well of the Porra wellfield, the vertical temperature profile established in February 2019 shows that the temperature is homogeneous over the entire depth of the permeable aquifer. The water temperature is quite distinct from that measured in the underlying aquitard layer (> 25 m depth) (Fig. 2). Temporal monitoring in the same observation well shows that seasonal temperature variations during the hydrological cycle are important ($\Delta T > 8$ °C) while the temperature is very stable in the underlying clayey layer (Fig. 3). Relying on the Roya water temperature estimated from the air data temperature, Ladouche et al. (2020) showed that the impulse response of the system which transforms the surface water temperature time series into the groundwater time series is characteristic of advective transfers. A shift comprised between 10 and 100 d is evaluated between the extrema of the surface water temperature curve and that of groundwater in different observations wells (Fig. 4). The CFCs and SF6 concentrations in groundwater confirmed that the aquifer is recharged with recent water. The chemical analysis and isotope study corroborated these interpretations of surface water – groundwater exchanges. It also highlighted the inputs of the Bevera River to the aquifer downstream the confluence, the Bevera water being characterized by high SO_4 and Sr contents (Fig. 5).

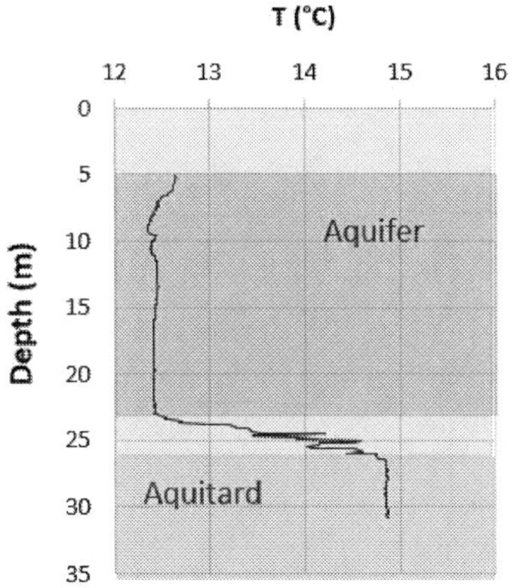

Figure 2. Temperature vertical evolution in the FR2 observation well (12 February 2019 logging data).

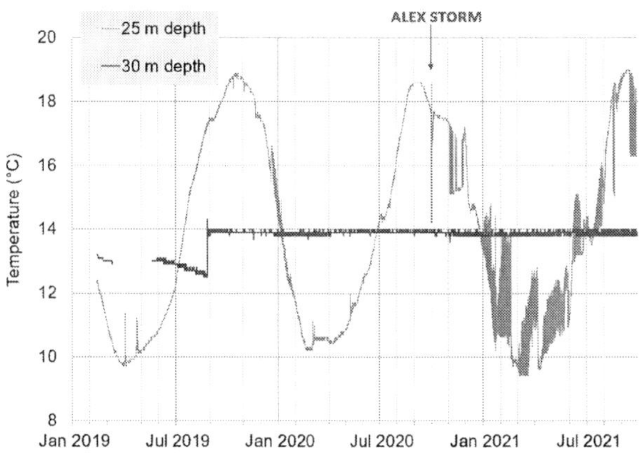

Figure 3. Groundwater temperature evolution at two different depths (data from FR2 observation well in the Porra wellfield).

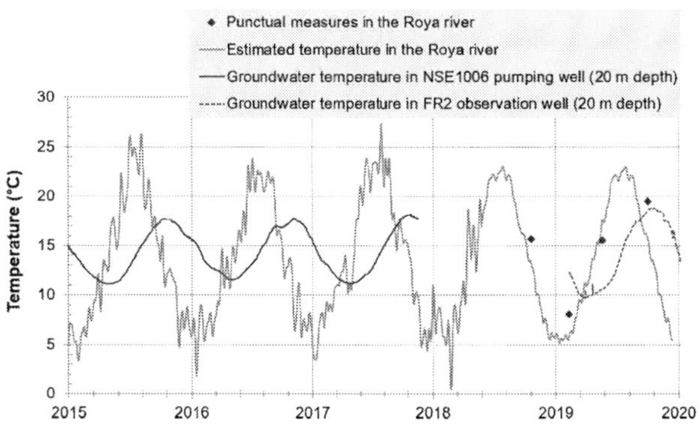

Figure 4. Surface water and groundwater temperature evolution with time.

The recharge study indicated that the aquifer recharge is comprised between 1400 and 1700 mm yr^{-1} and confirmed that the contribution of effective rainfall is low (it represents only 15 % to 20 % of the total recharge of the aquifer).

4.2 After the storm Alex

Following the storm Alex, piezometric measurements throughout the alluvial aquifer showed a continuous downward trend (Fig. 6, see piezometers location on Fig. 1). In order to avoid desaturation of the aquifer in the area of the wellfields, water authorities applied emergency measures as early as November 2020. Therefore, trenches were excavated along the riverbed to facilitate surface water infiltration. This produced rapid water level increases (Fig. 6). However, this impact was time-limited, and similar operations had to be done regularly (Genevier and Ibba, 2021).

It seems that the flood of the storm Alex has modified the preferential flow paths and main transfer patterns between the different reservoirs of the hydrosystem. As shown previously, before Alex, the aquifer temperature between 5 and 25 m depth exhibited an annual cycle directly related to the temperature of the Roya River water that infiltrates the aquifer. After the flood event, the groundwater temperature behaviour evolved in the transition zone. At 25 m depth, it seems to be also influenced by the temperature of the deeper layer (which remains stable, around 14 °C) with high frequency variations that are superimposed to the annual cycle (Fig. 3).

The observation point NSE1076 (depth 80 m, see location on Fig. 1) is located near the shoreline. It allows the monitoring of electrical conductivity of water at different depths in order to better understand the temporal evolution of marine intrusion. Measures show that the mineralization of the aquifer in the downstream zone has increased over the last 2 years (Fig. 7) due to the decrease of the hydraulic head in the aquifer.

5 Discussion

The changes observed in the Roya alluvial aquifer functioning after the storm Alex may be linked to several combined mechanisms which altered the capacity of the aquifer recharge by the river.

Firstly, during and after the flood event, an important volume of sediment has been transported by the Roya River from the upper valley to the alluvial plain and the Mediterranean sea, as shown by the Sentinel satellite images (e.g. on https://geoimage.cnes.fr/fr/les-consequences-de-la-tempetes-alex-sur-le-pourtour-mediterraneen-octobre-2020, last access: 10 May 2023). This lead to the clogging of the riverbed by finest particles, which strongly reduced its infiltration capabilities.

In addition, the morphology of the river has been modified with significant movements of the riverbed. Consequently, the relationships between surface water and groundwater

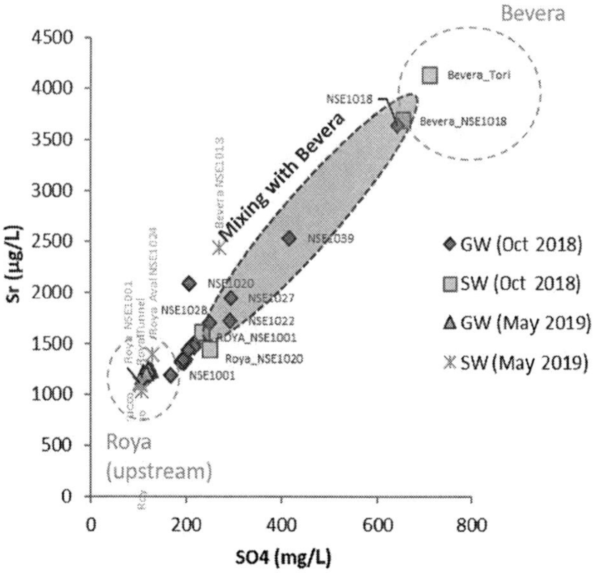

Figure 5. Chemical signatures of surface water (SW) and groundwater (GW) in the Roya plain.

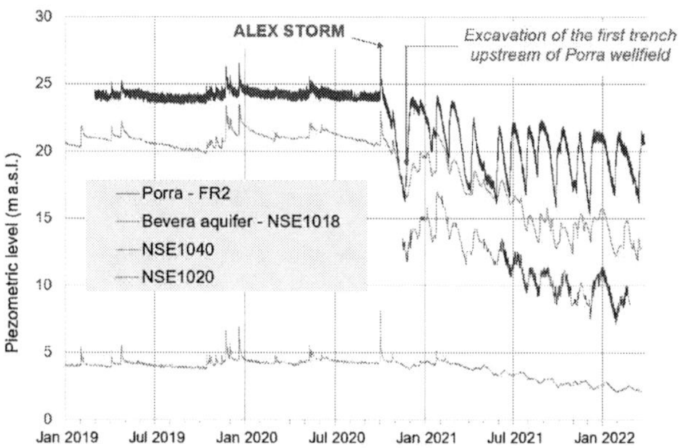

Figure 6. Water levels in the Roya alluvial aquifer since 2019.

have been impacted as the preferential flowpaths have been redrawn.

The upper alluvial part of the riverbed was washed out to the sea by the flood, modifying the sediment granulometry in the hyporheic zone. This constitutes another perturbation for superficial groundwater flows.

Changes in rock-water chemical interactions leading to mineral precipitation, and thus possible riverbed clogging, were investigated but no evidences of these processes was found yet (Gal and Genevier, 2022).

Finally, the flood from the storm Alex deeply modified the hydraulic equilibrium of the river – exploited aquifer system. The aquifer recharge upstream the wellfield is not sufficient any more to balance the withdrawal for drinking supply. This shortage might also be enhanced by the stop of the power-

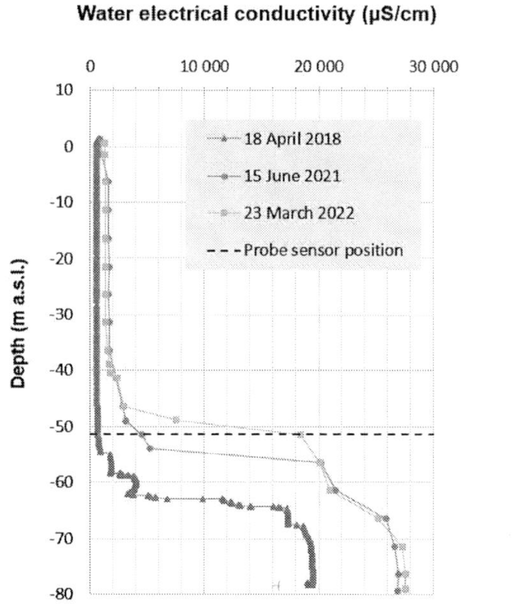

Figure 7. Water electrical conductivity profiles in a 100 m-deep well located near the coast line.

plant and thus the stop of its focused water release to the river just upstream the wellfield.

6 Conclusions

A monitoring network has been implemented over the alluvial plain of the Roya River since many years to highlight the functioning of the aquifer. It allowed understanding the important role of surface water in the recharge of the hydrogeological reservoirs. It proved to be notably useful to characterize the changes induced by the storm Alex in October 2020. Nevertheless, further research works would be needed to explore the links between sediment transport and water levels decreases, or the possible chemical modifications of surface water and groundwater induced by the flood.

Data availability. The data used in this study come from the Roya aquifer observatory, a numerical platform developed within the CONCERT-EAUX project which collects both French and Italian data from surface and groundwater monitoring networks. Data supporting the findings of this study are available upon request to the corresponding author.

Author contributions. MI, VBC and MG conducted the field investigation; SL, BL and BD analyzed the data; MG administrated the project; SL wrote the manuscript draft; SL, BL, BD, VBC and MG reviewed and edited the manuscript.

Competing interests. The contact author has declared that none of the authors has any competing interests.

Disclaimer. Publisher's note: Copernicus Publications remains neutral with regard to jurisdictional claims in published maps and institutional affiliations.

Special issue statement. This article is part of the special issue "IAHS2022 – Hydrological sciences in the Anthropocene: Variability and change across space, time, extremes, and interfaces". It is a result of the XIth Scientific Assembly of the International Association of Hydrological Sciences (IAHS 2022), Montpellier, France, 29 May–3 June 2022.

Acknowledgements. This research work has been mainly conducted in the framework of the CONCERT-EAUX project (2017–2020, https://www.concert-eaux.eu, last access: 9 May 2023)

Financial support. This research has been supported by the European Union (programme Interreg ALCOTRA), BRGM and the CARF intercommunal structure.

Review statement. This paper was edited by Christophe Cudennec and reviewed by three anonymous referees.

References

Chochon, R., Martin, N., Lebourg, T., and Vidal, M.: Analysis of Extreme Precipitation During the Mediterranean Event Associated with the Alex Storm in The Alpes-Maritimes: Atmospheric Mechanisms and Resulting Rainfall, in: Advances in Hydroinformatics, edited by: Gourbesville, P. and Caignaert, G., Springer Water, Springer Nature, Singapore, 397–418, https://doi.org/10.1007/978-981-19-1600-7_26, 2022.

Gal, F. and Genevier, M.: Champ captant de Porra sur la nappe de la Roya – Interprétation des analyses chimiques disponibles, Rapport BRGM/RP-71789-FR., 2022.

Genevier, M. and Ibba, M.: Appui scientifique et technique sur la problématique de colmatage des alluvions de la Vallée de la Roya aux abords du captage de Porra (Vintimille, Italie), Rapport BRGM/RC-70730-FR, 2021.

Genevier, M., Ibba, M., and Norie, A.: Chute anormale du niveau de la nappe de la Roya au droit du champ captant de Porra et aux alentours, Rapport BRGM/RP-70354-FR, 2020.

Healy, R. W. and Cook, P. G.: Using groundwater levels to estimate recharge, Hydrogeol. J., 10, 91–109, https://doi.org/10.1007/s10040-001-0178-0, 2002.

Ladouche, B., Bailly-Conte, V., Rivet, F., and Genevier, M.: Projet Interreg ALCOTRA " CONCERT-EAUX ". Eléments d'analyses géochimiques, isotopiques et thermiques de la nappe alluviale de la Roya Rapport final, BRGM/RP-71899-FR, 43 pp., 2020.

Lanini, S. and Caballero, Y.: ESPERE, a Tool for Multimethod Aquifer Recharge Estimation: What's New with Version 2, Groundwater, 59, 5–6, https://doi.org/10.1111/gwat.13049, 2021.

Lanini, S. and Wuillemier, A.: Projet Interreg ALCOTRA "CONCERT-EAUX". Synthèse hydrogéologique, étude de la recharge et modèle conceptuel de la plaine alluviale de la Roya, Rapport BRGM/RP-71900-FR, 68 pp., 2020.

Payrastre, O., Nicolle, P., Bonnifait, L., Brigode, P., Astagneau, P., Baise, A., Belleville, A., Bouamara, N., Bourgin, F., Breil, P., Brunet, P., Cerbelaud, A., Courapied, F., Devreux, L., Dreyfus, R., Gaume, E., Nomis, S., Poggio, J., Pons, F., Rabab, Y., and Sevrez, D.: Tempête Alex du 2 octobre 2020 dans les Alpes-Maritimes: une contribution de la communauté scientifique à l'estimation des débits de pointe des crues, LHB, 2082891, https://doi.org/10.1080/27678490.2022.2082891, 2022.

Vidal, J. P., Martin, E., Franchistéguy, L., Baillon, M., and Soubeyroux, J. M.: A 50-year high-resolution atmospheric reanalysis over France with the Safran system, Int. J. Climatol., 30, 1627–1644, 2010.

Proc. IAHS, 385, 281–290, 2024
https://doi.org/10.5194/piahs-385-281-2024

IAHS2022 – Variability and change across space, time, extremes, and interfaces

Spatially distributed calibration of a hydrological model with variational optimization constrained by physiographic maps for flash flood forecasting in France

Maxime Jay-Allemand[1], Julie Demargne[1], Pierre-André Garambois[2], Pierre Javelle[2], Igor Gejadze[3], François Colleoni[2], Didier Organde[1], Patrick Arnaud[2], and Catherine Fouchier[2]

[1]Hydris Hydrologie, Montferrier-sur-Lez, France
[2]INRAE, Aix-Marseille Univ., RECOVER, Aix-en-Provence, France
[3]INRAE, Montpellier Univ., GEAU, Montpellier, France

Correspondence: Pierre-André Garambois (pierre-andre.garambois@inrae.fr)

Received: 25 May 2022 – Revised: 6 April 2023 – Accepted: 21 April 2023 – Published: 18 April 2024

Abstract. This contribution presents a regionalization approach to estimate spatially distributed hydrologic parameters based on: (i) the SMASH (Spatially distributed Modelling and ASsimilation for Hydrology) hydrological modeling and assimilation platform (Jay-Allemand, 2020; Jay-Allemand et al., 2020) underlying the French national flash flood forecasting system Vigicrues Flash (Javelle et al., 2019); (ii) the variational assimilation algorithm from (Jay-Allemand et al., 2020), adapted to high dimensional inverse problems; (iii) spatial constraints added to the optimization problem, based on masks derived from physiographic maps (e.g., land cover, terrain slope); (iv) multi-site global optimization, which targets multiple independent watersheds. This method gives a regional estimation of the spatially distributed parameters over the whole modeled area. This study uses a distributed rainfall-runoff model with 4 parameters to calibrate, with a spatial resolution of $1 \times 1\,\mathrm{km}^2$ and a 15 min time step. Performances of the calibrated hydrological model and the parameters robustness are evaluated on two French study areas with 20 catchments in each, in spatio-temporal extrapolation based on cross-validation experiments over a 12-year period. Several spatial regularization strategies are tested to better constrain the high dimensional optimization problem. The model parameters are calibrated based on the Nash-Sutcliffe Efficiency (NSE) computed for multiple calibration basins in the study area. Results are discussed based on the Nash-Sutcliffe Efficiency and the Kling-Gupta Efficiency criteria obtained on calibration and validation catchments for two subperiods of 6 years. Further work aims to improve the global search of prior parameter sets and to better balance the adjoint sensitivity with respect to the spatial constraints resolution and catchment characteristics. This will ensure a better consistency of simulated fluxes variabilities and enhance the applicability of the regionalization method at higher spatial scales and over larger domains.

Keywords. UPH6; UPH19; UPH20; SDG13; distributed modeling; conceptualization; parameter optimization; ungauged basins; flash floods

1 Introduction

The estimation of storage and fluxes in surface hydrology is an essential scientific question related to major socio-economic issues as extreme floods and droughts forecasting, especially with the undergoing climate change. In order to better predict flash floods and reduce their potentially devastating impact, warning systems have been developed or are currently under development (Gourley et al., 2017; Javelle et al., 2016). In this context, advanced spatially distributed modeling tools are critically needed to generate reliable and skillful local forecasts. These models take into account the spatial variability of the catchment properties and of the forcing inputs (e.g., rainfall) to produce discharge predictions at ungauged locations. Nevertheless, hydrological modeling remains a challenging task because of limited observations of physical processes and modeling uncertainties, and internal storage fluxes are tinged with uncertainties.

Whatever their status and complexity, hydrological models are most often calibrated and validated using integrative discharge time series at the outlet of a catchment (e.g. Sebben et al., 2013). However, due to a potentially significant number of cells or subcatchments involved in distributed models, the calibration process has to deal with overparameterization and equifinality (uniqueness) issues. In particular, given the spatial sparsity of constraining discharge observations, hydrological modeling is faced with the challenge of producing predictions at ungauged locations based on the regionalization of the hydrological model parameters. Despite the overparameterization problem in spatially distributed modeling, the contribution of e.g. Jay-Allemand et al. (2020) and Jay-Allemand (2020) showed promising results for estimating the spatial variability of the distributed parameters within a given catchment using only downstream discharge observations. However, providing better spatial constrains on the estimated parameters patterns, within and outside the calibration catchments in a regionalization perspective, remains a challenge.

Different regionalization strategies have been proposed in hydrology. Some methods search for relations between the model parameters and physical descriptors (Seibert, 1999). Others methods explore regionalization schemes based on the spatial or physical proximity (e.g. Odry and Arnaud, 2017; Parajka et al., 2005). These methods have many requirements: (1) reliable parameters have to be identified on gauged catchment; (2) hydrological relations between gauged and ungauged catchments must exist; (3) model parameters should have a physical meaning. Unfortunately, these requirements are not always satisfied. To better constrain regionalization over large domains, promising regional calibration results of Samaniego et al. (2010), Poncelet (2016), Mizukami et al. (2017), and Beck et al. (2020) are based on the Multiscale Parameter Regionalization (MPR) upscaling and pre-regionalization scheme of Samaniego et al. (2010), which enables to account for various physiographic and climatic descriptors. It consists in imposing a mapping between fine scale descriptor maps and model parameter maps prior to regional optimization. Nevertheless, in all those studies, the mapping parameters are optimized as spatially uniform with classical optimization algorithms limited to low dimensional inverse problems.

In a regionalization perspective, we extend the distributed calibration approach developped by Jay-Allemand et al. (2020), which is adapted to high dimensional hydrological inverse problems. Our regional calibration method uses the following elements: (1) a distributed hydrological model; (2) a variational algorithm; (3) the adjoint of the forward model; (4) physical descriptors to constrain the spatial variability of the parameters; (5) a regularization function, required for ill-posed inverse problems (Tikhonov, 1963), adapted to constrain the distributed hydrological model with physical descriptors. Finally, the major difference with others regional calibration approaches lies in the fact that no prior guess based on hypothetical physical meanings of the parameters is required. The descriptors are only used to define classified physiographic maps to constrain the parameter patterns through the control setting or the regularization function.

2 Modeling approach

SMASH is a computational software framework dedicated to Spatially distributed Modeling and data ASsimilation for Hydrology at high resolution, capable to tackle high dimensional inverse problems with variational data assimilation. It underlies the French flash flood warning system called Vigicrues Flash (Javelle et al., 2019).

2.1 SMASH Model

Let $\Omega \subset \mathbb{R}^2$ be a 2D spatial domain, containing single to multipe nested and non-nested catchments, and $t > 0$ be the physical time. A regular lattice $\mathcal{T}_\Omega$ covers Ω and $D(x)$ is the drainage plan obtained from terrain elevation processing, with the only condition that a unique point in Ω has the highest drainage area, i.e. the outlet of the largest catchment in Ω. The hydrological model is a dynamic operator $\mathcal{M}$ mapping the observed input fields of rainfall and evapotranspiration, denoted $\boldsymbol{P}\left(x,t'\right)$ and $\boldsymbol{E}\left(x,t'\right)$, $\forall\left(x,t'\right) \in \Omega \times (0,t)$ onto the discharge field $Q(x,t)$ such that:

$$Q(x,t) = \mathcal{M}\left[\boldsymbol{P}\left(x,t'\right), \boldsymbol{E}\left(x,t'\right), \boldsymbol{h}(x,0), \boldsymbol{\theta}(x)\right] \quad \forall\left(x,t'\right) \in \Omega \times [0,t] \tag{1}$$

with $\boldsymbol{h}(x,t)$, the N_s-dimensional vector of model states (2D fields), and $\boldsymbol{\theta}$, the N_p-dimensional vector of model parameters (2D fields); $\boldsymbol{\theta}$ is also called the control vector in the optimization context.

In this study, we use a parsimonious 4-parameters gridded GR-like hydrological model schematized in Fig. 1. For a given pixel i of coordinates $x \in \Omega$, the proposed model includes three reservoirs $\mathcal{P}$, $\mathcal{T}_r$ and $\mathcal{T}_l$ of capacity c_p, c_{tr} and c_{tl}, to simulate respectively the production of runoff and its transfer within a cell (with two transfer reservoirs). Their stages are respectively denoted $\boldsymbol{h}_p$, $\boldsymbol{h}_{tr}$ and $\boldsymbol{h}_{tl}$. Given known flow directions, classically obtained from a Digital Elevation Model, the cell-to-cell routing is done with a nonlinear routing reservoir $\mathcal{R}$ of capacity c_r. The spatial resolution is set to $\Delta x = 1$ km and the simulation time step to $\Delta t = 15$ min corresponding to the space-time resolution of input rainfall and evaporation data, which are assumed to be constant over a given time step Δt. The flow operators, mostly consisting in first order ordinary differential equations, are analytically integrated over a time step, enabling a simple computation of the forward model defined in Eq. (1).

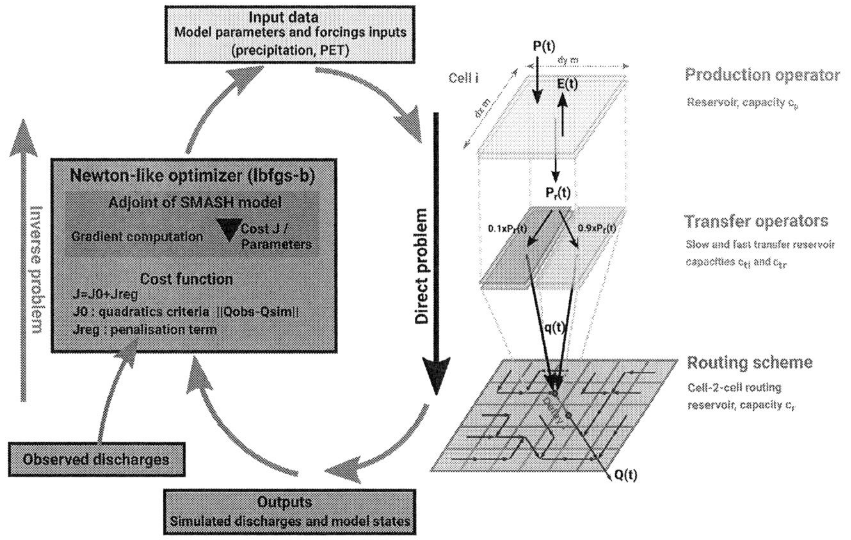

Figure 1. The distributed SMASH hydrological model and the variational algorithm involving the quasi-Newton optimizer lbfgs-b and the adjoint of the forward model.

2.2 Calibration algorithm

In order to better constrain discharge estimates at ungauged locations, the calibration aims to estimate spatially distributed parameters within the calibration catchments as well as outside them. Hence, the parameter vector $\boldsymbol{\theta}(x) \equiv \left(c_{\mathrm{p}}(x), c_{\mathrm{tr}}(x), c_{\mathrm{r}}(x), c_{\mathrm{tl}}(x)\right)^{\mathrm{T}}$ is of size $N_{\mathrm{c}} \times N_x$, where N_{c} denotes the number of parameters fields and N_x the total number of cells in Omega Ω, including cells in ungauged areas. Considering tens of cells or more over a given hydrological domain Ω with discharge observations being available only on a limited number of points (i.e. outlets of the gauged catchments), the regional calibration of a spatially distributed $\boldsymbol{\theta}$ is a high dimensional and difficult inverse problem.

The adjoint-based variational calibration algorithm developed by e.g. (Jay-Allemand, 2020; Jay-Allemand et al., 2020) and adapted to such hydrological inverse problems, is used in this study and summarized here and in Fig. 1. Given the simulated and observed discharges at gauged cells $x_k \in \Omega, k \in 1, .., N_{\mathrm{g}}$, respectively denoted $Q_k(t)$ and $Q_k^*(t)$, we define the objective function as:

$$J(\boldsymbol{\theta}) = J_{\mathrm{obs}}(\boldsymbol{\theta}) + \alpha J_{\mathrm{reg}}(\boldsymbol{\theta}) \tag{2}$$

where the observation cost function is $J_{\mathrm{obs}} = \sum_{k=1}^{N_{\mathrm{g}}} j_k^*$. In this study, j_k^* is classically defined as $1 - \mathrm{NSE}$, with NSE the Nash-Sutcliffe efficiency (Nash and Sutcliffe, 1970), a quadratic metric that measures the misfit between simulated and observed discharges. Note that the simulated discharge depends on the control vector $\boldsymbol{\theta}$ via the hydrological model (Eq. 1). The second term in Eq. (2) is a regularization term weighted by α.

Given the ill-posedness of the hydrological calibration problem, classified maps of different physiographic descriptors are introduced to spatially constrain the variability of model parameters fields. For each parameter field of $\boldsymbol{\theta}$, one classified physiographic map is considered to either reduce the control vector into semi-distributed controls or compute a regularization term by class as detailed hereafter.

In order to improve the obtained hydrologic parameter patterns, an additional regularization term is proposed to penalize parameter variability within each physiographic class. Consider for each of the $N_{\mathrm{c}} = 4$ parameters fields $\theta_{\mathrm{c} \in [1..N_{\mathrm{c}}]}(x) \in \boldsymbol{\theta}(x)$, a given physiographic map defining a partition of the catchment such that $\Omega_{\mathrm{c}} = \cup_{k=1}^{n_{\mathrm{p,c}}} \Omega_{\mathrm{c},k}$ with $n_{\mathrm{p,c}}$ the number of classes in the spatial partition of parameter c. We introduce the following regularization term:

$$J_{\mathrm{reg}} = \sum_{c=1}^{N_{\mathrm{c}}} \sum_{k=1}^{n_{\mathrm{p,c}}} \left(\sum_{i,j}^{i,j \in \Omega_{\mathrm{c}},k} \frac{1}{d_{i,j}} ||\theta_{\mathrm{c}}^i(x) - \theta_{\mathrm{c}}^j(x)||^2 \right) \tag{3}$$

where $d_{i,j}$ is the Euclidean distance between cells i and j. This function, inspired by the classic Tikhonov regularization (Tikhonov, 1963), enables a consistent regularization of the parameters spatial variability within the spatial partitioning Ω_{c} of each parameter. The weight α is chosen with the L-curve method (Hansen and O'Leary, 1993).

Given the hydrological model defined in Eq. (1), an optimal estimate $\hat{\boldsymbol{\theta}}$ of the model parameter set is obtained from the condition:

$$\hat{\boldsymbol{\theta}} = \underset{\boldsymbol{\theta}_{\min} \leq \boldsymbol{\theta} \leq \boldsymbol{\theta}_{\max}}{\arg\min} \; J(\boldsymbol{\theta}) \tag{4}$$

The optimization is performed with the L-BFGS-B algorithm adapted to high dimension (Zhu et al., 1994). This algorithm requires the gradient of the cost function with respect to the sought parameters, $\nabla_\theta J$, which is obtained by solving the adjoint model. The adjoint model has been generated with the automatic differentiation engine TAPENADE (Hascoet and Pascual, 2013) applied to the SMASH source code, which includes the new modeling components and regularization scheme, and has been validated with a standard gradient test. The first guess value $\bar{\boldsymbol{\theta}}$ needs to be defined as a starting point for the optimization process with the regularization term. $\bar{\boldsymbol{\theta}}$ is defined as a spatially uniform global optimum determined with a simple global minimization algorithm from a uniform parameter set, $\bar{\boldsymbol{\theta}}^*$ (Jay-Allemand et al., 2020). The steepest descent method summarized in (Edijatno, 1991) has been used in this study to determine this uniform first guess value. Note that the semi-distributed calibration of a semi-distributed conceptual model has been investigated in De Lavenne et al. (2019), using a sequential and low dimensional calibration approach involving a regularization strategy accounting for deviation to a priori parameter sets determined with catchment proximity. In the present work, a regularization term penalizing the spatial variability of parameters within physiographic catchment partitioning is introduced in a variational data assimilation algorithm adapted to high dimensional inverse problems; the cost function is global in space and time.

In order to investigate the effects of the proposed regularization term, the following calibration strategies are compared:

- a spatially uniform calibration (denoted U), where the control vector considered constant in space (hence $\boldsymbol{\theta}_{\mathrm{U}}(x) \equiv \bar{\boldsymbol{\theta}}$ reduced to 4 unknowns) is sought with the simple steepest descent algorithm from e.g. (Edijatno, 1991); this uniform calibrated parameter set $\hat{\boldsymbol{\theta}}_{\mathrm{U}}$ is used as the first guess value in the three subsequent calibration processes;
- a distributed calibration (denoted D), with a spatially distributed control $\boldsymbol{\theta}_{\mathrm{D}}(x)$, i.e. with free parameters in each cell within the gauged catchments used for calibration; the first guess values (from the uniform calibration) give parameter estimates for all grid cells outside of the gauged catchments;
- a semi-distributed calibration (denoted SD), with a semi-distributed control $\boldsymbol{\theta}_{\mathrm{SD}}(x)$ spatially constrained by classified physiographic maps over the whole domain, hence with one spatially uniform free parameter for each class of the physiographic mask relative to the model parameter;
- a distributed-regularized calibration (denoted DR), with a fully distributed control $\boldsymbol{\theta}_{\mathrm{DR}}(x)$ since the physiographic regularization from Eq. (3) (based on the same physiographic masks as in the SD calibration) is used in this case as a weak regionalization constrain over the whole domain.

In fact, in the optimization process, the observation cost function is only sensitive to the parameter values of the cells within the gauged catchments used for calibration. However, the regionalization effect stems from the proposed classified physiographic maps either through the semi-distributed control setup for the SD case or from the regularization term in the DR case.

3 Data and experimental setup

The calibration strategies are evaluated over two study areas with 20 gauged basins in each (see Fig. 2): (i) a Northern area in the Burgundy region, (ii) a Southern area in the Ardeche region. These 40 gauged basins, with catchment areas ranging from 25 to 768 km^2, have been split up in 2 subsets for cross-validation purposes: 28 basins for calibration and 12 basins for validation (5 nested and 7 neighboring catchments). The selection of catchments is based on the availability of long time series with high quality of observed flow and limited anthropogenic impacts. The choice of the localization of the calibration/validation and nested/disjoint catchments is arbitrary.

The SMASH model runs on a $1 \times 1\,\mathrm{km}^2$ grid resolution at a 15 min time step forced by: (i) observed rainfall grids based on hourly ANTILOPE J+1 radar-gauge rainfall reanalysis (Laurantin, 2013; Champeaux et al., 2009), which have been disaggregated using the temporal distribution of 15 min PANTHERE radar-gauge rainfall estimates available in real-time (both gridded products from Météo-France); (ii) PET estimates based on the Oudin formula and the temperature data from SAFRAN reanalysis produced by Météo-France on a $8 \times 8\,\mathrm{km}^2$.

Several combinations of physiographic descriptors were tested to spatially constrain the parameter calibration using land use, local slope, bedrock type and drainage areas. Results presented in this paper consider: (i) for the c_{p} production parameter, a 1 km^2 land use gridded mask with 3 classes derived from CORINE Land Cover (2012 version, https://land.copernicus.eu/pan-european/corine-land-cover, last access: 9 May 2023) (the 3 classes including artificial areas, open spaces with little or no vegetation and water bodies for class 1, agricultural areas for class 2, and forests for class 3); (ii) for the 3 other transfer and routing parameters (c_{tr}, c_{tl} and c_{r}), a 1 km^2 local slope grid mask with 4 classes (using the threshold values of 0.025, 0.5 and 1.8° for classes 1 to 4) from the Copernicus database (http://land.copernicus.eu/in-situ/eu-dem-derived-products/slope, last access: 9 May 2023). The numbers of classes was chosen so that each catchment is covered by at least two classes and both descriptors have more or less the same number of classes (3 and 4). We are aware that this choice affects the size of the control of the (SD) case.

The number of free parameters depends on the study area (Northern and Southern) and the optimization scheme. For

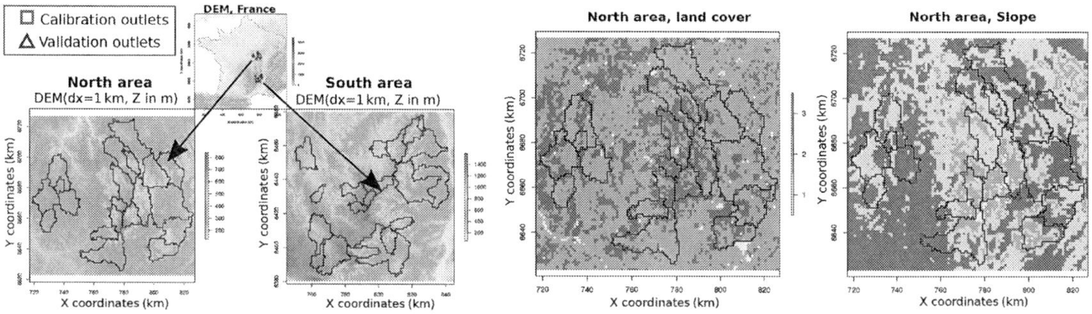

Figure 2. On the left side, catchment boundaries for the North and South study areas, with red squares for the calibration outlets (14 for each area) and blue triangles for the validation outlets (6 for each area). On the right side, the physiographic partitioning relative to the descriptors of land cover and slope for the Northern area.

the distributed case (D), the number of free parameters is equal to 12 328 (Northern) and 14 400 (Southern), which corresponds to the number of cells inside the gauged catchments times the number of hydrological operators in the forward model. Cells outside the gauged area are not sensitive to the cost function and are therefore not included in the optimization process. For the distributed-regularized case (DR), the number of free parameters is equal to 45 760 (Northern) and 40 000 (Southern), which corresponds to the total number of cells of the domain times the number of hydrological operators in the forward model. Thanks to the regularization function, all domain cells are sensitive to the cost function and are thus included in the optimization process. For the semi-distributed (SD) case, the use of classified physiographic maps for the reduction of the control leads to 15 free parameters, corresponding to the total number of descriptors classes used for all operators. For the uniform (U) case, the control contains only 4 parameters. Table 1 gives a summary description of the configurations of the four calibration experiments.

The parameter bounds for the uniform (U) and semi-distributed (SD) calibration are defined as follows: c_{p} and c_{tr} varying between 20 and 1000 mm, c_{tl} between 20 and 5000 mm, and c_{r} between 0.1 and 100 min. For the distributed (D) and distributed-regularized (DR) calibrations, the first guess value for each parameter has been used to define the upper bound as $\boldsymbol{\theta}_{\max} = \boldsymbol{\theta}(x_{\mathrm{b}}) + 3 \times \boldsymbol{\theta}(x_{\mathrm{b}})$, the lower bound being equal to 1.

The weight of the regularization, α, was estimated using the L-curve method (Hansen and O'Leary, 1993). The chosen α value varies in the interval of [7.0, 20.] depending on the experiments. It is interesting to mention that, at the end of the optimization, the relative weights $\frac{\alpha J_{\mathrm{reg}}(\hat{\theta})}{J_{\mathrm{obs}}(\hat{\theta})}$ remain within the range of [9.0, 12.0].

To focus the parameter calibration on high flow conditions, the optimization criterion (NSE) is calculated using only observed flow exceeding one fifth of the flood quantile relative to the 2-year return period. The evaluation scores are all computed with the same thresholding strategy, which leads to lower efficiency score values but a better description of the similarity between observed and simulated discharges for high flow conditions.

The evaluation of the model calibrations consists in analysing the model predictive performance (referred as MPP) using data not involved in calibration (Jay-Allemand et al., 2020). Therefore, the full set of observations $Q^*k(t), k = 1, \ldots, \mathrm{Ng}, t \in (0, T)$ is divided into two complementary subsets: a calibration subset and a validation subset. Since Q^* depends on k (defining the spatial distribution of sensors) and t, we distinguish the temporal validation and spatio-temporal validation, following the split sample test defined by Klemeš (1986). Two 6-year subperiods are defined: P1 from 1 August 2007 to 31 July 2013, and P2 from 1 August 2013 to 31 July 2019. Each subperiod (P1 or P2) is considered as the calibration period while the other period is the validation period. A model warm-up of one year (from August to July preceeding the calibration period) is performed before starting the simulations, one year being long enough to obtain stable model states for these catchments not strongly impacted by snow conditions.

If data from a station is used in calibration, the corresponding catchment is labelled as "calibration catchment"; otherwise it is a "validation catchment". Both the calibration quality and the MPP in validation are measured using the NSE criterion and the Kling–Gupta efficiency (KGE) criterion (Gupta et al., 2009). The NSE and KGE metrics are computed for high flow conditions only, on the calibration and validation subperiods. The gain in the NSE score is computed for each individual basin as a skill score to evaluate the relative improvement (in terms of NSE) of a given calibration in reference to the benchmark calibration (e.g., the uniform (U) case).

Table 1. Characteristics of the four calibration strategies, including the name of the experiment, the cost function, the optimization algorithm, the number of free parameters, the use of the physiographic descriptors, the value of the first guess and the notation of the calibrated parameter set.

Experiments	Cost function	Optimizer	Nb of free parameters	Descriptors	First guess	Calib. set
Uniform (U)	$J_{\text{obs}}(\boldsymbol{\theta})$	Steep-descent	4 (N, S)	No	$\overline{\boldsymbol{\theta}_{\text{U}}} = \bar{\theta}^*$	$\widehat{\boldsymbol{\theta}_{\text{U}}}$
Distributed (D)	$J_{\text{obs}}(\boldsymbol{\theta})$	lbfgs-b + adj.	12 328 (N), 14 400 (S)	No	$\bar{\boldsymbol{\theta}}_{\text{D}} = \hat{\boldsymbol{\theta}}_{\text{U}}$	$\hat{\boldsymbol{\theta}}_{\text{D}}$
Semi-Distributed (SD)	$J_{\text{obs}}(\boldsymbol{\theta})$	lbfgs-b + adj.	15 (N, S)	Yes	$\bar{\boldsymbol{\theta}}_{\text{SD}} = \hat{\boldsymbol{\theta}}_{\text{U}}$	$\hat{\boldsymbol{\theta}}_{\text{SD}}$
Distributed-Regularized (DR)	$J_{\text{obs}}(\boldsymbol{\theta}) + \alpha J_{\text{reg}}(\boldsymbol{\theta})$	lbfgs-b + adj.	45 760 (N), 40 000 (S)	Yes	$\bar{\boldsymbol{\theta}}_{\text{DR}} = \hat{\boldsymbol{\theta}}_{\text{U}}$	$\hat{\boldsymbol{\theta}}_{\text{DR}}$

4 Results and Discussions

4.1 Model predictive performances analysis

The distribution of the NSE and KGE values for all 28 calibration catchments (from both study areas) on the calibration and validation periods are plotted on Fig. 3, while results for the 12 validation catchments on the validation period (for spatiotemporal validation) are plotted on Fig. 4. Box-and-whisker plots represent the quartile values, the median value being indicated by the bar within the box and the whiskers extending from the smallest to the largest value.

Model performances at the calibration stations for the calibration subperiods (left plot of Fig. 3): the best model performances are obtained for the distributed and distributed-regularized calibrations, for which the NSE (and KGE) median values (for high flow conditions) are equal to respectively 0.8 (0.83) and 0.77 (0.83). The (fully) distributed calibration (D and DR) using the variational algorithm significantly improves the NSE and KGE scores at the calibration stations by 0.1 point on average compared to uniform calibration U. The semi-distributed calibration shows intermediate model performances due to the reduced control vector in space. The more flexible distributed-regularized calibration strategy (i.e., with more degrees of freedom) leads to better regional model performances in calibration compared to the semi-distributed method.

Model performances at the calibration stations for the validation subperiods (right plot of Fig. 3): the uniform parameters (U) strongly degrade the model predictive performances (MPP) from the calibration to the validation period, with a median NSE score reduced from 0.65 to 0.44. MPP are less degraded for the other experiments. Both fully distributed approaches (D and DR) outperform the semi-distributed (SD) calibration: for example, the distributed-regularized (DR) calibration improves the NSE score for 37 out of the 56 cases, with a median gain of 12 % (on all 56 cases). While these two approaches are over-parameterized, the distributed (D) parameter set obtains better scores in temporal validation at the calibration stations: in comparison with the SD calibration, the NSE score is improved for 38 out of the 56 cases, with a median gain in NSE of 14 % (on all 56 cases).

Model predictive performances at the validation stations for the validation period (Fig. 4): all four experiments degrade MPP in spatio-temporal validation: for example, the uniform parameter set (U) obtains a median NSE score of 0.32 on all 24 cases. The distributed calibration (D) uses the uniform parameter set outside of the calibration catchments, thus obtaining a similar median NSE score (0.33 on all 24 cases), the improvement being limited to the 5 validation catchments fully or partially nested within a calibration catchment (10 cases for the 2 validation subperiods). The semi-distributed (SD) and distributed-regularized (DR) calibrations slightly improve the MPP for 15 and 14 cases (over the 24 cases) in comparison to the uniform calibration (U), with a median NSE gain of 11 % and 4 % respectively. In comparison with the D calibration, the DR calibration leads to a NSE improvement for 5 of the 10 nested cases. The spatial constraints applied for the DR and SD experiments allow to estimate a spatially variable parameter set outside the calibration catchments, which should be preferred to the uniform parameter set (U).

4.2 Analysis of the parameter spatialisation

Calibrated parameter maps for period P1 and for the Northern study area are presented in Fig. 5, with spatial variability either inside the calibration (gauged) catchments for the distributed calibration case (D) or over the whole domain for the calibration cases SD and DR (thus for any ungauged catchments). The considered calibration strategies, involving different control and/or regularization setups in an equifinality context led to contrasted inferred parameter fields that are analyzed here.

The D and DR calibration experiments led to similar parameter maps since, for these methods, the initial first guess values play a key role during the optimization. Indeed the median and average values computed on the overall domain remain close to the first guess values. The estimation of the first guess is therefore critical to constraint the problem.

To globally analyse the differences between inferred parameter sets $\boldsymbol{\theta}_{\text{SD}}$ and $\boldsymbol{\theta}_{\text{DR}}$ obtained for periods P1 and P2 and catchment sets North and South, the relative difference to the uniform first guess $\boldsymbol{\theta}_{\text{U}}$ is introduced: $\Delta\theta_{\text{SD}}^{\Omega_k,p,z} = \left|(\hat{\boldsymbol{\theta}}_{\text{SD}}^{\Omega_k,p,z} - \boldsymbol{\theta}^*{}_{\text{U}})\right| / \boldsymbol{\theta}^*{}_{\text{U}}$ and $\Delta\theta_{\text{DR}}^{\Omega_k,p,z} = \left|(\hat{\boldsymbol{\theta}}_{\text{DR}}^{\Omega_k,p,z} - \boldsymbol{\theta}^*{}_{\text{U}})\right| / \boldsymbol{\theta}^*{}_{\text{U}}$; where $\hat{\boldsymbol{\theta}}_{\text{SD}}^{\Omega_k,p,z}$ and $\hat{\boldsymbol{\theta}}_{\text{DR}}^{\Omega_k,p,z}$ denote

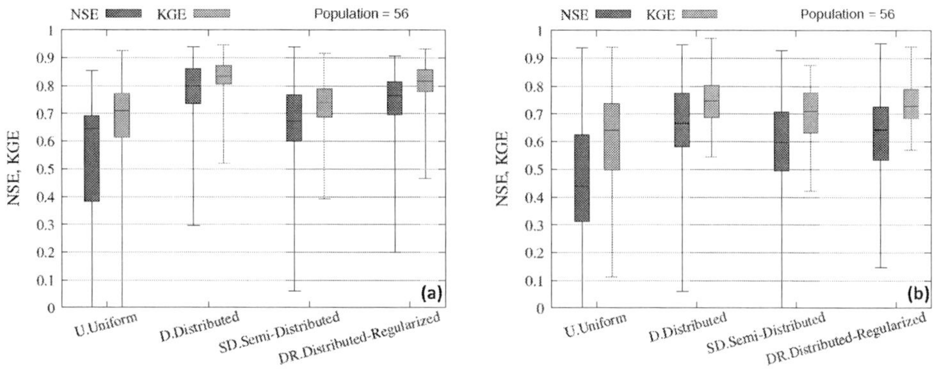

Figure 3. Model performances (NSE and KGE for high flow conditions) at the 28 calibration stations for the four experiments for the 2 calibration subperiods **(a)** and for the 2 validation subperiods **(b)**.

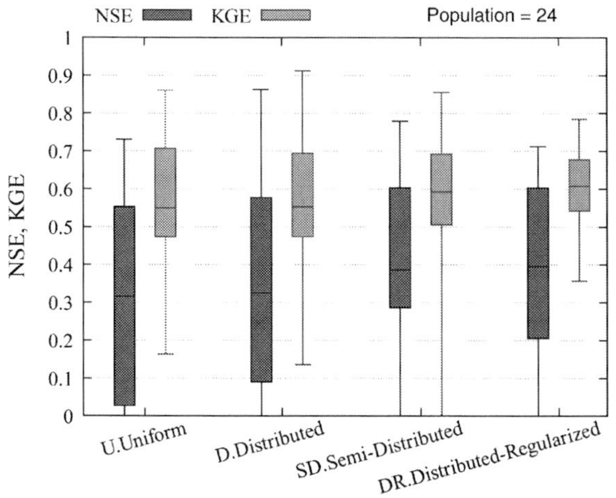

Figure 4. Model performances (NSE and KGE for high flow conditions) at the 12 validation stations for the four experiments and for the 2 validation subperiods.

the inferred parameters for each spatial partitioning $\Omega_k = \{\Omega_1, \ldots, \Omega_{N_k}\}$, period $p = \{\mathrm{P1}, \mathrm{P2}\}$ and study zone $z = \{\mathrm{North}, \mathrm{South}\}$, for the semi-distributed (SD) and distributed-regularized (DR) experiments. The corresponding global spatial averages over each zone, $\overline{\Delta\theta_{\mathrm{SD}}}$ and $\overline{\Delta\theta_{\mathrm{DR}}}$, and spatial variances, $\sigma(\Delta\theta_{\mathrm{SD}})$ and $\sigma(\Delta\theta_{\mathrm{DR}})$, are then computed (see Table 2). The comparison of the averaged relative differences $\overline{\Delta\theta_{\mathrm{SD}}} > \overline{\Delta\theta_{\mathrm{DR}}}$ (being equal to 2.06 and 0.24) indicate that the inferred SD parameter set $\hat{\boldsymbol{\theta}}_{\mathrm{SD}}$ strongly deviates from the first guess while the DR $\hat{\boldsymbol{\theta}}_{\mathrm{DR}}$ parameters with physiographic reguarization stay much closer. The comparison of the spatial variances $\sigma\,\Delta\theta_{\mathrm{SD}} > \sigma\,\Delta\theta_{\mathrm{DR}}$ (being equal to 74.43 and 0.21) indicates a much higher variablility for the semi-distributed calibration. These results can be seen in terms of deviation from the first guess and spatial variability of the inferred parameters on Fig. 3. The parameter sets from the semi-distributed calibration experiment (SD) are the most variable within the parameter bounds (i.e. higher σ values for the SD parameter maps). The resulting average and median values diverge from the first guess value. The variances of the parameters estimated on the overall domain are also higher in average.

When the parameter optimization targets multiple disjoint catchments, the uniform and semi-distributed calibration problems are under-parameterized. This could lead to unexpected optimal parameter sets when catchments exhibit different types of hydrological behavior. This issue occurred in the Southern study area, where the optimal first guess ($\boldsymbol{\theta}_{\mathrm{U}} = \overline{\boldsymbol{\theta}}$ being spatially uniform) for the slow transfer reservoir c_{tl} is equal to 2793 mm for P2 and to 281 mm for P1. For period P2, the semi-distributed calibration (SD) led to an average parameter value $\overline{c_{\mathrm{tl}}}$ equal to 283 mm; for the distributed calibration, the optimal parameter $\overline{c_{\mathrm{tl}}}$ could be lower than 500 mm for one catchment and larger than 3000 mm for two others. The uniform and semi-distributed optimization could be driven globally by the performances reached at only one or a few catchments, whereas data quality issues and/or model deficiencies could cause a performance loss for these catchments. Interestingly, the distributed calibration with physiographic regularization (DR) leads to spatially smoother inferred parameter fields.

5 Conclusions

This contribution presented a novel method to regionally calibrate the parameters of a distributed hydrological model using a variational algorithm. This methods relies on classified physiographic map(s) to impose spatial patterns to the parameters, either strongly (SD) via the definition of the control vector or weakly (D) via a regularization term added to the objective function. Both calibration strategies led to similar performances in spatio-temporal validation whereas the distributed-regularized method improved the efficiency scores for the calibration catchments. Both strategies have their own strengths and weaknesses.

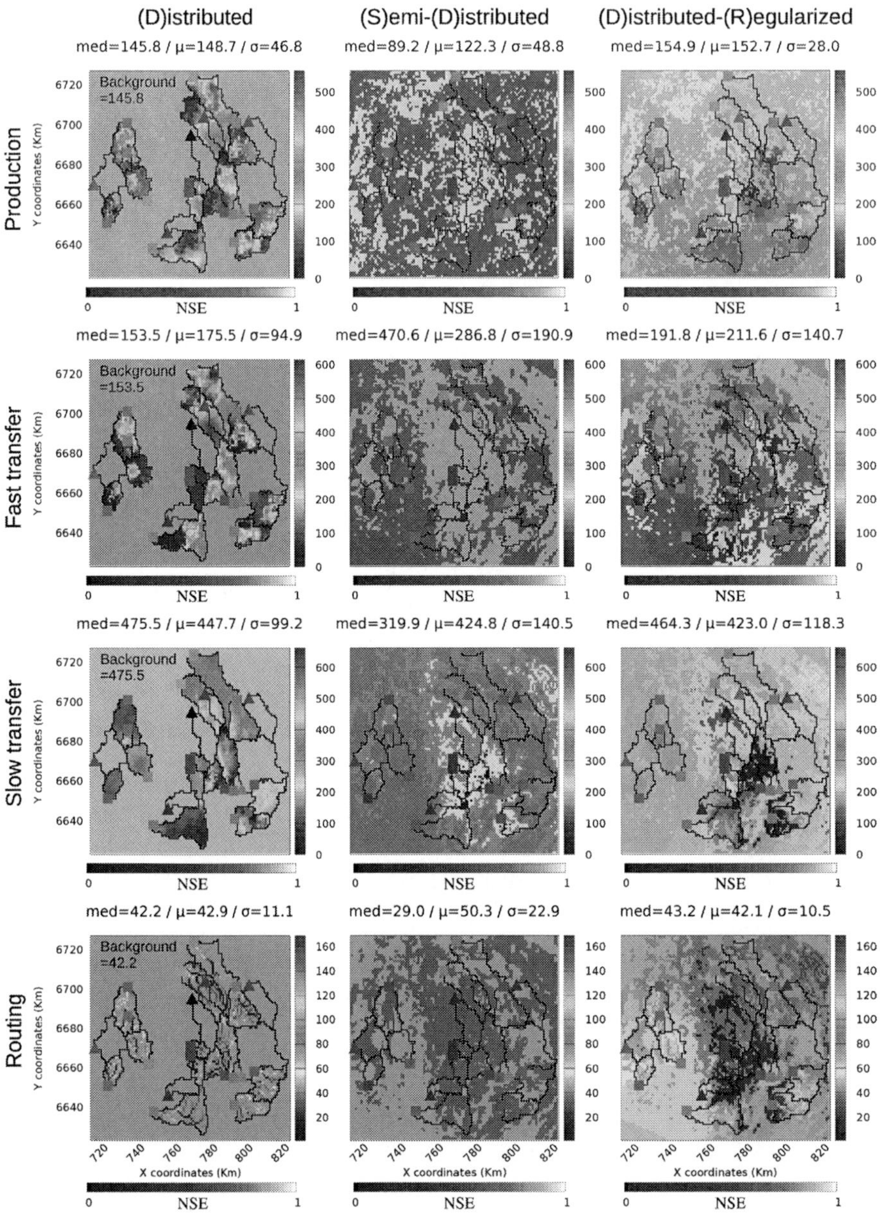

Figure 5. Optimal model parameters for the North study area and for period P1: parameter maps for the distributed, semi-distributed and distributed-regularized calibrations are plotted, while the first guess values from the uniform calibration is specified on the distributed calibration graphics (corresponding to the parameter value outside of the calibration catchments for that calibration experiment). The median, mean and standard deviation values are specified for each parameter map. Squares and triangles represent the NSE score for high flow conditions obtained on the calibration period P1 for the calibration and validation gauges respectively.

Table 2. Average $\overline{\Delta\theta}$ and variance $\sigma(\Delta\theta)$ values of the relative differences of the optimized model parameter set to the uniform first guess $\boldsymbol{\theta}_{\mathrm{U}}$, for the calibration experiments SD and DR.

	Semi-Distributed (SD)	Distributed-Regularized (DR)
$\overline{\Delta\theta}$	2.06	0.24
$\sigma(\Delta\theta)$	74.43	0.21

The (SD) strategy is the easiest approach to drastically reduce the size of the control vector. However, like the uniform calibration, it is very sensitive to the multi-site optimization function and local model deficiencies and data errors could potentially impact the calibration over the whole area. Moreover, this strategy is very dependent on the choice of the spatial partitioning (i.e, the physiographic descriptors and their number of classes), which should be further explored. It could be interesting to test a combination of these strategies, using the output of the (SD) method as a first guess for the (DR) approach. Finally as the (SD) calibration works similarly to a special case of the MPR approach, we plan to extend this work via the more flexible MPR framework. The fully distributed calibration approach using the physiographic regularization strategy (DR case) showed promising results since: (1) it maximises the model performances at the control stations (MPP similar to the fully distributed approach without regularization); (2) it gives an estimate of the parameters values for the overall area, thus for any ungauged catchments. These regional parameter estimates help to improve the discharge predictions at the pseudo-ungauged validation catchments (which were not used for calibration). One drawback could be the huge size of the control vector. However, the variational optimization using the adjoint model is an efficient method to optimize the parameter vector and the equifinality does not exclusively stem from its size (Jay-Allemand et al., 2020). In our opinion, the distributed-regularized approach is therefore more flexible and adequate for calibrating the parameters of a distributed hydrological model than the (SD) method. We believe that the regularization term can be further enhanced.

The spatio-temporal robustness of the calibrated parameters for (SD) and (DR) approach, is still to be improved: parameter maps are very different between the two calibration subperiods and the MPP at the validation stations are relatively poor compared to the MPP for the calibration stations. This lack of robustness has several causes. First the conceptual model includes parameters with limited physical meaning. Secondly, the optimization problem is overparameterized for the (DR) approach and the first guess values, which play a key role during the optimization, have to be improved using physical priors for example. Last, the lower calibration performance at validation catchments indicates a need to improve the model structure versatility to better adapt to different hydrologic behaviors.

Planned future work will focus on testing other regularizations with a physiographic distance metric and introducing hydrological units, as well as testing the forward model of upscaling and regionalization transfer functions of the MPR approach to upgrade the (SD) strategy. Moreover, a sensivity analysis during the first decisive iterations of the LBFGS-B descent could provide valuable information about the calibration process and its driving components. Finally, a statistical and spatial analysis of the optimized parameters should be performed to search for relations between model parameters and physiographic descriptors and select the most adequate descriptors.

Code and data availability. SMASH is planned to be released as an open-source software. A public online documentation has already been released at https://smash.recover.inrae.fr/index.html (Arnaud et al., 2023). Data sets can be provided on demand.

Author contributions. Authors performed the numerical simulations, analysis of the results and development of the SMASH model; they participated to the data and results analysis and contributed to the writing and reviewing of the article.

Competing interests. The contact author has declared that none of the authors has any competing interests.

Disclaimer. Publisher's note: Copernicus Publications remains neutral with regard to jurisdictional claims in published maps and institutional affiliations.

Special issue statement. This article is part of the special issue "IAHS2022 – Hydrological sciences in the Anthropocene: Variability and change across space, time, extremes, and interfaces". It is a result of the XIth Scientific Assembly of the International Association of Hydrological Sciences (IAHS 2022), Montpellier, France, 29 May–3 June 2022.

Acknowledgements. We acknowledge all the reviewers for their reviews and comments, which helped to improve the article. We thank all the partners who provided data and computational ressources.

Review statement. This paper was edited by Christophe Cudennec and reviewed by Alban de Lavenne and one anonymous referee.

References

Arnaud, P., Colleoni, F., Demargne, J., Ettalbi, M., Folton, N., Fouchier, C., Garambois, P., Gejadze, I., Godet, J., Haruna, A., Huynh Ngo Nghi, T., Javelle, P., Jay-Allemand, M., Organde, D., Paluszkiewicz, M., Pujol, K., Renard, B., Vigoureux, S., and Villenave, L.: The SMASH platform: Spatially distributed Modelling and ASsimilation for Hydrology, https://smash.recover.inrae.fr/index.html (last access: 12 May 2023), 2023.

Beck, H. E., Pan, M., Lin, P., Seibert, J., van Dijk, A. I. J. M., and Wood, E. F.: Global Fully Distributed Parameter Regionalization Based on Observed Streamflow From 4,229 Headwater Catchments, J. Geophys. Res.-Atmos., 125, e2019JD031485, https://doi.org/10.1029/2019JD031485, 2020.

Champeaux, J.-L., Dupuy, P., Laurantin, O., Soulan, I., Tabary, P., and Soubeyroux, J.-M.: Les mesures de précipitations et l'estimation des lames d'eau à Météo-France: état de l'art et perspectives, LHB, 5, 28–34, https://doi.org/10.1051/lhb/2009052, 2009.

De Lavenne, A., Andréassian, V., Thirel, G., Ramos, M.-H., and Perrin, C.: . A regularization approach to improve the sequential calibration of a semidistributed hydrological model, Water Resour. Res., 55, 8821–8839, https://doi.org/10.1029/2018WR024266, 2019.

Edijatno: Mise au point d'un modèle élémentaire pluie-débit au pas de temps journalier, PhD thesis, Université Louis Pasteur, ENGEES, Cemagref Antony, France, Strasbourg, 242 pp., https://webgr.inrae.fr/wp-content/uploads/2012/07/1991-EDIJATNO-THESE.pdf (last access: 9 May 2023), 1991.

Gourley, J. J., Flamig, Z. L., Vergara, H., Kirstetter, P.-E., Clark, R. A., Argyle, E., Arthur, A., Martinaitis, S., Terti, G., Erlingis, J. M., Hong, Y., and Howard, K. W.: The FLASH Project: Improving the Tools for Flash Flood Monitoring and Prediction across the United States, B. Am. Meteorol. Soc., 98, 361–372, https://doi.org/10.1175/BAMS-D-15-00247.1, 2017.

Gupta, H. V., Kling, H., Yilmaz, K. K., and Martinez, G. F.: Decomposition of the mean squared error and NSE performance criteria: Implications for improving hydrological modelling, J. Hydrol., 377, 80–91, 2009.

Hansen, P. C. and O'Leary, D. P.: The use of the L-curve in the regularization of discrete ill-posed problems, SIAM J. Sci. Comput., 14, 1487–1503, 1993.

Hascoet, L. and Pascual, V.: The Tapenade Automatic Differentiation tool: principles, model, and specification, ACM T. Math. Softw., 39, p. 43, https://doi.org/10.1145/2450153.2450158, 2013.

Javelle, P., Organde, D., Demargne, J., Saint-Martin, C., de Saint-Aubin, C., Garandeau, L., and Janet, B.: Setting up a French national flash flood warning system for ungauged catchments based on the AIGA method, in: 3rd European Conference on Flood Risk Management FLOODrisk 2016, vol. 7, p. 11, https://doi.org/10.1051/e3sconf/20160718010, 2016.

Javelle, P., Saint-Martin, C., Garandeau, L., and Janet, B.: Flash flood warnings: Recent achievements in France with the national Vigicrues Flash system, UNDRR GAR, https://www.undrr.org/publication/flash-flood-warnings-recent-achievements-france-national-vigicrues-flash-system (last access: 9 May 2023), 2019.

Jay-Allemand, M.: Estimation variationnelle des paramètres d'un modèle hydrologique, PhD thesis, Universite d'Aix-Marseille, https://www.theses.fr/2020AIXM0400/document (last access: 9 May 2023), 2020.

Jay-Allemand, M., Javelle, P., Gejadze, I., Arnaud, P., Malaterre, P.-O., Fine, J.-A., and Organde, D.: On the potential of variational calibration for a fully distributed hydrological model: application on a Mediterranean catchment, Hydrol. Earth Syst. Sci., 24, 5519–5538, https://doi.org/10.5194/hess-24-5519-2020, 2020.

Klemeš, V.: Operational testing of hydrological simulation models, Hydrolog. Sci. J., 31, 13–24, 1986.

Laurantin, O.: ANTILOPE: hourly rainfall analysis over France merging radar and rain gauges data, in: Proceedings of the 11th International Precipitation Conference, edited by: Leijnse, H. and Uijlenhoet, R., KNMI, Ede-Wageningen, the Netherlands, 30 June to 3 July 2013.

Mizukami, N., Clark, M. P., Newman, A. J., Wood, A. W., Gutmann, E. D., Nijssen, B., Rakovec, O., and Samaniego, L.: Towards seamless large-domain parameter estimation for hydrologic models, Water Resour. Res., 53, 8020–8040, https://doi.org/10.1002/2017WR020401, 2017.

Nash, J. and Sutcliffe, J.: River flow forecasting through conceptual models, Part I. A conceptual models discussion of principles, J. Hydrol., 10, 282–290, 1970.

Odry, J. and Arnaud, P.: Comparison of Flood Frequency Analysis Methods for Ungauged Catchments in France, Geosciences, 7, 88, https://doi.org/10.3390/geosciences7030088, 2017.

Parajka, J., Merz, R., and Blöschl, G.: A comparison of regionalisation methods for catchment model parameters, Hydrol. Earth Syst. Sci., 9, 157–171, https://doi.org/10.5194/hess-9-157-2005, 2005.

Poncelet, C.: Du bassin au paramètre : jusqu'où peut-on régionaliser un modèle hydrologique conceptuel?, Theses, Université Pierre et Marie Curie – Paris VI, https://tel.archives-ouvertes.fr/tel-01529196 (last access: 9 May 2023), 2016.

Samaniego, L., Kumar, R., and Attinger, S.: Multiscale parameter regionalization of a grid-based hydrologic model at the mesoscale, Water Resour. Res., 46, W05523, https://doi.org/10.1029/2008WR007327, 2010.

Sebben, M. L., Werner, A. D., Liggett, J. E., Partington, D., and Simmons, C. T.: On the testing of fully integrated surface–subsurface hydrological models, Hydrol. Process., 27, 1276–1285, https://doi.org/10.1002/hyp.9630, 2013.

Seibert, J.: Regionalisation of parameters for a conceptual rainfall-runoff model, Agr. Forest Meteorol., 98–99, 279–293, https://doi.org/10.1016/S0168-1923(99)00105-7, 1999.

Tikhonov, A. N.: On the regularization of ill-posed problems, in: Doklady Akademii Nauk, Russian Academy of Sciences, vol. 153, 49–52, 1963.

Zhu, C., Byrd, R. H., Lu, P., and Nocedal, J.: L-BFGS-B: a limited memory FORTRAN code for solving bound constrained optimization problems, EECS Department, Northwestern University, Evanston, IL, Technical Report No. NAM-11, https://doi.org/10.1145/279232.279236, 1994.

Proc. IAHS, 385, 291–296, 2024
https://doi.org/10.5194/piahs-385-291-2024

Open Access PIAHS

Estimating the water balance of a small reservoir in the Brazilian semiarid

Ályson B. S. Estácio, Leticia Vieira, Samiramisthais Linhares, Rafaela Alves, Renan Rocha, and Eduardo Martins

Department of Water Resources, Fundação Cearense de Meteorologia e Recursos Hídricos, Fortaleza, 60.115-221, Brazil

Correspondence: Ályson B. S. Estácio (alyson.estacio@funceme.br)

Received: 29 May 2022 – Revised: 13 April 2023 – Accepted: 21 April 2023 – Published: 18 April 2024

Abstract. Small communities in semi-arid regions usually suffer from a lack of data to enable effective water management. This is the case for many rural communities in Brazil that rely on small reservoirs. The aim of this study is to extract information on water storage and water flows of Lagoa Cercada, a small reservoir in the Brazilian semi-arid region. Monthly data series of water levels and water quality were combined with remote sensing information to evaluate water storage in the reservoir. Inflow, withdrawal, and infiltration were calculated using water and salinity balance in the reservoir. The calculated capacity of Lagoa Cercada reservoir was 1.14 million cubic meters. However, using satellite imagery instead of drone imagery resulted in a 51 % difference in the estimation of storage capacity. Additionally, Lagoa Cercada's water withdrawal has not returned to the same amount as before the drought.

Keywords. Water resources data; rural communities; UPH 16; UPH 18

1 Introduction

Small communities in semi-arid regions often face not only unreliable water availability but also a lack of sufficient and accurate hydrological data for water management (Jackson et al., 2019; Serrao et al., 2020). The absence of such data (e.g., stored water available, water volume consumption) can have a negative impact on water security and undermine the ability of communities to effectively manage their water resources (Paul et al., 2021; Alazard et al., 2015). This is the situation for many small reservoirs in the Brazilian semi-arid region (BSAR), which, despite being an important source of water for rural communities, are largely unmanaged.

Remote sensing has been a powerful tool to provide or complement information for water management (Wang and Xie, 2018; Garrido-Rubio et al., 2020). In the BSAR, it has been used for mapping weirs and water surfaces (FUNCEME, 2021). Remote sensing can be particularly useful for regions where on-site monitoring is not possible. However, the accuracy of different remote sensing techniques typically requires on-site data to be validated.

Only strategic reservoirs in BSAR have been consistently monitored. For instance, in the state of Ceará (in BSAR) 155 strategic reservoirs are monitored daily out of a total of 100 000 reservoirs. Then, most small reservoirs suffer from data scarcity.

Water balance is a powerful tool for obtaining relevant data for reservoir management (e.g. Arsiso et al., 2017).

Even when limited data is available, the use of appropriate techniques can provide useful estimates of water availability.

In this study, we evaluate the uncertainty of the storage curve, estimate the stored volumes, and water flows, including inflow, withdrawal, and infiltration, using a water balance combined with remote sensing techniques.

2 Study area

Lagoa Cercada is a reservoir located in a rural community in the municipality of Quixeramobim, in the state of Ceará, Brazil (see Fig. 1) That region is characterized by a semi-arid climate, with average annual rainfall of around 810 mm,

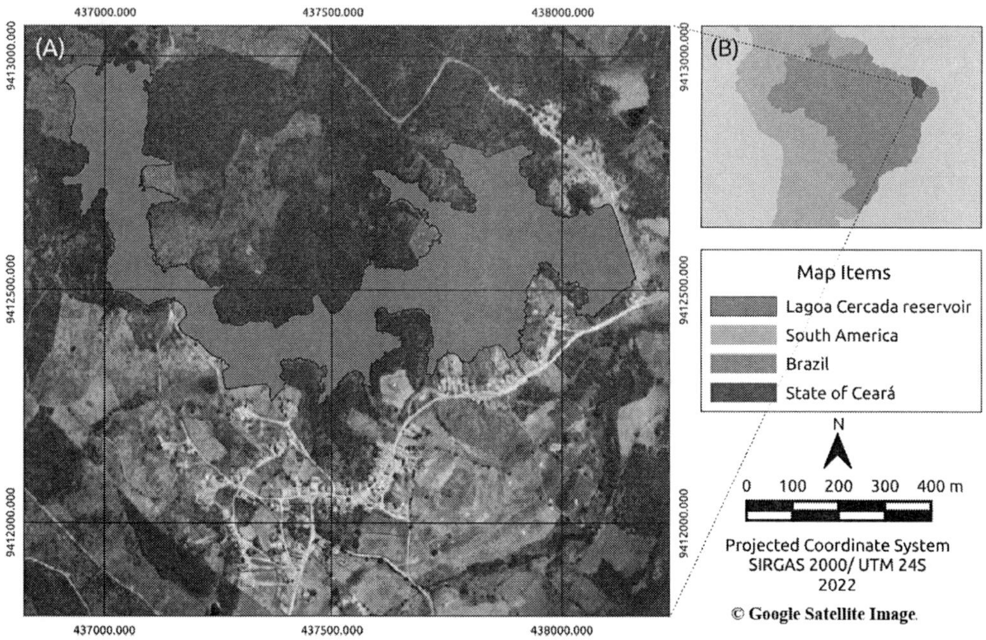

Figure 1. Study area. **(a)** Location of the study area. **(b)** Lagoa Cercada Community and reservoir. © Google Satellite Image.

dropping to less than 600 mm in the inland areas (Werner and Gerstengarbe, 2003). Actual evapotranspiration is about 78 % of annual precipitation (SUDENE, 1980). The rainy season occurs mainly between February and May, accounting for routhly 70 % of the total annual rainfall.

Lagoa Cercada is one of the 100 000 weirs mapped in Ceará, where only 155 strategic reservoirs are monitored daily. Although Lagoa Cercada is not considered a strategic reservoir, the 100-family community located near it uses its water for agriculture and household activities. Fundação Cearense de Meteorologia e Recursos Hídricos (FUNCEME) has been monitoring the water level and quality of Lagoa Cercada on a monthly basis since 2010.

3 Methods

In this study, remoting sensing is used to estimate the storage curve of Lagoa Cercada. This information was combined with monthly data of water level and water quality to perform the water and salinity balances and estimate inflow, withdrawal and infiltration.

The methodology was divided into four steps related to the estimation of (i) storage curve, (ii) stored volumes, (iii) reservoir inflow, and (iv) reservoir withdrawal and infiltration. The steps are described as follows.

3.1 Estimation of the storage curve

Two remote sensing techniques, drone imagery and a satellite product, were used to estimate the storage curve for the reservoir.

Drone images of the reservoir bottom were obtained from an aerial survey carried out in January 2022, when the reservoir was dry. Contour lines (0.2 m) were calculated from the drone images.

Additionally, we used the Digital Elevation Model (DEM) Alos Palsar 12.5 m, which is a satellite product, to directly calculate contour lines (1.0 m).

We considered the relative highs for both storage curves, and the lower high of each storage curve was referred to as high zero.

The differences between both estimations of storage curve were used to quantify the uncertainty of the volume estimations.

3.2 Stored volumes estimation

The stored volumes were calculated from water level measurements taken at the end of each month between September 2010 and December 2019. The storage curves, which were estimated in the previous step, was used to convert water level measurements into stored volume estimates. If the measured water level fell between two discrete levels of the storage curve (0.2 or 1.0 m), linear interpolation was used to estimate the volume. The use of different storage curves estimated from drone imagery or satellite products resulted in

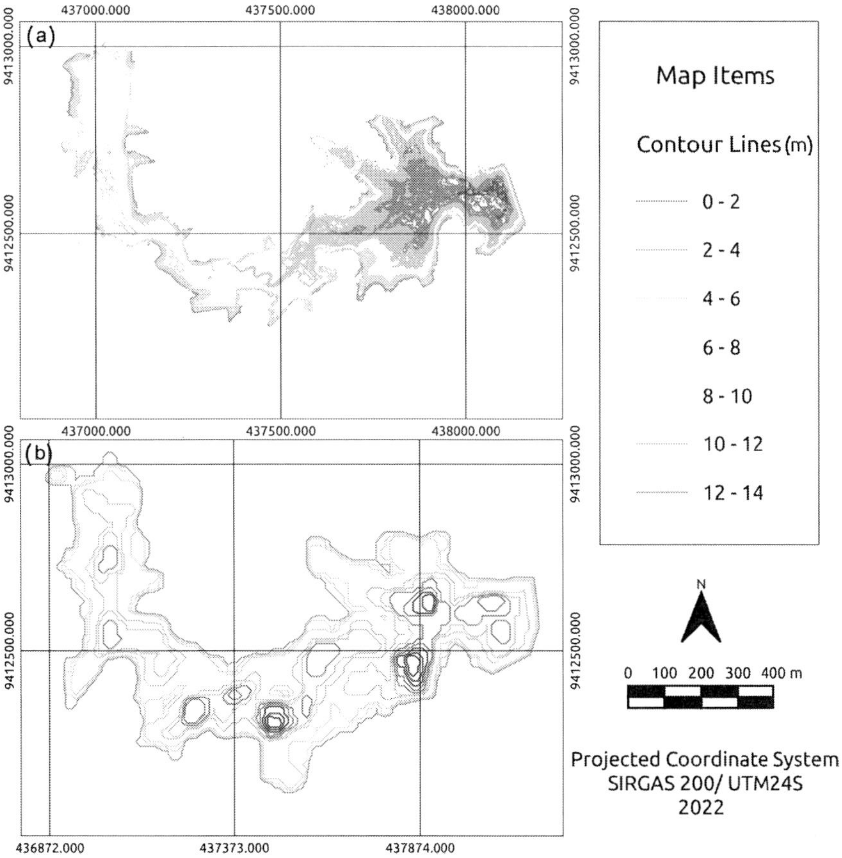

Figure 2. Contour lines from **(a)** drone imagery, and **(b)** DEM Alos Palsar satellite product.

different estimations of stored volumes. The difference between the stored volume estimations were assessed.

3.3 Water balance and inflow estimation

Monthly inflow volume in the reservoir was calculated using the water balance equation. The water balance between the end of month (i) and the end of the previous month ($i-1$) was represented by Eq. (1):

$$\Delta V_i = I_i - E_i - W_i \qquad (1)$$

where ΔV_i is the variation of stored volume during the period, I_i is the inflow volume, E_i is the volume of water evaporated, and W_i is the volume of water withdrawn or lost through lateral infiltration.

The terms I_i, E_i and W_i are incoming or outgoing volumes during the month and correspond numerically to the average flows at the monthly scale (i.e. in units of volume per month).

Water inflow is calculated from Eq. (1) only in the first semester of the hydrological year (December to May), when inflow is expected to occur due to direct precipitation over the reservoir, streamflow, and lateral inflow (superficial and underground). For the rest of the year, inflow is considered null since it is the dry season.

In addition, W_i can be ignored in the first semester. During this period, withdrawal is negligible because rain usually supplies the local demand for water. Rainfed agriculture is the dominant activity in the wet semester, while cisterns fed by rooftop runoff supply household consumption. Therefore, the community conserve Lagoa Cercada's reserve in the wet semester. Furthermore, lateral infiltration losses (i.e., groundwater leaving the reservoir) are also negligible in this period. In the wet period the lateral fluxes in the hydraulic basin tend to be towards the reservoir, i.e., in the opposite direction of infiltration losses. That incoming volume of groudwater is computed together with I_i. This direction of the underground fluxes is due to the shallow soils of the region. Because of that, groundwater levels surrounding the reservoir tend to be higher than the water level in the reservoir during the wet season.

Ignoring W_i in Eq. (1), monthly estimations of I_i were computed for the months of December to May, from December 2010 to December 2019, using stored volumes estima-

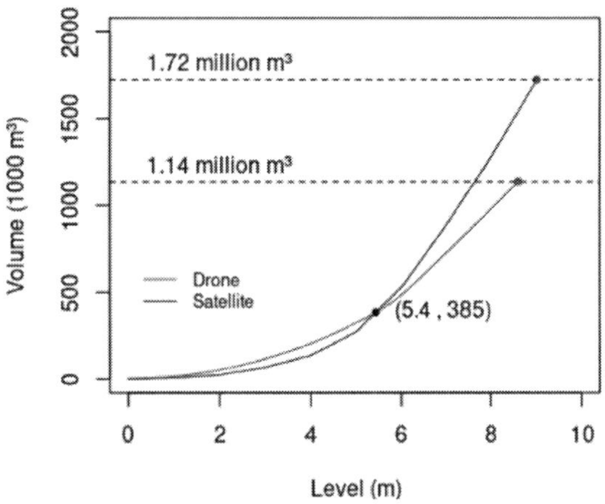

Figure 3. Storage curves estimated from remote sensing.

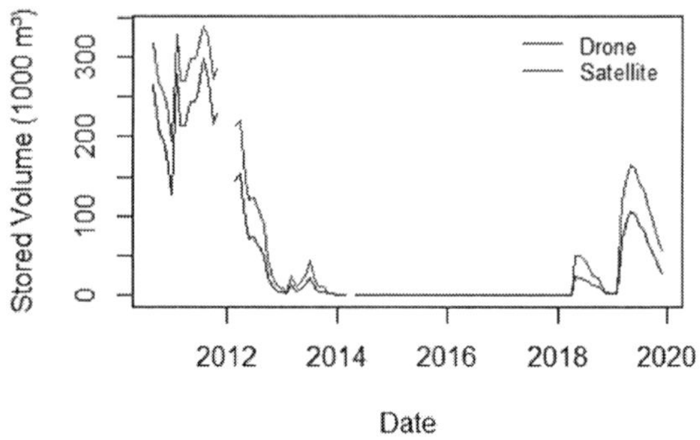

Figure 4. Time series of the stored volume in Lagoa Cercada reservoir.

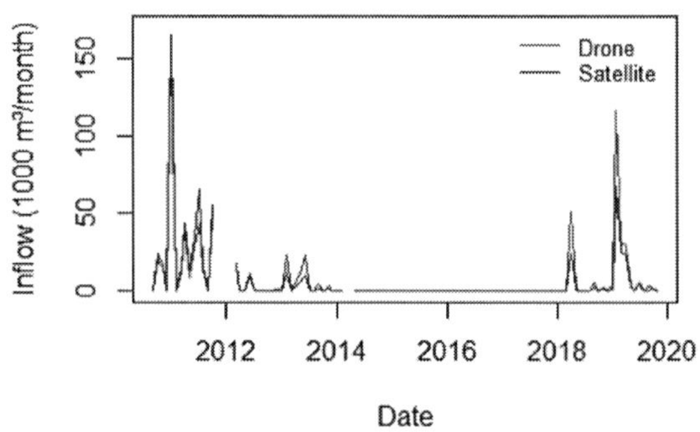

Figure 5. Estimated monthly inflow to Lagoa Cercada reservoir.

tions calculated in the previous step to calculate ΔV_i, and using an estimation of E_i.

E_i was estimated from the average monthly evaporation, calculated from the direct measures of an evaporimeter tar in the municipality of Quixeramobim.

Because of the different estimations of stored volumes (i.e. derived from the drone and the satellite storage curves), two estimations of inflow were obtained. We compared both.

3.4 Withdrawal and infiltration

Withdrawal and infiltration are both estimated in an aggregated way within W_i, using Eq. (1).

W_i is only calculated in the second semester of the hydrological year (June to November), when infiltration and withdrawal might be relevant, whereas I_i can be ignored during this period.

Then, to calculate W_i in the second semester, the estimations of ΔV_i, and E_i are sufficient, just as in the calculation of I_i in the first semester.

Nevertheless, we were able to calculate W_i independently of an external estimation of E_i, thanks to different effects of E_i and W_i on the reservoir salinity.

E_i is an outflow that preserve the salt mass in the reservoir, while W_i decrease the salt mass. Then, the salinity balance can be used to differentiate E_i and W_i.

In fact, W_i can be calculated by dividing the outgoing salt mass during the month by the average salinity of W_i in the period. Spatial variations of salinity across were not considered. Therefore, salinity of W_i for a specific month is considered equal to the salinity in the reservoir. Additionally, we considered that salinity is proportional to the electrical conductivity, which is also measured during with water level campaigns. Then, W_i was computed for the months June to November, from October 2010 to November 2019 using Eq. (2):

$$W_i = \frac{-V_i \times c_i + V_{i-1} \times c_{i-1}}{(c_i + c_{i-1})/2} \qquad (2)$$

where V_i and V_{i-1} are the storage volumes at the end of the month i and $(i-1)$, respectively; and c_i and c_{i-1} are the electrical conductivity at the same dates.

Two different results were obtained for W_i, which are related to the volumes estimated from drone and satellite storage curves.

4 Results and discussion

4.1 Storage curves

The contour lines obtained from drone imagery and satellite product (see Fig. 2) differ in precision, with the former being more precise at 0.2 m and the latter at 1.0 m. As a result, the storage curves shown in Fig. 3 also differ.

Since the storage curve derived from the drone imagery is considered more reliable due to the precision of its contour lines, the uncertainty associated with the use of the storage curve derived from the satellite DEM can be estimated by comparison.

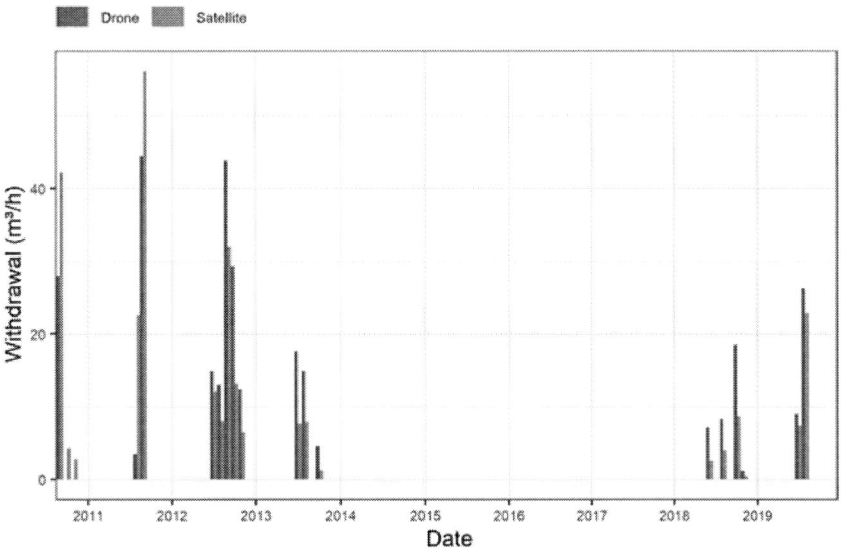

Figure 6. Estimated monthly W (withdrawal and infiltration) during the second semester of the hydrological year (June–November) for Lagoa Cercada reservoir.

The storage curve from satellite product underestimates lower volumes (i.e. volumes under 385 000 m^3) and overestimates higher one. The difference is smaller for low volumes.

The total capacity of Lagoa Cercada is estimated to be 1 140 000 m^3, according to the drone imagery.

However, the storage curve derived from the satellite DEM indicates a 51 % larger maximum volume (i.e. 1 720 000 m^3).

4.2 Stored volumes

The time series for the stored volume according to both storage curves is depicted in Fig. 4, with the discontinuity of the lines due to missing monitored data.

Both time series shows the rapid drying up of the reservoir in 2012 and the continuity of null volumes until 2018, corresponding to the recent drought in the state of Ceará (Pontes Filho et al., 2020).

The volumes derived from satellite storage curve are lower than those estimated from drone storage curve, which is used as reference due to its higher precision.

This result is due to the underestimation of volumes under 385 000 m^3 presented in the previous section. Throughout the analysis period, the volumes never exceeded that threshold.

Excluding the completed dry period 2014–2017, the mean absolute difference between drone- and satellite-derived volumes is 32 000 m^3. This represents only about 10 % of the higher volume in the assessed period.

4.3 Inflow

Figure 5 shows the estimated inflow series based on the two different storage curves. The difference between the estimations is related to the precision of the storage curves used to calculate the stored volumes for the water balance.

Since the storage curve derived from drone images is more precise, the inflows derived from it are considered as a reference in this analysis, as no inflow measurements are available.

The inflow calculated using the satellite storage curve underestimates the reference inflow until 2012 and overestimates it since 2013. The mean absolute difference between the two estimations was 3600 m^3 per month (i.e. 120 $m^3\,d^{-1}$), excluding the period 2012–2017, when there was no inflow.

4.4 Withdrawal and infiltration

In Fig. 6, the evolution of the withdrawal and infiltration (i.e. W_i) during the second semester is presented.

W_i was lower in the period 2018–2019 than in the period 2012–2013. Nevertheless, the stored volumes were similar in both periods (i.e. in the beginning the drought and after it). Using the hypothesis that infiltration losses are strongly related to the water level in the reservoir, one could state that the reduction of W_i is due to a reduction in withdrawal.

This result indicates that even if the reservoir volume has recovered to a level like the one observed in 2012–2013, the water consumption has not returned to the same amount. That means that water consumption may have been affected by the recent history of the local water system, which is an effect called hysteresis (Souza Filho et al., 2018).

5 Conclusions

Remote sensing and water balance techniques, combined with on-site data, were used to obtain information on the capacity of Lagoa Cercada reservoir and its water fluxes.

The main findings were:

- The capacity of Lagoa Cercada reservoir is 1.14 million m^3
- DEM Alos resulted in a 51 % difference in the estimation of storage capacity of Lagoa Cercada reservoir.
- Lagoa Cercada's water withdrawal has not returned to the same amount as before the drought.
- Drone imaging and water balance techniques can enable monthly reservoir monitoring to provide information on stored volume, inflow, withdrawal, and infiltration.

The framework proposed in this paper can be replicated in other reservoirs of similar size in the Brazilian semiarid region, which might be the main source of water for a local rural community, such as Lagoa Cercada reservoir.

Code and data availability. Code and data can be made available upon request directly to the authors.

Author contributions. The authors confirm contribution to the paper as follows: conceptualization, formal analysis, methodology, and writing: ÁBSE; aerial survey and data treatment: LV; data collection: SL and RA; supervision and validation: RR and EM.

Competing interests. The contact author has declared that none of the authors has any competing interests.

Disclaimer. Publisher's note: Copernicus Publications remains neutral with regard to jurisdictional claims in published maps and institutional affiliations.

Special issue statement. This article is part of the special issue "IAHS2022 – Hydrological sciences in the Anthropocene: Variability and change across space, time, extremes, and interfaces". It is a result of the XIth Scientific Assembly of the International Association of Hydrological Sciences (IAHS 2022), Montpellier, France, 29 May–3 June 2022.

Acknowledgements. The authors would like to thank Fundação Cearence de Meteorologia e Recursos Hídricos (FUNCEME), ARID as a researchers' network and French Agricultural Research Centre for International Development (CIRAD).

Review statement. This paper was edited by Christophe Cudennec and reviewed by two anonymous referees.

References

Alazard, M., Leduc, C., Travi, Y., Boulet, G., and Ben Salem, A.: Estimating evaporation in semi-arid areas facing data scarcity: Example of the El Haouareb dam (Merguellil catchment, Central Tunisia), J. Hydrol., 3, 265–284, https://doi.org/10.1016/j.ejrh.2014.11.007, 2015.

Arsiso, B. K., Tsidu, G. M., Stoffberg, G. H., and Tadesse, T.: Climate change and population growth impacts on surface water supply and demand of Addis Ababa, Ethiopia, Climate Risk Management, 18, 21–33, https://doi.org/10.1016/j.crm.2017.08.004, 2017.

Funceme: Mapeamento das barragens dos pequenos reservatórios d'água situados no Estado do Ceará, Technical report, Funceme, Fortaleza, 2021.

Garrido-Rubio, J., González-Piqueras, J., Campos, I., Osann, A., González-Gómez, L., and Calera, A.: Remote sensing-based soil water balance for irrigation water accounting at plot and water user association management scale, Agr. Water Manage., 238, 106236, https://doi.org/10.1016/j.agwat.2020.106236, 2020.

Jackson, M., Stewart, R., Fielding, K., Cochrane, J., and Beal, C.: Collaborating for Sustainable Water and Energy Management: Assessment and Categorisation of Indigenous Involvement in Remote Australian Communities, Sustainability-Basel, 11, 427–456, https://doi.org/10.3390/su11020427, 2019.

Paul, M., Rajib, A., Negahban-Azar, M., Shirmohammadi, A., and Srivastava, P.: Improved agricultural Water management in data-scarce semi-arid watersheds: Value of integrating remotely sensed leaf area index in hydrological modeling, Sci. Total Environ., 791, 148177, https://doi.org/10.1016/j.scitotenv.2021.148177, 2021.

Pontes Filho, J. D., Souza Filho, F. A., Martins, E. S. P. R., and Studart, T. M. C.: Copula-Based Multivariate Frequency Analysis of the 2012–2018 Drought in Northeast Brazil, Water-Sui, 12, 834–855, https://doi.org/10.3390/w12030834, 2020.

Serrao, L., Molinos-Senante, M., Bezzi, M, and Ragazzi, M.: Assessment of wastewater reuse potential for irrigation in rural semi-arid areas: the case study of Punitaqui, Chile, Clean Technol. Envir., 22, 1325–1338, https://doi.org/10.1007/s10098-020-01874-3, 2020.

Souza Filho, F. A., Formiga-Johnsson, R. M., de Carvalho Studart, T. M., and Abicalil, M. T.: From Drought to Water Security: Brazilian Experiences and Challenges, in: Water Resources Development and Management, Springer, Singapore, 233–265, https://doi.org/10.1007/978-981-10-7913-9_10, 2018.

Sudene: Plano de Aproveitamento Integrado de Recursos Hídricos do Nordeste do Brasil, Plan, Sundene, Recife, 1980.

Wang, X. and Xie, H.: A Review on Applications of Remote Sensing and Geographic Information Systems (GIS) in Water Resources and Flood Risk Management, Water-Sui, 10, 608–618, https://doi.org/10.3390/w10050608, 2018.

Werner, P. C. and Gerstengarbe, F. W.: The climate of Piauí and Ceará, in: Global Change and Regional Impacts, edited by: Gaiser, T., Krol, M., Frischkorn, H., and De Araújo, J. C., Springer, Berlin, Heidelberg 81–86, https://doi.org/10.1007/978-3-642-55659-3_5, 2003.

Proc. IAHS, 385, 297–303, 2024
https://doi.org/10.5194/piahs-385-297-2024

Open Access PIAHS

IAHS2022 – Past and future of open, inclusive, innovative, and society-interfacing approaches

Quantifying irrigation returns into a highly human managed wetland using remote sensing: The Primera de Palos freshwater lagoon (Spain)

Eva Contreras[1,2], Rafael Pimentel[1,2], Cristina Aguilar[1], Javier Aparicio[1,2], and María J. Polo[1,2]

[1]Fluvial Dynamics and Hydrology Research Group, Andalusian Institute for Earth System Research, University of Córdoba, 14071 Córdoba, Spain

[2]Department of Agronomy, Unit of Excellence María de Maeztu (DAUCO), University of Córdoba, Córdoba, Spain

Correspondence: Eva Contreras (econtreras@uco.es)

Received: 15 June 2022 – Revised: 30 March 2023 – Accepted: 21 April 2023 – Published: 18 April 2024

Abstract. In this paper, the Global Surface Water Explorer (GSWE) was combined with bathymetric and historical meteorological data to quantify water balance during the period 1984–2020 in the Primera de Palos freshwater lagoon (Southwest Spain). This allowed us, through a water balance approach, to estimate all water inputs and outputs to analyse the hydrological changes in the lagoon. The results showed high fluctuations with seasonal changes marked by the climatic regime during the first two decades of the study period. After this initial period, water extension remained stable above 70 % of the maximum lagoon extent. Thus, the natural hydrological regime of the lagoon was modified by water inputs from irrigation return, which are difficult to quantify. Thanks to a water balance approach, these irrigation returns were quantified as the closure term of the water balance. Three scenarios of irrigation return inputs can be defined, 4500, 1700, and 500 $m^3\,d^{-1}$, depending on the cropping season. The use of remote sensing combined with bathymetric and meteorological data can provide the knowledge to support better informed water-management decision-making, although it may have some limitations in dry periods related to image processing in border data.

Keywords. Remote sensing; irrigation returns; freshwater lagoon; human-managed system; UPH 16; Panta Rhei

1 Introduction

Wetlands are essential for ecology and environmental regulation (Westerberg et al., 2010; Newton et al., 2018). They also have a key role for the economy and social service functions (Ye and Sun, 2021; Ghanian et al., 2022). However, the natural hydrology of this kind of system is often modified by anthropogenic activities, such as drainage (Sica et al., 2016), mining and drastic modifications of the land uses in the upstream areas (Jia et al., 2013; Li et al., 2021). These changes in hydrological dynamics of these systems, for instance modifications of water fluxes partitioning, are especially relevant in wetlands where their upstream watersheds have a significant proportion of surface altered by these human activities (Merchán et al., 2015). In arid and semi-arid climates, wetlands often rely on inputs from irrigation return flows, which play an important role in water balances of hydrological watersheds (Scott et al., 2011; Merchán et al., 2013). To understand the changes over wetlands ecosystems and their spatiotemporal distribution is critical for the management, restoration, and conservation of these systems.

The development of remote sensing and GIS techniques have emerged as an alternative to traditional methods like field survey (usually laborious, time consuming and expensive) to account for changes. Open-source multispectral satellite remote sensing missions provide worldwide long-term time-series coverage of wetland evolution. There are an increasing number of studies providing evidence about the success of this kind of methods for spatiotemporal monitoring of changes in these habitats (Bortels et al., 2011; Blount et al., 2022; Chen et al., 2022). These techniques can be also used to analyse the effects of anthropogenic activities in the water bodies.

An example of a human-modified freshwater lagoon due to agricultural pressure is the Primera de Palos lagoon, which is located in a highly modified marshy environment in south-western Spain. In this study we aim to analyse the changes in the lagoon hydrological dynamics during the last thirty years narrowing down the ranges of the different water fluxes involved in the water balance and accounting specifically from irrigation returns. For that, remote sensing information in combination to meteorological data and the bathymetry of the lagoon were used. This work examines research question No 16 of the twenty-three unsolved problems in hydrology (UPH; Blöschl et al., 2019): "How can we use innovative technologies to measure surface and subsurface properties, states, and fluxes, at a range of spatial and temporal scales?"

2 Study site and available data

2.1 Primera de Palos freshwater lagoon

In this work we analysed the evolution of the Primera de Palos freshwater lagoon (Southwest Spain) (Fig. 1), a Ramsar site whose surrounding lands have been intensively modified by petrochemical industry and greenhouse strawberry crops. This water body is part of a complex wetland that has its origin in the closure of old valleys or riverbeds due to the advance of a coastal dune front that acts as a barrier for both, the inland water flowing to the sea and the sea water flowing into the lagoon avoiding saline intrusion (Mantecon et al., 1988). The contributing watershed, which is entirely dedicated to strawberry cultivation from 1995, is made up of a sandy substrate that allows irrigation water to drain into the wetland.

2.2 Data

We have used different sources of data in this study. On the one hand, since there are no historical series of flow measurements in the lagoon or in its contributing watershed, the Global Surface Water Explorer (GSWE), a remote sensing based virtual online machine was used to spatially quantify water surface during the historical period 1984–2020 (Pekel et al., 2016). The explorer provides extent and changes over time of the world's water extensions at 30 m spatial resolution. Since our study goes until 2022, Landsat information from the period 2020–2022 was used for deriving the lagoon extension using the Normalized Water Difference Index (NWDI, Gao, 1996). This information was combined with bathymetric data (provided by TYPMA S.L, a consultancy firm which participates in some management tasks in the wetland) and historical meteorological data from the weather station close to the lagoon with the longest dataserie – Palos de la Frontera (provided by the Spanish State Meteorological Agency).

Table 1. In situ measurements (outflow and water level) at the lagoon' outlet.

Date	Outflow ($L\,s^{-1}$)	Water level (cm)
12/05/2021	44.74	–
10/06/2021	28.58	–
21/06/2021	7.60	–
01/07/2021	1.36	–
08/07/2021	0.00	–
02/08/2021	0.00	−5
24/09/2021	–	−11
14/10/2021	–	−5
19/10/2021	–	0
26/10/2021	–	+9
02/11/2021	–	+25
22/11/2021	–	+33
18/01/2022	–	+23
14/02/2022	–	+26
17/02/2022	8.70	–
25/03/2022	222.00	+44
28/04/2022	37.38	+27
06/05/2022	45.89	+25
10/05/2022	–	+23
24/05/2022	22.46	+20
31/05/2022	24.67	+20
07/06/2022	16.83	+20

In addition, from May 2021 to March 2022, periodic measurements of streamflow and water level in the lagoon outlet (Fig. 1) were carried out with a weekly/biweekly frequency (Table 1). These observations will also allow us to establish a limit value of the water surface from which there is no outflow in the lagoon.

3 Methodology

A water balance approach was carried out to estimate the lagoon dynamics:

$$\frac{\Delta V}{\Delta t} = \text{inflows} - \text{outflows} = (P + Q_r + I_d + R) - (E + Q_{out})$$

where, Δt is the variation of time between consecutives dates with available satellite information, $\Delta V/\Delta t$ is the variation of volume in the lagoon in each time interval, P accounts for precipitation in the lagoon, Q_r is the direct runoff into the lagoon, E is the evapotranspiration from the lagoon, I_d represents the subsurface flow, R accounts for the irrigation returns and Q_{out} accounts for water outflow. Each term of the balance equation was computed as followed:

- $\Delta V/\Delta t$ is calculated as the difference between the water volumes between two consecutive images. For each date, the volume of the lagoon was derived using the water surface from the remote sensing information and the relationship between water surface – water volume –

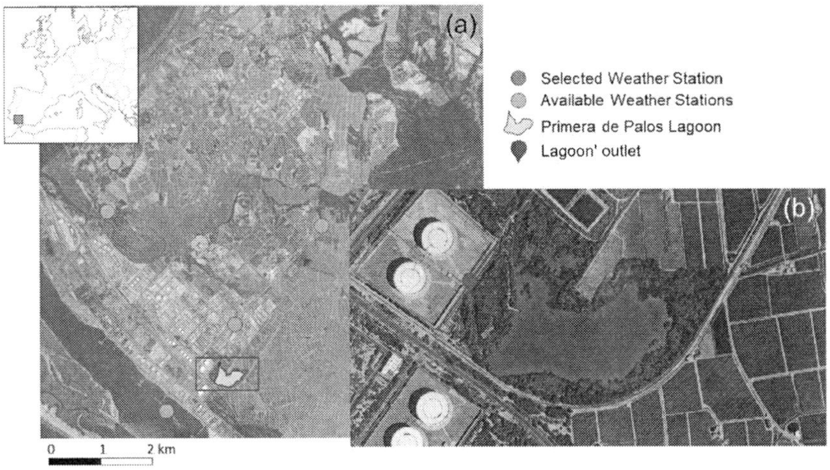

Figure 1. **(a)** Geographic location of the Primera de Palos lagoon in the Southwestern Spain. **(b)** Zoom to the Primera de Palos lagoon. © Google 2023 Primera de Palos freshwater lagoon. Available at: https://www.google.com/ (last access: 30 March 2023).

lagoon wet area – elevation, established using the bathymetric information.

- P was estimated as the product of the accumulated precipitation in each interval and the average area of the water surface of the lagoon between images.

- Q_r, was estimated as the product of the accumulated precipitation, which is assumed constant over the drainage area into the lagoon, this contributing area ($3.73\,km^2$) and a variable runoff coefficient (c) over time based on the main land use on this area. We assumed a $c = 0.31$ for no crops under plastic and $c = 0.50$ for crops under plastic.

- E, was estimated as the product of the potential evaporation accumulated between images and the average area of the water surface of the lagoon between these days. The potential evaporation on a daily scale was calculated from minimum, mean and maximum temperature data using the Hargreaves equation (Hargreaves and Samani, 1985). $I_d R$ and Q_{out} cannot be directly estimated. Therefore, they are grouped in one term and considered the closure term of the water balance.

To account for each of these individual terms we analysed specific situations in which some of the terms can be considered null.

- Situation A – Natural conditions: It assumes that there was no relevant irrigation contribution to the lagoon ($R = 0$) and a negligible value of the outflow ($Q_{out} = 0$), since the water levels in the lagoon were lower than the height of the outlet. The closure term becomes I_d in this case.

- Situation B – Modified conditions: Under this situation the closure term will be the sum of R, Q_{out} and I_d. However, analysing each of the water fluxes dynamics we can understand seasonal dynamics of this term. We assume that this period is representative of the current lagoon state in terms of water balance.

- Situation C – Year 2021–2022: We have measurements of Q_{out} during the last year. Then the closure term can be reduced to $R + I_d$. If we have obtained the I_d estimates for the situation A and that estimates are kept in time. We can provide a value for the irrigation returns into the lagoon.

4 Results

The analysis of the GSWE' water extension during the analysed period 1984–2020 (Fig. 2), showed high fluctuations during the first two decades of the study period. These fluctuations are clearly linked to the climatic regime, that is, the lagoon was emptier in dry years than in wet years. However, from 1995 linked to a boom in the use of the greenhouse for strawberry crops, the water extension in the lagoon did not show these high fluctuations. Since then, the water extension of the lagoon has always been above 70 % of the maximum lagoon extent ($121\,149\,km^2$). These results allowed us to verify the hypothesis of a modification in the lagoon dynamics due to the strawberry crop irrigation. Therefore, the period that represents Situation A – natural conditions goes from 1984 until 1995 and the period representing Situation B – modified conditions goes from 1996 until 2022.

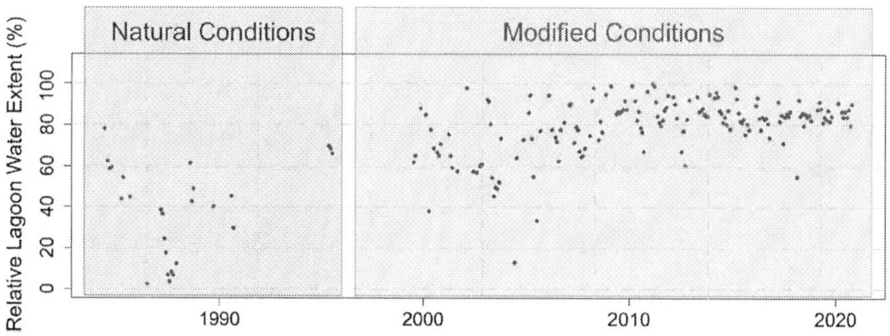

Figure 2. Evolution of the percentage of relative lagoon water surface with respect to the maximum extension value for the period 1984–2020.

4.1 Situation A – Natural conditions

For this first period 1984–1995, the number of satellite images was low, just 20 dates with available information (Table 2). This fact makes the Δt use in the balance too long, diluting the outcomes of the analysis. In addition, the images also coincide with very dry months, usually in summer, during the analysed period. Consequently, the water extensions in the lagoon were very reduced. This makes errors in the retrieving process to become higher. That is, most of the pixels identified as water were susceptible to be erroneous due to border effect. Previous studies (Blount et al., 2022) already pointed out that the main source of error in the satellite results is the dataset spatial resolution limits, thus, in this case, the error associated with the volume calculation could be relevant. For this reason, we decided to rule out the water balance estimates for this period. However, the images analysed showed that the water levels in the lagoon decrease during all year except for winter months (Table 2). Therefore, we can assume that there was no inflow into the lagoon and thus under natural conditions we can assume that $I_d = 0$ for this period. In addition, for these months, the water extent in the lagoon was always below 70 % of its maximum lagoon extent (Fig. 2).

4.2 Situation B – Modified conditions

Figure 3 shows the monthly evolution of the different terms of the water balance in the lagoon for the period 1996–2020. In this case the closure term refers to the sum of subsurface flow, irrigation return and outflow ($I_d + R - Q_{\text{out}}$). A negative value of this term indicates that outflows from the lagoon (Q_{out}) are greater than inflows (due to I_d and R) into the lagoon. These conditions were found from November to May (Fig. 4). Usually in this period, water extension in the lagoon presents higher values, always above 70 % of the maximum lagoon extent (Fig. 2). In addition, these months also coincide with the strawberry crops' irrigation period. On the other hand, results showed positive values of the closure term from June to October (Fig. 4). That implies that during these months Q_{out} is almost neglectable.

Table 2. Dates with available Global Water Surface Explorer (GSWE) information and percentage over the maximum lagoon water extension for the period 1984–1995. Bold cells indicate a decreasing change in water extension in the lagoon.

Date	%	Date	%
15/05/1984	78.12	**15/04/1987**	**24.48**
15/07/1984	**62.50**	**15/05/1987**	**17.71**
15/08/1984	**58.85**	**15/06/1987**	**6.77**
15/09/1984	59.38	**15/07/1987**	**3.65**
15/03/1985	43.75	**15/08/1987**	**8.33**
15/04/1985	**54.17**	**15/08/1990**	**45.31**
15/08/1985	**44.79**	**15/09/1990**	**26.69**
15/06/1986	2.60	**15/05/1995**	**69.79**
15/02/1987	38.54	**15/06/1995**	**68.75**
15/03/1987	**36.46**	**15/07/1995**	**66.15**

4.3 Situation C – Year 2021–2022

The Q_{out} measured in the outlet of the lagoon (Table 2) helped us to verify the outcomes obtained about seasonality in the lagoon. On the one hand, outflow dynamic follow the pattern previously defined. Q_{out} decreases from May to June, it is null from June to November and increases from November on. This last increase of Q_{out} follows two different rates (about $10\,\text{m}^3\,\text{s}^{-1}\,\text{d}^{-1}$ from November to March and about $500\,\text{m}^3\,\text{s}^{-1}\,\text{d}^{-1}$ from March to May). On the other hand, we have been able to verify the threshold that defines when the lagoon is draining out (70 % of the maximum lagoon extent).

Moreover, having Q_{out} values along this period allows us to reduce the closure term to the sum of $I_d + R$. Figure 4 shows the evolution of this term for the days when satellite images were available during the period with outflow measurements. According to our hypothesis of $I_d = 0$, except for winter, the closure term is equivalent to R.

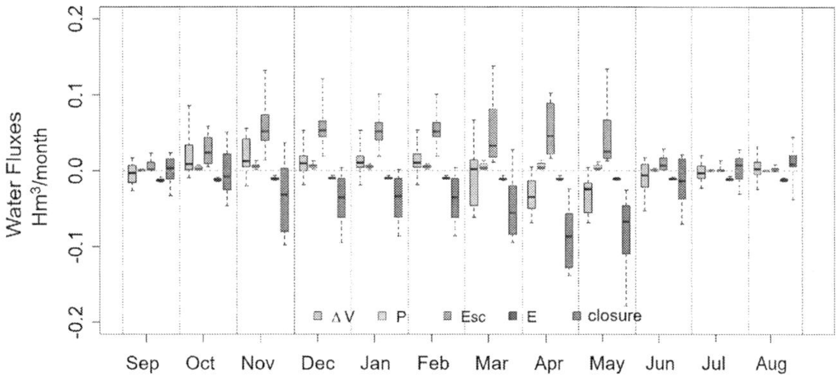

Figure 3. Monthly evolution of the different terms of the water balance in the Primera de Palos lagoon for the period 1996–2020.

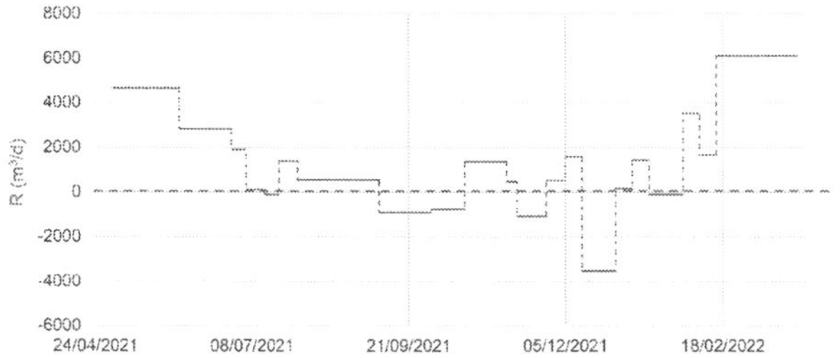

Figure 4. Evolution of the term R for the days in which there are satellite images during May 2021–March 2022.

Under these conditions, we were able to account for the irrigation return inputs into the lagoon establishing three possible scenarios: (i) high irrigation return inflow, about $4500\,m^3\,d^{-1}$ (in April); (ii) medium irrigation return inflow, about $1700\,m^3\,d^{-1}$ (July, October, and December); and low irrigation return inflow, about $500\,m^3\,d^{-1}$ (June and November).

For the higher contribution rate, assuming an irrigation period about $100\,d\,yr^{-1}$ and a water volume allocation for strawberries irrigation of $7000\,m^3\,ha^{-1}$, the irrigation volume allocated for strawberries in the watershed would be about $700\,mm\,yr^{-1}$, which means that a 15 % of the total irrigation volume is reaching the lagoon.

5 Conclusions

This study quantifies the hydrological dynamics of a highly human managed wetland in southwestern Spain. The case study of the Primera de Palos freshwater lagoon can be seen as a pilot area for testing a methodology to account for unmeasurable water fluxes in a lagoon water balance approach. The main findings of this work can be summarised:

- The remote sensing product used, GWSE, combined with local information, can be used to successfully evaluate hydrological behaviour in water bodies when historical hydrological data series are not available.

- The spatiotemporal resolution of GWSE has proven to be enough to capture seasonal dynamics of water fluxes in the analysed lagoon. Only, some limitations have been highlighted connected to low levels in the lagoon. This small water extent can be linked to a wrong water pixel identification. In this situation the use of higher resolution remote sensing data, as Sentinel-2 will be the solution.

- The methodology applied, identifying key situations in which some terms in the water balance equation can be neglected, allowed us to define the behaviour of the water surface for the period 1984–2020.

- Three scenarios of irrigation return were defined according to the season. The quantification of the magnitude of all water fluxes helped to not only reduce the uncertainty regarding the unmeasured water fluxes but also to establish the basis for a water balance model to better inform water-management decision-making.

Code availability. The software is available from the corresponding author, Eva Contreras, upon reasonable request.

Data availability. In relation to the data, two different Earth Observation (EO) products were used for this research: (1) Global Surface Water Explorer (Pekel et al., 2016) available at https://global-surface-water.appspot.com/download (last access: 1 June 2022) and (2) Landsat 8 scenes, available at https://earthexplorer.usgs.gov/ (USGS, 2022). Meteorological dataset, Palos de la Frontera station is available at https://www.aemet.es/es/datos_abiertos (AEMET, 2022). In situ measurements of outflow and water level are available in Table 1. Bathymetric information is not publicly available due to privacy restriction.

Author contributions. EC: Conceptualization, Data curation analysis, Formal analysis, Investigation, Methodology, Software, Visualization, Writing – original draft preparation.

RP: Conceptualization, Data curation, Formal analysis, Investigation, Methodology, Software, Visualization, Writing – review & editing.

CA: Methodology, Writing – review & editing.

JA: Data curation, Writing – review & editing.

MJP: Conceptualization, Funding acquisition, Supervision, Methodology, Writing – review & editing.

Competing interests. The contact author has declared that none of the authors has any competing interests.

Disclaimer. Publisher's note: Copernicus Publications remains neutral with regard to jurisdictional claims in published maps and institutional affiliations.

Special issue statement. This article is part of the special issue "IAHS2022 – Hydrological sciences in the Anthropocene: Past and future of open, inclusive, innovative, and society-interfacing approaches". It is a result of the XIth Scientific Assembly of the International Association of Hydrological Sciences (IAHS 2022), Montpellier, France, 29 May–3 June 2022.

Acknowledgements. Rafael Pimentel acknowledges fundings by the Juan de la Cierva In corporación Programme of the Spanish Ministry of Science and Innovation (IJC2018–038093-I). Javier Aparicio acknowledges fundings by the Programme Ayudas para contratos predoctorales para la formación de doctores of the Spanish Ministry of Science and Innovation (PRE2019–090493). Eva Contreras, Rafael Pimentel, Javier Aparicio and María J. Polo are members of DAUCO, Unit of Excellence ref. CEX2019–000968-M, with financial support from the Spanish Ministry of Science and Innovation, the Spanish State Research Agency, through the Severo Ochoa and María de Maeztu Program for Centers and Units of Excellence in R&D.

Financial support. This research has been supported by the environmental consulting company Tratamiento Proyectos Medioambientales, S.L (Analysis of the hydrological viability of the drainage actions in the Primera de Palos freshwater lagoon).

Review statement. This paper was edited by Christophe Cudennec and reviewed by two anonymous referees.

References

AEMET: AEMET Open Data, AEMET [data set], https://www.aemet.es/es/datos_abiertos, last access: 30 June 2022.

Blöschl, G. et al.: Twenty-three unsolved problems in hydrology (UPH) – a community perspective, Hydrol. Sci. J., 64, 1141–1158, https://doi.org/10.1080/02626667.2019.1620507, 2019.

Blount, T. R., Carrasco, A. R, Cristina, S., and Silvestri, S.: Exploring open-source multispectral satellite remote sensing as a tool to map long-term evolution of salt marsh shorelines, Estuar. Coast Shelf S., 266, 107664, https://doi.org/10.1016/j.ecss.2021.107664, 2022.

Bortels, L., Chan, J. C.-W., Merken, R., and Koedam, N.: Long-term monitoring of wetlands along the Western-Greek Bird Migration Route using Landsat and ASTER satellite images: Amvrakikos Gulf (Greece), J. Nat. Conserv., 19, 215–223, https://doi.org/10.1016/j.jnc.2011.01.004, 2011.

Chen, K., Cong, P., Qu, L., Liang, S., and Sun, Z.: Annual variation of the landscape pattern in the Liao River Delta wetland from 1976 to 2020, Ocean Coast. Manage., 224, 106175, https://doi.org/10.1016/j.ocecoaman.2022.106175, 2022.

Hargreaves, G. H. and Samani, Z. A.: Reference Crop Evapotranspiration from Temperature, Appl. Eng. Agric., 1, 96–99, https://doi.org/10.13031/2013.26773, 1985.

Gao, B.-C.: NDWI – A normalized difference water index for remote sensing of vegetation liquid water from space, Proc. SPIE, 58, 257–266, https://doi.org/10.1016/S0034-4257(96)00067-3, 1996.

Ghanian, M., Ghoochani, O. M., Noroozi, H., and Cotton, M.: Valuing wetland conservation: A contingent valuation analysis among Iranian beneficiaries, J. Nat. Conserv., 66, 126140, https://doi.org/10.1016/j.jnc.2022.126140, 2022.

Jia, Z., Wu, Z., Luo, W., Xi, W., Tang, S., Liu, W. L., and Fang S.: The impact of improving irrigation efficiency on wetland distribution in an agricultural landscape in the upper reaches of the Yellow River in China, Agr. Water Manage., 121, 54–61, https://doi.org/10.1016/j.agwat.2013.01.003, 2013.

Li, S., Wu, M., Jia, Z., Luo, W., Fei, L., and Li, J.: Study on drainage strategy of ditch wetland in semi-arid area under the influence of inflow from the upstream irrigation area, Agr. Water Manage., 248, 106792, https://doi.org/10.1016/j.agwat.2021.106792, 2021.

Mantecon-Gómez, R., Martín-Machuca, M., and Mediavilla-Laso, C.: Problemas de sobreexplotación e intrusión marina en los acuíferos costeros del litoral atlántico andaluz, 504–520, in: Tecnología de la intrusión de los acuíferos costeros, edited by: Fernández-Rubio, R., López Geta, J. A., and Ramos González, G., TIAC, 88, 654 pp., 1988.

Merchán, D., Causapé, J., and Abrahão, R.: Impact of irrigation implementation on hydrology and water quality in a small agricultural basin in Spain, Hydrolog. Sci. J., 58, 1400–1413, https://doi.org/10.1080/02626667.2013.829576, 2013.

Merchán, D., Causapé, J., Abrahão, R., and García-Garizábal, I.: Assessment of a newly implemented irrigated area (Lerma Basin, Spain) over a 10-year period. I: Water balances and irrigation performance, Agr. Water Manage., 158, 277–287, https://doi.org/10.1016/j.agwat.2015.04.016, 2015.

Newton, A., Brito, A. C., Icely, J. D., Derolez, V., Clara, I. Angus, S., Schernewski, G., Inácio, M., Lillebø, A. I., Sousa, A. I.,

Béjaoui, B., Solidoro, C., Tosic, M., Cañedo-Argüelles, M., Yamamuro, M., Reizopoulou, M., Tseng, S., Canu, H.-C., Roselli, D., Maanan, L., Cristina, S., Ruiz-Fernández, A. C., de Lima, R. F. d., Kjerfve, B., Rubio-Cisneros, N., Pérez-Ruzafa, A., Marcos, C., Pastres, R., Pranovi, F., Snoussi, M., Turpie, J., Tuchkovenko, Y., Dyack, B., Brookes, J., Povilanskas, R., and Khokhlov, V.: Assessing, quantifying and valuing the ecosystem services of coastal lagoons, J. Nat. Conserv., 44, 50–65, https://doi.org/10.1016/j.jnc.2018.02.009, 2018.

Pekel, J. F., Cottam, A., Gorelick, N., and Belward, A. S.: High-resolution mapping of global surface water and its long-term changes, Nature, 540, 418–422, https://doi.org/10.1038/nature20584, 2016 (data available at: https://global-surface-water.appspot.com/download, last access: 1 June 2022).

Scott, J., Rosen, M. R., Saito, L., and Decker, D. L.: The influence of irrigation water on the hydrology and lake water budgets of two small arid-climate lakes in Khorezm, Uzbekistan, J. Hydrol., 410, 114-125, https://doi.org/10.1016/j.jhydrol.2011.09.028, 2011.

Sica, Y. V., Quintana, R. D., Radeloff, V. C., and Gavier-Pizarro, G. I.: Wetland loss due to land use change in the Lower Paraná River Delta, Argentina, Sci. Total Environ., 568, 967–978, https://doi.org/10.1016/j.scitotenv.2016.04.200, 2016.

USGS: Landsat 8 scenes, USGS [data set], https://earthexplorer.usgs.gov/, last access: 1 June 2022.

Westerberg, V. H., Lifran, R., and Olsen, S. B.: To restore or not? A valuation of social and ecological functions of the Marais des Baux wetland in Southern France, Ecol. Econ., 69, 2383–2393, https://doi.org/10.1016/j.ecolecon.2010.07.005, 2010.

Ye, C. and Sun, F.: Development of a social value evaluation model for coastal wetlands, Ecol. Inform., 65, 101417, https://doi.org/10.1016/j.ecoinf.2021.101417, 2021.

Proc. IAHS, 385, 305–311, 2024
https://doi.org/10.5194/piahs-385-305-2024

Open Access PIAHS

Trend analysis of reference evapotranspiration and climate variables in the main hydrosystems of Senegal: Senegal, Gambia and Casamance River Basins

Papa Malick Ndiaye[1], **Ansoumana Bodian**[1], **Serigne Bassirou Diop**[1], **Lamine Diop**[2], **Alain Dezetter**[3], **Andrew Ogilvie**[4], **and Koffi Djaman**[5]

[1]Laboratoire Leïdi "Dynamique des Territoires et Développement", Université Gaston Berger (UGB), Saint-Louis BP 234, Senegal
[2]UFR S2ATA Sciences Agronomiques, de l'Aquaculture et des Technologies Alimentaires, Université Gaston Berger, Saint-Louis BP 234, Senegal
[3]HSM, Univ. Montpellier, IRD, CNRS, UFR Pharmacie, Bâtiment HYDROPOLIS, 15 avenue Charles Flahault, 34090 Montpellier, France
[4]UMR G-EAU, AgroParisTech, Cirad, University of Montpellier, IRD, INRAE, Montpellier SupAgro, 34196 Montpellier, France
[5]Department of Plant and Environmental Sciences, New Mexico State University, Agricultural Science Center at Farmington, P.O. Box 1018, Farmington, NM 87499, USA

Correspondence: Papa Malick Ndiaye (ndiaye.papa-malick@ugb.edu.sn)

Received: 24 May 2022 – Revised: 17 February 2023 – Accepted: 21 April 2023 – Published: 18 April 2024

Abstract. Analysis of reference evapotranspiration (ET_0) trends is essential for understanding the impacts of climate change on water resources. Thus, despite the continuous rise in temperatures, a decrease in evapotranspiration is noted in some parts of the world. This contrast is called the "evaporation paradox" and is thought to be related to the variation in wind speed, relative humidity and solar radiation. The objective of this work is to analyze the annual and seasonal trends of ET_0 and climate variables at the scale of the Senegal, Gambia and Casamance river basins. The reanalyze data of NASA/POWER are used over the period 1984–2019. Mann Kendall's test and Sen slope were used to analyze trends in ET_0 and climate variables. Results show that on annual scale, ET_0 increases significantly in 32 % of the Senegal basin and decreases in less than 1 % of it. In contrast, in the Casamance and Gambia basins, the annual ET_0 drops by 65 % and 18 %, respectively. On an annual scale, temperature and relative humidity show an increasing trend over all basins while wind speed and radiation decrease significantly. This confirms the existence of the "evaporation paradox" in the three basins. This phenomenon is explained by the increase in relative humidity and the decrease in wind speed and solar radiation.

Keywords. UPH1; SDG 6; Modelling; Senegal; Gambia; Casamance River Basins

1 Introduction

In a context of climate change, the spatiotemporal analysis of evapotranspiration is important for the management of water resources. Indeed, in semi-arid regions, 60 % to 80 % of rainfall is lost through evapotranspiration (Mjejra, 2015; Ndiaye et al., 2020a). In this regard, evapotranspiration is an important parameter for hydrological modelling, reservoir management, irrigation planning and scheduling, and climate change studies. In its latest assessment report, the GIEC (2021) continues to alert international opinion to the potential impacts of climate change on various sectors of activity. One of the consequences of global warming is the intensification of the hydrological cycle (Wilcox et al., 2018) which would result in an increase in evapotranspiration (Jiao and Wang, 2018).

However, in recent years, despite rising temperatures, studies (Peterson et al., 1995; Qin et al.. 2021) have shown a decreasing in evapotranspiration in certain regions of the world. This contrast between increasing temperatures and decreasing evapotranspiration highlights the concept of the "evaporation paradox" theorized since the 1990s (Peterson et al., 1995). The causes of the evaporation paradox are analyzed through the trends of climatic variables (temperatures, relative humidity, solar radiation and wind speed). Thus, according to Mekoya (2020), for the evaporation paradox to exist, the temperature must follow an upward trend. Others show that the drop in evapotranspiration is linked to the decreasing in wind speed (Han et al., 2012), solar radiation (Rodrick and Forquha, 2002) and the increase in relative humidity (Chattopadhyay and Hulme, 1997). The main reasons for this evaporation paradox are not yet well understood and research is continuing in this area. Ndiaye et al. (2020b) highlighted the evaporation paradox in the Senegal River basin over the period 1984–2017. This paradox is linked to the increase in relative humidity and the decrease in wind speed. This work aims to expand the study by Ndiaye et al. (2020b) on the other watersheds of Senegal (Gambia and Casamance) in order to analyze trends in ET_0 and climatic variables on the main hydrosystems of Senegal. These three hydrosystems are major challenges for the agricultural sector, drinking water supply, hydroelectric production in the countries that share these basins.

2 Material and methods

2.1 Study area

The main hydrosystems of Senegal are the basins of Senegal, Gambia and Casamance (Fig. 1). The Senegal and Gambia rivers have their source in Fouta Djallon at an altitude of 1150 m and drain areas of more than 300 000 km^2 and 77 100 km^2 respectively (Faye, 2018; Ndiaye et al., 2020a). The Senegal River is shared by Senegal, Guinea, Mauritania and Mali which form the Organization for the Development of the Senegal River (Organisation pour la Mise en Valeur du Fleuve Sénégal, OMVS, in french). The Gambia river basin is concerned by Senegal, Gambia and Guinea which constitute the Organization for the Development of the Gambia River (Organisation pour la Mise en Valeur du fleuve Gambie, OMVG in French). The Casamance basin extends over an area of 20 150 km^2 and is located almost entirely in Senegalese territory (Dacosta, 1989). From a climatic point of view, these three hydrosystems extend over three climatic zones: Guinean (average annual rainfall > 1500 mm); South Sudanian (1000 < rainfall < 1500 mm); North Sudanian (500 < rainfall < 1000 mm) and Sahelian (P rainfall 500 mm). The water resources of these three hydrosystems are used for the development of agriculture, hydroelectric production, drinking water supply, navigation and the maintenance of ecosystems. These water resources are therefore a major challenge for the socio-economic development of the riparian countries.

2.2 Data

In West Africa, climate data managed by national meteorological services are difficult to access due to their high acquisition cost (Bodian et al., 2020). Added to this is the low density of the observation network, which poses a problem of representativeness at the scale of the catchment area (Ndiaye et al., 2020a). It is for this reason that reanalyses from the NASA Langley Research Center (LaRC) POWER Project funded through the NASA Earth Science/Applied Science Program (https://power.larc.nasa.gov, last access: February 2022) are used as an alternative to observed data generally incomplete and inaccessible (Ndiaye et al., 2020a). The coordinates of 160 stations distributed across the three basins (Fig. 1) are used to extract climate variables on a daily scale over the period 1984–2019. The climatic variables collected are: temperatures (max and min), solar radiation, relative humidity and wind speed at 2 m from the ground.

2.3 Methods

The methodology adopted includes: (i) calculation of reference evapotranspiration, (ii) detection of the trend of ET_0 and climatic variables and (iii) sensitivity analysis of ET_0 to climatic variables.

Reference evapotranspiration (ET_0) was calculated by the Penman-Monteith method (Allen et al., 2005). Mann Kendall's test (Mann, 1945; Kendall, 1975) and Sen's slope (1968) were used for detecting trends in ET_0 and climatic variables. As a reminder, this test is based on two hypotheses: the null hypothesis (H0) which assumes an absence of trend and the alternative hypothesis (H1) which indicates the existence of a trend in the time series. The significance of the detected trend is determined by $\alpha = 0.05$. Before applying the Mann Kendall test, the autocorrelation was tested and the results showed a non-significant autocorrelation for more than 80 % of the stations selected. Therefore, the primary version of Mann Kendall's test was applied to the data. ET_0 sensitivity to climate variable was estimated by varying each variable individually by ±5 %, 10 %, 15 %, 20 % and 25 %, leaving the other variables constant. The sensitivity coefficient (S_{vi}) is calculated to determine the degree of influence of each climate variable on ET_0:

$$S_{vi} = \frac{\partial ET_0}{\partial_{vi}} \times \frac{vi}{ET_0} \quad (1)$$

Where S_{vi} is the sensitivity coefficient, ∂ET_0 variation of the reference evapotranspiration (ET_0) caused by the change of a variable (∂_{vi}), vi being the variable considered. A positive (negative) sensitivity coefficient indicates that the variable increases (decreases) the reference evapotranspiration. The higher the absolute value of the sensitivity coefficient, the greater the influence of the variable on ET_0 (Li et al., 2017).

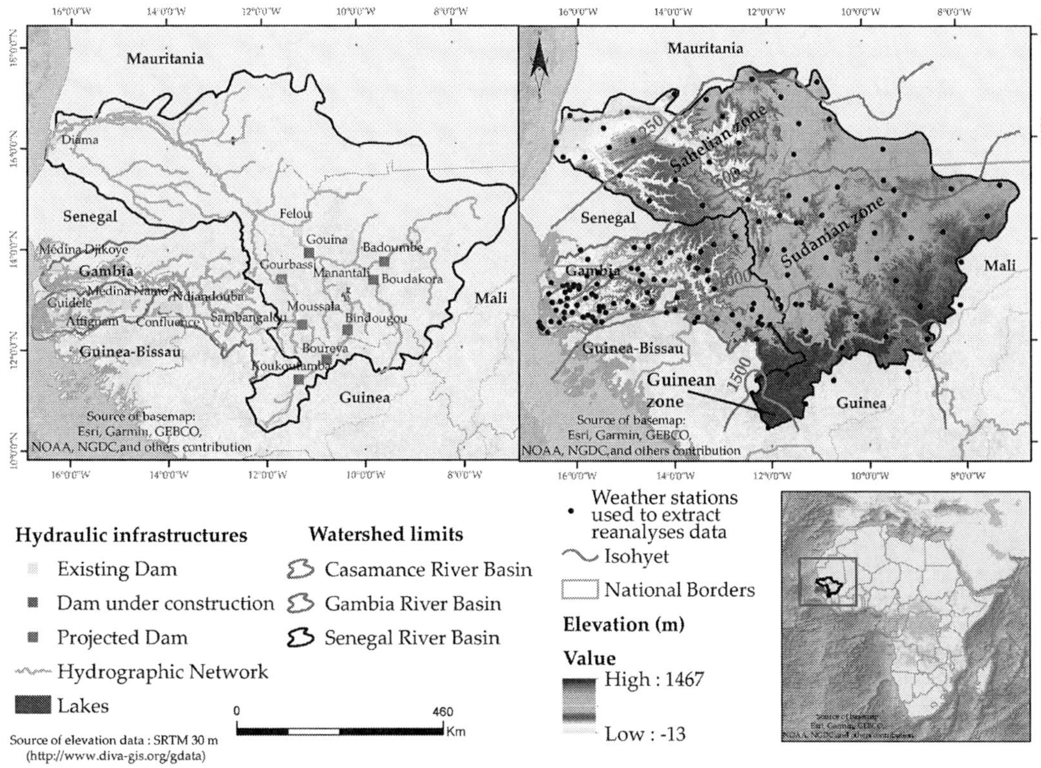

Figure 1. Location of the Senegal hydrosystems, hydraulic infrastructures, climate zones, hydrographic network and weather stations used to extract reanalyses data.

3 Results

3.1 Annual and seasonal trend of evapotranspiration and climate variables

Figure 2 shows the spatial repartition of ET_0 trend in the three basins and Fig. 3 gives the boxplot of Z of Mann Kendall's test. According to the figures a positive trend of ET_0 in annual and seasonal (dry season) is observed in the three basins. This positive trend is more marked in the Senegal River basin with an increase in ET_0 of 32.6 % on an annual scale and 49 % during the dry season. For the Gambia basins, ET_0 increases in about 2 % and 5 % of the basin on annual scale and during the dry season, respectively. During these same periods, a significant increase in ET_0 was noted in 4 % and 36 % of the Casamance river basin. A decreasing trend in ET_0 was also noted in the three hydrosystems, especially in the rainy season where the three respective basins exhibit a drop in ET_0 of 13 % (Senegal basin), 22 % (Gambia basin) and 99 % (Casamance basin) of their area. Over the entire Senegal River basin, the Sen's slope indicates an increase in ET_0 of 1.30 and 1.25 mm yr^{-1} on annual scale and during the dry season and a decrease of 1.13 mm yr^{-1} during the rainy season. For the Gambia basin, ET_0 increases on average by 0.24 mm yr^{-1} during the dry season and decreases by 0.54 to 0.66 mm yr^{-1}on annual scale and during the rainy season. For all periods, the ET_0 of the Casamance basin decreases from 0.15 to 1 mm yr^{-1}.

Figures 4 and 5 shows respectively the spatial repartition of the trend of climate variables of the three basins and the boxplot of Z at annual and seasonal scale. Thus, the average annual temperatures globally exhibit a significant increase in the basins of Senegal (T_{max} 0.023 °C yr^{-1} and T_{min} 0.021 °C yr^{-1}) and Gambia (T_{max} 0.021 °C yr^{-1} and T_{min} 0.044 °C yr^{-1}). On the other hand, for the Casamance basin, maximum temperatures decrease by 0.003 °C yr^{-1} and minimum temperatures increase by 0.012 °C yr^{-1}. Relative humidity increases significantly on annual scale and during the dry season for the three hydrosystems. This increase in relative humidity is on average 0.10 % yr^{-1} for the Senegal basin, 0.17 % yr^{-1} for the Gambia and 0.11 % yr^{-1} for the Casamance basin. During the rainy season, a non-significant increase in relative humidity (No change) was noted in the Casamance basin. Solar radiation drops significantly during the rainy season in the Casamance basin.

3.2 Sensitivity of ET_0 to climatic variables

The influence of climatic variables on ET_0 was determined by the sensitivity coefficient. The absolute value of this co-

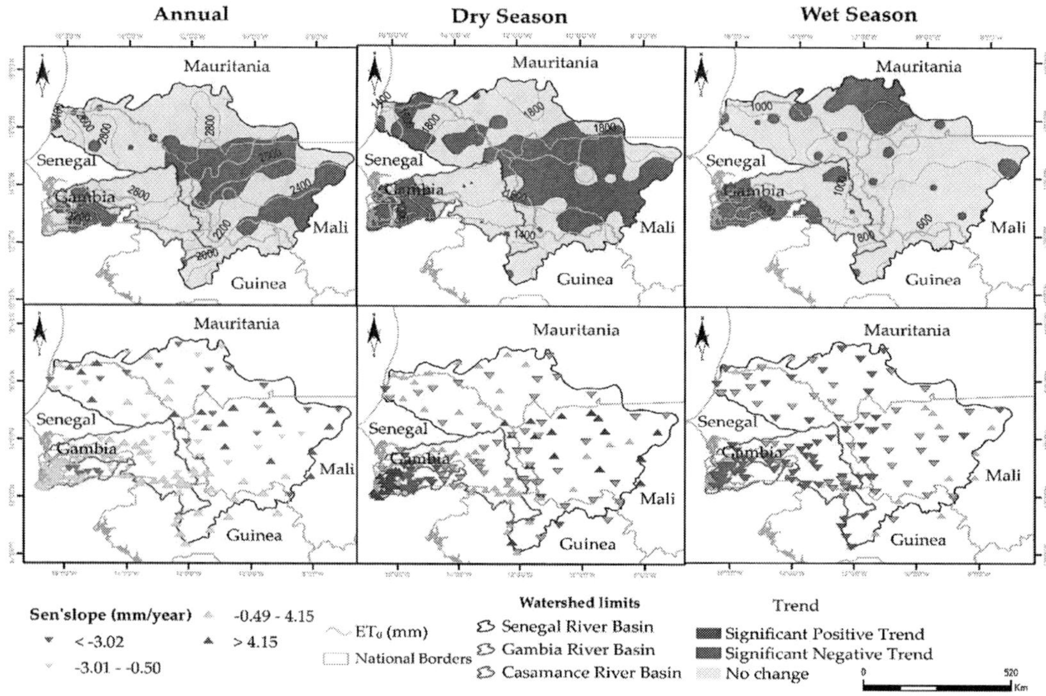

Figure 2. Trend of annual and seasonal reference evapotranspiration over the period 1984–2019.

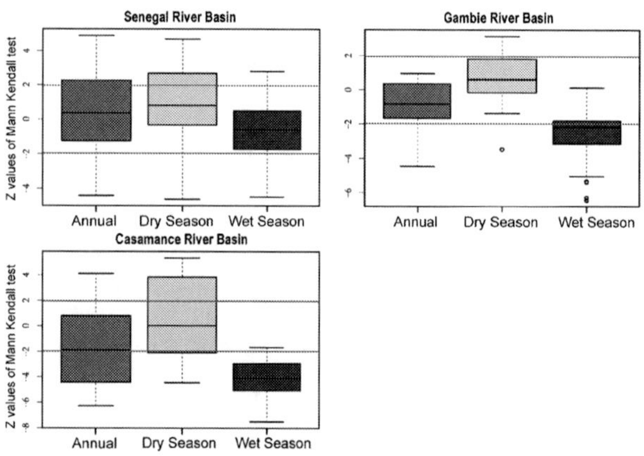

Figure 3. Z values of Mann Kendall test of reference evapotranspiration at annual and seasonal scale over the period 1984–2019 (red line is the significant level, $\alpha = 0.05$).

efficient made it possible to classify the variables according to their degree of influence on ET_0 (Fig. 6). Thus, for the Senegal River basin, evapotranspiration is more sensitive to variation in relative humidity, temperature and solar radiation. The influence of wind speed is minimal compared to the other variables. For the Gambia and Casamance basins, ET_0 is more sensitive to radiation and maximum temperature on annual scale and during the dry season. On the other hand, in the rainy season, solar radiation and relative humidity have more influence on ET_0 than the maximum temperature.

4 Discussion

Overall, on the three hydrosystems of Senegal, the ET_0 varies slightly. The maximum values of ET_0 are observed in the Senegal River basin which extends mainly over the Sahelian and Sudanian zones where the influence of temperatures is more significant. This explains the upward trend in ET_0 on annual scale and during the dry season. On the other hand, a downward trend in ET_0 was noted in the three hydrosystems, especially during the rainy season. This downward trend in evapotranspiration coincides with a significant increase in annual and seasonal temperatures. This highlights the concept of "evaporation paradox" in the three hydrosystems of Senegal. Because for Mekoya (2020) the decrease in ET_0 and the increase in temperature at the same time highlight the existence of evaporation paradox in an area. The decrease in ET_0 can be explained by the increase in relative humidity and the decrease in solar radiation and wind speed. In this study, the decrease in wind speed is observed in the three basins on annual scale and during the dry season. Solar radiation decreases significantly in the Gambia and Casamance basins. These results confirm those of Ndiaye et al. (2020b) who showed that the evaporation paradox is linked to the increase in relative humidity and the decrease in solar radia-

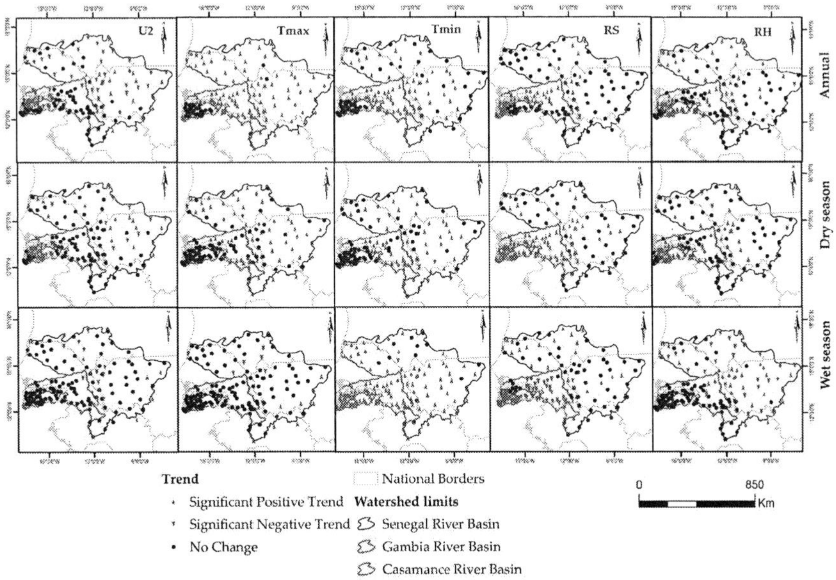

Figure 4. Spatial repartition of the trend of climatic variables on annual and seasonal scale over the period 1984–2019 (RH: relative humidity, T_{max}: maximum temperature, T_{min}: minimum temperature, R_s: Solar radiation and U_2: wind speed).

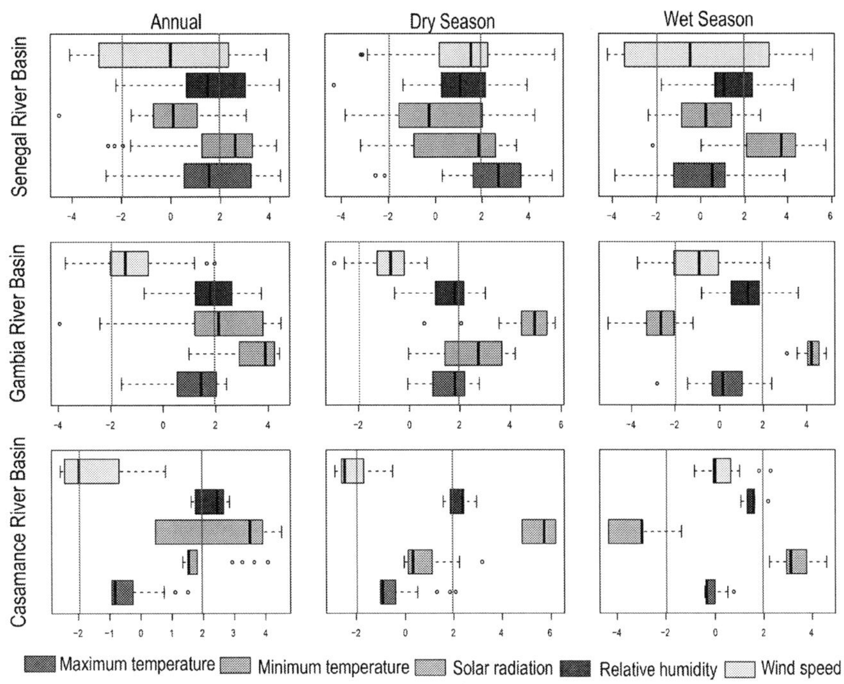

Figure 5. Z values of Mann Kendall test of the climate variables at annual and seasonal scale over the period 1984–2019 (red line is the significant level, $\alpha = 0.05$).

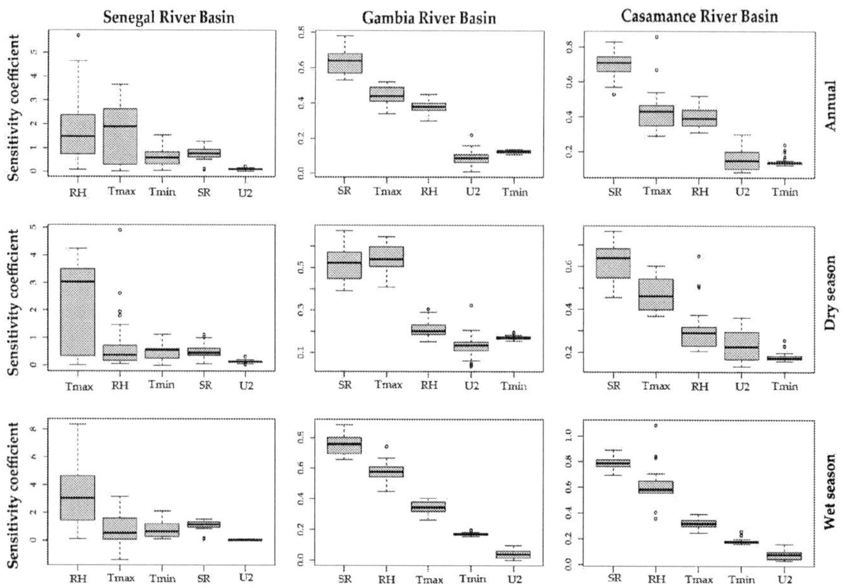

Figure 6. Mean sensitivity coefficient of climatic variables over the period 1984–2019 (RH: relative humidity, T_{max}: maximum temperature, T_{min}: minimum temperature, R_s: Solar radiation and U_2: wind speed).

tion and wind speed. Similar results were obtained by Qin et al. (2021) who noted that the evaporation paradox is caused by the decreasing in radiation and wind speed in the Huaibei region (China). The significant increase in ET_0 in the Sahelian and Sudanian zone can have a negative impact on water resources and vegetation. On the other hand, the downward trend in ET_0 may be an indicator of a return to wetter climatic conditions, which improved water availability in the various rivers.

5 Conclusions

In the three hydrosystems of Senegal, evapotranspiration follows both an upward and downward trend depending on the basins and the periods. The trend of rising temperatures and decreasing ET_0 highlights the evaporation paradox in the three hydrosystems of Senegal. This decrease is explained here by the increase in relative humidity and the decrease in radiation and wind speed. Sensitivity analysis reveals that reference evapotranspiration is more sensitive to relative humidity, maximum temperature and solar radiation. The influence of wind speed is not negligible, especially in the Gambia and Casamance basins. However, the reanalysis data used in this study as well as the temporal depth of the series used (1984–2019) may be a source of uncertainty in the results obtained. For a better understanding of ET_0 trend, it will be important to investigate the trend of rainfall and the change of landcover/land use in the main hydrosystem of Senegal.

Code availability. The software R used in this study is freely available (https://www.r-project.org/, last access: April 2023). The code is not publicly accessible. Please contact the authors if you are interested in this research code.

Data availability. The data of climate variables used in this study are available for public. They come from NASA Langley Research Center (LaRC) POWER Project funded through the NASA Earth Science/Applied Science Program (https://power.larc.nasa.gov/data-access-viewer/, NASA, 2022). Please contact the authors if you are interested in the evapotranspiration data used in this research.

Author contributions. Conceptualization, methodology, software and writing PMN, AB, SBD, review, editing and proofreading LD, AD, AO, and KD. All authors have read and agreed to the published version of the manuscript.

Competing interests. The contact author has declared that none of the authors has any competing interests.

Special issue statement. This article is part of the special issue "IAHS2022 – Hydrological sciences in the Anthropocene: Variability and change across space, time, extremes, and interfaces". It is a result of the XIth Scientific Assembly of the International Association of Hydrological Sciences (IAHS 2022), Montpellier, France, 29 May–3 June 2022.

Acknowledgements. The authors would like to thank ActNao for funding Bassirou Diop's Master's work, the results of which are presented in this document. They are also grateful to SYSTA for according the travel grant to Papa Malick Ndiaye to participate in the 11th Scientific Assembly of the IAHS from 29 May to 5 June 2022 in Montpellier, France.

Review statement. This paper was edited by Christophe Cudennec and reviewed by Jean-Marie Kileshye-Onema and one anonymous referee.

References

Allen, R. G., Walter, I. A., Elliot, R. L., Howell, T. A., Itenfisu, D., Jensen, M. E., and Snyder, R.: The ASCE standardized reference evapotranspiration equation, ASCE and American Society of Civil Engineers, 59 pp., https://xwww.mesonet.org/images/site/ASCE_Evapotranspiration_Formula.pdf (last access: March 2022), 2005.

Bodian, A., Diop, L., Panthou, G., Dacosta, H., Deme, A., Dezetter, A., Ndiaye, P. M., Diouf, I., and Vichel Th.: Recent Trend in Hydroclimatic Conditions in the Senegal River Basin, Water, 12, 436, https://doi.org/10.3390/w12020436, 2020.

Chattopadhyay, N. and Hulme, M.: Evaporation and potential evapotranspiration in India under conditions of recent and future climatic change, Agr. Forest Meteorol., 87, 55–74, 1997.

Dacosta, H.: Précipitations et écoulements sur le bassin de la Casamance. Dakar: ORSTOM, 278 pp. multigr. Th. 3e cycle: Géogr. Phys.: Hydrologie, Université Cheikh Anta Diop: Dakar, 1989/07/01, 1989.

Faye, C.: Caractérisation d'un bassin versant par l'analyse statistique des paramètres morphométriques: cas du bassin versant de la Gambie (bassin continental Guineo-Sénégalais), Revue Marocaine de Géomorphologie, 110–127, 2018.

GIEC: Synthèse du rapport AR6 du GIEC publié le 09/08/2021, 1–10, https://theshiftproject.org/article/climat-synthese-vulgarisee-6eme-rapport-giec/ (last access: March 2022), 2021.

Han, S., Xu, D., and Wang, S.: Decreasing potential evaporation trends in China from 1956 to 2005: Accelerated in regions with significant agricultural influence, Agr. Forest Meteorol., 154–155, 44–56, 2012.

Jiao, L. and Wang, D.: Climate Change, the Evaporation Paradox and Their Effects on Streamflow in Lijiang Watershed, Pol. J. Environ. Stud., 27, 2585–2591, https://doi.org/10.15244/pjoes/81290, 2018.

Kendall, M. G.: Rank Correlation Methods, 4th Edition, Charles Griffin, London, 1975.

Li, Y., Feng, A., Liu, W., Ma, X., and Dong, G.: Variation of Aridity Index and the Role of Climate Variables in the Southwest China, Water, 9, 743, https://doi.org/10.3390/w9100743, 2017.

Mann, H. B.: Non-parametric test against trend, Econometrika, 13, 245–259, 1945.

Mekoya, A.: Evaluation of Evaporation Paradox at Tharandt, Environ. Sci. Nat. Resour., 24, 209–214, 2020.

Mjejra, M.: Etude de l'évapotranspiration dans le bassin versant de la Mejerda (Tunisie): Apport de la télédétection satellitaire et des systèmes d'information géographiques, Thèse de doctorat en Géographie, Université de Rennes 2, Rennes. Disponible en ligne: https://tel.archives-ouvertes.fr/tel01281636/document (last access: February 2022), 2015.

NASA: Data Access Viewer, NASA [data set] https://power.larc.nasa.gov/data-access-viewer/, last access: February 2022.

Ndiaye, P. M., Bodian, A., Diop, L., Deme, A., Dezetter, A., and Djaman, K.: Evaluation and Calibration of Alternative Methods for Estimating Reference Evapotranspiration in the Senegal River Basin, Hydrology, 7, 24, https://doi.org/10.3390/hydrology7020024, 2020a.

Ndiaye, P. M., Bodian, A., Diop, L., Deme, A., Dezetter, A., Djaman, K., and Ogilvie, A.: Trend and Sensitivity Analysis of Reference Evapotranspiration in the Senegal River Basin Using NASA Meteorological Data, Water, 12, 1–19, https://doi.org/10.3390/w12071957, 2020b.

Peterson, T. C., Golubev, V. S., and Groisman, P. Y.: Evaporation losing its strength, Nature, 377, 687–688, 1995.

Qin, M., Zhang, Y., Wan, S., Yue, Y., Cheng, Y., and Zhang, B.: Impact of climate change on "evaporation paradox" in province of Jiangsu in southeastern China, PLOS ONE, 16, 16–26, 2021.

Roderick, M. L. and Farquhar, G. D.: The Cause of Decreased Pan Evaporation over the Past 50 Years, Science, 298, 1410–1411, 2002.

Sen, P. K.: Estimates of the regression coefficient based on Kendall's tau, J. Am. Stat. Assoc., 63, 1379–1389, 1968.

Wilcox, C., Vischela, T., Panthoua, G., Bodian, A., Blancheta, J., Descroix, L., Quantina, G., Casse, C., Tanimoune, B., and Kone, S.: Trends in hydrological extremes in the Senegal and Niger Rivers, J. Hydrol., 566, 531–545, 2018.

Proc. IAHS, 385, 313–317, 2024
https://doi.org/10.5194/piahs-385-313-2024

IAHS2022 – Past and future of open, inclusive, innovative, and society-interfacing approaches

Analysis of rainfed cereal-legume mixture cropping water productivity in Lebna catchment, Cap-Bon, Tunisia

Insaf Mekki[1], **Rim Zitouna-Chebbi**[2], **Salah Benyoussef**[3], **Aya Benabdelghaffar**[1,4], **Mariem Boukari**[2], **Frédéric Jacob**[5], **and Jean Albergel**[5]

[1]INRGREF, Carthage Univ., LR16INRAT05 L.STA, Ariana, Tunisia
[2]INRGREF, Carthage Univ., LR16INRGREF02 LRVENC, Ariana, Tunisia
[3]INRAT, Carthage Univ., LR16INRAT01 L.PAF, Ariana, Tunisia
[4]ESIM, Jendouba Univ., Jendouba, Tunisia
[5]LISAH, Univ Montpellier, INRAE, IRD, Institut Agro, Montpellier, France

Correspondence: Insaf Mekki (insaf.mekki.im@gmail.com)

Received: 16 June 2022 – Revised: 20 April 2023 – Accepted: 21 April 2023 – Published: 19 April 2024

Abstract. Under climate change conditions, optimizing water resources management in rainfed agricultural production systems requires the reasonable choice of crops. In this context, the adoption of crops diversification is promoted to increase the agricultural production and the added value per cubic meter of rain water (green water) used by crops. Contributing, therefore, to increase agricultural production and to preserve soil and water resources. The objective of this study is: (i) to identify mixed crops within agricultural fields and, (ii) to evaluate the biomass production and the water productivity in the Lebna watershed (Cap-Bon, Tunisia) using remote sensing and field measurements. The study area, covering 210 km^2, is characterized by the predominant of cereals, legumes and fodder cropping systems. The experiments allowed the quantification of crop evapotranspiration and the observed biomass production at the agricultural field plots. The use of the sentinel images and the observations at different agricultural fields allowed to produce NDVI maps. The results first confirmed a good correlation between biomass production and NDVI values. The exponential relationships showed a values of R^2 greater than 0.7. The use of sentinel images and GIS allowed to compute water productivity from field to watershed scale. The results revealed a considerable spatial variation in water productivity values for different crops. Compared to a single crop, the cereal-legume mixture cropping improved the water productivity. The maximum value with 9.07 $kg\,m^{-3}$ is observed for the mixture crops. The lowest value (0.12 to 2.40 $kg\,m^{-3}$) was obtained for the cereal crop. These results help to recommend adaptation measures in agricultural production systems to climate change.

Keywords. Biomass production; Water productivity; Sentinel 2

1 Introduction

Rainfed agriculture is mostly the primary source of food in the world; contributing to about 60 % of the total crop production (UNESCO, 2009). In Tunisia, rainfed agriculture occupies between 92 % of the utilized agricultural area and contributes to about 65 % of agricultural production (FAO, 2022). However, rainfed agriculture face climate change pressure that is expected to impact the crop yield and thus affecting the availability of food and interfering in various environmental factors such as water production and use efficiency. Hence, there is a need to explore reasonable choice of cropping system that increases the crop yield, optimize water resources management while ensuring a decent income for farmers in rainfed agriculture. In this context, the adoption of crops diversification is promoted to increase the agricultural production and the added value per cubic meter of rain water (green water) used by crops. Contributing, therefore, to increase agricultural production and to preserve soil and water resources. The objective of this study are: (i) to identify

Published by Copernicus Publications on behalf of the International Association of Hydrological Sciences.

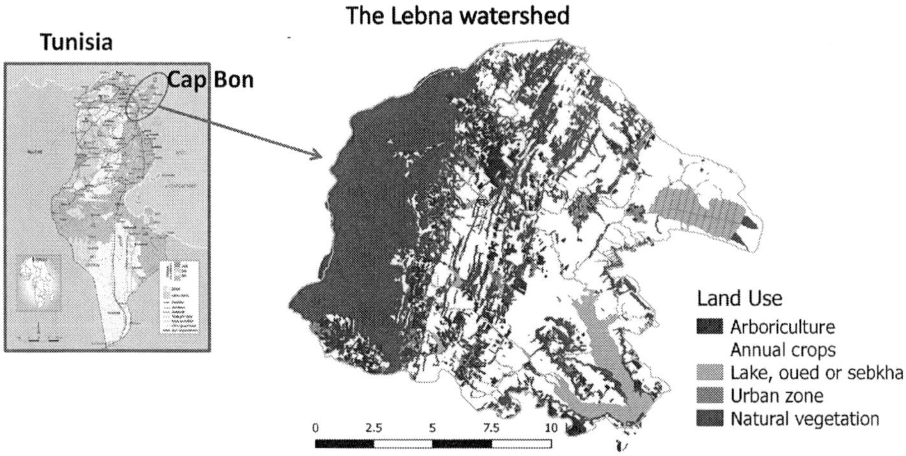

Figure 1. The Lebna watershed study area.

mixed cropping system within agricultural fields and, (ii) to evaluate the biomass production and the water productivity in the Lebna watershed (Cap-Bon, Tunisia) using remote sensing and field measurements.

2 Materials and methods

2.1 Study site

The study area was the Lebna watershed (210 km^2, 36°43′–36°53′ N, 10°40′–10°58′ E) located in the Nabeul Governorate (Fig. 1). The climate regime is at the boundary between subhumid and semiarid conditions (IAO, 2002). From downstream to upstream, the mean annual rainfall and the mean annual evapotranspiration (Penman- Monteith reference crop) range from 450 to 800 mm and from 1000 to 1500 mm, respectively (IAO, 2002; Zitouna-Chebbi et al., 2012). As described in Mekki et al. (2018a), agricultural systems are mainly based on rainfed mixed farming and livestock. Annual crop areas cover 67 % of the study area. Perennial crops (mainly olive trees) cover 8 % of the total area of the study area. Wheat is mainly rotated with legumes or spices to capture the benefits of nitrogen fixation (in the case of legumes) and/or to control pests or weeds. However, forage crops may sometimes replace legume or spice crops in biennial succession. Livestock husbandry includes cattle, sheep and goat breeding. Arable land (annual and perennial crops) accounts for 75 % of the study area, while natural and semi-natural areas account for one-fifth. The Lebna watershed is subject to soil erosion and subsequent reservoir siltation. Mekki et al. (2018b) have shown that evapotranspiration is the predominant factor influencing soil moisture dynamics in this area and that evapotranspiration differs significantly depending on the crops, cropping practices, soil properties and climatic conditions. Consequently, they assumed that it is possible to control the amount of green water and, in part, runoff and downstream water yield by adopting appropriate agricultural practices, including the spatiotemporal distribution of crops.

2.2 Data collection and analysis

The field surveys have been held during 2019/2020 growing season. The use of the sentinel images (Table 1) and the observations at different agricultural fields allowed to produce land use and NDVI maps (Fig. 2). The required georeferenced data were acquired by using existing data and classifying sentinel images. Data on the crops present in the fields during the growing season were obtained through field observations that have been used to train the supervised classification of the sentinel images series. The observed fields were distributed among the existing types of crops (217 fields for wheat, forage crops, 292 fields for spices and legumes, 317 fields for fodder crops among them 41 mixed cropping fields where 17 fields with cereal-legume crops). Five agricultural fields were selected to monitor the aboveground biomass for three types of crops: cereals (oats, barely, wheat), legume (faba bean), cereal-legume mixture (vetch-triticale mixture). Measurements of the aboveground biomass started from early January to mid-March, with around ten days' time step as shown in the Table 1. The spectral data were derived from a time series of multispectral sentinel images (Table 1) from which we determined the Vegetation Index (NDVI) which is the normalized difference of the red and the near infrared bands. The data processing was carried out with R software, we used the random forest library (Liaw and Wiener, 2002) for the classification of Sentinel 2 images and QGIS software for water productivity mapping. The calibration sample consisted of a subset of the observed fields. Six images taken between the end of December to the end of May were used for the supervised classification. The average prediction rate was 86 %. Once the classification is set

Table 1. Dates for ground measurements and Sentinel 2 images acquisition. The sentinel 2 images, scenes are provided by https://sentinels.copernicus.eu/web/sentinel/missions/sentinel-2/data-products (last access: 25 May 2020).

Ground measurements	15 January 2020; 5 February 2020; 13 February 2020; 27 February 2020; 3 March 2020; 10 March 2020
Sentinel-2A	27 November 2019; 7 December 2019; 25 February 2020; 5 May 2020; 25 May 2020
Sentinel-2B	10 February 2020; 1 March 2020

up, we determined the NDVI maps for the catchment and we established the biomass maps using the NDVI-biomass correlations.

The water productivity (WP) estimates the value obtained from each unit volume of water used in the production of a given production (Luan et al., 2018). The WP can be assessed considering the crop production in terms of vegetation biomass and/or grain production. The green water (rainfall) productivity (GWP) can be defined, where only effective rain "R_{Eff}" is consumed during the crop growth period (Luan et al., 2018). GWP ($\mathrm{kg\,m^{-3}}$) can be calculated at the watershed scale.

$$\mathrm{GWPgreen}\left(\mathrm{kg\,m^{-3}}\right) = Y/10 \times R_{\mathrm{Eff}} \tag{1}$$

Where Y is the total crop production and R_{Eff} is the effective rainfall (R) representing the total precipitation infiltrated in the soil for crop production. According to the observed measurements we estimated a runoff coefficient of 20 % and calculated the ($R_{\mathrm{Eff}} = R - 0.2 \cdot R$). The factor 10 is used to convert effective rain water depths (R_{Eff}) in millimeters into water volumes per land surface in $\mathrm{m^3\,ha^{-1}}$.

3 Results

The observed total fresh aboveground biomass between January 2020 and March 2020 for the different land use (cereals, legumes and mixture crops) shows variation within the watershed. The highest value was observed on mixture crops, happened in the end of March and reached a maximum of $34\,\mathrm{t\,ha^{-1}}$ in 2020. Figure 3a, b and c presents the correlation between the biomass production and the NDVI values for fababean, wheat and mixture crop respectively. The results show a good correlation between biomass production and NDVI values. The exponential relationships showed a values of R^2 greater than 0.7 as shown in Table 2.

Figure 4 presents the spatial variation of the GWP values for the different crops within the Lebna watershed. The results revealed a considerable spatial variation in water productivity values for different crops. Compared to a single crop, the cereal-legume mixture cropping seems to improve the water productivity in terms of biomass production. The maximum GWP value with $9.07\,\mathrm{kg\,m^{-3}}$ is observed for the mixture crops. The lowest GWP (0.12 to $2.40\,\mathrm{kg\,m^{-3}}$) was obtained for the cereal crop.

Table 2. The statistical indicators (R^2 and RMSE) of the correlation between the NDVI and the aboveground biomass.

	Cereals	Legume	Cereal–Legume mixture
R^2	0.86	0.72	0.72
RMSE	0.47	1.70	1.66

Figure 4 presents the spatial variation of the GWP values for the different crops within the Lebna watershed. The results revealed a considerable spatial variation in water productivity values for different crops. Compared to a single crop, the cereal-legume mixture cropping seems to improve the water productivity in terms of biomass production. The maximum GWP value with $9.07\,\mathrm{kg\,m^{-3}}$ is observed for the mixture crops. The lowest GWP (0.12 to $2.40\,\mathrm{kg\,m^{-3}}$) was obtained for the cereal crop.

The observed variation across the watershed could be ascribed to the soil properties, the crop management practices during the growing cycle and the crop rotation in relation to the available soil water to the crop use.

4 Conclusions

The green water (rainfall) productivity assessment in Lebna watershed along the 2019/2020 growing season was performed using Sentinel 2 satellite images and the aboveground observed biomass for the cereal-legume, mixture crops. The results first confirmed a good correlation between biomass production and NDVI values. Compared to a single crop, the cereal-legume mixture cropping improved the green water productivity. These results help to recommend measures to increase the adaptation of rainfed agricultural to climate change. Further studies may refine the quantification of local scale WP values. The results are currently to be validated using the high number of field observations and satellite images over several years and dates (three cropping cycles). Also, the measured crop evapotranspiration can be used instead of effective rainfall to calculate a more relevant GWP as shown in the equation above. Ongoing work coupling field experiments, remote sensing and agro-hydrological model aims to develop simulation and evaluate the agri-environmental impacts of the rainfed cereal-legume mixture cropping.

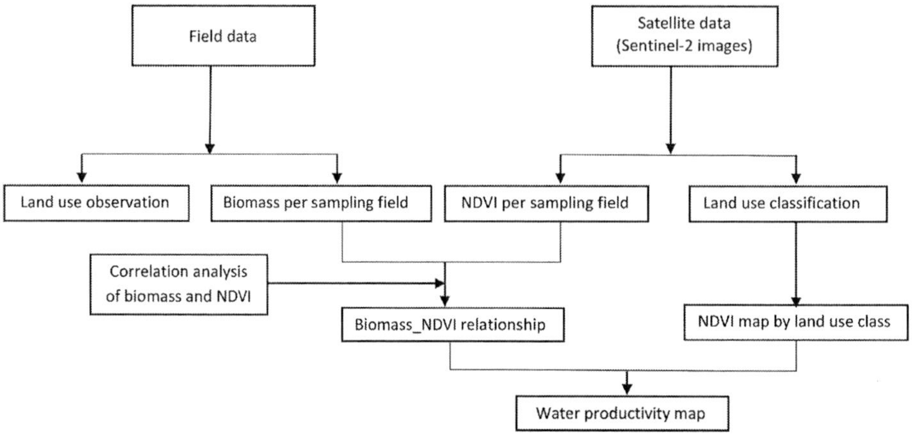

Figure 2. The flowchart of the methodology.

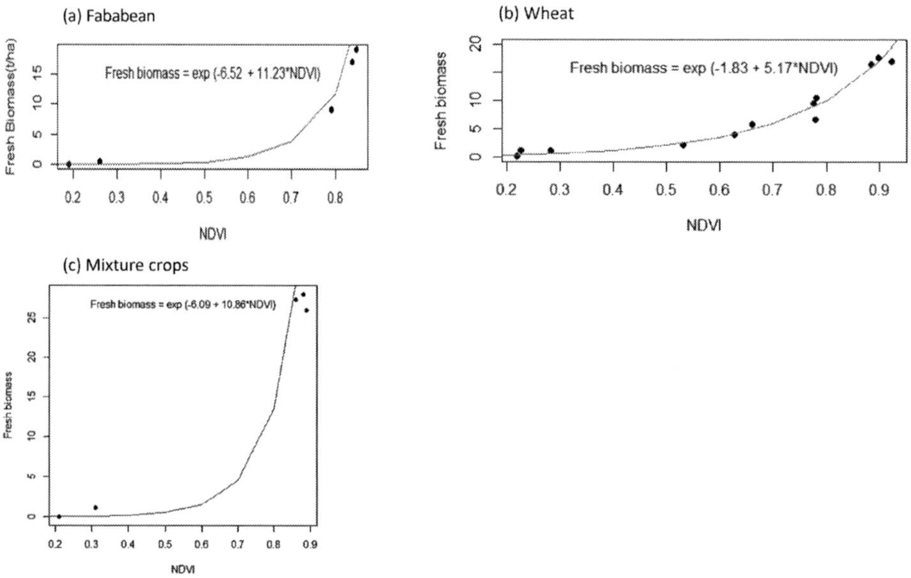

Figure 3. Correlation between biomass production and NDVI values for: **(a)** fababean, **(b)** wheat, and **(c)** mixture crops.

Code availability. The software code is randomForest (https://CRAN.R-project.org/doc/Rnews/, Liaw and Wiener, 2002).

Data availability. The underlying research data is currently not publicly available due to privacy. The sentinel 2 images, scenes are provided by https://sentinels.copernicus.eu/web/sentinel/missions/sentinel-2/data-products (Copernicus Open Access Hub, 2020).

Author contributions. Conceptualization, IM and RZC; investigation, MB and AB; methodology, IM, RZC; SB, MB and AB; writing–original draft, IM and RZC; review-editing, all authors; projects administration and funding acquisition, IM, RZC and FJ. All authors have read and agreed to the published version of the manuscript.

Competing interests. The contact author has declared that none of the authors has any competing interests.

Disclaimer. Publisher's note: Copernicus Publications remains neutral with regard to jurisdictional claims in published maps and institutional affiliations.

Special issue statement. This article is part of the special issue "IAHS2022 – Hydrological sciences in the Anthropocene: Past

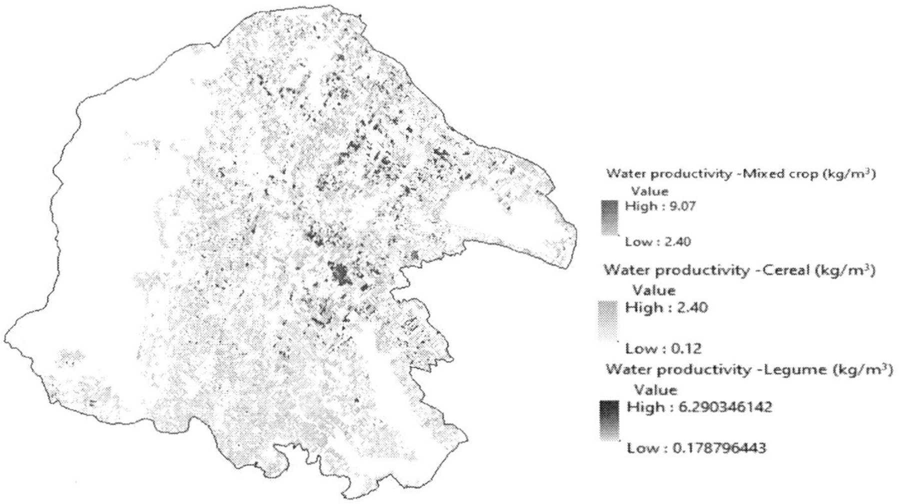

Figure 4. Map of the green water productivity (GWP) for wheat, legume and mixed crops for 2020–2021 for Lebna watershed.

and future of open, inclusive, innovative, and society-interfacing approaches". It is a result of the XIth Scientific Assembly of the International Association of Hydrological Sciences (IAHS 2022), Montpellier, France, 29 May–3 June 2022.

Acknowledgements. The fieldwork was performed within a memorandum of understanding between CRDA Nabeul and INRGREF. We gratefully acknowledge Imed Belouafi (CTV Haouaria), Tarek Ajmi, Rim Louati, Jihen Zrili (INRGREF) for participating in the fieldwork. The cooperation of the farmers in the Lebna region by providing access to their land is highly appreciated.

Financial support. This research was supported by the PRIMA ALTOS project (ANR-18-PRIM-0011-01) and by public funds received from the MESRS. This work was also supported by the H2020-MSCA-RISE-2018 ACCWA project and by funds received from MESRS, IRESA and IRD under the framework of LMI NAILA.

Review statement. This paper was edited by Christophe Cudennec and reviewed by two anonymous referees.

References

Copernicus Open Access Hub: Sentinel-2 archives, https://sentinels.copernicus.eu/web/sentinel/missions/sentinel-2/data-products, last access: 25 May 2020.

FAO: L'agriculture pluviale face aux changements climatiques en Afrique du Nord – Impact et perspective avec l'agroécologie, Tunis, https://doi.org/10.4060/cc0014fr, 2022.

IAO: Land resources of the Oued Lebna catchment (Tunisia), in: 20th Course Professional Master: Remote Sensing and Natural Resources Evaluation, Florence, Italie, Istituto Agronomico per l'Oltremare, p. 132, https://doi.org/10.13140/RG.2.2.10428.77449, 2020.

Luan, X., Wu, P., Sun, S., Wang, Y., and Gao, X.: Quantitative study of the crop production water footprint using the SWAT model, Ecol. Indic., 89, 1–10, 2018.

Mekki, I., Bailly, J. S., Jacob, F., Chebbi, H., Ajmi, T., Blanca, Y., Zairi, A., and Biarnès, A.: Impact of farmland fragmentation on rainfed crop allocation in Mediterranean landscapes: a case study of the Lebna watershed in cap bon, Tunisia, Land Use Policy, 75, 772–783, https://doi.org/10.1016/j.landusepol.2018.04.004, 2018a.

Mekki, I., Zitouna-Chebbi, R., Jacob, F., Ben, Mechlia N., Prévot, L., Albergel, J., and Voltz, M.: Impact of land use on soil water content in a hilly rainfed agrosystem: a case study in the cap bon peninsula in Tunisia, AGROFOR Int. J., 3, 64–75, https://doi.org/10.7251/AGRENG1801064M, 2018b.

Liaw, A. and Wiener, M.: Classification and Regression by randomForest, R News 2, 18–22, [code], https://CRAN.R-project.org/doc/Rnews/, 2002.

UNESCO: United Nations World Water Development Report 3: Water in a Changing World, Paris, UNESCO, and London, Earthscan, ISBN 978-92-3-104095-5, 978-1-84407-840-0, 978-89-8225-791-9, 2009.

Zitouna-Chebbi, R., Prévot, L., Jacob, F., Mougou, R., and Voltz, M.: Assessing the consistency of eddy covariance measurements under conditions of sloping topography within a hilly agricultural catchment, Agr. Forest Meteorol., 164, 123–135, 2012.

Proc. IAHS, 385, 319–326, 2024
https://doi.org/10.5194/piahs-385-319-2024

Performance assessment of daily GR conceptual rainfall-runoff models in the Upper Benue River (Cameroon) using airGR packages

Rodric Mérimé Nonki[1,2,3], **Ernest Amoussou**[3,4], **Raphael Muamba Tshimanga**[5], **Djan'na Koubodana Houteta**[6], **Domiho Japhet Kodja**[3], **Franck Eitel Kemgang Ghomsi**[7,8], and **André Lenouo**[9]

[1]Laboratory for Environmental Modeling and Atmospheric Physics (LEMAP), Department of Physics, Faculty of Sciences, University of Yaoundé 1, P.O. Box 812, Yaoundé, Cameroon
[2]West African Science Service Center on Climate Change and Adapted Land Use (WASCAL), WASCAL-Climate Change and Water Resources, University of Abomey – Calavi, P.O. Box 526, Cotonou, Benin
[3]Laboratory Pierre PAGNEY, Climate, Water, Ecosystem and Development (LACEEDE), University of Abomey – Calavi, P.O. Box 1122, Cotonou, Benin
[4]Department of Geography and Land Management, University of Parakou, P.O. Box 123, Parakou, Benin
[5]Congo Basin Water Resources Research Center (CRREBaC) and Department of Natural Resources Management, University of Kinshasa, Kinshasa, Democratic Republic of the Congo
[6]Climate Change Sciences Research Department, African Institute for Mathematical Sciences, Kigali, Rwanda
[7]Geodesy Laboratory, National Institute of Cartography, Yaoundé, Cameroon
[8]Nansen-Tutu Center for Marine Environmental Research, Department of Oceanography, University of Cape Town, Cape Town, South Africa
[9]Department of Physics, Faculty of Science, University of Douala, P.O. Box 24157, Douala, Cameroon

Correspondence: Rodric Mérimé Nonki (norome31@yahoo.fr)

Received: 15 June 2022 – Revised: 1 May 2023 – Accepted: 21 July 2023 – Published: 19 April 2024

Abstract. Many hydrological applications employ conceptual-lumped rainfall-runoff models to support water resource management techniques. Because of their ability to exploit minimal data and provide enough credible information, they provide an edge in data-scarce domains. The purpose of this research is to evaluate the feasibility of using multiple rainfall-runoff hydrologic models Génie Rural à 4, 5, 6 paramètres Journalier (GR4J, GR5J, and GR6J) in the Upper Benue River (UBR) in Northern Cameroon. The models were calibrated using Michel's calibration algorithm implemented in airGR packages, while considering different objective functions (NSE, KGE, composite criterion). The objective here was to find the optimum objective function that takes into account the various components of the hydrographs in this watershed and to assess the impact of the chosen objective function on parameter optimization. The results reveal that the model performance as well as the optimized parameters of the different GR models vary according to the selected objective function. According to the study, adopting the composite criterion as an objective function during model calibration improves model accuracy. The model's performance reveals that two of the three models (GR5J and GR6J) reproduced the discharge well in the considered catchment, with NSE and KGE greater than 0.62 during the model validation. This highlights how these two GR models can be applied to various water management concerns in the UBR.

Published by Copernicus Publications on behalf of the International Association of Hydrological Sciences.

Keywords. UPH 19; Open Science; Modelling; GR hydrological models, airGR packages

1 Introduction

Nowadays, hydrological models are essential and powerful tools for the efficient management of water resources. They are used to simulate the different hydrological processes occurring at the watershed scale, to prevent extreme events such as floods and droughts, and are now widely used to study the impacts of climate change and land cover on water resources at the global, regional and local scales (Tshimanga and Hughes, 2012; Nonki et al., 2019, 2021b; Lemaitre-Basset et al., 2021). A large number of hydrological models that vary from physical-based to conceptual models are currently available. But the choice between the distributed, semi-distributed and lumped-conceptual is very important for catchment hydrology. Given the greediness of semi-distributed and distributed models in terms of input data, conceptual-lumped rainfall-runoff models are the most commonly used especially in the context of data-scarce regions (Tegegne et al., 2017). They have the advantage that they can use limited data and generate sufficiently reliable information (Nonki et al., 2021c). They are also usually simple and relatively easy to implement.

They are GR (Génie Rural) models in terms of the number of extant and extensively used conceptual-lumped models (Perrin et al., 2009). The GR models were developed in the early 1980s by Claude Michel at Irstea, a French national applied research institute. Nowadays, there are numerous GR models that differ in terms of both time-step (annual, monthly, daily, and hourly) and the number of free model parameters. Because these models are rather parsimonious, end users can easily code them in their modeling environment. However, providing the models in a default implementation environment can make them easier to use and ensure their correct application. Coron et al. (2017) built the airGR packages in this context, which make the GR models available in R. They provided a GR simulation tool that is free, open source and easy to disseminate. Based on their accuracy to simulate discharge around the world and especially in the African catchments, this study aims to assess the feasibility of applying daily GR models in the Upper Benue River for future water resources management.

2 Materials and methods

2.1 Study area and data

The study is conducted in the Upper Benue River (UBR) located in northern Cameroon between latitudes 7 and 11° N, and Longitudes 12 and 16° E (Fig. 1). It is Cameroon's second largest river with a watershed of 64 000 km^2 at Garoua outlet and serves as the main tributary to the Niger River basin. This basin rises in the Adamaoua Plateau and constitutes the most socioeconomic economic impediment of the Northern region because of its potential for sustaining various water-related activities such as irrigated agriculture, hydroelectricity production and navigation. A tropical climate prevails in the basin with two main seasons: the rainy season lasts from May to October and the dry season from November to April. It's a unimodal rainfall region (maximum in August) with annual rainfall ranges between 900 and 1500 mm (Dassou et al., 2016). The prevailing vegetation type in the region is savanna, and elevation is characterized by three main hills: the Adamawa Plateau, and Mounts Alantika and Mandara.

Daily meteorological and hydrological data were used in this study to calibrate and validate the hydrological models. The meteorological data consisted of precipitation obtained from 25 stations and potential evapotranspiration (PET) computed by Penman's formula (Penman, 1948) from solar radiation, relative humidity, wind speed at 2 m, as well as minimum, maximum and average air temperatures from 4 stations located in the basin. These data were provided by the Direction of the National Meteorology of Cameroon (DNM). The daily measured discharge at the Garoua gauging stream station were obtained from the Système d'Information Environnementales sur les Ressources en Eaux en Afrique pour leur Modélisation (SIEREM) database of HydroSciences Montpellier (Boyer et al., 2006; http://www.hydrosciences.org/index.php/2020/08/26/bases-de-donnees/, last access: 1 May 2023).

2.2 Hydrological models description

Three daily time-step GR (Génie Rural) lumped-conceptual rainfall-runoff models were used: the GR model with four daily parameters (GR4J; Perrin et al., 2003), the GR model with five daily parameters (GR5J; Le Moine, 2008), and the GR model with six daily parameters (GR6J; Pushpalatha et al., 2011). To simulate discharge, these models use daily precipitation and PET as inputs. The GR5J is a modified version of the GR4J that includes an additional parameter (X_5). It is a threshold for inter-catchment exchange (–) which can take positive or negative values. While GR6J is an upgraded version of GR5J which incorporates an additional routing store (X_6: exponential store controlling parameter, mm) that allows more flexibility for low-flow modeling (Coron et al., 2017). Figure 2 depicts a schematic view of the various hydrological processes accounted for in the various models, where X_1 is the maximum capacity of the production store (mm), X_2 is the groundwater exchange coefficient (mm), X_3 is the maximum capacity of the routing store (mm) and X_4 is the base time of unit hydrograph (days).

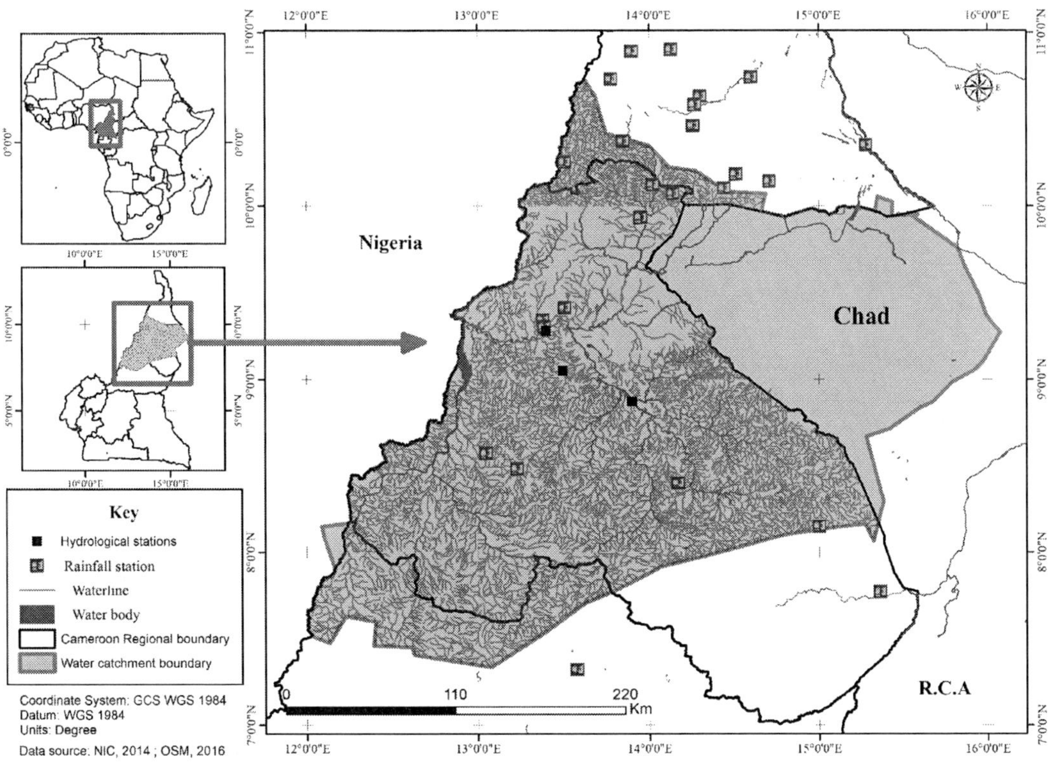

Figure 1. Study catchment: basin drainage area and rainfall and hydrological stations.

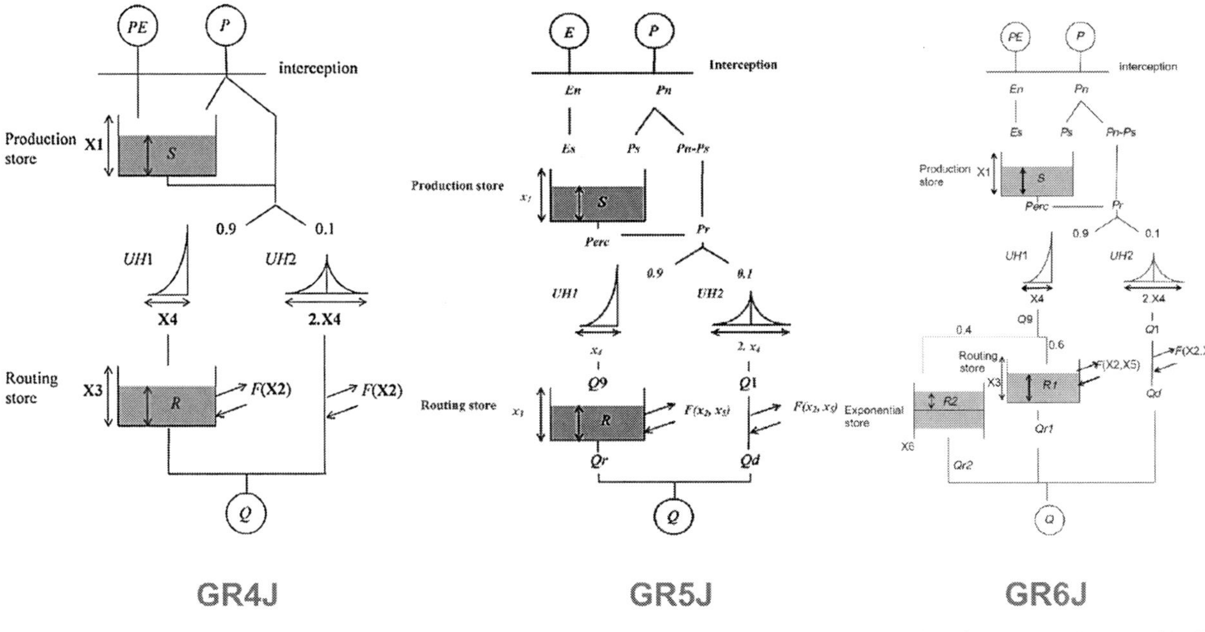

Figure 2. Schematic view of the different hydrological processes accounted for in the GR4J, GR5J and GR6J Conceptual-Lumped hydrological models.

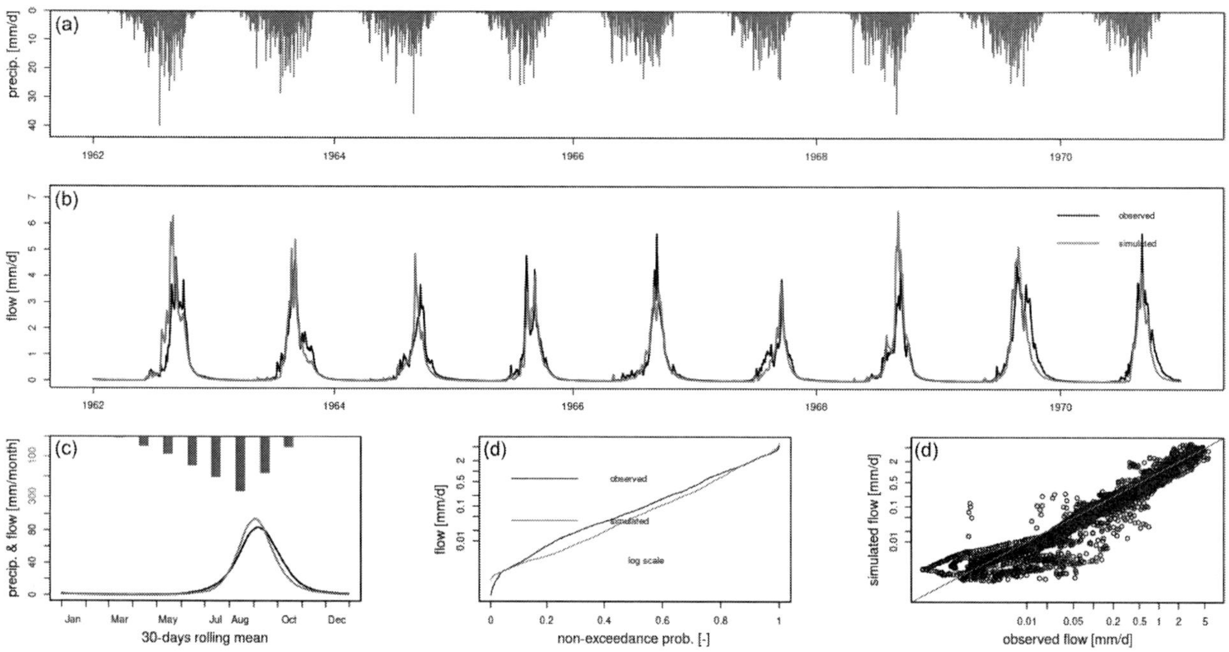

Figure 3. Observed precipitation at the UBR (upper panel), observed vs. simulated daily discharge at the UBR (Garoua hydrometric station) using the GR6J model (middle and lower panels) during the calibration (1961–1970).

2.3 Model calibration, validation and performance assessment

The split – Sample test (Klemeš, 1986) was used for model calibration and validation. Our data time-series were divided into two sub-periods: calibration from 1 January 1961 to 31 December 1970 and validation from 1 January 1971 to 31 December 1980. Calibration/validation was repeated twice, by swapping the two periods (the second half for calibration and first half for validation). By doing this, we can get rid of the dependency of model performance to the simulated period. The model was calibrated using Michel (1991)'s optimization algorithm. These model parameters were calibrated against three different objective function: the Nash and Sutcliffe Efficiency (NSE; Nash and Sutcliffe, 1970), the Kling–Gupta efficiency (KGE; Gupta et al., 2009) and the composite criterion (OF; Lemaitre-Basset et al., 2021) that allowed a fair compromise of both low and high flows. It's the average KGE value, calculated with discharge and inverse of discharge series. The formulation of the two first criteria can be found in Nonki et al. (2021a, c) and the last criterion is given in Eq. (1):

$$\mathrm{OF} = \frac{\mathrm{KGE}(Q) + \mathrm{KGE}(1/Q)}{2} \tag{1}$$

The model's performance was assessed by using both graphical (hydrographs of observed and simulated discharge and flow duration curves) and statistical criteria such as NSE with its different transformations (log and sqrt), KGE, root mean square error (RMSE) and Pearson correlation coefficient (r).

3 Results and analysis

3.1 Impact of selected objective function on optimized parameters and performance

Table 1 gives the optimized model parameters and the performance criteria obtained during the model calibration using different objective functions. The results reveal that the optimum model parameters for the three models differ depending on the criterion used. This result was rather predictable given that we trained the model for different purposes. We also notice that the use of a composite criterion as an objective function during model calibration is highly recommended due to its ability to account both low and high flows components (represented by NSE($\log Q$) and NSE($\sqrt{Q}$) respectively) compared to NSE and KGE in which the models fail to simulate low flows. This result corroborates with those of Garcia et al. (2017) which recommended using OF as an objective function to simulate low-flow indices with continuous conceptual rainfall–runoff models. Despite the fact that increasing model parameters number increase the model uncertainty, we found that increasing the number of model parameters improves model accuracy. This means that the addition of one process to the model improve its performance.

Table 1. Optimized parameters and performance criteria using different objective functions during the calibration period (1961–1970).

Models		GR4J			GR5J			GR6J		
Objective function		NSE	KGE	OF	NSE	KGE	OF	NSE	KGE	OF
Model parameters	X_1	120.76	59.4	162	257.96	175.92	0.03	5.35	72.9	0.01
	X_2	1.15	−4.21	−0.93	−0.12	−0.57	−2.66	−1.14	−0.52	−0.95
	X_3	16.20	268.57	22.75	34.38	130.32	239.20	251.52	101.67	170.59
	X_4	6.80	6.68	2.49	6.67	5.24	6.22	6.59	7.24	6.41
	X_5				1.0	0.367	0.09	0.20	0.22	0.01
	X_6							16.38	14.21	7.30
Performance criteria	NSE(Q)	0.84	0.86	0.36	0.84	0.83	0.82	0.87	0.85	0.80
	NSE($\sqrt{Q}$)	0.80	0.87	0.66	0.79	0.82	0.90	0.85	0.85	0.90
	NSE($\log Q$)	0.21	0.47	0.60	0.17	0.28	0.87	0.36	0.40	0.87
	KGE(Q)	0.87	0.93	0.63	0.89	0.91	0.90	0.86	0.92	0.88
	KGE($1/Q$)	−0.39	−0.31	0.17	−0.44	−0.40	0.45	−0.29	−0.50	0.43
	OF	0.24	0.31	0.40	0.23	0.26	0.67	0.29	0.29	0.65

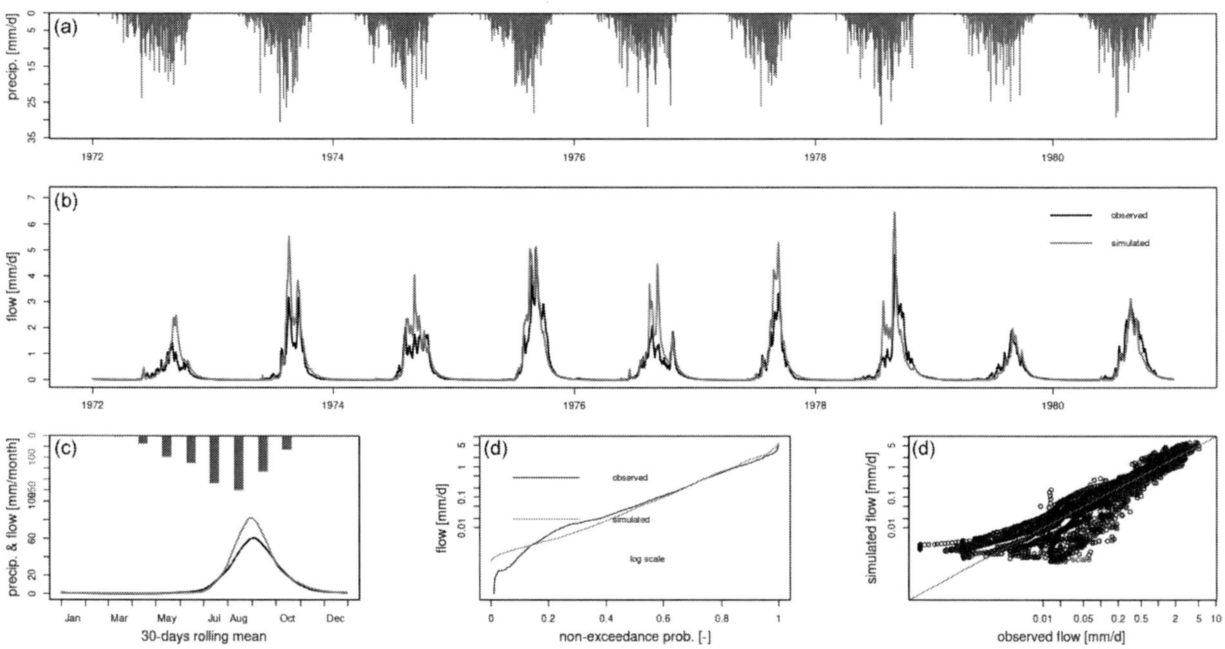

Figure 4. Same as Fig. 3 but during the validation period (1971–1980).

3.2 Models' evaluation results

To evaluate the model performance, we used the optimized model parameters obtained using composite criterion as objective function. Figures 3 and 4 show the comparison between the observed and simulated hydrograph as well as flow duration curves during the calibration and validation using the GR6J model. It's clear that the model captures the timing and magnitude of the discharge well in the UBR. In addition, the model well captured the annual cycle of discharge. However, we also notice that the model fails to simulate the low flows compared to high flows. The same results are achieved with GR5J although they differ slightly with GR4J (Figures not shown).

These findings are reinforced by multi-statistical criteria analysis shown in Table 2. Calibration and validation of the models within the same sub period reveal that the model performed better during model calibration. This result is not surprising as the model is calibrated to better depict the watershed's hydrologic conditions during the calibration period, which are never exactly the same throughout model validation (Merz et al., 2009). In addition, the models outperform during the first half of the calibration period and during the

Table 2. Statistical summary of the model performance of the three daily GR models during the calibration and validation periods at the Benue River (Garoua hydrometric station). The models were calibrated using OF as objective function.

Model		Period	NSE(Q)	NSE($\sqrt{Q}$)	NSE($\log Q$)	KGE(Q)	KGE($1/Q$)	OF	RMSE	R^2
GR4J	Calibration	1961–1970	0.36	0.66	0.60	0.63	0.18	0.40	0.76	0.73
		1971–1980	0.27	0.64	0.40	0.59	−0.03	0.28	0.61	0.73
		Mean	0.32	0.65	0.50	0.61	0.075	0.34	0.685	0.73
	Validation	1961–1970	0.40	0.61	0.24	0.54	−0.89	−0.18	0.74	0.71
		1971–1980	0.12	0.67	0.64	0.55	−0.23	0.16	0.67	0.71
		Mean	0.26	0.64	0.44	0.545	−0.56	−0.01	0.705	0.71
GR5J	Calibration	1961–1970	0.82	0.90	0.87	0.90	0.44	0.67	0.40	0.92
		1971–1980	0.75	0.75	0.36	0.81	0.00	0.40	0.36	0.89
		Mean	0.785	0.825	0.615	0.855	0.22	0.535	0.38	0.905
	Validation	1961–1970	0.79	0.76	0.20	0.72	−0.98	−0.13	0.44	0.90
		1971–1980	0.65	0.86	0.84	0.62	−0.09	0.26	0.43	0.90
		Mean	0.72	0.81	0.52	0.67	−0.535	0.065	0.435	0.90
GR6J	Calibration	1961–1970	0.80	0.90	0.87	0.88	0.43	0.65	0.42	0.91
		1971–1980	0.79	0.88	0.63	0.87	0.05	0.46	0.33	0.91
		Mean	0.795	0.89	0.75	0.875	0.24	0.555	0.375	0.91
	Validation	1961–1970	0.81	0.87	0.67	0.75	−0.89	−0.07	0.42	0.91
		1971–1980	0.66	0.87	0.80	0.65	−0.12	0.27	0.42	0.90
		Mean	0.735	0.87	0.735	0.70	−0.505	0.1	0.42	0.905

second half of the validation period. The GR5J and GR6J models clearly outperformed the GR4J model in the UBR. This can be explained by the fact that Northern Cameroon is a semi-arid region in Cameroon where many rivers dry up a few months after the rainy season, then the groundwater significantly contributes to the stream flow during the dry season. Therefore, the better the groundwater exchange is taken into account in the hydrological model, the better the accuracy is. An objective examination demonstrates that the GR6J model outperforms the other two models. This finding is consistent with earlier research (Flores et al., 2021; Tyralis et al., 2023).

4 Conclusions

The aim of this study was to assess the viability of applying daily GR Conceptual-Lumped models in the Upper Benue River for future water management challenges. We used the airGR packages in R, which give GR models. The Michel's algorithm implemented in these packages was used for model optimization using three different objective criteria, and the model's performance was evaluated using different statistical criteria. The results showed that the optimized parameters vary according to the selected objective function and suggested the use of composite criteria as an objective function during the model calibration. In addition, the GR5J and GR6J models outperformed the GR4J model. This demonstrates that adding one reservoir to the model for groundwater simulation outperforms the model.

Code and data availability. This study was conducted using airGR packages available online (https://CRAN.R-project.org/package=airGR, Coron et al., 2023). Meteorological and hydrological data are the properties of the Direction of the National Meteorology of Cameroon (DNM) and the SIEREM/HSM database respectively. Interested users can contact the corresponding author to have access to the data.

Author contributions. RMN: Funding acquisition, Conceptualization, Methodology, Software, Formal analysis, Validation, Writing – original draft. EA: Funding acquisition, Conceptualization, Software, Writing – original draft. RMT: Conceptualization, Methodology, Validation, Writing – review and editing. DKH: Data curation, Software, Writing – review and editing. DJK: Software, visualization, Writing – review and editing. FEKG: Investigation and editing. AL: Funding acquisition, editing and supervision.

Competing interests. The contact author has declared that none of the authors has any competing interests.

Special issue statement. This article is part of the special issue "IAHS2022 – Hydrological sciences in the Anthropocene: Variability and change across space, time, extremes, and interfaces". It is a result of the XIth Scientific Assembly of the International Association of Hydrological Sciences (IAHS 2022), Montpellier, France, 29 May–3 June 2022.

Acknowledgements. The authors would like to thank the airGR package developers and the data providers. We also thank the IAHS through the SYSTA award for funding three authors to attend this conference. We thank the Associate Editor and the reviewers, whose comments were important in improving the manuscript.

Financial support. This research has been supported by the IAHS (through the SYSTA award) and DAAD within the framework of the climapAfrica programme with funds of the Federal Ministry of Education and Research (grant no. 57610298).

Review statement. This paper was edited by Christophe Cudennec and reviewed by Alban de Lavenne and one anonymous referee.

References

Boyer, J. F., Dieulin, C., Rouche, N., Cres, A., Servat, E., Paturel, J. E., and Mahé, G.: SIEREM: an environmental information system for water resources, 5th World FRIEND Conference, La Havana, Cuba, November 2006, Climate Variability and Change – Hydrological Impacts IAHS Publ., 308, 19–25, 2006.

Coron, L., Thirel, G., Delaigue, O., Perrin, C., and Andréassian, V.: The suite of lumped GR hydrological models in an R package, Environ. Model. Soft., 94, 166–171, https://doi.org/10.1016/j.envsoft.2017.05.002, 2017.

Coron, L., Delaigue, O., Thirel, G., Dorchies, D., Perrin, C., and Michel, C.: airGR: Suite of GR Hydrological Models for Precipitation-Runoff Modelling, R Package Version 1.7.4, CRAN [code], https://CRAN.R-project.org/package=airGR (last access: 1 May 2023), 2023.

Dassou, E., Ombolo, A., Chouto, S., Mboudou, G., Essi, J., and Bineli, E.: Trends and geostatistical interpolation of spatio-temporal variability of precipitation in northern Cameroon, Am. J. Clim. Change, 5, 229–244, https://doi.org/10.4236/ajcc.2016.52020, 2016.

Flores, N., Rodríguez, R., Yépez, S., Osores, V., Rau, P., Rivera, D., and Balocchi, F.: Comparison of Three Daily Rainfall-Runoff Hydrological Models Using Four Evapotranspiration Models in Four Small Forested Watersheds with Different Land Cover in South-Central Chile, Water, 13, 3191, https://doi.org/10.3390/w13223191, 2021.

Garcia, F., Folton, N., and Oudin, L.: Which objective function to calibrate rainfall–runoff models for low-flow index simulations?, Hydrol. Sci. J., 62, 1149–1166, 10.1080/02626667.2017.1308511, 2017.

Gupta, H. V., Kling, H., Yilmaz, K. K., and Martinez, G. F.: Decomposition of the mean squared error and NSE performance criteria: implications for improving hydrological modelling, J. Hydrol., 377, 80–91, https://doi.org/10.1016/j.jhydrol.2009.08.003, 2009.

Klemeš, V.: Operational testing of hydrologic simulation models, Hydrol. Sci. J., 31, 13–24, https://doi.org/10.1080/02626668609491024, 1986.

Lemaitre-Basset, T., Collet, L., Thirel, G., Parajka, J., Evin, G., and Hingray, B.: Climate change impact and uncertainty analysis on hydrological extremes in a French Mediterranean catchment, Hydrol. Sci. J., 66, 888–903, 10.1080/02626667.2021.1895437, 2021.

Le Moine, N.: Le Bassin Versant De Surface Vu Par Le Souterrain: Une Voie d'amélioration Des Performances Et Du Réalisme Des Modèles Pluie-débit?, Dissertation, Université Pierre et Marie Curie, Paris, France, https://api.semanticscholar.org/CorpusID:160882984 (last access: 1 May 2023), 2008.

Merz, R., Parajka, J., and Blöschl, G.: Scale effects in conceptual hydrological modeling, Water Resour. Res., 45, W09405, https://doi.org/10.1029/2009WR007872, 2009.

Michel, C.: Hydrologie appliquée aux petits bassins ruraux, Hydrology handbook, Cemagref, Antony, France, 1991 (in French).

Nash, J. E. and Sutcliffe, J. V.: River flow forecasting through conceptual models. Part I – a discussion of principles, J. Hydrol., 10, 282–290, https://doi.org/10.1016/0022-1694(70)90255-6, 1970.

Nonki, R. M., Lenouo, A., Lennard, C. J., and Tchawoua, C.: Assessing climate change impacts on water resources in the Benue River Basin, Northern Cameroon, Environ. Earth Sci., 78, 606, https://doi.org/10.1007/s12665-019-8614-4, 2019.

Nonki, R. M., Lenouo, A., Lennard, C. J., Tshimanga, R. M., and Tchawoua, C.: Comparison between Dynamic and Static Sensitivity Analysis Approaches for Impact Assessment of Different Potential Evapotranspiration Methods on Hydrological Models' Performance, J. Hydrometeorol., 22, 345 2713–2730, https://doi.org/10.1175/JHM-D-20-0192.1, 2021a.

Nonki, R. M., Lenouo, A., Tchawoua, C., Lennard, C. J., and Amoussou, E.: Impact of climate change on hydropower potential of the Lagdo dam, Benue River Basin, Northern Cameroon, Proc. IAHS, 384, 337–342, https://doi.org/10.5194/piahs-384-337-2021, 2021b.

Nonki, R. M., Lenouo, A., Tshimanga, R. M., Donfack, F. C., and Tchawoua, C.: Performance assessment and uncertainty prediction of a daily time-step HBV-Light rainfall-runoff model for the Upper Benue River Basin, Northern Cameroon, J. Hydrol. Reg. Stud., 36, 100849, https://doi.org/10.1016/j.ejrh.2021.100849, 2021c.

Penman, H. L.: Natural evaporation from open water, bare soil and grass, Proc. Roy. Soc. A-Math. Phys., 193, 120–145, https://doi.org/10.1098/rspa.1948.0037, 1948.

Perrin, C., Michel, C., and Andréassian, V.: Improvement of a parsimonious model for streamflow simulation, J. Hydrol., 279, 275–289, https://doi.org/10.1016/S0022-1694(03)00225-7, 2003.

Perrin, C., Michel, C., and Andreassian, V.: A Set of Hydrological Models (Chapter 16), Environmental Hydraulics, edited by: Tanguy, J. M., 279, ISTE Ltd, John Wiley and Sons, Paris, 493–509. http://eu.wiley.com/WileyCDA/WileyTitle/productCd-1848211546.html (last access: 1 May 2023), 2009.

Pushpalatha, R., Perrin, C., Le Moine, N., Mathevet, T., and Andréassian, V.: A downward structural sensitivity analysis of hydrological models to improve low-flow simulation, J. Hydrol., 411, 66–76, https://doi.org/10.1016/j.jhydrol.2011.09.034, 2011.

Tegegne, G., Park, D. K., and Kim, Y. O.: Comparison of hydrological models for the assessment of water resources in a data-scarce region, the Upper Blue Nile river basin, J. Hydrol. Reg. Stud., 14, 49–66, https://doi.org/10.1016/j.ejrh.2017.10.002, 2017.

Tshimanga, R. M. and Hughes, D. A.: Climate change and impacts on the hydrology of the Congo Basin: the case of the northern sub-basins of the Oubangui and Sangha Rivers, Phys. Chem. Earth A/B/C, 50–52, 72–83, https://doi.org/10.1016/j.pce.2012.08.002, 2012.

Tyralis, H., Papacharalampous, G., and Khatami, S.: Expectile-based hydrological modelling for uncertainty estimation: Life after mean, J. Hydrol., 617, 128986, https://doi.org/10.1016/j.jhydrol.2022.128986, 2023.

Proc. IAHS, 385, 327–332, 2024
https://doi.org/10.5194/piahs-385-327-2024

Open Access
PIAHS

Application of Geochemical and Isotopic Tools to Investigate Water Recharge and Salinization in a Coastal Phreatic Aquifer Suffering Severe Natural and Anthropogenic Constraints: Case of the Mornag Aquifer, NE Tunisia

Mohsen Ben Alaya[1], **Jean-Denis Taupin**[2], **Mohamed Haythem Msaddek**[3], **Inès Ayari**[4], **Nicolas Patris**[2], **Najet Chaabene**[1], **Baya Toumi**[1], **and Fetheddine Melki**[5]

[1]LMU, National Institute of Research and Physical and Chemical Analysis – (INRAP), Technopole, 2020 Sidi Thabet, Ariana, Tunisia
[2]Hydrosciences, UMR 5569 (IRD, CNRS, UM), Montpellier, France
[3]LRME, Tunis-El Manar University, 1060 Tunis, Tunisia
[4]LBSGP, Tunis-El Manar University, 1060 Tunis, Tunisia
[5]L3G, Tunis-El Manar University, 1060 Tunis, Tunisia

Correspondence: Mohsen Ben Alaya (benalaya.mohsen@gmail.com)

Received: 29 May 2022 – Revised: 30 April 2023 – Accepted: 21 July 2023 – Published: 19 April 2024

Abstract. The determination of the origin of salinity in the superficial aquifer of Mornag (NE Tunisia), and the understanding of its hydrological and geochemical behaviours related to severe natural and anthropogenic constraints, were approached by the combined study of chemical elements and stable isotopes (^{2}H and ^{18}O). This study indicates that: (1) the high salinities of the superficial aquifer of Mornag are mainly explained by the dissolution/precipitation processes of evaporite minerals in the aquifer formation, (2) the present-day recharge during rainwater infiltration brings downward a high content of nitrates and other dissolved salts, (3) infiltration of untreated sewage from the main urban areas contaminates the aquifer, (4) two other sources of dissolved salts in groundwater exist, favoured by the intensive exploitation of the phreatic aquifer. The first one is due to mineralised water infiltration from Meliane Wadi where activities, mainly a cement factory, discharge their wastewater. Intrusion of marine saltwater is the second source of salinity caused by aquifer over-exploitation. This hypothesis is supported by the high chloride concentration ($>122\,\mathrm{m\,e.q.\,L^{-1}}$), Br/Cl ratios ($>1.8$ ‰) and the piezometric level lower than sea level. On the other hand, the artificial recharge with low mineralization waters by the Mejerda-Cap Bon Canal and the natural recharge in the valley of wadi El Hma, contribute to a dilution of groundwater. The freshwater/saltwater mixing causes geochemical interactions modifying the water chemistry: cationic exchange, precipitation phenomena. Isotopic tools (^{2}H and ^{18}O) show that the water of this aquifer system has been recently recharged by direct infiltration from the boundary and in the valley of wadis.

Keywords. SGD 6; field observation

1 Introduction

Groundwater resources of the Mediterranean coastal plains and more particularly the phreatic ones in the southern bank of the basin (Middle East and North Africa) show a qualitative and quantitative deterioration developing in time (Edmunds and Droubi, 1987; Custodio and Bruggeman, 1987; Richter and Kreitler, 1993; Vengosh and Rosenthal, 1994; Vengosh et al., 1999). This study concerns he Phreatic coastal Aquifer of Mornag is located in the Northeast of Tunisia (Fig. 1), where a semiarid Mediterranean climate prevails. The dry season is pronounced and this exacerbates the situation, given pumping rates higher usually coincides with the period of least recharge. Indeed, the coastal plain of Mornag hosts a large number (more than 1500) of wells with depths varying from a few meters to nearly 250 m. Some wells are used for drinking water, and little for industry, and around 97 % of the water is used for irrigation. depletion of the groundwater resources and degradation of their quality. The total water extraction in the These exploitation rates, which are required to meet the increasing demands of the current agricultural practices, exceed the natural replenishment of these basins and had led to basin is estimated by the local water authorities as 24.44 million cubic meters per year. On average, groundwater extraction in the basin exceeds recharge by an estimated 7 million cubic meters annually (DGRE, 2017). This over-pumping of the alluvial aquifer has resulted in water level declines ranging from 2.59 to 4.75 m yr^{-1} during the past three decades. The water quality is highly variable and in some areas reaches high salinity levels exceeding 5 g L^{-1}. Evaluation of the mechanisms that cause the degradation of the groundwater quality and water recharge in costal phreatic aquifer in Mornag basin is the focus of this work.

2 Study area

This study concerns the coastal phreatic aquifer of Mornag basin of northeast Tunisia (Fig. 1a). This region has a semi-arid climate, with extreme temperatures and rainfall variations with an average annual temperature and rainfall of 20 °C and 454 mm yr^{-1}, respectively. The age of geological formations cropping out in Mornag basin extends from the Triasic to Quaternary. The catchment area is composed of gypsum, halite, limestone, dolomite, sand, sandstone, clayey sand and conglomerates. A representative West–East cross section of the study area is shown in Fig. 1b. The costal phreatic aquifer is located in the Quaternary layer system formed by sand, clayey sand and conglomerates. The phreatic aquifer is characterized by moderate transmissivity ranging from 1 to 10^{-3} m^2 s^{-1}.

3 Sampling and analytical methods

Twenty-five groundwater samples from Mornag coastal phreatic aquifer were sampled in September–October 2016 from pumping wells (Fig. 1a). Several analyses, water temperature, pH, and electrical conductivity (EC) were carried out on-site. All samples were filtered directly in the field through 0.45 µm membrane filters, stored in high-density polyethylene bottles of 250 mL, and kept at 4 °C. Stable isotope composition of water samples were analysed at the Laboratory LAMA of HydroSciences Montpellier. Major chemistry and trace elements were determined by several methods (Titration, Ion Chromatography and flame photometry inductively at several Laboratorys of National Institute of Research and Physical-Chemical Analysis (INRAP-Tunisia). All samples showed an ionic unbalance smaller than ±5 % (Mandel and Shiftan, 1981).

4 Results and Discussion

4.1 Piezometry

Piezometric surveys of phreatic aquifer performed in September–October 2016 (Fig. 2a):

- The aquifer is recharged by direct infiltration and its higher limits is at Jebel Boukarnine in Southeast, Khlidia region, Jebel Tella and Rades Hill;
- The discharge limits coincide with the Mediterranean shore line,
- The groundwater flow is mainly toward the south-east and north-west to the middle of basin and may be locally disturbed by piezometric depressions (−5 m below sea level) due to the intensive exploitation,

The piezometric gradient varies from 7 ‰ to 12 ‰ in the upstream zone and 3.57 ‰ to 6 ‰ in the downstream zone.

4.2 Chemical Tracing

The measured physical and chemical parameters showed large spatial variations (Table 1). The electrical conductivity varied between 1748 and 11 530 µS cm^{-1} and the salinity between 1.15 and 8.53 g L^{-1} (Fig. 2b). The results of chemical analyses indicated most of the Mornag groundwater samples show similar water types: SO_4^{2-}–Cl^-–Ca^{2+} and Cl^-–SO_4^{2-}–Ca^{2+} (Fig. 3a).

The expanded Durov diagram was used (Fig. 3b) to identify processes and reaction paths such as mixing, ion exchange and dissolution affecting groundwater composition. Box 5, where a part of samples is present, is close to conservative mixing waters. However, the samples belong to Boxes 7 and 8, are characterized by mixing and reverse ion exchange affecting the composition of water sampled. A maximum increase in the salinity should produce water in box 9, where we see a mixture with seawater

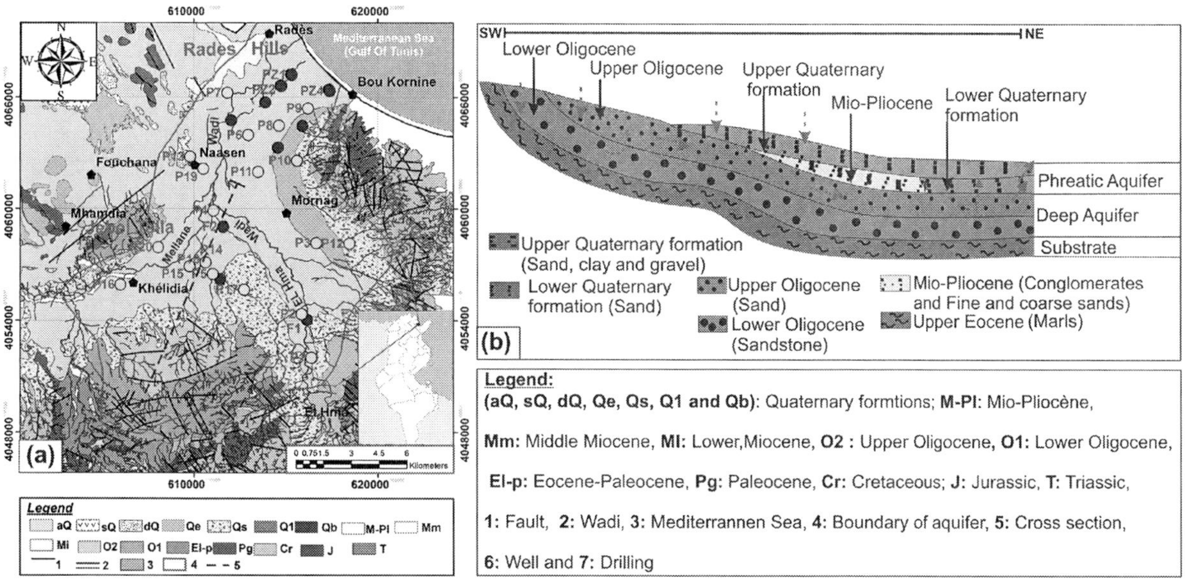

Figure 1. (**a**) Geological map of the Mornag basin (After Jauzein, 1957 and Bujulka et al., 1971); (**b**) A geological cross-section in the studied area.

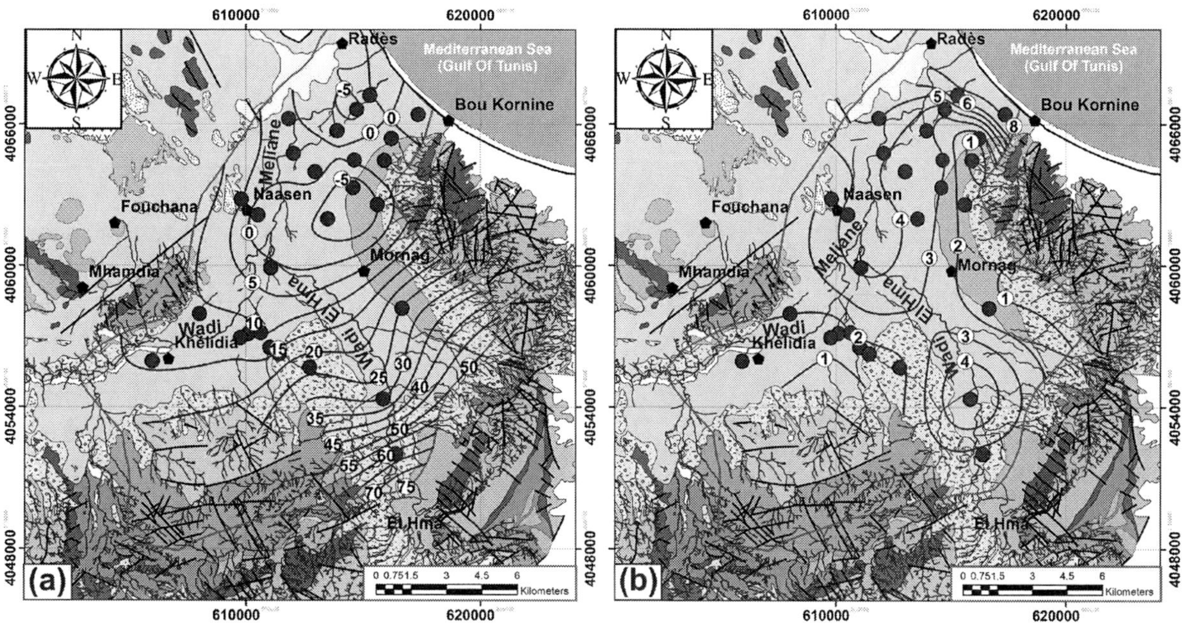

Figure 2. (**a**) Piezometric and (**b**) Salinity maps of coastal phreatic aquifer of Mornag.

Some samples present a depletion in Na^+ content with respect to Cl^- concentration, probably reflecting the cation exchange reactions leading to adsorption of Na^+ on clay minerals belonging to aquifer formations and simultaneous releasing of Ca^{2+} ions (Fig. 4a). On the other hand, the remaining part of the samples shows a more obvious loss of Ca^{2+} with respect to SO_4^{2-}. This may be the result of calcite precipitation controlled by gypsum dissolution which must be to maintain saturation or oversaturation. In fact, bicarbonates formed by CO_2 dissolution is balanced by Na^+ and Ca^{2+} released from clay minerals (Andrews et al., 1994). The cation exchange process was confirmed through the relation characterized by a slope of -1 (Fig. 4b) traced by the position

Table 1. Basic statistics of the various physico-chemical parameters measured.

	Min	Max	Standard deviation
T (°C)	17.40	22.70	3.74
pH	7.65	8.29	0.45
C.E ($mS\,cm^{-1}$)	1.56	11.53	7.05
TDS ($mg\,L^{-1}$)	1.15	8.53	5.22
Na^+	71.12	2042.40	1393.91
K^+	9.08	109.20	70.79
Ca^{2+}	128.00	780.00	461.03
Mg^{2+}	12.00	576.00	398.81
HCO_3^-	183.85	503.74	226.19
SO_4^{2-}	126.31	1672.30	1093.18
Cl^-	114.50	4331.00	2981.52
NO_3^-	2.16	90.75	62.65
Br^-	1.74	12.96	7.93
^2H (‰)	−34.5	−14.7	14.4
^{18}O (‰)	−5.90	−1.96	2.78

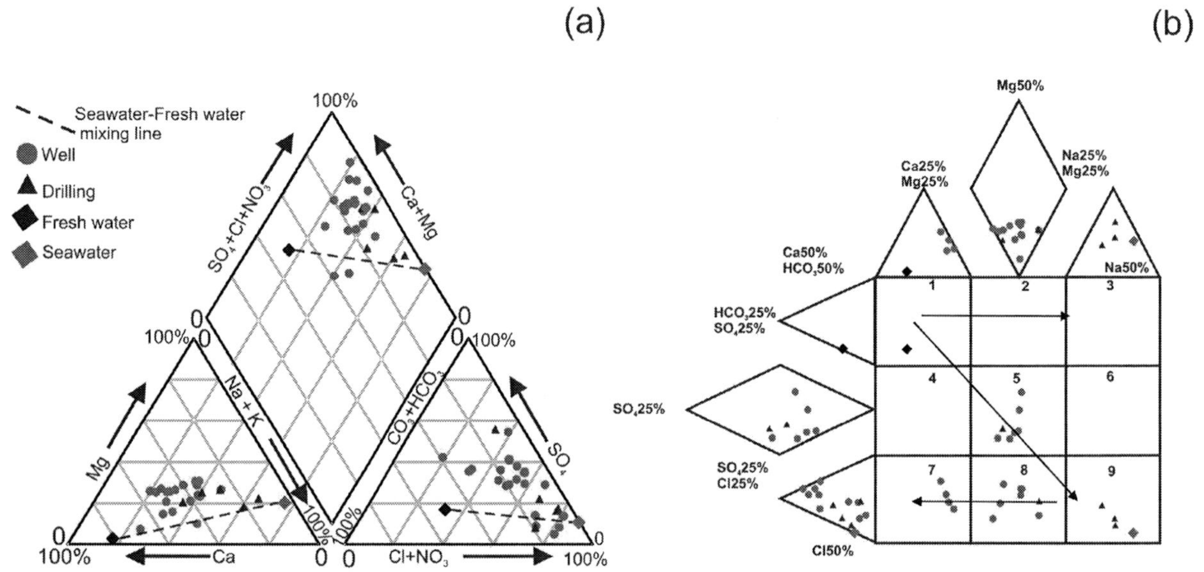

Figure 3. (a): Piper and **(b)**: expended Durov diagrams for the groundwater of the studied area.

of the samples (Garcia et al., 2001). Without exchange, all analytical points should lie close to the origin.

The majority of these waters sampled, are marked by relatively high chloride concentrations and the Br^-/Cl^- ratios varied between 0.45 ‰ and 8.97 ‰. Five wells show a Br^-/Cl^- ratio similar to that of seawater (1.3 ‰ to 2 ‰). Other waters with a high content (>1000 $mg\,L^{-1}$) and Br^-/Cl^- ranging from 0.45 ‰ to 1 ‰ could be linked to leaching of the Triassic salty evaporitic formations, outcropping in southern boundaries of basin. Waters with a relatively low chloride content (<1000 $mg\,L^{-1}$) and Br^-/Cl^- ranging from 2 ‰ and 8.97 ‰, according to the classification proposed by Alcalá and Custodio (2008) or Zabala et al. (2015), a contribution of atmospheric precipitation and agricultural activities on groundwater salinity. However, it can be suggested that the values observed in these wells can only be the consequence of intensive use of fertilizers in this part of the study area. An additional enrichment in Br^-/Cl^- is also found as a result of industrial pollution like this cement factory, discharge their wastewater in Meliane Wadi.

The excessive use of inorganic fertilizers and other substances (pesticides, fungicides, etc.) may introduce anthropogenic toxic elements into the groundwater. In Mornag coastal plain, nitrogen fertilizers are used for the cultiva-

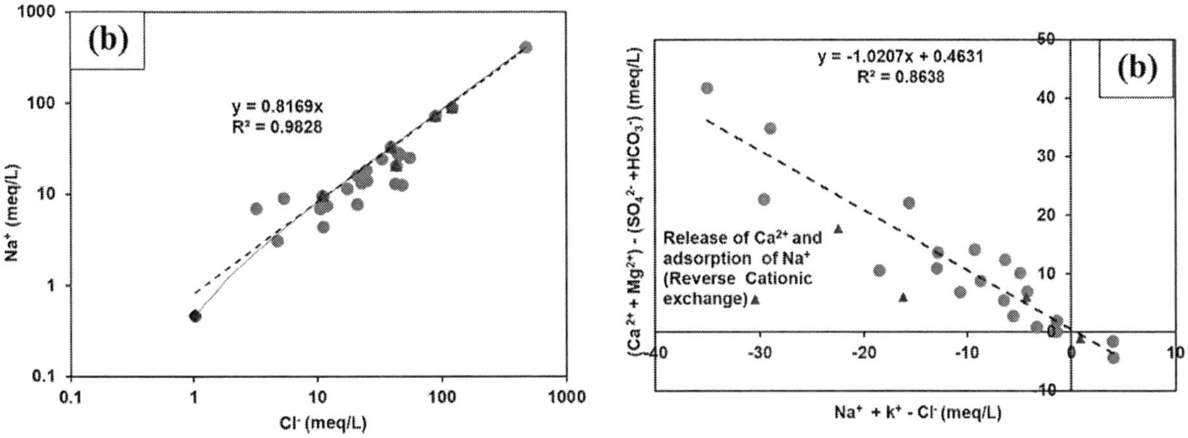

Figure 4. **(a)** Relationships between Cl^-/Na^+ **(a)** and Relationship between $[(Ca^{2+} + Mg^{2+}) - (HCO_3^- + SO_4^{2-})]$ and $(Na^+ + K^+ - Cl^-)$ reflecting the cation exchange processes.

tion of vegetable crops. Among all sampled water from the coastal phreatic aquifer of Mornag, 6 showed concentrations of nitrates ranging from 50.13 to 90.75 $mg\,L^{-1}$. Nitrate pollution affects shallow wells in the same way as deeper ones. This may be explained by a high permeability of the aquifer infiltration zone.

Stable isotopes composition of water samples from all wells collected in the phreatic coastal aquifer of Mornag suggests the presence of recently recharged water (Fig. 5a). Oxygen-18 and deuterium values in this area range, respectively, from −5.90 ‰ to −1.96 ‰ versus V-SMOW and from −34.4 ‰ to −14.7 ‰ versus V-SMOW. The mean isotopic values are, respectively, −5.04 ‰ and −29.3 ‰ versus V-SMOW for $\delta^{18}O$ and δ^2H, close to the values found in regional precipitations ($\delta^{18}O = -4.5$ ‰ and $\delta^2H = -25.1$ ‰) (Celle-Jeanton et al., 2001) in the humid season suggesting the groundwater was linked to current rainfall origin. The highest values come from the upstream part of the plain, near Hma dam; in this zone, the variability is high. Elsewhere, the isotopic composition of the aquifer is more homogeneous.

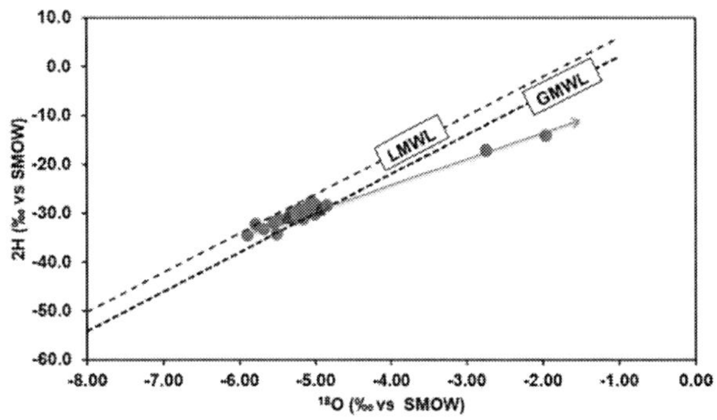

Figure 5. Stable isotope composition of coastal phreatic aquifer of Mornag (LMWL: Local Meteoric Water Line and GMWL: Global Meteoric Water Line).

5 Conclusion

A combination of piezometric data, major elements geochemistry, trace elements and stable isotopes has provided a preliminary comprehensive understanding of the hydrodynamic mode of coastal phreatic aquifer Mornag and has allowed to highlight mechanisms linked to observed mineralization and reduced water quality. The following conclusions can be drawn: (1) salinization processes of groundwater are due to washing away of the evaporate levels in aquifer formation and contamination from the surface by chlorides and nitrates, and (2) isotopic tracing (^{18}O and ^2H) provide a mean for understanding the specification and location of the groundwater recharge. The phreatic aquifer is very vulnerable and only the application of protective measures will allow a better management of this resource for future generations. This study proposes a conceptual model of flow circulation allowing in a second close step to present a numerical model which will take account possible scenarios of groundwater evolution.

Code availability. The code supporting the findings of this study is available from the corresponding author on request.

Data availability. All geochemistry data are not dealt with and Table 1 gives the mean of various physico-chemical parameters measured, the chemical data will be the subject of another paper.

Author contributions. MBA carried out the investigations and interpretations of this study in collaboration with scientific supervision of JDT and NP performed the stable isotope analyses. MBA prepared the manuscript with contributions from all co-authors.

Competing interests. The contact author has declared that none of the authors has any competing interests.

Disclaimer. Publisher's note: Copernicus Publications remains neutral with regard to jurisdictional claims in published maps and institutional affiliations.

Special issue statement. This article is part of the special issue "IAHS2022 – Hydrological sciences in the Anthropocene: Variability and change across space, time, extremes, and interfaces". It is a result of the XIth Scientific Assembly of the International Association of Hydrological Sciences (IAHS 2022), Montpellier, France, 29 May–3 June 2022.

Acknowledgements. The authors gratefully acknowledge the contributions of the staff members of Ben Arous Water Resources Division/Agriculture Ministry, for their help during field work. We also thank the technical staff of the Laboratory LAMA of Hydro-Sciences Montpellier and Laboratory of National Institute for Research and Physico-Chemical Analysis (INRAP) for their help and assistance during laboratory analyses.

Review statement. This paper was edited by Christophe Cudennec and reviewed by two anonymous referees.

References

Alcalá, F. J. and Custodio, E.: Using the Cl/Br ratio as a tracer to identify the origin of salinity in aquifers in Spain and Portugal, J. Hydrol., 359, 189–207, https://doi.org/10.1016/j.jhydrol.2008.06.028, 2008.

Andrews, J. N., Fontes, J. C., Aranyossy, J. F., Dodo, A., Edmunds, W. M., Joseph, A., and Travi, Y.: The evolution of alkaline aquifer of the Irhazer Plain, Niger, Water Resour. Res., 30, 45–61, https://doi.org/10.1029/93WR02226, 1994.

Bujalka, P., Johan, K., Krivy, M., Rakus, M., and Vacek, J.: Carte géologique de Grombalia à l'échelle 1 : 50 000, Service Géologique de Tunisie, 1971.

Celle-Jeanton, H., Zouari, K., Travi, Y., and Daoud, A.: Caractérisation isotopique des pluies en Tunisie, Essai de typologie dans la région de Sfax, C. R. Acad. Sci. Paris, 333(IIA), 625–631, https://doi.org/10.1016/S1251-8050(01)01671-8, 2001.

Custodio, E. and Bruggeman, G. A.: Groundwater problems in coastal areas, UNESCO, Studies and Reports in Hydrology, UNESCO Press, París, 45, 576 pp., ISBN 978-92-3-102415-3, 92-3-102415-9, 1987.

Direction Générale des Ressources en Eau (DGRE): Annuaire d'exploitation des nappes en Tunisie, Direction Générale des Ressources en Eau (DGRE), 2017.

Edmunds, W. M. and Droubi, A.: Groundwater salinity and environmentalchange, in: International Symposium on Isotope Techniques in the Study of Environmental Change, Keynote Paper, STI/PUB/1024, IAEA, Vienna, Austria, 503–518, ISSN: 0074-1884, 1997, 1987.

Garcia, M. G., Del Hidalgo, M., and Blesa, M. A.: Geochemistry of groundwater in the alluvial plain of Tucumàn province Argentina, Hydrogeol. J., 9, 597–610, https://doi.org/10.1007/s10040-001-0166-4, 2001.

Jauzein, A.: Carte géologique de Bir M'Cherga à l'échelle 1 : 50 000, Service Géologique de Tunisie, 1957.

Mandel, S. and Shiftan, Z. L.: Groundwater resources investigation and development, Academic Press Inc., New York, 274 pp., ISBN 9780323157827, 1981.

Richter, B. C. and Kreitler, C. W.: Geochemical Techniques for Identifying Sources of Groundwater Salinization, CRC Press, Boca Raton, FL, USA, 272 pp., ISBN 9781566700009, 1993.

Vengosh, A. and Rosenthal, A.: Saline groundwater in Israel: its bearing on the water crisis in the country, J. Hydrol., 156, 389–430, https://doi.org/10.1016/0022-1694(94)90087-6, 1994.

Vengosh, A., Spivack, A. J., Artzi, Y., and Ayalon, A.: Boron, strontium and oxygen isotopic and geochemical constraints for the origin of the salinity in groundwater from the Mediterranean Coast of Israel, Water Resour. Res, 35, 1877–1894, https://doi.org/10.1029/1999WR900024, 1999.

Zabala, M. E., Manzano, M., and Vives, L.: The origin of groundwater composition in the pampeano aquifer underlying the Del Azul Creek basin, Argentina, Sci. Total Environ., 518–519, 168–188, https://doi.org/10.1016/j.scitotenv.2015.02.065, 2015.

Proc. IAHS, 385, 333–337, 2024
https://doi.org/10.5194/piahs-385-333-2024

Open Access
PIAHS

Runoff uncertainty associated with global climate model chosen in regional climate modeling

Salah Basem Ajjur[1,2] **and Sami G. Al-Ghamdi**[1,3,4]

[1]Division of Sustainable Development, College of Science and Engineering, Hamad Bin Khalifa University, Qatar Foundation, Doha, Qatar
[2]Department of Earth, Environmental & Planetary Sciences, Brown University, Providence, RI, USA
[3]Environmental Science and Engineering Program, Biological and Environmental Science and Engineering Division, King Abdullah University of Science and Technology (KAUST), Thuwal 23955-6900, Saudi Arabia
[4]KAUST Climate and Livability Initiative, King Abdullah University of Science and Technology (KAUST), Thuwal 23955-6900, Saudi Arabia

Correspondence: Sami G. Al-Ghamdi (sami.alghamdi@kaust.edu.sa)

Received: 21 May 2022 – Revised: 7 May 2023 – Accepted: 21 July 2023 – Published: 19 April 2024

Abstract. Uncertain climatic projections result in uncertain runoff predictions, which lead to unreliable water resources management and floods mitigation measures. This study investigates runoff uncertainty associated with Global Climate Models (GCMs) chosen as boundary conditions in regional climate models (RCMs). To this end, the study projected surface air temperature, precipitation, wind speed, and potential evapotranspiration, over Qatar, from the Massachusetts Institute of Technology (MIT) RCM driven by three GCMs: CCSM4, MPI-ESM, and NorESM. Historical projections were made during 1976–2005 while future projections were made during 2071–2100 under the business-as-usual Representative Concentration Pathways (RCP8.5). The study then inputs climatic parameters in addition to topographic and groundwater data, in a physically based water balance model to compare runoff simulations. During historical and future periods, climatic projections differed among the RCM runs, especially precipitation and potential evapotranspiration. Uncertainty in climatic projections caused significant uncertainty in runoff estimations. Runoff estimations varied from 12.2 to 45 $\mathrm{Mm^3\,yr^{-1}}$, with an average value of 24.4 $\mathrm{Mm^3\,yr^{-1}}$ during 1976–2005, and from 12.3 to 52.4 $\mathrm{Mm^3\,yr^{-1}}$, with an average value of 32.4 $\mathrm{Mm^3\,yr^{-1}}$ during 2071–2100. All RCM runs agreed on future runoff increases. Since runoff is a main cause of floods and alleviating flood risk can be hindered if climatic projections are incompetent, the study's findings emphasize on narrowing the uncertainty in GCM projections. The study also encourages Qatar authorities to implement managed aquifer recharge projects like rainwater harvesting to alleviate floods risk.

Keywords. Floods; modeling uncertainty; climate models; RCMs; Qatar

1 Introduction

Accurate estimation of the hydrologic cycle components is essential for water resources management and climate change mitigation and adaptation strategies (Pozdniakov et al., 2020; Xie et al., 2018; Anibas et al., 2016). Previous literature, therefore, tried to identify and narrow the uncertainty while estimating hydrologic cycle components. Among these components, runoff uncertainty is a big concern. Surface runoff cannot be measured on a large scale. Overestimate runoff may lead to significant losses in water requirements, whereas underestimated runoff can lead to severe damaging floods (Xie et al., 2018; Moiwo et al., 2010; Ahmed et al., 2018). Therefore, significant efforts have been devoted to exploring runoff uncertainty and enabling getting reliable estimations. The focus was on surface parameters and groundwater inputs (Xie et al., 2018; Martos-Rosillo et al., 2015). Though, a key source of runoff uncertainty might be climatic inputs. Accordingly, this study quantifies and compares runoff simulations obtained from implementing different climatic inputs but same topographic and groundwater conditions in a constructed calibrated water balance model (i.e., WetSpass) in Qatar. WetSpass is a common, reliable physically based model that utilizes the water balance concept at grid points to simulate surface runoff, groundwater recharge, and actual evapotranspiration (Ajjur and Mogheir, 2012; Mogheir and Ajjur, 2013; Batelaan and De Smedt, 2007; Moiwo et al., 2010). WetSpass relates runoff to precipitation amount and intensity, interception fraction, and soil infiltration capacity. The interception fraction varies based on the land cover and represents a constant ratio of annual precipitation.

This study obtained Qatar's topographic and groundwater data from the Ministry of Municipality and Environment. The study uses the Massachusetts Institute of Technology (MIT) Regional Climate Model (RCM) for climatic inputs. The MIT RCM has proven its reliability over southwest Asia at a 25-km spatial resolution (Pal and Eltahir, 2016). It uses three Global Climate Models (GCMs) as atmospheric boundary conditions. Table 1 tabulates information on these GCMs and their horizontal resolutions. The three GCMs were selected based on objective criteria of their representation of regional water bodies (read further details on these criteria in Pal and Eltahir, 2016). For the purpose of this study, the spatial resolution of the MIT RCM has been improved, over Qatar, from 25 to 12.5 km, and the MIT RCM has been run under the three GCMs boundary conditions, producing three runs' outputs during historical and future climates.

Case study description

Qatar is a country in the eastern Arabian Peninsula, encompassing an area of 11 650 km^2. The elevation changes from zero at coasts to up to 92 m above mean sea level in the southwestern parts (Ajjur et al., 2021). There is has no surface water, and potable and industrial water needs are filled through seawater desalination (Ajjur and Baalousha, 2021; Ajjur and Al-Ghamdi, 2022). The climate is hyper-arid, with an average long-term maximum temperature of 35 °C and a long-term annual rainfall of 77 mm. Qatar, and the Arabian Peninsula in general, aridity and warming trend are expected to increase through the 21st century (Ajjur and Al-Ghamdi, 2021a, b, c; Al-Humaiqani and Al-Ghamdi, 2022; Serdar et al., 2022), with a possible link to global teleconnection patterns (Ajjur and Al-Ghamdi, 2021d; Ajjur and Riffi, 2020).

There are three main aquifers in Qatar distributed in the northern, southern, and southwest areas. Among these aquifers, the northern aquifer gets higher recharge from rainfall, and hence has better quality. During the last three decades, Qatar aquifers have been recently overexploited due to population growth, urbanization, and climate change impacts, which intensified some managed aquifer recharge initiatives to improve aquifers' quality and storage (Ajjur and Baalousha, 2021; Ajjur and Mogheir, 2019).

2 Data and Methods

The study methodology includes three steps to investigate runoff uncertainty associated with driven GCM chosen in RCM. First, historical and future climatic parameters were projected from the MIT RCMs. The projected parameters included average air temperature, precipitation, wind speed, and potential evapotranspiration. The historical period was considered 1976–2005, while future period was regarded as the end-21st century (2071–2100) under the business-as-usual greenhouse gas emission Representative Concentration Pathways (RCP8.5). The RCP8.5 denotes 8.5 Watt per m^2 of radiative forcing in 2100 relative to the pre-industrial period (1850–1900). Mean monthly values of temperature and wind speed were computed to get seasonal averages, while precipitation and potential evapotranspiration data were aggregated to get seasonal sum during both historical and future periods. Summer was represented between April and September, whereas the months between October and March represented winter. Second, topographic, groundwater, slope, and land use maps were prepared, and, together with climatic maps, all components were applied in WetSpass. Land use and soil type maps were connected with WetSpass as attribute tables. WetSpass analysis was performed for summer and winter. Seasonal runoffs were then summed to obtain yearly rates. Third, runoff volumes were approximated using GIS spatial analyst tools, and runoff direction was investigated in each run.

3 Results and Discussions

This section first describes how climatic parameters, projected from the three RCM runs, varied during the historical and future periods. It then shows how annual runoff simulations differed according to the variations in these climatic inputs.

3.1 Historical and future climate in Qatar

Figure 1 illustrates the differences among the three RCMs when projecting the mean climate over Qatar during the historical and future periods. In both periods, RCM runs projected similar temperature and wind speed values but different precipitation and potential evapotranspiration. For instance, mean annual precipitation varied among RCM runs from 26 to 64 mm (24 to 58 mm) during the historical (future) period. Similarly, annual potential evapotranspiration varied between 91 and 189 mm during the historical period and between 125 and 276 mm during the future period. These results agree with previous literature (Ajjur and Al-Ghamdi, 2021e, f) and can be attributed to two reasons: (1) there is a poor representation of rainfall, and potential evapotranspiration in GCMs over the Arabian Peninsula (Ajjur and Al-Ghamdi, 2021a), and (2) uncertainty in temperature and pre-

Table 1. Information on the three GCMs used as the boundary condition for the MIT RCM.

RCM runs	Diving GCM model	GCM Institution	Spatial resolution (lat by lon)	
			Atmospheric	Oceanic
1	Community Climate System Model V4 (CCSM4)	National Center for Atmospheric Research, USA	0.9° × 1.25°	1.11° × 0.27°–0.54°
2	Max-Planck Institute Earth System Model (MPI-ESM)	Max Planck Institute for Meteorology, Germany	T63/1.875°	0.4° × 0.4°
3	Norwegian community Earth System Model (NorESM)	Norwegian Climate Centre, Norway	1.875° × 2.5°	1° × 1°

cipitation projections led to potential evapotranspiration uncertainty.

The RCMs agreed on temperature increases of 3 to 3.6 °C and potential evapotranspiration increases of 33 to 86 $mm^3 yr^{-1}$, under RCP8.5, relative to the historical period. However, RCMs disagree on the future trend of precipitation (compare blue boxes in Fig. 1a and b). While CCSM4 projected a small increase in precipitation, both RCMs (MPI-ESM and NorESM) projected precipitation decreases. The future evolution of wind speed was negligible in all RCM runs. On average, the MIT RCM projected an overall increase of 3.28 °C in average air temperature and an increase of 64 mm in annual potential evapotranspiration under RCP8.5, relative to the historical period.

3.2 Runoff simulation

The previous section showed substantial uncertainty among RCM runs when projecting Qatar's climate during historical and future periods. It is still unclear whether this uncertainty could lead to uncertainty in runoff simulations. To investigate this, WetSpass analysis was performed, and runoff simulations obtained from the RCM runs were compared during the historical and future periods (Fig. 2). Runoff values varied up to 22 $mm^3 yr^{-1}$ in all RCM runs, except the MPI-ESM where maximum runoff hit 63 $mm^3 yr^{-1}$ in the historical period and 72 $mm^3 yr^{-1}$ in the future period (see eastern parts in Fig. 2b and e). Worth notably that the MPI-ESM projected higher precipitation than other runs, producing larger values of runoff. Some spatial uniformity is found between runoff simulations. Urban areas in the metropolitan Doha, represented by the eastern coastal part, received higher runoff since these areas are mainly composed of impervious surfaces. Impervious surfaces hinder groundwater recharge, forcing excess precipitation to flow at the land surface (Batelaan and De Smedt, 2007; Moiwo et al., 2010).

To quantify the variations in runoff simulations, spatial analysis tools were used to compute runoff volumes in all

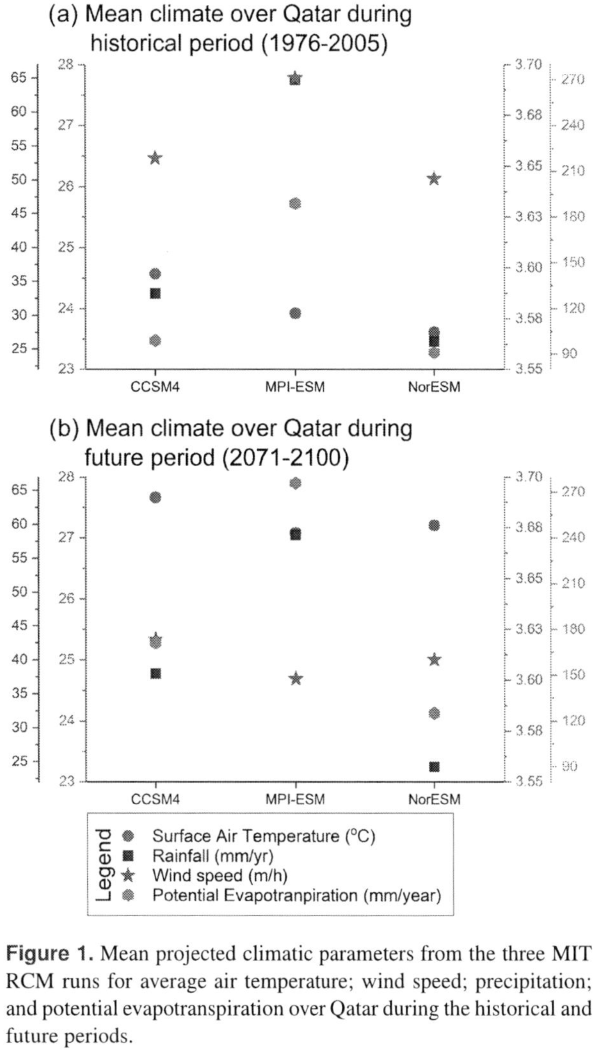

Figure 1. Mean projected climatic parameters from the three MIT RCM runs for average air temperature; wind speed; precipitation; and potential evapotranspiration over Qatar during the historical and future periods.

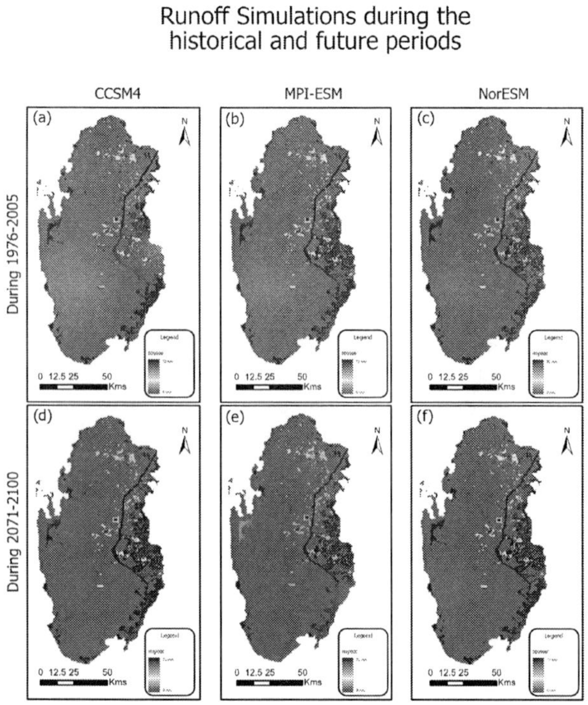

Figure 2. Annual runoff simulations during the historical **(a–c)** and end-21st century **(d–f)** periods simulated using RCM runs inputs. Future simulations are under RCP8.5.

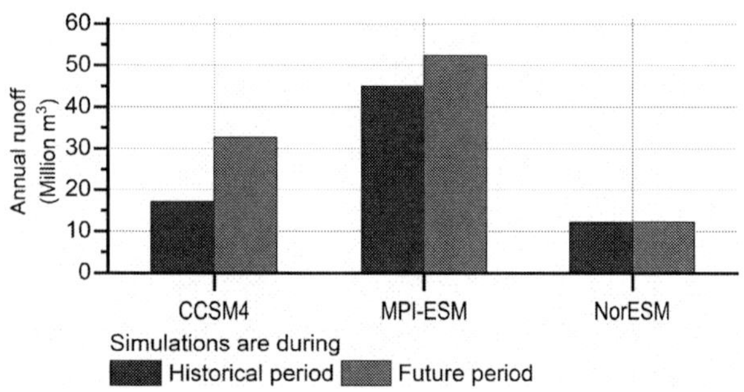

Figure 3. Annual runoff volumes, in Million m^3, simulated using RCM runs inputs during the historical period (1976–2005) and the end-21st century (2071–2100) under RCP8.5.

runs. Figure 3 depicts Qatar runoff volumes simulated by WetSpass using RCM runs inputs. The simulated runoff ranged from 12.2 to 45 Mm^3 yr^{-1}, with an average value of 24.8 Mm^3 yr^{-1} during the historical period. Future simulations ranged from 12.3 to 52.4 Mm^3 yr^{-1}, with an average value of 32.4 Mm^3 yr^{-1}. All runs agreed on future runoff increases, however minimal increase was simulated under NorESM. During 2071–2100, runoff volumes increased by 91 % in CCSM4 and by 16 % in MPI-ESM relative to 1976–2005.

4 Conclusions and recommendations

The presented study used climatic inputs from the same RCM version that three different GCMs drive. It then follows on previous work by investigating the uncertainty resulting from the driving GCM model. Runoff estimations simulated using these inputs showed substantial changes, ranging from 12.2 to 45 Mm^3 yr^{-1} during 1976–2005 and from 12.3 to 52.4 Mm^3 yr^{-1} during 2071–2100. Qatar is a hyper-arid area receives a minimal rainfall of 77 mm^3 yr^{-1} on average. Therefore, the uncertainty in runoff simulations matters. Such uncertainty can hinder characterizing the magnitude and the direction of current and future runoff, posing questions on water resources management and climate change adaptation and mitigation measures. Therefore, it is paramount that researchers in the climate community devote serious efforts to narrow the uncertainty in GCMs.

Regarding Qatar, local authorities are highly encouraged to implement managed aquifer recharge systems like rainwater harvesting. This is an optimal way to benefit from runoff and alleviate floods risk on people, assets, and the built-up environment.

Data availability. Topographic and groundwater data are available under reasonable request from the Qatar Ministry of Municipality and Environment.

Author contributions. SA: Conceptualization, Methodology, Data curation, Formal analysis, Validation, Visualization, Writing-original draft, and Writing-review and editing. SGAG: Conceptualization, Funding acquisition, Project administration, and Writing—review and editing.

Competing interests. The contact author has declared that neither of the authors has any competing interests.

Disclaimer. Publisher's note: Copernicus Publications remains neutral with regard to jurisdictional claims in published maps and institutional affiliations.

Special issue statement. This article is part of the special issue "IAHS2022 – Hydrological sciences in the Anthropocene: Variability and change across space, time, extremes, and interfaces". It is a result of the XIth Scientific Assembly of the International Association of Hydrological Sciences (IAHS 2022), Montpellier, France, 29 May–3 June 2022.

Acknowledgements. The authors thank Elfatih Eltahir from Massachusetts Institute of Technology and Jeremy Pal from Loyola Marymount University for providing MIT climatic data. This work was made possible through a National Priorities Research Program

(NPRP) grant (NPRP12S-0212-190073) from the Qatar National Research Fund (QNRF), a member of Qatar Foundation (QF) to Sami Al-Ghamdi and a PhD scholarship from Hamad Bin Khalifa University (HBKU), a member of QF, to Salah Ajjur.

Financial support. This research has been supported by the National Priorities Research Program from the Qatar National Research Fund (QNRF), a member of Qatar Foundation (QF) (grant no. NPRP12S-0212-190073).

Review statement. This paper was edited by Christophe Cudennec and reviewed by two anonymous referees.

References

Ahmed, F., Moors, E., Khan, M. S. A., Warner, J., and Terwisscha van Scheltinga, C.: Tipping points in adaptation to urban flooding under climate change and urban growth: The case of the Dhaka megacity, Land Use Policy, 79, 496–506, https://doi.org/10.1016/j.landusepol.2018.05.051, 2018.

Ajjur, S. B. and Al-Ghamdi, S. G.: Evapotranspiration and water availability response to climate change in the Middle East and North Africa, Clim. Change, 166, 1–19, https://doi.org/10.1007/s10584-021-03122-z, 2021a.

Ajjur, S. B. and Al-Ghamdi, S. G.: Seventy-year disruption of seasons characteristics in the Arabian Peninsula, Int. J. Climatol., 41, 5920–5937, https://doi.org/10.1002/joc.7160, 2021b.

Ajjur, S. B. and Al-Ghamdi, S. G.: Global Hotspots for Future Absolute Temperature Extremes From CMIP6 Models, Earth Space Sci., 8, e2021EA001817, https://doi.org/10.1029/2021EA001817, 2021c.

Ajjur, S. B. and Al-Ghamdi, S. G.: Variation in Seasonal Precipitation over Gaza (Palestine) and Its Sensitivity to Teleconnection Patterns, Water, 13, 667, https://doi.org/10.3390/w13050667, 2021d.

Ajjur, S. B. and Al-Ghamdi, S. G.: Evapotranspiration and water availability response to climate change in the Middle East and North Africa, Clim. Change, 166, 1–19, https://doi.org/10.1007/s10584-021-03122-z, 2021e.

Ajjur, S. B. and Al-Ghamdi, S. G.: Seventy-year disruption of seasons characteristics in the Arabian Peninsula, Int. J. Climatol., 41, 5920–5937, https://doi.org/10.1002/joc.7160, 2021f.

Ajjur, S. B. and Al-Ghamdi, S. G.: Towards sustainable energy, water and food security in Qatar under climate change and anthropogenic stresses, Energy Reports, 8, 514–518, https://doi.org/10.1016/j.egyr.2022.02.099, 2022.

Ajjur, S. B. and Baalousha, H. M.: A Review on Implementing Managed Aquifer Recharge in the Middle East and North Africa Region: Methods, Progress, and Challenges, Water Int., 46, 578–604, https://doi.org/10.1080/02508060.2021.1889192, 2021.

Ajjur, S. B. and Mogheir, Y.: Effects of Climate Change on Groundwater Resources, The 4th International Engineering Conference-Towards engineering of 21st century, Gaza, Palestine, 2012.

Ajjur, S. B. and Mogheir, Y. K.: Identification of intrinsic suitable sites in Gaza Strip for the application of artificial groundwater recharge using a geographic information system multicriteria decision analysis, Journal of Multi-Criteria Decision Analysis, 27, 255–265, https://doi.org/10.1002/mcda.1701, 2019.

Ajjur, S. B. and Riffi, M. I.: Analysis of the observed trends in daily extreme precipitation indices in Gaza Strip during 1974–2016, Int. J. Climatol., 40, 6189–6200, https://doi.org/10.1002/joc.6576, 2020.

Ajjur, S. B., Al-Ghamdi, S. G., and Baalousha, H. M.: Sustainable Development of Qatar Aquifers Under Global Warming Impact, Int. J. Global Warm., 25, 323–338, https://doi.org/10.1504/IJGW.2021.119003, 2021.

Al-Humaiqani, M. M. and Al-Ghamdi, S. G.: The built environment resilience qualities to climate change impact: Concepts, frameworks, and directions for future research, Sustain. Cities Soc., 80, 103797, https://doi.org/10.1016/j.scs.2022.103797, 2022.

Anibas, C., Schneidewind, U., Vandersteen, G., Joris, I., Seuntjens, P., and Batelaan, O.: From streambed temperature measurements to spatial-temporal flux quantification: using the LPML method to study groundwater-surface water interaction, Hydrol. Process., 30, 203–216, https://doi.org/10.1002/hyp.10588, 2016.

Batelaan, O. and De Smedt, F.: GIS-based recharge estimation by coupling surface–subsurface water balances, J. Hydrol., 337, 337–355, https://doi.org/10.1016/j.jhydrol.2007.02.001, 2007.

Martos-Rosillo, S., González-Ramón, A., Jiménez-Gavilán, P., Andreo, B., Durán, J. J., and Mancera, E.: Review on groundwater recharge in carbonate aquifers from SW Mediterranean (Betic Cordillera, S Spain), Environ. Earth Sci., 74, 7571–7581, https://doi.org/10.1007/s12665-015-4673-3, 2015.

Mogheir, Y. and Ajjur, S.: Effects of climate change on groundwater resources (Gaza strip case study), Int. J. Sustain. Energy Environ., 1, 136–149, 2013.

Moiwo, J. P., Yang, Y., Li, H., Han, S., and Yang, Y.: Impact of water resource exploitation on the hydrology and water storage in Baiyangdian Lake, Hydrol. Process., 24, 3026–3039, https://doi.org/10.1002/hyp.7716, 2010.

Pal, J. S. and Eltahir, E. A. B.: Future temperature in southwest Asia projected to exceed a threshold for human adaptability, Nature Clim. Change, 6, 197–200, https://doi.org/10.1038/nclimate2833, 2016.

Pozdniakov, S. P., Vasilevsky, P. Y., Grinevskiy, S. O., Lekhov, V. A., Sizov, N. E., and Wang, P.: Variability in spatial–temporal recharge under the observed and projected climate: A site-specific simulation in the black soil region of Russia, J. Hydrol., 590, 125247, https://doi.org/10.1016/j.jhydrol.2020.125247, 2020.

Serdar, M. Z., Koç, M., and Al-Ghamdi, S. G.: Urban Transportation Networks Resilience: Indicators, Disturbances, and Assessment Methods, Sustain. Cities Soc., 76, 103452, https://doi.org/10.1016/j.scs.2021.103452, 2022.

Xie, Y., Cook, P. G., Simmons, C. T., Partington, D., Crosbie, R., and Batelaan, O.: Uncertainty of groundwater recharge estimated from a water and energy balance model, J. Hydrol., 561, 1081–1093, https://doi.org/10.1016/j.jhydrol.2017.08.010, 2018.

Proc. IAHS, 385, 339–344, 2024
https://doi.org/10.5194/piahs-385-339-2024

Open Access
PIAHS

Influence of data spatial resolution in water resources management for oak-savanna distinctive vegetation patches

Ana Andreu[1,2], **Elisabet Carpintero**[3], **María J. Muñoz-Gomez**[4], **Ángel Blázquez-Carrasco**[4], and **María P. González-Dugo**[4]

[1]Fluvial Dynamics and Hydrology Research Group, Andalusian Institute for Earth System Research, University of Córdoba, 14071, Córdoba, Spain
[2]Dept. Agronomy, Unidad de Excelencia María de Maeztu (DAUCO), University of Córdoba, 14071 Córdoba, Spain
[3]Instituto de Agricultura Sostenible (IAS), Consejo Superior de Investigaciones Científicas (CSIC), Alameda del Obispo s/n, 14004 Córdoba, Spain
[4]Instituto Andaluz de Investigación y Formación Agraria y Pesquera, IFAPA, Centro Alameda del Obispo. Avd. Menéndez Pidal s/n, 14071 Córdoba, Spain

Correspondence: Ana Andreu (ana.andreu@uco.es)

Received: 20 May 2022 – Revised: 22 April 2023 – Accepted: 21 July 2023 – Published: 19 April 2024

Abstract. In Mediterranean oak-savannas (known as *dehesas* in Spain), it is important to better understand the interactions between vegetation structure and local climate regulation at scales relevant to farm management and the dominant hydrological regime. This study evaluated the water use patterns of *dehesa* vegetation patches (open grasslands, lowland grasslands with high soil moisture, tree + grass, and riparian vegetation), estimating actual evapotranspiration (ET). We used different models, previously validated in the area, that integrate remotely sensed data. They apply (a) a soil water balance (Kc-FAO56), (b) a surface energy balance (ALEXI/DisALEXI, and SEBS), and (c) a sharpening algorithm (STARFM), obtaining products at multiple spatial resolutions (30 m, 1 km, 5 km). The conceptual and operational differences between the methodologies reinforce the idea of a combined application of models. We demonstrated the need for high spatial and temporal resolution for on-farm livestock management due to the importance of the grasslands layer. This scale is crucial to determine the grass's emergence/drying cycle, which is key for livestock feeding planning. In humid/denser areas that provide essential ecosystem services (e.g., refuge, pasture rotation), transpiration rates are higher throughout the year and were underestimated when coarser spatial scale data was used. Over the typical system (grass with dispersed trees), the ET maps at low spatial resolution reflected the water use trends, and all models correlated well. Higher differences were found when comparing the models' performance over open grasslands.

Keywords. Remote sensing; Mediterranean oak-savannas; soil water balance; surface energy balance; UPH 5; Panta Rhei

1 Introduction

Dehesas, Spanish semiarid oak-savannas, are agroforestry ecosystems distributed in Mediterranean-type climatic areas, thus periodically subject to water scarcity conditions. They play a key role in rural development, basing their production on multiple uses (e.g., livestock, agriculture, hunting). Their importance as biodiversity reservoirs is recognized internationally (Habitats Directive 92/43/EEC and Natura 2000 Network). Although *dehesas* are adapted to highly variable climatic conditions and dry periods during the summer, the increase in drought intensity, duration, and frequency, changes in agricultural practices, pests, diseases, and other socioeconomic and environmental factors contribute to their degradation.

Moreover, global change alters climate patterns by increasing interannual variability and the incidence of extreme and torrential events. The vegetation cover largely determines the interaction between climate processes in semiarid agroforestry systems. Therefore, vegetation management is crucial not only to preserve the biodiversity and profitability of *dehesas* but to design strategies to reduce their vulnerability.

Oak-savannas have a complex structure with multiple layers of vegetation that differ in phenology, physiology, and function, contributing differently to the ecosystem fluxes. This structure plays a crucial role in savannas' resilience, allowing them to efficiently convect sensible heat and to maintain vegetation temperature within the adequate range (Johnston et al., 2021), providing food and heat comfort to animals.

In these systems, the variability of vegetation distribution creates different spaces that influence ecological processes. The diversity of spaces is altered and maintained through agricultural practices. For example, (1) tree dispersion allows radiation to reach the downstory, but the trees act as a screen, decreasing the light underneath, and animals use their shade in summer. (2) During day time, grasses have an efficient cooling effect when atmospheric evaporative demand is high. During the hot season, a dry grass layer also reduces surface radiometric temperature (Johnston et al., 2022). (3) Trees influence grass communities by augmenting habitat heterogeneity and increasing the total species richness. Altering the structural environment (e.g., regeneration of trees and shrubs or modification of grass communities) can be valuable in conserving resources while enhancing carbon fixation or animal health.

As a first step, we want to better understand the interactions between vegetation structure and water consumption dynamics at scales relevant to farm management and the dominant hydrological regime.

In this study, we analyze the patterns of water use (evapotranspiration – ET) of different spaces of the *dehesa*, (A) open grasslands, (B) lowland grasslands with high soil moisture, (C) tree + grass, and (D) riparian vegetation, estimated with different model schemes that integrate data obtained by remote sensors at different scales (5 km, 1 km, 30 m). Our objective was to evaluate the possibilities offered by the different spatial resolutions and methodologies for the management of this ecosystem's vegetated spaces.

2 Materials and methods

The study was conducted in an experimental *dehesa* farm located in the Sierra de Cardeña & Montoro Natural Park in southern Spain (Santa Clotilde, 38°12′ N, 4°17′ W, 736 m a.s.l., with two eddy covariance towers measuring water and carbon fluxes, Fig. 1). More information about the installation in Andreu et al. (2018) and Carpintero et al. (2020). The area has Mediterranean climate, with strong seasonality, moderately cold winters and long, hot, and dry summers. Precipitation is highly variable (annual average = 895 mm from 1990 to 2015), mainly concentrated in spring and autumn. The vegetation consists primarily of widely spaced *Quercus ilex* L. combined with a herbaceous stratum composed of annual grasslands (which emerge after the autumn rains and dry up in late spring).

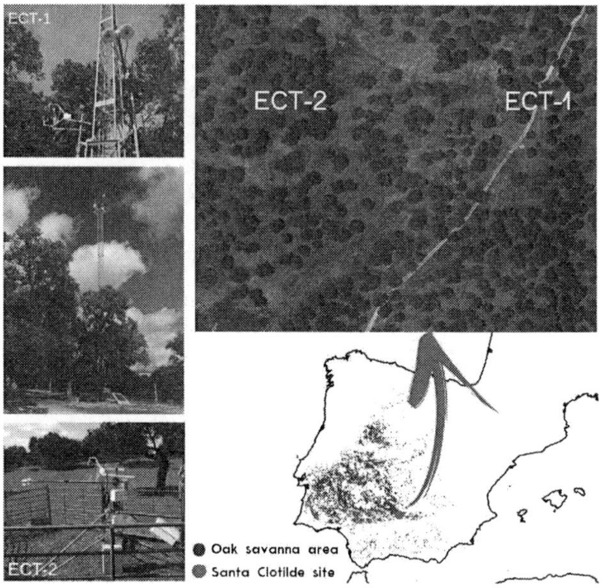

Figure 1. Dehesa farm study site and the location of the two eddy covariance towers over the area (ECT-1 tree + grass system, which is zone C, and the ECT-2 over grass). Aerial image from © Google Maps.

ET patterns were analyzed in 4 zones (Fig. 2). Zone A (1700 m^2) is a grassland with high temporal and spatial variability. Zone B (2400 m^2) is a grassland located in a land depression that accumulates water most of the year and maintains higher humidity. Zone C (4300 m^2) is a tree + grass structure (25 % tree fractional cover) standard of *dehesas* (the area surrounding ECT-1). Zone D (3300 m^2) is located near a stream with riparian vegetation (trees and an underlayer of evergreen scrubs).

Models and earth observation data

The models, previously validated in the study area (using ECT-1 data for all models and ECT-2 data for Kc-ETo due to footprint) with good results (Carpintero et al., 2020, 2021; González-Dugo et al., 2021) and applied over 2013–2015, are:

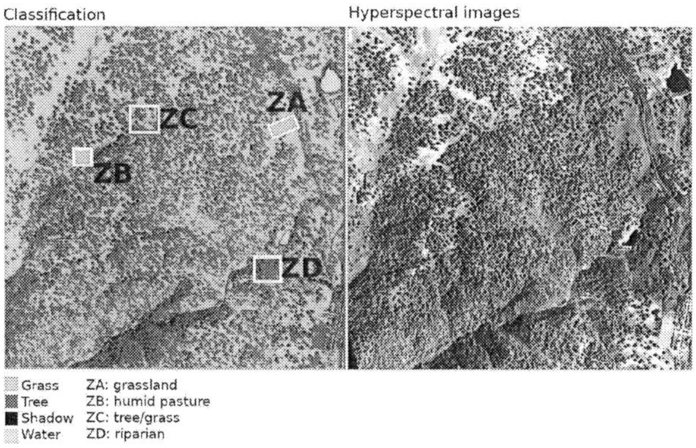

Figure 2. Classification map used as input in the VI-ETo model and the different zones (white squares). Hyperspectral image from 27 August 2015 (Hector Nieto and Pablo Zarco-Tejada).

1. An adaptation of the Kc-ETo method of FAO56 (Allen et al., 1998) to assimilate remote data, the VI-ETo model (González-Dugo et al., 2008), which estimates ET with a high spatial resolution (30 m) from visible and near-infrared Landsat 8 data. This model requires the calibration of vegetation parameters and information on soil parameters. To refine the characterization of these parameters that define the water balance (e.g., roots depth), we performed a digital classification during the dry season using an airborne hyperspectral image with a high spatial resolution (1 m).

2. A surface energy balance to estimate ET from thermal sensors, the ALEXI-DisALEXI model (Anderson et al., 1997), was used with MODIS and Landsat 7/8 images. Both types of ET maps were combined using the STARFM data fusion technique (Gao et al., 2006) that provides daily ET estimates at 30 m. With this method, we produced maps of daily ET at 1 km and 30 m without the need for local calibration.

3. We used the SEBS model (Surface Energy Balance System, Su et al., 2002), which does not need local calibration, to provide the regional approach integrating remotely sensed thermal data of 5 km spatial resolution and monthly temporal frequency. Due to the coarse resolution, it is impossible to differentiate between the zones, as all are located inside a 5 km × 5 km pixel.

3 Results and Discussion

In Fig. 2, we present the classification map used to refine the vegetation and soil parameters that influence the water balance of the VI-ETo. Model inputs such as the root depth or the basal crop coefficient under full vegetation cover greatly influence the overall balance because they determine the control volume. From a previous analysis (Andreu et al., 2022), when general average values were selected (e.g., root depth = 2 m, instead of $\text{root}_{\text{tree}} = 4$ m and $\text{root}_{\text{grass}} = 1$ m), the pattern of the ET estimated with VI-ETo integrating Landsat images (30 m) was similar to the ET pattern estimated with coarser resolutions (DisALEXI MODIS model with 1 km). In this case, we used a detailed classification, in which we assumed shadows as part of the herbaceous layer, obtaining a 51 % of grass and 47 % of tree cover for the study area. Depending on the zone selected, tree fractional cover ranges between 0 (zones A and B) and almost full coverage (zone D).

Figure 3 shows the ET estimates for the zones in which the results of the selected models are available on (a) daily (VI-ETo, ALEXI-DisALEXI, and STARFM) or (b) monthly (including SEBS) temporal resolutions. It can be observed that the spatial scale of meters (VI-ETo and STARFM) better captures heterogeneity, as for the herbaceous senescence date in zone A. Since there are mixed pixels at the scale of SEBS and DisALEXI MODIS, multiple vegetation covers contribute to the estimates resulting in more similar ET rates in all zones. Daily temporal resolutions allowed for strong rain events to be shown as ET peaks.

With models needing calibration, the input parameters have to be refined to take advantage of the spatial resolution provided by missions such as Sentinel 2. For example, root depth, permanent wilting point, or soil texture greatly influence the water storage capacity and, thus, the evapotranspiration flux. When different canopies (trees, shrubs, and grasses) were present, the classification needed to be more detailed (Fig. 3a, zone D: Scrubs).

In zone A (Fig. 3a and b), the date of grass senescence was earlier (in late June) when applying the STARFM and VI-ETo models, reaching $0\,\text{mm}\,\text{d}^{-1}$ during summer. MODIS-DisALEXI produced minimum ET rates from August. During the 2013 summer, ET maintained values $\sim 1\,\text{mm}\,\text{d}^{-1}$. MODIS-DisALEXI presented higher values than the other models in the spring production peaks except for SEBS (Fig. 3b). Thus, using coarser resolutions may imply an overestimation of the accumulated ET and the correspondent biomass production in *dehesa* pasture areas.

In zone B (Fig. 3a), the slope of the spring ET decline is less steep because of the soil moisture conditions, maintaining a residual ET rate during 2013 with all models and during 2014 with VI-ETo. ET time series at metres or kilometres are similar in this zone since wet grasslands maintained their transpiration longer during summer. These areas with higher production capacity are essential for managing grazing rotations.

Zone C represents the most common *dehesa* structure (an herbaceous layer with dispersed trees). Here the spatial resolutions gave similar results due to the homogeneity of this patch. In this case, the information provided at low resolution is helpful for management at regional or basin scales.

In zone D, ET estimates at m resolution suggest groundwater use by vegetation during the summer, with ET remain-

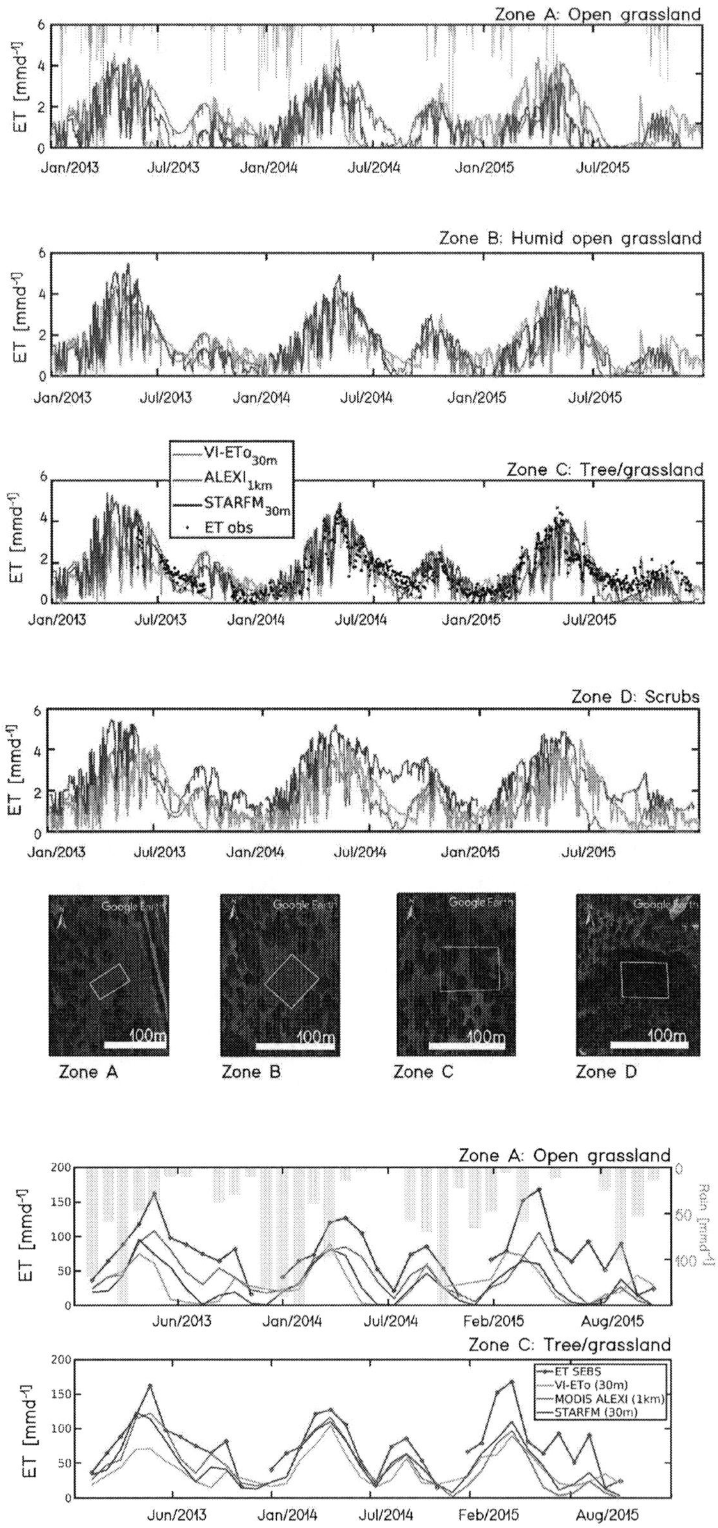

Figure 3. Upper panel: daily ET estimated for all zones (A, B, C, and D) using VI-ETo (30 m), DisALEXI (1 km), and STARFM (30 m) models. In zone C, daily ET measured by the ECT-1 is plotted (ET observations are the black dots). Aerial image from © Google Earth. Lower panel: monthly ET estimated for zones A and C, using SEBS (5 km), VI-ETo (30 m), DisALEXI (1 km), and STARFM (30 m) models.

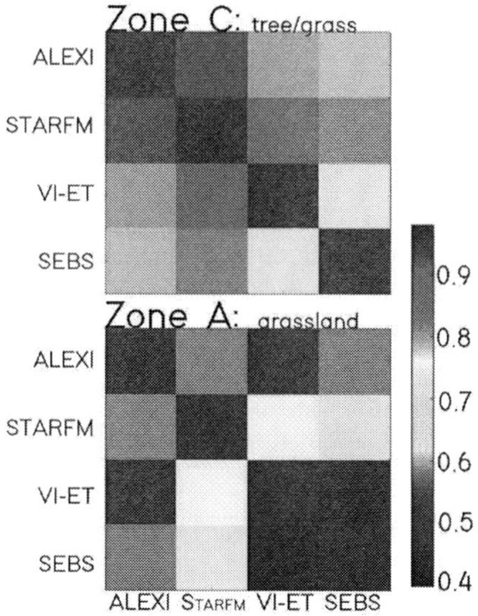

Figure 4. Model correlation for zone C: tree + grass and zone A: Open grassland.

ing above 1 mm d^{-1}. At 1 km, surrounding grasslands were incorporated, and ET values were reduced. This layer is the most heterogeneous concerning the different models' behavior, as we need to include the scrub's characteristics in the formulation. These scrub species are of nutritional interest for domestic livestock and game and can act as a refuge due to the conditions under the canopy.

As shown in Fig. 3b, the different models' monthly ET averages on zone C present a better agreement, except for the low VI-ETo 2013 spring peak and SEBS's usual overestimation of ET rates.

For regional purposes, when the *dehesa* is considered a homogeneous landscape, for example, for vulnerable areas zonification or drought early warning systems, all models performed similarly and correlated well (Fig. 4 zone C), showing the same trends. Nevertheless, the differences between them were higher for the management of the patches within the ecosystem (Fig. 4 zone A) .

4 Conclusions

The results showed that, for the *dehesa* ecosystem, ET estimations at a low resolution might be sufficient for management at the basin level. Information at a higher resolution may help identify areas with different water storage capacities or vegetation growth. For livestock management, this information can support grazing rotations and delimitation of areas containing fragile vegetation or high biodiversity. Considering the methodologies used, combining models with conceptual and operational differences may improve the outcomes.

Code availability. The SEBS code can be downloaded from the GitHub repository (https://github.com/TSEBS/SEBS_Spain, last access: 5 February 2021; Chen et al., 2019). A description of the regional Atmosphere-Land Exchange Inverse (ALEXI) and the DisALEXI model (Anderson et al., 1997) code is provided in the Algorithm Theoretical Basis Document, and the code may be distributed on request from the authors (ATBD: https://lpdaac.usgs.gov/documents/340/ECO4ESIALEXIU_ATBD_V1.pdf; Anderson et al., 2018). ALEXI and DisALEXI use the Two-Source Energy Balance model, which can be downloaded from the GitHub repository (https://github.com/hectornieto/pyTSEB, Nieto, 2016). Kc-FAO56 VI model may be distributed on request to the author (Carpintero et al., 2020; https://doi.org/10.3390/w12051418). STARFM code can be downloaded from the USDA dataset repository (https://data.nal.usda.gov/dataset/starfm, last access: June 2021; Gao et al., 2006).

Data availability. Data from the Santa Clotilde site may be distributed on request to the principal investigator of the experimental site (María P. González-Dugo, mariap.gonzalez.d@juntadeandalucia.es).

Author contributions. AA and MPGD conceptualized and designed the paper. All authors collect field data. EC, MPGD, and AA performed the simulations. AA prepared the manuscript with contributions from all co-authors.

Competing interests. The contact author has declared that none of the authors has any competing interests.

Special issue statement. This article is part of the special issue "IAHS2022 – Hydrological sciences in the Anthropocene: Past and future of open, inclusive, innovative, and society-interfacing approaches". It is a result of the XIth Scientific Assembly of the International Association of Hydrological Sciences (IAHS 2022), Montpellier, France, 29 May–3 June 2022.

Acknowledgements. The authors thank Hector Nieto and Pablo Zarco-Tejada for using the high-resolution image from 27 August 2015 over Santa Clotilde.

Financial support. Ana Andreu has received funding from the Ministry of Science and Innovation and the State Research Agency of Spain through the Severo Ochoa and María de Maeztu Program for Units of Excellence in R&D&I (Ref. CEX2019-000968-M) and the SIERRA 1265875-R project, with the economic collaboration of European Funds for Rural Development and the Office of Economy, Knowledge, Business, and University of the Junta de Andalucía. Additional funding was provided by PID2019-107693RR-C22 (MCIU/AEI/FEDER, EU) project and EO-AFRICA R&D Tactic research project.

Review statement. This paper was edited by Christophe Cudennec and reviewed by two anonymous referees.

References

Allen, R. G., Pereira, L. S., Raes, D., and Smith, M.: Crop ET-Guidelines for Computing Crop Water Requirements-FAO, Irrigation and Drainage Paper 56, FAO, Rome, Italy, ISBN 92-5-104219-5, 1998.

Anderson, M. C., Norman, J. M., Diak, G. R., Kustas, W. P., and Mecikalski, J. R.: A two-source time-integrated model for estimating surface fluxes using thermal infrared remote sensing, Remote Sens. Environ., 60, 195–216, https://doi.org/10.1016/S0034-4257(96)00215-5, 1997.

Anderson, M. C., Fisher, J. B., and Hook, S. J.: Level-4 Evaporative Stress Index, Algorithm Theoretical Basis Document, JPL Publication D-94648, 16 pp., https://lpdaac.usgs.gov/documents/340/ECO4ESIALEXIU_ATBD_V1.pdf (last access: 1 June 2021), 2018.

Andreu, A., Kustas, W. P., Polo, M. J., Carrara, A., and González-Dugo, M. P.: Modeling Surface Energy Fluxes over a Dehesa Ecosystem Using TSEB model I, Remote Sens.-Basel, 10, 567, https://doi.org/10.3390/rs10040567, 2018.

Andreu, A., Carpintero, E., and González-Dugo, M. P.: Influencia de la resolución espacial en la gestión de los recursos hídricos en zonas distintivas de la dehesa, in: Proceedings of XIX Congreso de la Asociación Española de Teledetección, edited by: Ruiz, L. A., Estornell, J., González-Audícana, M., and Álvarez-Mozos, J., Pamplona, Spain, 29 June–1 July 2022, eISBN: 978-84-9769-383-7, 2022.

Carpintero, E., Andreu, A., Gómez-Giráldez, P.J., Blázquez, Á., and González-Dugo, M. P.: Remote-Sensing-Based Water Balance for Monitoring of Evapotranspiration and Water Stress of a Mediterranean Oak–Grass Savanna, Water, 12, 1418, https://doi.org/10.3390/w12051418, 2020.

Carpintero, E., Anderson, M. C., Andreu, A., Hain, C., Gao, F., Kustas, W. P., and González-Dugo, M. P.: Estimating Evapotranspiration of Mediterranean Oak Savanna at Multiple Temporal and Spatial Resolutions, Remote Sens.-Basel, 13, 3701, https://doi.org/10.3390/rs13183701, 2021.

Chen, X., Su, Z., Ma, Y., and Middleton, E. M.: Optimization of a remote sensing energy balance method over different canopy applied at global scale, Agr. Forest Meteorol., 279, 107633, https://doi.org/10.1016/j.agrformet.2019.107633, 2019 (code available at: https://github.com/TSEBS/SEBS_Spain, last access: 5 February 2021).

Gao, F., Masek, J., Schwaller, M., and Hall, F.: On the blending of the Landsat and MODIS surface reflectance: Predicting daily Landsat surface reflectance, IEEE T. Geosci. Remote, 44, 2207–2218, https://doi.org/10.1109/TGRS.2006.872081, 2006 (data available at: https://data.nal.usda.gov/dataset/starfm, last access: June 2021).

González-Dugo, M. P. and Mateos, L.: Spectral veg. indices for benchmarking water productivity of irrigated cotton and sugarbeet crops, Agr. Water Manage., 95, 48–58, https://doi.org/10.1016/j.agwat.2007.09.001, 2008.

González-Dugo, M. P., Chen, X., Andreu, A., Carpintero, E., Gómez-Giraldez, P. J., Carrara, A., and Su, Z.: Long-term water stress and drought assessment of Mediterranean oak savanna vegetation using thermal remote sensing, Hydrol. Earth Syst. Sci., 25, 755–768, https://doi.org/10.5194/hess-25-755-2021, 2021.

Johnston, M., Andreu, A., Verfaillie, J., Baldocchi, D., Gonzalez-Dugo, M. P., and Moorcroft, P.: Measuring surface temperatures in a woodland savanna: Opportunities and challenges of thermal imaging in an open-canopy ecosystem, Agr. Forest Meteorol., 310, 108484, https://doi.org/10.1016/j.agrformet.2021.108484, 2021.

Johnston, M. R., Andreu, A., Verfaillie, J., Baldocchi, D., and Moorcroft, P. R.: What lies beneath, Remote Sens. Environ., 274, 112950, https://doi.org/10.1016/j.rse.2022.112950, 2022.

Nieto, H.: hectornieto/pyTSEB, GitHub [code], https://github.com/hectornieto/pyTSEB (last access: July 2011), 2016.

Su, Z.: The Surface Energy Balance System (SEBS) for estimation of turbulent heat fluxes, Hydrol. Earth Syst. Sci., 6, 85–100, https://doi.org/10.5194/hess-6-85-2002, 2002.

Proc. IAHS, 385, 345–349, 2024
https://doi.org/10.5194/piahs-385-345-2024

Recharge and dynamics of the Tamanrasset alluvial aquifer (Algerian Sahara)

Samia Hadj-Said[1], **Aziez Zeddouri**[1], **Jean-Denis Taupin**[2], **Nicolas Patris**[2], **and Christian Leduc**[3]

[1]RSPGA, University Kasdi Merbah, Ouargla, Algeria
[2]HSM, University Montpellier, IRD, CNRS, Montpellier, France
[3]IRD, UMR G-EAU, Montpellier, France

Correspondence: Samia Hadj-Said (hadjsaidsamia@gmail.com)

Received: 15 June 2022 – Revised: 3 May 2023 – Accepted: 21 July 2023 – Published: 19 April 2024

Abstract. Population growth and agricultural intensification are ever more demanding in water resources, and threaten their quality and quantity, especially in semi-arid and arid areas. The Sahara desert is typical of these multiple constraints and rapid changes making sustainable management of water resources a major issue. The wide regional aquifers have nearly fossil water. In contrast, alluvial aquifers associated with small wadis are recharged nearly each year by the rare floods. In the case of Wadi Tamanrasset, in southern Algeria, the alluvial aquifer is exploited intensively because of its accessibility and its good chemical quality. The physical conditions commanding flow and recharge of groundwater, and especially the transfer from surface to groundwater, were studied by combining hydrodynamics and isotopic tracers (^{18}O, ^{2}H). Monthly monitoring of groundwater level in around thirty wells was complemented by automatic recorders with hourly time steps since March 2016 in three wells, along a 600 m transect perpendicular to the wadi bed. The slow response of the water table is visible only two months after the first flood; its maximum rise was 1 m in 2016. Isotopic analyses of rainwater on a daily time scale, of the wadi water on a 1 h time scale during runoff and of groundwater (two complete campaigns in the dry and wet seasons and a specific monitoring of the transect every month) have provided additional information on surface-groundwater transfer.

Keywords. UPH13; Field observation; Arid areas; Hydrodynamics; Stable isotopes; Surface-groundwater interaction

1 Introduction

The last periods of drought that have affected the Mediterranean region and the economic and demographic development in spite of the arid climate have increased the solicitation of groundwater resources in the Sahara. Groundwater, which is now very actively exploited, is often vulnerable both qualitatively and quantitatively. Future exploitation plans should therefore be preceded by a better identification of these fragile water resources.

Oxygen-18 and deuterium are excellent tracers of the water circulation, since they are not subject to fixation processes that affect even the best other chemical tracers. They give information on all stages of the water cycle, from the atmospheric vapor to groundwater. The isotopic fractionation depends mostly on temperature and environmental factors conditioning the various processes, especially condensation and evaporation (Shamsuddin et al., 2018).

Groundwater geochemistry in the Sahara was studied by various works (e.g. Saighi et al., 2001; Guendouz et al., 2003; Moulla et al., 2005) but the groundwater/river relationship is rarely investigated. This issue is the main purpose of this work.

Most of the water used in the city of Tamanrasset is groundwater, with a special interest of inhabitants for small alluvial aquifers because of their good quality. Groundwater is recharged by occasional floods of wadis whose upstream morphology is marked by steep slopes and narrow valleys.

In Tamanrasset, the wadi/aquifer relationship is still poorly known. The study of the stable isotopes of water allowed to identify the origin of the input signal to aquifers, to estimate the recharge altitudes and to define the relationships between groundwater and surface water. The analysis of piezometric fluctuations complemented this study of transfers between surface water and groundwater.

2 Study area

2.1 Geographic and climatic context

The study area (Fig. 1) is located in southern Algeria, between 22°10′ and 22°60′ N and between 5°32′ and 5°39′ E. It extends over an area of 106 km^2 and is the downstream part of the much larger Wadi Tamanrasset catchment (about 500 km^2). In this arid climate, rain may originate from the Sudan-Saharan monsoon whose extreme northern extension limit is the south of the Tademaït plateau, or from a few Mediterranean rain systems. Maximum values of rainfall are recorded at the Assekrem station (elevation of about 2700 m). Rain usually occurs between May and September (monsoon rain) but sporadic rains exist in winter and spring (Mediterranean or North Atlantic rainfall).

In 2016 the annual rainfall was 150 mm in Assekrem and 99 mm in Tamanrasset. The average annual temperature in Assekrem was 11.9 °C, and the hottest months were June and July (24.3 °C), while the coldest months were December and January (6.0 °C). In Tamanrasset, the average annual temperature was 23.4 °C.

2.2 Hydrological and geological context

The elevation of the study area varies mostly between 1350 and 1500 m. The upstream part of the Tamanrasset catchment consists of basaltic plateaus (near Assekrem, at a mean elevation of 2000 m a.s.l.) while the downstream sub-catchment is mostly made of granites.

The Tamanrasset wadi starts on the South side of the Atakor, in the region of Assekrem (highest peak at 2726 m a.s.l.), and then flows down along a SSW direction. It has nearly no tributaries on its left bank and very few on its right. The longitudinal profile of the wadi is quite regular, in spite of small discontinuities due to the presence of recent volcanic flows and crystalline sills. The wadi sometimes flows in very narrow gorges.

Upstream of these crystalline thresholds, the wadi may spread out and form small alluvial accumulations. The Tamanrasset plain is the largest of them.

A gauging test in the city of Tamanrasset during the flood of 11 May 2014 gave flow values between 0.5 and 5.24 $m^3\,s^{-1}$ (Tarenga et al., 2015).

The main geological formations of the study area are the unaltered basement, alterites and the recent alluvium. The basement is made of granite, gneiss and micaschist, with frequent diaclases. The depth of altered bedrock varies significantly, depending on fracturation density and local uplift of harder rocks.

The aquifer is a notch in the gneissic and granitic basement. The water resource is mainly stored in coarse alluvial deposits resulting from the erosion of the Atakor massif. The geophysical study carried out in the area (Ben Elkheir et al., 2017; Zeddouri et al., 2018) identified four layers: dry coarse alluvium (gravel, pebbles and cobbles) at the surface and in sub-surface; medium to fine alluvium that may also contain groundwater; altered bedrock; and fresh bedrock, which is very resistant and can be met at a depth of up to 40 m. The hydrogeological system is composed of five consecutive alluvial basins, separated by sills corresponding to outcrops of the gneissic basement.

These alluviums, which offer a renewable resource irregularly fed by the few floods, have an average thickness of 21 m that decreases from upstream to downstream. The piezometric maps for March and September 2016 show a general flow that coincides with the course of the wadi.

3 Materials and methods

In order to study groundwater recharge and its relationship with the wadi, an isotopic monitoring was carried out:

- in precipitation: 16 rainfall events were sampled in 2016 at the Assekrem station (23.3° N; 5.6° E), among which one was discarded because of imperfect storage, and seven events in Tamanrasset (22.8° N, 5.5° E; elevation of 1376 m a.s.l.),
- in the wadi: 31 samples from the four floods that occurred in 2016; each flood was sampled every 30 min,
- in groundwater: 60 samples.

Oxygen 18 and deuterium were measured by mass spectrometry (IRMS Isoprime) at the LAMA laboratory of Hydro-Sciences Montpellier (HSM). Results are expressed as delta values relative to the international standard VSMOW (Vienna Standard Mean Ocean Water) with errors of ±0.06 ‰ and ±0.8 ‰ for $\delta^{18}O$ and $\delta^{2}H$, respectively.

The water table was monitored in three wells (P12, P7, P4) equipped with piezometric recorders at a hourly time step along a 600 m transect perpendicular to the wadi bed (Fig. 1).

4 Results and discussion

4.1 Isotopic content of rainfall

The stable isotope content of rainfall water samples collected at the Assekrem station varied between −4.65 ‰ and +3.37 ‰ for $\delta^{18}O$, and −24.8 ‰ and +24.6 ‰ for $\delta^{2}H$. The deuterium excess varied between −7.6 ‰ and +16.6 ‰. Four rainfall events, with d-excess $< +8$ ‰ and relatively enriched isotope values, showed a clear evaporation process, either during rainfall or before sampling but in general the events show no evaporation process during the fall under the

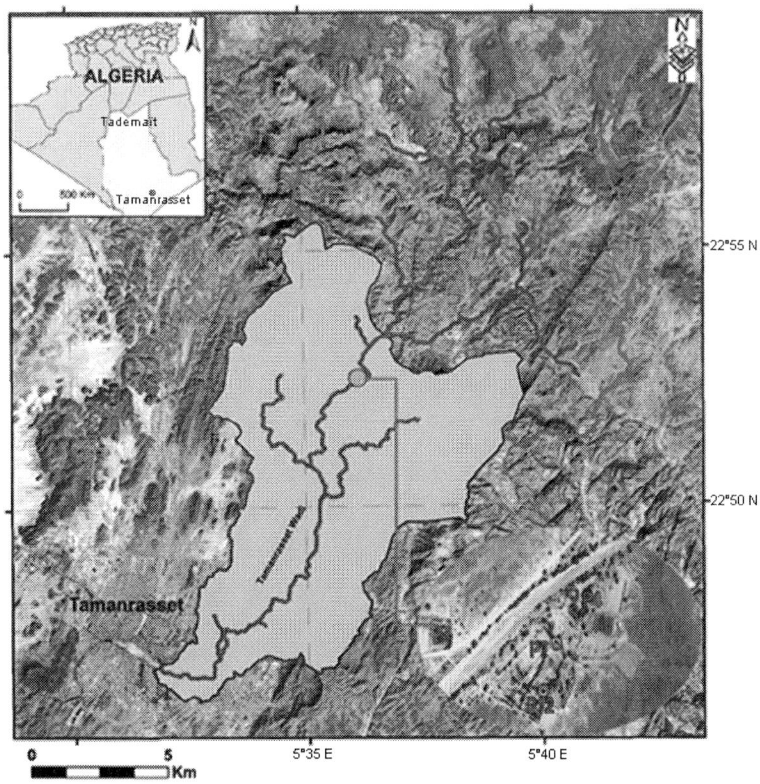

Figure 1. The study area (light blue), downstream part of the Wadi Tamanrasset catchment (orange contour), and the transect location.

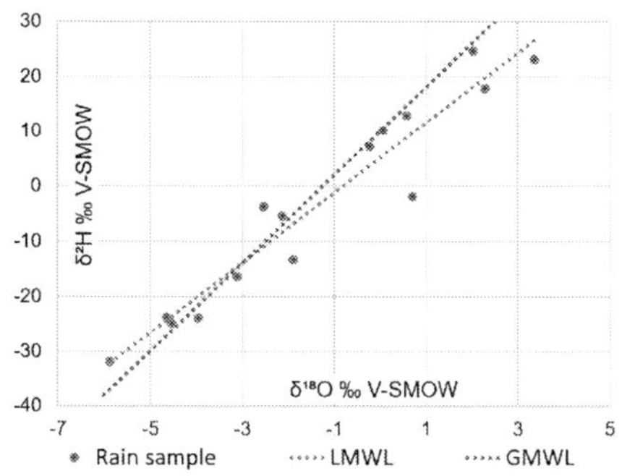

Figure 2. Deuterium vs. oxygen-18 relationship of Assekrem rainwater in 2016 (Global and Local Meteoric Water Lines).

cloud, despite the small quantity sampled. The precipitation-weighted mean values were −3.65 ‰ for δ^{18}O and −18.9 ‰ for δ^{2}H (Table 1). The isotopic meteoric line at the Assekrem station is: $\delta^{2}\mathrm{H} = 4.2\,\delta^{18}\mathrm{O} + 1.2$ (Fig. 2). Similar results were obtained by Saighi et al. (2001).

In the seven Tamanrasset events, the stable isotope of rainwater varied between −2.95 ‰ and +4.02 ‰ for δ^{18}O and between −14.0 ‰ and +13.3 ‰ for δ^{2}H. The deuterium excess varied between −23.8 ‰ and +9.7 ‰. The annual precipitation-weighted mean in Tamanrasset was +0.13 ‰ for δ^{18}O and +1.2 ‰ for δ^{2}H.

Table 1. Isotopic content of rainfall events at the Assekrem station in 2016 (ND = no data).

Date (dd/mm/yyyy)	δ^{18}O	δ^{2}H	D excess	P (mm)
22/05/2016	−5.87	−31.9	+15.1	27.9
24/05/2016	+0.71	−1.9	−7.6	2.0
27/06/2016	−0.23	+7.2	+9.0	2.6
30/06/2016	+0.06	+10.2	+9.7	1.9
20/07/2016	+2.03	+24.6	+8.4	0.6
21/07/2016	−2.53	−3.7	+16.6	2.8
22/07/2016	−3.12	−16.3	+8.7	7.7
12/08/2016	+3.37	+23.2	−3.8	1.8
15/08/2016	−2.13	−5.3	+11.7	9.8
23/08/2016	−4.65	−23.7	+13.4	0.1
26/08/2016	+0.58	+12.9	+8.3	0.5
27/08/2016	−3.96	−24.0	+7.7	27.7
28/08/2016	−4.54	−24.8	+11.5	1.5
10/09/2016	−1.90	−13.3	+1.8	ND
27/09/2016	+2.29	+17.8	−0.5	1.2

Table 2. Comparison of $\delta^{18}O$ of rain and floods events.

Flood date (dd/mm/yyyy)	$\delta^{18}O$ Tamanrasset	$\delta^{18}O$ Assekrem	$\delta^{18}O$ Wadi Mean	$\delta^{18}O$ Wadi Min / Max
17/07/2016	+0.27	–	+0.31	+0.02/+0.7
21/07/2016	−1.43	−2.53	−2.85	−3.43/−1.96
10/08/2016	+3.05	+3.37	−3.5	−3.72/−3.36
29/08/2016	−0.35	+0.58	−3.0	−3.46/−2.13

The Tamanrasset meteoric line is marked by a greater evaporation process in four events: $\delta^{2}H = 3.22\,\delta^{18}O + 0.5$ but its representativeness is limited by the weak number of events sampled.

The isotopic data of rainwater showed temporal variations. Precipitation at the beginning and end of the rainy season was more enriched than the heavier rainfall in the middle of the rainy season (July to August) which had more depleted values. Such evolution can be attributed to the saturation state of the atmosphere at the time of precipitation, and the mass effect and low atmospheric temperature during strong thunderstorms.

The variation of the annual weighted mean $\delta^{18}O$ in precipitation highlights the altitude effect that leads to a depletion of heavy isotopes in precipitation due to lower condensation temperatures.

4.2 Isotopic content of the wadi water

In the wadi water, $\delta^{18}O$ ranged from −3.72 ‰ to +0.70 ‰ and $\delta^{2}H$ from −21 ‰ to +10 ‰. Isotope variations during the flood (3 to 24 h) were rather limited, showing a good mixing of rainfall in the wadi water. Floods are generated in the highest part of mountainous zone, except maybe for the 17 July flood where no rainfall was measured concurrently in the Assekrem station. The isotopic value in the wadi for this event was close to the isotopic value of rain at the Tamanrasset station. However, only one flood event (21 July) can be directly linked with Assekrem in isotope terms, indicating that the wadi is fed by the whole catchment with its own isotope variability due to altitude, to condensation levels and origin of the convective rainfall. No evaporation process marked the floods during the wet season (July and August) (Table 2). Thus, it can be concluded that the surface water of the Wadi Tamanrasset basin is strongly affected by precipitation at higher elevations.

4.3 Isotopic content of groundwater

In groundwater collected in March 2016 $\delta^{18}O$ ranged from −3.79 ‰ to −1.55 ‰, and $\delta^{2}H$ from −24.8 ‰ to −7.5 ‰. In groundwater sampled in August 2016 $\delta^{18}O$ ranged from −4.19 ‰ to −1.64 ‰ and $\delta^{2}H$ varied between −23.7 ‰ and −6.4 ‰. The isotopic composition of groundwater sampled

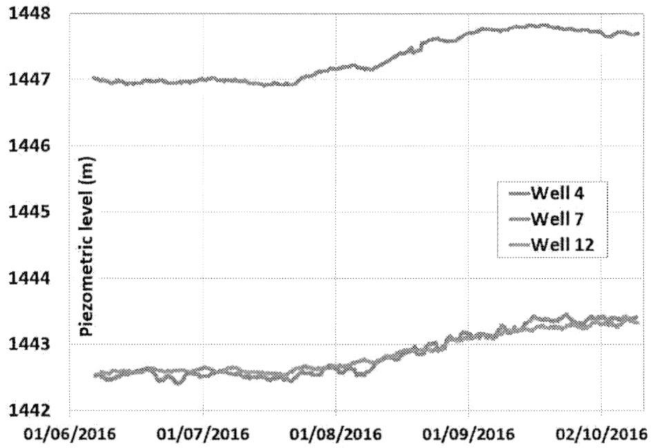

Figure 3. Variation of the groundwater level along a transect (June/October 2016).

in March 2016 was similar to wadi water indicating that the source of groundwater is related to the wadi water. The isotopic composition of groundwater varies, depending on the isotopic signature of the recharge water. Since recharge needs some time to reach the water table, the isotopic composition in March would be affected by the various previous events. In the high-water season, the signatures in groundwater are more depleted than those collected in the low water season corresponding to the flooding of the wadi originating in the Assekrem rain.

4.4 Interactions between groundwater and wadi

Contrary to expectation, the three wells equipped with automatic recorders did not show individual signals of flood but rose progressively by about 1 m with a phase shift of two months after the first flood and a stabilization one month after the last flood (Fig. 3). The delayed response of the flood effect on the piezometric level may be due to (i) the important thickness of the unsaturated zone which can reach 15 m, (ii) the large amount of clays in the alluvium according to the granulometric analyses and the measurements of electrical resistivity (Zeddouri et al., 2018). The annual recharge for 2016 was estimated at about 21 hm^3.

5 Conclusion

This study aimed at determining the main components of the recharge of the alluvial aquifer and especially the wadi-alluvium connection, with the help of stable isotopes and piezometric measurements. Isotopic content of rainwater varied seasonally, with more enriched events in the core of the thunderstorms season. The annual weighted means of ^{18}O in precipitation in the two stations of Assekrem and Tamanrasset highlighted the effect of altitude. The isotopic signature in groundwater was similar to that of the wadi floods originating from the rainfall in Assekrem. Automatic recorders of the groundwater level showed a slow velocity of the wadi/-groundwater transfer during the two months of September and October making the level of the groundwater progressively rise by over 1 m. This slow response of the water table is explained by the low permeability of the unsaturated zone because of clays in the alluvium and by its thickness (up to 15 m).

Data availability. A more complete formulation of data used in this paper may be requested from the first author.

Author contributions. SHS and AZ contributed to the field activities; JDT and NP contributed to the isotopic analyses; all coauthors contributed to the data interpretation and paper writing.

Competing interests. The contact author has declared that none of the authors has any competing interests.

Disclaimer. Publisher's note: Copernicus Publications remains neutral with regard to jurisdictional claims in published maps and institutional affiliations.

Special issue statement. This article is part of the special issue "IAHS2022 – Hydrological sciences in the Anthropocene: Variability and change across space, time, extremes, and interfaces". It is a result of the XIth Scientific Assembly of the International Association of Hydrological Sciences (IAHS 2022), Montpellier, France, 29 May–3 June 2022.

Acknowledgements. We thank the LAMA laboratory of Hydrosciences Montpellier (France) for the isotopic analyses.

Financial support. This research has been supported by the French-Algerian PHC TASSILI funding (grant no. 16MDU961) led by IRD/University of Montpellier (France) and the RSPGA laboratory of University Kasdi Merbah-Ouargla (Algeria).

Review statement. This paper was edited by Christophe Cudennec and reviewed by Mohammad Merheb and one anonymous referee.

References

Ben Elkheir, A., Zeddouri, A., Hadj-Said, S., Satouh, A., Taupin, J.-D., Leduc, C., and Patris, N.: Configuration du système aquifère de la région de Tamanrasset (Sud de l'Algérie) : application des méthodes géophysiques à l'établissement d'un schéma conceptuel, J. Int. Sci. Technique Eau Env., 2, 21–24, 2017.

Guendouz, A., Moulla, A. S., Edmunds, W. M., Zouari, K., Shand, P., and Mamou, A.: Hydrogeochemical and isotopic evolution of water in the Complexe Terminal aquifer in the Algerian Sahara, Hydrogeol. J., 11, 483–495, https://doi.org/10.1007/s10040-003-0263-7, 2003.

Moulla, A. S., Guendouz, A., Reghis, Z., Chaid, Z., Larbes, M., Ouarezki, S., and Rezka, A.: Un aperçu de quelques résultats d'hydrologie isotopique importants obtenus au Sahara algérien, in: Colloque International sur les Ressources en Eau au Sahara, 13 December 2005, Ouargla, Algérie, Agence de Bassin Hydrographique du Sahara, 12 pp., 2005.

Saighi, O., Michelot, J. L., and Filly, A.: Isotopic characteristics of meteoric water and groundwater in Ahaggar massif (Central Sahara), International Atomic Energy Agency, Vienna, Austria, TEC-DOC no. 1207, 7–25, https://inis.iaea.org/collection/NCLCollectionStore/_Public/32/018/32018317.pdf?r=1 (last access: 25 August 2023), 2001.

Shamsuddin, M. K. N., Sulaiman, W. N. A., Ramli, M. F., Kusin, F. M., and Samuding, K.: Assessments of seasonal groundwater recharge and discharge using environmental stable isotopes at Lower Muda River Basin, Malaysia, Appl. Water Sci., 8, 120, https://doi.org/10.1007/s13201-018-0767-x, 2018.

Tarenga, L., Hadj-Said, S., Zeddouri, A., Ben Elkheir, A., and Taupin, J. D.: Alluvial groundwater a vulnerable resource in arid zone, First International Conference on the Hydrology of Large African Rivers, Hammamet, Tunisia, 2015.

Zeddouri, A., Ben Elkheir, A., Hadj-Said, S., Taupin, J. D., Leduc, C., and Patris, N.: Analysis and interpretation of electrical resistivity tomography data of alluvial aquifer of Tamanrasset Southern Algeria, in: Technologies and materials for renewable energy, environment and sustainability, edited by: Salam, C. T., Aillerie, M., Papageorgas, P., Haider, A. J., Vokas, G., and Jabur, A. R., TMREES18 conference in Beyrouth, Lebanon, 1 February 2018, AIP Conference Proceedings 1968, AIP Publ., New York, USA, 020024-1-020024-8, ISBN 978-0-7354-1675-8, 2018.

Proc. IAHS, 385, 351–357, 2024
https://doi.org/10.5194/piahs-385-351-2024

IAHS2022 – Variability and change across space, time, extremes, and interfaces

Impacts of climate variations and human influences on groundwater recharge in the semi-arid Trarza aquifer (SW Mauritania)

Ahmed Salem Mohamed[1], **Christian Leduc**[2], **Sidi Yeslem**[3], **Guinka Tagarinska**[4], and **Salek Aichete**[4]

[1]ESP (Ecole Supérieure Polytechnique), BP 4303 Cité Cadres – Sebkha, Nouakchott, Mauritania
[2]IRD, UMR G-EAU, BP 5095, 34196 Montpellier CEDEX 5, France
[3]Autorité de Régulation Multisectorielle, Nouakchott, Mauritania
[4]SNDE, BP 796, Nouakchott, Mauritania

Correspondence: Ahmed Salem Mohamed (ahmedsalemmd@gmail.com)

Received: 15 June 2022 – Revised: 3 May 2023 – Accepted: 21 July 2023 – Published: 19 April 2024

Abstract. Hydrological processes are rapidly changing in semi-arid areas under the double constraint of climate change and anthropization. Such changes were investigated at different scales of time and space in the Trarza aquifer (SW Mauritania), a flat region of about 40 000 km^2 with a low rainfall (100 to 250 $mm\,a^{-1}$). All old sources of hydrogeological information were exploited in complement with the more recent surveys. The geochemical information shows the importance of Quaternary eustatic variations in the present groundwater mineralization. In spite of very great variations of rainfall, the piezometric stability over the last 50 years shows that infiltration (about 4 $mm\,a^{-1}$) did not vary significantly. Since 1989, the big Diama dam has raised the mean level of the Senegal River, which brings an additional inflow to the alluvium and to the Trarza aquifer (about 34 $hm^3\,a^{-1}$). Groundwater recharge from the river and rainfall infiltration over the whole area have the same order of magnitude. In the Nouakchott area, the new supply of water to the capital (45 $hm^3\,a^{-1}$) has led to a significant groundwater rise and the submersion of several urban districts.

Keywords. UPH3; UPH13; Panta Rhei; Field observation; Mauritania; Groundwater recharge; Surface-groundwater interaction; Temporal changes

1 Introduction

The climate change is often evoked as a major threat on water resources in the next decades, especially in semi-arid areas. In fact, most of these regions are already exposed to multiple human activities that deeply modify the water cycle and these changes undermine the fragile balance between resources and uses that has existed for centuries (e.g. Scanlon et al., 2006; Leduc et al., 2017).

In Africa, the Sahelian region was marked by the catastrophic droughts of the 1970s and 1980s (e.g. Balme et al., 2006). In contrast with this violent phenomenon, the ever increasing demographic pressure and its heavy consequences on land use and land cover is an even more important driver of change, but it has acted continuously and progressively over the long term (e.g. Leblanc et al., 2008). In the same time, the multiplication of large and small hydraulic works, linked to the development of irrigation and the greater control of floods, modifies the flow and storage of both surface water and groundwater. Research on the evolution of water resources must therefore be carried out at multiple spatial and temporal scales. In the Sahel, wide areas with little infrastructure, a small amount of heterogeneous and sparse data do not represent the most favourable context. The Trarza aquifer in south-western Mauritania is emblematic of these difficulties where multiple influences must be unravelled among anthropization and climatic fluctuations of the last few decades, but also older paleo-environmental changes. These different drivers have considerably impacted the groundwater quantity and quality. The analysis of all the available data, recent and old, shows that beyond the apparent regional stability of the water table, water flows have changed considerably since 50 years.

2 The study area

The Trarza region extends over 40 000 km^2 in SW Mauritania and is limited by the Senegal River, the Atlantic Ocean and the Mauritanides metamorphic range (Fig. 1). In this very flat region, the succession of Quaternary dunes and interdune depressions is the main relief figure. The short rainy season (June to September) and the long dry season are typical of the Sahelian climate, as well as the S-N rainfall gradient (about 250 mm a^{-1} in the Senegal River valley and only 100 mm a^{-1} in Nouakchott). Like everywhere in semi-arid areas, rainfall is very irregular in space and over time: the annual average may fall in one or two rainy events (as in the case of Rosso in 2010) and, at another scale of time, the whole Sahelian region was marked by a long and severe drought in the 1970s and 1980s. The phreatic aquifer is the highest in a multi-layered aquifer system of the Senegal-Mauritania sedimentary basin. Its Tertiary and Quaternary sediments are alternating lacustrine, eolian and marine deposits. The rural population is less than 300 000 inhabitants, to be compared with more one million people living now in Nouakchott, the capital city that was created from nearly nothing at the end of the 1950s. Until 2011, the water supply for Nouakchott was taken from the Trarza aquifer by pumping in the Idini area. Water for the city is now transferred from the Senegal River by a 180 km pipe.

The mean depth to the water table is about 30 m, but it varies between a few metres close to the sea and the Senegal River and more than 70 m in the north-eastern part. The thickness of the saturated zone varies between 10 and 50 m, with a mean of 20 m. The volume of water stored in the aquifer is estimated at 100 km^3.

3 Data and methods

All available hydrogeological data since the first measurements in 1959 were collected and exploited. This data set is very heterogeneous, with a few big campaigns like in 1959 (Paloc, 1962), 1973 (unpublished) and 2011 (unpublished), and many sparse and discontinuous point observations. In a few sites, a specific survey provides a much denser information in space and/or over time. The richest example is the Idini area where the national water company SNDE has accumulated about 1000 measurements since 1972 in the 30 km^2 of the pumping site.

These different measurements consist mostly of groundwater levels and, secondarily, electrical conductivity of groundwater. Analyses of major ions and isotopes and estimations of hydrodynamic characteristics are much rarer.

In spite of its heterogeneity, this data set enables the differentiation of the multiple drivers acting on the Trarza phreatic aquifer at various scales in space and over time, from the local pumping to the climatic variations and marine transgressions of the Quaternary.

4 Results

4.1 Long-term regional stability

The comparison of measurements over 60 years requires that wells are indisputably identified and not exposed to significant perturbations like a local pumping or a distant big hydraulic work. Depending on the data subsets that are compared, the number of common points varies between 13 and 97. The main result is the absence of a clear trend over decades. For instance, in the most copious comparison (1973 vs. 2011), rising wells are as numerous as decreasing wells and their geographical distribution has no visible pattern. This overall stability is a significant help: isolated measurements can be used for complementing the major field campaigns and their merging provides a very robust piezometric map (Fig. 1).

4.2 Rainfall infiltration in the interdune depressions

After clearing of possible disturbances by local pumping, a seasonal fluctuation of the groundwater level is frequently observed in interdune depressions during the rainy season (between 0.2 and 1 m in most cases).The continuous measurement in the Nkek village (Fig. 2) illustrates the phenomenon, in complete coherence with other continuous observations and measurements taken before and after the rainy season in 36 wells in 2010 and 49 wells in 2015.

In spite of the low annual rainfall, this recharge seems present over the whole aquifer. Nine measurements of ^{14}C (Mohamed, 2012) and 17 measurements of ^{3}H (IAEA, 2011, 2017) revealed highly variable mixing of recent and older waters (0–6 TU; 0–9 ka) and confirmed the existence of a modern contribution everywhere in the aquifer.

Using the water table fluctuation method (WTF) and assuming an effective porosity of 8 %, the median value of recharge was estimated at 1.5 % of the total rainfall in the north of the study area, and 2.1 % in the south, which represents a regional recharge by rainfall of about 4 mm a^{-1}.

4.3 Recharge from the Senegal River

In the whole Trarza region, the highest groundwater levels are met along the Senegal River. Before the construction of the Diama dam, at 23 km from the river mouth, and the Manantali dam, in Mali, at the end of the 1980s, the annual flood of the river was very variable in duration and amplitude. Exchanges between the river and alluvium reversed seasonally (Illy, 1973). In the low flow period, sea water could reach 250 km up the valley, which obviously impacted the groundwater quality. Since the construction of the two dams, the

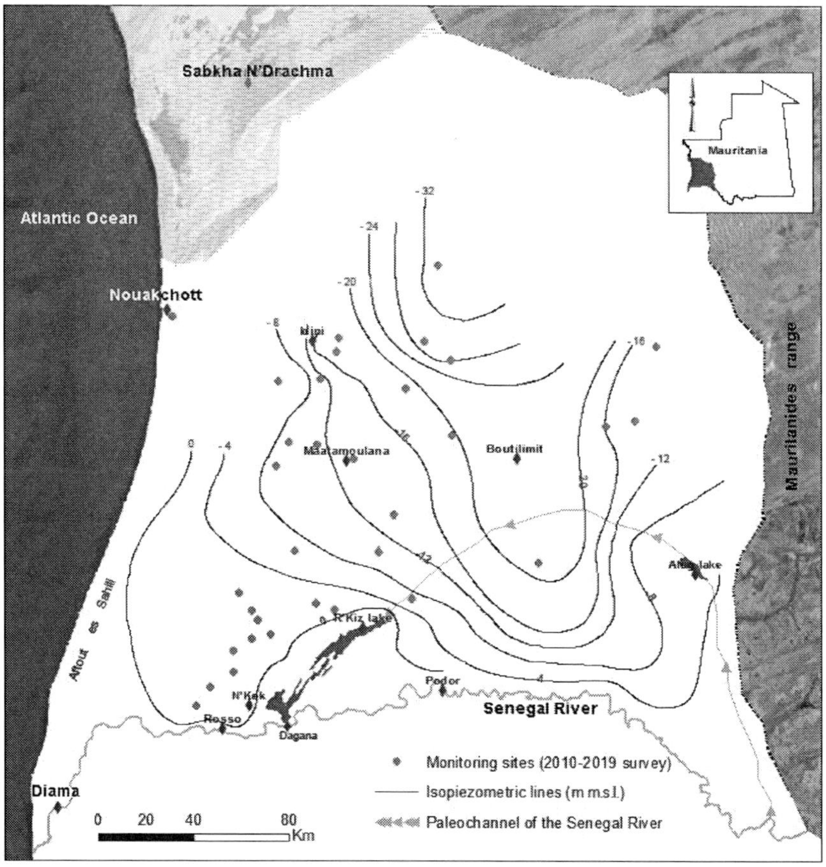

Figure 1. Extent of the Trarza phreatic aquifer and its piezometric contour lines drawn from the 2010–2020 survey.

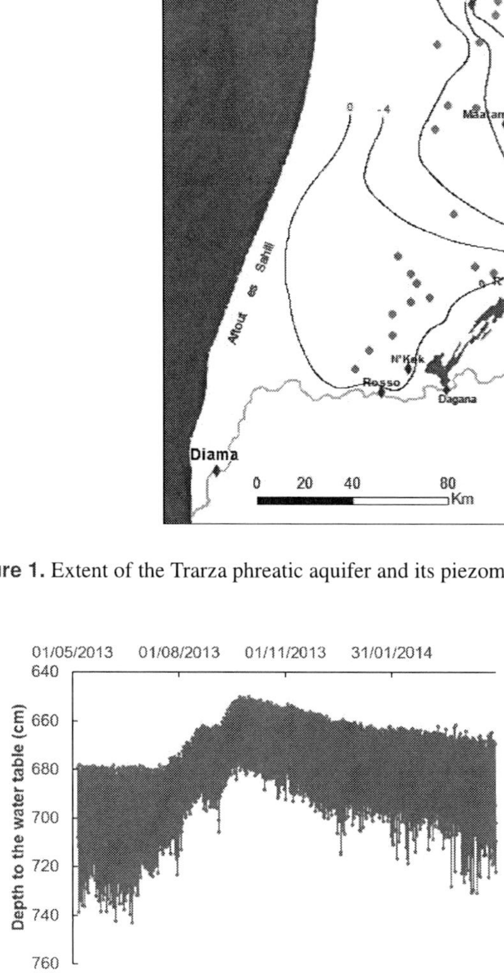

Figure 2. Fluctuation of the groundwater level in a well at the foot of a dune: seasonal recharge by rain infiltration and daily perturbation by pumping.

river level is kept much higher (in Rosso, the minimum river level has risen by 2 m). The river control induces a permanent recharge to the Trarza phreatic aquifer, and the rise of groundwater levels on both banks of the river is up to 1 m (Mohamed et al., 2014; Ngom et al., 2016). This recharge flow is estimated at about $110\,\mathrm{hm^3\,a^{-1}}$, a hydraulic gradient of 1 ‰ and a median transmissivity of $5.5 \times 10^{-3}\,\mathrm{m^2\,s^{-1}}$ in the neighbouring aquifer. Uncertainty is still large but independent calculations in surface water and groundwater converge.

4.4 Urban recharge in Nouakchott

In its natural state, groundwater of the Nouakchott area was naturally brackish and not used for human consumption. It was, and still is, very close to the ground surface and hydraulically constrained by the sea and the strong evaporation uptake. Population in the capital city changed from 5000 inhabitants in 1960 to more than one million. Most of domestic waste water is discharged in individual septic tanks and recharges the phreatic aquifer. Since 2011, the water supply provided from the Senegal River has tripled the quantity of water distributed in the capital (about $0.12\,\mathrm{hm^3\,d^{-1}}$ presently). Increased leaks from the water distribution network and increased waste water release have increased the aquifer recharge and the water table has risen by up to 0.6 m (Mohamed et al., 2017). Large urban districts are now submerged and the phenomenon is limited only by a constant pumping in multiple sites of the city (Fig. 3).

4.5 Intense pumping in Idini

After 40 years of progressive increase of pumping, the maximum drawdown observed in the piezometers spread over the 30 km^2 of the Idini pumping site was 7.7 m. Because of the strong reduction in pumping since 2011, the groundwater level has partially recovered (up to 3 m). The volume of dewatered aquifer was about 16 hm^3 in 2011, which is very small in comparison with the total of pumped volumes over four decades (0.4 km^3). This limited drawdown is consistent with the local hydrodynamic characteristics (median transmissivity of $5.5 \times 10^{-3}\,m^2\,s^{-1}$).

Whatever the pumping rate, there has been no overall degradation of the groundwater quality in Idini over time. Only three boreholes among the 46 wells in operation have experienced a significant change in mineralization since 1972 (from 300 to 1210 $mg\,L^{-1}$ for the most important variation) while in all others, no fluctuations exceeded 10 % in 50 years.

5 Discussion

5.1 Regional groundwater flows

The uneven distribution of information, with a smaller density in the north and east of the Trarza aquifer, is a difficulty but not an obstacle to the understanding of regional hydrogeology.

The highest groundwater levels are observed close to the coast and in the Senegal River valley, and the water table progressively decreases to the northeast. This reminds the natural piezometric closed depressions that are frequently observed in the Sahelian region (e.g. Favreau et al., 2002). The singularity of a dissymmetrical shape can be explained by the contact with the impervious metamorphic Mauritanides rocks.

5.2 Long-term stability and recharge

Like all other Sahelian regions, Trarza was subject to the very severe droughts of the 1970s and 1980s and to rapid demographic changes but, surprisingly, there is no obvious trace of these major events in groundwater while many other Sahelian aquifers experimented important changes (e.g. Leduc et al., 2001). The stability of the water table over 60 years is an indicator of a low renewal rate. Out of the Nouakchott area, recharge is due to rainfall infiltration and inflow from the Senegal River, estimated at about 0.16 and 0.11 km^3a^{-1} respectively. Even if uncertainty remains large on the two figures, the order of magnitude is certain. Their sum is less than 3 ‰ of the water stored in the Trarza aquifer, which is coherent with the small impact of recent events. The net recharge is probably even smaller than 3 ‰ as a significant part of this inflow may go back to the atmosphere by evapotranspiration in the first months after infiltration.

The low value of rain infiltration, between 1 % and 5 % of annual rainfall, agrees with other estimates of recharge in semi-arid areas. Under the natural conditions that previously existed in south-western Niger, Le Gal La Salle et al. (2001) proposed a value of 5 $mm\,a^{-1}$ (i.e., 1 % of annual rainfall of about 560 mm). The same ratio was also calculated by Leduc et al. (2000) in the drier north-western Lake Chad basin. Beyond the Sahelian belt, the proposed value also agrees with the wide compilation of studies in semi-arid regions by Scanlon et al. (2006), who concluded that recharge usually varies between 0.1 % and 5 % of annual rainfall.

A first independent confirmation of the low renewal rate comes from the stability of the piezometric closed depression. In continental areas, the only acceptable explanation for such shapes is evapotranspiration uptake, which is necessarily very low when groundwater lies at some tens of metres below the ground surface. An important recharge would rapidly erase the depression.

Another confirmation comes from the presence of seawater in the aquifer, at tens of kilometres of the present coastline (Mohamed et al., 2014). These relicts of the last transgression episodes of Quaternary would have disappeared by dilution in case of an important annual recharge.

5.3 Groundwater memory

Transgression episodes have been essential drivers of groundwater dynamics in the past (Elouard et al., 1972; Miskovsky et al., 1991). They covered a very large extent and lasted a long time and their impact is still visible locally in the aquifer mineralization (high salinity, Br / Cl ratio close to the marine ratio), even very far from the present coastline (Fig. 4). Traces of more local and/or shorter hydrological events of the past are identified with greater difficulty.

In 1950, an exceptional flood of the Senegal River spread floodwater throughout the Aftout es Sahili, a long topographic depression parallel to the coastline, at a distance of 1 to 3 km from the sea. It brought water from the Senegal River right up to Nouakchott, 190 km north of the river delta. There was no hydrogeological measurements at that time but the presence of modern fresh water in the coastal brackish groundwater is a good indicator of an exceptional recharge (Mohamed, 2012). This fresh water may result from the 1950 flood only or from the accumulation of several similar older floods that occurred a very few times per century. Their impact was limited to a narrow coastal strip but the amount of water entering the aquifer may have been as important locally as the rain infiltration for 30 or 50 years.

Further into the past, the bed of the Senegal River has moved significantly over time. A former river bed is still visible farther north in Mauritania in the form of Lake Aleg and Lake R'kiz, then farther south in Senegal with Lake Guiers

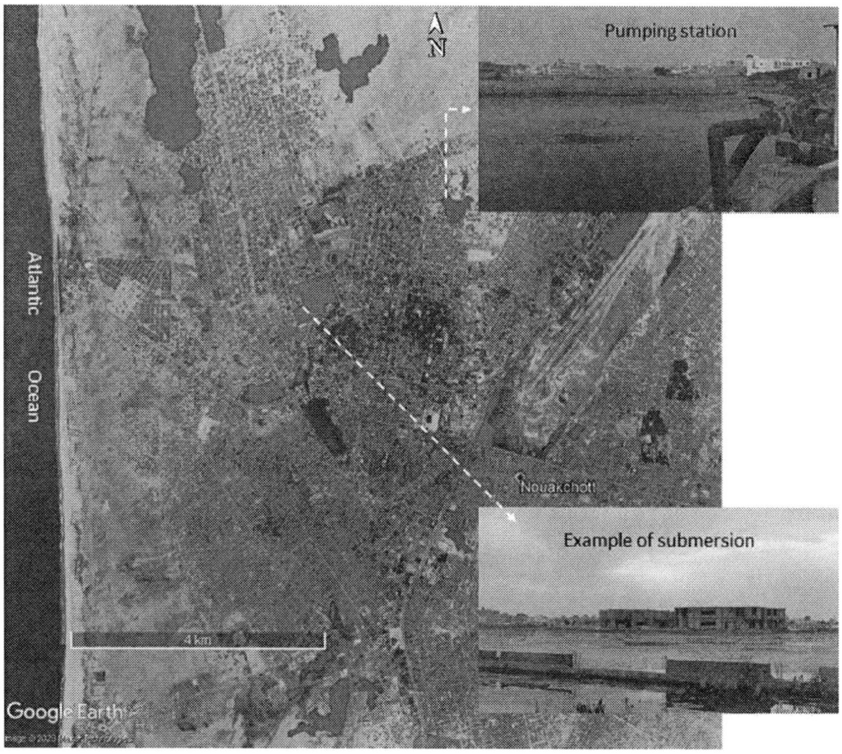

Figure 3. Areas submerged by the groundwater rise in Nouakchott in light blue (background © Google Earth).

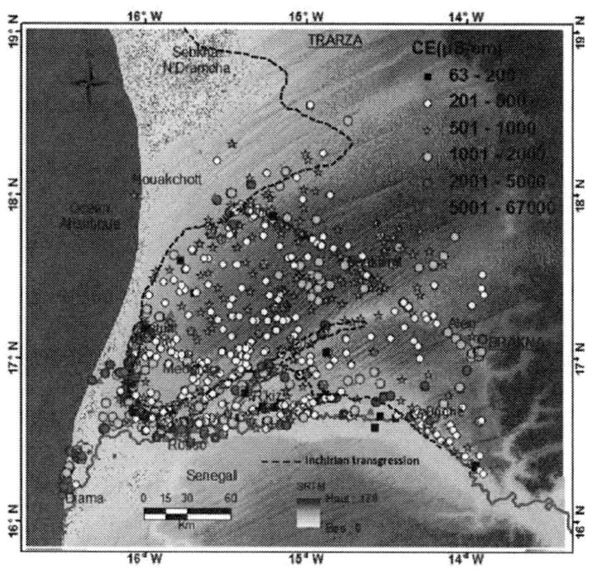

Figure 4. Groundwater salinity in the Trarza aquifer (Mohamed, 2012; complemented with the data set of the present study). The black dashed line limits the maximum of Quaternary transgressions.

(Fig. 1). The maximum distance to the north from the present bed is about 70 km. Each year Lake R'kiz is filled up by a distributary of the Senegal River, which follows the ancient bed and recharges the surrounding aquifer with a preferential flow along the former bed.

The long history of the Trarza aquifer implies that the contemporary observations have to be interpreted in connection with present drivers and also in reference to much older events.

5.4 Anthropogenic perturbations in Nouakchott

Increasing groundwater recharge in Nouakchott obviously has considerable social, sanitary and geotechnical consequences. The process started 10 years ago and will probably continue over the next decades. Even partially polluted by domestic uses, this artificial recharge will gradually reduce the groundwater salinity, which was originally very high in this area.

As a counterpoint to the geochemical improvement, this artificial recharge is compensated by a permanent pumping to limit the extension of flooded areas. Part of this water is discharged into the sea, the rest is discharged over the ground surface away from the city, which only shifts the problem. In this very flat city, partly below sea level and with a very insufficient wastewater collection system, a more sustainable alternative is still to be found.

Another important issue in the Nouakchott area is the weakening of the coastal dune bar by uncontrolled sand extraction and by harbour dykes modifying the littoral sediment transport (Mohamed et al., 2017). The rise of the sea level and the increased number of storms expected as a consequence of the global warming will lead to more frequent intrusion of seawater in the coastal area. The same threat concerns two other sites of new big harbours along the Mauritanian coast.

5.5 Anthropogenic perturbations in the Senegal River valley

Thirty years after Diama dam was built, groundwater can be considered as having reached a new equilibrium with the raised base level of the river. Our present estimate of annual inflow from the Senegal River of about 110 hm^3 is a net recharge to the Trarza aquifer, lower than the total sum of fluxes exchanged each year in both directions during the annual flood and the following recession. It is consistent with the hydrological surveys of the Senegal River (e.g. Dieng, 1999). In the most recent synthesis, Bader (2015) estimated the river loss between Bakel (630 km upstream of Diama) and Diama at a mean value of 725 $hm^3\ a^{-1}$, to be divided into direct evaporation, evapotranspiration in the riparian vegetation and groundwater recharge through the two banks.

The Senegal delta area downstream of Diama is very fragile hydraulically and sedimentarily, depending on oceanic and continental conditions (e.g. Barusseau et al., 1998) and human interventions could significantly impact it. A dramatic example is the artificial breach that was opened in October 2003 in the Langue de Barbarie, which is a narrow strip of land between the Senegal River and the Atlantic Ocean, downstream of Diama. The breach was dug to prevent a disastrous flood of the Senegal River, with an initial width of a few metres that rapidly reached 5 km and changed the freshwater-salt water balance in the whole area. In a less violent evolution, irrigation expands along the Senegal valley (34 700 ha in 2015) and will even increase with the construction of a new canal to the Aftout es Sahili coastal depression, transferring water from the Senegal River (55 km northwards) to a new 16 000 ha irrigated area. These new hydraulic conditions will progressively modify both the level and quality of groundwater, northwards and eastwards respectively, as already observed ion the southern bank of the Senegal River (e.g. Diaw et al., 2019).

6 Conclusion

Like many other semi-arid areas, South-west Mauritania depends on groundwater for supplying water to rural communities, but hydrogeological information is still scarce. This may be easily explained by the size of the country, the small and poor population, and the sparse hydraulic infrastructure. The lack of any information before the 1950s and the limited amount of data acquired since could limit the identification of major physical processes. In fact, the large extent of the aquifer and the small renewal rate give it a great inertia, which makes possible to concatenate heterogeneous data sets. This inertia allows scientists to go back into the past and find traces of exceptional events that took place a few decades, centuries or thousands of years ago.

In contrast with many other semi-arid aquifers, the Trarza aquifer does not appear exposed to high risks in the near future. In quantitative terms, the water supply to Nouakchott from the Senegal River has considerably reduced the volume of water pumped from the aquifer and the higher river level in the Senegal River has increased the recharge from the river. In qualitative terms, the control of the Senegal River now avoids the marine intrusion that used to come up every year and brings an additional input of freshwater to the aquifer. At the local scale of Nouakchott, the massive introduction of fresh water in the aquifer reduces the groundwater salinity.

Numerous decisions concerning water management in Mauritania are now taken without a clear idea of the hydrological context and their possible consequences. The relative groundwater inertia and the absence of an immediate big threat cannot justify that observation networks are not seen as a necessity by authorities while they undertake new large-scale hydraulic infrastructures.

Data availability. Older data were exploited from numerous original technical reports, mostly unpublished. Data are not freely available but may be requested from the first author.

Author contributions. The paper was mostly written by ASM and CL. All authors contributed to the collection and interpretation of recent data.

Competing interests. The contact author has declared that none of the authors has any competing interests.

Disclaimer. Publisher's note: Copernicus Publications remains neutral with regard to jurisdictional claims in published maps and institutional affiliations.

Special issue statement. This article is part of the special issue "IAHS2022 – Hydrological sciences in the Anthropocene: Variability and change across space, time, extremes, and interfaces". It is a result of the XIth Scientific Assembly of the International Association of Hydrological Sciences (IAHS 2022), Montpellier, France, 29 May–3 June 2022.

Review statement. This paper was edited by Christophe Cudennec and reviewed by Mohammad Merheb and one anonymous referee.

References

Bader, J. C.: Monographie hydrologique du fleuve Sénégal. De l'origine des mesures jusqu'à 2011, IRD Editions, ISBN 978-2-7099-1885-5, 2015.

Balme, M., Lebel, T., and Amani, A.: Années sèches et années humides au Sahel: quo vadimus?, Hydrol. Sci. J., 51, 254–271, 2006.

Barusseau, J. P., Ba, M., Descamps, C., Diop, E. S., Diouf, B., Kane, A., Saos, J. L., and Soumaré, A.: Morphological and sedimentological changes in the Senegal River estuary after the construction of the Diama dam, J. Afr. Earth Sci., 2, 317–326, 1998.

Diaw, M., Mall, I., Leblanc, M., Faye, S., and Travi, Y.: Quantitative estimation of recharge potentialities of shallow aquifers in Senegal River Delta hydrosystem, American J. Water Sci. Engineering, 5, 47–61, 2019.

Dieng, B.: Impacts des barrages de Diama et Manantali sur les nappes de la vallée du fleuve Sénégal. Sud Sci. Technologies. 3, 31–38, 1999.

Elouard, P., Faure, H., and Hébrard, L.: Quaternaire du littoral mauritanien entre Nouakchott et Port-Etienne (18°–21° latitude Nord), ASEQUA, 23, 15–24, 1972.

Favreau G., Leduc, C., Marlin, C., and Guero, A.: Une dépression piézométrique naturelle en hausse au Sahel (Sud-Ouest du Niger), C.R. Geosci., 334, 395–401, 2002.

IAEA (International Atomic Energy Agency): Studies on groundwater in Mauritania, Water & Environment news no. 28, IAEA, Vienna, Austria, ISSN 1020-7120, 2011.

IAEA (International Atomic Energy Agency): Integrated and sustainable management of shared aquifer systems and basins of the Sahel Region, The Senegalo-Mauritanian basin. Report of the RAF/7/011 project, Vienna, Austria, 43 pp., https://www.iaea.org/sites/default/files/raf7011_senegalo-mauritanian_basin.pdf (last access: 1 July 2023), 2017.

Illy, P.: Etude hydrogéologique de la vallée du fleuve Sénégal, Technical report for "Projet hydro-agricole du bassin du fleuve Sénégal", OMVS technical report, Dakar, Senegal, 163 pp., 1973.

Leblanc, M. J., Favreau, G., Massuel, S., Tweed, S. O., Loireau, M., and Cappelaere, B.: Land clearance and hydrological change in the Sahel: SW Niger, Global Planet. Change, 61, 135–150, 2008.

Leduc, C., Sabljak, S., Taupin, J. D., Marlin, C., and Favreau, G.: Estimation de la recharge de la nappe quaternaire dans le Nord-Ouest du bassin du lac Tchad (Niger oriental) à partir de mesures isotopiques, C.R. Acad. Sci. Paris, 330, 355–361, 2000.

Leduc, C., Favreau, G., and Schroeter, P.: Long-term rise in a Sahelian water-table: the Continental Terminal in South-West Niger, J. Hydrol., 243, 43–54, 2001.

Leduc, C., Pulido-Bosch, A., and Remini, B.: Anthropization of groundwater resources in the Mediterranean region: processes and challenges, Hydrogeol. J., 25, 1529–1547, 2017.

Le Gal La Salle, C., Marlin, C., Leduc, C., Taupin, J. D., Massault, M., and Favreau, G.: Renewal rate estimation of groundwater based on radioactive tracers (^{3}H, ^{14}C) in an unconfined aquifer in a semi-arid area, Iullemeden Basin, Niger, J. Hydrol., 254, 145–156, 2001.

Miskovsky, J. C., Saos, J. L., and Bonkel, A.: Paléohydrologie et paléoclimatologie des bassins endoréiques du sud de la Mauritanie depuis 18 000 B.P., Bull. Soc. Géol. France, 162, 1141–1150, 1991.

Mohamed, A. S.: Approches hydrodynamique et géochimique de la recharge de la nappe du Trarza (SO de la Mauritanie), PhD thesis, University Paris-Sud 11, France, https://theses.hal.science/tel-00800604 (last access: 1 July 2023), 2012.

Mohamed, A. S., Marlin, C., Leduc, C., and Jiddou, M.: Means of groundwater recharge in a semi-arid area: example of the Trarza aquifer (southwest Mauritania), Hydrol. Sci. J., 59, 1046–1062, 2014.

Mohamed, A. S., Leduc, C., Marlin, C., Wagué, O., and Sidi Cheikh, M. A.: Impacts of climate change and anthropization on groundwater resources in the Nouakchott urban area (coastal Mauritania), C.R. Geosci., 349, 280–289, 2017.

Ngom, F. D., Tweed, S., Bader, J. C., Saos, J. L., Malou, R., Leduc, C., and Leblanc, M.: Rapid evolution of water resources in the Senegal delta, Global Planet. Change, 144, 34–47, 2016.

Paloc, H.: Carte hydrogéologique du bassin sud-ouest mauritanien au 1 / 500 000 et notice explicative d'après les travaux d'Elouard P. (1959), Technical report, BRGM, Dakar, 1962.

Scanlon, B. R., Keese, K. E., Flint, A. L., Flint, L. E., Gaye, C. B., Edmunds, W. M., and Simmers, I.: Global synthesis of groundwater recharge in semiarid and arid regions, Hydrol. Process., 20, 3335–3370, 2006.

Proc. IAHS, 385, 359–364, 2024
https://doi.org/10.5194/piahs-385-359-2024

Open Access PIAHS

Investigation of drought and flooding areas in coastal countries of West Africa in the context of global warming

Domiho Japhet Kodja[1], Gandomè Mayeul Leger Davy Quenum[3], Houteta Djan'na Koubodana[4,5], Ernest Amoussou[1,2], Isaiah Sewanu Akoteyon[7], Arsène Sègla Josué Akognongbé[1], Mahougnon Fidèle Ahéhéhinnou Yêdo[1], Gil Mahé[6], Jean-Emmanuel Paturel[6], Expédit Wilfrid Vissin[1], and Constant Houndénou[1]

[1]Laboratoire Pierre Pagney, Climat, Eaux, Ecosystèmes et Développement (LACEEDE), Université d'Abomey Calavi (Bénin), 03 BP 1112 Cotonou, Bénin
[2]Département de Géographie et Aménagement du Territoire, Université of Parakou, Parakou, Bénin
[3]Laboratory of Applied Hydrology (LHA), National Water Institute (INE), 01 BP 4521 Cotonou, Benin
[4]African Institute for Mathematical Sciences (AIMS), Kigali, Rwanda
[5]Institut Togolais de Recherche Agronomique, ITRA, Lomé, Togo
[6]UMR 5599 HydroSciences Montpellier, CNRS, IMT, IRD, Montpellier, France
[7]Department of Geography and Planning Lagos State University, PMB 0001 LASU Post Office, Ojo, Lagos, Nigeria

Correspondence: Domiho Japhet Kodja (japhdom@gmail.com)

Received: 15 June 2022 – Revised: 10 May 2023 – Accepted: 21 July 2023 – Published: 19 April 2024

Abstract. This study investigated drought and flooding changes in West Africa between 1983–2012 and projected near future (2025–2054) periods. The datasets used are the CHIRTS and CHIRPS-2 for observed reanalysis and five (05) models of ISIMIP2b for Shared Socio-economic Pathways (SSP1.2-6 and SSP5-8.5). Extremely and very wet days total precipitation (R95pTOT; R99pTOT) and Standardized Precipitation Evapotranspiration Index (SPEI) were employed to investigate floods and drought spatial distribution using Sen Slope trend analysis method. The results showed that there is a variability in the spatial distribution of extreme indices with an upward and downward trend of dry and wet rainfall periods in West Africa in both historical and projected periods. This observation suggests that the study area is faced with rainfall variability marked by extreme events. A further examination on the spatial and temporal distribution of flood occurrence showed that more flood events were observed in the Gulf of Guinea and Savannah countries, followed by an increase in uniform spatial distribution and moderate wet days both under SSP1.2.6, and SSP 5.8.5. In addition, result showed that an upward trend in wet periods can cause the occurrence of extreme events associated with floods in the context of global warming. However, with these scenarios negative changes are not excluded in the East, the Sahel and some western part of the Gulf of Guinea in the study area for the SSP5.8.5 scenario. Thus, the results revealed that the spatio-temporal variability of extreme rainfall can have repercussions on the hydrological functioning of watersheds, water availability and water-dependent activities.

Keywords. Flood; drought; extreme precipitation; SPEI; global warming; West Africa

1 Introduction

In recent decades, West Africa has been faced with a high frequency and intensity of extreme hydro-meteorological events which are mostly detrimental to socio-economic and environmental issues (Quenum et al., 2019; Ajayi et al., 2020). Thus, with the effects of global warming, precipitation regimes are not only affected, but also an uneven distribution of precipitation is observed with repercussions on the hydrodynamic functioning of watersheds and the quality of water resources within the in ecosystems (Quenum et al., 2019). The Inter-Governmental Panel on Climate Change (IPCC) predicts changes in temperature and precipitation regimes with a high degree of confidence in the increase in the frequency of extreme events (Pedersen et al., 2022). This frequency is linked to the increase in the intensity and duration of storms, cold or heat waves, the increase in the extent of regions affected by drought and unexpected floods accompanied by uneven consequences distributed for many regions. According to Obahoundjè et al. (2017), the rainfall recession can reduce the availability of water resources with exacerbation on agricultural land, industry, energy, food security and river flow regimes. As for the occurrence of extreme rainfall, it can cause the destruction of lives and properties of communities, most of which are already vulnerable to extreme weather and climate conditions with threats of terrorism (Odoulami et al., 2019). Faced with this situation, the analysis of the spatio-temporal distribution of rainfall as well as that of these trends arouse the curiosity of this study not only to contribute to the improvement and integrated management of water resources but also for policy formulation, taking into account the impact of climate change on socio-economic and environmental developmental programs with a view to integrating it into the Sustainable Development Goals (SDGs). The analysis of various characteristics and trends of precipitation is an essential task to improve the utilization of water resources. To do this, several works have frequently used trend analysis in the context of rainfall or even hydrological variability based on approaches such as linear regression, Sen slope, Spearman's rho and trend tests by Mann-Kendall (Quenum et al., 2019). However, this study conducted statistical analyses of precipitation data using the Sen's slope that is basically applied to identify the magnitude of trend in a data series which are not serially auto-correlated (Şen, 2015).

The study therefore seek to investigate the extreme events of drought and flooding inherent to rainfall in coastal countries in West Africa in the context of global warming.

1.1 The study area

The study area (Fig. 1) is located between latitudes 0–20° N and longitudes 20° E–20° N which covers the Sahel, the Sudano-Sahelian and the Guinea coast regions. The study focused much more on this last region with specific analyses on the coastal countries (Senegal, Côte d'Ivoire, Benin, Togo and Nigeria) which have watersheds with hydrological potential for energy and agricultural purposes.

Table 1. Type of the scenarios.

Types of Scenarios	Description
SSP1-2.6	More gradual decline in emissions than RCP2.6 and a higher starting point
SSP2-4.5	Higher starting point and slightly slower decline than RCP4.5
SSP4-6.0	Quite different from RCP6.0, with CO_2 emissions peaking and declining
SSP3-7.0	Middle of the range of baseline results produced by energy system models.
SSP5-8.5	Substantially higher CO_2 emissions than RCP8.5

Source: Adapted from CMIP 5 and CMIP6 Scenario comparison

2 Data sources and methods

2.1 Data sources

For this study we used the daily observed reanalysis covering (1983–2012) of CHIRTS (T_{min} and T_{max}) and CHIRPS-2 (Rainfall) for 0.05° × 0.05° resolution to evaluate the accuracy of the climate models (https://data.chc.ucsb.edu/products/CHIRPS-2.0/, last access: 7 September 2023) CHIRTS and CHIRPs-2 data have been validated with observation over West Africa by previous studies (Atiah et al., 2020; Dembélé et al., 2016). The ensemble mean of the five models of ISIMIP2b (GFDL, IPSL, MPI, MRI, UKESM1, daily 0.5° × 0.5°) of (pr, tasmin, tasmax) over the historical period spanning 1983 to 2012 and near future 2025–2054 with the scenarios SSP 1.2.6 and SSP 5.8.5 were merged to analyse drought and flooding areas in coastal countries of West Africa (Table 1) https://www.isimip.org/gettingstarted/isimip2b-bias-adjustment/ (last access: 7 September 2023).

We used minimum and maximum temperature to compute the Potential Evapotranspiration (PET) in the study area.

2.2 Methods

2.2.1 Characterization of flood events

For the characterization of areas or land prone to drought or flooding in the coastal countries of West Africa in the context of global warming, certain indices were developed by the Ex-

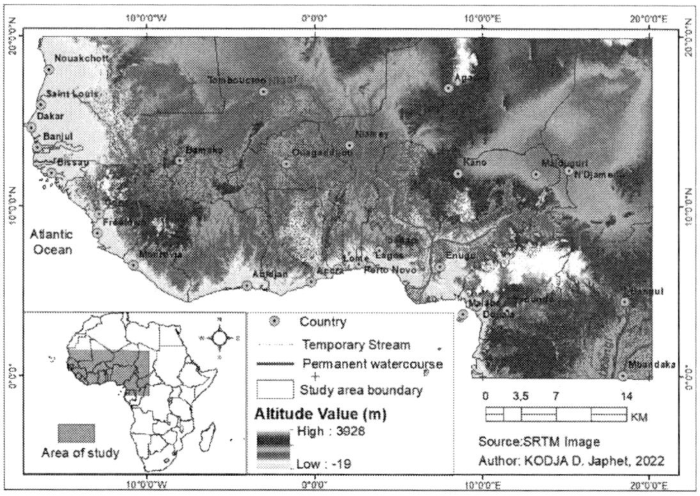

Figure 1. Geographical location of the study area.

pert Team on Climate Change Detection Indices (ETCCDI) were employed. Detailed descriptions of these indices are available on the ETCCDI website (http://cccma.seos.uvic.ca/ETCCDMI/software.shtml, last access: 7 September 2023). In the present study, two (02) indices were calculated and they are based on the extreme rainfall characteristic of floods.

- *R95pTOT.* Annual total PRCP when $RR > 95p$. Let RR_{wj} be the daily precipitation amount on a wet day w ($RR \geq 1.0$ mm) in period I and let $RR_{wn}95$ be the 95th percentile of precipitation on wet days. If W represents the number of wet days in the period. R95pTOT were calculated by Eq. (1):

$$R95_{pj} = \sum_{w=1}^{w} RR_{wj} \text{ with } RR_{wj} > RR_{wn}95 \quad (1)$$

- *R99pTOT.* Annual total PRCP when $RR > 99p$: Let RR_{wj} be the daily precipitation amount on a wet day w ($RR \geq 1.0$ mm) in period i and let $RR_{wn}99$ be the 99th percentile of precipitation on wet days. If W represents the number of wet days in the period. R99pTOT were calculated by Eq. (2):

$$R99_{pj} = \sum_{w=1}^{w} RR_{wj} \text{ with } RR_{wj} > RR_{wn}995 \quad (2)$$

2.2.2 Standardized Precipitation and Evapotranspiration Index (SPEI)

The normalized precipitation evapotranspiration index (SPEI) was highlighted in 2010 and has been the subject of several studies in climatology, hydrology, water availability and in research relating to agrarian systems (N'guessan Bi et al., 2020). The normalized precipitation evapotranspiration index (SPEI) focus on the difference between precipitation and potential evaporation instead of just precipitation (Vicente-Serrano et al., 2010). Thus, given its specificity in measuring the severity of drought, analyzed drought variability and also of extreme wetness (Allen et al., 2011), it was adopted in this research. The normalized precipitation evapotranspiration index were calculated for 1, 3, 6, 9 and 12 months on each grid point for each dataset (Quenum et al., 2019). However for this study, the results presented are those of the SPEI generated over 12 months because the latter provide a representation of long-term precipitation patterns such as those calculated over 12 months and contribute to the detection of hydrological impacts (OMM and GWP, 2016). The calculation method used in this research is presented in Table 2.

Table 2. Classification of SPEI values.

Periods	Indicators	SPEI Values
Wet periods	Extremely wet	$SPEI \geq 2$
	Severalty/very wet	$1.5 \leq SPEI < 1.99$
	Moderately wet	$1 \leq SPEI < 1.49$
Near-Normal periods	Near-Normal	$-0.99 < SPEI < 0.99$
Drought periods	Moderately dry	$-1.49 < SPEI \leq -1$
	Severalty/very dry	$-1.99 < SPEI \leq -1.5$
	Extremely dry	$SPEI \leq -2$

Source: Adapted from Vicente-Serrano et al. (2010).

The computation of potential evapotranspiration is used to calculate the normalized Precipitation-Evapotranspiration Index (SPEI). For that, Thornthwaite equation was used in this research.

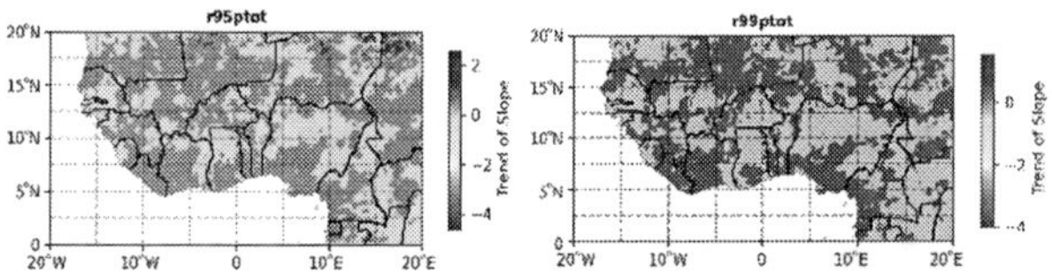

Figure 2. Trends of the slope in the precipitation on very and extremely wet period during 1983–2012.

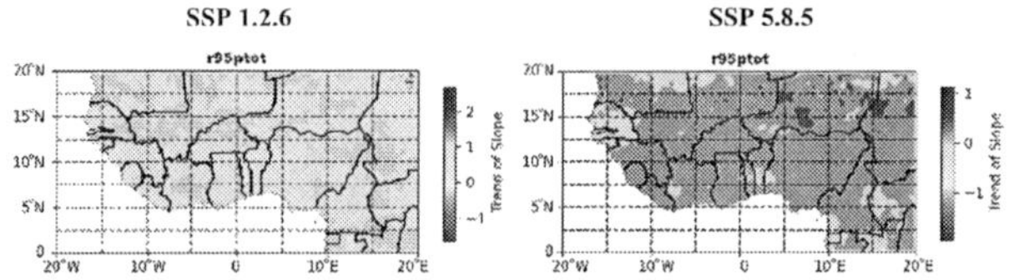

Figure 3. Trend of Slope on very wet period in near future period (2025–2054) with scenario SSP 1.2.6 and SSP 5.8.5.

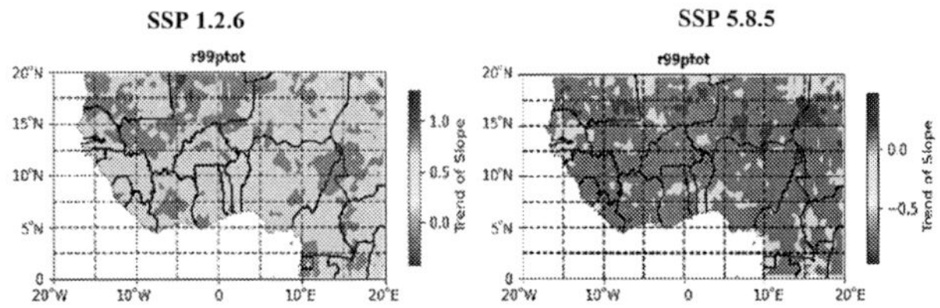

Figure 4. Trend of Slope of the Scenarios SSP 1.2.6 and SSP 5.8.5 of wet period from SPEI for 12 months on 2025–2054.

2.2.3 Method of trend analysis

Trend analysis of rainfall time series includes determination of increasing and decreasing trend and magnitude of trend and its statistical significance" (Quenum et al., 2019). Then the Sen Slope trend was employed for this research. In addition these indicators were calculated on historical (1983–2012) and near future periods (2025–2054) of precipitation in West Africa.

3 Results and discussion

3.1 Spatial trend analysis of precipitation on very wet by R95pTOT and extremely wet R99pTOT under historical period (1983–2012)

The spatial trend of the extremely wet period under 1983–2012 (Fig. 2), reveals on the one part that the very wet period (R95pTOT) trend increase (between 0 to 0.5) over the entire West Africa domain characterized by a positive slope. It is important to note that this increase is slightly small in Southern and concern Benin, Togo, Ghana, Ivory Coast during the historical periods. This is also true in Sahel countries.

As for the trend of extremely wet (R99pTOT) period, a slight increase was observed (between 0 to 1) over the entire West Africa domain characterized by a positive slope. For this result, it is clear that there will be a decreasing trend of R99pTOT along the Guinea Coast (North Benin, Togo and Ghana). A similar trend was also observed in other countries located in the Sahara and Sahel such as Burkina Faso and Mali as well as Northern Cameroon and the lac Tchad.

3.2 Spatial trend analysis of precipitation on very wet by R95pTOT and extremely wet R99pTOT near future period (2025–2054) with scenario SSP 1.2.6 and SSP 5.8.5

Trend in very wet period characterize by R95pTOT (Fig. 3) shows a slight decrease and increase trend during the SSP 1.2.6 scenario in the near future while the results reveals a clear increasing trend over the entire West Africa domain under SSP 5.8.5 scenario characterized by a positive slope.

The analysis of the results by R99pTOT (Fig. 4) shows a slight decrease trend during the SSP1.2.6 scenario in the near future while we noted a clear increasing trend over the entire West Africa, basically in the gulf of Guinea under SSP 5.8.5 scenario characterized by a positive slope (Fig. 4).

3.3 Analysis of trend in the wet and drought period with SPEI for 12 months in the historical period (1983–2012)

From near normal to extremely wet (Fig. 5), the spatial analysis of trend by SPEI under 12 months, show a slight decrease (between 0 to −2) and an increase (between 0 to 2) trend during the historical which varies from one country to another. Such a situation can contribute to the occurrence of floods.

Concerning the characterization of drought (Fig 6), it was observed that from moderate to extremely dry, the spatial analysis of trend by SPEI shows a slight increase (between 0 to 1) and decrease (between 0 to −2) trend during the historical area; which suggests that there is an upward trend of wet rain events reported on the analysis of Fig. 3.

This situation described above may have repercussions both on water-dependent activities and on the hydrological and hydroelectric dynamics of the West African watersheds.

The results show that from normal to extremely wet, the spatial analysis of trend by SPEI show a slight decrease (between 0 to −2) and increase (between 0 to 2) which varies from one country to another in near future (2025–2054) under the SSP 1.2.6 (Fig. 7).

According to the analysis of Fig. 8, it should be noted that the same scenario under SSP 1.2.6 was observed but the wet period under SSP 5.8.5 will decrease slightly more than the SSP 1.2.6 scenario.

Based on this result, we inferred that the data from the SSP 1.2.6 scenarios are less pessimistic than those of SSP 5.8.5 and can be used to characterize wet rainfall events.

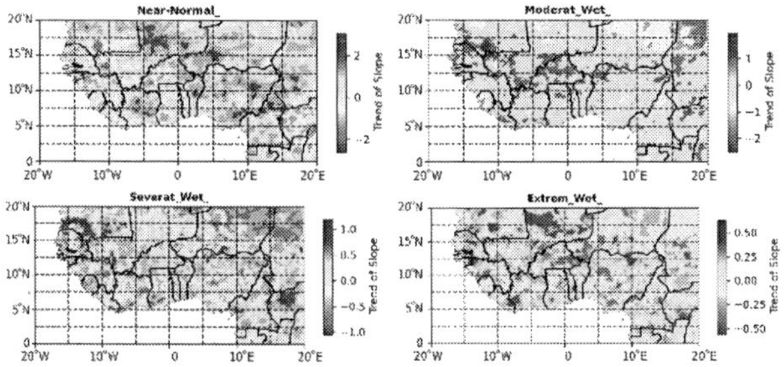

Figure 5. Trend of the slope of historical wet from SPEI for 12 months under 1983–2012.

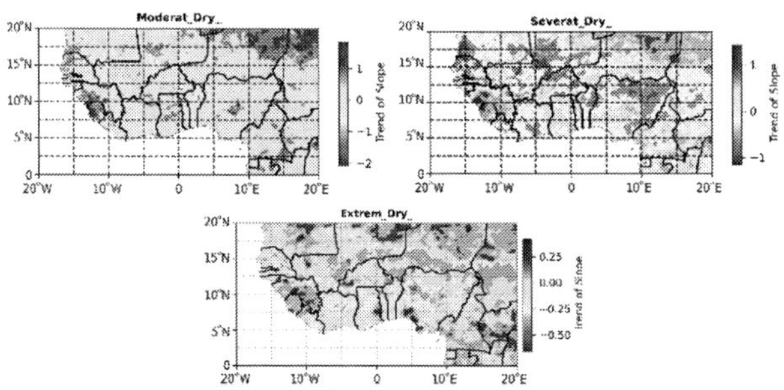

Figure 6. Trend of the slope of historical drought from SPEI for 12 months under 1983–2012.

During the near future period (2025–2054) with scenario SSP 1.2.6, the results (Fig. 9), depict moderate to extremely dry, while the spatial trend by the SPEI show a slight increase of dry period in costal country in the Gulf of Guinea and significance increase in the Sahel under the SSP 1.2.6.

This situation with SSP 1.2.6 scenario in the Gulf of Guinea may be due to the influence of other physical factors such as oceanic effects that contributed to rainfall and the regulation of global warming.

The results of Fig. 10, revealed the same situation as scenario SSP 1.2.6. However, the drought with SSP 5.8.5 will be slightly more severe than the 1.2.6 scenario that was characterized by the drought period. We opined that the data of the SSP 5.8.5 scenario are more pessimistic than those of the SSP 1.2.6 scenario in terms of global warming.

4 Conclusions

The study confirmed that there is variability in the spatial distribution of rainfall indices with an upward and downward trend of dry and wet periods in the study area.

This situation is much more remarkable in the countries of the Gulf of Guinea and can however cause extreme rainfall events associated with floods or dry spells and consequently repercussions on water-dependent activities and the hydrological dynamics of rivers. Considering the data series studied, the upward trends in rainfall confirmed the return period of rainfall reported in many studies over West African countries. However, the upward trend is weak. In conclusion, the countries of the Gulf of Guinea and the Sahelian border to the north of the study area are the wettest. In addition, the upward or downward trend varies from one scenario to another. The data analysis of the different scenarios SSP 1.2.6 and 5.8.5 deserve to be deepened in future studies.

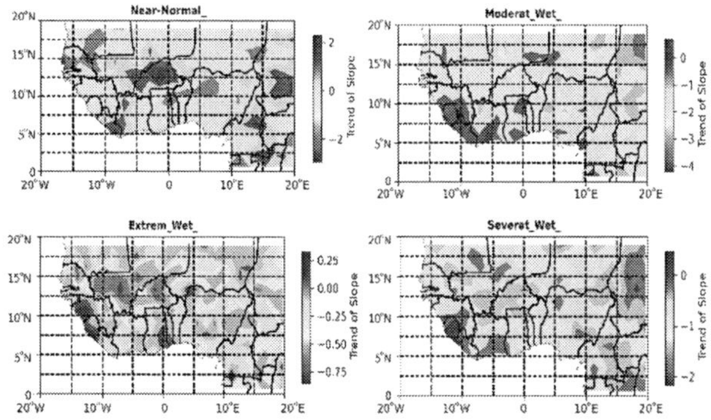

Figure 7. Trend of the in the wet period from SPEI for 12 months in near future period (2025–2054) with scenario SSP 1.2.6.

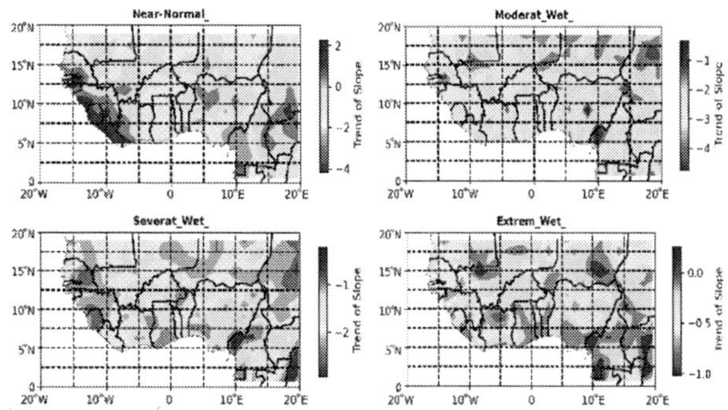

Figure 8. Trend of the in the wet period with SPEI for 12 months in near future period (2025–2054) with scenario SSP 1.2.6.

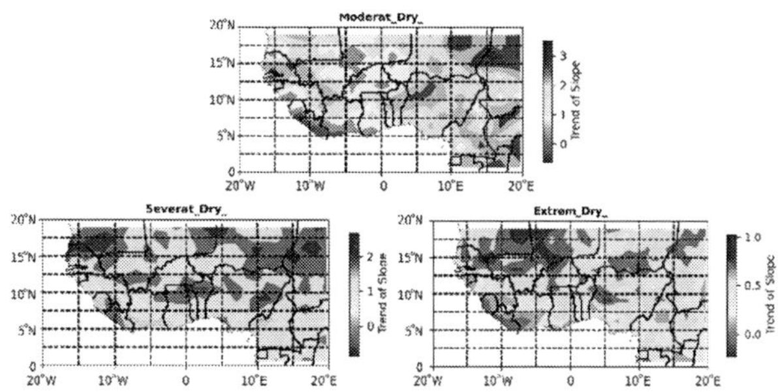

Figure 9. Trend of the in the drought period with SPEI for 12 months in near future period (2025–2054) with scenario SSP 1.2.6.

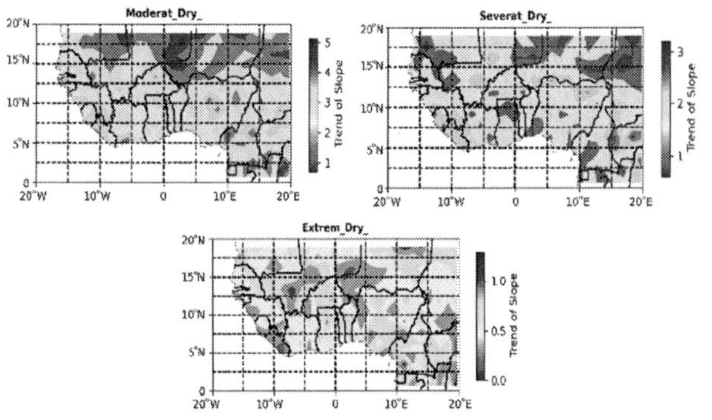

Figure 10. Trend of the in the drought period with SPEI for 12 months in near future period (2025–2054) with scenario SSP 5.8.5.

Code and data availability. The CHIRTS, CHIRPS-2 observation reanalysis and climate model data are available online at https://data.chc.ucsb.edu/products/CHIRPS-2.0/ (last access: 7 September 2023) and https://doi.org/10.5281/zenodo.1069050 (Lange, 2017).

Author contributions. DJK: Conceptualization, Methodology, Software, Data curation, Formal analysis, Validation, Writing – original draft. GMLDQ and HDK: Conceptualization, formal analysis review, editing the original draft. EA, ISA, ASJA, MFAY, GM, JEP, EWV, and CH: Investigation editing, review and supervision.

Competing interests. The contact author has declared that none of the authors has any competing interests.

Disclaimer. Publisher's note: Copernicus Publications remains neutral with regard to jurisdictional claims in published maps and institutional affiliations.

Special issue statement. This article is part of the special issue "IAHS2022 – Hydrological sciences in the Anthropocene: Variability and change across space, time, extremes, and interfaces". It is a result of the XIth Scientific Assembly of the International Association of Hydrological Sciences (IAHS 2022), Montpellier, France, 29 May–3 June 2022.

Acknowledgements. Authors would like to thank the IAHS SYSTA award funding to participate at IAHS 2022 Scientific Assembly, Montpellier, France, 2022. We also thank the anonymous reviewers of this work for their contribution to improving the quality of this study.

Review statement. This paper was edited by Christophe Cudennec and reviewed by two anonymous referees.

References

Ajayi, V. O. and Ilori, O. W.: Projected drought events over West Africa using RCA4 regional climate model, Earth Syst. Environ., 4, 329–348, https://doi.org/10.1007/s41748-020-00174-6, 2020.

Allen, K. J., Ogden, J., Buckley, B. M., Cook, E. R., and Baker P. J.: The potential to reconstruct broadscale climate indices associated with southeast Australian droughts from Athrotaxis species, Tasmania, Clim. Dynam., 37, 1799–1821, https://doi.org/10.1007/s00382-011-1011-7, 2011.

Atiah, W. A., Amekudzi, L. K., Aryee, J. N. A., Preko, K., and Danuor, S. K.: Validation of satellite and merged rainfall data over Ghana, West Africa, Atmosphere, 11, 859, https://doi.org/10.3390/atmos11080859, 2020.

Dembélé, M. and Zwart, S. J.: Evaluation and comparison of satellite-based rainfall products in Burkina Faso, West Africa, Int. J. Remote Sens., 37, 3995–4014, https://doi.org/10.1080/01431161.2016.1207258, 2016.

Lange, S.: ISIMIP2b Bias-Correction Code, Zenodo [code], https://doi.org/10.5281/zenodo.1069050, 2017.

N'guessan Bi, V. N. G., Adjakpa, T., Allechy, F., Ta, M. Y., Assa, Y., and Affian, K.: Characterization of the drought by the SPI and SPEI indices in the west center of Côte d'Ivoire case of the lobo basin, LARHYSS J., 43, 23–39, 2020.

Obahoundjè, S., Ofosu, E. A., Akpoti, K., and Kabobah, A. T.: Land Use and Land Cover Changes under Climate Uncertainty: Modelling the Impacts on Hydropower Production in Western Africa, Hydrology, 4, 2, https://doi.org/10.3390/hydrology4010002, 2017.

Odoulami, R. C., Abiodun, B. J., and Ajayi, A. E.: Modelling the potential impacts of afforestation on extreme precipitation over West Africa, Clim. Dynam., 52, 2185–2198 https://doi.org/10.1007/s00382-018-4248-6, 2019.

OMM and GWP: Handbook of Drought Indicators and Indices, Integrated Drought Management Program, Integrated Drought Management Program Tools and Guidelines Series 2, edited by: Svoboda, M. and Fuchs, B. A., Geneva, ISBN 978-92-63-11173-9/ISBN 978-91-87823-24-4, 2016.

Pedersen, J. T. S., van Vuuren, D., Gupta, J., Santos, F. D., Edmonds, J., and Swart, R.: IPCC emission scenarios: How did critiques affect their quality and relevance 1990–2022, Global Environ. Chang., 75, 102538, https://doi.org/10.1016/j.gloenvcha.2022.102538, 2022.

Quenum, G. M. L., Klutse, N. A., Dieng, D., Laux, P., Arnault, J., Kodja, J., and Oguntunde, P. G.: Identification of potential drought areas in West Africa under climate change and variability, Earth Syst. Environ., 3, 429–444, https://doi.org/10.1007/s41748-019-00133-w, 2019.

Şen, Z.: Innovative trend significance test and applications, Theor. Appl. Climatol., 127, 939–947, https://doi.org/10.1007/s00704-015-1681-x, 2015.

Vicente-Serrano, S. M., Begueria, S., and Lopez-Moreno, J. I.: A Multiscalar Drought Index Sensitive to Global Warming: The Standardized Precipitation Evapotranspiration Index, J. Climate, 23, 1696–1718, https://doi.org/10.1175/2009JCLI2909.1, 2010.

Proc. IAHS, 385, 365–370, 2024
https://doi.org/10.5194/piahs-385-365-2024

Analyse en ondelettes des séries temporelles aux stations de pluies et débits dans le bassin versant de Tortiya amont (Nord de la Côte d'Ivoire)

Marc Auriol Amalaman[1,5], **Gil Mahé**[2], **Béh Ibrahim Diomande**[3], **Armand Zamblé Tra Bi**[3], **Nathalie Rouché**[2], **Zeineddine Nouaceur**[4], **and Benoit Laignel**[5]

[1]Département de géographie, Université Peleforo Gon Coulibaly, Korhogo, Côte d'Ivoire
[2]Hydrosciences Montpellier, University Montpellier, CNRS, IRD, IMT Alès, Montpellier, France
[3]Département de géographie, UFR CMS, Université Alassane Ouattara, Bouaké, Côte d'Ivoire
[4]Université de Rouen Normandie, UMR IDÉES CNRS 6226, Mont – Saint-Aignan CEDEX, France
[5]Université de Rouen Normandie, UMR CNRS 6143 M2C, Mont – Saint-Aignan CEDEX, France

Correspondence: Marc Auriol Amalaman (marcauriol@outlook.com, marcamal@upgc.edu.ci)

Received: 15 June 2022 – Revised: 11 May 2023 – Accepted: 20 July 2023 – Published: 19 April 2024

Résumé. L'objectif de cette étude est d'analyser les liens entre les indices climatiques et la variabilité des séries de précipitations et de débits. Afin de mieux comprendre la non stationnarité des différentes stations, les données de débits et de pluviométries utilisées concernent la station de Tortiya (1960–1996). Les indices climatiques couplés à ces séries sont l'indice NAO (North Atlantic Oscillation) et l'ENSO (El Niño – Southern Oscillation) sur la même période d'étude. La méthodologie a consisté à appliquer l'analyse et la cohérence en ondelettes sur les séries temporelles. Ces méthodes ont mis en évidence les modes de variabilité dans les séries chronologiques : le mode infra annuel, le mode annuel et le mode interannuel (1–2 ans ; 2–4 ans ; 4–8 ans). D'une part, les résultats montrent que la variabilité du signal est expliquée dans les hautes fréquences (6 mois à 1 an) dans les différentes séries chronologiques. À cette fréquence, c'est le mode annuel (1 an) qui enregistre toute la variabilité du signal comprise entre 30 % et 70 %. D'autre part, ce travail présente aussi des signaux sur d'autres fréquences et périodes assez faibles. L'analyse en ondelette a révélé que le signal dominant est très largement significatif au niveau du cycle annuel. Par ailleurs, l'usage de la cohérence en ondelettes entre les indices climatiques (ENSO, NAO) et les précipitations, indiquent une forte influence du NAO sur les séries pluviométriques et de débits.

Mots clés. Ondelettes continues ; cohérence en ondelettes ; pluie-débit ; bassin versant Tortiya Amont ; Côte d'Ivoire

1 Introduction

L'Afrique a été au centre de plusieurs études à cause des effets du changement climatique. L'impact des indices climatiques sur la précipitation et le cycle hydrologique constitue une question fondamentale dans plusieurs débats scientifiques. À cet effet, de nombreux auteurs tels que (Tra Bi et al., 2021 ; Zamrane et al., 2021 ; Khedimallah et al., 2020 ; Nouaceur, 2020 ; Benoit et al., 2013 ; Dieppois et al., 2010) ont mis en relation les variations du régime hydro-pluviométrique et les indices climatiques. Toutefois, ce lien entre les paramètres hydroclimatiques et les indices climatiques suscite maintes préoccupations. Pour certains, la variabilité des précipitations est liée à l'Oscillation Nord Atlantique (NAO : North Atlantic Oscillation) ; (Zamrane et al., 2021 ; Laignel et al., 2013). Tandis que d'autres expliquent cette variabilité par l'indice El Nino (Bogning et al., 2021). De manière générale, la variabilité du climat global à court terme est associée à des phases de couplage de phénomènes océaniques et atmosphériques dont El Niño (ENSO) et le NAO, l'Oscillation Nord Atlantique (Laignel et al., 2014). En Côte d'Ivoire, plusieurs travaux menés sur le bassin versant de Tortiya tels que (Anzoumanan et al., 2019 ; Soro et al., 2013) ont montré les changements survenus dans les séries hydroclimatiques à partir d'une analyse empirique. L'objectif de cette étude est de mettre en évidence les liens entre les indices climatiques ainsi que les paramètres hydrologiques et climatiques du bassin versant de Tortiya amont par usage de l'analyse spectrale.

1.1 Présentation de la zone d'étude

Le bassin versant de Tortiya amont est un sous-bassin du bassin versant du Bandama Blanc. Il est situé au nord de la Côte d'Ivoire, entre les longitudes 5 et 6°20′ Ouest et les latitudes 8°40′ et 10°20′ Nord, dans les départements de Korhogo, Katiola, Niellé, Tengréla et Boundiali (Fig. 1). Il couvre une superficie estimée à 14 500 km^2 et est caractérisé dans son ensemble par des surfaces aplanies dont l'altitude varie entre 300 et 400 m. Toutefois, cette monotonie du plateau est interrompue par des dômes granitiques à l'allure d'inselbergs (Soro et al., 2013). Il est drainé par le Bandama et ses principaux affluents (Solomougou, Lopkoho, Yoréloro, Lafigué, Badénou, etc). Le climat du bassin versant est de type soudanien (régime tropical de transition) à deux saisons contrastées : une saison des pluies de mai à octobre et une saison sèche de novembre à avril avec une pluviométrie moyenne annuelle évaluée à 1230 mm (période 1950–2000). Le régime hydrologique est calqué sur celui des précipitations. On y distingue une période de faibles écoulements (étiage) allant de novembre à avril et une période de forts écoulements (crue) allant de mai à octobre. La crue la plus importante est enregistrée en septembre avec un débit de 300 $m^3\,s^{-1}$ (Soro et al., 2013).

2 Données et méthodes

2.1 Données

Les données pluviométriques et de débits utilisées dans cette étude proviennent de la station de Tortiya (1960–1996). Ces données de pluviométries ont été obtenues à la SODEXAM (*Société d'Exploitation pour le Développement Aéroportuaire et Météorologique de la Côte d'Ivoire*). Quant aux débits, ils résultent de la direction de l'hydrologie. En outre, les indices climatiques ont été téléchargés sur le site de NOAA : https://psl.noaa.gov/data/climateindices/list/ (la date du dernier accès : 15 May 2022). Ces indices climatiques (ENSO et NAO) permettent d'évaluer leur influence sur l'évolution de la pluviométrie et du débit dans le bassin versant. S'agissant de l'image Aster de type SRTM, elle provient du site internet « USGS Earth Explorer » https://earthexplorer.usgs.gov/ (la date du dernier accès : 20 January 2022). Elle a été utile pour réaliser la cartographie du réseau hydrographique, la carte des altitudes et délimiter le bassin versant. Enfin, l'analyse spectrale a été faite à l'aide du logiciel libre R sous la version 4.1.1. Enfin, les traitements cartographiques ont été effectués sur le logiciel ArcGIS 10.8.

2.2 Méthodes

L'analyse en ondelettes (CWT) est une technique mathématique très utile pour l'analyse numérique et la manipulation de signaux multidimensionnels et discrets. Elle vise à identifier et quantifier les caractéristiques temporelles des principales composantes spectrales de la série temporelle. Ici, les spectres de puissance des ondelettes ont été calculés à l'aide de la transformée en ondelettes de Morlet pour l'analyse du signal. L'ondelette de Morlet introduit un ensemble de fonctions sous forme de petites ondes créées par des dilatations et translations à partir d'une simple fonction génératrice. En pratique, l'ondelette de Morlet est définie comme une onde harmonique dont la fréquence est multipliée par une fenêtre temporelle gaussienne (Zamrane et al., 2021). L'objectif est de percevoir si ces fluctuations sont aléatoires ou suivent un modèle cyclique. Les modes de variabilité détectés ont ensuite été comparés à ceux caractérisant les indices climatiques.

3 Résultats

3.1 Caractérisation de la variabilité pluviométrique dans le bassin versant

L'analyse de la variabilité de la série pluviométrique dans le bassin versant de Tortiya amont indique différents modes de variabilités. D'emblée, la Fig. 2 présente plusieurs modes de variabilités tant dans les hautes que dans les basses fréquences. Dans les hautes fréquences, le cycle saisonnier (6 mois à 1 an) présente une grande énergie de 1960 à 1980 au détriment du reste de la série. À ce niveau, cette variabilité à 6 mois est estimée à 27.55 % de la variabilité saisonnière. S'agissant du cycle annuel (1 an), il indique une significativité tout au long de la série. Ce mode de variabilité représente 67.07 % de la variabilité du signal. Il ressort de cette analyse que les hautes fréquences représentent le total de l'énergie avec 94.62 % de variabilité. Concernant de la variabilité dans les basses fréquences (>1 an), elle indique un signal estimé à 4.03 %. À ce stade, un signal est plus indiqué au mode (1–2 ans et 2–4 ans) avant la grande sécheresse survenue dans la décennie 1970.

3.2 Caractérisation de la variabilité hydrométrique à la station de Tortiya

Pour cerner le contenu fréquentiel au niveau de la station hydrométrique de Tortiya, l'analyse en ondelettes continues a été appliquée à celle-ci (Fig. 3). L'analyse saisonnière (6 mois–1 an) indique 37 % de la variabilité du signal. Celle-ci (variabilité) est assez significative avant la grande sécheresse survenue dans la décennie 1970. À ce titre, le constat indique une reprise de cette variabilité hydrométrique entre 1975 et 1983. Une significativité est présentée en 1975 et en 1980. Au niveau du mode annuel (1 an), quand bien même

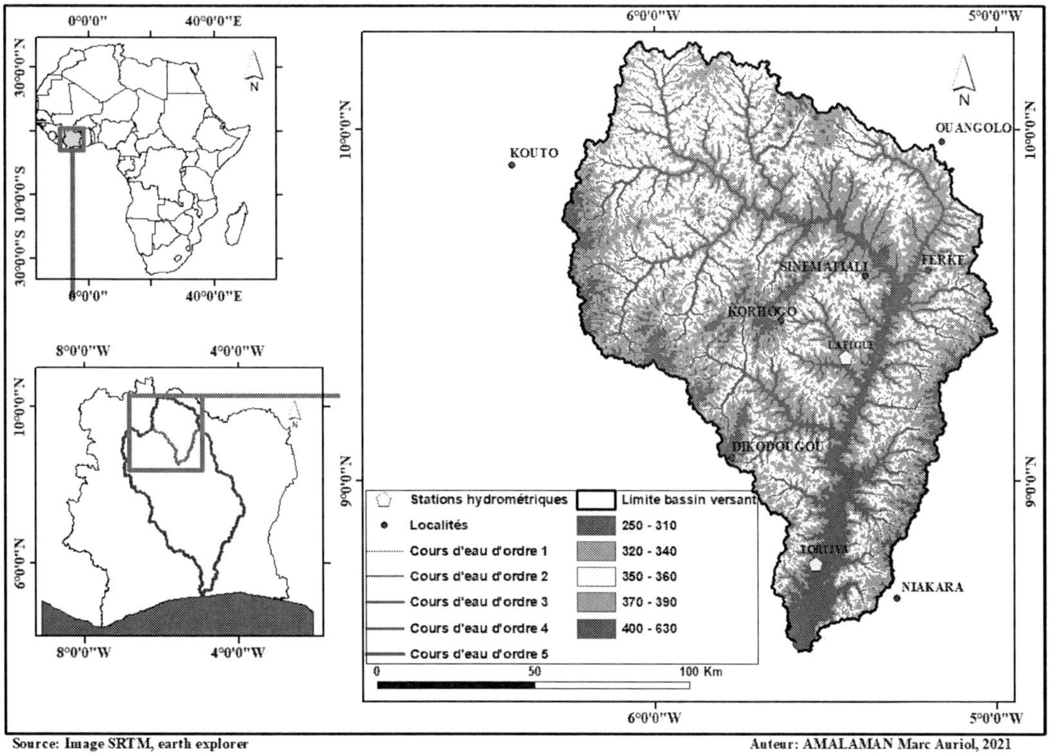

Figure 1. Localisation de la zone d'étude.

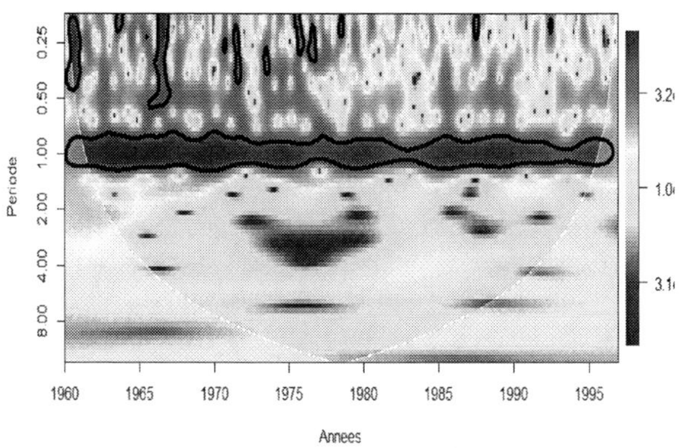

Figure 2. Analyse en ondelettes continues de la pluviométrie à la station de Tortiya.

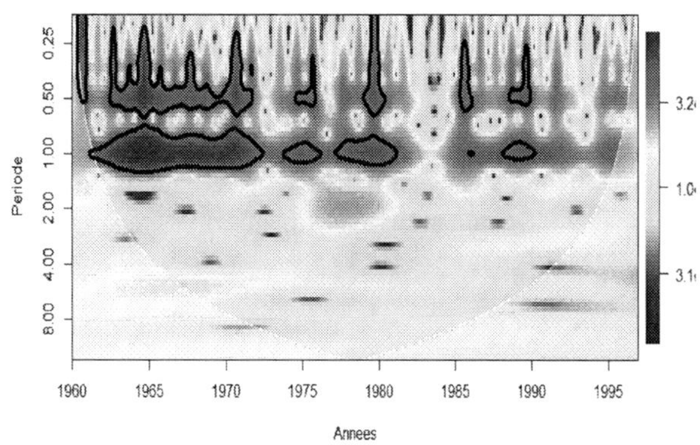

Figure 3. Analyse en ondelettes continues du débit à la station de Tortiya.

la fréquence des variances est estimée à 50.58 %, la tendance représente les mêmes réalités que celle du mode saisonnier. La somme des variabilités est donc plus représentative dans les hautes fréquences (87.58 %) que dans les basses fréquences (6.02 %). Il en ressort que la variabilité s'est ressentie au mode (1–2 ans) entre 1975 et 1980. S'agissant de la périodicité (4–8 ans), la variabilité est visible dans la décennie 1960.

4 Analyse de la cohérence en ondelettes entre les indices climatiques et la variabilité des précipitations et du débit

La compréhension de l'impact des indices climatiques (NAO et ENSO) sur la précipitation et le débit amène à faire une analyse basée sur la cohérence en ondelettes. Il s'agit d'étudier l'impact de ces indices sur la pluviométrie d'une part, et sur le débit d'autre part.

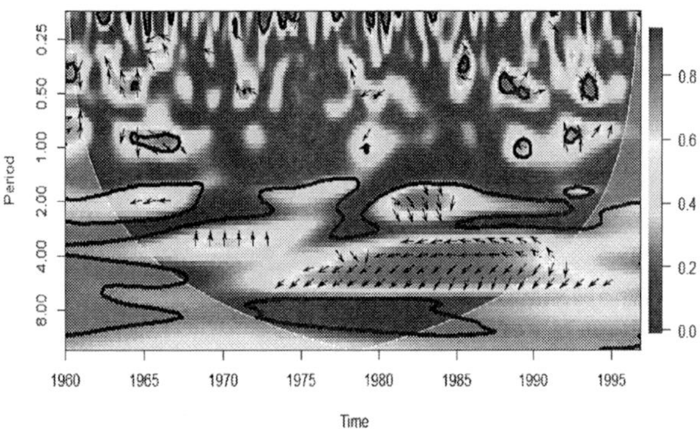

Figure 4. Cohérence en ondelettes entre l'indice ENSO et la pluviométrie à Tortiya.

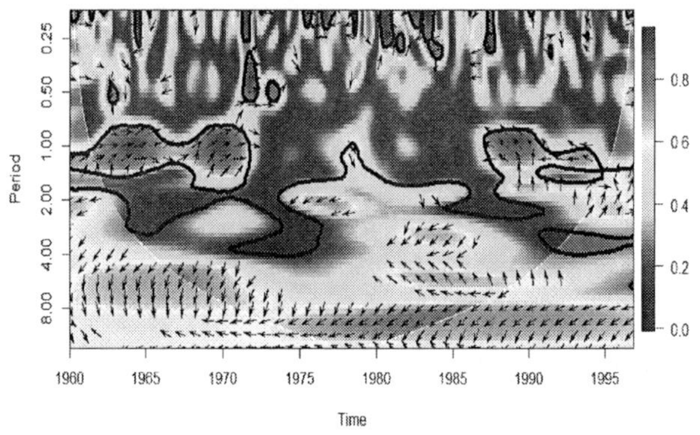

Figure 5. Cohérence en ondelettes entre l'indice NAO et la pluviométrie à Tortiya.

4.1 Analyse de la cohérence en ondelettes entre les indices climatiques et la précipitation

4.1.1 Relations entre la pluviométrie et l'indice ENSO

Le spectre en ondelettes continues présente une cohérence entre l'indice ENSO et la pluviométrie dans les basses fréquences (Fig. 4). Cette cohérence est plus marquée au niveau des périodicités (2–4 ans et 4–8 ans). Une cohérence en quart de phase entre l'indice ENSO et la pluviométrie à Tortiya est observée à la périodicité (2–4 ans) entre 1980 et 1985. Au niveau de la fréquence (4–8 ans), une antiphase est remarquée dès la décennie 1970 à la décennie 1990.

4.1.2 Relations entre la pluviométrie et l'indice NAO

À la station de Tortiya, la Fig. 5 présente de manière générale une forte influence de l'indice NAO sur la pluviométrie. Cet impact est perçu tant au niveau annuel (1 an) qu'au niveau pluriannuel (2–4 ans et 4–8 ans). En effet, au niveau de la bande d'énergie annuelle (1 an), une évolution paire entre la pluie et le NAO est constatée de 1960 à 1970. C'est l'effet contraire (une antiphase) qui s'est produit de 1990 jusqu'à la fin de la série chronologique. À la fréquence (2–4 ans), cette évolution en antiphase s'observe sur toute la périodicité. Quant à l'échelle (4–8 ans), elle souligne une forte influence du NAO sur la pluviométrie en antiphase sur toute la série. Au-delà de la fréquence (8 ans), cette cohérence en phase est constatée sur la majorité de la série malgré un faible signal entre 1960 et 1966.

4.2 Analyse de la cohérence en ondelettes entre les indices climatiques (ENSO – NAO) et le débit à la station de Tortiya

Cette analyse consistera à mettre en relation les différents indices climatiques (ENSO et NAO) et les données de débits à la station de Tortiya.

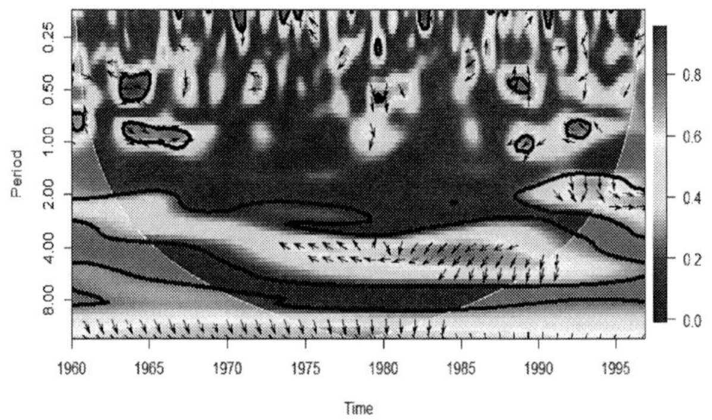

Figure 6. Cohérence en ondelettes entre l'indice ENSO et le débit à Tortiya amont.

4.2.1 Relations entre le débit et l'indice ENSO

L'impact de l'indice ENSO sur le débit est assez faible de manière générale. Dans les hautes fréquences, ce sont les spots de cohérence en antiphase qui sont visibles dans la décennie 1960 et 1990. C'est le cycle interannuelle (2–4 ans ; >8 ans) qui présente un signal assez marqué. La fréquence du signal (2 ans) indique une cohérence en phase à la fin décennie 1990. Par contre, dans la période (2–4 ans), une cohérence en antiphase de 1980 à 1991 est constaté. Sur la même fréquence de signal, une cohérence en quart de phase est affichée de 1974 à 1980 par le biais de la Fig. 6.

La présente étude s'est attelée à indiquer l'impact de l'indice ENSO sur le débit à la station de Tortiya. Il convient donc, dans l'analyse qui suit, de montrer l'influence du NAO sur le débit dans ladite station.

4.2.2 Relations entre le débit et l'indice NAO

L'impact de l'indice NAO sur le débit à la station de Tortiya amont est perceptible aux cycles annuels et pluriannuels. À l'échelle annuelle (1 an), une forte cohérence en phase est constatée durant la décennie 1960. S'agissant de la décennie

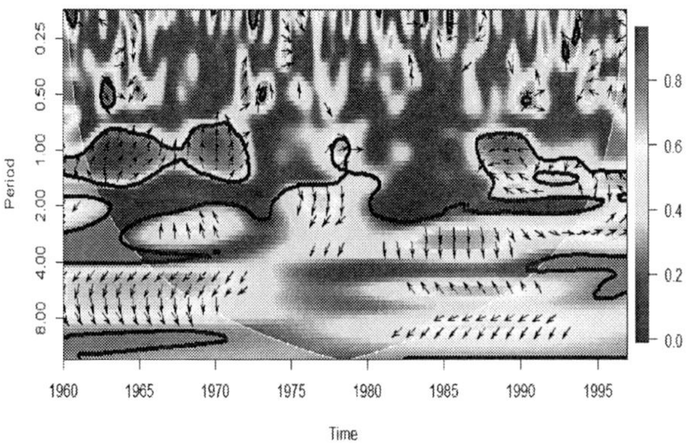

Figure 7. Cohérence en ondelettes entre l'indice NAO et le débit à Tortiya.

1970, c'est plutôt une cohérence en quart de phase qui s'observe. Cette situation s'explique par les modifications des conditions climatiques qui annoncent la grande sécheresse. Toutefois, à la fin de la série, celle de la décennie 1990, le constat montre une opposition entre les données, plus marquée de 1988 à 1991. La perte de signal enregistrée se situe entre 1972 et 1988. Quant à l'échelle pluriannuelle (2–4 ans et 4–8 ans), une évolution en quart de phase se fait remarquer dans ces deux échelles. À la fréquence (2–4 ans), la cohérence en phase se situe entre 1965 et 1970 avant la grande sécheresse. À partir de 1975, une évolution en quart de phase est indiquée jusqu'en 1990. Partant de cette date, une opposition de phase est visible durant la décennie 1990. Enfin, à l'échelle (4–8 ans), une cohérence en antiphase assez prononcée est observée de 1960 à 1970. Après une perte de signal de 1970 à 1983, c'est une antiphase qui continue de 1984 à 1993 comme le démontre la Fig. 7.

5 Discussions et conclusion

L'usage de l'analyse spectrale dans cette étude a permis de montrer les différents modes de variabilités dans les données hydroclimatiques du bassin versant de Tortiya. D'une part, à partir de l'analyse en ondelettes continues, Cette étude souligne que les modes de variabilités situés dans les hautes fréquences indiquent un plus fort taux de signal que ceux des basses fréquences. Il s'agit du cycle saisonnier (6 mois) et du cycle annuel (1 an). À cette périodicité, la variance du signal peut atteindre plus de 90 % au détriment de celle des basses fréquences estimée à environ 10 %. À ce titre, les résultats de l'étude de Tra Bi et al. (2021) abondent dans le même sens. Ils confirment que la variabilité hydrologique à Fêtêkro, dans les hautes fréquences, enregistre toute la variance du signal. À ce niveau, le signal est estimé à 93.73 %. S'agissant de l'analyse de la cohérence en ondelettes, cette analyse indique une influence assez marquée par les indices ENSO et NAO sur les données pluviométriques et hydrométriques. Cet impact se voit tant dans la périodicité (1 an) qu'à l'échelle située à partir de 2 ans. S'inscrivant dans le même sens, Zamrane et al. (2021) observent que l'indice NAO a une influence sur la précipitation et le ruissellement dans le bassin du Moulouya et Tensift. Selon eux, la contribution au débit varie entre 61 % et 74 %. Les précipitations, pour leur part, peuvent être estimées à 70 %. Quant à l'étude de Bogning et al. (2021) dans le bassin du fleuve Ogooué (ORB), les résultats ont montré le lien entre la variabilité pluviométrique et l'indice ENSO. L'analyse de la cohérence en ondelettes des deux séries temporelles confirme qu'il existe des liens significatifs entre l'indice ENSO et les précipitations dans l'ORB. Ils insinuent que ce lien est visible autour de la fréquence (6 mois) et (1 an). En somme, cette étude indique une forte influence de l'indice NAO sur les données de pluie et de débits à la station de Tortiya.

Disponibilité du code. The software used to analyse and process the data is free (R). It can be downloaded from the CRAN website at https://cran.rproject.org/bin/windows/base/ (R Core Team, 2021). The ArcGis software was used in the continental and coastal morphodynamics laboratory (M2C) at the University of Rouen Normandy.

Disponibilité des données. The various links proposed above in the methodology of this work give access to the sites from which the data were downloaded. These are the climate index data and satellite images. Flow data was provided by the hydrology department and rainfall data was made possible by SODEXAM in Côte d'Ivoire.

Collaborateurs. La conception de ce travail vient de MAA. La collecte, le traitement et l'analyse des données ont été effectués par MAA. Le projet a été relu et corrigé par GM. L'apport de l'analyse spectrale vient de GM, BL et ZN. L'analyse des données collectées et la relecture du document ont été faites par IBD, AZTB et NR.

Intérêts concurrents. The contact author has declared that none of the authors has any competing interests.

Déclaration du numéro spécial. This article is part of the special issue "IAHS2022 – Hydrological sciences in the Anthropocene : Variability and change across space, time, extremes, and interfaces". It is a result of the XIth Scientific Assembly of the International Association of Hydrological Sciences (IAHS 2022), Montpellier, France, 29 May–3 June 2022.

Remerciements. The authors would like to thank the various laboratories (M2C ; IDEES ; Laboraddys ; Hydrosciences Montpellier), the Hydrology Directorate and the Cooperation and Cultural Action Department of the French Embassy in Côte d'Ivoire for their support.

Financement. This research has been supported by the French Embassy's Department of Cooperation and Cultural Action.

Contrôle par les pairs. This paper was edited by Christophe Cudennec and reviewed by Abdoulaye Faty and Mohammad Merheb.

Références

Anzoumanan, K., Koffi, Y. B., Kouassi, A. M., Kouakou, B. D., and Seydou, D. : Impact des Évolutions Climatiques sur les Ressources en eau des Petits Bassins en Afrique Sub-Saharienne : Application au Bassin Versant du Bandama à Tortiya (Nord Côte d'Ivoire), ESJ, 15, https://doi.org/10.19044/esj.2019.v15n9p84, 2019.

Benoit, L., Zeineddine, N., Imen, T., Hiba, J., Habib, A., and Ellouz, M. : North Atlantic Oscillation and Rainfall Variability on the Southern Coast of the Mediterranean 203–210, https://www.proquest.com/docview/1611002688 (la date du dernier accès : July 2023), 2014.

Bogning, S., Frappart, F., Mahé, G., Paris, A., Onguene, R., Blarel, F., Niño, F., Etame, J., et Braun, J.-J. : Investigating links between rainfall variations in the Ogooué River basin and ENSO in the Pacific Ocean over the period 1940–1999, Proc. IAHS, 384, 181–186, https://doi.org/10.5194/piahs-384-181-2021, 2021.

Dieppois, B., Durand, A., Fournier, M., Massei, N., Sebag, D., et Bouba, H. : Variabilité des précipitations au sahel Central et Recherche du forçage climatique par analyse du signal : la station de Maïne-Soroa (SE Niger) entre 1950 et 2005. PANGEA, Géosciences et Développement, 47/48, juin/décembre 2010.

Khedimallah, A., Meddi, M., et Mahé, G. : Characterization of the interannual variability of precipitation and runoff in the Cheliff and Medjerda basins (Algeria), J. Earth Syst. Sci., 129, 134, https://doi.org/10.1007/s12040-020-01385-1, 2020.

Laignel, B., Nouaceur, Z., Turki, Imen., Jemai, H., Habib, A., et Ellouz, M. : North atlantic oscillation and rainfall variability on the southern coast of the Mediterranean, in : Conference : Air and water components of the environment, January, https://www.proquest.com/docview/1611002688 (la date du dernier accès : July 2023), 2014.

Nouaceur, Z. : La reprise des pluies et la recrudescence des inondations en Afrique de l'Ouest sahélienne, Physio-Géo. Géographie physique et environnement, 15, 89–109, https://doi.org/10.4000/physio-geo.10966, 2020.

R Core Team : R : A language and environment for statistical computing. R Foundation for Statistical Computing, Vienna, Austria, https://www.R-project.org/ (la date du dernier accès : January 2022), 2021.

Soro, T. D., Kouakou, B. D., Kouassi, E. A., Soro, G., Kouassi, A. M., Kouadio, K. E., Yéi, M.-S. O., et Soro, N. : Hydroclimatologie et dynamique de l'occupation du sol du bassin versant du Haut Bandama à Tortiya (Nord de la Côte d'Ivoire), VertigO – la revue électronique en sciences de l'environnement, 13, 3, https://doi.org/10.4000/vertigo.14468, 2013.

Tra Bi, Z. A., M'Baye, K. A. C., et Amalaman, M. A. : Variabilité hydrologique du bassin versant du N'zi dans la région du Gbêkê (centre de la Côte d'Ivoire), Actes de colloque international de Géographie, 10–12 Juin 2021, in Géovision, Hors – série no. 2, Décembre 2021.

Zamrane, Z., Mahé, G., et Laftouhi, N. E. : Wavelet Analysis of Rainfall and Runoff Multidecadal Time Series on Large River Basins in Western North Africa, Water, 13, 3243, https://doi.org/10.3390/w13223243, 2021.

Proc. IAHS, 385, 371–376, 2024
https://doi.org/10.5194/piahs-385-371-2024

Open Access PIAHS

Technical, economic and social rehabilitation of old canals to cope with global change: the case of the Neste Canal (France)

Patrice Garin[1], Marielle Montginoul[1], Daniel Lepercq[2], and Pascal Chisne[2]

[1]INRAE UMR G-EAU Université de Montpellier, Montpellier, 34196, France
[2]CACG, Tarbes, 65004, France

Correspondence: Marielle Montginoul (marielle.montginoul@inrae.fr)

Received: 27 May 2022 – Revised: 4 May 2023 – Accepted: 21 July 2023 – Published: 19 April 2024

Abstract. Many canals were built during the 19th century to satisfy multiple uses, which have since highly changed, calling into questions about their function. This article assumes that these old hydraulic works can help the territories to adapt, if reforms of their hydraulic, economic and institutional management are carried out at the same time. It illustrates this assumption and its consequences with the Neste Canal (in South France). The evolution of the multiple uses and the decrease in the flows derived over the last 70 years are described conducting to a structural imbalance of its economic model. Its future depends on the political recognition of its contribution to the minimum water flows of the rivers of Gascony, the introduction of a payment for this ecological function, and changes in the hydraulic regulation system to satisfy this last function previously managed as a hydraulic constraint.

Keywords. UPH 18, UPH 22, SDG 6; inter basin water transfers in France, economic rehabilitation

1 Introduction

Interbassin water transfer (IBWT) is defined as "the transfer of water from one geographically distinct river catchment, or basin to another, or from one river reach to another" (Davies et al., 1992). The first large transfer canals, in Mesopotamia, date back to 2500 BC, but they grew in number especially between 1970 and 1990 (Rollason et al., 2021). In France, many were built in the 19th century for multiple-use purposes (irrigation, hydropower, navigation, drinking water, dilution of domestic pollution). With increasing water demands and decreasing resources, water transfers could account for 25 % of the world's water withdrawals by 2025 (Gupta and van der Zaag, 2008). The environmental, social, political and economic consequences of recent and future transfers are being critically analysed (Gao et al., 2021; Rollason et al., 2021). The modernization of old transfer infrastructure to adapt them to current demands also comes into question. After more than a century of functioning, they have profoundly modified aquatic ecosystems, sometimes described as "hybrid" (Crifasi, 2005), and have historically embedded in the relationships of populations to water and to these developed landscapes. However, climate change and the growing pressure on water resources are destabilizing the governance of these old structures. Water rights, historical uses and the maintenance of these "hybrid" ecosystems may be challenged (Wiener et al., 2008). In this article, we illustrate how the economic model on which the sustainable management of these old transfer structures is based is also altered by these changes in uses, resource availability and societal expectations. The illustration will focus on the Neste Canal, operational since 1862, in the south-west of France, in the foothills of the Pyrenees. In the first part, we present the context of this study and the methods used. The next part deals with the history of this canal, its operation and the evolution of its multiple uses. It emphasizes on environmental objectives, translated into management constraints. And it details flow sharing between uses. The third part describes the current economic model and its difficulties to cover the sustainable cost of the water transfer system. The fourth part discusses the recognition to supply water to hybrid ecosystems as a priority use, involving then flow support pricing and a governance reform to ensure the water transfer system's sustainability.

2 Background and methodology for the Neste Interbasin Transfer Study

The IBWT is based on the 28 km long Neste Canal, taking water from the River Neste at Sarrancolin, a tributary of the Garonne, at up to $14\,m^3\,s^{-1}$ (Fig. 1). It feeds 17 rivers in Gascony, almost at their source, directly from the canal, or via a network of 200 km of secondary earth canals and 15 dams ($78\,Mm^3$) in the foothills, which store water diverted by the canal in winter. Upstream of the Neste River, $48\,Mm^3$ kept in 4 hydroelectric dams are mobilized to feed the canal when the river flow at Sarrancolin is insufficient to ensure the IBWT and the minimum environmental flow in the Neste River ($4\,m^3\,s^{-1}$) (Fig. 2). This complex and interconnected hydraulic system constitutes the Neste system. Prior to the canal, the rivers of Gascony were intermittent in their upstream part. In the absence of an exploitable aquifer, drinking water was taken from these rivers, which were highly polluted by domestic wastewater, causing serious epidemics, a major argument for the construction of a transfer canal to ensure a cleanliness flow. Navigability at the confluence of these rivers with the Garonne was the other major justification for the building of this transfer canal by the State (Fernandez and Trottier, 2012) along with the operation of wheat mills. Industrial uses (cooling, agro-industry), irrigation (by gravity on the upstream slopes, by sprinkler from river pumping after 1970), and small hydroelectricity (on some river weirs) were very secondary. They became increasingly significant with the building of hydroelectric dams upstream on the Neste at the beginning of the 20th century, and the piedmont reservoirs from 1950 to 2010. The canal and most of the piedmont reservoirs belong to the State, which granted the concession to the "Compagnie d'Aménagement des Coteaux de Gascogne" (CACG) since 1990. The article is the result of an audit entrusted to INRAE during the transfer of infrastructure ownership to the two concerned Regions. This audit focused on the physical condition of the infrastructure, the water use dynamics, the sustainable cost of the Neste System, and the threats and opportunities of inter-basin transfers in the light of climate change and societal expectations (Garin et al., 2019a, b). The study methods combined document analyses (internal to the CACG, external study reports), visits to works, audits of accounts and customer database, semi-directive surveys of 40 local water stakeholders (to estimate the perceived benefits of the Neste system and the management issues) and 20 irrigators (to identify irrigation practices, water-saving potential and the economic value of irrigation).

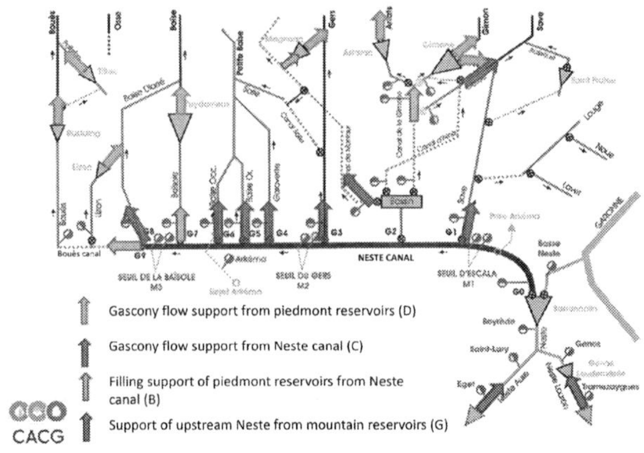

Figure 1. Scheme for the operation and distribution of water diverted from the Neste inter-basin transfer canal (CACG, 2018).

3 Brief history and hydraulic principles of the Neste System

3.1 An IBWT regulated by upstream since their inception

A decree published in 1909 defined the canal's diversion right – with an obligation of a minimum flow of $4\,m^3\,s^{-1}$ remaining in the river Neste – and the rules for sharing the water between the rivers re-fed by the canal. These rules took into account the uses: flow water support for drinking water and sanitation along rivers, several hundred hectares of irrigated meadows from the canal's gullies, and river navigation over 30 to 60 km downstream. This decree is currently still in application, even if uses have highly changed. The hydraulic operating rules have become more complex with the management of the kept volume in the upstream hydroelectric dams and the integration of the piedmont reservoirs. The upstream allocation is used during the low-water period (June to October) as soon as the flow of the Neste at Sarrancolin is less than $7\,m^3\,s^{-1}$ ($3\,m^3\,s^{-1}$ minimum for the canal $+\,4\,m^3\,s^{-1}$ for the in-stream flow of the Neste River downstream of the catchment). The piedmont reservoirs are partly filled by the transfer canal during low water ($43\,Mm^3\,yr^{-1}$ on average, with a maximum flow of $6\,m^3\,s^{-1}$, the rest comes from the watershed flow – $61\,Mm^3\,yr^{-1}$ on average). During low water, the transfer canal and the emptying of the piedmont reservoirs are managed together to satisfy the uses, under environmental constraints. As mentioned earlier, uses have changed over a century. The supply of $14\,Mm^3\,yr^{-1}$ of raw water for drinking water utilities now supplies 300 000 inhabitants. This urban demand has been relatively stable over the last 20 years, with the fall of individual consumption and the improved efficiency of the networks compensating for the increase in population. Industrial water now concerns only two companies, directly taking $7.3\,Mm^3$ from the canal for cooling. Their discharge into a small tributary of the Garonne is not used for

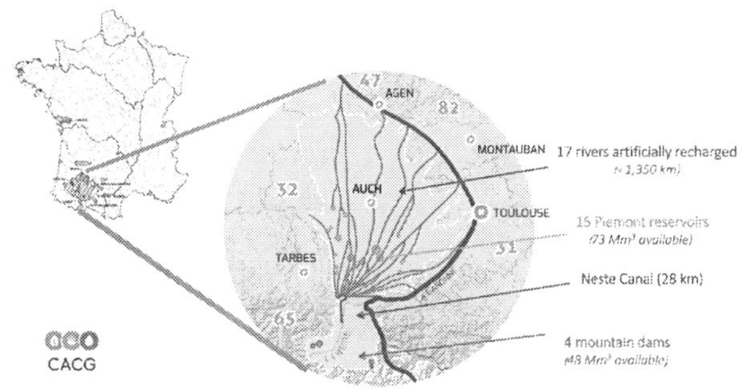

Figure 2. Simplified map of the Gascony rivers recharged by the transfer of water from the river Neste (CACG, 2018).

other purposes in the territory. Drinking water and industrial water represent a continuous flow of about 0.7 $m^3\,s^{-1}$.

3.2 Irrigation, the key issue for the last 50 years

Since the 1970s, irrigation from pumping in the recharged rivers has developed substantially (65 000 irrigable hectares). Irrigators sign an annual contract with CACG, tacitly renewed, mentioning the maximum flow that they take. This subscribed flow entitles them to a quota with two components (1000 $m^3\,L\,s^{-1}$ on June + 3000 $m^3\,L\,s^{-1}$ from July to October, during the peak of the low water support). Since the end of the 1990s, two declarations of the index of individual meters (June, October) make it possible to check consumption and to invoice penalties in the event of overruns (less than 1 % per year). Smart meters have been deployed for the past five years at all abstraction points to provide a precise map of the abstractions and to improve the regulation of the canal and the reservoirs. This allowance of 4000 $m^3\,L\,s^{-1}$ can be revised according to the hydrological situation evaluated each spring by the Neste commission. This commission brings together CACG, State services and Water Agency, as well as representatives of all water uses, but irrigators form the majority. This power struggle explains why 32 $m^3\,s^{-1}$ and 130 Mm^3 are subscribed for irrigation, to ensure access to water without constraint, even if the effective consumption is much lower: indeed, for the past twenty years, irrigation has withdrawn between 30 and 85 Mm^3 per year depending on the climate (56 Mm^3 on average for 2015-2020). The crop diversification (soya, arboriculture, legumes) does not compensate for the drop in water demand caused by the decline of irrigated corn.

3.3 Downstream regulation, more recently, to comply with environmental regulations

The environmental constraints of this IBWT have been sharpened and strengthened over time on the resupplied rivers. Each piedmont reservoir must respect an in-stream flow at its outlet. The resupplied rivers have also been

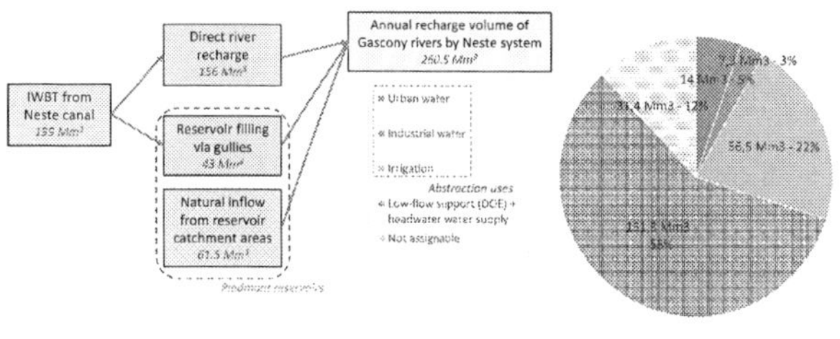

Figure 3. Annual water flow balance in the Neste system and uses (in millions of $m^3\,yr^{-1}$) (Garin et al., 2019a).

given an objective low-water flow (DOE "Débit Objectif d'Etiage") at their confluence with the Garonne, like most non-intermittent rivers in France. The DOE is calculated to allow good water status to be achieved downstream of the measurement point and to satisfy all uses on average 8 years out of 10. The DOE is checked a posteriori to see whether it has been met, i.e. whether the lowest average flow of 10 consecutive days has been maintained above 80 % of its value. Other threshold values are used to trigger restrictions on use. The DOEs of the 10 main resupplied rivers have become one of the major set points for hydraulic regulation downstream of the Neste system. At the head of the basins, the cleanliness objectives have led to raising the target flows on the canal intakes and in the gullies, in order to dilute effluents from the nearest treatment plants. The system must thus combine a complex upstream and downstream regulation with different abstraction contexts depending on the season. The hydrological regimes of the resupplied rivers are inverted: from June to October, flows are higher than in spring and autumn, to compensate agricultural abstractions in addition to satisfying the maximum demand for cleanliness during the tourist period. After several decades of this hydrological regime, the aquatic ecosystems and the riverside vegetation have been deeply modified. Human activities depending on these hybrid ecosystems have then developed: recreational fishing, river recreational activities and boating along the banks and reservoirs, and micro-hydroelectricity. The living conditions have thus been profoundly changed. All local water stakeholders interviewed are aware of the territory's dependence on the Neste system. Most of them are attached to its long-term maintenance in order to cope with climate change and the expected decrease in natural flows. However, some environmental associations question the justification for maintaining these hybrid ecosystems. But they find it difficult to propose an alternative vision for the re-supply of water to the territory through this inter-basin transfer. However, despite this general awareness of the territory's dependence on the Neste system, most of the stakeholders interviewed are convinced that the transfer primarily serves irrigation, reflecting the composition of the Neste Commission. They consider that the water not withdrawn constitutes a minor part of the hydrological balance of the Neste system. However, this view is questionable: withdrawals only represent 30 % of the vol-

Table 1. Average revenue of the Neste system and estimated costs (Operation and Maintenance – O&M, and Renovation) (in EUR thousands) following two cost assignment scenarios according to the types of use (Garin et al., 2019b). In italic, the part of costs taken by low-flow support for each scenario.

	Receipts	Cost scenario 1 (2024 value)			Cost scenario 2 (2024 value)		
	(2011–2017)	O&M costs	Renovation costs	Total costs	O&M costs	Renovation costs	Total costs
Irrigation (rivers)	2766	1086	689	1775	2190	2405	4595
Urban water	1000	473	242	715	952	842	1794
Industrial water	300	298	129	427	601	451	1052
Low-flow support		*1887*	*2638*	*4525*			*0*
Total	4066	3744	3698	7442	3744	3698	7442

umes managed (transfers and reservoirs), while environmental regulations require 58 % (cleanliness of the headwaters of the basin and low water levels downstream) (see Fig. 3 which presents the hydrological balance, breaking down the share of water withdrawn and the support for the flows of the resupplied rivers). The inefficiency of the system can thus be estimated at 12% (volumes not allocated to an objective). Considering the hydrological impacts of the various climate scenarios for the year 2050 (Lamblin et al., 2015) and assuming that the objective low-water flows remain unchanged as well as drinking water needs, the volumes available for agriculture would be reduced to zero. The sustainability of the system is therefore problematic, especially as today the contribution of the various uses to the system financing does not reflect their water needs.

4 Economic model and sustainable cost of the Neste System

In order to guarantee the long-term water supply, the manager must cover the sustainable cost, i.e. the sum of the costs of renewal, maintenance and operation, as defined by Tardieu (1999). This cost, whose structure is mainly fixed as in all the network industries, must effectively be "*supported by the current beneficiaries who will hand over the assets in good condition to future generations*" (Tardieu, 1999).

The audit (Garin et al., 2019b) attempted to estimate it. The two parallel methods applied (current international standards and the state of the networks) gave similar estimates, evaluating the need to collect at EUR 3.7 million per year to ensure the sustainability of the infrastructure, in addition to operation and maintenance costs also evaluated at around EUR 3.7 million.

However, today only withdrawers pay for water, with revenues not exceeding EUR 4 million. Irrigation provides the greater part (70 %), with a fixed rate per litre subscribed, independent of the volumes consumed. As 90 % of the subscribed flows for irrigation are signed up, there is little hope of additional revenue without a radical change in the pricing system.

This reform is needed to better reflect the sustainable cost of the system, to encourage water savings and improve the equity of the effort requested between all uses. To do so, two contrasted scenarios have been built. In the first, the sustainable cost is broken down by use, based on the costs of specific services and by allocating non-allocable costs, according to the annual volumes required by each (Table 1). Compared to current average revenues per use, it leads to a doubling of the water average price for these uses. In the second scenario, following an alternative equity approach, the financial effort has to be assumed by the industry (50 % increase) and by the beneficiaries of low water support. The questions of the implementation of the low-water support service payment and its level are currently the key issues (between EUR 0 and EUR 4525 million per year).

5 Discussion: introducing low water support pricing and reforming the Governance

The Neste system is an illustrative case of these old interbasin water transfers, carried out for multiple, interconnected, and intended (irrigation, industry, drinking water, navigation, sanitation and low-water support) or indirect (tourism, landscape, hybrid ecosystems) uses, which evolve over time. They thus profoundly shape the territories they serve (Rollason et al., 2021). However, these structures require costly long-term maintenance, whose financial impact is often underestimated during the first few decades, when the primary aim is to recover the heavy initial investments made by the end users.

When the time comes for costly renovations, the governance system and the economic model are destabilized: how to finance them and how to justify this new expenditure? Beneficiaries have become accustomed to not paying for services rendered, directly or indirectly, at their sustainable costs. By raising the problem of equitable sharing of these costs, the actors concerned are questioning the nature, importance and distribution of the benefits within the territory served, and the public interest in this transfer. In addition, the representations of these old structures, their economic, social and environmental functions today, and their justifications are being called into question, as was the case when they were built.

A related question concerns the effects of basins supplying the water, which could also valorize it, especially from a climate change perspective, even if this aspect was not covered in the terms of reference of the Neste System audit. Indeed, it is possible to note that 182 Mm^3 return to the Garonne out of the 199 Mm^3 taken from the Neste, one of its tributaries (cf. Fig. 3) but 150 km downstream, which deprives them of the corresponding flows. All these issues highlight the need to rethink the whole governance and sustainable cost sharing of the Neste system, which is currently underway. It is very likely that this will involve a change in the political balance of power between withdrawal and non-withdrawal users. The new governance will need to reflect better the priorities of existing uses and the contribution levels of direct and indirect beneficiaries. This renewed governance will have to be even more robust than the previous one to regulate the sharing of water and its costs, in the context of climate change and water demand increased by rising temperatures.

Code and data availability. The raw data used in this article is confidential, being for the most part a company secret.

Author contributions. PG and MM analysed the case of the Neste Canal, in close collaboration with DL and PC. PC calculated and presented hydraulic elements of the Neste System (especially Figs. 1 to 3); DL described the history of the system and calculated economic elements; PG and MM wrote the manuscript with contribution from DL and PC; MM, with the help of PG and DL, reviewed and edited the manuscript.

Competing interests. The authors declare that they have no conflict of interest. At the time of the study and the writing of the article, Daniel Lepercq and Pascal Chisne belonged to the company that operates the Neste system.

Disclaimer. Publisher's note: Copernicus Publications remains neutral with regard to jurisdictional claims in published maps and institutional affiliations.

Special issue statement. This article is part of the special issue "IAHS2022 – Hydrological sciences in the Anthropocene: Past and future of open, inclusive, innovative, and society-interfacing approaches". It is a result of the XIth Scientific Assembly of the International Association of Hydrological Sciences (IAHS 2022), Montpellier, France, 29 May–3 June 2022.

Acknowledgements. We thank François Brelle who did an external (independent) estimation of costs.

Review statement. This paper was edited by Christophe Cudennec and reviewed by Mohammad Merheb and one anonymous referee.

References

CACG: Le système Neste. Une concession d'Etat vitale pour un territoire de 8400 km^2. Présentation et cartographie, Communication document, 11 pp., 2018.

Crifasi, R. R.: Reflections in a Stock Pond: Are Anthropogenically Derived Freshwater Ecosystems Natural, Artificial, or Something Else?, Environ. Manage., 36, 625–639, https://doi.org/10.1007/s00267-004-0147-1, 2005.

Davies, B. R., Thoms, M., and Meador, M.: An assessment of the ecological impacts of inter-basin water transfers, and their threats to river basin integrity and conservation, Aquat. Conserv., 2, 325–349, 1992.

Fernandez, S. and Trottier, J.: Chapitre 9. La longue construction du débit d'objectif d'étiage: l'odyssée d'une métamorphose (la gestion des cours d'eau du bassin Adour-Garonne), edited by: Papy, F., Nouveaux rapports à la nature dans les campagnes, 153–167, https://doi.org/10.3917/quae.papy.2012.01.0153, 2012.

Gao, T., Liu, H., Sun, Y., and Zhang, E.: Sustaining environmental flows in water-deficient rivers via interbasin hydropower transfer, Hydrol. Process., 35, e14027, https://doi.org/10.1002/hyp.14027, 2021.

Garin, P., Montginoul, M., Bouarfa, S., Dorchies, D., Schneider -Guerin, L., Malaterre, P.-O., Peyras, L., Wittling-Serra, C., Brelle, F., and Tripiana, V.: Etat des lieux, analyse AFOM et perspectives d'évolution des concessions hydrauliques d'état en gestion CACG. Partie 1 : Etat du patrimoine hydraulique et de ses usages, INRAE – UMR G-Eau, Montpellier, 108 pp., https://hal-inrae-fr.ezproxy.u-pec.fr/hal-03925845 (last access: 29 August 2023), 2019a.

Garin, P., Montginoul, M., Guérin-Schneider, L., Wittling, C. S., Brelle, F., and Ferroudji, A. R.: État des lieux, analyse AFOM et perspectives d'évolution des concessions hydrauliques d'Etat en gestion CACG. Partie 2 : Synthèse des études prospectives et des attentes des acteurs, report, INRAE – UMR G-Eau, 88 pp., https://hal.inrae.fr/hal-03925854 (last access: 29 August 2023), 2019b.

Gupta, J. and van der Zaag, P.: Interbasin water transfers and integrated water resources management: Where engineering, science and politics interlock, Phys. Chem. Earth, Parts A/B/C, 33, 28–40, 2008.

Lamblin, V., Arama, Y., Goulard, F., Lhuissier, L., and Sauquet, E.: Garonne 2050. Un exercice de prospective participative sur la gestion de l'eau du bassin de la Garonne, Futuribles, 407, 57–65, 2015.

Rollason, E., Sinha, P., and Bracken, L. J.: Interbasin water transfer in a changing world: A new conceptual model, Prog. Phys. Geog. Earth Environ., 46, 371–397, https://doi.org/10.1177/03091333211065004, 2021.

Tardieu, H.: La valeur de l'eau en agriculture irriguée: une information économique nécessaire pour mieux réguler la gestion de l'eau et des productions agricoles dans un marché ouvert, edited by: CIID, Congrès de la CIID, 19, 13 pp., https://www.oieau.fr/eaudoc/system/files/documents/44/223188/223188_doc.pdf (last access: 29 August 2023), 1999.

Wiener, J., D., Dwire, K. A., Skagen, S. K., Crifasi, R. R., and Yates, D.: Riparian ecosystem consequences of water redistribution along the Colorado Front Range, Water Resour. Impact, 10, 18–21, 2008.

Proc. IAHS, 385, 377–385, 2024
https://doi.org/10.5194/piahs-385-377-2024

Open Access PIAHS

IAHS2022 – Past and future of open, inclusive, innovative, and society-interfacing approaches

On the Growing Socioeconomic Vulnerability of Southern Mediterranean Coastal Lagoons

Abderraouf Hzami[1], **Oula Amrouni**[1], **Essam Heggy**[2,3], **Gil Mahé**[4], **and Hechmi Missaoui**[1]

[1]National Institute of Marine Sciences and Technologies, University of Carthage, Tunis, Tunisia
[2]Viterbi School of Engineering, University of Southern California, Los Angeles, CA 90089-1112, USA
[3]Jet Propulsion Laboratory, California Institute of Technology, Pasadena, CA, USA
[4]HydroSciences Laboratory, IRD, CNRS, University of Montpellier, Montpellier, France

Correspondence: Abderraouf Hzami (abderraouf.hzami@instm.rnrt.tn)

Received: 15 June 2022 – Revised: 23 August 2023 – Accepted: 28 August 2023 – Published: 19 April 2024

Abstract. Due to pronounced hydroclimatic fluctuations and rapid urban growth, coastal lagoonal systems in the Southern Mediterranean semi-arid areas undergo pronounced shoreline retreats and increased coastal floodings. The socioeconomic impacts of these alarming coastal changes remain poorly uncharacterized. To address this deficiency, we calculate the Socioeconomic Vulnerability Index (SVI) based on the decadal changes in shoreline retreat and land use occupation considering the study case of the coastal lagoon of Ghar El Melh in Tunisia which is representative of several lagoonal systems in the southern Mediterranean basin. We first monitor the diachronic shoreline changes from 1882 to 2016 using aerial and orbital photogrammetric scenes combined with topographic and bathymetric maps. Using the Digital Shoreline Analysis System (DSAS) and historic bathymetric records, we then measure the changes in shoreline position from 1882 to 2016. Secondly, the above is correlated with the Land Use/Land Cover (LULC) assessment using a supervised classification algorithm from multiple Landsat 5-8 orbital photogrammetric scenes. We assess the SVI using five socioeconomic parameters: (1) population density, (2) land use change, (3) road network, and (4) settlements. Our results reveal the occurrence of severe coastal erosion with a maximum Net Shoreline Movement of -1580 m (± 6 m) between 1882 and 2016. Moreover, the changes in land use from 1985 to 2021 reveal the extension in urban growth around the lagoon, rising from 5.4 % to 13.5 %. Whereas the vegetation coverage is decreasing by 9 % over the same period. The resulting SVI map reveals that 85 % of the lagoon area is characterized by high to very high vulnerability. This high coastal vulnerability is found to be primarily due to the extensive change in land occupation over the last few decades and is aggravated by the increase in hydroclimatic extremes.

Keywords. Coastal lagoon; sediment transport; vulnerability; monitoring; socioeconomic indicators; shoreline retreat

1 Introduction

Coastal sandy barriers change their morphology and composition due to several forcing processes, such as marine submersion and storm surges (Cooper et al., 2018). The frequency, intensity, and geomorphology of these coastal barriers influence how they react to hydroclimatic disturbances (Thieler and Young, 1991). Coastal lagoons are also unique ecosystems where various socioeconomic activities, such as fisheries, tourism, aquaculture, maritime shipping, and urban development, can occur. The ecological function of these lagoons must be carefully studied and assessed in the specific context of each of these water bodies having their own characteristics (Pérez-Ruzafa et al., 2010). Due to the rapid changes that have occurred in recent decades, lagoon habitats are of concern because of the negative impact on the structure and dynamics of their biodiversity. Some of the changes are due to coastal development for tourist amenities, while others are due to changes in agricultural practices in the watershed, such as increased inputs of agricultural waste and nitrogen into the lagoons (Gaaloul et al., 2022).

The Mediterranean region is considered a climate change "hotspot" (Lionello and Scarascia, 2018; Pausas and Millán, 2019). Coastal and inland soil erosions, as well as floodings, are widely observed in North Africa as a result of hydroclimatic changes associated with global warming (Salhi et al., 2023). Coastal lagoons in North Africa are unique ecosystems that are often damaged by human activities (Fouad et al., 2023). The Tunisian coast is 2290 km long, including the shores of coastal lagoons (445 km) and islands (457 km). This coast is threatened by coastal erosion and marine submersion related to sea level rise and anthropogenic factors (Oueslati, 2015; Amrouni et al., 2019b; Hzami et al., 2021). The coastal area of Ghar El Melh, like all the coasts of the country, is densely impacted by urban growth and socioeconomic development (Amrouni et al., 2019a; Hzami et al., 2021). But the vulnerability of the Ghar El Melh coast is also aggravated by the strong reduction of sediment supply by the Medjerda River since the construction of the Sidi Salem dam in 1981 (Louati et al., 2014; Oueslati et al., 2015). Indeed, recent studies have demonstrated that the coarser fraction of fluvial sediments transported by the Medjerda River is stored in the dam and does not reach the coast of Ghar El Melh (Kotti et al., 2018; Benmoussa et al., 2019). Due to this, the sand of the Ghar El Melh shores is no more renewed, leading to the rapid reduction of the shoreline width between the sea and the lagoon (Hzami et al., 2018). These studies were based on regional data (Gulf of Tunis), with low spatial resolution data (>30 m).

The shoreline retreat has already impacted local socioeconomic activities by reduction of agricultural and tourist areas (Samàali, 2011), but there are no studies quantifying the level of socioeconomic impacts. Eventually, the ecohydrological local system is negatively impacted, although the Ghar El Melh lagoon is protected by the Ramsar Convention since 2007 (Abulhawa and Cummings, 2017). Developing a specific vulnerability method adapted to local socioeconomic variables is essential to give local and regional authorities some figures to define the location and the intensity of the vulnerability better and more precisely along the shore. To understand these local environmental issues of the Ghar El Melh coastal lagoon, we use high-resolution remote sensing and GIS techniques, to quantify the level of socioeconomic vulnerability of the Ghar El Melh coastal area, based on a vulnerability index (Mahapatra et al., 2015). The study also aims to increase our knowledge to better protect the ecosystem.

2 Study area

The Ghar El Melh lagoon is a Mediterranean water body, situated in North-Eastern Tunisia (Fig. 1), on the western shores of the Gulf of Tunis (37°06′–37°10′ N, 10°08′–10°15′ E). This lagoon is a vestigial part of the Gulf of the Utique Sea (Delile et al., 2015). The sandy coastal barrier separating the lagoon from the sea was interrupted by a channel named "Boughaz", 85 m wide and 2.5 m deep (Moussa et al., 2005). The lagoon covers an area of 28.5 km^2 and within an average depth of 0.8 m.

The Medjerda River produced the principal inflows of the Ghar El Melh Lagoon until the 19th century. Since 1973, was naturally derived 18 km southward of the Ghar El Melh lagoon (Paskoff, 1985).

The lagoon climate setting is Mediterranean, with medium rainfall (540 mm yr^{-1}) and an evapotranspiration rate of about 1450 mm yr^{-1}. The seasonal variations in air temperature, with medium values ranging from 12 to 27 °C in winter and summer, respectively (INM, 2020).

3 Methodology

3.1 Data description and sources

In this study, we used time-series remote sensing data to assess the shoreline evolution of land occupation and socioeconomic vulnerability of the Ghar El Melh coast.

For the orbital satellite scenes, we use Landsat (1985), Spot (1988; 1999), Sentinel 2B (2016), and Sentinel 2A (2021) which have been acquired during the dry summer seasons from June to August (Table 1). These scenes are available in the archives of the U.S. Geological Survey (USGS). To perform long-term shoreline analysis (1882–2016), we also use photo-interpretation scenes such as aerial photography (1974), topographic maps (1936), and marine maps (1882) (Table 1).

3.2 Land Cover/Land Use (LCLU) assessment

We used the supervised algorithmic method to generate Land Use maps using ENVI software. We applied a geometric, atmospheric, and radiometric correction to all multispectral images in Ghar El Melh between 1985 and 2021 to improve the visibility and quality of land use classifications. The region of interest (ROI) is determined from the pre-processed images and then processed to identify different terrain classes to better understand the evolution of the identified terrain classes: continental water, urban spaces, vegetations, bare soils, and sand deposits for the entire study area. For the supervised classification, we used the maximum likelihood algorithm (MacQueen, 1967).

By comparing the ROI class results generated in ENVI 5 with reference data obtained during a field campaign in the research area in April 2021, the accuracy of the Landsat 5 (1985) and Sentinel (2021) classifications (Table 1) was determined using the confusion matrix approach. Our five land cover classes, continental water, urban space, vegetation, bare soil, and sand deposit are based on statistical data on the temporal evolution of the region. For post-treatment classification algorithms, the errors of the classification methods are determined using the kappa indices and the accuracy of

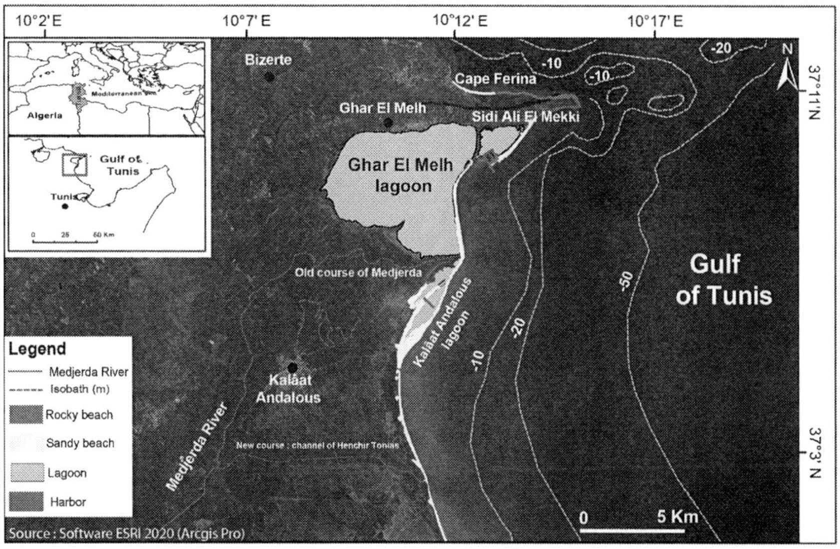

Figure 1. Location and characteristics of Ghar El Melh lagoon and coast, Gulf of Tunis.

Table 1. Spatial scenes and thematic maps used to assess the socioeconomic vulnerability in the Ghar El Melh (Tunisia).

Spatial database	Date (Scale)	Resolution (m)	Error (m)	Source
Topographic maps	1936 (1/50 000) 1974 (1/25 000)	2	±0.18	Tunisian Office of Cartography (OTC)
Aerial photography	1974 (1/25 000)	1	±0.1	
Marine maps	1882 (1/100 000)	4	±0.28	
Satellite scenes	Landsat 5 (1985) Spot 1 (1988) Spot 4 (1999) Sentinel 2A (2016) Sentinel 2B (2021)	30 20 20 10 10	±0.12	United State Geological Survey (USGS) https://earthexplorer.usgs.gov/ (last access: June 2021)

the two land use images of the confusion matrix (Amrouni et al., 2019b).

3.3 Socioeconomic vulnerability index

The Socioeconomic Vulnerability Index (SVI) was calculated by combining four quantitative socioeconomic variables: (1) population density, (2) land use, (3) road network, and (4) settlement, using ArcGIS software (Table 2). These variables are not exhaustive but relevant to the study area's social vulnerability. The weightings for SVI are derived using Forman and Gass (2001) Analytical Hierarchical Process (AHP) approach. The indices for the dry area of the Ghar El Melh lagoon are calculated using a mesh with each element measuring 150 m by 350 m (Hzami et al., 2021).

The AHP was created by Saaty (2000) and is used to establish the necessary weighing factors using a priority matrix. To begin, all variables are compared pairwise, and the matrix is completed by assigning a relative dominating value between 1 and 9. The priority values for socioeconomic variables were calculated, with population density at 60 % and coastal land use at 25 %.

The socioeconomic factors for the lagoon are derived from published data and international sources. ArcGIS 10.2 was used to create the input data for our coastal vulnerability analysis. All raster data is converted to vector format using the WGS82 projection for mapping purposes.

After identifying the socioeconomic parameters of the entire Ghar El Melh lagoon coast from the source data, the variables of each factor are arranged in GIS and all parameters have been assigned corresponding numerical values from 1 (very low) to 5 (very high) for each segment of the study area (Thieler and Hammar-Klose, 1999).

Table 2. Ranges of the Socioeconomic Vulnerability Index, SVI (McLaughlin et al., 2002).

Variables	Score				
Socioeconomic	Very low 1	Low 2	Moderate 3	High 4	Very high 5
Population density (people km^{-2})	<100	100–200	200–400	400–600	>600
Coastal land use	Water bodies, marsh, and moor, sparsely vegetated areas	Bare land	Water/wetland, grassland	Forest, farmland	Urban space
Road network	>1500	1000–1500	500–1000	250–500	<250
Settlement	No settlement	Village	Small town	Large town	City

3.4 Shoreline dynamic

There are several methods to quantify the shoreline retreat on sandy coasts. The first approach was developed by Kriebel (1982). This model estimates the potential shoreline retreat rate in relation to the return period of a high-water level (FEMA, 2003). The second method is based on the development and using numerical models. The XBeach and Telemac models are the most used (Roelvink et al., 2009; Pender et al., 2015). However, these models did not use high-resolution spatial data (satellite scenes, aerial photography, etc.).

Finally, the third model is the Digital Shoreline Analysis System (DSAS). It consists in quantifying the impact of storms by measuring, using transects perpendicular to the shoreline, the distance between the pre-storm and post-storm shoreline which have been previously digitized in a database of aerial photographs. This method relies on a high-resolution database of photogrammetric and thematic scenes to assess short, medium, and long-term historical shoreline trends (Thieler et al., 2009; Amrouni et al., 2019b). The high use of the DSAS tool is due to its easy integration with Esri ArcGIS software. This combination helps calculate the distance between each transect and the baseline and provides a simple interface for statistical analysis of shoreline changes (Mishra et al., 2020).

In the present study, we use the DSAS approach implemented using datasets from Spot (1988 and 1999) and Sentinel 2B (2016), aerial photography (1974), topographic maps (1936), and marine maps to perform long-term shoreline monitoring (1882) (Table 1). Orthogonal transects of 1000 m are used to quantify the shoreline changes which have occurred along the sandy barrier of the Ghar El Melh. The statistics of the computed rate-of-change enable us to create time series in GIS that depict the temporal evolution of the shoreline's geometries and positions over the last century. We use two statistical outputs from the DSAS model in our analysis: (1) the Net Shoreline Movement (NSM), which shows the total reported separation between the oldest (1882) and most recent (2016) shoreline positions; and (2) the End Point Rate (EPR), which shows the annual rate of shoreline retreat. Finally, using Morgan et al. (1997) and Van der Wal and Pye's (2003) techniques, the inaccuracy in estimating shoreline change rates is measured.

4 Results and discussion

4.1 Shoreline evolution of Ghar El Melh coastline

The evolution of the coastline of Ghar El Melh during the 20th century is mainly influenced by the action of the sea, the river discharge of the Medjerda River, and the maritime anthropogenic developments along the coastal ridge of the Sabkha Sidi Ali El Mekki (Amrouni et al., 2020).

The subdivision of the mean time scale was carried out according to the dynamics of the littoral system at the multi-decadal scale. Two periods can be individualized:

- Period 1882–1974: This period concerns the natural dynamics of the coastline.
- Period 1974–2016: This is the period when human activities started with the deviation of the Medjerda River in 1973 and the building of the new port of Ghar El Melh in 1974.

a. *Natural evolution.* Ghar El Melh beach is bordered in the north by the fishing port and in the south by the old mouth of the Medjerda River (Fig. 2). From 1882 to 1974, the sandy barrier of Ghar El Melh underwent a natural progradation with accretion rates EPR between $+1.43 \pm 0.18$ and $3.85 \pm 0.16\,m\,yr^{-1}$.

b. *Anthropogenic evolution.* After the construction of the fishing harbor in 1974, the northern part of the sandy barrier experienced intense erosion between 1974 and 2016, with an average EPR rate of $-2 \pm 0.20\,m\,yr^{-1}$. The highest erosion rate observed in this region is an NSM of $-19.61 \pm 0.22\,m\,yr^{-1}$, between 1999 and 2016 (Fig. 2).

This trend has resulted in the erosion of much of the cultivated land along the sandy barrier of Ghar El Melh Lagoon, as well as the destruction of several urban structures along this beach. A similar study using Landsat scenes along this zone shows that the maximum shoreline retreat is $-17 \pm 2.4\,m\,yr^{-1}$ (Saïdi et al., 2014). This severe erosion is

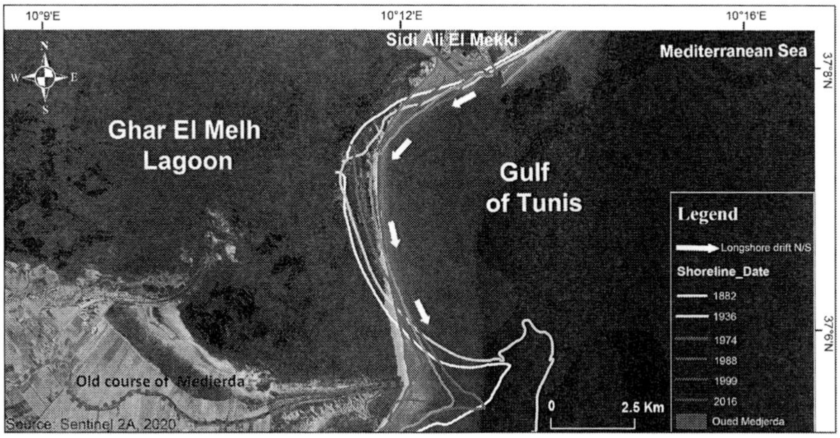

Figure 2. The shoreline movement trend along the beaches of the Ghar El Melh coastline from 1882 to 2016.

largely due to the change in the morphology of the alluvial coastal plain and the reduction in the sediment supply of the Medjerda River (Paskoff, 1994; Benmoussa et al., 2019). After the great flood of 1973, 18 km south of the Ghar El Melh lagoon was naturally drained (Fig. 1). According to Amrouni al. (2023), the low sedimentation in the highly vulnerable coast, will lead to a near coastal submersion of the sandy barrier of Ghar El Melh lagoon.

The sediment transport vectors observed at the lagoon coast show a dominant N/S direction (Beji, 2019). This longitudinal coastal transit causes the reduction of the sediment budget of the coast in coarse load and the erosion of the shoreline (Amrouni et al., 2014, 2019a).

4.2 The Land Use/Land Cover changes

The supervised classification of both Landsat (1985) and Sentinel 2A (2021) images into 6 classes is based on the "maximum likelihood" method. This method provides the resolution closest to nature, as described in Fig. 3, based on the natural colors of the different classes. Based on the distribution of these colors, we notice a decrease in vegetation cover and an increase in urban space between 1985 and 2021. The statistical class analysis for the supervised classification of the two Landsat (1985) and Sentinel 2A (2021) images is shown in Fig. 3. According to the diagram of the areas of the land occupation units of Ghar El Melh in 1985 and 2021 (Fig. 4), we note:

- Reduction of the vegetation class by 9 %;
- Increase in urban space by 8 %;
- Increase in the surfaces of bare land by 5 %;
- Reduction of agricultural activities by 5 %.

The identification of land cover changes between 1985 and 2020 shows that the natural space is significantly reduced by the rapid expansion of cities in the surrounding area. Hydraulic management of the river network reduces sediment and water flow into the depression, resulting in a reduction in the water balance and an increase in the evaporation rate into the low-lying watershed, which is here covered by the Sabkha environment. The anthropogenic factors are also enhanced by the climatic aridification of the North African regions.

Evaluation of the supervised classification methods for the two sets of satellite scenes using the confusion matrices shows that the kappa indices are 0.90 for Landsat 5 (1985) and 0.93 for Sentinel 2A (2021). The accuracy of this classification is $\sim 95\,\%$, indicating maps with high reliability.

4.3 Socioeconomic vulnerability

Mapping of SVI values reveals that 55 % of the lagoon coastline experienced a very high vulnerability (scores 4 and 5), 30 % is at high vulnerability (score 3), and 15 % is classified as moderately vulnerable (scores 1 and 2) (Fig. 5). SVI levels are significantly higher in areas with rapid urbanization. The most vulnerable areas are in the low-lying deltaic plain of the Medjerda River. The extent of these coastal vulnerabilities results in significant degradation of coastal aquifers and consequent deterioration of crop productivity, threatening food security and leading to migration out-flows (Hzami et al., 2021).

5 Conclusion

Our study suggests that the high socioeconomic coastal vulnerability area observed in the Ghar El Melh lagoon is primarily due to the rapid and extensive urbanization occurring over the last few decades. SVI values reveal that 85 % of the Ghar El Melh lagoon and its associated coast are characterized by high to very high vulnerability. Additionally,

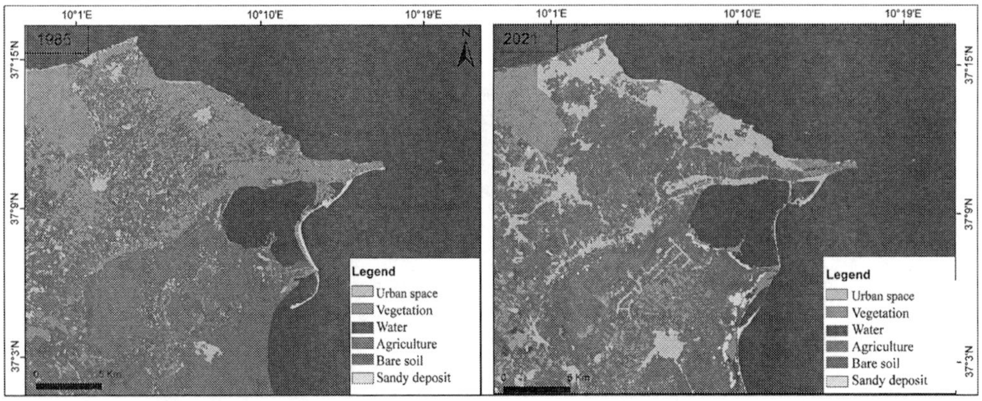

Figure 3. Detection of changes in LULC from the supervised classification between 1985 and 2021 from Landsat 5 (1985) and Sentinel 2A (2021) scenes.

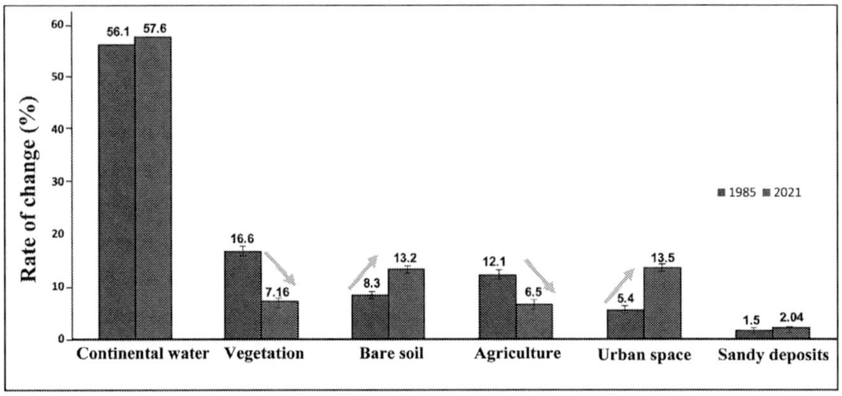

Figure 4. The statistical temporal evolution of LULC classes in the lagoon between 1985 and 2021.

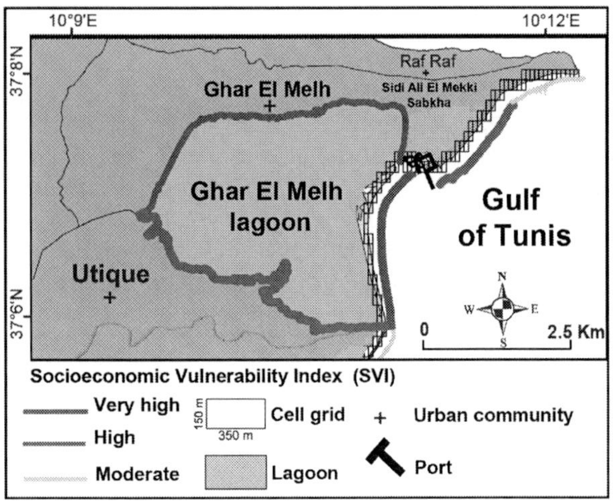

Figure 5. Vulnerability scale of socioeconomic variables SVI for the Ghar El Melh Lagoon, Gulf of Tunis.

the temporal analysis of the aerial and orbital photogrammetric scenes suggests that $\sim 70\,\%$ of the Ghar El Melh coast's sandy beaches are currently eroding at an average EPR rate of $-2\,\mathrm{m\,yr^{-1}}$. Anthropogenic factors, such as population density and land occupation, are found to be the main drivers of the observed shoreline retreat and the associated socioeconomic vulnerability in Ghar El Melh. The above results from decreased vegetation cover, water budget, sand deposit, and the increase in salt flats and urban spaces. These anthropogenic factors further aggravate the impacts of hydroclimatic extremes widely observed in the region over the last decade. Restoring a substantial part of the vegetation coverage and stabilizing the sandy barriers with the natural-based solution can reduce the vulnerability in the mid and long terms, but comprehensive urban planning that considers the recent hydroclimatic changes in this area need to be urgently implemented to mitigate the short-term impacts.

Code availability. The data processing was done by ArcGIS Software. The type of my license is "GIS Professional Advanced" provided by the university of Soutern California (USC) https://uscssi.maps.arcgis.com/home/index.html (last access: September 2023) subscription code for my account: Abderraoufhzami_USCSSI.

Data availability. The USGS Earth Explorer gives a quick way to download free satellite imagery for our paper (https://www.usgs.gov/centers/eros/science/usgs-eros-archive-sentinel-2, USGS, 2018). All the data used in this analysis are openly available as listed in Table 1.

Author contributions. Conceptualization, AH; methodology, AH, OA and EH; writing – original draft preparation: AH, data acquisition, and analysis: OA; writing – review and editing, AH, OA, EH and GM; supervision, OA and EH; project administration, HM; funding acquisition, GM and EH. The published version of the manuscript has been approved by all authors.

Competing interests. The contact author has declared that none of the authors has any competing interests.

Disclaimer. Publisher's note: Copernicus Publications remains neutral with regard to jurisdictional claims in published maps and institutional affiliations.

Special issue statement. This article is part of the special issue "IAHS2022 – Hydrological sciences in the Anthropocene: Past and future of open, inclusive, innovative, and society-interfacing approaches". It is a result of the XIth Scientific Assembly of the International Association of Hydrological Sciences (IAHS 2022), Montpellier, France, 29 May–3 June 2022.

Acknowledgements. The authors would like to thank all researchers who contributed to the implementation of the scientific tasks of this work. The authors also thank the Copernicus editorial team for the review that helped improve this paper.

Financial support. This research has been supported by the National Institute of Marine Sciences and Technologies in Tunisia, University of Carthage. This work was developed under the tasks of the PHC-UTIQUE-ECOBAM (grant no. 21G1002) project and the Arid and Water Research Center at the University of Southern California. Postdoc Abderraouf Hzami received funding under support from the Zumberge Research and Innovation Fund of the University of Southern California.

Review statement. This paper was edited by Christophe Cudennec and reviewed by Ernest Amoussou and one anonymous referee.

References

Abulhawa, T. and Cummings, T.: Rapid cultural inventories of wetlands in Arab states including Ramsar Sites and World Heritage Properties: Building greater understanding of cultural values and practices as a contribution to conservation success, Arab Regional Centre for World Heritage and Ramsar Convention Secretariat, https://www.ramsar.org/rapid-cultural-inventories-wetlands-arab-states (last access: September 2023), 2017.

Amrouni, O., Hermassi, T., Abdeljaouad, S., and Messaoudi, S.: Contribution of Grain–size Trend to Sediment of a Microtidal Beach. Case of the Gulf of Tunis Bay (Cape Ferina–Cape Gammarth, Tunisia), Research Journal of Environmental Sciences, 8, 161–177, https://doi.org/10.1007/978-3-319-70548-4_495, 2014.

Amrouni, O., Sánchez, A., Khélifi, N., Ben Moussa, T., Chiarella, D., Mahé, G., Abdeljaouad, S., and McLaren, P.: Sensitivity assessment of the deltaic coast of Medjerda based on fine grained sediment dynamics, Gulf of Tunis, Western Mediterranean, J. Coast. Conserv., 23, 571–587, https://doi.org/10.1007/s11852-019-00687-x, 2019a.

Amrouni, O., Hzami, A., and Heggy, E.: Photogrammetric assessment of shoreline retreat in North Africa: Anthropogenic and natural drivers, ISPRS J. Photogram. Remote Sens. 157C, 73–92, 2019b.

Amrouni, O., Beji, Y., Heggy, E., and Sánchez, A.: Coastal Sedimentary dynamic of wetland ecosystem of semi-arid areas in north Africa, in: 3rd Conference of the Arabian Journal of Geosciences, Sousse, Tunisia, 2–5 November 2020, http://www.cajg.org/index.php (last access: September 2023), 2020.

Amrouni, O., Ibrahim, H.S, Hzami, A., Missaoui, H., Mahe, G., and Heggy, E.: Eroding coastal sandy barriers under changing aeolian flux in the Gulf of Tunis, Bull. Inst. Natl. Sci. Technol. Mer., 48, 37–48, https://www.instm-bulletin.tn/index.php/bulletin/article/view/298, last access: September 2023.

Beji, Y.: Dynamique sédimentaire du littoral de Ghar El Melah: Evolution naturelle et impacts anthropiques, Université de Tunis, Faculté des sciences de Tunis, 109 pp., http://hdl.handle.net/1834/41840 (last access: September 2023), 2019.

Benmoussa, B. T., Amrouni, O., Hzami, A., Dezileau, L., Mahe, G., Condomines, M., and Saadi, A.: Progradation and retrogradation of the Medjerda delta during the 20th century (Tunisia, western Mediterranean). Comptes Rendus Geosciences, 351, 340–350, https://doi.org/10.1016/J.Crte.2018.10.004, 2019.

Cooper, J. A. G., Green, A. N., and Loureiro, C.: Geological constraints on mesoscale coastal barrier behaviour, Global Planet. Change, 168, 15–34, https://doi.org/10.1016/j.gloplacha.2018.06.006, 2018.

Delile, H., Abichou, A., Gadhoum, A., Goiran, J. P., Pleuger, E., Monchambert, J. Y., Wilson, A., Fentress, E., Quinn, J., Ben Jerbania, I., and Ghozzi, F.: The geoarchaeology of Utica, Tunisia: the paleogeography of the Mejerda Delta and hypotheses concerning the location of the ancient harbor, Geoarchaeology, 30, 291–306, https://doi.org/10.1002/gea.21514, 2015.

Federal Emergency Management Agency (FEMA): Guidelines and Specifications for Flood Hazard Mapping Partners, Appendix D: Guidance for Coastal Flooding Analyses and Mapping, Prepared under FEMA's Flood Hazard Mapping Program, https://www.fema.gov/sites/default/files/nepa/ (last access: September 2023), 2003.

Forman, E. and Gass, S.: The Analytic Hierarchy Process – An Exposition, Oper. Res., 49, 469–486, 2001.

Fouad, S. S., Heggy, E., and Weilacher.: Waterways transformation in the vulnerable port city of Alexandria, Cities, 141, 104426, https://doi.org/10.1016/j.cities.2023.104426, 2023.

Gaaloul, N., Amrouni, O., Heggy, E., Douss, N., Hzami, A., Khélifi, N., Bejaoui, B., and Sánchez, A.: Impacts of water stress on lagoonal ecosystem degradation in semi-arid coastal areas, Mar. Pollut. Bull., 179, 113445, https://doi.org/10.1016/j.marpolbul.2022.113445, 2022.

Hzami, A., Amrouni, O., Romanescu, G., Constantin Stoleriu, C., Mihu-Pintilie, A., and Saâdi, A.: Satellite images survey for the identification of the coastal sedimentary system changes and associated vulnerability along the western bay of the Gulf of Tunis (northern Africa), Proc. IAHS, 377, 83–89, https://doi.org/10.5194/piahs-377-83-2018, 2018.

Hzami, A., Heggy, E., Amrouni, O., Mahé, G., Maanan, M., and Abdeljaouad, S.: Alarming coastal vulnerability of the deltaic and sandy beaches of North Africa, Sci. Rep., 11, 2320, https://doi.org/10.1038/s41598-020-77926-x, 2021.

Institut National de la Météorologie (INM): Données numériques (brutes), https://www.meteo.tn/fr/donnees-climatiques, last access: June 2020.

Kotti, F., Dezileau, L., Mahé, G., Habaieb, H., Bentkaya, M., Dieulin, C., and Amrouni, O.: Etude de l'impact des barrages sur la réduction des transports sédimentaires jusqu'à la mer par approche paléohydrologique dans la basse vallée de la Medjerda, Proc. IAHS, 377, 67–76, https://doi.org/10.5194/piahs-377-67-2018, 2018.

Kriebel, D. L.: Beach and Dune Response to Hurricanes, unpublished M.S. Thesis, Department of Civil Engineering, University of Delaware, Newark, DE, 1982.

Lionello, P. and Scarascia, L.: The relation between climate change in the Mediterranean region and global warming, Reg. Environ. Chang., 18, 1481–1493, 2018.

Louati, M., Saïdi, H., and Zargouni, F.: Shoreline change assessment using remote sensing and GIS techniques: a case study of the Medjerda delta coast, Tunisia, Arab. J. Geosci., 8, 4239–4255, https://doi.org/10.1007/s12517-014-1472-1, 2014.

MacQueen, J. B.: Some Methods for classification and Analysis of Multivariate Observations, in: Proceedings of 5th Berkeley Symposium on Mathematical Statistics and Probability, Berkeley, University of California Press, 1, 281–297, http://projecteuclid.org/euclid.bsmsp/1200512992 (last access: September 2023), 1967.

Mahapatra, M., Ramakrishnan, R., and Rajawat, A.: Coastal vulnerability assessment using analytical hierarchical process for South Gujarat coast, India, Nat. Hazards, 76, 139–159, 2015.

McLaughlin, S., McKenna, J., and Cooper J. A. G.: Socio-economic data in coastal vulnerability indices: constraints and opportunities, J. Coast. Res., 36, 487–497, 2002.

Mishra, M., Sudarsan, D., Kar, D., Naik, A. K., DAS, P. P., Santos, C. A., and da Silva, R. M.: The development and Research trend of using DSAS tool for shoreline change analysis: A scientometric analysis, Urban Environ. Eng., 14, 69–77, https://doi.org/10.4090/juee.2020.v14n1.069077, 2020.

Morgan, A., Larson, R., and Gorman, L.: Monitoring the coastal environment: part III: geophysical research methods, J. Coastal Res., 13, 1064–1085, 1997.

Moussa, M., Baccar, L., and Ben Khemis, R.: La lagune de Ghar El Melh: Diagnostic écologique et perspectives d'aménagement hydraulique, Revue des Sciences de l'Eau, 18, 13–26, 2005.

Oueslati, A., El Aroui, O., and Sahtout, N.: Sur la grande vulnérabilité du lido du complexe lagunaire de Ghar El Melh et de ses terres humides (Tunisie septentrionale). Érosion, risque de maritimisation et menaces sur le terroir original Ramli, Méditerranée, 125, 65–73, 2015.

Paskoff, R.: Les plages de la Tunisie, Editec, Caen, 198 pp., https://journals.flvc.org/jcr/article/view/77401 (last access: September 2023), 1985.

Paskoff, R.: Les littoraux. Impact des aménagements sur leur évolution, Paris, Masson, https://www.persee.fr/doc/rga_0035-1121_1985_num_73_3_2607_t1_0345_0000_1 (last access: September 2023), 1994.

Pausas, J. G. and Millán, M. M.: Greening and Browning in a Climate Change Hotspot: The Mediterranean Basin, Bioscience, 69, 143–151, 2019.

Pender, D., Callaghan, D. P., and Karunarathna, H.: An evaluation of methods available for quantifying extreme beach erosion, Journal of Ocean Engineering and Marine Energy, 1, 31–43, https://doi.org/10.1007/s40722-014-0003-1, 2015.

Pérez-Ruzafa, A., Marcos, C., and Pérez-Ruzafa, I. M.: Mediterranean coastal lagoons in an ecosystem and aquatic resources management context, Phys. Chem. Earth, 36, 160–166, https://doi.org/10.1016/j.pce.2010.04.013, 2010.

Roelvink, D., Reniers, A., van Dongeren, A., van Thiel de Vries, J., McCall, R., and Lescinski J.: Modelling storm impacts on beaches, dunes and barrier islands, Coast. Eng., 56, 1133–1152, https://doi.org/10.1016/j.coastaleng.2009.08.006, 2009.

Saaty, T. L.: Fundamentals of decision-making and priority theory, Pittsburgh, RWS Publications, ISBN 0962031763, https://books.google.tn/books?id=wct10TlbbIUC&hl=fr&source=gbs_book_other_versions (last access: September 2023), 2000.

Saïdi, H., Souissi, R., Louati, M., and Zargouni, F.: Morphologic changes and sedimentary budgets along a Mediterranean coastline with a sand spit: case of the littoral fringe Sidi Ali El Mekki-Gammarth (NE Tunisia), Rend. Fis. Acc. Lincei, 25, 393–401, https://doi.org/10.1007/s12210-014-0314-0, 2014.

Salhi, A., El Hasnaoui, Y., Pérez Cutillas, P., and Heggy E.: Soil erosion and hydroclimatic hazards in major African port cities: the case study of Tangier, Sci. Rep., 13, 13158, https://doi.org/10.1038/s41598-023-40135-3, 2023.

Samàali, H.: Etude de l'évolution de l'occupation et de l'utilisation du sol dans le delta de Medjerda par télédétection et systèmes d'informations géographiques, PhD thesis, University of Tunis, Fac. Sci. Hum. Soc. Geography., 376 pp., https://hal.archives-ouvertes.fr/tel-00612952/ (last access: September 2023), 2011.

Thieler, E. R. and Hammar–Klose, E. S.: National assessment of coastal vulnerability to future sea–level rise: preliminary results for the U.S. Atlantic coast, U.S. Geological Survey, Open-File Report, 99-593, https://doi.org/10.3133/ofr99593, 1999.

Thieler, E. R., Himmelstoss, E. A., Zichichi, J. L., and Ergul, A.: The Digital Shoreline Analysis System (DSAS) Version

4.0-An ArcGIS Extension for Calculating Shoreline Change, Open-File Report, US Geological Survey Report, 2008-1278, https://doi.org/10.3133/ofr20081278, 2009.

Thieler, E. R. and Young, R. S.: Quantitative Evaluation of Coastal Geomorphological Changes in South Carolina after Hurricane Hugo, J. Coast. Res., 8, 187–200, 1991.

USGS: USGS EROS Archive – Sentinel-2, USGS [data set], https://www.usgs.gov/centers/eros/science/usgs-eros-archive-sentinel-2 (last access: September 2023), 2018.

Van der Wal, D. and Pye, K.: The use of historical bathymetric charts in a GIS to assess morphological change in estuaries, Geogr. J., 169, 21–31, 2003.

Proc. IAHS, 385, 387–391, 2024
https://doi.org/10.5194/piahs-385-387-2024

Open Access PIAHS

Quantifying snow sublimation by Eddy covariance measurements on the High Atlas Mountain of Marrakech at Tazaghart plateau, Morocco

Lahoucine Hanich[1], **Ouiaam Lahnik**[1], **Simon Gascoin**[2], **Adnane Chakir**[3], **and Vincent Simonneaux**[2]

[1]L3G Laboratory, Earth Sciences Department, Faculty of Sciences and Techniques, Cadi Ayyad University, Marrakech, Morocco
[2]CNRS/CNES/IRD/UPS, Centre d'Études Spatiales de la Biosphère (CESBIO), University of Toulouse, Toulouse, France
[3]Faculty of Sciences Semlalia, Cadi Ayyad University, Marrakech, Morocco

Correspondence: Lahoucine Hanich (l.hanich@uca.ma)

Received: 24 May 2022 – Revised: 24 April 2023 – Accepted: 28 August 2023 – Published: 19 April 2024

Abstract. The High Atlas Mountains are a valuable source of water supply to lowland areas, in the Tensift watershed around Marrakech and play an important role in the hydrologic processes and water balance. The solid part of precipitation in these areas is important and forms a seasonal snowpack where water is stored during winter. Therefore, an accurate assessment of sublimation is essential in the overall hydrological cycle. In this context, we estimated the sublimation on the Tazaghart plateau located in the High Atlas of Marrakech, using a dataset measured with the Eddy covariance system (EC) for a period from September 2020 to January 2021. The temporal variation in measured sublimation illustrated a diurnal cycle with a peak in the early afternoon and the highest relative snow sublimation fluxes occurred with subzero temperatures and low atmospheric humidity. The average daily sublimation rate measured was 0.41 mm d^{-1}. Measured sublimation accounted for 42 % and 40 % of snow ablation, based on the energy and water balance respectively. The estimated percentages are comparable, which supports reliable measurements and provides important validation of the Eddy covariance method. This study shows how the Eddy covariance method can be useful for direct measurements of winter water loss due to the sublimation of snowpacks in rather inaccessible regions where measurements are relatively rare.

Keywords. UPH 16; SDG 06; Snow sublimation; Eddy covariance; Tazaghart plateau; High Atlas

1 Introduction

Mountain areas are often a valuable source of water supply in semi-arid watersheds and play an important role in the hydrologic processes and water balance. Precipitation in high-altitude areas is important and forms a seasonal snowpack where water is stored during winter (Hanich et al., 2022). The High Atlas Mountains in Central Morocco, where a large part of precipitation falls as snow, are the country's most vital water supply and contribute greatly to socioeconomic development. Despite the importance of snow in the High Atlas Mountains, the processes controlling snowpack evolution in this region have not been thoroughly researched. According to a prior study, high altitude circumstances in the High Atlas, i.e., low air pressure, high wind speed, and dry air, should favour sublimation, where sublimation is defined as a loss of water resources from the snowpack to the atmosphere. However, sublimation rates in the High Atlas are unknown. Therefore, a better knowledge of snow sublimation's contribution to the annual water balance of mountain watersheds is required.

Many studies have been conducted in arid and semi-arid region to investigate the spatial and temporal variability of snow sublimation (Liston and Elder, 2006; Ault et al., 2006; Bosilovich, 2006; Dietz et al., 2015), but few have been based on point measurements in mountainous terrain. Previous studies in the High Atlas region have estimated snowmelt

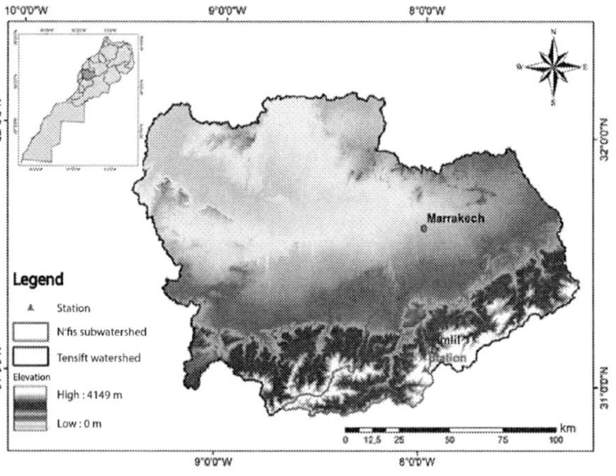

Figure 1. Geographic situation of the Study area and the location of Tazaghart's Eddy covariance station.

Figure 2. Photo of the Tazaghart Plateau automatic weather station, including the Eddy covariance system.

and sublimation rates using modelling approaches (Boudhar et al., 2016; Baba et al., 2018). Therefore, an accurate assessment of sublimation is essential in the overall hydrological cycle. In this context, an eddy-covariance flux station was installed on the Tazaghart plateau, a wide flat area located at 3850 m a.s.l. in the High Atlas Mountain of Morocco, allowing continuous acquisitions between September 2020 and January 2021, with a timestep of 30 min, in order to estimate evapotranspiration and sublimation processes.

2 Study area

This research focuses on the Tazaghart plateau, which is located in the N'fis watershed of the High Atlas of Marrakech, as depicted in Fig. 1. The plateau covers roughly 1.3 km^3 and has a mean altitude of 3850 m. It is located at 31°3′40.3″ N and of 7°58′1.5″ W. The plateau has a typical continental climate, with cold, dry weather and high radiation. The Tazaghart plateau, with its peculiar climatic conditions and topography, is an ideal place for conducting Eddy covariance measurements, which are noted for their homogeneity at the plateau level.

3 Materials and methods

Meteorological data were collected between September 2020 and January 2021 using an automatic weather station (AWS) located on the Tazaghart plateau (Fig. 2). The eddy covariance method is a direct measurement of the turbulent flux density of a scalar (temperature, momentum, water vapor concentration) across the wind streamlines (Reba et al., 2012). The method analyzes high-frequency datasets of wind and scalar values and produces fluxes of its properties.

3.1 Energy balance

In our study, we aimed to determine the balance between all the energy processes that affect the snowpack during a specific time interval. To achieve this, we utilized the following equation:

$$\mathrm{Rn} - G = H + \mathrm{LE} + F, \tag{1}$$

that incorporates various energy fluxes, including net radiation flux (Rn), soil heat flux (G), sensible heat flux (H), latent heat flux bound to sublimation (LE), and energy available for melting the snow (F). We focused on a time interval from 09:00 to 17:00 UTC when the net energy is positive.

3.2 Sublimation measurements

The snow sublimation flux is calculated according to the following equation:

$$E_{\mathrm{subl}} = \mathrm{LE}/\mathrm{Lvi} \tag{2}$$

Where Lvi = 2.838106 J kg^{-1} at 0 °C is the latent heat of sublimation of the ice and LE is the latent heat flux.

3.3 Sublimation rate

The sublimation and melt rates are calculated according to the relationship:

$$\text{Sublimation rate (\%)} = \text{sublimation (mm)/snow ablation} \cdot 100 \tag{3}$$

$$\text{Snowmelt rate (\%)} = \text{snowmelt (mm)/snow ablation} \cdot 100 \tag{4}$$

Neglecting the transport of snow by the wind and the refreezing of liquid water in the snowpack, the total ablation of snow is estimated either from the water balance:

$$\text{Ablation} = \text{precipitation} - \text{residue} \tag{5}$$

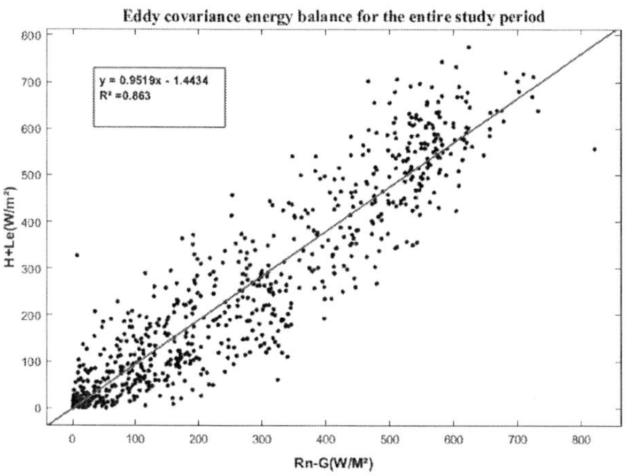

Figure 3. Eddy covariance energy balance closure [comparison between the sum of turbulent fluxes $[H + \mathrm{Le}]$ and measured available energy $[\mathrm{Rn} - G]$ for valid complete days measured at Tazaghart station during all measure periods.

or from the energy balance:

$$\text{Ablation} = \text{snowmelt} + \text{sublimation} \quad (6)$$

The remaining snow residue (R) calculated from following equation:

$$\Sigma P = \Sigma S + \Sigma F + R, \quad (7)$$

Where P is precipitation, S is sublimation, and F is melting.

The residual is calculated by converting the snow height into its water equivalent (SWE Snow water Equivalent), according to the relation:

$$\text{Snow depth(mm)} \cdot \text{Average snow density} = \text{SWE (mm)} \quad (8)$$

Where the average snow density is estimated to be $300\,\mathrm{kg\,m^{-3}}$ (for fresh snow), based on Boudhar et al. (2009) measurements. This value is for fresh snow as the snow does not stay long and melts.

4 Results and discussions

The verification of the closure of the energy balance is crucial to ensure the accuracy of turbulent flow measurements. Generally, we ensure that the sum of these flows closely matches the available energy ($\mathrm{Rn} - G$). Our analysis, as shown in Fig. 3, indicates an energy deficit of 5 % with a correlation coefficient (R) of 0.86, which is considered an adequate closure term in the application of this method. Overall, our results provide valuable insights into the energy balance of the snowpack and highlight the importance of considering all energy processes when evaluating it.

The daily variations of sublimation and snowmelt over the Tazaghrat plateau are reported in Fig. 4. The graph shows

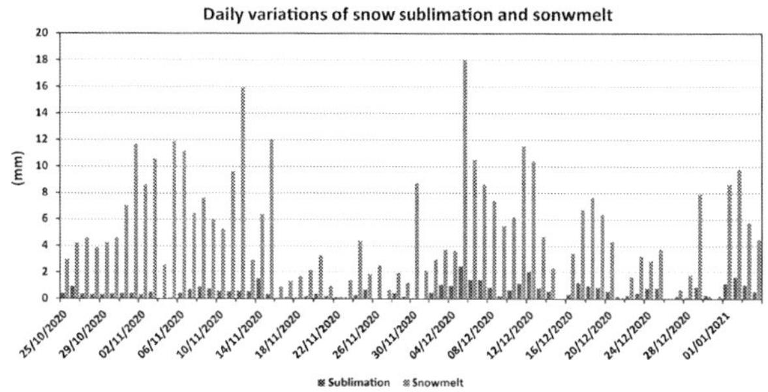

Figure 4. Daily variations of sublimation and snowmelt (mm) at Tazaghart station.

that the average daily rate of sublimation was $0.56\,\mathrm{mm\,d^{-1}}$, while the average daily rate of snowmelt was $5.14\,\mathrm{mm\,d^{-1}}$. The maximum amplitude of sublimation was observed to be $2.5\,\mathrm{mm\,d^{-1}}$, and the maximum amplitude of snowmelt was $18\,\mathrm{mm\,d^{-1}}$. These results can be explained by various factors. The rate of sublimation is influenced by temperature, humidity, and wind speed. Higher temperatures and lower humidity levels can increase the rate of sublimation, while stronger winds can enhance the transport of water vapor away from the snow surface, resulting in increased sublimation. On the other hand, the rate of snowmelt is mainly driven by temperature, solar radiation, and precipitation. Warmer temperatures and increased solar radiation can enhance snowmelt, while precipitation can either increase or decrease the rate of snowmelt depending on the form (rain or snow).

During our observation period, we identified three distinct phases, each with different dominant factors affecting sublimation rates. These phases included bare soil in September, a period dominated by snowmelt in October after the first snowfall, and finally, after two other snowfall events, a period dominated by sublimation in December (Fig. 5).

The comparison between the periods dominated by snowmelt or sublimation suggests that the sublimation period, characterized by strong wind speeds, low air temperatures, and high incident radiation, produced higher rates of sublimation and total sublimation than the snowmelt period. The rates of sublimation were 0.72 and $0.43\,\mathrm{mm\,d^{-1}}$ for the sublimation-dominated period and snowmelt-dominated period, respectively (Fig. 4).

To better understand the influence of meteorological variables on sublimation rates, we analyzed sublimation measurements in conjunction with different meteorological factors (Fig. 5). Our findings indicate that the highest relative snow sublimation fluxes occurred during subzero temperatures, low atmospheric humidity, and high wind speeds. In order to further investigate the role of wind speed in sublimation rates, we included it as a third variable in the scatter plots of meteorological variables (Fig. 6). Our results show that low sublimation values were recorded at subzero air

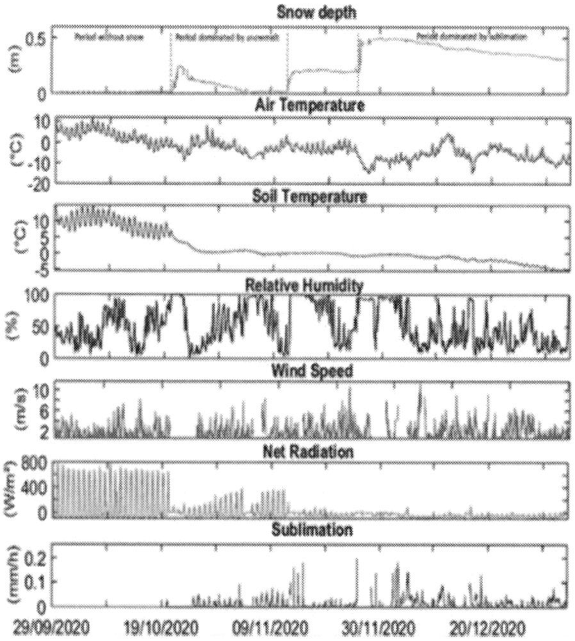

Figure 5. Sublimation measurements with different meteorological variables at an hourly time step.

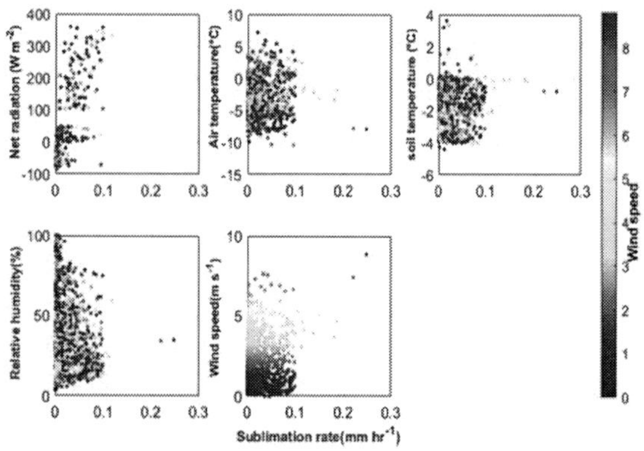

Figure 6. Scatter plots of meteorological variables against sublimation rate, observed at Tazaghart station. The color of the data points refers to the observed wind.

temperatures and low wind speeds. The highest daily sublimation values of approximately 0.3 mm h^{-1} were obtained when wind speeds were high (around 11 m s^{-1}) and air temperatures were near zero. However, warm air temperatures resulted in a decrease in snow sublimation, even with high wind speeds. Additionally, high humidity levels resulted in lower vapor pressure gradients near the surface, which subsequently led to lower sublimation rates. Overall, our results suggest that moderate to high wind speeds are necessary for significant sublimation to occur. These findings provide valuable insights into the complex interplay between meteorological variables and sublimation rates in the study area.

Table 1. Sublimation and melt rates throughout the study period.

	Sublimation Rate (%)	Melting Rate (%)	Ablation (mm)
From the water balance	40	54	104
From the energy balance	42	58	97

Based on the energy and water balance, calculated at a timestep of 30 min, the measured sublimation accounted for 42 % and 40 % of snow ablation respectively (Table 1). The estimated percentages are comparable, which supports the reliable measurements and provides validation of the Eddy covariance method. Using a modelling method, Chaponnière et al. (2005) and Boudhar et al. (2009) calculated sublimation in the northern High Atlas and discovered that it ranges between 16 % and 38 % of the total annual solid precipitation. Using a modelling method, Schulz and de Jong (2004) calculated that the drier southern slopes of the High Atlas sublimate around 45 % of the total precipitation.

5 Conclusion

The article describes a study conducted in the Tazaghart plateau in the High Atlas Mountains of Morocco to estimate sublimation and evapotranspiration processes. The study utilized an eddy-covariance flux station to estimate the energy balance and sublimation rates between September 2020 and January 2021. The preliminary data analyses show a diurnal sublimation cycle with a peak in the early afternoon and the highest relative snow sublimation fluxes that occurred with subzero temperatures, low atmospheric humidity, and high wind speeds. The study found that the Tazaghart plateau area is ideal for eddy covariance measurements, and the daily rate of sublimation was 0.56 mm d^{-1}, while the daily rate of snowmelt was 5.14 mm d^{-1}. Measured sublimation accounted for 42 % and 40 % of snow ablation, based on the energy and water balance respectively. The results show the ability of the Eddy covariance technic to quantify the direct measurements of winter water loss due to the sublimation of snowpack, especially in higher areas where measurements are relatively difficult and rare. It provides valuable insights into the hydrological processes in mountain areas and highlights the importance of accurately assessing sublimation in the overall hydrological cycle which can help inform water management strategies in semi-arid watersheds. Unfortunately, the altitude of the plateau, the very severe climatic conditions, and the difficulty of access limit the maintenance of this monitoring system on a permanent basis in these areas.

Data availability. All raw data can be provided by the corresponding authors upon request.

Author contributions. LH, VS, and SG planned the campaign; LH, VS, and AC performed the measurements; LH, OL, SG, and AC analyzed the data; LH and OL wrote the manuscript draft; SG, VS, and AC reviewed and edited the manuscript.

Competing interests. The contact author has declared that none of the authors has any competing interests.

Disclaimer. Publisher's note: Copernicus Publications remains neutral with regard to jurisdictional claims in published maps and institutional affiliations.

Special issue statement. This article is part of the special issue "IAHS2022 – Hydrological sciences in the Anthropocene: Variability and change across space, time, extremes, and interfaces". It is a result of the XIth Scientific Assembly of the International Association of Hydrological Sciences (IAHS 2022), Montpellier, France, 29 May–3 June 2022.

Review statement. This paper was edited by Christophe Cudennec and reviewed by two anonymous referees.

References

Ault, T. W., Czajkowski, K. P., Benko, T., Coss, J., Struble, J., Spongberg, A., Templin, M., and Gross, C.: Validation of the MODIS snow product and cloud mask using student and NWS cooperative station observations in the Lower Great Lakes Region, Remote Sens. Environ., 105, 341–353, https://doi.org/10.1016/j.rse.2006.07.004, 2006.

Baba, M. W., Gascoin, S., and Hanich, L.: Assimilation of Sentinel-2 data into a snowpack model in the High Atlas of Morocco, Remote Sens., 10, 1–23, https://doi.org/10.3390/rs10121982, 2018.

Bosilovich, M. G.: A comparison of MODIS land surface temperature with in situ observations, Geophys. Res. Lett., 33, 1–5, https://doi.org/10.1029/2006GL027519, 2006.

Boudhar, A., Hanich, L., Boulet, G., Duchemin, B., Berjamy, B., and Chehbouni, A.: Evaluation of the Snowmelt Runoff model in the Moroccan High Atlas Mountains using two snow-cover estimates, Hydrol. Sci. J., 54, 1094–1113, https://doi.org/10.1623/hysj.54.6.1094, 2009.

Boudhar, A., Boulet, G., Hanich, L., Sicart, J. E., and Chehbouni, A.: Energy fluxes and melt rate of a seasonal snow cover in the Moroccan High Atlas, Hydrol. Sci. J., 61, 931–943, https://doi.org/10.1080/02626667.2014.965173, 2016.

Chaponnière, A., Maisongrande, P., Duchemin, B., Hanich, L., Boulet, G., Escadafal, R., and Elouaddat, S.: A combined high and low spatial resolution approach for mapping snow covered areas in the Atlas mountains, Int. J. Remote Sens., 26, 2755–2777, https://doi.org/10.1080/01431160500117758, 2005.

Dietz, A. J., Kuenzer, C., and Dech, S.: Global SnowPack: a new set of snow cover parameters for studying status and dynamics of the planetary snow cover extent, Remote Sens. Lett., 6, 844–853, https://doi.org/10.1080/2150704X.2015.1084551, 2015.

Hanich, L., Chehbouni, A., Gascoin, S., Boudhar, A., Jarlan, L., Tramblay, Y., Boulet, G., Marchane, A., Baba, M. W., Kinnard, C., Simonneaux, V., Fakir, Y., Bouchaou, L., Leblanc, M., Le Page, M., Bouamri, H., Er-Raki, S., and Khabba, S.: Snow hydrology in the Moroccan Atlas Mountains, J. Hydrol. Reg. Stud., 42, 101101, https://doi.org/10.1016/J.EJRH.2022.101101, 2022.

Liston, G. E. and Elder, K.: A distributed snow-evolution modeling system (snowmodel), J. Hydrometeorol., 7, 1259–1276, https://doi.org/10.1175/JHM548.1, 2006.

Reba, M. L., Pomeroy, J., and Danny Marks, D.: Estimating surface sublimation losses from snowpacks in a mountain catchment using eddy covariance and turbulent transfer calculations, Hydrol. Process., 26, 3699–3711, https://doi.org/10.1002/hyp.8372, 2012.

Schulz, O. and de Jong, C.: Snowmelt and sublimation: field experiments and modelling in the High Atlas Mountains of Morocco, Hydrol. Earth Syst. Sci., 8, 1076–1089, https://doi.org/10.5194/hess-8-1076-2004, 2004.

Proc. IAHS, 385, 393–398, 2024
https://doi.org/10.5194/piahs-385-393-2024

Open Access PIAHS

Multidisciplinary approach to understand the salinization of fractured crystalline aquifers in semi-arid region

Marjorie Beate Kreis[1,2,3,6], Jean-Denis Taupin[3], Nicolas Patris[3], Patrick Lachassagne[3], Virginie Vergnaud-Ayraud[4], Julien Daniel Pierre Burte[5], Christian Leduc[5], and Eduardo Sávio Passos Rodrigues Martins[6]

[1]Doctoral School GAIA ED 584, University of Montpellier, Montpellier, France
[2]Dep. of Civil Engineering, Federal Univ. of Ceará, Fortaleza, Brazil
[3]HSM, Univ Montpellier, CNRS, IRD, Montpellier, France
[4]Geosciences Rennes, Univ. Rennes, CNRS, OSUR- Plateforme Condate Eau, Rennes, France
[5]G-EAU, AgroParisTech, CIRAD, IRD, INRAE, Montpellier SupAgro, Montpellier, France
[6]Research Institute of Meteorology and Water Resources (FUNCEME), Fortaleza CE, Brazil

Correspondence: Marjorie Beate Kreis (marjorie.kreis@hotmail.fr)

Received: 16 May 2022 – Revised: 13 April 2023 – Accepted: 28 August 2023 – Published: 19 April 2024

Abstract. In semi-arid areas, groundwater (GW) represents the most reliable resource for water supply. In the semi-arid region of Ceará (Northeastern Brazil), GW of the crystalline fractured aquifers has a salinity higher than expected in this type of environment. This study implemented a multidisciplinary approach based on hydrodynamic (piezometric) and hydrogeochemical (^{18}O, ^{2}H, ^{3}H, ^{14}C, CFC, SF_6, Electrical Conductivity, major ions) data to understand the salinization processes. Results demonstrate that GW is generally recent and recharged by meteoric waters mainly through indirect infiltration from evaporated surface water. The study suggests that GW, originally bicarbonated, becomes progressively enriched in chloride due to the dissolution and leaching of salts, during the wet season, which come from meteoric inputs and have precipitated in the unsaturated zone and pond sediments during dryer periods.

Keywords. SDG6; Field observation

1 Introduction

In crystalline aquifers, groundwater (GW) resources are generally scarce and may be subject to salinization, notably in semi-arid areas (Rabemanana et al., 2005; Jalali, 2007; Sreedevi et al., 2021). In semi-arid Northeastern Brazil (NEB), the processes at the origin of the high and heterogeneous GW salinity observed in the crystalline basement are still controversial. In Ceará (a state in semi-arid NEB), almost 75 % of the area consists of Precambrian basement rock (Santiago et al., 2001; Osorio, 2018). However, the hydrogeological functioning of the crystalline aquifers remains unclear despite the importance of this resource to cope with the lack of water in rural areas, especially during prolonged drought. Given the complexity of fractured medias, this study used a multidisciplinary approach to understand the hydrogeological functioning and hydrochemical processes that lead to GW salinization through the implementation of a monitoring network of the electrical conductivity (EC) and the piezometric levels, a multi-tracer sampling (^{18}O, ^{2}H, ^{3}H, ^{14}C, CFC, SF_6) and the analysis of major ions.

2 Study area

2.1 Geographic and climatic context

The study area is included in the Banabuiú watershed (19 647 km^2), in the center of the Ceará at approximately 180 km of the oceanic coast. Investigations were carried out in four sub-basins of the Banabuiú watershed: Forquilha (214 km^2), Pirabibú (127 km^2), Vista Alegre (550 km^2) and Ibicuitinga (286 km^2), located between the latitudes 4°54′38″ S and 5°29′08″ S, and the longitudes 38°27′38″ W

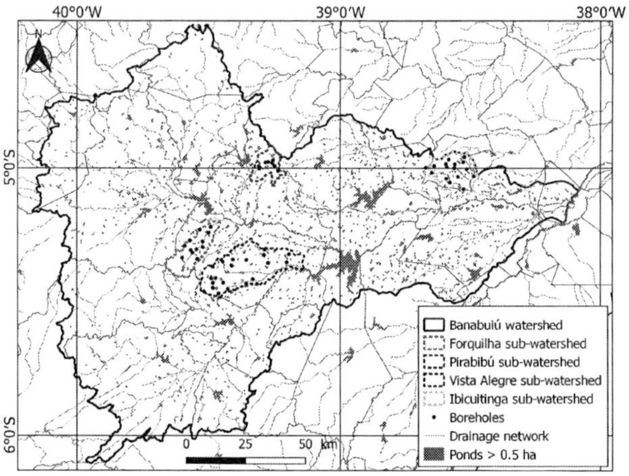

Figure 1. Map of the study area.

and 39°39′55″ W (Fig. 1). These are rural areas of low human density, inserted in the Caatinga biome, whose socioeconomic development is mainly based on livestock and agriculture (MME, 1998).

The semi-arid climate of the study zone is characterized by water deficiency due to low precipitation levels ($< 800\,\mathrm{mm\,yr^{-1}}$), high evapotranspiration (PET) rates ($> 1000\,\mathrm{mm\,yr^{-1}}$), irregular rainfall distribution and recurrent drought periods (Burte, 2008; Martins and Júnior, 2017; Pontes Filho et al., 2020). The wet season, which extends from December to July, concentrates around 75 % of the precipitation over 4 months of the year (from February to May) in the study area. Moreover, these four months of rain are characterized by intense rainfall events (Costa and Herrera, 2016). The annual mean temperature is 27.2 °C, with monthly mean temperatures ranging between 25.7 and 28.5 °C. The monthly mean relative humidity varies from 41.6 % to 85.9 %, with an annual mean of 61.0 % (data from hydrometric INMET Station no. 82586, 1981–2010 period).

2.2 Hydrological context

In the crystalline zones of the study area, surface runoff is of Hortonian type due to the low thickness (< 1 m) and the low permeability of soils (Burte, 2008). In combination with the meteorological characteristics, all Ceara rivers are naturally intermittent (Santiago et al., 2001; Burte, 2008; Ribeiro Neto et al., 2022). The Banabuiú watershed is characterized by a large number of small and big dams, with more than 12 820 reservoirs mapped out by the FUNCEME (with an area greater than 0.5 ha). These reservoirs are subject to high evaporation losses and, consequently, to salinization and eutrophication problems (Santiago et al., 2001).

2.3 Geological and hydrogeological context

The Banabuiú watershed is characterized by the outcropping of the crystalline basement over 96.5 % of its surface. These rocks are represented by various types of gneisses and migmatites of Precambrian age, associated with plutonic and metaplutonic rocks of predominantly granitic composition (INESP, 2009). The remaining 3.5 % of the area are composed of Tertiary and Quaternary sediments (probably only Pleistocene according to Cavalcanti and Cavalcante, 2014) and colluvial-eluvial Quaternary covers. All of these Cenozoic sediments are interpreted as fluvial-lacustrine deposits (Da Silva et al., 2003).

The hydrogeological knowledge of the Ceará State is still very limited (Funceme, 2007). The presence of aquifers in the crystalline basement is related to the degree of alteration and fracturing of the rock, considering their very low primary porosity and permeability (Wyns, 2004; Lachassagne et al., 2021). In the study region, the median depth of water wells is 60 m and the level of the water table is generally around 10 m below the ground surface (Da Silva et al., 2007). Wells present low flow rates (1 to $3\,\mathrm{m^3\,h^{-1}}$). GW of the crystalline aquifers of Ceará may present very high EC values for this type of environment. No regional flow is observed according to Santiago et al. (2000), which means that the natural outlet of the crystalline aquifers is locally formed by the river or the alluvium (or a spring), at the scale of small sub-basins, unless evapotranspiration creates endorheic conditions.

3 Methodology and methods

Investigations were developed on two main axes. On one side, the study of the subterranean hydrodynamics thanks to the implementation of a piezometric monitoring network and the realization of a multi-tracers sampling (^{18}O, ^{2}H, ^{3}H, ^{14}C, CFC, SF_6). On the other side, the study of the hydrochemical composition and evolution of GW thanks to the establishment of an EC monitoring network and the analysis of major ions. Studied borewells are distributed in the crystalline basement of the four sub-watersheds. All sampled borewells were pumped before sampling until stabilization of pH, temperature and EC parameters. Field parameters were measured in situ using an Aquaprobe® AP-700 multiparameter instrument.

3.1 Piezometric monitoring network

Monthly and hourly piezometric data were collected from March 2018 to December 2019 over a network of 56 boreholes, and from November 2016 to December 2019 in four of these boreholes, respectively.

3.2 Stable isotopes (^{18}O, ^{2}H) analyses

The water stable isotopes ($\delta^{18}O$, $\delta^{2}H$) were measured on GW from 75 borewells (137 samples) sampled between August 2016 and August 2019. Analyses were performed in the water stable isotope analysis Laboratory (LAMA) of Hydrosciences Montpellier (France). Results are expressed in δ ‰ vs V-SMOW (Vienna Standard Mean Ocean Water) as defined by Craig (1961). Analytical errors are ±0.08 ‰ and ±0.8 ‰ for $\delta^{18}O$ and $\delta^{2}H$, respectively.

3.3 Multi-tracer dating (^{3}H, ^{14}C, CFC, SF_6) analyses

GW samples were taken from 3 borewells in November 2009 and from 7 borewells in February 2018 for ^{14}C and ^{3}H analyses, and from 10 borewells in June 2019 for CFC and SF_6 analyses. ^{14}C and ^{3}H analyses were carried out at the laboratory of Avignon University (France) and are respectively expressed as pMC (percent modern carbon) and TU (tritium unit). Analytical errors are about 0.5 pMC and 0.3 TU, respectively. CFC and SF_6 analyses were performed at the Plateforme Condate Eau of Rennes University (France) and are within a precision of 3 % for CFC and 5 % for SF_6.

3.4 Hydrochemical analyses (EC, major ions)

An EC monitoring was carried out in conjunction with the monthly piezometric monitoring to evaluate the salinity dynamics of GW. Major ions sampling campaigns were conducted in December 2017 (dry season) and June 2018 (wet season) in 27 and 38 borewells, respectively. Major ions analyses were performed in the GECOQ laboratory of the CAGECE in Fortaleza (Brazil), in accordance with methodologies recommended by APHA (2012).

4 Results and discussions

4.1 Hydrodynamic characterization

4.1.1 Piezometric monitoring

Piezometric monitoring evidenced that GW levels are relatively shallow, between 0.4 and 21.4 m deep. A contrasted temporal variation of piezometric levels was observed, with annual variations varying from 0.2 to 5.9 m depending on the boreholes. Seasonal recharge occurs during the rainy season, between February and May. Hourly piezometric monitoring showed a strong reactivity of GW levels (< 24 h) to significant rainy events (> 50–60 mm), but a decrease of the level is almost never observed following the event. Instead, the levels gradually increase throughout the rainy season. Analysis of the variations in GW levels suggests that the seasonal recharge depends on the distribution and amounts of daily rainfall and is favoured by the storage of surface runoff in ponds.

4.1.2 Stable isotopes monitoring (^{18}O, ^{2}H)

$\delta^{18}O$ and $\delta^{2}H$ values of GW samples showed a strong spatio-temporal variability and ranged between −4.04 ‰ and +1.75 ‰ and between −23.3 ‰ and +8.0 ‰, respectively, with median values of -2.52 ± 1.08 ‰ for $\delta^{18}O$ and -14.0 ± 5.7 ‰ for $\delta^{2}H$. GW presents low values of deuterium excess, with a median value of $+6.1 \pm 3.6$ ‰. The relationship between $\delta^{18}O$ and $\delta^{2}H$ of GW is expressed by Eq. (1):

$$\delta^{2}H = 5.09 \times \delta^{18}O - 1.22 \; (r^2 = 0.9), \tag{1}$$

The slope of 5.09 of Eq. (1) is lower than the slope of the local meteoric water line (LMWL) defined in Kreis et al. (2020) and the global meteoric water line (GMWL), and clearly indicates that meteoric waters undergo significant evaporation before their infiltration (Chatton et al., 2016). Moreover, Eq. (1) is very similar to the linear regressions obtained by Matsui (1978) and by Stolf et al. (1977), cited in Matsui (1978) for surface waters in the NEB, which are respectively of $\delta^{2}H = 4.6 \times \delta^{18}O - 0.5$ ($r^2 = 0.97$) and $\delta^{2}H = 4.75 \times \delta^{18}O - 1.3$. Therefore, the strong similarity between the linear regressions of GW and surface water highlights that part of GW recharge occurs through a mixing process involving surface water marked by evaporation. This hypothesis is consistent with the numerous dams, lagoons and temporary drainage networks existing in the study area, and is supported by the GW level survey.

4.1.3 Multi-tracer dating (^{14}C, ^{3}H, CFC, SF_6)

The assessment of GW residence time through a multi-tracer approach showed that the apparent age of the GW varies from a few decades to hundreds of years. It also underlined the large contribution of post-1950s water in the recharge processes with ^{14}C values generally higher than 100 pMC and measurable ^{3}H values, but highlighted the complexity of flow path organization in the fractured crystalline system. Indeed, CFC and SF_6 analyses, best interpreted by a binary mixing model (BMM), suggested the influence of water mixing between a recent pole (15 % to 85 % of the contributions) corresponding to a present-day recharge (fast vertical flows) and an older pole of longer transit time (interpreted as slow horizontal transfers). The BMM is based on the mixing of two piston flow water bodies. Contrary to the classical BMM which allows mixing between young (current CFC values) and old (CFC free) reservoirs, Condate Eau Plateform (Rennes University) tests the date allowing mixing with more or less young and more or less old waters. The result given is the best combination of mixing result (age and proportion) obtained for each tracers. The uncertainty in the model results depends on the adequacy between the results given by the different tracers and does not exceed 20 % and 10 years. Results of the multi-tracer dating are synthesized in Table 1.

Table 1. Results of the multi-tracer dating. FOR = Forquilha, PIR = Pirabibú, VA = Vista Alegre, IBI = Ibicuitinga.

Id. Samples	Sampling year	^{14}C (pMC)	^{3}H (UT)	CFC and SF_6 (Binary mixing model)	
				Recent pole (%)	Old pole (%)
PIR 44bis	2019	–	–	–	–
PIR 45	2019	–	–	85	15
FOR P7	2009	126.7	0.7	–	–
PIR P15	2009	124.4	1	–	–
VA Zuca	2009	119.5	0.9	–	–
PIR P16	2018	105.3	0.7	–	–
FOR 144	2018/9	103.6	0.6	65	35
VA 132	2018/9	110.1	≤ 0.7	45	55
PIR 52	2018/9	105.6	≤ 0.5	35	65
FOR 151	2018/9	98.3	≤ 0.7	50	50
IBI 89	2019	–	–	40	60
VA 123	2019	–	–	25	75
IBI 96	2018/9	73.2	≤ 0.4	20	80
IBI 152	2019	–	–	15	85
IBI 160	2018	91.8	≤ 0.3	–	–

4.2 Hydrogeochemical characterization

4.2.1 Electrical conductivity (EC) monitoring

The EC monitoring highlighted a strong spatial heterogeneity of the crystalline media, with EC values ranging between 886 and 19 310 $\mu S\,cm^{-1}$ (median value of 2850 $\mu S\,cm^{-1}$) and no salinity gradient. During the wet season, most boreholes show a temporary and fast increase of EC which suggests a leaching of salts stored in sediments, soil or in the unsaturated zone (USZ) during drier periods.

4.2.2 Major ions sampling

More than 75 % of the samples presented a mixed chloride or sodium chloride facies. However, the less salty waters (EC $< 1600\,\mu S\,cm^{-1}$) presented a bicarbonated facies ($< 10\,\%$ of the samples), whereas they tended towards a chloride facies with the increase in salinity. The chemical sampling before and after the rainy season showed a displacement of the points from the sodium pole towards the mixed pole, which suggests, in association with the observed increase in EC, a phenomenon of dissolution of calcium and/or magnesium salts (as calcite or dolomite). The strong correlation between EC and chloride ($r^2 = 0.9$) or cations ($0.7 < r^2 < 0.8$) and the absence of correlation between EC and bicarbonate, sulphate or nitrate ($r^2 < 0.1$) indicates that the origin of high salinity is not related to the water/rock interaction or the anthropic pollution.

4.3 Origin of groundwater salinization

Among the processes that lead to salinization of GW, marine intrusions and the presence of marine evaporitic rocks are ruled out considering the distance from the study zone to the oceanic coast and the absence of primary evaporitic terms in the Precambrian crystalline basement of the region. Moreover, the hypothesis of a past marine transgression is incompatible with the measured residence time of water. Mineral hydrolysis of crystalline rocks may explain part of the water mineralization, but not the predominance of chloride ion or the high salinization of GW. Considering that the chlorides and the GW salinity are not related to anthropic pollution, it results that the chloride source is external to the aquifer and derives from meteoric inputs (in the form of wet or dry depositions). Furthermore, the chemical study suggests that the GW, originally bicarbonated, becomes progressively enriched in chloride due to the dissolution and leaching of salts that were precipitated in the USZ and pond sediments during the dry season owing to the strong evapotranspiration rates, as observed in Madagascar (Rabemanana et al., 2005) or India (Sreedevi et al., 2021), for example. The precipitation of salts on the surface is favoured by alternating dry and wet periods (Sreedevi et al., 2021). The hypothesis of leaching of salts is supported by the observation of evaporite minerals (calcite, dolomite, gypsum, halite and probably bischofite) on the banks and sediments of the dams in the study area (Laraque, 1991; Araújo, 2017) and the presence of evaporated water in the aquifers.

5 Conclusions

This multidisciplinary study used hydrogeochemical and isotopic approaches, associated with piezometric data, to understand the hydrogeological functioning and the origin of the salinity of a complex crystalline system. Generally, the contrasted temporal and spatial variations of piezometric levels or hydrochemical composition of GW highlighted the strong heterogeneity and compartmentalization of the crystalline aquifers, as well as the influence of local processes. The combination of results allows to identify that GW recharge occurs through both direct and indirect infiltration, with a significant proportion of evaporated surface water. The GW residence time is relatively short (significant supply of post-1950s water) and the flow paths organization is mainly composed of rapid vertical flows. In these aquifers, the salinity is interpreted as originating from rainfall. The evapo(transpi)ration of small rainfall events or surface waters during the dry periods favors the concentration of atmospheric salts on the surface. This study suggests that the GW becomes progressively salinized due to the dissolution and leaching of evapoconcentrated salts precipitated in the unsaturated zone and pond sediments, given the high evapotranspiration in the partially endorheic context.

Data availability. The research data presented in this study can be found in Kreis (2021), accessible at: https://www.theses.fr/2021MONTG031.

Author contributions. MBK carried out the investigations and interpretations of this study under the scientific supervision of JDT, CL, PL and ESPRM. JDPB carried out the 2009 sampling campaign for the measurement of ^{14}C and ^{3}H. NP and VVA performed respectively the stable isotope and the CFC/SF_6 analyses. MBK prepared the manuscript with contributions from all co-authors.

Competing interests. The contact author has declared that none of the authors has any competing interests.

Disclaimer. Publisher's note: Copernicus Publications remains neutral with regard to jurisdictional claims in published maps and institutional affiliations.

Special issue statement. This article is part of the special issue "IAHS2022 – Hydrological sciences in the Anthropocene: Variability and change across space, time, extremes, and interfaces". It is a result of the XIth Scientific Assembly of the International Association of Hydrological Sciences (IAHS 2022), Montpellier, France, 29 May–3 June 2022.

Acknowledgements. This work is part of the multidisciplinary French-Brazilian cooperation between IRD and FUNCEME. The authors would thank the LAMA laboratory and the FUNCEME for all the financial and technical support.

Financial support. This research has been supported by internal funding from Research Institutes of authors (LAMA laboratory from Hydrosciences Montpellier and FUNCEME).

Review statement. This paper was edited by Christophe Cudennec and reviewed by three anonymous referees.

References

American Public Health Association (APHA): Standard methods for the examination of water and wastewater, 22th edition, edited by: Rice, E. W., Baird, R. B., Eaton, A. D., and Clesceri, L. S., American Public Health Association (APHA), American Water Works Association (AWWA) and Water Environment Federation (WEF), Wahington, DC, 1100 pp., 2012.

Araújo, A.: Mecanismos dominantes na salinização de água em rochas anisotrópicas com base em razões iônicas e número fuzzy, em área do sertão central no Ceará, PhD Thesis, Federal university of Ceará, 188 pp., http://repositorio.ufc.br/handle/riufc/53426, 2017.

Burte, J.: Os pequenos aquiferos aluviais nas areas cristalinas semi-aridas: Funcionamento e estrategias de gestao, Estudo de caso no Nordeste brasileiro, PhD Thesis, University of Montpellier, https://agritrop.cirad.fr/547686/ (last access: 28 September 2023), 2008.

Cavalcanti, J. A. D. and Cavalcante, J. C.: Evolução geológica, in: Geodiversidade do estado do Ceará, CPRM, 19–34, https://rigeo.sgb.gov.br/handle/doc/16726, 2014.

Chatton E., Aquilina L., Petelet-Giraud E., Cary, L., Bertrand, G., Labasque, T., Hirata, R., Martins, V., Montenegro, S., Vergnaud, V., Aurouet, A., and Kloppmann, W.: Glacial recharge, salinization and anthropogenic contamination in the coastal aquifers of Recife (Brazil), Sci. Total Environ., 569, 1114–1125, https://doi.org/10.1016/j.scitotenv.2016.06.180, 2016.

Costa, C. S. B. and Herrera, O. B.: Halophytic life in Brazilian Salt Flats: Biodiversity, Uses and Threats, in: Sabkha Ecosystems, edited by: Khan, M. A., Boër, B., Özturk, M., Clüsener-Godt, M., Gul, B., and Breckle, S. W., Volume V, The Americas (Vol. 48), Springer, https://doi.org/10.1007/978-3-319-27093-7_2, 2016.

Craig, H.: Standard for reporting concentrations of deuterium and oxygen-18 in natural waters, Science, 133, 1833–1834, 1961.

Da Silva, A. J. P., Lopes, R. C. L., Vasconcelos, A. M., and Bahia, R. B. C.: Capítulo II – Bacias sedimentares paleozoicas e meso-cenozoicas interiores, in: Geologia, Tectônica e Recursos Minerais do Brasil, CPRM: Brasília, ISBN 85-230-0790-3, 55–85, 2003.

Da Silva, F. J. A., de Araújo, A. L., and de Souza, R. O.: Águas subterrâneas no Ceará – poços instalados e salinidade, Revista Tecnologia, 28, 136–159, 2007.

Funceme: Mapeamento e Avaliação do Potencial Hídrico Subterrâneo dos Aluviões em Zonas SemiÁridas Utilizando Técnicas de Sensoriamento Remoto e SIG, Fortaleza: Governo do Estado do Ceará, http://www.funceme.br/wp-content/uploads/2019/02/Aluviao.pdf (last access: 30 April 2022), 2007.

INESP: Caderno regional da sub-bacia do Banabuiú, Vol.2, Fortaleza, Instituto Nacional de Ensino e Pesquisa (INESP), ISBN 978-85-87764-99-7, 2009.

Jalali, M.: Salinization of groundwater in arid and semi-arid zones: an example from Tajarak, western Iran, Environ. Geol., 52, 1133–1149, https://doi.org/10.1007/s00254-006-0551-3, 2007.

Kreis, M.: Origine et dynamique de salinité des aquifères fracturés cristallins de la région semi-aride du Ceará, Brésil, PhD

Thesis, University of Montpellier, Federal University of Ceará, NNT: 2021MONTG03, https://www.theses.fr/2021MONTG031 (last access: 23 May 2022), 2021.

Kreis, M., Taupin, J. D., Patris, N., and Martins, E. S. P. R.: Isotopic characterisation and dating of groundwater recharge mechanisms in crystalline fractured aquifers: example of the semi-arid Banabuiú watershed (Brazil), Isot. Environ. Healt. S., 56, 418–430, https://doi.org/10.1080/10256016.2020.1797275, 2020.

Lachassagne, P., Dewandel, B., and Wyns, R.: Review: Hydrogeology of weathered crystalline/hard-rock aquifers–guidelines for the operational survey and management of their groundwater resources, Hydrogeol. J., 29, 2561–2594, https://doi.org/10.1007/s10040-021-02339-7, 2021.

Laraque, A.: Comportements hydrochimiques des" açudes" du Nordeste brésilien semi-aride: évolutions et prévisions pour un usage en irrigation, PhD Thesis, University of Montpellier 2, 446 pp., https://www.theses.fr/1991MON20003 (last access: 23 May 2022), 1991.

Martins, E. S. P. R. and Júnior, F. D. C. V.: O clima da Região Nordeste entre 2009 e 2017: Monitoramento e previsão, Parcerias Estratégicas, 22, 63–79, 2017.

Matsui, E.: Origem e dinâmica de salinização da água do nordeste brasileiro – Bacia do rio Pajeú/PE, PhD Thesis, USP: Piracicaba/SP, 1978.

Ministério de Minas e Energia (MME): Programa de recenseamento de fontes de abastecimento por água subterrânea no Estado do Ceará. Diagnóstico do município de Quixeramobim, CPRM: Fortaleza, 15 pp., https://rigeo.sgb.gov.br/handle/doc/16611, 1998.

Osorio, D. B.: Hidrogeologia do semiárido Cearense, M.S. Thesis, University of São Paulo, https://doi.org/10.11606/D.44.2019.tde-30052019-131248, 2018.

Pontes Filho, J. D., Souza Filho, F. D. A., Martins, E. S. P. R., and Studart, T. M. D. C.: CopulaBased Multivariate Frequency Analysis of the 2012–2018 Drought in Northeast Brazil, Water, 12, 834, https://doi.org/10.3390/w12030834, 2020.

Rabemanana, V., Violette, S., De Marsily, G., Robain, H., Deffontaines, B., Andrieux, P., Bensimon, M., and Parriaux, A: Origin of the high variability of water mineral content in the bedrock aquifers of Southern Madagascar, J. Hydrol., 310, 143–156, https://doi.org/10.1016/j.jhydrol.2004.11.025, 2005.

Ribeiro Neto, G. G., Melsen, L. A., Martins, E. S. P. R., Walker, D. W., and van Oel, P. R.: Drought Cycle Analysis to evaluate the influence of a Dense Network of small reservoirs on drought evolution, Water Resour. Res., 58, https://doi.org/10.1029/2021WR030799, 2022.

Santiago, M. F., Frischkorn, H., Neto, P. S., and Filho, M. J.: The recharge mechanisms in an alluvial aquifer zone in northeast Brazil, Groundwater, 39, 18–23, https://doi.org/10.1111/j.1745-6584.2001.tb00347.x, 2001.

Santiago, M. M. F., Frischkorn, H., and Mendes Filho, J.: Mecanismos de salinização em águas do Ceará, Rio Grande do Norte e Piauí. Águas Subterrâneas, Anais do XI Congresso Brasileiro de Águas Subterrâneas, 16 pp., https://aguassubterraneas.abas.org/asubterraneas/article/view/23481 (last access: 28 September 2023), 2000.

Sreedevi, P. D., Sreekanth, P. D., and Reddy, D. V.: Influence of hydrological and hydrogeological factors on inland groundwater salinity in a hard rock aquifer, south India, J. Earth Syst. Sci., 130, 215, https://doi.org/10.1007/s12040-021-01715-x, 2021.

Stolf, R., Leal, J. M., Fritz, P., and Salati, E.: Water budget of a dam in the semi-arid Northeast of Brazil based on oxygen-18 and chlorine contents, in: Isotopes in lake studies, Proceedings Series, 29 August–2 September 1977, Vienna International Atomic Energy Agency No. ST1/PUB/511, 57–66, Vienna, 1979.

Wyns, R., Baltassat, J. M., Lachassagne, P., Legtchenko, A., Vairon, J., and Mathieu, F.: Application of Proton Magnetic Resonance Soundings to groundwater reserve mapping in weathered basement rocks (Brittany, France), B. Soc. Geol. Fr., 175, 21–34, https://doi.org/10.2113/175.1.21, 2004.

Proc. IAHS, 385, 399–406, 2024
https://doi.org/10.5194/piahs-385-399-2024

Open Access PIAHS

Bayesian inference of synthetic daily rating curves by coupling Chebyshev Polynomials and the GR4J model

Rafael Navas[1,3], Pablo Gamazo[1], and R. Willem Vervoort[2]

[1]Hidrología e Hidráulica, Departamento del Agua, CENUR – Litoral Norte,
Universidad de la República, Salto 50000, Uruguay
[2]ARC ITTC in Data Analytics for Resources and Environments & School of Life and Environmental Sciences,
The University of Sydney, Sydney, NSW 2006, Australia
[3]Programa Nacional de Investigación en Producción y Sustentabilidad Ambiental,
Instituto Nacional de Investigación Agropecuaria, La Estanzuela 70000, Uruguay

Correspondence: Rafael Navas (rafaelnavas23@gmail.com)

Received: 13 May 2022 – Revised: 30 May 2023 – Accepted: 28 August 2023 – Published: 19 April 2024

Abstract. In fluvial dynamics studies, there are instances where it becomes necessary to estimate the daily discharge of a river in locations where only one instantaneous level record is available per day. In such cases, there may be no rating curve, or one that is unreliable, making it difficult to make accurate discharge estimates. A daily rating curve would be an estimate of the daily discharge of a river, from a single instantaneous stage level. This work proposes to estimate synthetic (non-gauged) daily rating curves from nearby gauged locations using a rainfall-runoff model. A rainfall-runoff model (GR4J) is coupled with an instantaneous/stage–daily/discharge relationship based on third order Chebyshev polynomials. The parameters in the joint daily rating curve and rainfall-runoff model are optimized and uncertainty is quantified with Bayesian inference and the Delayed Rejection Adaptive Metropolis algorithm assuming model residuals to be normally distributed $N(0,\sigma)$. A case study with four gauging sites in New South Wales, Australia, and periods with no changes in the stage-discharge relationship were selected. The method is implemented four times across the gauging sites, where three sites are assumed gauged and one site is assumed to have only instantaneous water level records. The results of this methodology can help provide a more comprehensive understanding of the hydrological functioning of systems, where only one instantaneous stage level per day is available. This is particularly useful in situations where historical observations or satellite altimetry data in rivers is used to estimate daily flows.

Keywords. UPH 16; UPH 17; modelling; new monitoring

1 Introduction

Runoff time-series give valuable information for water resources. In practice, hydrometric stations measure only river stage. Runoff time series are estimated through the rating curve (RC), which is the relationship between paired instantaneous stage and discharge measurements (gaugings). RCs usually are approximated with a power law (WMO, 2010), but alternative approaches are polynomial regression, splines and fuzzy regression (Fenton, 2018; Jalbert et al., 2011; McMahon and Peel, 2019). Despite the plethora of techniques, building the RC is still complex since it depends on the availability of gaugings, variable hydraulic conditions, and understanding the physical process governing the stage-discharge relationship (Le Coz, 2012).

Piecewise relationships (Lee et al., 2010; Reitan and Petersen-Øverleir, 2009), hysteresis, backwater effects (Dottori et al., 2009; Hidayat et al., 2011; Kim et al., 2021; Petersen-Øverleir, 2006), changes in the river bed over time (Bhandari et al., 2023; Morlot et al., 2014), aquatic vegetation dynamics (Perret et al., 2021), or lack of gaugings (Lang et al., 2010; Pedersen et al., 2019; Reistad et al., 2007) are some limitations for building a RC. This is particularly true at low or high stages, where, for operational or safety reasons, gaugings are absent and extrapolation techniques are needed (Di Baldassarre and Claps, 2011; Lang et al., 2010; Reistad et al., 2007). Some example extrapolation techniques are conveyance slope, areal comparison of peak discharges, flood rooting, step backwater and hydraulic modelling (WMO, 2010). All these methods are based on the availability of instantaneous discharges and most of them require additional field information.

The quality of the rating curve data and approximation defines the uncertainty of the observed discharge and affects the performance of the hydrological model as well as the optimization procedure (McMillan et al., 2010; Sellami et al., 2013). But what happens if we take advantage of this link to estimate the rating curve? Perhaps due to the complexity of the problem, only a few authors have investigated this. For example: Sikorska and Renard (2017) re-calibrate the rating curve using a Bayesian framework which couples the rating curve with the hydrological model outputs, and therefore consider structural and parametric uncertainties in the discharge prediction. Jian et al. (2017) use only water levels and a hydrological model to make discharge predictions. They calibrate the model using Spearman Rank correlation and the inverse rating curve. Equifinality is crucial in these kinds of problems since the parameters of the hydrological model could compensate for the errors associated with the parametric uncertainty of the rating curve (Lima et al., 2019).

The aim of this work is to take advantage of the link between stream stage observations and hydrological model outputs to develop a framework to estimate daily discharge at sites with incomplete rating curves. The daily rating curve represents the relationship between an instantaneous water level within a day and the mean daily discharge. The concept is introduced to address the problem of scaling between instantaneous water level observations and discharge simulations. The size of the basin affects the sub-daily variability of discharge (Blöschl and Sivapalan, 1995), affecting the estimated mean daily discharge and the magnitude of estimated discharges is further influenced by the nonlinearity of the rating curve (Kiang et al., 2018). These factors add uncertainties to discharge computation. The working hypothesis is that parametric uncertainties from a daily rating curve model and a hydrological model can be quantified using Bayesian Inference, resulting in an estimate of the daily rating curve based on sparse or limited data. For example, this approach could be used to estimate rating curves in a cross-catchment approach where at some locations only one instantaneous level data per day is available, for example, based on satellite altimetry of rivers (Kittel et al., 2021).

2 Study area and dataset

The study area is located in the headwater of the Lachlan River, which flows from the Great Dividing Range into Wyangala Dam (a tributary of the Murray-Darling river system, New South Wales, Australia). Average annual rainfall is around 1100 mm; monthly precipitation being generally uniform throughout the year. Annual evaporation varies from 900 to 1200 mm with strong seasonal behaviour between winter and summer. Groundwater is governed by fractured rocks and topography.

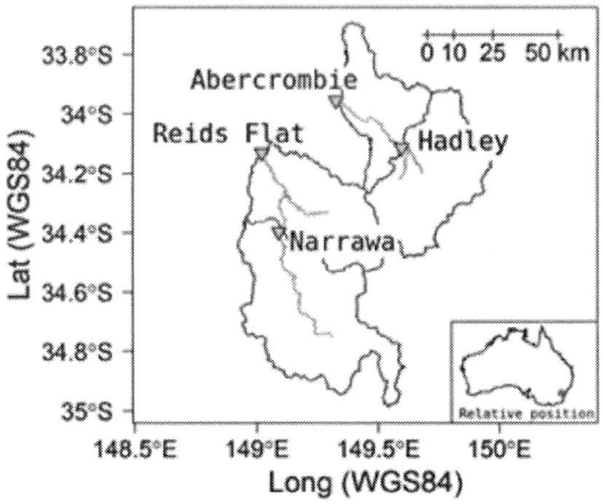

Figure 1. Gauging sites of Abercrombie and Lachlan rivers: Lachlan River at Reids Flat (3742 km^2), Lachlan River at Narrawa (2256 km^2), Abercrombie River at Abercrombie (2636 km^2), and Abercrombie River at Hadley (1635 km^2).

Figure 1 shows the location of the gauging sites. The stage discharge relationships at the gauging sites are well known and have relatively dense data sets (Water Data Online, http://www.bom.gov.au/waterdata/, last access: 14 September 2023). The simulation period chosen is from 2008 to 2013 to maximize the data availability with no significant changes in the stage-discharge relationship. Generally, the scatter of percentage differences between instantaneous discharge and daily discharge increases as daily discharge increases (Fig. 2). These differences are normally distributed with a zero mean being more homogeneous as basin size increases (Fig. 2). Climatological forcing using mean areal daily values of potential evapotranspiration and precipitation for each basin was generated using the gridded SILO – Australian climate database (Scientific Information for Land Owners, https://www.longpaddock.qld.gov.au/silo/, last access: 11 September 2023), which provides values at an approximate 5 km grid scale. Potential evapotranspiration is calculated by the FAO56 Penman–Monteith formula. Further information about the interpolation techniques used in SILO can be found in Jeffrey et al. (2001).

3 Method

The daily rating curve is the relationship between mean daily discharge (Q) and an arbitrary instantaneous water level within the day (H). It differs from the classical definition of

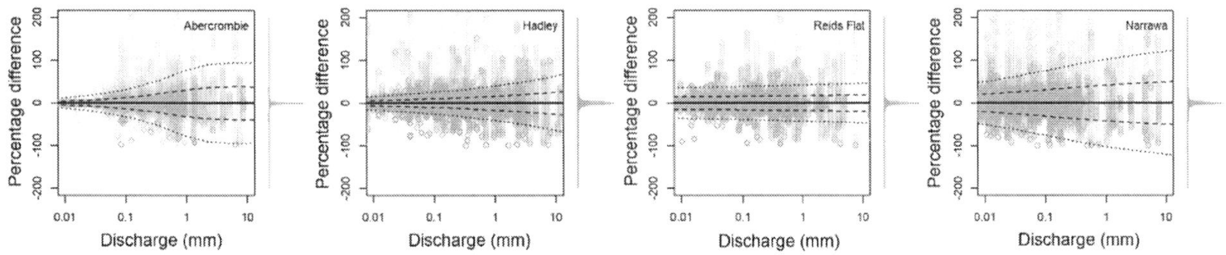

Figure 2. Percentage difference of reported 15 min discharge and mean daily discharge as a function of daily discharge (lines: Generalised Additive Models for Location Scale and Shape, continuous line: 50th percentile, dashed line: 25–75th percentiles, pointed line: 5–95th percentiles).

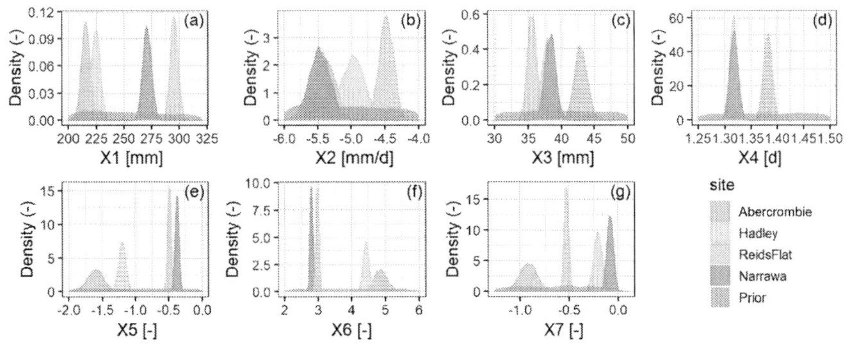

Figure 3. Prior and posterior density functions of parameters of the GR4J **(a, b, c, d)** and daily rating curve models **(e, f, g)** for the four hypothetical instantaneous water level sites.

RC since the latter works with paired instantaneous values of stages and discharges. We propose a daily rating curve model coupled with a rainfall-runoff model to obtain flow estimates at sites where only one instantaneous water level value is available per day. This work uses a cross catchment verification where a single rainfall-runoff model is fitted across the four sites (Fig. 1), and where one of the sites is assumed to have only instantaneous water level records, this is called the test site. Next, the daily discharge on the test site with only water level records is derived from the hydrological model to estimate the daily rating curve model. At the same time, the coupled models are optimized by Bayesian inference.

3.1 Rainfall–Runoff Model

The rainfall-runoff model simulates discharges at the 4 sites. It is based on the GR4J model (Perrin et al., 2003). GR4J is used since it is a simple lumped rainfall-runoff model with 4 parameters (X_1: the production storage capacity; X_2: the groundwater exchange coefficient; X_3: one day ahead maximum capacity of the routing storage; and X_4: the time base of unit hydrograph). Furthermore, the GR4J model is available in several programming languages and packages like Fortran or R with a very low computational cost (Andrews et al., 2011; Coron et al., 2017).

The model requires the daily time series of precipitation and potential evapotranspiration as input variables. The implementation of the GR4J uses the same set of parameters for all sites, and the model predicting different gauging sites only differs in the climatological forcing. This approach considers that the characteristics of basins are similar, and the transfer of parameters across the different basins can be performed with a minor loss of prediction skill, which is, of course, a strong assumption that we aim to handle with the Bayesian Inference.

A general formulation of the rainfall-runoff model follows Eq. (1):

$$Q_u = \mathrm{GR4J}(\phi_u, X_1, X_2, X_3, X_4), \quad (1)$$

where Q is the discharge (mm d^{-1}), the subscript u is the gauging site, and GR4J() is the rainfall-runoff model as function of the climatological forcing denoted by ϕ (precipitation and potential evapotranspiration in mm d^{-1}), and X_1, X_2, X_3, X_4 the GR4J parameters.

3.2 Daily rating curve model

The daily rating curve model is built from instantaneous observed water levels and daily simulated discharges using GR4J. This preliminary approach assumes that the stage-discharge relationship could have up to one possible changing point caused by a change in the hydraulic controls or flows above the bank full stage. This assumption is represented using 3rd order Chebyshev polynomials (Eq. 2). Chebyshev polynomials instead of other models overcome some problems of the automatic generation of rating curves, being a computationally efficient alternative (Fenton, 2018) with a reduced number of parameters.

$$\log(Q_{u=\mathrm{test}}) = X_5 + X_6 \cos\left(\mathrm{acos}(h^*_{u=\mathrm{test}})\right) + X_7 \cos\left(2 \times \mathrm{acos}(h^*_{u=\mathrm{test}})\right) \quad (2)$$

where $u = \text{test}$ is the instantaneous water level test site, X_5, X_6, X_7 are the 3rd order Chebyshev coefficients and h^* is a transformation which rescales the stage between -1 to 1 (McMahon and Peel, 2019).

3.3 Bayesian Inference

Bayesian inference is often used for parameter optimization and uncertainty estimation. The method estimates the probability density function of the parameters of the model (known as a posterior distribution) by using a likelihood function and prior distributions of parameters and the Bayes theorem.

In this preliminary implementation, the likelihood function assumes that residual errors are Gaussian, homoscedastic and independent:

$$L = \sum_u \left(-\frac{n}{2} \log\log\left(2\pi\sigma_u^2\right) + \sum_u \frac{\epsilon_u^2}{2\sigma_u^2} \right) \quad (3)$$

where L is likelihood function, σ the variance of residuals, and ϵ the model residuals as $\epsilon = Q_{\mathrm{obs}} - Q_{\mathrm{sim}}$. On the full gauged sites, Q_{obs} is given by the daily observed discharges downloaded from the Bureau of Meteorology of Australia. At the water level test site, Q_{obs} is imposed by the daily rating curve Eq. (2).

The Bayesian Inference of the coupled models is using the Delayed Rejection Adaptive Metropolis algorithm (DRAM). This technique finds an ensemble of parameters values that represent parameter distributions and uncertainties. The implementation is using the package FME in the R environment (Soetaert and Petzoldt, 2010). Prior distributions of parameters have been defined by Gaussian probability distributions for parameters X_1 and X_2, and non-informative uniform distributions for parameters X_3 to X_7 as well as σ.

4 Results and discussion

4.1 Application of the Bayesian Inference

The prior and posterior distributions of parameters resulting from applying the DRAM algorithm across the instantaneous water level test sites are shown in Fig. 3. Posterior distributions of the parameters differ between the test cases, where each test case represents a different gauging station with only daily instantaneous values. Differences might be caused by the assumption that all basins are parameterized with the same set of parameters, ignoring catchment differences. Depending on which catchment is the test case, this would change the parameter distributions. A clear demonstration of this effect is shown by the time base of the unit hydrograph (X_4, Fig. 3d). The parameter X_4 is related to the size, shape and slope of the basin. Posterior distributions of X_4 are similar for test cases which have similar basin characteristics, such as Abocrombie and Reids Flat. However not too much physical interpretation should be given to the parameters of GR4J since the parameters cannot always be related to the physical characteristics of the basins (Narbondo et al., 2020). In contrast, the daily rating curve model shows no similarities between parameters across gauging sites.

Despite being a parsimonious hydrological model the X_1, X_2, X_3 parameters of GR4J were correlated (Fig. 4). This has been highlighted before for GR4J (Yang et al., 2018; Arsenault and Brissette, 2014; Qi et al., 2020). The rating curve model also showed a high correlation between the parameters (Fig. 4), indicating a possible overparameterization that could be due to the choice of the degree of the Chebyshev polynomials. Here 3rd degree polynomials were included to allow for changes in the stage-discharge relationship. However, there is no interaction between the parameters of the rating curve and the parameters of the hydrological model (Fig. 4). This could be due to the effect introduced by the coupled scheme, which incorporates discharge observations from neighbouring basins, reducing the problem of equifinality in the coupling of the rating curve and hydrologic model, compared to similar work (Lima et al., 2019).

4.2 Daily rating curve estimation, potential applications and limitations

Two simulations of the daily discharge have been generated for the sites where only daily water level data are available. The first simulation is generated by the rainfall-runoff model, and the second by the daily rating curve model. Figure 5 shows the residuals of both simulations as a function of observed discharges. Two models overestimate the low flow with the GR4J estimations being less biased. Additionally, for medium to high flows, the overestimation decreases and the scatter of residuals of the daily rating curve is lower than for GR4J.

Daily rating curves tend to overestimate/underestimate the flow at low/high stages (Fig. 6). Overall change points in the estimated rating curves are relatively smooth, which would point to potential overparameterization discussed in Sect. 4.1. An interesting finding of this work is that the upper part of the daily rating curve has similarities with the instantaneous rating curve. In contrast, the lower part of the daily rating curve has non-negligible errors that could come from the simplistic structure of the rainfall-runoff model (Flores

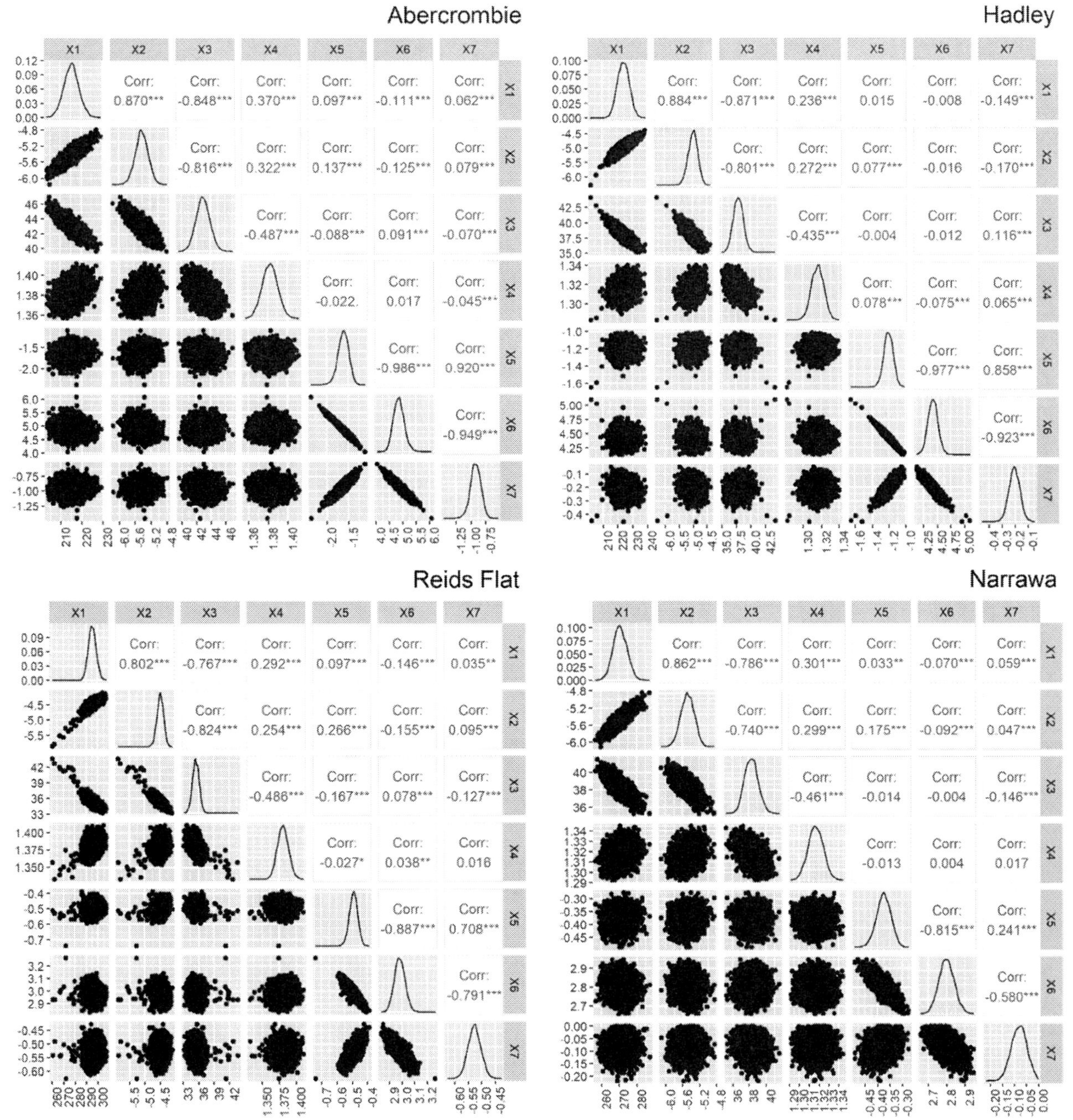

Figure 4. Correlation matrix, scatter plots and posterior density function of parameters.

et al., 2021). Also, it could be related to the structure of the differences between instantaneous discharge and daily discharge (Fig. 2); It is noted that the best fit is obtained for Reids Flat, which is the larger basin with the lowest sub-daily discharge variability.

The daily rating curve is a new concept that aims to establish a relationship between the stage and discharges in a different time scale. This addresses the problem of scaling between instantaneous water level observations and discharge simulations of a hydrological model which delivers mean discharges over a time step (a time aggregation), which is often forgotten. Although Fig. 6 compares the daily rating curve with actual instantaneous gaugings, it is important to remember that the daily approach has a different purpose and

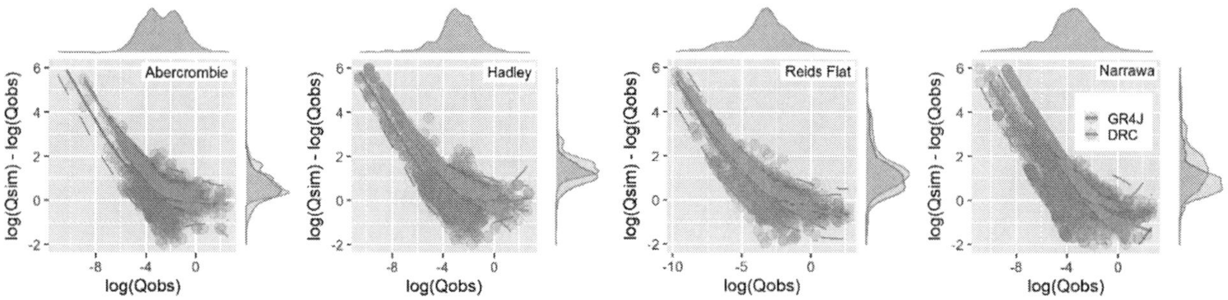

Figure 5. GR4J and Daily Rating Curve (DRC) residuals as a function of daily discharges at the water level site and probability density function of log discharges (top) and log residuals (right).

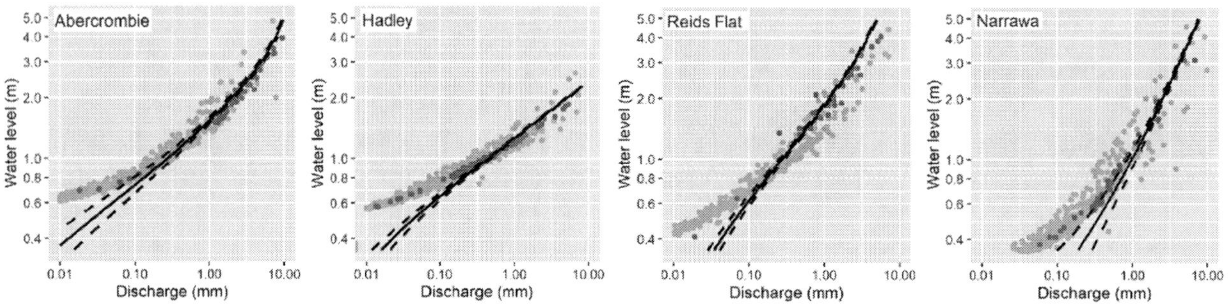

Figure 6. Daily rating curve estimation (black line) with the 95 % credible interval (dashed lines), observed daily discharges (grey dots), and observed instantaneous discharges (red dots).

potential use that should not be confused with the classic instantaneous rating curve. We believe that the main potential of this approach could be data assimilation at sites with only water level records, such as from satellite altimetry. The presented approach also has some limitations which require further research. Temporal changes in the river geometry could be one of the most important constraints (Bhandari et al., 2023; Morlot et al., 2014), for this reason, the study period was limited to a period of time without significant changes in the stage-discharge relationship. Other problems, such as hysteresis or backwater, could occur in other study areas. In these cases, autoregressive models or more elegant formulations of the rating curve model could be included in the framework (Petersen-Øverleir, 2006).

5 Conclusions

This work introduces a framework to estimate daily rating curves at partly gauged sites. The concept of daily rating curves differs from the classical approach since it uses daily values of discharges rather than instantaneous values. Bayesian optimization of the model parameters results in a significant overestimation of the flow at the low stages and moderate underestimation at the high stages. For medium stages, the daily rating curves do not differ greatly from the instantaneous rating curves. These results suggest that the daily rating curves have the potential to be used to estimate flows from stage levels under average conditions, which should be evaluated by extending the analysis to a broader set of river basins.

Code availability. FME R package version 1.3.6.2 (URL: https://CRAN.R-project.org/package=FME, last access: 14 September 2023, https://doi.org/10.18637/jss.v033.i03, Soetaert and Petzoldt, 2010) was used for the Bayesian Inference. Other codes are available from the corresponding author on request.

Data availability. The discharge and water level dataset are available at http://www.bom.gov.au/waterdata/ (Bureau of Meteorology, 2023). The climate dataset is available at https://www.longpaddock.qld.gov.au/silo/ (Queensland Government, 2023).

Author contributions. RN: data curation, methodology, software, visualization, writing – original draft preparation. PG: Resources. RWV: conceptualization, supervision, writing – review & editing.

Competing interests. The contact author has declared that none of the authors has any competing interests.

Disclaimer. Publisher's note: Copernicus Publications remains neutral with regard to jurisdictional claims in published maps and institutional affiliations.

Special issue statement. This article is part of the special issue "IAHS2022 – Hydrological sciences in the Anthropocene: Variability and change across space, time, extremes, and interfaces". It is a result of the XIth Scientific Assembly of the International Association of Hydrological Sciences (IAHS 2022), Montpellier, France, 29 May–3 June 2022.

Acknowledgements. We sincerely appreciate all the valuable comments and suggestions the reviewers made on the original draft. Their efforts contributed to significantly improving the manuscript.

Financial support. This research has been supported by the Instituto Nacional de Investigación Agropecuaria (grant no. FPTA-341) and the Comisión Sectorial de Investigación Científica (grant no. I+D 2018-ID108).

Review statement. This paper was edited by Christophe Cudennec and reviewed by two anonymous referees.

References

Andrews, F. T., Croke, B. F. W., and Jakeman, A. J.: An open software environment for hydrological model assessment and development, Environ. Modell. Softw., 26, 1171–1185, https://doi.org/10.1016/j.envsoft.2011.04.006, 2011.

Arsenault, R. and Brissette, F. P.: Continuous streamflow prediction in ungauged basins: The effects of equifinality and parameter set selection on uncertainty in regionalization approaches, Water Resour. Res., 50, 6135–6153, https://doi.org/10.1002/2013WR014898, 2014.

Bhandari, B., Markert, K., Mishra, V., Markert, A., and Griffin, R.: Investigation of Data-Driven Rating Curve (DDRC) Approach, Water, 15, 604, https://doi.org/10.3390/w15030604, 2023.

Blöschl, G. and Sivapalan, M.: Scale issues in hydrological modelling: A review, Hydrol. Process., 9, 251–290, https://doi.org/10.1002/hyp.3360090305, 1995.

Bureau of Meteorology: Water Data Online, Water Information, Bureau of Meteorology [data set], http://www.bom.gov.au/waterdata/, last access: 14 September 2023.

Coron, L., Thirel, G., Delaigue, O., Perrin, C., and Andréassian, V.: The suite of lumped GR hydrological models in an R package, Environ. Modell. Softw., 94, 166–171, https://doi.org/10.1016/j.envsoft.2017.05.002, 2017.

Di Baldassarre, G. and Claps, P.: A hydraulic study on the applicability of flood rating curves, Hydrol. Res., 42, 10–19, https://doi.org/10.2166/nh.2010.098, 2011.

Dottori, F., Martina, M. L. V., and Todini, E.: A dynamic rating curve approach to indirect discharge measurement, Hydrol. Earth Syst. Sci., 13, 847–863, https://doi.org/10.5194/hess-13-847-2009, 2009.

Fenton, J. D.: On the generation of stream rating curves, J. Hydrol., 564, 748–757, https://doi.org/10.1016/j.jhydrol.2018.07.025, 2018.

Flores, N., Rodríguez, R., Yépez, S., Osores, V., Rau, P., Rivera, D., and Balocchi, F.: Comparison of Three Daily Rainfall-Runoff Hydrological Models Using Four Evapotranspiration Models in Four Small Forested Watersheds with Different Land Cover in South-Central Chile, Water, 13, 3191, https://doi.org/10.3390/w13223191, 2021.

Hidayat, H., Vermeulen, B., Sassi, M. G., Torfs, P. J. J. F., and Hoitink, A. J. F.: Discharge estimation in a backwater affected meandering river, Hydrol. Earth Syst. Sci., 15, 2717–2728, https://doi.org/10.5194/hess-15-2717-2011, 2011.

Jalbert, J., Mathevet, T., and Favre, A.-C.: Temporal uncertainty estimation of discharges from rating curves using a variographic analysis, J. Hydrol., 397, 83–92, https://doi.org/10.1016/j.jhydrol.2010.11.031, 2011.

Jeffrey, S. J., Carter, J. O., Moodie, K. B., Beswick, A. R.: Using Spatial Interpolation to Construct a Comprehensive Archive of Australian Climate Data, Environ. Modell. Softw., 16, 309–330, https://doi.org/10.1016/S1364-8152(01)00008-1, 2001.

Jian, J., Ryu, D., Costelloe, J. F., and Su, C.-H.: Towards hydrological model calibration using river level measurements, J. Hydrol. Reg. Stud., 10, 95–109, https://doi.org/10.1016/j.ejrh.2016.12.085, 2017.

Kiang, J. E., Gazoorian, C., McMillan, H., Coxon, G., Le Coz, J., Westerberg, I. K., Belleville, A., Sevrez, D., Sikorska, A. E., Petersen-Øverleir, A., Reitan, T., Freer, J., Renard, B., Mansanarez, V., and Mason, R.: A Comparison of Methods for Streamflow Uncertainty Estimation, Water Resour. Res., 54, 7149–7176, https://doi.org/10.1029/2018WR022708, 2018.

Kim, Y., Oh, S., Lee, S., Byun, J., and An, H.: Application of Stage-Fall-Discharge Rating Curves to a Reservoir Based on Acoustic Doppler Velocity Meter Measurement Data, Water, 13, 2443, https://doi.org/10.3390/w13172443, 2021.

Kittel, C. M. M., Jiang, L., Tøttrup, C., and Bauer-Gottwein, P.: Sentinel-3 radar altimetry for river monitoring – a catchment-scale evaluation of satellite water surface elevation from Sentinel-3A and Sentinel-3B, Hydrol. Earth Syst. Sci., 25, 333–357, https://doi.org/10.5194/hess-25-333-2021, 2021.

Lang, M., Pobanz, K., Renard, B., Renouf, E., and Sauquet, E.: Extrapolation of rating curves by hydraulic modelling, with application to flood frequency analysis, Hydrolog. Sci. J., 55, 883–898, https://doi.org/10.1080/02626667.2010.504186, 2010.

Le Coz, J.: A literature review of methods for estimating the uncertainty associated with stage-discharge relations, WMO, Rep. PO6a, 21, https://citeseerx.ist.psu.edu/viewdoc/download?doi=10.1.1.400.8656&rep=rep1&type=pdf (last access: 14 September 2023), 2012.

Lee, W. S., Lee, K. S., Kim, S. U., and Chung, E.-S.: The Development of Rating Curve Considering Variance Function Using Pseudo-likelihood Estimation Method, Water Resour. Manage., 24, 321–348, https://doi.org/10.1007/s11269-009-9448-8, 2010.

Lima, F. N., Fernandes, W., and Nascimento, N.: Joint calibration of a hydrological model and rating curve parameters for simulation of flash flood in urban areas, Rev. Bras. de Recur. Hidr., 24, https://doi.org/10.1590/2318-0331.241920180066, 2019.

McMahon, T. A. and Peel, M. C.: Uncertainty in stage–discharge rating curves: application to Australian Hydrologic

Reference Stations data, Hydrolog. Sci. J., 64, 255–275, https://doi.org/10.1080/02626667.2019.1577555, 2019.

McMillan, H., Freer, J., Pappenberger, F., Krueger, T., and Clark, M.: Impacts of uncertain river flow data on rainfall-runoff model calibration and discharge predictions, Hydrol. Process., 24, 1270–1284, https://doi.org/10.1002/hyp.7587, 2010.

Morlot, T., Perret, C., Favre, A.-C., and Jalbert, J.: Dynamic rating curve assessment for hydrometric stations and computation of the associated uncertainties: Quality and station management indicators, J. Hydrol., 517, 173–186, https://doi.org/10.1016/j.jhydrol.2014.05.007, 2014.

Narbondo, S., Gorgoglione, A., Crisci, M., and Chreties, C.: Enhancing Physical Similarity Approach to Predict Runoff in Ungauged Watersheds in Sub-Tropical Regions, Water, 12, 528, https://doi.org/10.3390/w12020528, 2020.

Petersen-Øverleir, A.: Modelling stage – discharge relationships affected by hysteresis using the Jones formula and nonlinear regression, Hydrolog. Sci. J., 51, 365–388, https://doi.org/10.1623/hysj.51.3.365, 2006.

Pedersen, Ø., Aberle, J., and Rüther, N.: Hydraulic scale modelling of the rating curve for a gauging station with challenging geometry, Hydrol. Res., 50, 825–836, https://doi.org/10.2166/nh.2019.044, 2019.

Perret, E., Renard, B., and Le Coz, J.: A Rating Curve Model Accounting for Cyclic Stage-Discharge Shifts due to Seasonal Aquatic Vegetation, Water Resour. Res., 57, e2020WR027745, https://doi.org/10.1029/2020WR027745, 2021.

Perrin, C., Michel, C., and Andréassian, V.: Improvement of a parsimonious model for streamflow simulation, J. Hydrol., 279, 275–289, https://doi.org/10.1016/S0022-1694(03)00225-7, 2003.

Qi, W., Chen, J., Li, L., Xu, C., Li, J., Xiang, Y., and Zhang, S.: A framework to regionalize conceptual model parameters for global hydrological modeling, Hydrol. Earth Syst. Sci. Discuss. [preprint], https://doi.org/10.5194/hess-2020-127, 2020.

Queensland Government: SILO – Australian climate data from 1889 to yesterday, Queensland Government [data set], https://www.longpaddock.qld.gov.au/silo/, last access: 14 September 2023.

Reistad, K., Petersen-Øverleir, A., and Bogetveit, L.: Setting up rating curves using HEC-RAS, VANN, Journal of the Norwegian Water Association, 3, 20–30, 2007.

Reitan, T. and Petersen-Øverleir, A.: Bayesian methods for estimating multi-segment discharge rating curves, Stoch. Environ. Res. Risk Assess., 23, 627–642, https://doi.org/10.1007/s00477-008-0248-0, 2009.

Sellami, H., La Jeunesse, I., Benabdallah, S., and Vanclooster, M.: Parameter and rating curve uncertainty propagation analysis of the SWAT model for two small Mediterranean catchments, Hydrolog. Sci. J., 58, 1635–1657, https://doi.org/10.1080/02626667.2013.837222, 2013.

Sikorska, A. E. and Renard, B.: Calibrating a hydrological model in stage space to account for rating curve uncertainties: general framework and key challenges, Adv. Water Resour., 105, 51–66, https://doi.org/10.1016/j.advwatres.2017.04.011, 2017.

Soetaert, K. and Petzoldt, T.: Inverse Modelling, Sensitivity and Monte Carlo Analysis in R Using Package FME, J. Stat. Soft. [software], 33, 1–28, https://doi.org/10.18637/jss.v033.i03, 2010.

WMO: Manual on stream gauging, World Meteorological Organization, Geneva, 2 pp., https://library.wmo.int/viewer/35848?medianame=1044_Vol_I_en_#page=1&viewer=picture&o=&n=0&q=, 2010.

Yang, X., Magnusson, J., Rizzi, J., and Xu, C.-Y.: Runoff prediction in ungauged catchments in Norway: comparison of regionalization approaches, Hydrol. Res., 49, 487–505, https://doi.org/10.2166/nh.2017.071, 2018.

Proc. IAHS, 385, 407–413, 2024
https://doi.org/10.5194/piahs-385-407-2024

Qualitative flood risk assessment for road and railway infrastructures: the experience of the MOVIDA project

Natasha Petruccelli[1], Luca Mantecchini[1], Alice Gallazzi[2], Daniela Molinari[2], Mohammed Hammouti[3], Marco Zazzeri[3], Simone Sterlacchini[3], Francesco Ballio[2], Armando Brath[1], and Alessio Domeneghetti[1]

[1]Department of Civil, Chemical, Environmental and Materials Engineering (DICAM), Alma Mater Studiorum – Università di Bologna, Bologna, 40136, Italia
[2]Department of Civil and Environmental Engineering (DICA), Politecnico di Milano, Milano, 20133, Italia
[3]Italian National Research Council, Institute of Environmental Geology and Geoengineering (CNR-IGAG), Milano, 20131, Italia

Correspondence: Natasha Petruccelli (natasha.petruccelli@unibo.it)

Received: 12 May 2022 – Revised: 24 July 2023 – Accepted: 28 August 2023 – Published: 19 April 2024

Abstract. The Po River District Authority promoted the MOVIDA project with the aim to define appropriate methodologies for flood risk assessment and being compliant with the European Floods Directive (Directive 2007/60/EC). A dedicated Open Source Geographic Information System (i.e. QGIS geoprocessing modules) has been developed for mapping the expected damages in all areas at significant risk in the Po District (Northern Italy), considering five categories of exposed elements (population, infrastructures, economic activities, environmental and cultural heritage, and na-tech sites). Focusing on road and railway infrastructures, the methodology proposed within the project adopts information coming from different data sources (Regional Geoportals, Open Street Map, etc.) and allows to qualitatively estimate the potential risk associated with a flood event. Different risk classes (High, Medium, Low and Null) are assigned in relation to roads category (i.e., Highways, Main, Secondary, Service, Other) or railways type (High-Speed train or not), thus considering both the relevance of the infrastructure itself (as well as its topographical characteristics: e.g. tunnel, bridge, etc.) and the magnitude of the expected event (i.e., hazard). The definition of the risk matrix led to the estimation of the lengths of the sections exposed to different risk levels, which is useful to support the definition of potential mitigation measures and support the competent bodies in the organization of the rescue.

Keywords. Flood risk; infrastructures; road; railway; risk levels; expected impact

1 Introduction

Damages of hydrological origin represent about 37 % of all-natural disasters related ones (geophysical, meteorological, biological; CRED and UNDRR, 2021). In particular, flood represents one of the most frequent and widespread type of natural risk globally (Adhikari et al., 2010). In the last 20 years, the number of floods has more than doubled compared to the previous two decades and their further increase worldwide is expected due to climate change (Tanoue et al., 2016; Blöschl et al., 2019), likely causing a progressive rise in damage because of the greater concentration of the population in the most exposed areas (Ferreira et al., 2018) and to economic growth. The development of flood risk management and mitigation policies, promoted at EU level by the Floods Directive (Directive 2007/60/EC), pushed the entire scientific community towards the definition of methodologies capable to assess flood-related damages to the different exposed assets and support the Cost-Benefit Analysis (CBA) of mitigation measures. With the intent of upgrading the flood risk management plan for the Po River District (the largest basin in Italy), the Po River Basin Authority (AdBPo) funded the MOVIDA project (MOdello per la Valutazione Integrata del Danno Alluvionale – MOdel for the integrated assessment of flood damage) to assess flood-related damage to several elements (i.e., population, infrastructures, economic activities, environmental and cultural heritage, and na-tech sites) in all areas classified at significant risk of inundation (APSFR). This paper briefly summarizes the procedure developed for the qualitative assessment of the flood risk to road and railway infrastructures.

The transport system plays a fundamental role in the economic and social development of a country; a physical failure of the network or a traffic interruption, caused by the presence of water on the infrastructure, can lead to significant socio-economic losses (Kellermann et al., 2015; Zhang and Alipour, 2021). In the literature, there are several studies attempting to quantify the direct and indirect damages suffered by infrastructures and vehicles in circulation; however, in most of the cases a detailed description of the exposure and vulnerability of the network is still hampered by lack of information. This study reports expert-based (for road) and data-driven (for railway) approaches developed to sustain a large exposure classification of linear infrastructures. Referring to the road and railway network of one region of the Po River District (i.e., the Emilia-Romagna region), the scope was the attribution of risk classes to exposed infrastructure sections, based on their structural and functional characteristics (e.g. type of traction, traffic flows, etc.), as well as on inundation characteristics. This enables the identification of the most critical sections of the network, which is essential to strengthening the system's resilience and adaptability to natural disasters, such as floods.

2 Case study and database

The analysis concerns the regional territory of Emilia-Romagna falling within the Hydrographic District of the Po River. 7 APSFR (out of 22 identified at the district level by the Flood Risk Management Plan – FRMP) are in the study area (as shown in Fig. 1), with an extent of about 3500 km^2 corresponding to 16 % of the total regional area.

For each APSFR, the FRMP provides hazard maps (i.e., inundation depth; spatial resolution of 5–10 m) for three probability scenarios, according to Floods Directive: Low Probability Hazard (LPH – 500-year return period), Medium Probability Hazard (MPH – 200-year return period) and High Probability Hazard (HPH – 50-year return period). These maps, superimposed on the graphs of the transport network, identify the sections exposed to potential inundations. Concerning the transportation network, a detailed database containing the physical and functional characteristics of the infrastructures was built. For both roads and railways, reference has been made to Open Street Map (OSM), whose information has been integrated with that coming from the Regional Geoportal (https://geoportale.regione.emilia-romagna.it, last access: March 2021). Roads were classified by order of importance (e.g., Highway, Main, Secondary, Service or Dirt, Other) and in the light of topological characteristics: roads in elevation (bridge, viaduct, overpass) and tunnels (underpass). For railways, the Integrated Regional Transport Plan (Piano Regionale Integrato dei Trasporti, 2021) was also consulted to associate the number of tracks, the category and the type of traction to each section of the network. Data regarding the number and train typology in transit along a given railway arch in a specific period of time were also collected. The latter data were used to quantify the load on the network by two different types of users: passengers and freight. In particular, traffic data relating to any midweek working day were collected, as regards passenger flows, while data concerning freight train traffic were acquired over a weekly period. A real traffic value was therefore associated with each arch. In addition, the OSM information layer was intersected with that of the buildings of the Regional Topographical database to evaluate the presence of any railway station exposed to flood risk.

3 Methodology

Functional damages associated with the operability of a given infrastructure are in many cases of greater importance than physical damages (Pregnolato et al., 2017). In this work we propose a methodology for the assessment, in qualitative terms (i.e., null, low, medium, high), of the functional risk associated to the network about a specific flood event. Despite the singularities of the systems, the approach developed for road and railway infrastructures is similar and involves the definition of a risk matrix resulting from the combination of two sub-matrices: (i) a first one representative of the impact of the flood event in terms of slowdown or interruption of vehicular transit (depending on the hazardous variables; i.e. hydraulic load or water depth); a second one (ii) indicating the uniqueness of the performance services offered by the infrastructure, which depends on the type of flooded section. The overall risk matrix is obtained by carrying out a weighted average of the aforementioned matrices and consulting the existing literature (Kramer et al., 2016; Pregnolato et al., 2017; Kellermann et al., 2015) for the definition of the different hazardous thresholds. Specifically, those cases are foreseen as regards road infrastructures: (i) no risk to road sections with average water depth and average water velocity equal to zero, as well as for roads at high altitude; (ii) a high risk on the most critical sections in case of flooding (i.e., tunnels); (iii) a variable risk on the remaining sections, depending on the hazard and vulnerability (as Table 1 depicts). For railways, the definition of the risk in Table 2 matrix is more complex. Unlike roads, two possible scenarios have been distinguished: (A) the railway line is close to the flooding front, but not directly affected by water; this causes trains to slow down and, in the case of significant loads or water depth, the activation of the monitoring service due to risk of erosion; (B) the railway line is surmounted with possible damage to the power supply system and possible tran-

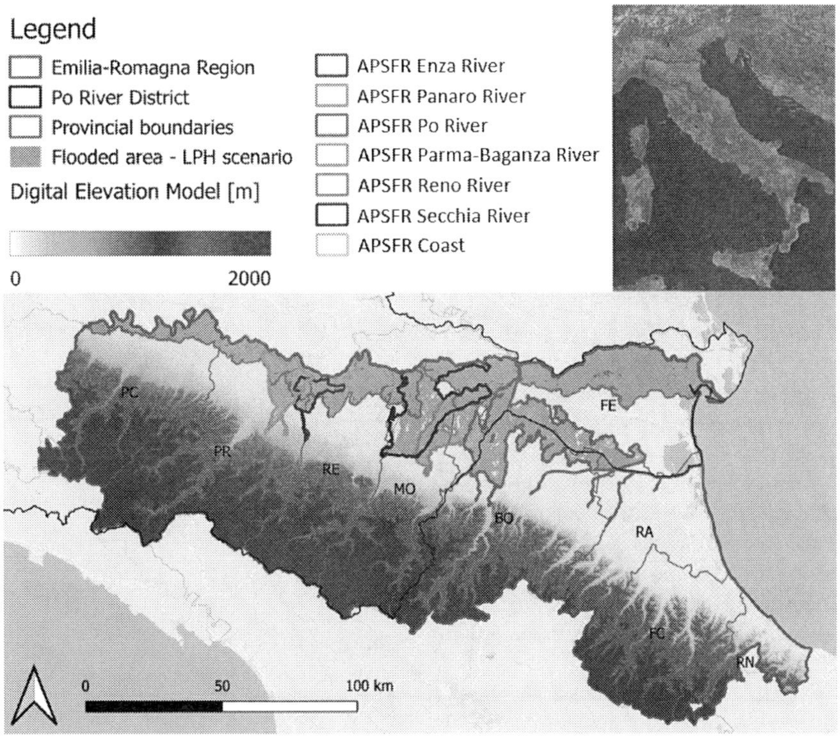

Figure 1. Study area and regional APSFRs.

Table 1. Risk matrix for road infrastructures.

Hazard classes	Bridge	Other	Service and/or Dirt	Secondary	Main	Highway	Tunnel
h=0 m and v=0 m/s	Null	Null	Null	Null	Null	Null	Null
0<h<0.15 m	Null	Low	Low	Low	Medium	High	High
h≥0.15 m	Null	Medium	Medium	Medium	High	High	High

Table 2. Risk matrix for railway infrastructures.

Hazard classes	I	II	III	IV
h=0 m and v=0 m/s	Null	Null	Null	Null
0<hA<0.2 m	Null	Null	Null	Null
hA≥0.2 m	Null	Low	Medium	High
0<hB<0.1 m	Null	Low	Medium	High
hB≥0.1 m	Low	Medium	High	High

sit interruption caused by the compromised stability of the roadbed and embankment.

Rail journeys are linked to the stations of origin and destination; thus, it is possible to build the arch-paths incidence matrix associating the traffic values to each railway arch. This information, together with the structural and functional characteristics of the network, reveals the importance of the railway section and, therefore, how severe the consequences are expected in case of its interruption. Attribution of relevance class to each arch (from I to IV, associate with higher relevance) was carried out through a multi-criteria analysis by weighing the characteristics of each arch: type of traction, category and number of trains in transit. In particular, as regards passenger transport, a weight (w_{trac}) of 0.5 has been assigned to diesel traction, to reduce the contribution of the minor lines where this technology is still present, while electric traction has been assigned a unit value. To the four present arch categories (Regional, National 1-track, National 2-track and High-Speed line), weights (w_{cat}) between 0.25 (for regional lines) and 1 (for High-Speed lines) have been attributed. The weight of train traffic on each network arch (w_{flux}) was evaluated by normalizing this variable among maximum and minimum values observed on the overall network, which result on a weight that varies continuously between 0 and 1. With this approach, each arch of the network was assessed by adopting a specific index (p_{equi}), expressed in terms of the number of equivalent passengers, estimated using the following expression:

$$p_{\mathrm{equi}} = n_{\mathrm{pas}}[1 + (w_{\mathrm{trac}} \cdot w_{\mathrm{cat}} \cdot w_{\mathrm{flux}})] \quad (1)$$

where n_{pas} represents the number of passengers travelling along each arch, calculated in relation to the type of train in transit and its transport capacity. Four different relevance classes have been identified based on p_{equi} (as illustrated in Fig. 2 in the top panel). A similar approach was used for freight transport (depicted in Fig. 2 in the bottom panel), according to the values assumed by the index f_{equi} evaluated as:

$$f_{equi} = tr_{equ}[1 + (w_{trac} \cdot w_{cat})] \quad (2)$$

where the number of trains in transit was transformed into the number of equivalent trains (tr_{equ}) considering the type of freight train. A maximum value of 3 has been assigned to Combined Trains – CT (international freight trains used for the intermodal transport of trucks, semi-trailers etc.), a minimum of 1 to the Terminal Trains – TT (freight trains in terminal service to and from a shunting area to railway yard). The relative weight of the type of traction is unchanged with respect to the previous case, while, for the classes inherent to category, the weight attributed to the High-Speed lines has been changed (set equal to 0.25), as these are of marginal importance for the freight transit.

The methodology developed for the qualitative assessment of functional damage to railway infrastructures also includes the identification of impacted railway stations; it was decided to assign the maximum class to all stations in the flooded area.

4 Results and discussion

The implemented GIS procedures makes possible the evaluation of road and railway sections most exposed to flood risk according to the identified classes. The risk matrices shown in Tables 1 and 2 were applied considering the hazard maps provided by the FRMP and the relevance classes attributed to roads and railways. From Fig. 2 it can be seen that the constructed index depicts different relevance classes to a given railway arch in relation to the type of transport considered (passengers or freight), which is representative of the current condition. Concerning passenger transport, the Firenze-Milano line, the only High-Speed line in the region, falls into the highest class, while the historic regional lines (e.g., Ancona-Piacenza and Bologna-Padova), reach class III having a significant passengers flow. The intra-regional connections, mainly concentrated in the provinces of Ravenna and Ferrara, all fall into the lowest relevance classes. The overall risk picture changes in case of considering freight transport: lines with the highest relevance index are those that connect the main cities with national ports or cross the border; the highest number of freight trains of international importance, such as European Combined Trains (ECT) and International Combined Trains (ICT), transits over these lines. Furthermore, it can be observed that lines strategic for passenger transport have a minor, if not null, impact on freight

Table 3. Risk associated with an MPH event for road infrastructures.

APSFR	Null [km]	Low [km]	Medium [km]	High [km]
Enza	3	18	244	20
Panaro	7	48	1126	120
Po	46	52	5507	644
Parma-Baganza	8	41	772	78
Reno	14	63	1320	140
Secchia	13	93	2153	249

(such as the High-Speed line and intra-regional diesel traction lines which, providing only passenger service, have been neglected in the risk analysis for the displacement of freight). The application of the risk matrices made possible to evaluate, for each of the three hazard scenarios, the effect on the road and railway infrastructures within the 7 APSFR of the Po District in the Emilia-Romagna region. Tables 3 and 4 show the length of roads and railways impacted by an event with an intermediate probability of occurrence (MPH).

Concerning the reported hazard scenario (MPH, although the same applies to other events), all APSFRs have a significant extension of medium-risk roads: this is due to the greater presence of secondary and service or dirt roads within each area, compared to the main roads and highways (as shown in the Fig. 3 for the APSFR of Secchia River).

Highways sections severely impacted are: the A1, which crosses the APSFR of Secchia River for 7 km, Parma-Baganza for 13 km and Enza for 3 km; A13 that crosses the APSFR Reno for 14 km and the A21 that intersects the APSFR of Po River in the Emilia-Romagna region.

As regards passenger transport on railway lines, most of the lines resulted associated with high and low risk classes for all APSFR. The first case is attributable to the High-Speed line, for a total length of 45 km, and to the Bologna-Piacenza line, parallel to the previous one, which is exposed for a length of 32 km. The significant extent of low risk classes is due to railway sections not directly affected by floods but rather located nearby the inundation edge (see Table 2). The overall risk for freight transport appears lower: arches relevant in this case are not severely impacted by inundations.

5 Conclusions

The developed methodology allows for qualitatively assess the functional risk for transport infrastructures in case of flooding. The evaluation considers water depth impacts on slowing down or interrupting the transit, as well as the ability of the network to guarantee an adequate level of redundancy, calculated according to the type of infrastructure affected by the flood. The procedure is applied for roads and for railways, for the latter distinguishing the case of person

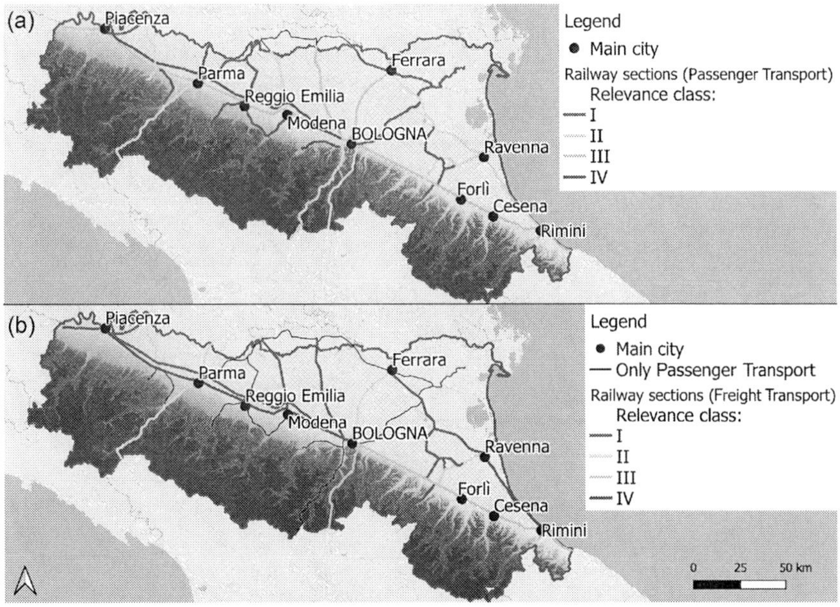

Figure 2. Relevance classes of railway sections used for the transport of passengers **(a)** and freight **(b)**.

Table 4. Risk associated with an MPH event for railway infrastructures (*P* refers to passenger transport and *F* to freight transport).

APSFR	Null [km]		Low [km]		Medium [km]		High [km]		Railway stations [nr]
	P	*F*	*P*	*F*	*P*	*F*	*P*	*F*	
Enza	0	0	14	7	3	0	5	0	1
Panaro	0	0	0	12	19	0	12	0	1
Po	0	0	75	149	13	0	24	0	17
Parma-Baganza	0	0	22	33	0	0	16	0	2
Reno	0	0	20	65	9	0	15	0	1
Secchia	0	0	37	22	8	0	29	0	2
Coast	0	0	12	12	0	0	0	0	0

and freight transport. The risk is assessed at the scale of the single exposed section or station (microscale).

As regards the procedure implemented for road infrastructures, a limit of the present investigation concerns high altitude sections (i.e., bridges, viaducts): comparison of deck level and water depth is not carried out, associating these sections to the null risk class, regardless the stability of the road infrastructure affected by the flood and the freeboard.

As regards the railway infrastructures, the current limitations concern: the definition of the arches, which are represented continuously from station to station, ignoring the presence of logistics stops; stations are considered as a punctual element, without distinguishing them in terms of importance and dimensions. For both infrastructures, only a qualitative estimate of the risk was made through the use of an "expert-based" approach. A proper validation of risk attribution was not possible at this stage due to the lack of extensive data relating to the impacts of past flood events, as well as by the limited involvement of infrastructures managers, which should be considered in the validation of categorical approaches, as the one proposed here. That said, the method refers to reasonable considerations and literature findings (literature data and experimental observations were used to evaluate the impact of water depth on the speed of road and rail vehicles), which we believe make the approach scientific and technically sounding. Nevertheless, the validation of methodology certainly represents one of the challenges that we intend to overcome in the future developments. Also, the study will be extended considering physical damages to infrastructures, in terms of costs of reconstruction/restoration, as well indirect damages (i.e., the impact on the travel times of passengers or the temporary stop of freight convoys). These are rather complex analyses that require in-depth knowledge, such as the geometric characteristics of the

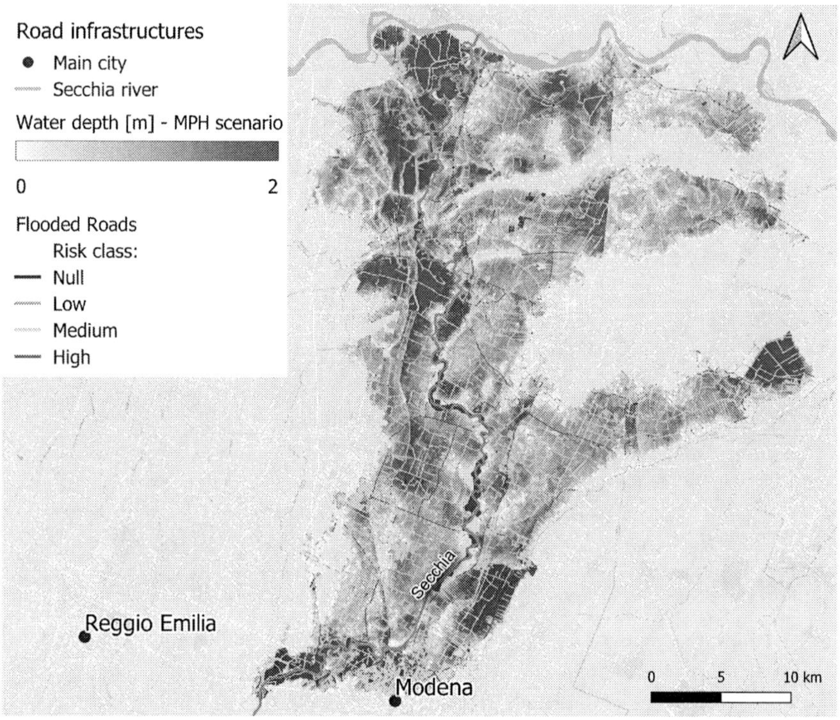

Figure 3. Inundation extent according to MPH scenario and road risk classification for the APSFR of Secchia River.

infrastructure, on which the cost items significantly depend. However, the presented analysis of the flood risk on transport infrastructures constitutes an added value for the operators of the sector in terms of efficient management of the network, promoting targeted planning and implementation of measures aimed at reducing the risk.

Code availability. The methodology was developed in a GIS environment (https://download.qgis.org/downloads/, QGIS, 2023); latest version released on 16 January 2021. The tool can be provided to scientists for reasonable research purposes upon request to the corresponding author.

Data availability. Data of transport networks refers to OSM (https://download.geofabrik.de/europe/italy/nord-est.html, OpenStreetMap contributors, 2020). Hazard maps can be downloaded from the national geo portal website (http://www.pcn.minambiente.it/mattm/direttiva-alluvioni/, Po River District Authority, 2021). Additional data are available from the website of the National Institute of Statistics (ISTAT) at this link: https://www.istat.it/it/archivio/104317 (ISTAT, 2011).

Author contributions. The work was carried out thanks to the participation of all authors. NP was responsible for defining the methodology, analyzing the data and writing the manuscript, AD gave a significant contribution to the analysis of the data and writing the manuscript, which was also reviewed by AB. LM, however, provided detailed information on the characteristics of transport networks. DM, AG and FB took care of the direction and coordination of the works, while MH, MZ and SS allowed the implementation of the procedure in a GIS environment and the definition of a structured database.

Competing interests. The contact author has declared that none of the authors has any competing interests.

Disclaimer. Publisher's note: Copernicus Publications remains neutral with regard to jurisdictional claims in published maps and institutional affiliations.

Special issue statement. This article is part of the special issue "IAHS2022 – Hydrological sciences in the Anthropocene: Variability and change across space, time, extremes, and interfaces". It is a result of the XIth Scientific Assembly of the International Association of Hydrological Sciences (IAHS 2022), Montpellier, France, 29 May–3 June 2022.

Acknowledgements. This work was supported by the Po River Basin District Authority for the update of the flood risk management plan, released in 2021.

Financial support. The research has been supported by the Po River Basin District Authority (research agreement ex art.15 L. N. 241/90 – Metodologie e applicazioni per l'aggiornamento delle mappe di danno alluvionale relativamente alla revisione del PGRA – "PGRA-Danni").

Review statement. This paper was edited by Christophe Cudennec and reviewed by Eric Gaume and one anonymous referee.

References

Adhikari, P., Hong, Y., Douglas, K., Kirschbaum, D., Gourley, J., Adler, R., and Brakenridge, R.: A digitized global flood inventory (1998–2008): compilation and preliminary results, Nat. Hazards, 55, 405–422, 2010.

Blöschl, G., Hall, J., Viglione, A., Perdigão, R., Parajka, J., Merz, B., Lun, D., Arheimer, B., Aronica, G. T., Bilibashi, A., Boháč, M., Bonacci, O., Borga, M., Čanjevac, I., Castellarin, A., Chirico, G., Claps, P., Frolova, N., Ganora, D. Gorbachova, L., Gül, A., Hannaford, J., Harrigan, S., Kireeva, M., Kiss, A., Kjeldsen, T., Kohnová, S., Koskela, J., Ledvinka, O., Macdonald, N., Guirguinova, M. M., Mediero, L., Merz, R., Molnar, P., Montanari, A., Murphy, C., Osuch, M., Ovcharuk, V., Radevski, I., Salinas, J., Sauquet, E., Šraj, M., Szolgay, J., Volpi, E., Wilson, D., Zaimi, K., and Živković, N.: Changing climate both increases and decreases European river floods, Nature, 573, 108–111, https://doi.org/10.1038/s41586-019-1495-6, 2019.

Centre for Research on the Epidemiology of Disasters, in collaboration with United Nations Office for Disaster Risk Reduction (CRED and UNDRR): The Non-Covid year in disasters, CRED, Brussels, Belgium, https://www.cred.be/publications (last access: December 2022), 2021.

Ferreira, C., Walsh, R., and Ferreira, A.: Degradation in urban areas, Current Opinion in Environmental Science & Health, 5, 19–25, https://doi.org/10.1016/j.coesh.2018.04.001, 2018.

ISTAT: Territorial bases – census sections, ISTAT [data set], https://www.istat.it/it/archivio/104317 (last access: November 2021), 2011.

Kellermann, P., Schöbel, A., Kundela, G., and Thieken, A. H.: Estimating flood damage to railway infrastructure – the case study of the March River flood in 2006 at the Austrian Northern Railway, Nat. Hazards Earth Syst. Sci., 15, 2485–2496, https://doi.org/10.5194/nhess-15-2485-2015, 2015.

Kramer, M., Terheiden, K., and Wieprecht, S.: Safety criteria for the trafficability of inundated roads in urban floodings, Int. J. Disast. Risk Re., 17, 77–84, https://doi.org/10.1016/j.ijdrr.2016.04.003, 2016.

OpenStreetMap contributors: Nord-Est, Geofabrik [data set], https://download.geofabrik.de/europe/italy/nord-est.html (last access: March 2021), 2020.

Piano Regionale Integrato dei Trasporti (PRIT 2025): Emilia-Romagna region, https://mobilita.regione.emilia-romagna.it/prit-piano-regionale-integrato-dei-trasporti/sezioni/prit-2025-elaborati-tecnici (last access: March 2022), 2021.

Po River District Authority: Flood Risk Management Plans, Geoportale Nazionale [data set], http://www.pcn.minambiente.it/mattm/direttiva-alluvioni/ (last access: March 2022), 2021.

Pregnolato, M., Alistair, F., Wilkinson, S., and Dawson, R.: The impact of flooding on road transport: A depth-disruption function, Transport. Res. D-Tr. E., 55, 67–81, https://doi.org/10.1016/j.trd.2017.06.020, 2017.

QGIS: v. 3.10.14, QGIS [code], https://download.qgis.org/downloads/, last access: January 2023.

Tanoue, M., Hirabayashi, Y., and Ikeuchi, H.: Global-scale river flood vulnerability in the last 50 years, Scientific Reports, 6, 36021, https://doi.org/10.1038/srep36021, 2016.

Zhang, N. and Alipour, A.: A multi-step assessment framework for optimization of flood mitigation strategies in transportation networks, Int. J. Disast. Risk Re., 63, 102439, https://doi.org/10.1016/j.ijdrr.2021.102439, 2021.

Proc. IAHS, 385, 415–421, 2024
https://doi.org/10.5194/piahs-385-415-2024

Evaluation and calibration of alternative methods for estimating reference evapotranspiration in the main hydrosystems of Senegal: Senegal, Gambia and Casamance River Basins

Ansoumana Bodian[1], **Papa Malick Ndiaye**[1], **Serigne Bassirou Diop**[1], **Lamine Diop**[2], **Alain Dezetter**[3], **Andrew Ogilvie**[4], **and Koffi Djaman**[5]

[1]Laboratoire Leïdi "Dynamique des Territoires et Développement", Université Gaston Berger (UGB), Saint-Louis BP 234, Senegal
[2]Laboratoire des Sciences Biologiques, Agronomiques, Alimentaires et de Modélisation des Systèmes Complexes (LaBAAM)Alimentaires, Université Gaston Berger, Saint-Louis BP 234, Senegal
[3]HydroSciences Montpellier, Univ. Montpellier, IRD, CNRS, UFR Pharmacie, Bâtiment HYDROPOLIS, 15 avenue Charles Flahault, 34090 Montpellier, France
[4]UMR G-EAU, AgroParisTech, Cirad, University of Montpellier, IRD, INRAE, Montpellier SupAgro, 34196 Montpellier, France
[5]Department of Plant and Environmental Sciences, New Mexico State University, Agricultural Science Center at Farmington, P.O. Box 1018, Farmington, NM 87499, USA

Correspondence: Ansoumana Bodian (ansoumana.bodian@ugb.edu.sn)

Received: 24 May 2022 – Revised: 26 July 2023 – Accepted: 28 August 2023 – Published: 19 April 2024

Abstract. Reference evapotranspiration (ET_0) is an essential parameter for hydrological modeling, irrigation planning and for studying the impacts of climate change on water resources. The Penman–Monteith method is recommended for estimating ET_0 under all climatic conditions, but its application is limited in regions where data is difficult to access. The alternative is to use methods that incorporate fewer climatic parameters. The objective of this work is to evaluate twenty alternative methods using few climate parameters of estimating ET_0 and to calibrate/validate the best ones in order to adapt them to the climatic context of the main hydrosystems of Senegal: Senegal, Gambia and Casamance river basins. The KGE and the PBIAS were used to evaluate the performance of the methods. The results show that after calibration/validation the methods of Trabert, Hargreaves, Hargreaves & Samani, Trajkovic and Oudin are the best on for the estimation of the reference evapotranspiration in the Senegal, Gambia and Casamance River Basins.

Keywords. UPH1; SDG6; Reference evapotranspiration; Senegal; Gambia; Casamance River Basins

1 Introduction

The estimation of evapotranspiration is a major difficulty in the context of developing countries. This is generally explained by the difficulty of accessing the climatic data necessary for its estimation and the low density of measurement stations. To overcome this, reanalysis data can be an alternative to often incomplete observed data. These data give the opportunity to calculate the evapotranspiration with several methods. The latter vary according to the climatic parameters that they integrate. A distinction is made between aerodynamic methods according to which evapotranspiration is proportional to wind speed and saturation deficit (Dalton, 1802), methods based on temperature (Hargreaves, 1975) and solar radiation (Makking, 1957) and those based on which combine several climatic variables (Penman, 1948; Monteith, 1965). These methods make it possible to estimate the reference evapotranspiration (ET_0) and their performance varies according to the geographical areas (Djaman et al., 2015). Therefore, it is important to evaluate them before using them in a specific context. Moreover, even after evaluation, these methods must be calibrated in order to adapt them to the local climatic context other than the original climates under which these models were developed. Several studies around the world (Bogawski and Bednorz, 2014; Valipour, 2015; Djaman et al., 2015; Ndiaye et al., 2020) have focused on the evaluation of different ET_0 estimation methods. In the Senegal River Basin, Ndiaye et al. (2020) evaluated the performance of twenty alternative methods for estimating ET_0. Their results showed that the best methods retained after calibration/validation: Valiantzas 2, Trabert, Valiantzas 3 and Hargreaves and Samani. This work aims to expand the study by Ndiaye et al. (2020) on the other watersheds of Senegal (Gambia and Casamance) in order to evaluate alternative methods using few climate parameters of estimating ET_0 on the main hydrosystems of Senegal.

2 Material and methods

2.1 Study area

The Senegal, Gambia and Casamance river basin constitute the main hydrosystems of the Senegalese territory (Fig. 1). The Senegal and Gambia rivers have their source in the Fouta Djallon at an altitude of 1150 m and drain respective areas of more than 300 000 and 77 100 km^2, respectively (Ndiaye et al., 2020). The Senegal river basin extends into Senegal, Guinea, Mauritania and Mali and is managed by the Organization for the Development of the Senegal River. The Gambia river covers parts of Senegal, Gambia and Guinea together with Guinea-Bissau which constitute the Organization for the Development of the Gambia River. The Casamance basin extends over an area of 20 150 km^2 and is located almost entirely in Senegalese territory (Dacosta, 1989). From a climatic point of view, these three hydrosystems extend over three climatic zones: Guinean (average annual rainfall > 1500 mm); South Soudanian (1000 < rainfall < 1500 mm); North Sudanian (500 < rainfall < 1000 mm) and Sahelian (rainfall < 500 mm). The water resources of these three hydrosystems are used for the development of agriculture, hydroelectric production, drinking water supply, navigation and the maintenance of ecosystems.

2.2 Data

In West Africa, meteorological data managed by national meteorological services are difficult to access due to their high acquisition costs (Bodian et al., 2020). Additionally, the low density of the observation networks limits their spatial representativeness at the scale of the catchment area (Ndiaye et al., 2020). For this reason, reanalyses from the NASA Langley Research Center (LaRC) POWER Project funded through the NASA Earth Science/Applied Science Program (https://power.larc.nasa.gov, last access: February 2022) are used as an alternative (Ndiaye et al., 2020). The coordinates of 160 (64, 58 and 38 for Senegal, Gambia and Casamance river basins, respectively) stations distributed across the three basins (Fig. 1) are used to extract climate variables on a daily scale over the period 1984–2019. The climatic variables collected are: maximum and minimum air temperatures, solar radiation, air relative humidity and wind speed at 2 m above ground.

2.3 Methods

2.3.1 Estimation method of reference evapotranspiration

The reference method used is that of FAO56-PM which is given by the following formula:

$$ET_{0_{FAO56-PM}} = \frac{0.408\Delta(R_n - G) + \frac{\gamma Cn}{T+273.3}u2(es - ea)}{\Delta + \gamma(1 + Cdu2)} \quad (1)$$

Where $ET_{0_{FAO56-PM}}$ is the reference evapotranspiration ($mm\,d^{-1}$), R_n: net radiation on the crop surface ($MJ\,m^{-2}\,d^{-1}$), G is the heat flux density of the soil ($MJ\,m^{-2}\,d^{-1}$) which is ignored on a daily scale, T is the average daily air temperature at a height of 2 m (°C), Cn and Cd are constant values which change according to the scale of time used (on a daily scale Cn and Cd are 900 and 0.34 respectively), $u2$ is the wind speed at a height of 2 m ($m\,s^{-1}$), es is the saturated vapor pressure (kPa) , ea is real vapor pressure (kPa), (es-ea) is the saturation deficit (in kPa) Δ is the vapor pressure slope curve ($kPa\,°C^{-1}$) and γ is the psychrometric constant psychometric ($kPa\,°C^{-1}$).

The list of the twenty alternative methods used in this work as well as their mathematical formulations are detailed in Ndiaye et al. (2020). The Table 1 give only the names of methods used and their categories.

The calibration of the best methods was done by the gradient reduction method (Bogawski and Bednorz, 2014). This seeks to change the constant values of the original methods to minimize estimation errors. For this, the series is divided into two part as recommended by Xu and Singh (2001): 2/3 of the series (1984–2005) for calibration and 1/3 (2006–2019) for validation. The calibrated methods were re-evaluated using the same criteria as for the evaluation with different time-frame dataset.

2.3.2 Evaluation criteria

Kling Gupta Efficiency (KGE, Gupta et al., 2009) and percentage bias (PBIAS) are used (Table 2) to assess the performance of alternative methods. The KGE includes both the correlation coefficient (r), the biases (β) and the variability (γ) and makes it possible to determine the degree of concordance between the reference ET_0 and that of the alternative methods. The PBIAS indicates errors as a percentage. A

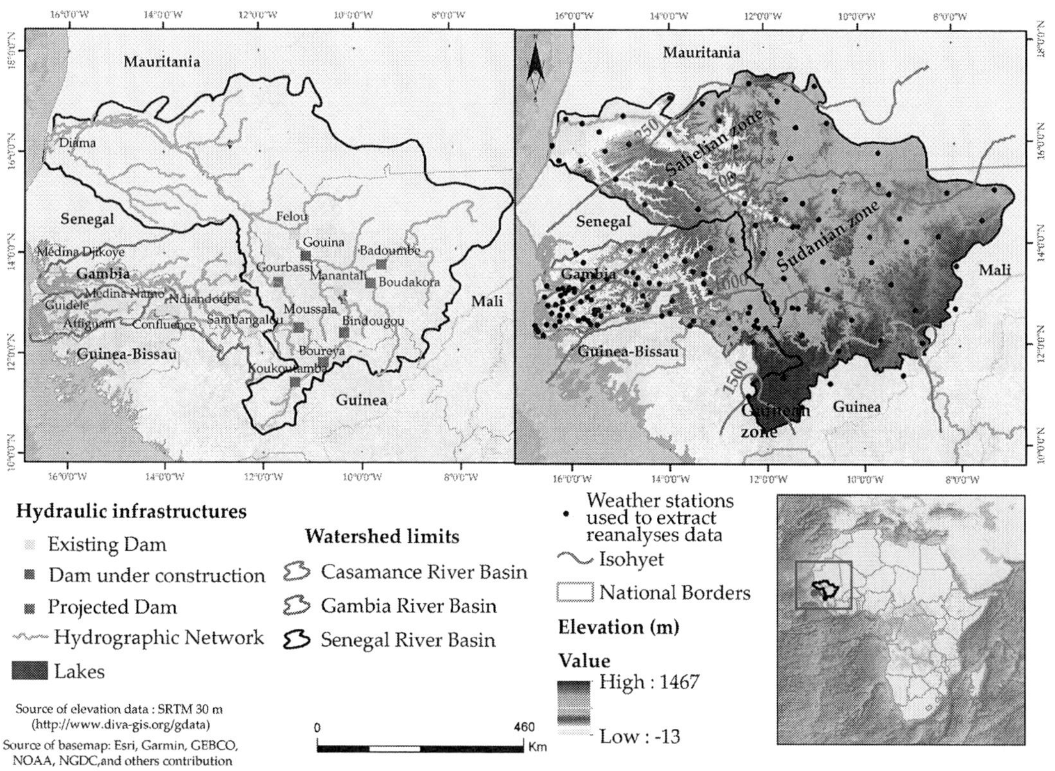

Figure 1. Location of the three Senegalese hydrosystems, hydraulic infrastructures, climate zones, hydrographic network and weather stations used to extract reanalyses data.

Table 1. List of the twenty alternative methods evaluated against FAO56-PM.

Categories	References	Abbreviation	Categories	References	Abbreviation
	Dalton (1802)	DN (2)		Makking (1957)	MK (12)
	Trabert (1896)	TRB (3)		Jensen et Haise (1963)	JH (13)
Aerodynamic	Penman (1948)	PNM (4)	Radiation	Priestley and Taylor (1972)	PT (14)
	Rohwer (1931)	RW (5)		Abtew (1996)	AB (15)
	Mahringer (1970)	MHR (6)		Oudin (2005)	OD (16)
	Hargreaves (1975)	HG (7)		Penman (1963)	PNM (17)
	Hargreaves and Samani (1985)	HS (8)		Doorenbos and Pruitt (1977)	DP (18)
Temperature	Trajkovic and Stojnic (2007)	TRA (9)	Combinatory	Valiantzas 1 (Valiantzas, 2013)	Val 1 (19)
	Droogers and Allen (2012)	DA (10)		Valiantzas 2 (Valiantzas, 2013)	Val 2 (20)
	Heydari and Heydari (2014)	HH (11)		Valiantzas 3 (Valiantzas, 2013)	Val 3 (21)

positive/negative PBIAS indicates an overestimation/underestimation of ET_0 by the alternative method.

3 Results and discussion

3.1 Evaluation of twenty alternative methods against FAO56-PM

Figure 2 gives the synthesis of the KGE of the twenty alternative methods compared to the FAO56-PM (Penman–Monteith). Because of the objective of this work, only the results of methods integrating few climatic variables are presented. The performance of these methods varies slightly depending on the river basins. For the Senegal River basin, the five best methods are those of Trabert, Valiantzas 3, Jensen & Haise and Hagreaves & Samani. For the Gambia River basin, in addition to the methods of Valiantzas (2 and 3), those of Oudin, Droogers & Allen and Hargreaves & Samani are the most robust for estimating ET_0. The KGE values of these methods are generally greater than 0.60 and estima-

Table 2. Evaluation criterea, range, optimal values.

Criteria	Formula	Range	Optimal values	
PBIAS	$\left[\frac{\frac{1}{n}\sum_{i=1}^{n}\left(ET0_{\text{alt}}-ET0_{\text{FAO56-PM}}\right)}{\frac{1}{n}\sum_{i=1}^{n}\left(ET0_{\text{alt}}\right)}\right]\cdot 100$	$-\infty, +\infty$	0	(22)
KGE	$1-\sqrt{(r-1)^2+(\beta-1)^2+(\alpha-1)^2}$	$-\infty, 1$	1	(23)

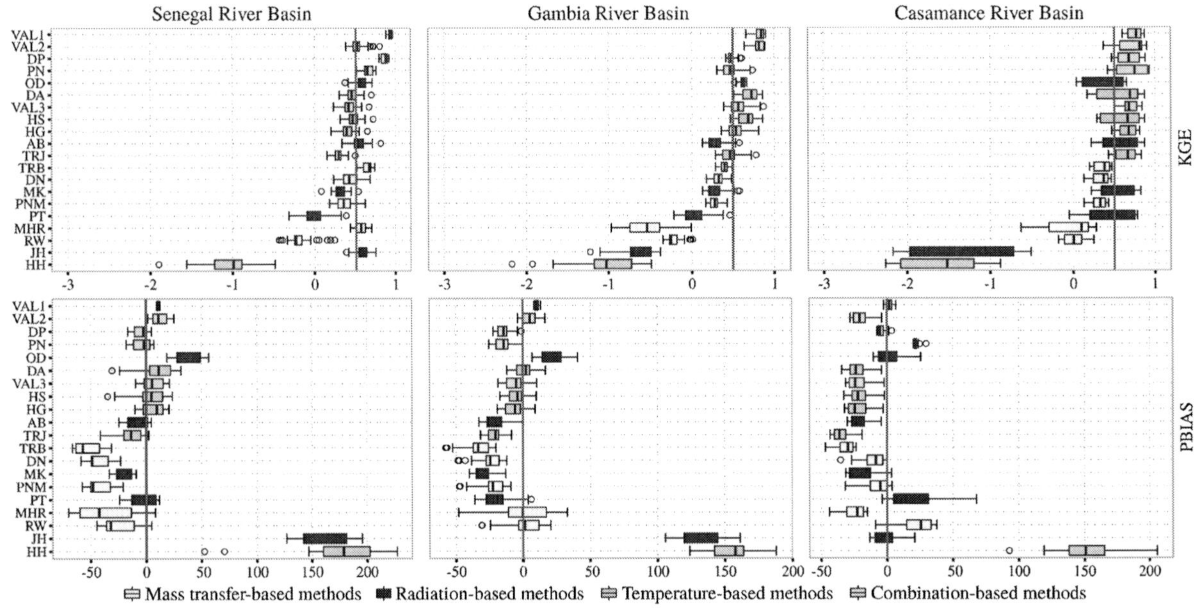

Figure 2. Performance of the methods according to the KGE (The red line in each figure represents the mean values of each KGE).

tion errors of −4 % to 23 %. In the Casamance River basin, the temperature-based methods of Hargreaves (HG) and Trajkovic give the best results with KGE of nearly 0.88 and low estimation errors ranging from −4.6 % to 11 %.

3.2 Calibration and validation of the best methods

The choice of methods to be calibrated is guided by the performance criteria and the number of climatic variables they include. On the basis of these criteria, five methods were selected in each basin for calibration. In the Senegal River basin, the mean KGE values of Trabert and Hargreaves & Samani methods increased globally by 24 % to 40 %. Trabert estimation errors were reduced by 97 % after calibration. In the Gambia River basin, the KGE values of the calibrated methods increased by 12 % to 28 %. The estimation errors of the VAL2 and DA methods decreased by 37 % and 23 %, respectively, after calibration in the basin. In the Casamance River basin, calibration slightly improved the performance of the methods. For example, the KGE values of the VAL2, HG and TRJ methods increased by 7 %, 5 % and 2 % respectively (Fig. 3 and Table 3). Figure 4 gives the spatial distribution of the PBIAS, which shows that the calibrated methods are effective in all the river basins. The Trabert method overestimates ET_0 over almost the entire Senegal River basin while the combinatory (VAL2, VAL3), radiation-based (Oudin) and temperature-based (Hargreaves) methods underestimate it in the Gambia basin. In the Casamance River basin, ET_0 is overestimated on the coastal fringe and underestimated on the mainland. Given the similarity of certain results and based on the number of climatic variables, the best methods retained in this study are summarized in Table 2. Thus, the calibration has improved the performance of the methods for estimating ET_0 in the three basins. The performance of these methods was noted by Ndiaye et al. (2020) in the Senegal River Basin and Djaman et al. (2015) in the Senegal River Valley and Delta.

4 Conclusion

This study provides information on the choice of simple methods for estimating reference evapotranspiration in the main hydrosystems of Senegal. The results show that the Trabert method can be used in the Senegal River basin

Table 3. Formula of methods before and after calibration.

Reference	Before calibration	After calibration
Senegal River Basin		
Hargreaves and Samani (1985)	$ET0_{HS} = 0.408 \times 0.0023(T + 17.8) \times (T_{max} - T_{min})^{0.5} \times Ra$	$ET0_{HScal} = 0.408 \times 00031(T + 17.8) \times (T_{max} - T_{min})^{0.5} \times Ra$
Trabert (1896)	$ET0_{TRB} = 0.3075\sqrt{u2}(es - ea)$	$ET0_{TRBcal} = 2770\sqrt{u2}(es - ea)$
Valiantzas-2 (Valiantzas, 2013)	$ET0_{VAL2} = 0.0393 R_s\sqrt{T + 9.5} - 0.19 R_s^{0.6} \times \phi^{0.15} + 0.078(T + 20) \times (1 - Hr/100)$	$ET0_{VAL2cal} = 0027 Rs\sqrt{T + 9.5} - 0.19 R_s^{0.6} \times \phi^{0.15} + 0159(T + 20) \times (1 - Hr/100)$
Gambia River Basin		
Hargreaves and Samani (1985)	$ET0_{HS} = 0.408 \times 0.0023(T + 17.8) \times (T_{max} - T_{min})^{0.5} \times Ra$	$ET0_{HScal} = 0.408 \times 000299(T + 2.22) \times (T_{max} - T_{min})^{0.577} \times Ra$
Oudin (2005)	$ET0_{OD} = R_s(T + 5)/100$	$ET0_{ODcal} = R_s(T - 10)/676$
Casamance River Basin		
Hargreaves (1975)	$ET0_{HG} = 0.0135 \times 0.408 R_s(T + 17.8)$	$ET0_{HGcal} = 0024 \times 0.408 R_s(T - 3)$
Trajkovic and Stojnic (2007)	$ET0_{TRJ} = 0.0023 \times 0.408 \times (T + 17.8) \times (T_{max} - T_{min})^{0.424} \times Ra$	$ET0_{TRJcal} = 0.0035 \times 0.263 \times (T + 16.63) \times (T_{max} - T_{min})^{0.495} \times Ra$

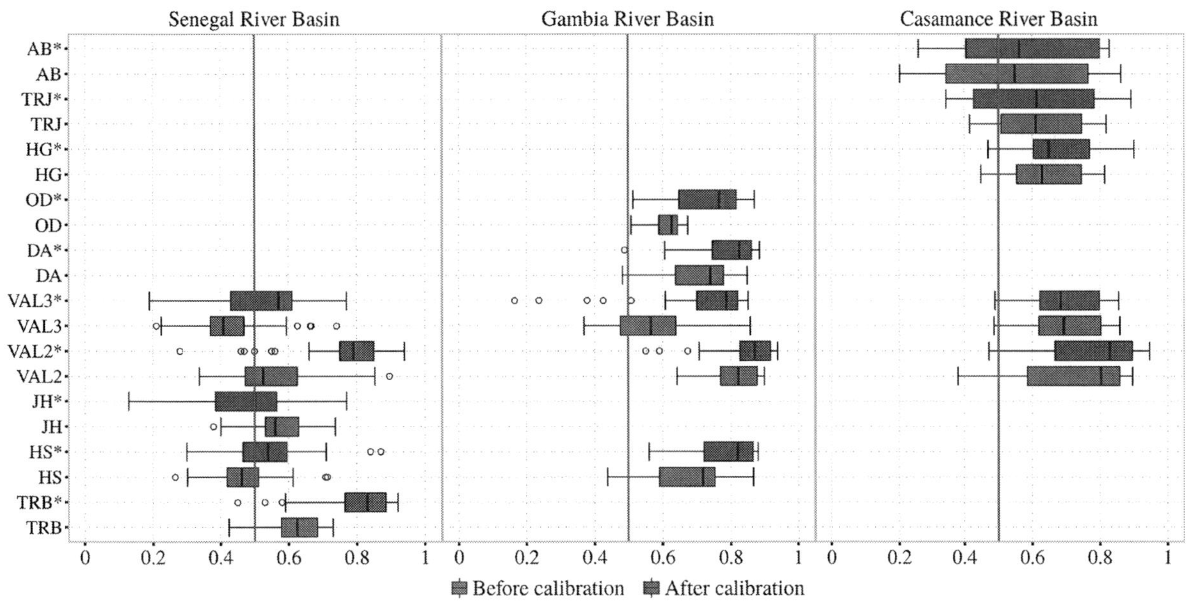

Figure 3. KGE values before and after calibration.

when temperature and wind speed data are available. The Valiantzas 2 method is applicable to all three basins. For the Gambia and Casamance river basins, temperature-based (Hargreaves, Hargreaves & Samani and Trajkovic) and radiation-based (Oudin) methods can be used when only temperature or solar radiation data is available. This information can be useful for water managers and irrigation development. However, the reanalyses data used can constitute a limit of this work.

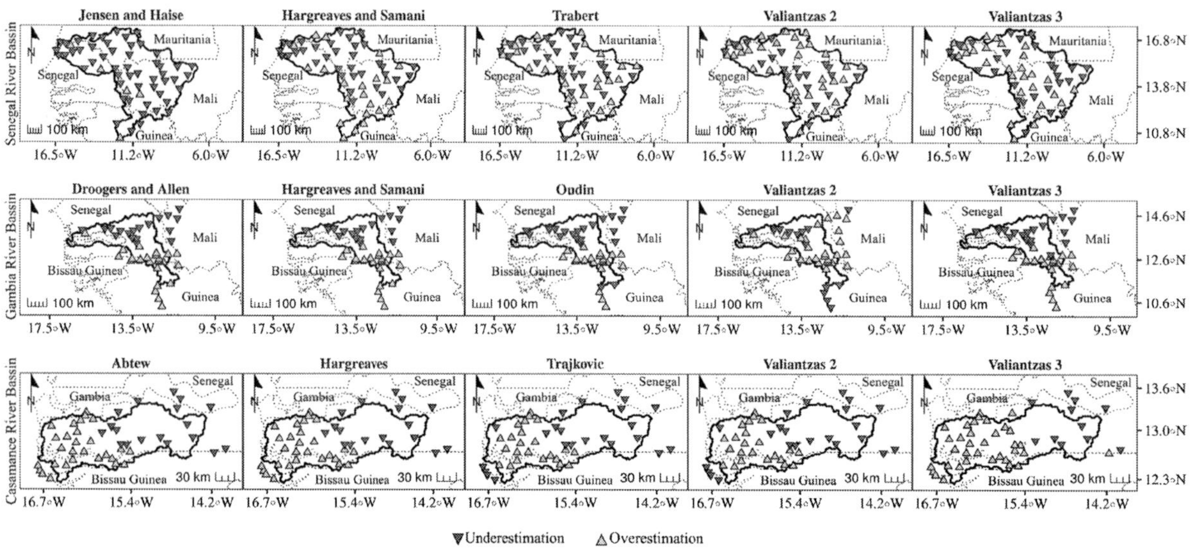

Figure 4. Spatial repartition of PBIAS according to calibrating methods.

Code availability. The software R used in this study is freely available (https://www.r-project.org/, last access: February 2022; R Development Team, 2022). The code is not publicly accessible. Please contact the authors if you are interested in this research code.

Data availability. The data of climate variables used in this study are available for public. They come from NASA Langley Research Center (LaRC) POWER Project funded through the NASA Earth Science/Applied Science Program (https://power.larc.nasa.gov/data-access-viewer/, last access: February 2022; NASA, 2022). Please contact the authors if you are interested in the evapotranspiration data used in this research.

Author contributions. Conceptualization, methodology, software and writing: AB, PMN, SBD. Review, editing and proofreading: AD, LD, AO, and KD. All authors have read and agreed to the published version of the manuscript.

Competing interests. The contact author has declared that none of the authors has any competing interests.

Disclaimer. Publisher's note: Copernicus Publications remains neutral with regard to jurisdictional claims in published maps and institutional affiliations.

Special issue statement. This article is part of the special issue "IAHS2022 – Hydrological sciences in the Anthropocene: Variability and change across space, time, extremes, and interfaces". It is a result of the XIth Scientific Assembly of the International Association of Hydrological Sciences (IAHS 2022), Montpellier, France, 29 May–3 June 2022.

Acknowledgements. The authors would like to thank ActNAO for funding Bassirou Diop's Master's work, the results of which are presented in this document. They are also grateful to SYSTA for according the travel grant to Papa Malick Ndiaye to participate in the 11th Scientific Assembly of the IAHS from 29 May to 5 June 2022 in Montpellier, France.

Review statement. This paper was edited by Christophe Cudennec and reviewed by Abdoulaye Faty and one anonymous referee.

References

Abtew, W.: Evapotranspiration measurement and modeling for three wetland systems in South Florida, Water Resour. B., 32, 465–473, 1996.

Bodian, A., Diop, L., Panthou, G., Dacosta, H., Deme, A., Dezetter, A., Ndiaye, P. M., Diouf, I., and Vichel, Th.: Recent Trend in Hydroclimatic Conditions in the Senegal River Basin, Water, 12, 436, https://doi.org/10.3390/w12020436, 2020.

Bogawski, P. and Bednorz, E.: Comparison and validation of selected evapotranspiration models for conditions in Poland (Central Europe), Water Resour. Manag., 28, 5021–5038, https://doi.org/10.1007/s11269-014-0787-8, 2014.

Dacosta, H.: Précipitations et écoulements sur le bassin de la Casamance. Dakar: ORSTOM, 278 p. multigr. Th. 3e cycle: Géogr. Phys., Hydrologie, Université Cheikh Anta Diop: Dakar, 1989.

Dalton, J.: Experimental essays on the constitution of mixed gases; on the force of steam of vapor from waters and other liquids in different temperatures, both in a Torricellian vacuum and in air on evaporation and on the expansion of gases by heat, Memoirs of the Manchester Literary and Philosophical Society, 5, 535–602, 1802.

Djaman, K., Balde, A. B., Sow, A., Muller, B., Irmak, S., Ndiaye, M. K., and Saito, K.: Evaluation of sixteen reference evapotranspiration methods under sahelian conditions in Senegal River Valley, J. Hydrol. Reg. Stud., 3, 139–159, https://doi.org/10.1016/j.ejrh.2015.02.002, 2015.

Doorenbos, J. and Pruitt, W. O.: Guidelines for Predicting Crop Water Requirements, FAO Irrigation and Drainag, Paper No. 24, FAO, Rome, Italy, 1977.

Droogers, P. and Allen, R. G.: Estimating Reference Evapotranspiration under Inaccurate Data Conditions, Irrig. Drain. Syst., 16, 33–45, https://doi.org/10.1023/A:1015508322413, 2002.

Gupta, H. V. K., Yilmaz, G. K. K., and Martinez, G. F.: Decomposition of the mean squared error and NSE performance criteria: Implications for improving hydrological modeling, J. Hydrol., 377, 80–91, 2009.

Hargreaves, G. H.: Moisture availability and crop production. Transactions of the ASAE, 18, 980–984, 1975.

Hargreaves, G. H. and Samani, Z. A.: Reference Crop Evapotranspiration from Temperature, Am. Soc. Agric. Eng., 1, 96–99, 1985

Heydari, M. M. and Heydari, M.: Evaluation of Pan Coefficient Equations for Estimating Reference Crop Evapotranspiration in the Arid Region, Archives of Agronomy and Soil Science, 60, 715–731, https://doi.org/10.1080/03650340.2013.830286, 2014.

Jensen, M. E. and Haise, H. R.: Estimating evapotranspiration from solar radiation, J. Irrig. Drain. Div., 89, 15–41, 1963.

Mahringer, W.: Verdunstungsstudien and Neusiedler See, Archiv für Meteorologie Geophysik und Bioklimatologie, série B, 18, 1–20, 1970.

Makking, G. F.: Testing the Penman formula by means of lysimeters, J. Inst. Water Eng., 11, 277–288, 1957.

Monteith, J. L.: Evaporation and Environment. 19th Symposia of the Society for Experimental Biology, University Press, Cambridge, 19, 205–234, 1965.

NASA: POWER Data Access Viewer, NASA [data set], https://power.larc.nasa.gov/data-access-viewer/, last access: February 2022, 2022.

Ndiaye, P. M., Bodian, A., Diop, L., Deme, A., Dezetter, A., and Djaman, K.: Evaluation and Calibration of Alternative Methods for Estimating Reference Evapotranspiration in the Senegal River Basin, Hydrology, 7, 24, https://doi.org/10.3390/hydrology7020024, 2020.

Oudin, L.: Recherche d'un Modèle D'évapotranspiration Potentielle Pertinent Comme Entrée d'un Modèle Pluie-Débit Global (Search for a Relevant Potential Evapotranspiration Model as Input for a Global RainFlow Model), PhD Thesis, L'école National de Génie Rural, des Eaux et Des Forêts (ENGREF), 2005, Paris, France, 496 pp., https://pastel.archivesouvertes.fr/file/index/docid/499816/filename/memoire.pdf (last access: March 2022), 2005 (in French).

Penman, H. L.: Natural evaporation from open water, bare soil and grass, P. Roy. Meteorol. Soc., 193, 120–145, 1948.

Penman, H. L.: Vegetation and hydrology, Tech. Comm. No. 53, Commonwealth Bureau of Soils, Harpenden, England, 125 pp., 1963.

Priestley, C. H. B. and Taylor, R. J.: On the assessment of surface heat flux and evaporation using large scale parameters, Mon. Weather Rev., 100, 81–92, 1972.

R Development Team: The R Project for Statistical Computing, R Development Team [software], https://www.r-project.org/, last access: February 2022.

Rohwer, C.: Evaporation from Free Water Surfaces, Technical Bulletin, 271, US Department of Agriculture, Washington DC, 1931.

Trabert, W.: Neue beobachtungen über verdampfungsgeschwindigkeiten, Meteorol. Z., 13, 261–263, 1896.

Trajkovic, S. and Stojvic, V.: Effect of wind speed on accuracy of Turc method in humid climate, Architecture and civil engineering, 5, 107–113, 2007.

Valiantzas, J.: Simple ET0 of Penman's Equation without wind/or Humidity Data. II: Comparisons Reduced Set-FAO and other methodologies, Am. Soc. Civ. Eng., 139, 9–19, https://doi.org/10.1061/(ASCE)IR.19434774.0000502, 2013.

Valipour, M.: Calibration of mass transfer-based methods to predict reference crop evapotranspiration, Application Water Science, 1, 1–11, 2015.

Xu, C. Y. and Singh, V. P.: Evaluation and generalization of temperature-based methods for calculating evaporation, Hydrol. Process., 14, 305–319, 2001.

Proc. IAHS, 385, 423–427, 2024
https://doi.org/10.5194/piahs-385-423-2024

IAHS2022 – Past and future of open, inclusive, innovative, and society-interfacing approaches

Assessing impacts of irrigation on flows frequency downstream of an irrigated agricultural system by the SWAT model

Andrés Saracho, Rafael Navas, Pablo Gamazo, and Elena Alvareda

Departamento del Agua, Centro Universitario Regional Litoral Norte,
Universidad de la República, Salto, Uruguay

Correspondence: Andrés Saracho (andressaracho.23@gmail.com)

Received: 26 May 2022 – Revised: 8 August 2023 – Accepted: 28 August 2023 – Published: 19 April 2024

Abstract. There are multiple conflicts between agronomic uses and environmental conservation. Environmental flows must be respected on basins with human interventions. The water regulation of Uruguay establishes a value of environmental flow per month for watercourses. For each month, the value corresponds to a 60 % probability of excess of the daily flows. However, the return flow of irrigation systems is not considered by current national regulations. The objective of this work is to assess the effect of irrigation downstream of reservoirs of agricultural basins. For this purpose, surface and subsurface water flow generation, and interactions between atmosphere, plant, water, and soil are quantified with the SWAT model. The model was implemented for a small basin in the north of Uruguay (Tala catchment, $120\,km^2$). The main crops are rice, soybeans, and corn, which are irrigated by pivots and border irrigation, where water is conducted by long dug channels. The water source is a reservoir with $1200\,Hm^3$ capacity, which is constructed with an earth dam and placed over vertisols soils. Water volumes in the reservoir and flows at the outlet of the catchment were validated with a moving window. Once the model was calibrated, it was used to simulate 30 years of irrigated agriculture and compared with simulations without human intervention. Results show the effect of return flow on the quantity and frequency of water fluxes through the basin. Flow duration curves show that the magnitude of low flows increases downstream on the irrigated system. On the other hand, high flows are attenuated by the storage capacity of the reservoir. The outcomes of this work could support new policies and water regulations, as they show that changes in the frequency of flows are associated with irrigation operations.

Keywords. UPH; UPH 19; Field observations; Modelling; SWAT; Irrigation return flows

1 Introduction

Water and soil resources support agricultural activity, which allows the economic development of the territory where it is carried out. However, some agricultural practices have effects on the flow frequency of flows downstream productive areas. To achieve sustainable production systems, it is necessary to minimize the impacts associated with each activity without compromising the performance of the system (Moltz et al., 2020). Irrigation return flows affect the flow regime of downstream rivers, being a key hydrological process that needs to be studied to achieve sustainable production (Song et al., 2020). For that purpose, hydrological modeling is a useful tool, especially on watersheds with scarce data (Gosain et al., 2005). The Soil Water Assessment Tool (SWAT) (Arnold et al., 2012) has been widely used to simulate the environmental impacts of land use, and land management practices. Irrigation return flows have also been simulated with SWAT, where the features of the model have allowed quantifying the dependence of return flows on different types of irrigation (Gosain et al., 2005). An approach to understanding the irrigation return flows was proposed by Kim et al. (2009) by means of SWAT modeling. These authors developed operating rules for reservoirs considering downstream environmental flow releases.

Uruguay is mainly an agricultural country with intensive use of water and soil resources, where assessing environmental conservation through hydrological modeling has been an attractive alternative (Mer et al., 2020). Models for better resource management, measuring economic benefits, and preserving water and environmental quality have been developed (Souto et al., 2021). National water regulation establishes reservoir operation using the probabilistic approach (AÑADIR REFERENCIA). Reservoirs must operate to guarantee an outflow greater than the flow with a 60 % probability of exceedance in the corresponding month. This approach gives a practical solution to estimate the water flow needed to preserve ecosystem services. However, local conditions such as irrigation or land use are not taken into account, even though they could have a significant effect on return flows (Ferencz and Tidwell, 2022).

The aim of this study is to assess the effect of irrigation downstream of a reservoir of a small agricultural basin in Uruguay. For that purpose, different scenarios will be modeled with the SWAT model. Additionally, to compare the effect of irrigation on flows with respect to the pristine condition, the flow frequency index (FFI) will be introduced and applied in several downstream sites of the irrigated system (analysis nodes).

2 Area of study and dataset

The study basin is located in the north of Uruguay in the department of Salto (Fig. 1). It has a humid subtropical climate (Cfa) according to the Köppen climate classification. Mean annual cumulative precipitation is 1430 mm with a slight seasonality, with mean daily temperatures around 10–15 °C for the winter and 20–30 °C for the summer. Soils are mainly brunisols and vertisols. The basin has an area of 120 km^2 and has been dedicated to rice, soybean, corn, and sorghum production since 1995 (Fig. 2). Water is supplied to crops through 40 km of canals from a reservoir with a volume of 12 Hm^3. This configuration makes the system efficient in soil use and conservation thanks to the agricultural rotations employed but inefficient in water use due to infiltration losses.

The basin is instrumented with a hydrometric station at the outlet, a water level station in the reservoir, and a complete climatological station (temperature, precipitation, wind, humidity, radiation). The simultaneous records of level, flow, and climatology are available for 2019–2020 agronomic year, in which the reservoir level is only available during the summer. The hydrometric station records instantaneous water levels and flow every 5 min. Then, the daily flow is calculated from instantaneous values.

3 Modelling scheme

SWAT is a semi-distributed model designed to predict the impact on water resources of various land management practices. It has a large number of parameters that serves to adjust biophysical processes in agricultural watersheds (e.g. runoff, infiltration, water storage, routing, crop yield, sediment transport, and nutrient cycles). To identify the important parameters, for the hydrological model, a sensitivity analysis was performed using the Fourier Amplitude Sensitivity Test (Reusser et al., 2011) . Then, a moving window calibration/validation routine was implemented for the 2019–2020 agronomic year. The routine uses a 7-month window for calibration and a 3-month window for validation. The windows are then scrolled to generate 8 independent calibrations and validations.

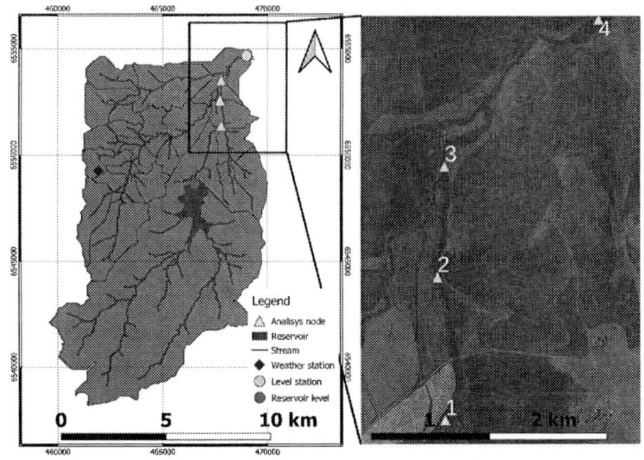

Figure 1. Tala basin and study area showing the location of the analysis nodes (Digital Orthoimagery: IDEUY, 2023).

Parameter optimization in each window is performed using Generalized Likelihood Uncertainty Estimation (GLUE) (Beven and Binley, 1992) with an objective function that quantifies the similarity of simulations and observations of flows and volume of water stored in the reservoir (Eq. 1).

$$\mathrm{OF} = 0.2\mathrm{KGE}_{\mathrm{res}} + 0.8\mathrm{KGE}_{\mathrm{reach}}, \tag{1}$$

where $\mathrm{KGE}_{\mathrm{res}}$ and $\mathrm{KGE}_{\mathrm{reach}}$ are Kling-Gupta efficiency (Gupta et al., 2009) for reservoir volume and river flow respectively. The coefficients 0.2 and 0.8 are weights assigned to reduce the effect of missing water levels on the reservoir, since during the winter water levels are not collected. These weights give priority to flow at the outlet, since it is the dataset with fewer missing values.

The impacts of irrigation were assessed with the Flow Frequency Index (FFI). The FFI index is introduced in this work to compare the frequency of flows of 2 scenarios. It is calculated as the ratio $Q_{\mathrm{i}}/Q_{\mathrm{o}}$, where Q is the flow rate for a certain frequency of occurrence, the subscript "i" is used for the irrigation case, and the subscript "o" for the basin without human intervention. To apply FFI, the simulation period was extended to 30 years. Climatological forcing was built with the data collected in a nearby climatological station. The

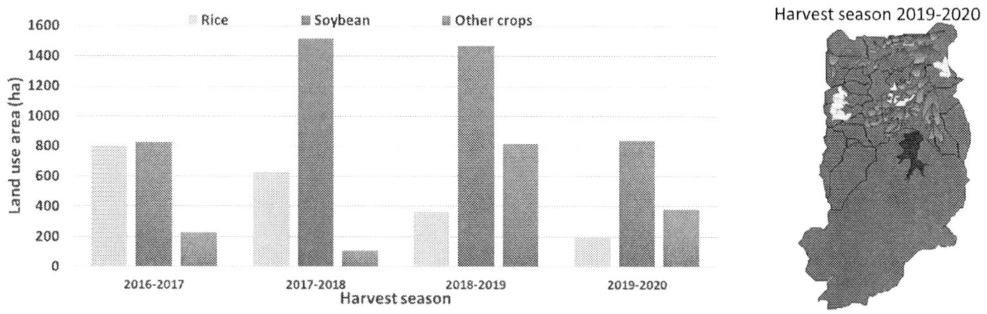

Figure 2. Areas of the different land use of the farms within the Tala watershed. The map shows the distribution of crops in the 2019–2020 harvest.

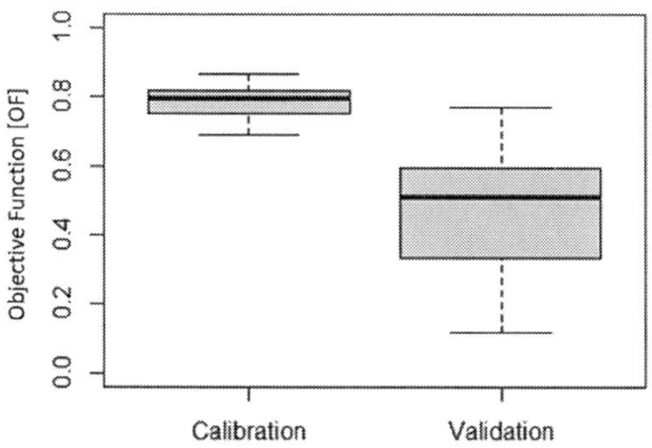

Figure 3. Objective Function values for the calibration and validation of the model.

pristine basin was simulated by eliminating the reservoir and incorporating natural herbaceous as the unique land use.

4 Results and discussion

The moving window calibration/validation showed objective function values between 0.6 and 0.8 for calibration (Fig. 3). The hydrographs show that good predictions are obtained for low flows and underestimates for high flows (Fig. 4). The slight degradation of the objective function between the calibration and validation periods may be due to the difference in size between the two windows as well as the climatic differences between the two periods (Ekmekcioğlu et al., 2022).

Figure 4 shows the simulated volumes of the irrigation reservoir are highly sensitive to the calibration/validation windows. One notable example of this sensitivity is evident when comparing the results obtained during the September–October–November validation window (Fig. 4a and b) with the November–December–January validation window (Fig. 4c and d). In the first period, a better fit is observed for reservoir volumes where more observed data is available for calibration (Fig. 4b). In contrast, the second period shows a slight degradation of the model fit, however, it still captures the depletion of the reservoir (Fig. 4c). On the other hand, the hydrographs, do not show significant changes (Fig. 4a and c).

The FFI was applied at four points downstream of the irrigated system. Values lower than 1 mean flow reduction for a given frequency. Figure 5 shows that the impact is different at each site. For example, a reduction of flows is obtained near the reservoir (Fig. 5a, b), where the return flow is supposed to be small as there are no large irrigated lands upstream. The reduction for higher non-exceedance probabilities (high flows) is explained by the storage effect of the reservoir, while the reduction for lower non-exceedance probabilities (low flows) is due to the detour and transfer of water to other irrigated lands. The impact is quite similar in winter/summer with small differences that may be due to the change in land use from natural grassland to agricultural rotations. In contrast, an increase of low flows is obtained in the analysis nodes 3 and 4 (Fig. 5c, d). The reason for these results is that the irrigation water losses are incorporated into the channel as return flows. The contribution of return flows to the main channel is greater for nodes 3–4 than nodes for 1–2 because nodes 3–4 have larger irrigated lands.

In future works, there are several aspects that can be addressed to enhance the study's findings. Firstly, extending the observation period would provide a more comprehensive dataset for model training and validation, ultimately leading to improved prediction accuracy. Secondly, a crucial area for improvement lies in refining the representation of groundwater-surface water interaction within the SWAT model. Currently, the model employs a simplified approach, which may not fully capture the complexities of this relationship. Considering a coupling with more sophisticated models would be beneficial in such cases, allowing for a more robust and accurate analysis of this interaction (Guzman et al., 2015).

5 Conclusions

This work evaluated the impact of irrigation downstream of reservoirs in agricultural basins. For this purpose, the SWAT model was used and the FFI was proposed. It was found that

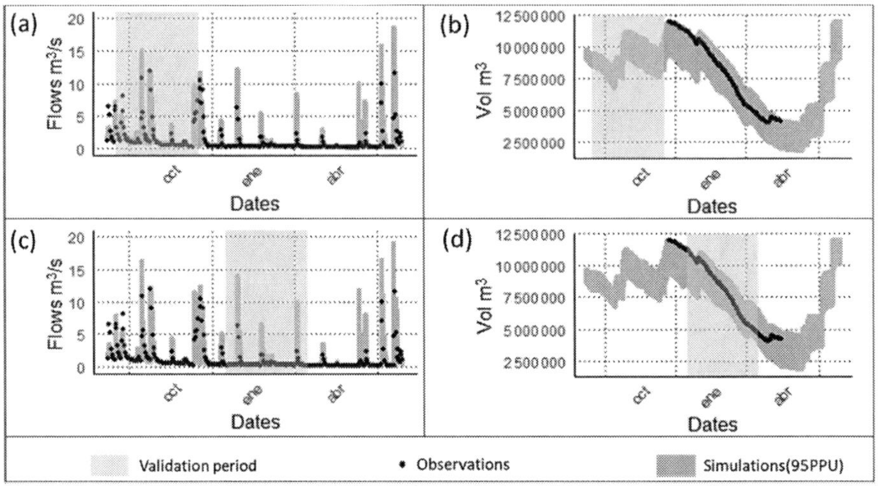

Figure 4. Comparison of two moving window calibration and validation periods for flow **(a, c)** and reservoir volume **(b, d)**.

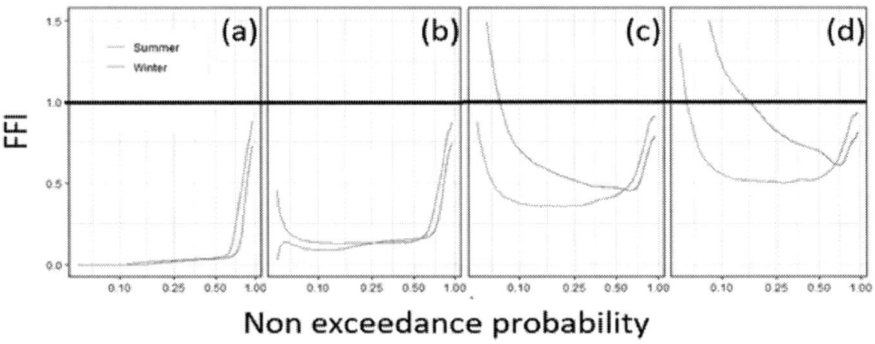

Figure 5. Flow Frequency Index as a function of non-exceedance probability on sites 1 **(a)**, 2 **(b)**, 3 **(c)** and 4 **(d)** (analysis node).

the impact of irrigation due to return flows is not homogeneous across the stream and also varies between seasons. The results of the model indicated that although the general effect of the operation of the reservoir is the reduction of the flow in the channel, this effect can be partially reversed under certain conditions: particularly during the dry season and in sections with significant accumulation of irrigation return flows.

The spatio-temporal heterogeneity of return flows limits the application of the probabilistic approach to define optimal reservoir operations since local conditions (e.g., irrigation type, land use, soil type) are not considered. The results of this work can provide new tools for defining policies and water regulations. This is extremely relevant given the worrying scenarios of increasingly accentuated droughts linked to climate change.

Code availability. The model is available at https://doi.org/10.5281/ZENODO.5708713 (Saracho, 2021).

Data availability. Observed flow, volume and climate data are available from the corresponding author on request.

Author contributions. AS: working on data processing, methodology, evaluation of the results, writing the paper; RN: conceptualization, methodology, supervision; PG: resources and review; AE: Evaluation of the results and discussion, review and editing.

Competing interests. The contact author has declared that none of the authors has any competing interests.

Disclaimer. Publisher's note: Copernicus Publications remains neutral with regard to jurisdictional claims made in the text, published maps, institutional affiliations, or any other geographical representation in this paper. While Copernicus Publications makes every effort to include appropriate place names, the final responsibility lies with the authors.

Special issue statement. This article is part of the special issue "IAHS2022 – Hydrological sciences in the Anthropocene: Past and future of open, inclusive, innovative, and society-interfacing approaches". It is a result of the XIth Scientific Assembly of the International Association of Hydrological Sciences (IAHS 2022), Montpellier, France, 29 May–3 June 2022.

Acknowledgements. We sincerely thank the Junco establishment, with special appreciation to Bernardo Bocking and Santiago Bandeira, for their invaluable assistance and provision of essential information for our study.

Financial support. This research has been supported by the Instituto Nacional de Investigación Agropecuaria (grant no. FPTA-341) and the Comisión Sectorial de Investigación Científica (grant no. I+D 2018-ID108).

Review statement. This paper was edited by Christophe Cudennec and reviewed by two anonymous referees.

References

Arnold, J. G., Moriasi, D. N., Gassman, P. W., Abbaspour, K. C., White, M. J., Srinivasan, R., Santhi, C., Harmel, R. D., van Griensven, A., Van Liew, M. W., Kannan, N., and Jha, M. K.: SWAT: Model Use, Calibration, and Validation, T. ASABE, 55, 1491–1508, https://doi.org/10.13031/2013.42256, 2012.

Beven, K. and Binley, A.: The future of distributed models: Model calibration and uncertainty prediction, Hydrol. Process., 6, 279–298, https://doi.org/10.1002/hyp.3360060305, 1992.

Ekmekcioğlu, Ö., Demirel, M. C., and Booij, M. J.: Effect of data length, spin-up period and spatial model resolution on fully distributed hydrological model calibration in the Moselle basin, Hydrolog. Sci. J., 67, 759–772, https://doi.org/10.1080/02626667.2022.2046754, 2022.

Ferencz, S. B. and Tidwell, V. C.: Physical Controls on Irrigation Return Flow Contributions to Stream Flow in Irrigated Alluvial Valleys, Front. Water, 4, 828099, https://doi.org/10.3389/frwa.2022.828099, 2022.

Gosain, A. K., Rao, S., Srinivasan, R., and Reddy, N. G.: Return-flow assessment for irrigation command in the Palleru river basin using SWAT model, Hydrol. Process., 19, 673–682, https://doi.org/10.1002/hyp.5622, 2005.

Gupta, H. V., Kling, H., Yilmaz, K. K., and Martinez, G. F.: Decomposition of the mean squared error and NSE performance criteria: Implications for improving hydrological modelling, J. Hydrol., 377, 80–91, https://doi.org/10.1016/j.jhydrol.2009.08.003, 2009.

Guzman, J. A., Moriasi, D. N., Gowda, P. H., Steiner, J. L., Starks, P. J., Arnold, J. G., and Srinivasan, R.: A model integration framework for linking SWAT and MODFLOW, Environ. Model. Softw., 73, 103–116, https://doi.org/10.1016/j.envsoft.2015.08.011, 2015.

IDEUY: https://visualizador.ide.uy/ideuy/core/load_public_project/ideuy/, last access: 19 April 2023.

Kim, H. K., Jang, T. I., Im, S. J., and Park, S. W.: Estimation of irrigation return flow from paddy fields considering the soil moisture, Agr. Water Manage., 96, 875–882, https://doi.org/10.1016/j.agwat.2008.11.009, 2009.

Mer, F., Baethgen, W., and Vervoort, R. W.: Building trust in SWAT model scenarios through a multi-institutional approach in Uruguay, Socio-Environmental Systems Modelling, 2, 17892–17892, https://doi.org/10.18174/sesmo.2020a17892, 2020.

Moltz, H. L. N., Wallace, C. W., Sharifi, E., and Bencala, K.: Integrating Sustainable Water Resource Management and Land Use Decision-Making, Water, 12, 2282, https://doi.org/10.3390/w12082282, 2020.

Reusser, D. E., Buytaert, W., and Zehe, E.: Temporal dynamics of model parameter sensitivity for computationally expensive models with the Fourier amplitude sensitivity test, Water Resour. Res., 47, W07551, https://doi.org/10.1029/2010WR009947, 2011.

Saracho, A.: AndresSaracho/SWATtala: v1.0.0, Zenodo [code], https://doi.org/10.5281/ZENODO.5708713, 2021.

Song, J.-H., Her, Y., Hwang, S., and Kang, M.-S.: Uncertainty in Irrigation Return Flow Estimation: Comparing Conceptual and Physically-Based Parameterization Approaches, Water, 12, 1125, https://doi.org/10.3390/w12041125, 2020.

Souto, A., Carriquiry, M. A., and Rosas, J. F.: Assessing the Impact of Agricultural Intensification on Water Pollution: An Integrated Model Assessment of the San Salvador Basin in Uruguay, 2021 Annual Meeting, 1–3 August, Austin, Texas, https://doi.org/10.22004/ag.econ.314037, 2021.

Proc. IAHS, 385, 429–433, 2024
https://doi.org/10.5194/piahs-385-429-2024

Open Access PIAHS

Exploring ways to improve agricultural water management in two Mediterranean irrigated systems: promises of wireless low-tech sensor networks

Paul Vandôme[1], Crystele Leauthaud[1], Simon Moinard[2], Insaf Mekki[3], Abdelaziz Zairi[3], François Charron[1], Julien Leconte[1], Intissar Ferchichi[3], Tarek Ajmi[3], and Gilles Belaud[1]

[1]G-EAU, AgroParisTech, Cirad, IRD, INRAE, L'Institut Agro, Univ. Montpellier, Montpellier, France
[2]ITAP, Univ. Montpellier, INRAE, L'Institut Agro, 2 Place Pierre Viala, 34060 Montpellier, France
[3]Institut National de Recherches en Génie Rural, Eaux et Forêts (INRGREF), Univ. Carthage, LR16INRAT05 L.STA, Tunis, Tunisie

Correspondence: Paul Vandôme (paul.vandome@gmail.com)

Received: 29 May 2022 – Revised: 1 September 2023 – Accepted: 25 September 2023 – Published: 19 April 2024

Abstract. Unsustainable use of water resources and climate change will exacerbate the already existing tensions on resources, especially in the Mediterranean context. Despite investments in modern, economically and energetically costly equipment, the performance of irrigated agriculture remains below expectations, notably because of inappropriate irrigation practices, due to insufficient knowledge of irrigation actual need and limited use of decision support tools. Access to information at an unprecedented level, via easily accessible low-cost and low-tech sensors, may be a major lever for better identifying achievable performance gains, at different spatial and temporal scales, and for guiding stakeholders towards more sustainable practices. To explore this hypothesis, we have worked on the emergence of such technologies within two Mediterranean irrigated systems (Provence, France, and Cap Bon, Tunisia) facing major water use efficiency issues. Interviews were conducted on each site in order to identify main local needs and constraints that limit sustainable water management, and potential levers to improve irrigation performance. Innovative technological systems (water sensors, automation, Internet of Things networks) have been developed in response and tested in field through a participatory approach. The technologies were then designed to be low energy, low-tech and low-cost, based on the hypothesis that the lack of accessibility – investment and maintenance costs, system readability – of existing equipment was a brake to the dissemination of innovations in the agricultural sector. We believe that the adoption of such technologies could contribute to improve irrigated systems sustainability by playing on several dimensions: promote suitable and sparing water use by improving decision-making; help maintain agricultural production by decreasing input costs; improving water users' working conditions. Generally, accompanying the transition towards more sustainable practices, by providing to the stakeholders keys for better understanding of their system. The performance gains achievable with these innovations, heeding their inherent weaknesses (eg. lower robustness and accuracy), and the potential impacts of their adoption at a larger scale remain to be assessed in an integrated way.

Published by Copernicus Publications on behalf of the International Association of Hydrological Sciences.

Keywords. UPH 16; Irrigation; Mediterranean agriculture; Participatory design; IoT

1 Introduction

The challenge of food security in a context of climate change and demographic growth brings irrigated agriculture – leading consumer of fresh water and source of more than 40 % of the world's agricultural production – under the spotlight (FAO, 2020). Within irrigated territories, pressure on water resources may causes allocation conflicts between agricultural, ecological, industrial and domestic uses. Major investments have been made to improve the performance of irrigated systems, from plot level to catchment scale. Measuring equipment (flow meters, soil moisture probes, water level sensors ...) is among the solutions deployed to support decision-making or irrigation systems automation. However, these tools are still under-used, especially at farm scale and in the global South. The digital boom and recent development of new technologies such as on-board electronics and Internet of Things (IoT) expand the possibilities for better understanding and management of agricultural resources (OCDE, 2019). For irrigated agriculture, these advances offer the opportunity of real-time water flows measurements through wireless sensor networks (WSN) deployed within territories. Many studies have shown that irrigation monitoring is technically feasible by means of systems based on different kinds of sensors, measured parameters and communication networks (Hamami and Nassereddine, 2020). Work has been done on diverse agronomic conditions and some studies have shown that performance gains are achievable with the help of such tools (Liao et al., 2021; Bwambale et al., 2022). However, to the best of our knowledge few studies have gone beyond the purely conceptual and technological aspects and actually tried to match such innovations with agrarian contexts. This article describes first the paths that led to the development of technological innovations within two irrigated territories and second provides an insight of the innovations that were implemented.

2 Methods and study sites

2.1 Shaping innovation to match territories

The aim of the first phase of the process is to understand the agrarian contexts and identify main issues related to agricultural water management. An agrarian diagnosis is carried out on two irrigated territories. Data is collected through a field survey campaign. To understand the main trajectories of local agriculture, open interviews with key resource persons and retired farmers are conducted. Around thirty semi-directive technical and economic interviews are conducted with farmers on each case study site. Besides we performed interviews with network operators and heads of water user associations, institutions, producers' and environmental associations. The analysis of these survey data allows (i) to build a typology of the farming and cropping systems composing the irrigated territories; (ii) to identify the performance gaps; (iii) a first insight into the innovation potential in the area, i.e. the identification of practices or tools improving performances and the inclination of stakeholders for innovation (access to education, digital technology, awareness of the stakes, investment capacity, etc).

At this stage, two cases may be encountered: innovation is existing on the territory, some "pilot" stakeholders (suppliers, farmers, managers) have developed expertise and tools to meet local needs. In this case, the objective is to support the existing innovation by asking why its adoption is limited for the moment, and by what means these solutions can be improved to encourage their dissemination. In the second case, the survey does not reveal any particular local innovation, despite the identification of strong water management issues. In this situation, the method consists of thinking up and designing new tools with the water users.

The design of innovations is based on a participatory approach, which aims to involve stakeholders throughout the process (Fig. 1). The development of the first prototype responds to a set of specifications dictated by the survey results, based on the needs and constraints formulated by the stakeholders. The tool is then introduced to potential users in the form of demonstration workshops, and then evolves regularly based on their feedback. Within the territories, work spaces for the design and maintenance of the systems (fablabs), also providing a place for interaction on the topic, are set up. Field trials carried out at "pilot" farmers' plots are an opportunity to test the technologies in real conditions, while user feedback contributes to the continuous improvement of the solutions developed.

2.2 Case-study sites

To explore our hypotheses, we experimented the envisaged method within territories known for the water management challenges they are facing (pressure on the water resource, agro-economic constraints). It was important that the study sites show a diversity of agrarian contexts (cropping and farming systems, access to the resource) while being representative of the key issues of irrigation in the Mediterranean basin.

2.2.1 The Crau plain, South of France

The Crau plain is located in the South-East of France, on the Mediterranean coast. The area has no natural river, but a network of irrigation canals developed over 500 years distributes the resource across the plain. The plain covers a vast groundwater table, the hydrodynamic functioning of which is highly dependent on anthropogenic activities (recharge and withdrawals). Water management in the Crau plain lies on a fragile balance between the different water uses. Three main

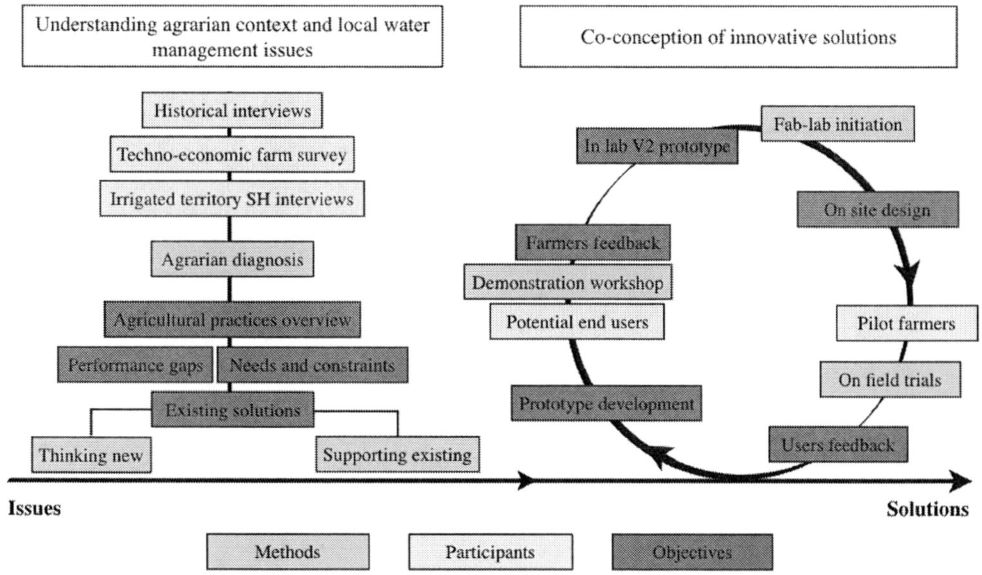

Figure 1. Shaping innovation to match territories. SH: stakeholders.

productions require irrigation in the area: Crau hay, whose gravity fed irrigation is supplied by the collectively managed surface network, arboriculture and market gardening, which mostly use groundwater to feed localized irrigation systems.

2.2.2 The Haouaria plain, North of Tunisia

The Haouaria plain is located at the northern tip of the Tunisian Cap Bon, governorate of Nabeul. The water resource of the plain is exclusively groundwater (agricultural, domestic and industrial water). The water tables, both deep and superficial, are heavily exploited by the numerous surface wells and boreholes. Access to agricultural water may organized within public irrigated schemes through water users' associations, or private through individual pumping. The main irrigation system is drip irrigation, both for tree crops (citrus, olive) and market gardening (potato, tomato, chilli, groundnut, etc.).

3 Results

3.1 Innovation pathways

In the Crau plain, the hay cropping system and the associated irrigation practices crystallise the issues and conflicts surrounding agricultural water management. Gravity irrigation is constraining in terms of workload and difficult to control. Many factors are at the root of performance gaps with strong consequences on the volumes withdrawn. Farmers and managers emphasise the lack of information that would be required to improve their practices (water advancement along the plot, irrigation events duration, water level in the canals, etc.). The search for solutions is active within the territory and several actors have contributed to the emergence of technological innovations (e.g. water gates controlled by timers, water detection systems). However, these innovations have not been widely disseminated throughout the territory, for technical and practical, economic or social reasons.

In the Haouaria plain, drip irrigation of vegetable crops on sandy and draining soils is problematic. Significant differences in water productivity are observed between users, who account for the high equipment cost without succeeding in reducing the volumes applied and the associated fees. The lack of information and the variable flow rates delivered by the network make irrigation monitoring complex. On citrus farming systems, the use of water tensiometers has been observed to adjust the irrigation schedule and reduce percolation losses, but the tool is hardly adopted due to its lack of technical (direct reading of the tension) and economic accessibility.

3.2 Overview of the innovation basket

The innovation process has led to the development of different kinds of technologies: (i) the "Pilowtech", a capacitive based soil moisture sensor for drip irrigation monitoring; (ii) the "WatAr sensor", which aims to provide information about surface water advancement for border irrigation and (iii) the "Flowter" sensor, for the monitoring of gravity based irrigation schedule (Fig. 2).

The technologies are designed to follow common properties, in response to common concerns raised on both study sites. They are low-tech, meaning easy to maintain and to reproduce by users. They are low-cost, so they may be

Figure 2. Example of innovations developed: WatAr sensor **(a)**, Flowter sensor **(b)**, Pilowtech sensor **(c)**.

used by any farming system. They are wireless and low-power, so they are autonomous without the need for regular visits. Developed systems are all based on open-source micro-controllers and LoRaWAN communication protocol. The data reading (historic and real-time) is made available to each user on an internet platform developed under Grafana cloud. A threshold-based alert system is set up through a smartphone messaging application. Besides collective features, the innovations differ relatively to the territories they have been designed for. The solutions developed in the Crau plain provide binary information about surface water flows. WatAr sensors localised downstream on the plot are alerting the farmer when time has come to operate the water gate and irrigate the next plot. Flowter sensors, placed in the tertiary channels, allow to monitor and record surface irrigation schedule (frequency and duration) at plot scale. On the other hand, the innovation that emerged from the plain of Haouaria allows double depth continuous soil moisture measurement and data is readable in various formats (moisture content percentage, graphic display of soil water storage level) both online and on-site.

4 Conclusions

The fieldwork carried out on the two study sites made it possible to identify needs relating to agricultural water management that are specific to each irrigated area. The methodology of the innovation process was adapted to the agrarian contexts. Regular exchanges with stakeholders led to the design of an innovation basket reflecting the inherent needs and constraints of the territories. The wireless water sensors developed meet common objectives and constraints for both irrigated systems: they represent technically and economically accessible means of real time and continuous flow measurement. The plasticity of on-board electronics and IoT systems then allows solutions to be adapted locally. Innovations are not static and continue to change as they are used in the field and feedback are made. Thus, recent technological advances in these fields presage progress in agricultural water management, provided that innovation is adapted to the territories and not the other way around. The impacts of these technologies adoption on the irrigated agriculture performances remain to be assessed. An integrated multi-criteria and multi-scale approach should help determine the achievable performance gains without neglecting potential undesirable effects. It would also be interesting to study the process of adoption of these tools in the longer term and on a larger scale: their re-appropriation by users, and the probable "bricolage" that would make them change over time.

Data availability. Data will be made available on request.

Author contributions. This paper is part of PV PhD work. CL, GB, IM and AZ are the PhD supervisory team and provide conceptual, technical and editorial support. SM, FC, JL, IF and TA contributed to resources and investigation.

Competing interests. The contact author has declared that none of the authors has any competing interests.

Disclaimer. Publisher's note: Copernicus Publications remains neutral with regard to jurisdictional claims made in the text, published maps, institutional affiliations, or any other geographical representation in this paper. While Copernicus Publications makes every effort to include appropriate place names, the final responsibility lies with the authors.

Special issue statement. This article is part of the special issue "IAHS2022 – Hydrological sciences in the Anthropocene: Past and future of open, inclusive, innovative, and society-interfacing approaches". It is a result of the XIth Scientific Assembly of the International Association of Hydrological Sciences (IAHS 2022), Montpellier, France, 29 May–3 June 2022.

Acknowledgements. This paper is part of the first author PhD project, funded by the CIRAD and the Occitanie region. The work is carried out as part of the European programme HubIS. We gratefully acknowledge CTV Haouaria, GDA Chraf, ASA Haute Crau and the Domaine du Merle for participating in the project. The cooperation of the farmers in the Chraf irrigated scheme and across the Crau plain was highly appreciated.

Financial support. This research has been supported by the EU's PRIMA program 2019 (grant no. ANR-19-P026–0006–02).

Review statement. This paper was edited by Christophe Cudennec and reviewed by one anonymous referee.

References

Bwambale, E., Abagale, F. K., and Anornu, G. K.: Smart irrigation monitoring and control strategies for improving water use efficiency in precision agriculture: A review, Agr. Water Manage., 260, 107324, https://doi.org/10.1016/j.agwat.2021.107324, 2022.

FAO: The State of Food and Agriculture, https://doi.org/10.4060/cb1447en, 2020.

Hamami, L. and Nassereddine, B.: Application of wireless sensor networks in the field of irrigation: A review, Comput. Electron. Agr., 179, 105782, https://doi.org/10.1016/j.compag.2020.105782, 2020.

Liao, R., Zhang, S., Zhang, X., Wang, M., Wu, H., and Zhangzhong, L.: Development of smart irrigation systems based on real-time soil moisture data in a greenhouse: Proof of concept, Agr. Water Manage., 245, 106632, https://doi.org/10.1016/j.agwat.2020.106632, 2021.

OCDE: Digital Opportunities for Better Agricultural Policies, https://doi.org/10.1787/571a0812-en, 2019.

Proc. IAHS, 385, 435–441, 2024
https://doi.org/10.5194/piahs-385-435-2024

Comparison of one-site vs. multi-sites calibration schemes for hydrological modelling of nested catchments in the West African Sahel

Yetchékpo Patrick Gbohoui[1,2,3,4], Roland Yonaba[1], Tazen Fowé[1], Bernadin Elégbédé Manou[4], Taofic Bacharou[4,6], Yvon-Carmen Hountondji[5], Ernest Amoussou[7], Luc O. Sintondji[3], Jean-Emmanuel Paturel[2], Harouna Karambiri[1], and Hamma Yacouba[1]

[1]Laboratoire Eaux, Hydro-Systèmes et Agriculture (LEHSA), Institut International d'Ingénierie de l'Eau et de l'Environnement (2iE), Rue de la Science, Ouagadougou 01 01-BP-594, Burkina Faso
[2]HydroSciences Montpellier, Univ Montpellier, CNRS, IMT, IRD, Montpellier, France
[3]Laboratoire d'Hydraulique et de Maîtrise de l'Eau, Institut National de l'Eau, Université d'Abomey-Calavi, 01 B.P. 526 Recette Principale Cotonou 01, Benin
[4]Laboratoire des Sciences et Techniques de l'Eau et de l'Environnement (LSTEE), Institut National de l'Eau, Université d'Abomey-Calavi, 01 BP 2009 Cotonou, Benin
[5]Department of Natural Resources Management, Faculty of Agronomy, University of Parakou, Parakou, Benin
[6]Laboratoire d'Energétique et de Mécanique Appliquée (LEMA), Université d'Abomey-Calavi, 01 BP 2009 Cotonou, Benin
[7]Département de Géographie et Aménagement du Territoire (DGAT), Université de Parakou, Parakou, Benin

Correspondence: Yetchékpo Patrick Gbohoui (patrickgbohoui@gmail.com)

Received: 27 May 2022 – Revised: 14 September 2023 – Accepted: 25 September 2023 – Published: 19 April 2024

Abstract. Runoff simulation in highly anthropized catchments is complex, but essential for water management, especially in poorly gauged and data-scarce hydrosystems of the West African Sahel. This study aims to evaluate the effect of different calibration schemes on runoff simulation. The physically-based and semi-distributed hydrological SWAT (Soil and Water Assessment Tool) model is used to simulate daily runoff in the Nakanbé catchment at Wayen station in Burkina Faso (in the West African Sahel) over the period 2006–2012. Four (4) hydrometric stations (Dombré, Rambo, Ramsa and Wayen) gauging 4 nested catchments (ranging from 1060 to 21 178 km^2 in size) are considered. The added value of the consideration of nested catchments is assessed through the following 3 calibration schemes: one-site (OS) at the entire catchment outlet (Wayen); multi-sites with nested sub-catchments (MS1); and multi-sites without considering nested sub-catchments (MS2). The results indicate that OS and MS2 schemes perform well (KGE > 0.7, |PBIAS| < 3 %), with MS2 scheme being superior (KGE, PBIAS). However, the MS1 scheme (KGE = 0.68; PBIAS = −22.9 %) performed worse in comparison to the traditional OS scheme. The comparison of the three modelling schemes provides evidence that accounting for nested sub-catchments does not necessarily improve the quality of rainfall-runoff simulations. Yet, multi-site calibration should be favoured when catchments are not nested.

Published by Copernicus Publications on behalf of the International Association of Hydrological Sciences.

Keywords. UPH6; Modelling; Nested catchments; Sahel

1 Introduction

The combined effects of climate and environmental changes increase the complexity of runoff modelling. In highly anthropized catchments, for example, natural hydrological behaviour is altered, affecting the ability of hydrological models to simulate runoff, especially if detailed and accurate data is not available (Zhang et al., 2013; Abbaspour, 2015; Zanin et al., 2018). Physically-based and distributed/semi-distributed models are known to be suitable for rainfall-runoff modelling in such contexts (Zanin et al., 2018). However, these models generally involve large number of parameters. The calibration of distributed/semi-distributed and physically-based models remains a challenging task.

Calibration is usually performed at the outlet of a single catchment and does not always represents the hydrological processes in the different river sub-catchments (Zanin et al., 2018). Multi-sites/multi-variables calibration approaches are therefore often applied to obtain realistic optimized parameter values (Cao et al., 2006; Zanin et al., 2018). These approaches often improve the simulation performance of hydrological models.

The Soil and Water Assessment Tool (SWAT, https://swat.tamu.edu/, last access: 13 October 2023) model is a physically-based and semi-distributed agro-hydrological modelling tool that allows the implementation of multi-sites modelling. The SWAT model is widely applied around the world and in Africa (Tan et al., 2020; Akoko et al., 2021).

The West African Sahel is known for the fragility of its hydrological system, for which a wide range of questions remains unsolved (Blöschl et al., 2019). Indeed, the hydrological behaviour of some West African Sahel catchments has been strongly affected by global change (Descroix et al., 2018; Gbohoui et al., 2021; Yonaba et al., 2021a). Volta is one of the most important rivers in West African and has its source in the Sahelian part of Burkina Faso (Nakanbé River). The Volta catchment is highly anthropized and shelters several dams for hydroelectric or agricultural purposes. Previous studies on modelling the runoffs of this river were mainly devoted to its downstream outlets (Schuol et al., 2008; Sood et al., 2013; Poméon et al., 2018). Furthermore, these studies did not analyse the effects of considering nested or non-nested catchments in the multi-sites modelling process.

This study focuses on the headwater sub-catchments and the Wayen station, whose gauged catchments covers the entire Sahelian part of the Volta catchment. It aims to assess various calibration schemes, namely one-site calibration (OS, at the entire catchment outlet) and different multi-sites (MS) calibration schemes using the SWAT model and 4 nested catchments (ranging in size from 1060 to 21 178 km^2) in the West African Sahel.

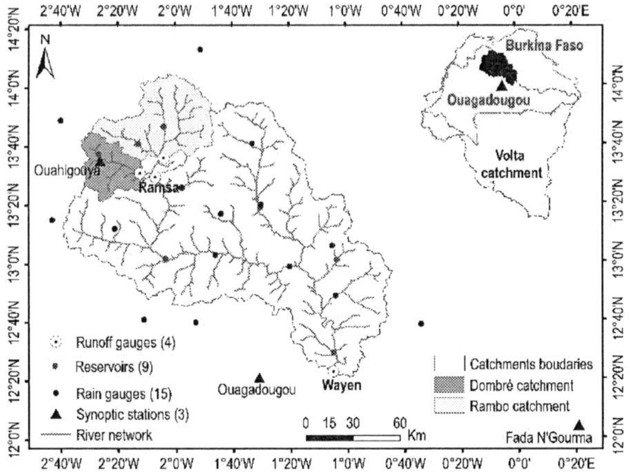

Figure 1. Study area (headwater catchments of the White Volta) with drainage network, reservoirs, synoptic stations, rain gauges and runoff gauges (data source: IGB and DGRE).

2 Material and methods

2.1 Study area

The study area covers the headwater sub-catchment of the Nakanbé, located between 0°40′ and 2°40′ W longitudes and 12°20′ and 14°00′ N latitudes. The relief of the catchment is generally flat with an extremely low average longitudinal slope (Slope $< 0.2\,\%$).

The study area (Fig. 1) is a combination of four nested catchments: Rambo (2471 km^2), Dombré (1060 km^2), Ramsa (3826 km^2), and Wayen (21 178 km^2). The Rambo and Dombré sub-catchments are nested within the Ramsa sub-catchment, while Ramsa itself lies within the larger Wayen catchment. The Wayen catchment shelters several natural and artificial reservoirs of various sizes, the most important of which are shown in Fig. 1.

2.2 Data

2.2.1 Hydrometeorological data

To evaluate the OS and MS calibrations, 4 hydrometric stations (runoff gauges) are selected: Rambo, Dombré, Ramsa and Wayen (whose location is presented in Fig. 1). The daily runoff data available at these stations covers the period 1965–2018. These stations were operational at different dates and their records are filled with several gaps. Therefore, a common gap-free period for the 4 runoff gauges is used in this study, which is the period 2006–2012. The data was obtained from a national agency (in French, "*Direction Générale des Ressources en Eau*", DGRE) in Burkina Faso.

The characteristics of the 9 main reservoirs (shown in Fig. 1) in the catchment were obtained from the DGRE. Water withdrawals data at Ziga dam (close to Wayen runoff

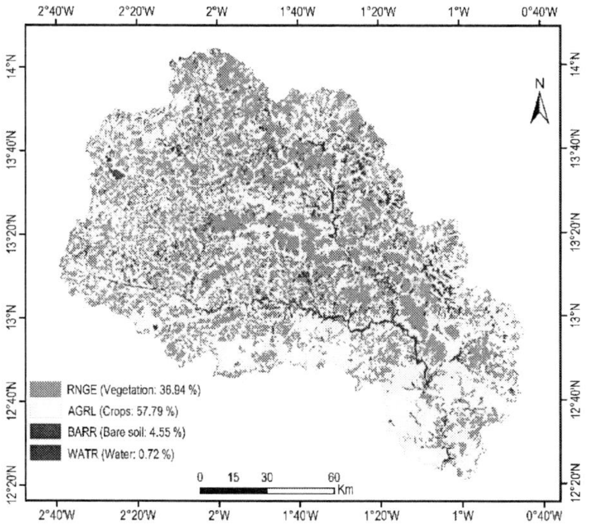

Figure 2. Land use and land cover map (data source: DGRE). RNGE, AGRL, BARR and WATR are identification codes for the LULC types implemented in SWAT.

gauge in Fig. 1) was provided by the national office in charge of water and sanitation (in French "*Office National de l'Eau et de l'Assainissement*", ONEA) in Burkina Faso.

The meteorological data for the period 2006–2012 were provided at the daily time step and included precipitation, temperature (minimum and maximum), solar radiation, wind speed, and relative humidity. The data was collected from the national meteorology agency (ANAM-BF) in Burkina Faso for 15 rain gauges and 3 synoptic stations (Fig. 1).

2.2.2 Ancillary data

The Digital Elevation Model (DEM) was downloaded from the Japan Aerospace Exploration Agency (JAXA) website (https://www.eorc.jaxa.jp/ALOS/en/aw3d30/index.htm, last access: 30 June 2022). It has a spatial resolution of 30 m × 30 m.

The Food and Agriculture Organization (FAO) 1/5 000 000 digital soil map (http://www.fao.org/soils-portal/soil-survey/cartes-historiques-et-bases-de-donnees-des-sols/carte-faounesco-des-sols-du-monde/fr/, last access: 30 June 2022) was used to generate the soil data needed for SWAT. In terms of hydraulic characteristics, the catchment basements are crystalline, which limits infiltration.

The land use and land cover (LULC) data were obtained from the national geography institute (in French, "*Institut Géographique du Burkina*", IGB), and was further reclassified into 4 units: water, bare soil, cultivated areas and vegetation (Fig. 2). These units were deemed reliable to cover the variability and physical processes in Sahelian landscapes (Yonaba et al., 2021b; Paturel et al., 2016).

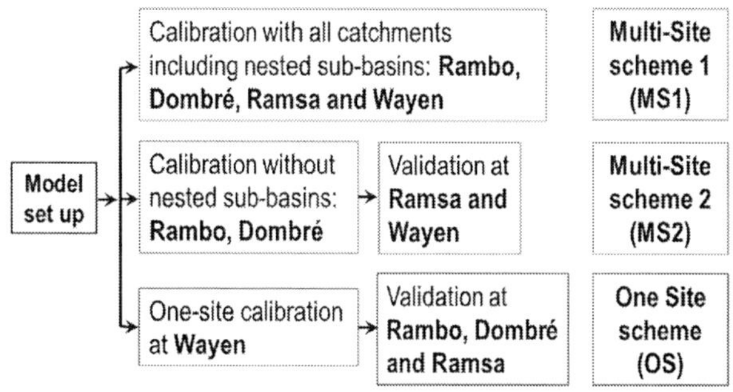

Figure 3. Calibration/Validation schemes.

2.3 Model setup

The model setup is carried out through the ArcSWAT 2012.10_5.21 interface. SWAT simulates hydrology at the scale of element areas known as Hydrological Response Units (HRUs). HRUs are created from the overlay of subcatchment contours, LULC units, soil types, and slope classes. The model outputs are estimated at the HRU scale and summed up to obtain processes outputs at the subcatchment scale, further routed at the entire catchment outlet (Neitsch et al., 2011).

Surface runoff is simulated through the Soil Conservation Service (SCS) Curve Number (CN) method, and the Muskingum method is further used for routing through the channel network system. Potential evapotranspiration is estimated through the Penman–Monteith formula.

In this study, the model was set up over the period 1965–2018 with a warm-up period of 3 years (1965–1968). The common data period (2006–2012) for the 4 runoff gauges is used to implement the modelling schemes.

2.4 Model calibration and validation

Based on a literature review of SWAT model applications in West African Sahel (Schuol et al., 2008; Sood et al., 2013; Adjei et al., 2015; Awotwi et al., 2015; Akpoti et al., 2016; Poméon et al., 2018; Yonaba et al., 2021a), a set of 33 parameters is initially screened for sensitivity analysis Thus, 17 parameters (presented in Table 1) were selected and considered for calibration and validation.

The calibration/validation process was performed under SWAT-CUP with the Sequential Uncertainty Fitting (SUFI2) algorithm, which combines optimization and uncertainty analysis (Abbaspour, 2015). The objective function used was the Kling-Gupta Efficiency (KGE) criterion with an acceptability threshold of 0.5 (Thiemig et al., 2013). The under-/overestimation of the model was assessed with the bias percentage (PBIAS) criterion. These criteria were defined in the SWAT-CUP user manual (Abbaspour, 2015).

The effects of multi-sites modelling compared to traditional modelling at the catchment outlet are evaluated

Table 1. SWAT parameters for model calibration.

Name	Definition
CN_2	Curve number for moisture condition II
OV_N/ CH_N2	Manning's n value for overland flow/ main channel
SOL_AWC	Available water capacity of the soil layer
SOL_K	Saturated hydraulic conductivity
SOL_Z	Depth from soil surface to bottom of layer
SOL_BD	Moist bulk density
CH_K1/ CH_K2	Effective hydraulic conductivity in tributary/main channel alluvium
MSK_CO1/ MSK_CO2	Calibration coefficient of Muskingum method
ESCO	Soil evaporation compensation factor
EPCO	Plant uptake compensation factor
REVAPMN	Threshold depth of water in the shallow aquifer for "revap" or percolation to the deep aquifer to occur
GW_REVAP	Groundwater "revap" coefficient
ALPHA_BF	Baseflow alpha factor
RCHRG_DP	Deep aquifer percolation fraction

through three schemes presented in Fig. 3: one-site (OS) at the entire catchment outlet (Wayen); multi-sites calibration with nested sub-catchments (MS1) and multi-sites without nested sub-catchments (MS2). The modelling process in SWAT is carried out through successive iterations. In this study, each modelling scheme consisted of 3 iterations of 500 simulations: 2 iterations for parameters calibration et 1 iteration for runoff simulation.

A comparative study of the results of the three modelling schemes allowed us to evaluate the effects of the one-site and the different multi-sites schemes on surface runoff simulation at the Wayen outlet. The results of the MS1 scheme (including nested catchments) compared to those of the MS2 scheme (non-nested catchments) allowed us to assess the consideration of the upstream-downstream hydraulic connections between nested catchments in the calibration process.

Table 2. Performance of the three modelling schemes at Wayen outlet.

Criterion	OS	MS1	MS2
KGE	0.73	0.68	0.75
PBIAS (%)	−1.30	−22.9	−1.70

3 Results and discussion

3.1 One-Site vs. Multi-Sites (nested or not) modelling

The performance of the different modelling schemes for the Wayen runoff gauge is shown in Table 2. Further details on the simulation results can be found in Gbohoui (2021).

The results indicated that OS (one-site calibration) and MS2 (multi-site without nested sub-catchments) schemes are good to very good (KGE > 0.7; $|\mathrm{PBIAS}| < 3\,\%$) at Wayen station. Previous attempts to model the runoff at Wayen station using the SWAT model (Sood et al., 2013; Poméon et al., 2018) resulted in weaker performance (KGE ≤ 0.5). The improved simulation performance attained in our study should be related to the effort to integrate some characteristics of the main water reservoirs. However, the availability of accurate data on soil types and catchment management might have further helped improving the accuracy of the simulations, especially at intermediate points within the catchment.

The MS2 scheme provided the optimal runoff simulation results at Wayen (KGE = 0.75). This suggests that the multi-sites calibration scheme without nested sub-catchments could therefore be applied to obtain regionalized parameter values for the West African Sahel region from small catchments distributed across all climatic zones.

The MS1 scheme (KGE = 0.68; PBIAS = −22.9 %) performed less than the OS scheme (KGE = 0.73; PBIAS = −1.30 %). Thus, it can be seen that the multi-sites scheme with nested sub-catchments did not improve significantly surface runoff simulation at the catchment outlet (Wayen), while considering the hydraulic connection between nested sub-catchments should have improved the modelling performance. The underperformance of the so-called multi-sites modelling scheme with nested sub-catchments is also reported in previous studies (Migliaccio and Chaubey, 2007; Zanin et al., 2018). The underlying explanation might be related in the operating mode of the optimisation algorithms. In multi-sites calibration, the optimisation algorithms apply ratios to the outputs of each catchment to obtain the final result (Migliaccio and Chaubey, 2007). However, the hydrological response at the outlet of a downstream catchment considers the upstream sub-catchments. Therefore, in the case of nested sub-catchments,

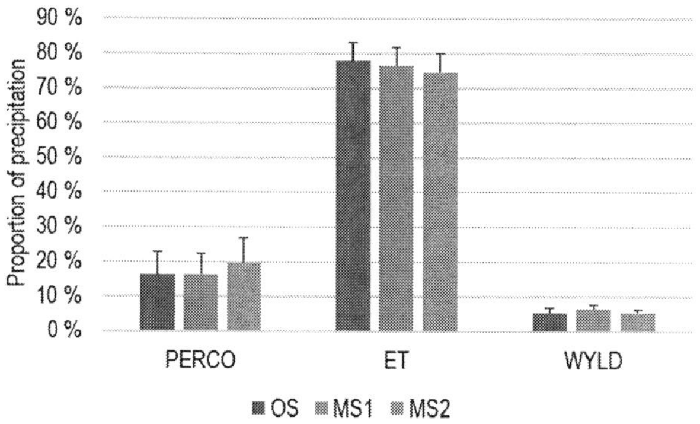

Figure 4. Water balance of the White Volta catchment at Wayen over the period 2006–2012. PERCO: Percolation, WYLD: Water yield, ET: Actual Evapotranspiration.

such upstream sub-catchments are considered several times, leading to an overall bias in the final result.

3.2 Water balance of the Nakanbé catchment at Wayen station

The three modelling schemes led to similar results for the water balance of the Wayen catchment. Figure 4 shows the repartition of rainfall in the different components of the water balance. Over the period 2006–2012, more than 70 % of the rainfall is lost as actual evapotranspiration. The high evapotranspiration values observed in the Wayen catchment reflects the overall behaviour of the West African Sahel catchments. In such context, evapotranspiration is the most important term of the water balance (Derive, 2003).

4 Conclusion

In this study, we implemented several modelling schemes to capture the hydrological response of the Nakanbé catchment at Wayen station (21 178 km^2). To this end, the SWAT model is applied with one-site, multi-sites with and without nested sub-catchments calibration schemes. The results showed that the multi-sites without nested sub-catchments scheme (MS2) performs well overall optimal (KGE = 0.75 and PBIAS = −1.70 %).

The SWAT agro-hydrological model is widely used due to its performance and wide range of applicability. However, this study findings suggest that multi-sites modelling under SWAT does not consider the upstream-downstream hydrological connection in nested sub-catchment during calibration. Therefore, integrating such hydrological connectivity in the model functioning is suggested to further improve the performance of the model, especially when applied to such contexts.

Data availability. The data used are the property of national organizations in Burkina Faso: Direction Générale des Ressources en Eau (DGRE), Agence Nationale de la Météorologie (ANAM).

Author contributions. PYG: Conceptualization, Data curation, Methodology, Formal analysis, Software, Writing – original draft preparation, Writing-review and editing. RY and TF: Methodology, Formal analysis, Software, Writing-original draft preparation, Writing – review and editing. BEM, TB, YCH, EA and LOS: Writing – original draft preparation, Writing – review and editing. JEP, HK and HY: Conceptualization, Methodology, Formal analysis, Funding acquisition, Project administration, Writing original draft preparation, Writing-review and editing.

Competing interests. The contact author has declared that none of the authors has any competing interests.

Special issue statement. This article is part of the special issue "IAHS2022 – Hydrological sciences in the Anthropocene: Variability and change across space, time, extremes, and interfaces". It is a result of the XIth Scientific Assembly of the International Association of Hydrological Sciences (IAHS 2022), Montpellier, France, 29 May–3 June 2022.

Financial support. Participation in the conference was paid for by SYSTA (SIVAPALAN Young Scientists Travel Award). The research has been supported by the African Development Bank Group (Project no. P-Z1-IA0-013, grant no. 2100155032824).

Review statement. This paper was edited by Christophe Cudennec and reviewed by Domiho Japhet Kodja and one anonymous referee.

References

Abbaspour, K. C.: SWAT calibration and uncertainty programs – A User Manual, eawag, 17–66, 2015.

Adjei, K. A., Ren, L., Appiah-Adjei, E. K., and Odai, S. N.: Application of satellite-derived rainfall for hydrological modelling in the data-scarce Black Volta trans-boundary basin, Hydrol. Res., 46, 777–791, https://doi.org/10.2166/nh.2014.111, 2015.

Akoko, G., Le, T. H., Gomi, T., and Kato, T.: A Review of SWAT Model Application in Africa, Water, 13, 1313, https://doi.org/10.3390/w13091313, 2021.

Akpoti, K., Antwi, E. O., and Kabo-bah, A. T.: Impacts of Rainfall Variability, Land Use and Land Cover Change on Stream Flow of the Black Volta Basin, West Africa, Hydrology, 3, 26, https://doi.org/10.3390/hydrology3030026, 2016.

Awotwi, A., Yeboah, F., and Kumi, M.: Assessing the impact of land cover changes on water balance components of White Volta Basin in West Africa, Water Environ. J., 29, 259–267, https://doi.org/10.1111/wej.12100, 2015.

Blöschl, G., Bierkens, M. F., Chambel, A., Cudennec, C., Destouni, G., Fiori, A., Kirchner, J. W., McDonnell, J. J., Savenije, H. H., Sivapalan, M., Stumpp, C., Toth, E., Volpi, E., Carr, G., Lupton, C., Salinas, J., Széles, B., Viglione, A., Aksoy, H., Allen, S. T., Amin, A., Andréassian, V., Arheimer, B., Aryal, S. K., Baker, V., Bardsley, E., Barendrecht, M. H., Bartosova, A., Batelaan, O., Berghuijs, W. R., Beven, K., Blume, T., Bogaard, T., Borges de Amorim, P., Böttcher, M. E., Boulet, G., Breinl, K., Brilly, M., Brocca, L., Buytaert, W., Castellarin, A., Castelletti, A., Chen, X., Chen, Y., Chen, Y., Chifflard, P., Claps, P., Clark, M. P., Collins, A. L., Croke, B., Dathe, A., David, P. C., de Barros, F. P. J., de Rooij, G., Di Baldassarre, G., Driscoll, J. M., Duethmann, D., Dwivedi, R., Eris, E., Farmer, W. H., Feiccabrino, J., Ferguson, G., Ferrari, E., Ferraris, S., Fersch, B., Finger, D., Foglia, L., Fowler, K., Gartsman, B., Gascoin, S., Gaume, E., Gelfan, A., Geris, J., Gharari, S., Gleeson, T., Glendell, M., Bevacqua, A. G., González-Dugo, M. P., Grimaldi, S., Gupta, A. B., Guse, B., Han, D., Hannah, D., Harpold, A., Haun, S., Heal, K., Helfricht, K., Herrnegger, M., Hipsey, M., Hlaváčiková, H., Hohmann, C., Holko, L., Hopkinson, C., Hrachowitz, M., Illangasekare, T. H., Inam, A., Innocente, C., Istanbulluoglu, E., Jarihani, B., Kalantari, Z., Kalvans, A., Khanal, S., Khatami, S., Kiesel, J., Kirkby, M., Knoben, W., Kochanek, K., Kohnová, S., Kolechkina, A., Krause, S., Kreamer, D., Kreibich, H., Kunstmann, H., Lange, H., Liberato, M. L. R., Lindquist, E., Link, T., Liu, J., Loucks, D. P., Luce, C., Mahé, G., Makarieva, O., Malard, J., Mashtayeva, S., Maskey, S., Mas-Pla, J., Mavrova-Guirguinova, M., Mazzoleni, M., Mernild, S., Misstear, B. D., Montanari, A., Müller-Thomy, H., Nabizadeh, A., Nardi, F., Neale, C., Nesterova, N., Nurtaev, B., Odongo, V. O., Panda, S., Pande, S., Pang, Z., Papacharalampous, G., Perrin, C., Pfister, L., Pimentel, R., Polo, M. J., Post, D., Sierra, C. P., Ramos, M.-H., Renner, M., Reynolds, J. E., Ridolfi, E., Rigon, R., Riva, M., Robertson, D. E., Rosso, R., Roy, T., Sá, J. H. M., Salvadori, G., Sandells, M., Schaefli, B., Schumann, A., Scolobig, A., Seibert, J., Servat, E., Shafiei, M., Sharma, A., Sidibe, M., Sidle, R. C., Skaugen, T., Smith, H., Spiessl, S. M., Stein, L., Steinsland, I., Strasser, U., Su, B., Szolgay, J., Tarboton, D., Tauro, F., Thirel, G., Tian, F., Tong, R., Tussupova, K., Tyralis, H., Uijlenhoet, R., Beek, R. V., Ent, R. J. V. , Ploeg, M. V. D., Loon, A. F. V., Meerveld, I. V., Oel. P. R. V., Vidal, J.-P., Freyberg, J. V., Vorogushyn, S., Wachniew, P., Wade, A. J., Ward, P., Westerberg, I. K., White, C., Wood, E. F., Woods, R., Xu, Z., Yilmaz, K. K., and Zhang, Y.: Twenty-three unsolved problems in hydrology (UPH) – a community perspective, Hydrol. Sci. J., 64, 1141–1158, https://doi.org/10.1080/02626667.2019.1620507, 2019.

Cao, W., Bowden, W. B., Davie, T., and Fenemor, A.: Multi-variable and multi-site calibration and validation of SWAT in a large mountainous catchment with high spatial variability, Hydrol. Process., 20, 1057–1073, https://doi.org/10.1002/hyp.5933, 2006.

Derive, G.: Estimation de l'évapotranspiration en région sahélienne: synthèse des connaissances et évaluation de modélisations (SISVAT, RITCHIE): application à la zone d'HAPEX-Sahel (Niger), PhD Thesis, Institut National Polytechnique De Grenoble, 189 pp., https://www.theses.fr/2003INPG0052 (last access: 14 October 2023), 2003.

Descroix, L., Guichard, F., Grippa, M., Lambert, L., Panthou, G., Mahé, G., Gal, L., Dardel, C., Quantin, G., Kergoat, L., Bouaïta, Y., Hiernaux, P., Vischel, T., Pellarin, T., Faty, B., Wilcox, C., Abdou, M. M., Mamadou, I., Vandervaere, J.-P., Diongue-Niang, A., Ndiaye, O., Sané, Y., Dacosta, H., Gosset, M., Cassé, C., Sultan, B., Barry, A., Amogu, O., Nka Nnomo, B., Barry, A., and Paturel J.-E.: Evolution of Surface Hydrology in the Sahelo-Sudanian Strip: An Updated Review, Water, 10, 748, https://doi.org/10.3390/w10060748, 2018.

Gbohoui, P.: Modélisation hydrologique de bassins emboîtes du Nakanbé au Burkina Faso dans un contexte de changement global (PhD Thesis), Université de Montpellier ; Institut International d'Ingénierie de l'Eau et de l'Environnement (2iE), 303 pp., https://hal-enpc.archives-ouvertes.fr/tel-03541807 (last access: 14 October 2023), 2021.

Gbohoui, Y. P., Paturel, J.-E., Fowe, T., Mounirou, L. A., Yonaba, R., Karambiri, H., and Yacouba, H.: Impacts of climate and environmental changes on water resources: A multi-scale study based on Nakanbé nested watersheds in West African Sahel, J. Hydrol.-Reg. Stud., 35, 100828, https://doi.org/10.1016/j.ejrh.2021.100828, 2021.

Migliaccio, K. W. and Chaubey, I.: Comment on Cao, W., Bowden, B. W., Davie, T., and Fenemor, A. 2006. "Multi-variable and multi-site calibration and validation of SWAT in a large mountainous catchment with high spatial variability", Hydrol. Process., 20, 1057–1073, Hydrol. Process., 21, 3226–3228, https://doi.org/10.1002/hyp.6491, 2007.

Neitsch, S. L., Arnold, J. G., Kiniry, J. R., and Williams, J. R.: Soil and water assessment tool theoretical documentation version 2009, Texas Water Resources Institute, 2011.

Paturel, J. E., Mahé, G., Diello, P., Barbier, B., Dezetter, A., Dieulin, C., Karambiri, H., Yacouba, H., and Maiga, A.: Using land cover changes and demographic data to improve hydrological modeling in the Sahel, Hydrol. Process., 31, 811–824, https://doi.org/10.1002/hyp.11057, 2016.

Poméon, T., Diekkrüger, B., Springer, A., Kusche, J., and Eicker, A.: Multi-Objective Validation of SWAT for Sparsely-Gauged West African River Basins – A Remote Sensing Approach, Water, 10, 451, https://doi.org/10.3390/w10040451, 2018.

Schuol, J., Abbaspour, K. C., Srinivasan, R., and Yang, H.: Estimation of freshwater availability in the West African sub-continent using the SWAT hydrologic model, J. Hydrol., 352, 30–49, 2008.

Sood, A., Muthuwatta, L., and McCartney, M.: A SWAT evaluation of the effect of climate change on the hydrology of the Volta River basin, Water Int. 38, 297–311, https://doi.org/10.1080/02508060.2013.792404, 2013.

Tan, M. L., Gassman, P. W., Yang, X., and Haywood, J.: A review of SWAT applications, performance and future needs for simulation of hydro-climatic extremes, Adv. Water Resour., 143, 103662, https://doi.org/10.1016/j.advwatres.2020.103662, 2020.

Thiemig, V., Rojas, R., Zambrano-Bigiarini, M., and De Roo, A.: Hydrological evaluation of satellite-based rainfall estimates over the Volta and Baro-Akobo Basin, J. Hydrol., 499, 324–338, https://doi.org/10.1016/j.jhydrol.2013.07.012, 2013.

Yonaba, R., Biaou, A. C., Koïta, M., Tazen, F., Mounirou, L. A., Zouré, C. O., Queloz, P., Karambiri, H., and Yacouba, H.: A dynamic land use/land cover input helps in picturing the Sahelian paradox: Assessing variability and attribution of changes in surface runoff in a Sahelian watershed, Sci. Total Environ., 757, 143792, https://doi.org/10.1016/j.scitotenv.2020.143792, 2021a.

Yonaba, R., Koïta, M., Mounirou, L. A., Tazen, F., Queloz, P., Biaou, A. C., Niang, D., Zouré, C., Karambiri, H., and Yacouba, H.: Spatial and transient modelling of land use/land cover (LULC) dynamics in a Sahelian landscape under semi-arid climate in northern Burkina Faso, Land Use Policy, 103, 105305, https://doi.org/10.1016/j.landusepol.2021.105305, 2021b.

Zanin, P. R., Bonuma, N. B., and Corseuil, C. W.: Hydrosedimentological modeling with SWAT using multi-site calibration in nested basins with reservoirs, RBRH, 23, 26, https://doi.org/10.1590/2318-0331.231820170153, 2018.

Zhang, Y., Xia, J., Shao, Q., and Zhai, X.: Water quantity and quality simulation by improved SWAT in highly regulated Huai River Basin of China, Stoch. Environ. Res. Risk A., 27, 11–27, https://doi.org/10.1007/s00477-011-0546-9, 2013.

Proc. IAHS, 385, 443–448, 2024
https://doi.org/10.5194/piahs-385-443-2024

Open Access PIAHS

Assessing the Accuracy of Multiple Classification Algorithms Combining Sentinel-1 and Sentinel-2 for the Citrus Crop Classification and spatialization of the Actual Evapotranspiration Obtained from Flux Tower Eddy Covariance: Case Study of Cap Bon, Tunisia

Amal Chakhar[1], Rim Zitouna-Chebbi[2], David Hernández-López[1], Rocío Ballesteros[1], Imen Mahjoub[3], and Miguel A. Moreno[1]

[1]Institute of Regional Development, University of Castilla-La Mancha, 02071 Albacete, Spain
[2]Institut National de Recherches en Génie Rural Eaux et Forêts, Université de Carthage, LR16INRGREF02 LRVENC, Rue Hédi Karray, 2080, Ariana, Tunisia
[3]Centre Technique des Agrumes, Université de Carthage, LR16INRGREF02 LRVENC, 8099, Béni Khalled, Tunisia

Correspondence: Amal Chakhar (amal.chakhar@uclm.es)

Received: 25 May 2022 – Revised: 27 January 2023 – Accepted: 30 March 2023 – Published: 19 April 2024

Abstract. Land use and water resources are closely linked. Every single type of land use has a different influence on the hydrologic cycle, consequently impacting the people and the natural resources. The use of advanced technologies, for example monitoring the agricultural resources with remote sensing, offers the possibility to assess the water demand, to know the total cultivated area with the precise distribution of crops and enables the regularly acquisition of data distributed in space and time. The citrus sub-sector is of paramount importance in the Tunisian agricultural sector. The Cap Bon region has the main production area with 75 % of the total citrus area. The possibility of classifying citrus crops is important for water resource management at regional scale and for economic stability. Given the socio-economic importance of the citrus sector in the Cap Bon region, it is very important to have accurate estimation of the total area of citrus plots in this region. Therefore, the main objectives of this current work are:

- To integrate multitemporal synthetic aperture radar SAR data, Sentinel-1, and optical data Sentinel-2, together to determine the best machine learning algorithm that allowed obtaining the most accurate citrus crop classification in the region.
- To study and analyze the temporal signatures of the Normalized Difference Vegetation Index (NDVI) of the classified crops, mainly the citrus, with the purpose to provide the maximum amount of information that allow the differentiation between the crops.
- To study the potential relation between NDVI and Actual Evapotranspiration (ET_a) fluxes measured with the eddy covariance method for a citrus orchard to extrapolate the eddy tower measurements to greater scales.

To achieve these objectives, we evaluated the performance of 22 nonparametric classifiers during the period September 2020–June 2021. Additionally, ET measured by the eddy covariance method was available for the same period, so we tried to find the potential relation between NDVI and Actual Evapotranspiration (ET_a). The results revealed that the best performing classifier is the Support Vector Machine SVM with an accuracy around 91 %. Consequently, our results provided a significant contribution to the citrus classification in the Cap Bon region but can be further improved. Also, the obtained results highlighted the potential to extrapolate accurate ET estimation to larger scales using the vegetation index obtained from Sentinel-2 data.

Published by Copernicus Publications on behalf of the International Association of Hydrological Sciences.

Keywords. Sentinel-1; Sentinel-2; crop classification; citrus crop; Actual Evapotranspiration; Cap Bon

1 Introduction

The citrus sub-sector is of paramount importance in the Tunisian agricultural sector. The Cap Bon region is considered to be the main production area of citrus with 75 % of the total area (Zekri and Laajimi, 2001). Given the socio-economic importance of the citrus sector in the Cap Bon region, it is of principal importance to make an estimate of the total area of citrus plots and their spatial distribution, especially with the absence of up-to-date agricultural census data and their spatial distribution. The information obtained from this type of work will allow the estimation of water requirements of citrus orchards and the improvement of irrigation scheduling methodologies which need to be more precise as water and energy supplies became scarce. Citrus monitoring with remote sensing can provide an affordable and effective method to support field inspection efforts. SAR data can offer valuable information that can enhance optical remote sensing data and can be particularly beneficial for crop classification application.

Various studies have shown that satellite derived canopy cover and Vegetation Indices (VI's) have strong correlation with the crop coefficient K_c (Mateos et al., 2013; Er-Raki et al., 2007). The VI's such as, Normalized Difference Vegetation Index (NDVI), Soil Adjusted Vegetation Index (SAVI) and Transformed Soil Adjusted Vegetation Index (TSAVI) are widely used for of actual crop evapotranspiration (ET_a) estimation (Glenn et al., 2011). Observation of these parameters over regional scale can be possible with high temporal resolution specially with actual satellite remote sensing products such Sentinel-2.

Therefore, the main objectives of this current work are:

- integrate multitemporal synthetic aperture radar SAR data, Sentinel-1, and optical data Sentinel-2, together to determine the best machine learning algorithm that allowed obtaining the most accurate citrus crop classification in the region.
- study and analyze the temporal signatures of the NDVI of the classified crops, mainly the citrus, with the purpose to provide the maximum amount of information that allow the differentiation between the crops.
- study the potential relation between NDVI and observed ET_a on a citrus orchard to extrapolate the eddy tower measurements to greater scales.

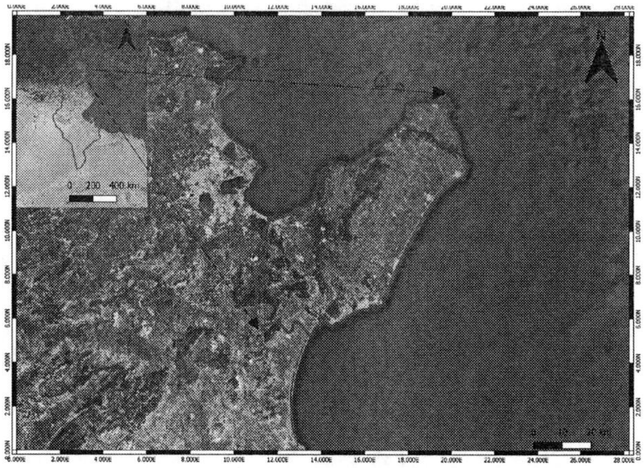

Figure 1. Cap Bon Tunisia: The study area (caption from © Google Maps).

2 Materials and methods

2.1 Presentation of the study area

This study was carried out on the scale of the entire Cap Bon region (Fig. 1). The Cap Bon region covers 2840 km^2. As for the climate, it varies from subhumid to semi-arid, and the average annual amounts of rain in the region vary between 390 and 630 mm. The study area is characterized by variable evaporation in time. The estimated total annual evapotranspiration exceeds 1100 mm (Hamouda, 2008). The dominant agricultural products are citrus, vineyards, spices, and vegetable culture (basically tomato, pepper, and strawberry).

2.2 Field truth data collection

2.2.1 Land use

During this study, we conducted land visits and interviews with some farmers, we collected the coordinates of the locations of citrus plots with a portable GPS. In total we made three visits, in 18, 25 and 31 August 2021. In addition, most of acquired truth data have been carried out based on a laborious interpretation of the Google Earth Map based on two main criteria: First criterion: The windbreaks surround the citrus orchards. Second criterion: the planting density of citrus fruits which is usually of the order of 3 out of 5 and 5 out of 6. Finally, we were able to delimit 1857 plots. Based on the distribution of major crops in the region, the following vegetable coverage classes were selected: citrus and open field.

2.2.2 Actual evapotranspiration

Long term observation of ET_a on citrus in the Cap Bon region have been acquired. A flux tower measuring ET_a has been installed since 2004. Measurements are still in progress. Description of the orchard, instruments and calculations methods are in Zitouna-Chebbi et al. (2017).

2.3 Remote sensing data: Sentinel-1 and Sentinel-2 dataset

Currently the new cloud solution the Google Earth Engine GEE is receiving an increasing attention as a new settlement to search, extract and download Earth Observation EO data in a way that was not possible before (rapid access to large spatio-temporal data). Therefore, we decided to use GEE to download the Sentinel-1 and Sentinel-2 dataset for the period September 2019–June 2021. For the Sentinel-2 we used a command to calculate and download the NDVI scenes directly instead of downloading the red and NIR bands and calculate the NDVI for each available date. As for the Sentinel-1 data, we downloaded the VV, VH and angle of incidence bands which were processed by the GEE default processing streams.

2.4 Classification methodology

One of the objectives of this study is to evaluate the effectiveness of the incorporation of Sentinel-1 time series data with optical remote sensing data of Sentinel-2 for citrus classification. We performed the evaluation of the performance of the 22 non-parametric classifiers with the integration of SAR data, to classify citrus trees. We applied the developed methodology (Chakhar et al., 2021) to Cap Bon case study. Therefore, following we present the adapted methodology:

- *Data collection.* We downloaded optical (Sentinel-2) and SAR (Sentinel-1) data during the period September 2019–June 2021 from GEE platform. Concerning the ground truth data, were collected by visual interpretation using very high resolution (VHR) Google Earth imagery and land visits.

- *Data preparation.* In pursuit of the preprocessing process, we extracted two features from the Sentinel-1 data set: VH, VV. We computed the mean of these features at the level of plot (plot-based approach) for each available date of Sentinel-1 and Sentinel-2 time series.

- *Assessing the accuracy of multiple classification algorithms.* In this step set of 22 classification algorithms was evaluated including decision trees, discriminant analysis, support vector machines, nearest neighbor and ensemble classifiers. These algorithms are the most frequent algorithms used in the literature. The Classification Learner application of Matlab® was utilized with the aim of calibrating and validating the different algorithms. The Classification Learner app can be found in the latest versions of the Statistics and Machine Learning toolbox of Matlab. Given that one of the critical issues for the use of SAR time series is the choice of the features used for the classification, we decided to evaluate VV and VH SAR features. The approach is about the combination of these feature with NDVI. Three classification scenarios based on the selection of the input feature were then applied for all the available dates:

  - The first scenario: only the optical feature was used as input.

  - The second scenario: VV and the VH channels derived from SAR data.

  - The third scenario: NDVI, the VV and VH channels.

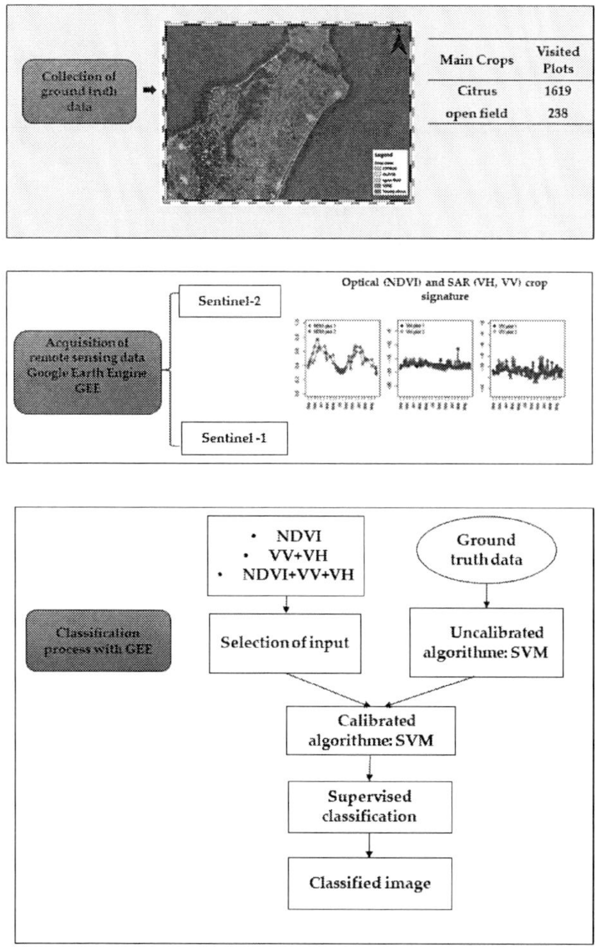

Figure 2. Classification methodology scheme (caption from © Google Maps).

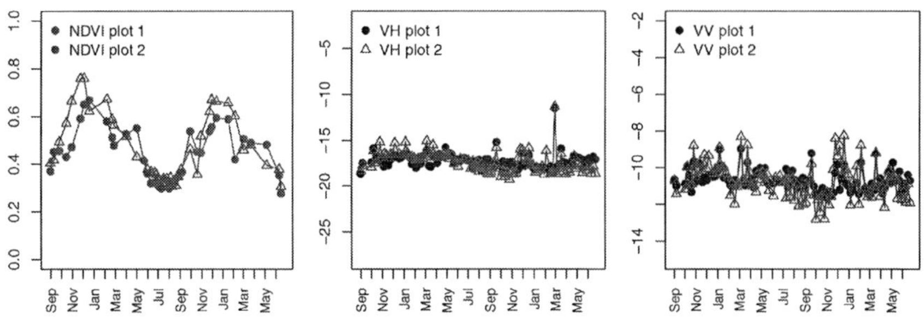

Figure 3. NDVI, VV and VH time series of two selected citrus plots.

Table 1. Overall Accuracy (OA) (expressed in (%)) of the different classification methods with the available information.

	Decision Trees			Discriminant Analysis		SVM					
	M1	M2	M3	M4	M5	M6	M7	M8	M9	M10	M11
NDVI	84.4	87.4	87.9	90.9	89.1	90.8	91.5	88.7	88.2	90.4	88
VV+VH	87.1	90.2	91.4	91.7	fail	91.9	90.6	89.8	87.2	91.5	90.3
NDVI+VV+VH	88	90.5	90	91.5	fail	91.9	91.9	90.5	87.1	91.4	90.4
	Nearest Neighbor Classifiers					Ensemble Classifiers					
	M12	M13	M14	M15	M16	M17	M18	M19	M20	M21	M22
NDVI	89.3	89.8	89.2	88.9	89.7	90.3	90.5	90.1	90.4	91.4	81.2
VV+VH	86.9	90.8	89.3	90.3	90.9	91.1	91.5	91.5	91.8	88.5	85.6
NDVI+VV+VH	87.9	90.7	89.4	89.7	90.6	91.3	91.8	90.8	91.1	88.3	84.4

Classification process with GEE: The proposed classification method is presented in Fig. 2. The applied methodology was entirely based on the use of the GEE cloud platform environment. The method started with the collection of the ground truth data. The second step consisted in downloading the available dataset Sentinel-1 and -2 corresponding for the study area and the required period from the GEE catalogue and building a consistent time-series. Afterward the evaluation of the accuracy of 22 classification algorithms was done and the identification of the most accurate classifiers for all the scenarios was performed and revealed the outperformance of SVM. Finally, the classification process was accomplished with the SVM algorithm within the GEE platform, and we have tried it with the three scenarios (NDVI, VV+VH, NDVI+VV+VH).

3 Results

3.1 Analysis of Temporal Signatures of citrus

Figure 3 shows NDVI, VV, VH time series for two selected plots for the period September 2019–June 2021. The graph showed the minimum and maximum NDVI founded values, respectively, in Summer (June–August) and in Autumn–Winter (January–February/October–December), the same statement was made in Vanella et al. (2020). The time-series of NDVI of these two selected plots have demonstrated a standard annual shape (Vanella et al., 2020; Sawant et al., 2016) originating from the sum of two contributions (i) the stable vegetation dynamic of citrus trees and (ii) the presence of grassed soil. Days of minimum NDVI were in the summer months (June–August) justified by low canopy increase due to peak of water stress (Sawant et al., 2016).

3.2 Evaluation of Classification Methods for Crop Classification

The results of using only Sentinel-2 information (NDVI), only Sentinel-1 information (VV+VH) and a combination of both sources of information (NDVI+VV+VH) were obtained for the evaluated classification methods. A comparison between the 22 classification methods with the different combinations of available information is shown in Table 1. Among the tested classifications algorithms and for the three scenarios, the best algorithm is the Support Vector Machine SVM, with an Overall Accuracy (OA) = 91.5 % for the first scenario, OA = 91.9 % for the second scenario and OA = 91.9 % for the third scenario.

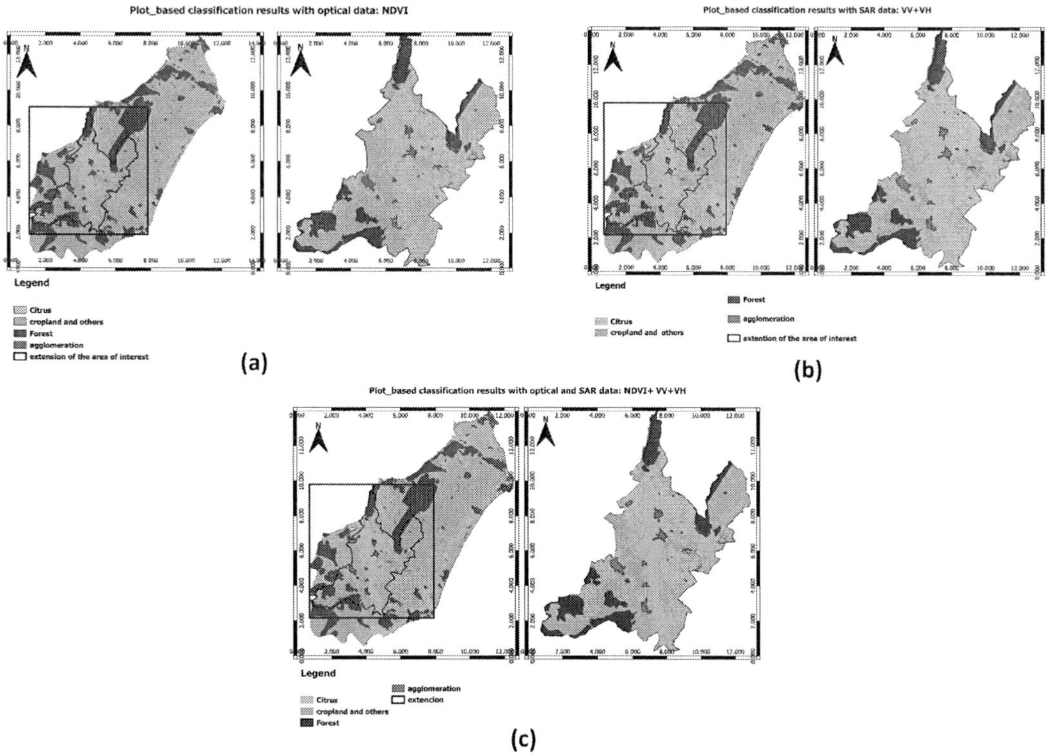

Figure 4. Cap Bon citrus maps obtained by the SVM classifiers for each adopted scenarios: **(a)** with NDVI as input data, **(b)** with VV+VH as input and **(c)** with NDVI+VV+VH as input data.

3.3 Evaluation of Classification results

Figure 4 shows the classification maps of the three adopted scenarios using the SVM classifiers. These classification maps did not offer high OA. The first scenario provided an OA = 67 %, while the second scenario supplied a slight high OA than the first scenario equal to 69 % and the third scenario had an OA = 73 %. However overly all the classification results overestimate the total area of citrus plots.

3.4 Relation between NDVI and ET_a measured with the eddy.

With the objective to extrapolate ET_a to larger scales using the vegetation index obtained from Sentinel-2 data, we studied, for the same period of classification 2019–2021 the potential relation between NDVI, and ET_a fluxes measured with the eddy covariance method installed in an experimental citrus orchard located in the study area. Figure 5 showed that linear relationships between NDVI and ET_a with a high coefficient of determination $R^2 = 0.89$ indicating a promising result that can be extrapolated to larger scale.

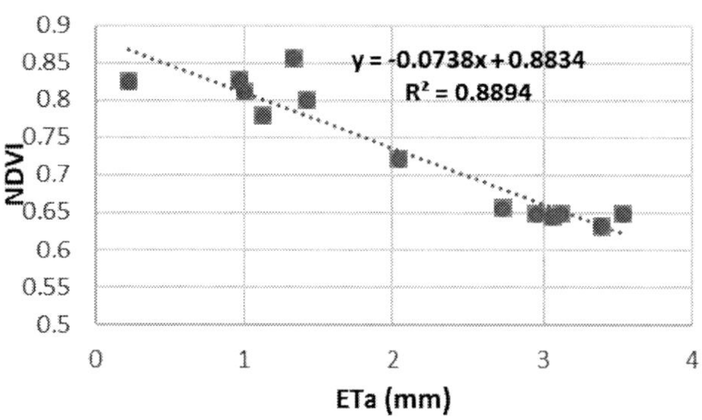

Figure 5. Vegetation index NDVI–ET_a relationship of an experimental plot in the study.

4 Conclusions

Image classification results were obtained with GEE for the three adopted scenario (NDVI; VV+VH; NDVI+VV+VH). We found that for all these classification scenarios there is an overestimation of the total area of citrus plots. Therefore, we think that these classification results can be further improved because GEE does not offer the option to work with all the time-series of the input data, instead it reduces the selected dataset to a single image containing each image's information to perform the classification. Consequently, for future classification tasks we are going to try to use all the time-series dataset by adapting the Datacubes methodology combined with the specific Local Nested Grid. Concerning the relation between NDVI and ET_a fluxes measured with the eddy covariance indicates promising result that can be extrapolated to larger scale.

Data availability. The underlying research data is currently not publicly available due to privacy. The Sentinel-1 and Sentinel-2 images, scenes are provided by https://dataspace.copernicus.eu/ (Copernicus Open Access Hub, 2023).

Author contributions. Conceptualization, DHL, MAM and RZC; data curation, AC; formal analysis, AC, DHL and MAM; funding acquisition, DHL, RB and MAM; investigation, AC, DHL and IM; methodology, AC, DHL and MAM; project administration, RB; resources, RB and MAM; software, DHL, and MAM; supervision, DHL and MAM; visualization, AC; writing – original draft, AC; writing – review and editing, AC and MAM. All authors have read and agreed to the published version of the manuscript.

Competing interests. The contact author has declared that none of the authors has any competing interests.

Disclaimer. Publisher's note: Copernicus Publications remains neutral with regard to jurisdictional claims made in the text, published maps, institutional affiliations, or any other geographical representation in this paper. While Copernicus Publications makes every effort to include appropriate place names, the final responsibility lies with the authors.

Special issue statement. This article is part of the special issue "IAHS2022 – Hydrological sciences in the Anthropocene: Past and future of open, inclusive, innovative, and society-interfacing approaches". It is a result of the XIth Scientific Assembly of the International Association of Hydrological Sciences (IAHS 2022), Montpellier, France, 29 May–3 June 2022.

Acknowledgements. This work was supported by H2020-MSCA-RISE-2018 ACCWA project and Water JPI 2018 FLUXMED project.

Financial support. This research has been supported by the University of Castilla–La Mancha for funding a Ph.D. grant for Amal Chakhar ("Predoctoral contract for research staff in training within the framework of the I+D+i plan, has been co-financed by the European Social Fund ESF within the 2014–2020 Operational Plan of Castila-La Mancha").

Review statement. This paper was edited by Christophe Cudennec and reviewed by two anonymous referees.

References

Chakhar, A., Hernández-López, D., Ballesteros, R., and Moreno, M. A.: Improving the Accuracy of Multiple Algorithms for Crop Classification by Integrating Sentinel-1 Observations with Sentinel-2 Data, Remote Sens., 13, 1–21, https://doi.org/10.3390/rs13020243, 2021.

Copernicus Open Access Hub: Sentinel-2 archives, https://sentinels.copernicus.eu/web/sentinel/missions/sentinel-2/data-products, last access: 12 December 2023.

Er-Raki, S., Chehbouni, A., Guemouria, N., Duchemin, B., Ezzahar, J., and Hadria, R.: Combining FAO-56 Model and Ground-Based Remote Sensing to Estimate Water Consumptions of Wheat Crops in a Semi-Arid Region, Agr. Water Manage., 87, 41–54, https://doi.org/10.1016/j.agwat.2006.02.004, 2007.

Glenn, E. P., Christopher, M. U. N., Hunsaker, D. J., and Nagler, P. L.: Vegetation Index-Based Crop Coefficients to Estimate Evapotranspiration by Remote Sensing in Agricultural and Natural Ecosystems, Hydrol. Process., 25, 4050–4062, https://doi.org/10.1002/hyp.8392, 2011.

Hamouda, M. F. B.: Approche Hydrogéologique et Isotopique Des Systèmes Aquifères Côtiers Du Cap Bon: Cas Des Nappes de La Côte Orientale et d'El Haouaria, Tunisie, PhD, INAT, 2008.

Mateos, L., González-Dugo, M. P., Testi, L., and Villalobos, F. J.: Monitoring Evapotranspiration of Irrigated Crops Using Crop Coefficients Derived from Time Series of Satellite Images. I. Method Validation, Agr. Water Manage., 125, 81–91, https://doi.org/10.1016/j.agwat.2012.11.005, 2013.

Sawant, S. A., Chakraborty, M., Suradhaniwar, S., Adinarayana, J., and Durbha, S. S.: TIME SERIES ANALYSIS OF REMOTE SENSING OBSERVATIONS FOR CITRUS CROP GROWTH STAGE AND EVAPOTRANSPIRATION ESTIMATION, Int. Arch. Photogramm. Remote Sens. Spatial Inf. Sci., XLI-B8, 1037–1042, https://doi.org/10.5194/isprs-archives-XLI-B8-1037-2016, 2016.

Vanella, D., Consoli, S., Ramírez-Cuesta, J. M., and Tessitori, M.: Suitability of the MODIS-NDVI Time-Series for a Posteriori Evaluation of the Citrus Tristeza Virus Epidemic, Remote Sens., 12, 12, https://doi.org/10.3390/rs12121965, 2020.

Zekri, S. and Laajimi, A.: Etude de La Compétitivité Du Sous-Secteur Agrumicole En Tunisie", Cahier Opt. Médit, January 2001, 9–16, https://www.researchgate.net/publication/235980822_Etude_de_la_Competitivite_du_Sous-secteur_Agrumicole_en_Tunisie (last access: 12 December 2023), 2001.

Zitouna-Chebbi, R., Mahjoub, I., Mekki, I., and Ben Mechlia, N.: Comparing evapotranspiration rates estimated from atmospheric flux, soil water balance and fao56 method of a small orange orchard in Tunisia, Acta Hortic., 1150, 23–30, https://doi.org/10.17660/ActaHortic.2017.1150.4, 2017.

Proc. IAHS, 385, 449–455, 2024
https://doi.org/10.5194/piahs-385-449-2024

Continental fishery and risk of contamination of the halieutic ecosystems in the gbaga lagoon channel (Benin-Togo coastal zone)

Mahougnon Fidèle Ahéhéhinnou Yêdo[1], **Ernest Amoussou**[2], **Djidjoho Mathieu Maurice Ahouansou**[3], **Domiho Japhet Kodja**[1], **and Sourou Henri Totin Vodounon**[2]

[1]Pierre Pagney Laboratory: Climate, Water, Ecosystems and Development (LACEEDE), University of Abomey-Calavi, Department of Geography and Land Management, Benin, BP 1338 Abomey-Calavi, Benin
[2]Pierre Pagney Laboratory: Climate, Water, Ecosystems and Development (LACEEDE), University of Parakou, Department of Geography and Land Management, Benin, BP 123 Parakou, Benin
[3]Laboratoire of Hydrology and Water Management (LHME), University of Abomey-Calavi/ School of Environmental Management, Faculty of Agronomic Sciences, Benin, 01 B.P526 Cotonou, Benin

Correspondence: Mahougnon Fidèle Ahéhéhinnou Yêdo (fideleyedo@gmail.com)

Received: 28 May 2022 – Revised: 5 December 2023 – Accepted: 13 December 2023 – Published: 19 April 2024

Abstract. The coastal zones are areas where a very high number of lagoon-marine species reproduce, grow or transit, which are most often the object of halieutic exploitation. The present study aims to analyze the risk of contamination of fisheries ecosystems and its consequences on the evolution of aquatic resources in the coastal zone, more precisely from southwest Benin to southeast Togo. The methodology adopted is based on the collection and processing of raw data such as: hydrometric data, granulometric data, climatological data, data from fauna and flora inventories, completed by those from in situ measurements and those from laboratory analyses made from sediment samples and some species caught during fishing activities during the different hydrological seasons. The analysis of the results revealed that the pollution of the lagoon channel in Cd, As, Pb and some major ions such as Cl^-, NO_3^- and PO_4^{3-} are responsible for its eutrophication. Furthermore, organic matter and high concentrations of SS lead to the covering of aquatic environments, particularly spawning grounds and the smothering of eggs through the reduction of essential metabolic exchanges (oxygen and nutrient supply, evacuation of metabolic waste) through the reduction of the concentration of dissolved oxygen within the substrate. Trace element analysis in the flesh of Captain (Polydactylus quadrifilis), white grouper (Epinephelus aeneus), tilefish (Tilapia zillii), and white catfish samples revealed their trace element contamination levels. Behavioral effects (abandonment of shelter areas and avoidance reaction); sublethal effects (short-term reduction in feeding rate; minor physiological stress; increase in coughing frequency; increase in the breathing rhythm; moderate physiological stress; major; delayed hatching; reduced number of fish; mortality) were recorded. According to 89 % of the fishermen, the decrease of the fished fish species is registered during the dry periods, when the channel has a high level of contaminating heavy metals and major ions.

Keywords. Lagoon ecosystems; conservation; aquatic life

1 Introduction

The coasts ecosystems are more and more affected by human activities. Industrialization and urban, agricultural or tourist development in these areas are the main causes of the increase in pressure on these environments (Lougbégnon and Houessou, 2007). Indeed, the geographical location of the Gbaga hydrosystem makes it a defluent of human discharges from the Mono River (Ahéhéhinnou Yêdo, 2020a). Certainly, this reservoir of receptacle of liquid and solid flows, exposes its ecosystems to high risks of pollution (Hodabalo et al., 2014). Thus, the Gbaga lagoon system is characterized by its concentration of heavy metals and major ions (Ahéhéhinnou Yêdo, 2020b). The ichthyological fauna associated with this hydrosystem is therefore not immune to this contamination. Faced with the objectives of the MAB-UNESCO (Man and the Biosphere) program, this study aims to analyze the risk of pollution of the Gbaga lagoon channel and the consequences of the contamination of its fisheries ecosystems on their health.

2 Study area

The study area is part of the Benin-Togolese coastal plain that covers a small part of Togo and Benin, and is located in the lower Mono River valley. It is located between parallels 6°15′ and 6°20′ N latitude and meridians 1°35′ and 1°46′ E longitude (Fig. 1). It is a wetland straddling the togolese and beninese territories with an area of approximately 7272.69 ha (Ahéhéhinnou Yêdo, 2020a). Being a flood bed of the Mono River, it contained seventy-six species of fish belonging to thirteen different families (Lougbégnon and Hesou, 2007). This area belongs to the West Complex Ramsar Site 1017 (Benin) and the Coastal Wetlands Ramsar Site of Togo (Site 1722). On the Togo side, the lagoon channel communicates with the Atlantic Ocean by describing a wide meander giving birth to the Aného lagoon.

3 Datas and methods

The analyses of the rate of organic matter and the rate of total organic carbon; of metallic traces elements; major ions; organic matter and dissolved organic carbon contained in the sediments and the four species of fish studied were carried out at the Laboratory of Quality Control of Water and Food (LCQEA) of the National Direction of Public Health (DNSP); at the Laboratory Central for the Control of Food Safety (LCSSA) of the MAEP and at the Laboratory of Soil, Water and Environmental Research (LSSEE) of the INRAB on 45 representatives samples of mud. They concern ETM (Cd, Pb, As); major ions (Cl^-, NO_3^- and PO_4^{3-}); DO and COT. These data allowed us to determine the degree of pollution of the channel and to analyze the consequences that this has on the biosphere of the Gbaga Channel. The data were obtained from analyses conducted in the LCQEA and LCSSA (Laboratory Central for the Control of Food Safety) laboratories of the MAEP using samples taken during high and low water periods.

The bioaccumulation factors of the fish studied were estimated from field data using the formula of Thomann et al. (1992): BSF = C_(b,L)/C_(s,oc); with : C_(b,L) = concentration of the substance in the organism (b) normalized to lipids;

C_(s,oc) = concentration of toxicant in silt normalized to total organic carbon. This bioaccumulation method allowed for a perspective analysis of the risk of channel eutrophication on the fish species studied. These data were processed using R software (Version 4.01). The SEPOCA analysis model was used to highlight the weaknesses in terms of failure to apply the environmental protection laws in force, the strengths in terms of success, the opportunities and the obstacles for a better appreciation of the ichthyological population of the Gbaga lagoon channel.

4 Results

4.1 Variation of the ecological monitoring parameters of the Gbaga lagoon channel during the different hydrological periods

The pollution of the lagoon channel can be evaluated by the variation of metallics traces elements and major ions on the one hand and that of dissolved oxygen on the other hand during the different hydrological periods.

4.1.1 Variation of trace metal elements and major ions during different hydrological periods

The Gbaga lagoon channel is subject to several anthropogenic pressures such as the discharge of pollutants. Toxic chemicals attach to or are absorbed by sediment particles, then transported and deposited in the bottom of the Gbaga channel resulting in eutrophication (Fig. 2).

From the analysis of Fig. 2, it is to be understood that the pollutants of the Gbaga channel mud have a very significant variation during the hydrological periods (high and low water periods) especially for trace elements Pb, As and major ions Cl^-, NO_3^-, PO_4^3; on the other hand, the pollutants Cd, MO and TOC have a not very significant variation during the high and low water periods. During low water periods, Pb concentrations in the channel exceed the legal concentration and vary between 159.99 and 273.64 $mg\,kg^{-1}$ with an average of 203.85 $mg\,kg^{-1}$. This explains why the ecosystems of the channel are subject to high levels of lead pollution during this period. However, during high water periods, Pb concentrations are relatively consistent with the very legal norm and vary between 16.57 and 58.32 $mg\,kg^{-1}$ es-

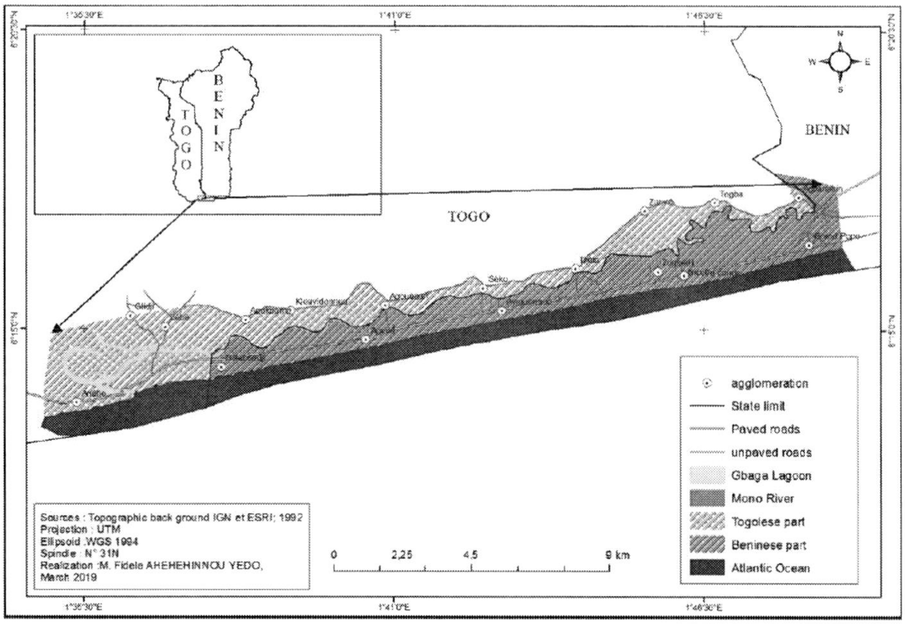

Figure 1. Geographic location of the Gbaga lagoon channel.

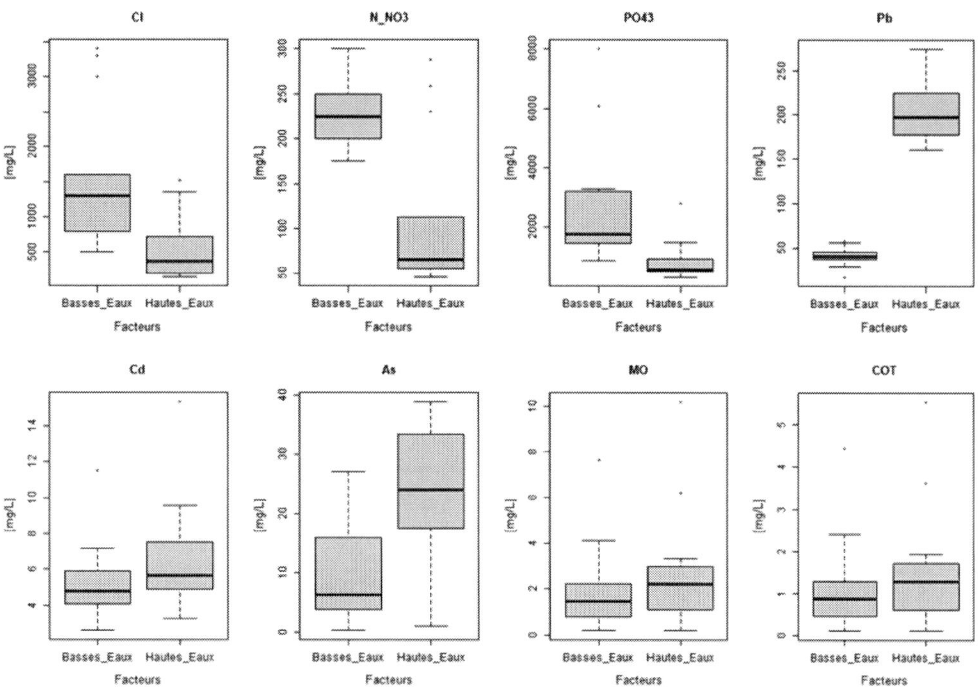

Figure 2. Temporal variation of available pollutants in the Gbaga Channel silt during high and low water periods. Source: Laboratory data processing, 2020.

Table 1. Average TME content in fresh fish caught in the Gbaga Channel.

ETM	Unity	Method used	White catfish	Grouper	Tilapia	Captain	Standards
Lead (Pb)	$mg\,kg^{-1}$	Chloroform extraction	±0.271	±0.226	±0.213	±0.177	0.2
Cadmium (Cd)	$mg\,kg^{-1}$	Chloroform extraction	±0.045	±0.0484	±0.026	±0.026	0.05
Arsenic (As)	$mg\,kg^{-1}$	Pyridine	±0.002	±0.002	±0.002	±0.002	0.1

Source: LCSSA data analysis, June 2020.

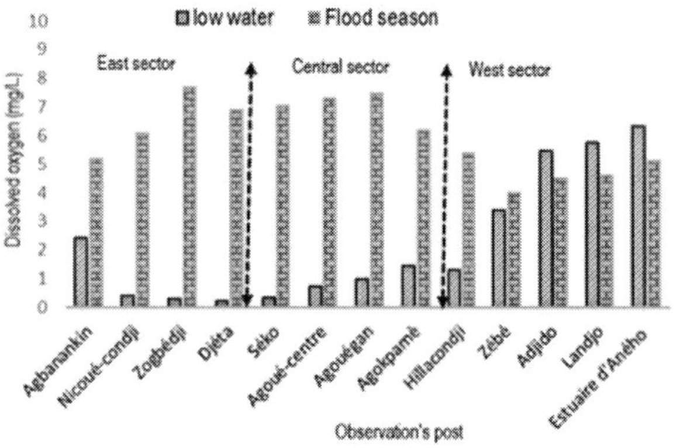

Figure 3. Temporal variation of dissolved oxygen in the Gbaga Channel during high and low water periods. Source: Field campaigns, September 2019 and January 2020.

Figure 4. Illustrates the behavior of crabs in the face of respiratory problems. Photo 1: Behavioral changes in crabs at Nicouécondji Shot: Ahéhéhinnou Yêdo, September 2020.

pecially in its western sector. As for Arsenic concentrations, they fluctuate between 1.4 and 27.2 $mg\,kg^{-1}$ with an average of 11.32 $mg\,kg^{-1}$ during the high water period. In fact, the lowest concentrations recorded are for Cadmium, which are lower than 2.613 $mg\,kg^{-1}$, whatever the hydrological period.

4.1.2 Dissolved oxygen variation in the Gbaga hydrosystem

Dissolved oxygen concentrations characterize and differentiate the sectors and hydrological periods of the channel. The following Fig. 3 presents the variation of dissolved oxygen in the Gbaga hydrosystem.

From the analysis of Fig. 3, it appears that the eastern and central sectors are very poor in dissolved oxygen during high water periods with an average of 0.76 $mg\,L^{-1}$.

While the western sector is very rich in dissolved oxygen with an average of 5.23 $mg\,L^{-1}$. On the other hand, the western sector is never devoid of dissolved oxygen, regardless of the hydrological period, just like the Agbanankin observation station. Variations in dissolved oxygen are dependent on exchanges with oceanic and continental waters and with the atmosphere.

It is to be understood that in the face of dissolved oxygen deficits in the aquatic environment, the crabs come to inhale oxygen on the hydrophytes, in order to survive during periods of high water.

4.2 Contamination of fish species with traces metals

Trace element analysis in the flesh of Captain (Polydactylus quadrifilis), white grouper (Epinephelus aeneus), Tilapia zillii, and white catfish allowed us to determine the average level of trace element contamination of these fish in the Gbaga Channel (Table 1).

Only the captain is safe from the effects that Pb produces in the flesh of fish. But there is a very strong alert of the effects that Cd produces in the Grouper. But these four endangered species are still safe from the effects of As in the waters of the Gbaga Channel.

The fish fauna is vulnerable to the phenomena of bioaccumulation and biomagnification of pollutants (Fig. 5).

The bioaccumulation estimate reveals that tilapia are the species most vulnerable to as bioaccumulation risk and groupers are also the least vulnerable to as bioaccumulation risk in the Gbaga Channel.

As bioaccumulation risk and groupers are also the least vulnerable to As bioaccumulation risk in the Gbaga Channel. It appears that only the captain is safe from the effects of Pb in the flesh of the fish. But we note a very strong alert of the effects that Cd produces in the Grouper. These species of fish being on the verge of extinction in the Gbaga channel, one can therefore deduce that beyond these concentrations, the species still present will end up having an emblaved reproduction.

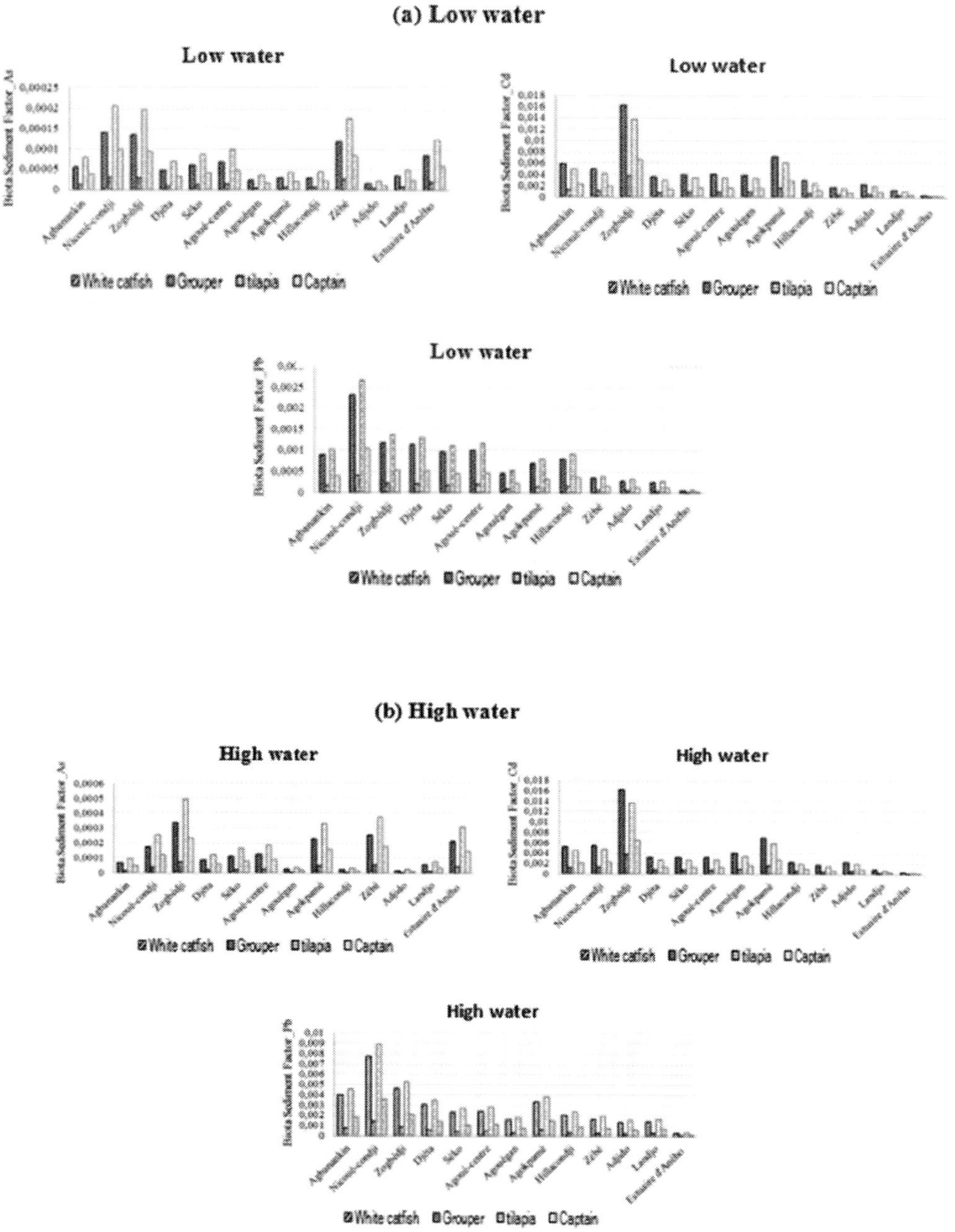

Figure 5. Risk of bioaccumulation of ETM in fish flesh during high and low water periods in the Gbaga Channel. Source: Laboratory data processing, 2020.

5 Discussion

At the end of this study, the Gbaga lagoon channel contains in its silt, a significant concentration of trace metal elements and major ions. These results obtained are in agreement with those obtained from the studies of Shaker (2016), who had stated that the lowest concentrations of heavy metals were recorded for Cd with less than $3\,mg\,kg^{-1}$, regardless of the season and sampling location. It is also worth noting that the silt of the Gbaga channel is less polluted with heavy metals than that of Lake Ahémé in Benin (Adéchina, 2018). This could be explained by the topographic aspect of the Gbaga Channel whose waters flow unlike those of Lake Ahémé whose waters are stagnant. This also corroborates the results obtained from the studies of Moussa Moumouni Djermakoye (2005) who demonstrated that the mineralization of the silt is strong, but the lowest values were recorded during the flood, due to the phenomenon of water dilution.

In addition, the pollution of the hydrosystems are the basis of the contamination of the ichthyological fauna and the constraints that hinder their blooming and development. These results obtained reinforce the results obtained from studies conducted on Lake Ahémé by Hounkpè (2017). These manifestations which are more frequent favor the transformation of water into brown or cloudy water with proliferation of hydrophytes such as algae which ecologically unbalance the Gbaga channel through their strong consumption of dissolved oxygen. The manifestations of this imbalance are complicated by excessive phosphate (PO_4^{3-}) and nitrogen (NO_3^-) inputs.

6 Conclusions

This study shows that eutrophication of the channel is occurring due to pollution by Cd, As, Pb and certain major ions such as Cl^-, NO_3^- and PO_4^{3-}. High concentrations of organic matter and suspended solids cause aquatic habitats to cover over, particularly spawning grounds, and suffocate eggs by reducing essential metabolic exchanges (supply of oxygen and nutrients, evacuation of metabolic waste) through lower concentrations of dissolved oxygen in the substrate. This zone is made up of highly fragile ecosystems, and every species found there is fortuitously threatened and endangered.

Code availability. No licensed data processing software was used for this article. For its writing, we used Excèl and the R software (Version 4.01 free).

Data availability. The data related to TME (Cd, Pb, As); major ions (Cl^-, NO_3^- and PO_4^{3-}), MO and TOC come from the analyses made on the samples of mud and fish of the channel in the laboratories LCQEA and LCSSA of the MAEP and LSSEE of the INRAB; taken during the periods of high water and low water. The results of these analyses are delivered on behalf of MFAY in 2020 by the heads of these institutes and are available at their levels and not online. The socio-economic data were obtained through our own investigations of the Gbaga Channel sectors and direct observations. In order to appreciate the evolution of the populations of these fish species, these data were completed by data from previous work carried out by the NGOs: Cooperative de Solidarité Partenê Group (COSOL PG – NGO); Actions d'aide Humanitaire pour le Développement (AHD NGO); the NGO Pro-Sol and the Association FAH-Gbaga (Federation of Associations of the Gbaga riverine villages); the NGO Borne Fonden, the NGO OFEDI and the NGO Action Plus, ECO-ECOLO.

Author contributions. MFAY designed the study, collected the samples that were analyzed by the different laboratories mentioned above, analyzed and processed the data, and wrote the manuscript. MMA developed the model and performed the simulations. EA provided the scientific orientations for this study. MMA, DJK, SHTV contributed to the improvement of the adopted methodology, the analysis of the data and the proofreading of this work.

Competing interests. The contact author has declared that none of the authors has any competing interests.

Disclaimer. Publisher's note: Copernicus Publications remains neutral with regard to jurisdictional claims made in the text, published maps, institutional affiliations, or any other geographical representation in this paper. While Copernicus Publications makes every effort to include appropriate place names, the final responsibility lies with the authors.

Special issue statement. This article is part of the special issue "IAHS2022 – Hydrological sciences in the Anthropocene: Variability and change across space, time, extremes, and interfaces". It is a result of the XIth Scientific Assembly of the International Association of Hydrological Sciences (IAHS 2022), Montpellier, France, 29 May–3 June 2022.

Acknowledgements. The authors thank the anonymous reviewers who contributed significantly to the scientific quality of this article by devoting their time to its critical evaluation.

Review statement. This paper was edited by Christophe Cudennec and reviewed by two anonymous referees.

References

Adéchina, A. R. M. A.: Influence of paleoclimatic, geological and anthropogenic factors on the environment of the Ouémé Delta (Benin), PhD Thesis in Hydrology and Integrated Water Resources Management, Université of Abomey-Calavi, Bénin, 200 pp., 2018.

Ahehehinnou Yêdo, M. F., Amoussou, E., Allagbe, B. S. Y., and Vissin, E. W.: Dynamique Spatio-Temporelle de l'Occupation des Berges de la Lagune Gbaga (Afrique de l'Ouest), International Journal of Progressive Sciences and Technologies, ISSN 2509-0119, 22, 235–243, 2020a.

Ahehehinnou Yêdo, M. F.: Dynamique éco-sédimentaire du chenal lagunaire Gbaga dans le contexte de la variabilité climatique en Afrique de l'Ouest, Thèse de doctorat 799-2020/EDP-ECD, de l'université d'Abomey-Calavi, Bénin, 280 pp., 2020b.

Hodabalo, D. S., Ouro-Sama, K., Gnandi, K., Afiademanyo, K. M., and Bowessidjaou, E. J.: Assessment and health risks of heavy metal bioaccumulation in fish species from the Togolese lagoon system, VertigO, 14, https://doi.org/10.4000/vertigo.15093, 2014.

Hounkpe, J. B.: Morpho-sedimentary environment and trace element geochemistry of water and sediments of Lake Ahémé in southern Benin, PhD thesis in Earth and Universe Sciences, Environmental Geosciences option, University of Abomey-Calavi, Benin, 200 pp., 2017.

Lougbegnon, T. O. and Houessou, L. G.: Projet transfrontalier d'appui à la gestion participative des ressources de la chenal Gbaga au sud-est du Togo et au sud-ouest du Benin, Rapport pour les études forestières et aménagement participatif des mangroves, December, 36 pp., 2007.

Moussa Moumouni Djermakoye, H.: Physico-chemical and bacteriological characteristics and impact on surface water and groundwater, Doctor of Pharmacy thesis, Faculty of Medicine, Pharmacy and Odonto-Stomatology, University of Bamako, 135 pp., 2005.

Shaker, A. A. A.: Impact of trace metals on the environment and the contribution of flow cytometry to the study of the functioning of the Étueffont landfill lagoons (Belfort region, France), Doctoral thesis in science and technology from the University of Franche-Comté, 339 pp., 2016.

Thomann, V. R., Connelly, J. P., and Thomas, F. P.: An equilibrium model of organic chemical accumulation in aquatic food webs with sediment interactions," Environ. Tox. Chem., 11, 615–629, 1992.

Proc. IAHS, 385, 457–463, 2024
https://doi.org/10.5194/piahs-385-457-2024

Open Access PIAHS

IAHS2022 – Past and future of open, inclusive, innovative, and society-interfacing approaches

SAR, SARin, RDSAR and FF-SAR Altimetry Processing on Demand for Cryosat-2, Sentinel-3 & Sentinel-6 at ESA's Altimetry Virtual Lab

Jérôme Benveniste[1], Salvatore Dinardo[2], Luciana Fenoglio-Marc[3], Christopher Buchhaupt[4], Michele Scagliola[5,a], Marcello Passaro[6], Karina Nielsen[7], Marco Restano[8], Américo Ambrózio[9], Giovanni Sabatino[10], Carla Orrù[10], and Beniamino Abis[10]

[1]European Space Agency-ESRIN, Via Galileo Galilei, 00044 Frascati, Italy
[2]Collecte Localisation Satellites, Toulouse, France
[3]Institute of Geodesy and Geoinformation, University of Bonn, Germany
[4]Cooperative Institute for Satellite Earth System Studies, University of Maryland, College Park, MD, USA
[5]Aresys Srl, Via Flumendosa 16, 20132 Milan, Italy
[6]DGFI, Technische Universität München, Munich, Germany
[7]DTU Space, Technical University of Denmark, Elektrovej 328, 2800 Kongens Lyngby, Denmark
[8]SERCO/ESRIN, Largo Galileo Galilei, Frascati, Italy
[9]DEIMOS/ESRIN, Largo Galileo Galilei, Frascati, Italy
[10]Progressive Systems/ESRIN, Largo Galileo Galilei, Frascati, Italy
[a]now at: RHEA/ESRIN, Largo Galileo Galilei, Frascati, Italy

Correspondence: Jérôme Benveniste (jerome@benveniste.space)

Received: 15 June 2022 – Accepted: 16 November 2023 – Published: 19 April 2024

Abstract. This paper presents the RDSAR, SAR/SARin & FF-SAR altimetry processors available in the ESA Altimetry Virtual Lab (AVL) hosted on the EarthConsole® platform. An overview on processors and features as well as preliminary analyses using AVL output data are reported to demonstrate the quality of the ESA Altimetry Virtual Lab altimetry services. Soon additional processors from ESA research contracts will be added to the AVL portfolio to continue providing innovative solutions to the radar altimetry community.

Keywords. UPH16; SDG6.5; Water Level; Synthetic Aperture Radar Altimetry; FF-SAR; SARvatore; SAMOSA+

1 Introduction

Since 1970s radar altimetry plays a key role in supporting research activities on oceanography and climate change. In the last decade, with the advent of improved SAR & FF-SAR processing schemes (Raney, 1998; Egido and Smith, 2017) and advanced retrackers such as those of the SAMOSA family (Ray et al., 2015), radar altimetry demonstrated its capability to provide useful data in challenging areas such as polar regions, the coastal zone and in the inland water domain.

The European Space Agency (ESA) research contracts in the field of radar altimetry are constantly innovating processing baselines providing prototyped products tailored for specific areas of interest. The smooth provisioning of these innovative products to researchers is of upmost importance to improve official altimetry products from ESA's Cryosat-2, Sentinel-3 and Sentinel-6 missions.

This motivated ESA to support the development of a dedicated virtual space to allow users processing on-line and on-demand altimetry data with advanced customizable processing baselines not yet deployed in ESA official ground segments.

The EarthConsole® Altimetry Virtual Lab (AVL), funded by ESA, aims at providing a virtual space to (1) Support the Altimetry community in the development & operation of new Earth Observation applications and (2) Foster collaboration by leveraging on knowledge-sharing tools.

The Altimetry Virtual Lab has been developed on the new EarthConsole® platform (https://earthconsole.eu/, last access: 24 December 2023) and hosts the SARvatore (SAR Versatile Altimetric TOolkit for Research & Exploitation) family of processors which was previously available in the ESA Grid Processing On-Demand (G-POD) environment.

All Altimetry Virtual Lab services are available via the Network of Resources (NoR), an ESA initiative aiming at facilitating the use of cloud environments. Research, development, or pre-commercial projects may be eligible to receive a voucher to exploit the EarthConsole® Altimetry Virtual Lab services free of charge or at very competitive prices.

The scope of this presentation is to feature and provide an update on the SARvatore family of altimetry services portfolio for the exploitation of CryoSat-2, Sentinel-3 & Sentinel-6 data from L1A (FBR) data products up to RDSAR, SAR/SARin & FF-SAR Level-2 geophysical data products.

2 Implementation and validation of SAR altimeter processing algorithms

The Altimetry Virtual Lab is a virtual space hosted on the EarthConsole® platform (http://www.earthconsole.eu, last access: 24 December 2023) and designed on the needs of the altimetry community to support the development and operations of new Earth Observation applications, and foster collaboration by leveraging on knowledge-sharing tools.

To achieve these objectives, the Altimetry Virtual Lab offers:

- *P-PRO (Parallel Processing) and P-PRO ON DEMAND services.* to perform bulk or on-demand processing with the SARvatore (SAR Versatile Altimetric Toolkit for Research & Exploitation) family of processors;
- *G-BOX (Algorithm Development and Execution Environment) service.* to develop and test new algorithms and perform post-processing analyses on a virtual machine (VM) with fast access to the dataset offered by the Copernicus Data and Information Access Services – DIAS. The VM comes with the most commonly used software for altimetry data analysis already installed: Broadview Radar Altimetry Toolbox (BRAT) (http://www.altimetry.info/toolbox/, last access: 24 December 2023), Panoply (https://www.giss.nasa.gov/tools/panoply/download/, last access: 24 December 2023), Python and SNAP (https://step.esa.int/main/download/snap-download/, last access: 24 December 2023);
- *I-APP (Application Integration) service.* to receive expert support for the integration of new processors which are not currently available in the EarthConsole® catalogue;
- *I-SHARE service.* to take advantage from a set of tools to network with colleagues, keep up with the latest news and publications and share results from the altimetry domain.

Through the *P-PRO & P-PRO ON DEMAND services*, users can access the SARvatore (SAR Versatile Altimetric TOolkit for Research & Exploitation) family of processors which was previously available in the ESA Grid Processing On-Demand (G-POD) environment.

The AVL graphical interface (Fig. 1) allows users to: (1) select, in all the services, a geographical area of interest within the time-frame related to the L1A (FBR) & L1b data products availability in the service catalogue, (2) set specific options to tailor the processing for the specific area of interest (e.g. open ocean, coastal zone, inland water, sea ice...).

After the task submission, users can follow, in real time, the status of the processing. The output data products are generated in standard NetCDF format, therefore being compatible with the multi-mission BRAT toolbox and typical tools.

At present, the following processors are part of the portfolio:

- The *SARvatore (SAR Versatile Altimetric TOolkit for Research & Exploitation) for CryoSat-2 and Sentinel-3 services* developed by the Altimetry Team in the R&D division at ESA-ESRIN. The SARvatore services are versatile and allow the users to customize and adapt the processing at L1b & L2 according to their specific requirements by setting a list of configurable options (ESA, 2013). The scope is to provide users with specific processing options not available in the operational processing chains (e.g. range walk correction, stack subsetting, extended receiving window, zero padding, high-posting rate and burst weighting at L1b & SAMOSA+, SAMOSA++ and ALES+ SAR retrackers at L2).
- The *TUDaBo SAR-RDSAR (Technical University Darmstadt – University Bonn SAR-Reduced SAR)* for CryoSat-2 and Sentinel-3 service (Buchhaupt et al., 2018; Buchhaupt, 2019). It allows users to generate reduced SAR, unfocused SAR & LRMC data. Several configurable L1b & L2 processing options and retrackers (BMLE3, SINC2, TALES, SINCS) are available. The processor will be extended during an additional activity related to the ESA HYDROCOASTAL Project (https://www.satoc.eu/projects/hydrocoastal/, last access: 24 December 2023) to account in the open ocean for the vertical motion of the wave particles (VMWP) in unfocused SAR (retracker SINCS-OV) and in a simplified form of the fully focused SAR called here Low Resolution Range Cell Migration Correction-Focused (LRMC-F).
- The *ALES + SAR for CryoSat-2 and Sentinel-3 service.* It allows users to process official L1b data and produces L2 NetCDF products by applying the empirical

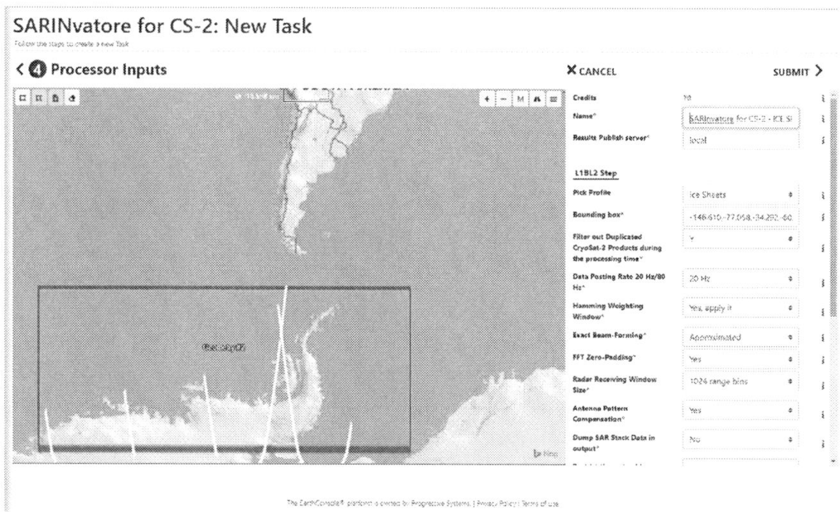

Figure 1. AVL graphical interface showing the selected area of interest (red box) and the selectable processing options (image credit: https://earthconsole.eu/, last access: 24 December 2023).

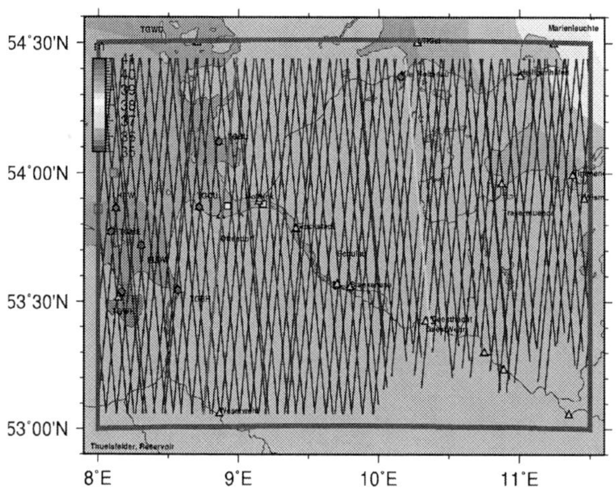

Figure 2. CryoSat-2 ground-tracks in open sea and in Elbe estuary and river.

ALES+ SAR sub-waveform retracker (Müller et al., 2020; Passaro et al., 2021), including a dedicated SSB solution, developed by DGFI-TUM in the frame of the ESA Sea Level CCI (http://www.esa-sealevel-cci.org/, last access: 24 December 2023) & BALTIC+ SEAL Projects (http://balticseal.eu/, last access: 24 December 2023).

- The *Aresys Fully Focused SAR for CryoSat-2, Sentinel-3 & Sentinel-6 services*. They provide the capability to produce FF-SAR L1b products thanks to the Aresys 2D transformed frequency domain AREALT-FF1 processor prototype (Guccione et al., 2018). Output products also include geophysical corrections and threshold peak (Davis, 1997) & ALES-like empirical sub-waveform retracker estimates (ALES + FF-SAR).

In the next section, some analyses in the inland water domain are presented to demonstrate the potential of the SARvatore processors. A full description of the adopted processing options can be found here: https://earthconsole.eu/knowledge-base-2/processors-manuals/ (last access: 24 December 2023).

3 Analyses of water level in the inland water domain

3.1 Elbe estuary

This case study analyses the CryoSat-2 data provided by the AVL Aresys FF-SAR processor over the Elbe estuary and river (Fig. 2). The time period from January 2019 and February 2020 includes a full CryoSat-2 geodetic orbit (ground track repeat cycle: 369 d). The selected setting for FFSAR is: multi-look posting rate 200 Hz ($\sim$ 35 m separation between two consecutive estimates), zero padding applied (set to 2), bandwidth factor 0.5 ($\sim$ 1 m resolution in the along-track direction), Threshold retracker (Davis, 1997). The goal is to investigate the quality of the FF-SAR data.

The river Elbe has about 300–400 m width between Geerstacht and Neue Darchau and is larger in the estuarine part which is affected by ocean tides. It ends in the German Bight at Ottendorf. The studied section is a downstream reach of $\sim$ 150 km long. From an along-track comparison in coastal and open ocean of FFSAR with EarthConsole® (former GPOD) SARVatore and TUDaBo unfocused SAR data (Fig. 3) at 20 Hz posting rate ($\sim$ 300 m separation between two consecutive estimates), FF-SAR data look noisier

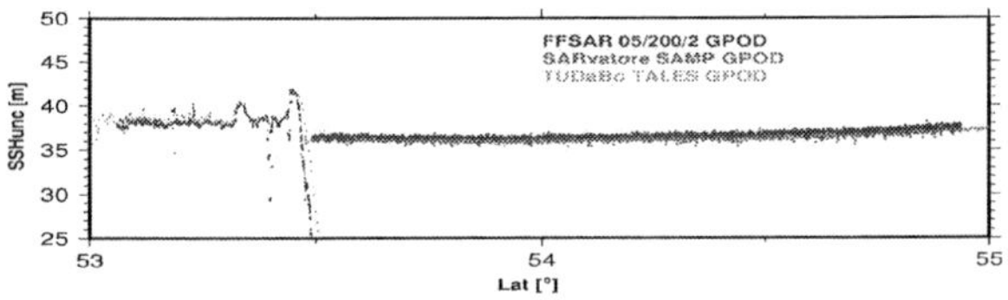

Figure 3. Uncorrected heights in open sea from FF-SAR (blue), unfocused SAR (red) and reduced SAR (violet) along open sea ground-track location.

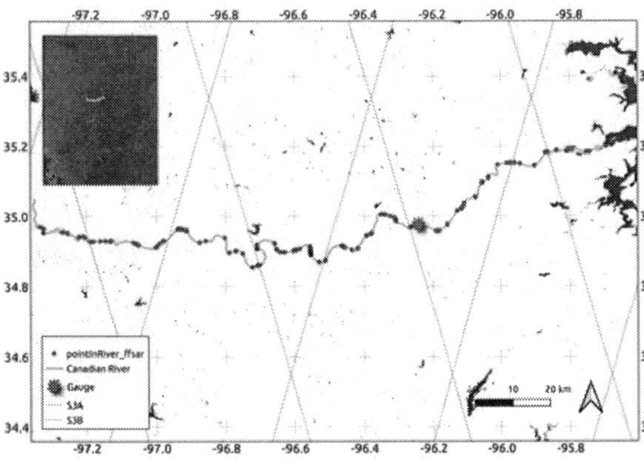

Figure 4. Study area (aerial image: © Google Earth 2019).

than the other products. This is expected considering the improved resolution in along-track (∼ 1 m) w.r.t. conventional unfocused SAR (330 m resolution in the along-track direction).

In the estuarine intertidal and in river parts FF-SAR provides a smooth surface and more data due to its higher sampling.

3.2 Canadian River in the Mississippi basin

In this case study, CryoSat-2 data processed with the AVL Aresys FF-SAR processor over the Canadian River in the Mississippi basin were evaluated. This river has a width of a few hundred meters. The FF-SAR derived water levels were evaluated using ICESat-2 laser altimetry data and gauge data from the USGS (https://maps.waterdata.usgs.gov/mapper/index.html, last access: 24 December 2023).

In the study we considered water levels based on different retrackers including ALES+ FF-SAR and threshold retrackers. To enhance the amount of data we selected a posting rate of 500 Hz (∼ 14 m separation between two consecutive estimates) and the bandwidth of 0.5 (∼ 1 m resolution in the along-track direction). To compare the altimetry data to the gauge data we need to create a water level time series at the location of the gauge. We use the CryoSat-2 FF-SAR data and ICESat-2 data to compile a time series.

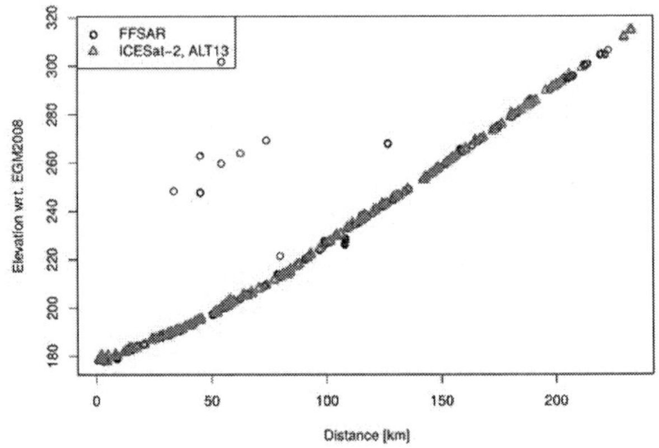

Figure 5. FF-SAR and ICESat-2 water levels plotted against distance along the reach.

The time series is constructed using data available along a reach of approximately 200 km (Figs. 4 and 5).

Firstly, the data is projected to the centreline of the river. Providing the data as a function of space and time, the data is then combined via a state-space model (Nielsen et al., 2022).

As seen in Fig. 6, we can identify the main water level signal. We find RMSE values of approximately 30 cm and a correlation of 0.6. The result is similar for all the retrackers.

3.3 Lake Bracciano

Lake Bracciano is of volcanic origin, located 32 km northwest of Rome. It is one of the major lakes of Italy and has a circular perimeter of approximately 32 km and a diameter of 7 km. Its in flow is from precipitation only as there are no inflowing rivers. As the lake serves as a drinking water reservoir for the city of Rome, it has been under control since 1986 to avoid the pollution of its waters.

Because of a drought that has hit Italy in 2017 and for the intensive water drainage by the Regional Water Supply Agency (ACEA), Lake Bracciano has suffered a significant drop of the water level between March and December 2017, as registered as well by in-situ measurements (https://braccianosmartlake.com, last access: 24 December 2023).

Since Sentinel-3A Orbit Number 213 over flies Lake Bracciano and considering the satellite's 27 d repeat period, water level variations detected by the SRAL altimeter can be regularly compared to insitu measurements to measure the performance of the instrument and the data processing algorithm estimating the radar range over such a very small lake.

The time series of the Bracciano Lake water level variation since April 2016 to March 2019 as measured by Sentinel-3A is shown in Fig. 7 (red points). The data have been processed in the AVL SARvatore for Sentinel-3 service and retracked using the SAMOSA++ retracker (Dinardo et al., 2017). The insitu measurements provided by Bracciano Smart Lake

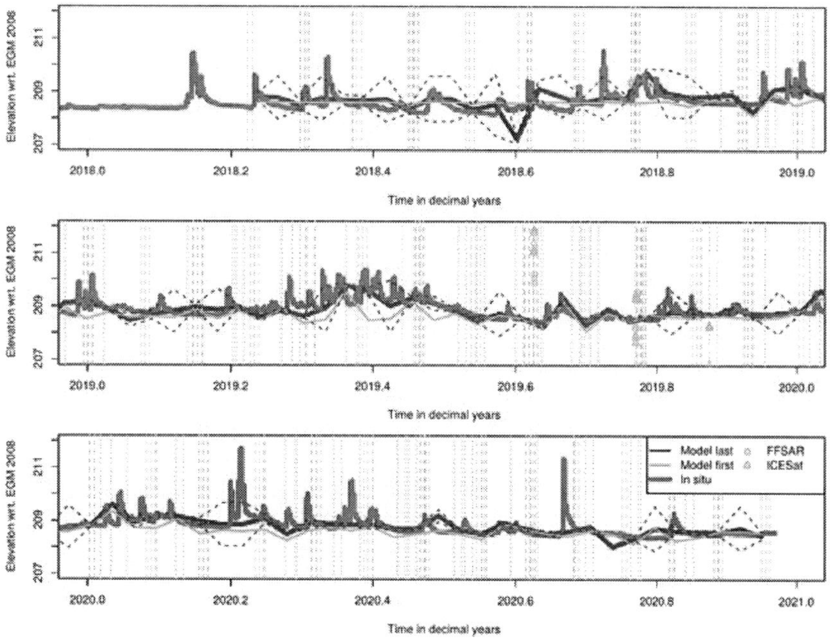

Figure 6. The reconstructed water level time series (blue) compared to the gauge water levels (red). The grey lines indicate the times where data is available.

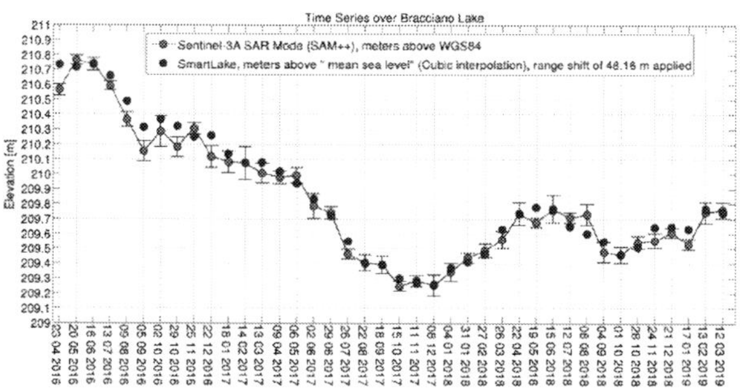

Figure 7. The time series of the Bracciano Lake water level variation from April 2016 to March 2019 as measured by Sentinel-3A.

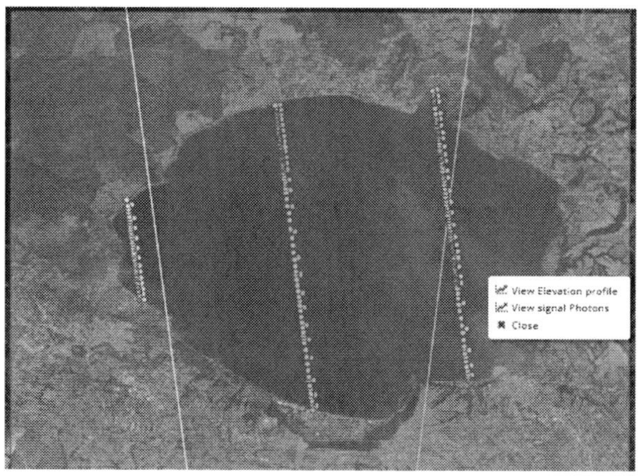

Figure 8. Position of ICESat-2 tracks (crossing the lake) and estimates from its beams (see points inside the lake) over Lake Bracciano. Only in the optimal case measurements from all 6 beams are available (image credit: https://openaltimetry.org/data/icesat2/, last access: 24 December 2023).

web platform (https://braccianosmartlake.com, last access: 24 December 2023) are depicted in blue.

The standard of the difference (STDD) between altimetry and in-situ is about 6 cm whereas the error bar average size is around 4 cm. Both in-situ and altimetry data have recorded the drop of the lake water level occurred during summer 2017. The lake is then slowly recovering the pre-drought level.

The ICESat-2 laser altimeter is also collecting measurements over lake Bracciano with its 6 beams (three pairs of strong and weak beams, Fig. 8). The repeat cycle of the mission is of 91 d. This laser altimeter generates individual laser footprints of nearly 14 m (in diameter). The ICESat-2 ATL13 products (available at https://openaltimetry.org/data/icesat2/, last access: 24 December 2023) are specialized geophysical data products that give along-track and near-shore water surface height distribution within the water masks.

Heights of surface water level is computed from strong beams (gt1r, gt2r, and gt3r) and weak beams (gt1l, gt2l, and gt3l). Data have been filtered to remove outliers using a typ-

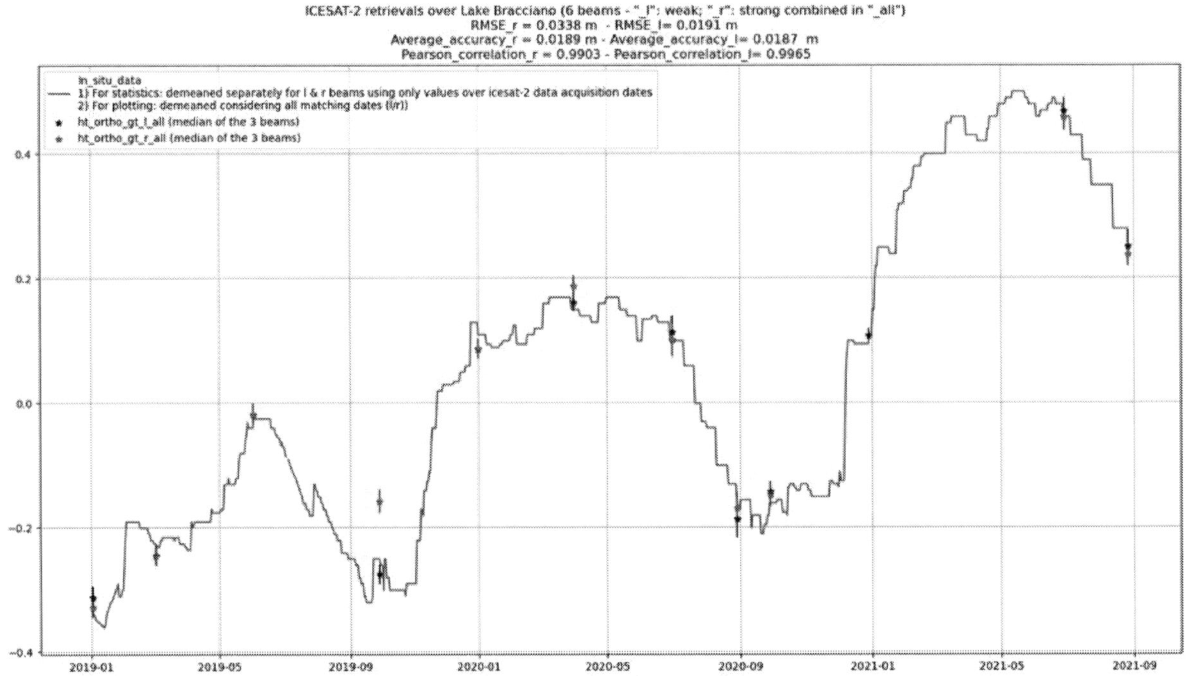

Figure 9. Time series of the Bracciano Lake water level variation from January 2019 to September 2021 as measured by filtered ICESat-2 ATL13 data. In the title and legend "r" refers to strong beams and "l" to weak beams.

ical mean ±2· SD criterion and quality flags (set as in Xu et al., 2021).

Then, at each date, the median of all available weak/strong beam estimates is taken and kept separated. In this way, there will be a "weak" orthrometric height time series (ht_ortho_gt_l_all) and a "strong" orthometric height time series (ht_ortho_gt_r_all). These can be seen in Fig. 9 along with the in-situ data time series.

The accuracy is taken as the standard deviation of the orthometric height time series.

Comparing to in-situ data, results in terms of RMSE are both at cm level (the "weak" time series performs better, RMSE is 1.9 vs. 3.3 cm, probably because of a non-negligible outlier in the 'strong' time series around September 2019). Weak beams perform better also in providing one additional estimate around January 2021 which is missing in the strong beam time series. The accuracy is below 2 cm and the Pearson correlation is above 99 % for both time series.

As expected from the higher resolution of ICESat-2, associated with footprints not contaminated by the presence of land, the reported RMSE are lower than the one indicated for the unfocused SAR SAMOSA++ Sentinel-3 estimates in the first analysis of this subsection.

The unavailability of ICESat-2 data is severe in 2021 and this is probably due to the cloud coverage, which is along with the 91 d repeat cycle, the main limitation in using ICESat-2 data in the inland water domain.

As ICESat-2 data are not available through the ESA Altimetry Virtual Lab, the presented analysis aimed at providing an example of possible synergies between radar and laser altimetry.

4 Conclusions

This paper presented the ESA Altimetry Virtual Lab available on the EarthConsole® platform reporting some preliminary inland water altimetry analyses.

Soon additional processors from ESA research contracts will be added to the AVL portfolio to continue providing innovative solutions to the radar altimetry community.

Code and data availability. Some training material and a brochure including all the information needed to access the services are available at: https://earthconsole.eu/knowledge-base/ (section "Earth Console Library"; Earth Console, 2021).

SAR and Fully-Focused SAR algorithms are freely available in Python through the ESA DeDop suite (https://dedop.org/; ESA, 2017) and the SMAP processor (https://github.com/cls-obsnadir-dev/SMAP-FFSAR; Guccione, 2018), respectively.

Author contributions. The conceptualization was done by JB and MR. JB ensured the funding acquisition, the project management and supervision. The methodology and software were developed, validated and reviewed by SD, LFM, CB, MS, MP, KN, AA, MR and JB. GS, BA and CO implemented the Altimetry Virtual Lab in the EarthConsole®. The original draft was prepared by MR. The review and editing were done by all authors.

Competing interests. The contact author has declared that none of the authors has any competing interests.

Disclaimer. Publisher's note: Copernicus Publications remains neutral with regard to jurisdictional claims made in the text, published maps, institutional affiliations, or any other geographical representation in this paper. While Copernicus Publications makes every effort to include appropriate place names, the final responsibility lies with the authors.

Special issue statement. This article is part of the special issue "IAHS2022 – Hydrological sciences in the Anthropocene: Past and future of open, inclusive, innovative, and society-interfacing approaches". It is a result of the XIth Scientific Assembly of the International Association of Hydrological Sciences (IAHS 2022), Montpellier, France, 29 May–3 June 2022.

Acknowledgements. The ESA Altimetry Virtual Laboratory, hosted in the EarthConsole®, is funded by the European Space Agency via the contract number 4000135127/21/I-DT-lr.

Review statement. This paper was edited by Christophe Cudennec and reviewed by two anonymous referees.

References

Buchhaupt, C.: Model Improvement for SAR Altimetry, PhD Thesis, University of Darmstadt, https://tuprints.ulb.tu-darmstadt.de/9015/ (last access: 24 December 2023), 2019.

Buchhaupt, C., Fenoglio-Marc, L., Dinardo, S., Scharroo, R., and Becker, M.: A fast convolution-based waveform model for conventional and unfocused SAR altimetry, Adv. Space Res., 62, 1445–1463, https://doi.org/10.1016/j.asr.2017.11.039, 2018.

Davis, C. H.: A robust threshold retracking algorithm for measuring ice-sheet surface elevation change from satellite radar altimeters, IEEE T. Geosci. Remote, 35, 974–979, https://doi.org/10.1109/36.602540, 1997.

Dinardo, S., Fenoglio, L., Buchhaupt, C., Becker, M., Scharroo, R., Fernandes, J. M., and Benveniste, J.: Coastal SAR and PLRM altimetry in German Bight and West Baltic Sea, Adv. Space Res., 62, 1371–1404, https://doi.org/10.1016/j.asr.2017.12.018, 2017.

Earth Console: Homepage, https://earthconsole.eu/knowledge-base/ (last access: 24 December 2023), 2021.

Egido, A. and Smith, W. H. F.: Fully Focused SAR Altimetry: Theory and Applications, IEEE T. Geosci. Remote, 55, 392–406, https://doi.org/10.1109/TGRS.2016.2607122, 2017.

ESA: Guidelines for the SAR (Delay-Doppler) L1b Processing, https://earth.esa.int/eogateway/documents/20142/37627/Guidelines-for-the-SAR-Delay-Doppler-L1b-Processing.pdf (last access: 24 December 2023), 2013.

ESA: The DeDop toolbox, ESA [code], https://dedop.org/ (last access: 24 December 2023), 2017.

Guccione, P., Scagliola, M., and Giudici, D.: 2D Frequency Domain Fully Focused SAR Processing for High PRF Radar Altimeters, Remote Sens.-Basel, 10, 1943, https://doi.org/10.3390/rs10121943, 2018 (code available at: https://github.com/cls-obsnadir-dev/SMAP-FFSAR, last access: 24 December 2023).

Müller, F., Passaro, M., Dettmering, D., and Baltic+ SEAL: Algorithm Theoretical Baseline Document (ATBD), Version 2.1. Technical report delivered under the Baltic+ SEAL project, https://doi.org/10.5270/esa.BalticSEAL.ATBDV2.1, 2020.

Nielsen, K., Zakharova, E., Tarpanelli, A., Andersen, O. B., and Benveniste, J.: River levels from multi mission altimetry, a statistical approach, Remote Sens. Environ., 270, 112876, https://doi.org/10.1016/j.rse.2021.112876, 2022.

Passaro, M., Müller, F. L., Oelsmann, J., Rautiainen, L., Dettmering, D., Hart-Davis, M. G., Abulaitijiang, A., Andersen, O. B., Høyer, J. L., Madsen, K. S., Ringgaard, I. M., Särkkä, J., Scarrott, R., Schwatke, C., Seitz, F., Tuomi, L., Restano, M., and Benveniste, J.: Absolute Baltic Sea Level Trends in the Satellite Altimetry Era: A Revisit, Front. Mar. Sci., 8, 647607, https://doi.org/10.3389/fmars.2021.647607, 2021.

Raney, R. K.: The delay/Doppler radar altimeter, IEEE T. Geosci. Remote, 36, 1578–1588, https://doi.org/10.1109/36.718861, 1998.

Ray, C., Martin-Puig, C., Clarizia, M. P., Ruffini, G., Dinardo, S., Gommenginger, C., and Benveniste, J.: SAR Altimeter Backscattered Waveform Model, IEEE T. Geosci. Remote, 53, 911–919, https://doi.org/10.1109/TGRS.2014.2330423, 2015.

Xu, N., Zheng, H., Mą, Y., Yang, J., Liu, X., and Wang, X.: Global Estimation and Assessment of Monthly Lake/Reservoir Water Level Changes Using ICESat-2 ATL13 Products, Remote Sens.-Basel, 13, 2744, https://doi.org/10.3390/rs13142744, 2021.

Proc. IAHS, 385, 465–470, 2024
https://doi.org/10.5194/piahs-385-465-2024

Open Access PIAHS

IAHS2022 – Variability and change across space, time, extremes, and interfaces

Analyzing the Effects of Climate Change for the Water Tower of Mesopotamia, Türkiye

Gokcen Uysal, Yusuf Ogulcan Dogan, Huseyin Soykan Civelek, Ali Arda Sorman, and Aynur Sensoy

Department of Civil Engineering, Eskisehir Technical University, Eskisehir, 26555, Türkiye

Correspondence: Gokcen Uysal (gokcenuysal@eskisehir.edu.tr)

Received: 14 June 2022 – Revised: 27 December 2023 – Accepted: 26 January 2024 – Published: 19 April 2024

Abstract. This study aims to analyze future snowmelt runoff and snow extent for the Upper Euphrates Basin (41 109 km^2) which is the most upstream basin of the water tower Mesopotamia and located in the Eastern mountainous part of Türkiye. The future characterization of snowmelt runoff and snow water equivalent is investigated taking the regional climate model projections into account through hydrological model applications. The basin is divided into two major sub-basins Karasu Basin and Murat Basin. The conceptual hydrological models, HBV and HEC-HMS have been utilized to establish a rainfall-runoff relationship. The models are calibrated and validated with observed daily total precipitation and average temperature data for the 1980–2011 period. Future projections are selected for the 2025–2100 periods using EU-CORDEX data with various global circulation models. MPI-ESM-LR and CNRM-CM5 data sets and RCP4.5 and RCP8.5 emission scenarios are used for future projections. The results indicate a decrease in snow cover extent and an increase in snowmelt runoff. The overall assessment will help us to understand how these changes will affect the operations of water resources systems (downstream reservoirs) in terms of flood control, energy production, irrigation, water supply, etc.

Keywords. Hydrological modelling; Climate change; Snowmelt; Euphrates River Basin; UPH 2

1 Introduction

Global climate change greatly affects climatological and hydrological elements like temperature, precipitation regime, groundwater levels and hydrological response (Gleick, 1989). Although floods and droughts are/will be the main expected consequences in the future, also it is expected that global change will have different effects on each region (Tabari, 2020). Therefore, the assessment of potential impacts of climate change on regional water resources is receiving growing attention from various researchers. The Mediterranean Basin has been identified as one of the world's most sensitive regions to climate change consequences (URL-1, 2022). Fujihara et al. (2008) explored the potential impacts of climate change on the Seyhan River Basin in Türkiye and concluded an increase in temperature and a reduction in precipitation for future projections. Bozkurt and Sen (2013) investigated future climate change in the Euphrates–Tigris Basin using dynamically downscaled outputs of different Global Circulation Models (GCMs) and emissions scenario simulations for two 30 year periods, 2041–2070 and 2071–2099. According to their analyses, significant decreases in the annual surface runoff is expected.

GCMs are generally used in large-scale climate change impact studies and play a key role in recognizing future global climatic changes (Fujihara et al., 2008). The model outputs necessitate a proper hydrological modeling background to produce future projections. The eastern part of Türkiye is covered with full or partially snow during winter and large rivers such as Euphrates and Tigris are mainly fed by the snowmelt. This study aims to project future runoff and snow components in the snow-dominated basins of the Euphrates River basin by integrating climate model data sets and two well-known conceptual hydrological models.

2 Data and Methodology

This section briefly describes the data used for the set-up and the structures of hydrological modelling approaches together with the climate data and their assessment for the future projections. Hydrological modelling is accomplished with two conceptual models: HBV-Light and HEC-HMS. After calibrating and validating models, GCM data set based runs are employed to simulate the historical and future changes. The changes over the reference period are calculated for runoff and snow components.

2.1 Hydro-Meteorological and Climate Data

Türkiye has a complex topography and variable climate. Two upstream snow-dominated catchments (Fig. 1) are selected; Karasu Basin (15 330 km^2) and Murat Basin (25 779 km^2) having the elevation ranges between 860–3565 and 840–4045 m, respectively. Annual total precipitation is 429.3 and 593.9 mm yr^{-1}, annual average temperature is 9.1 and 8.2 °C and annual average discharge is 153.5 and 239.0 $m^3\,s^{-1}$ between 1980-2011 for Karasu and Murat, respectively. These basins play an important role in downstream water structures. The region is also the concern of various national and international projects (Şensoy and Uysal, 2012; Uysal et al., 2016; Şorman et al., 2019; Uysal and Şorman, 2021).

Historical simulations and projections of future changes in precipitation and temperature are provided by the EURO-CORDEX platform (URL-2, 2021). MPI-ESM-LR and CNRM-CM5 GCMs data sets having RCP4.5 and RCP8.5 concentration scenarios with 112.5 km spatial resolution are selected in the study. Data over the basin are derived from Regional Climate Model (RCM) ModeRegCM4.3.4 to provide a series of high-resolution inputs (20 km resolution). Linear bias correction is applied for precipitation and temperature data sets. Stationary climate simulations are used for reference (1971–2000) and future (2025–2100) periods. The future projection period is evaluated under 2025–2050, 2051–2075, and 2076–2100 periods.

2.2 HBV Hydrological Model

The Swedish Meteorological and Hydrological Institute (SMHI) developed HBV as a conceptual and continuous hydrological model (Bergström, 1995). The HBV-Light version (Seibert and Vis, 2012) was utilized in this study as it offers a user-friendly graphical interface and an automatic calibration opportunity. The HBV model distributes precipitation, temperature, and potential evapotranspiration inputs based on elevation and vegetation zones. The model calculates snow and soil moisture layers based on zones, as well as runoff and routing processes for the entire basin, to mimic daily flow.

2.3 HEC-HMS Hydrological Model

HEC-HMS (Hydrologic Modeling System) model is developed by the U.S. Army Corps of Engineers Hydrologic Engineering Center (USACE) and is designed to simulate the rainfall-runoff process of dendritic watershed systems as continuous or event-based and to be applicable in a wide range of geographic areas (Scharffenberg and Fleming, 2000). Inputs are precipitation, temperature, and potential evapotranspiration. In this study, soil moisture accounting (SMA), linear reservoir and Clark unit hydrograph method are chosen respectively as a loss, baseflow and transform methods. SMA module considers continuous situation-based soil moisture variation which is carried out by calculating the amount of water stored in and moving through vegetation, soil surface, profile and groundwater layers. Linear reservoir method is more compatible with SMA.

2.4 Assessment Metrics

Nash-Sutcliffe Efficiency (NSE), Killing-Gupta Efficiency (KGE) and Pearson correlation coefficient (r) are used to test hydrological model performances (Eqs. 1–4):

$$\mathrm{NSE} = 1 - \frac{\sum (x-y)^2}{\sum (x-\overline{x})^2} \tag{1}$$

$$\mathrm{KGE} = 1 - \sqrt{(r-1)^2 + (a-1)^2 + (b-1)^2} \tag{2}$$

$$a = \frac{S_y}{S_x}; \quad b = \frac{\overline{y}}{x} \tag{3}$$

$$r = \frac{n\sum xy - \left(\sum x\right)\left(\sum y\right)}{\sqrt{\left[nx^2 - \left(\sum x\right)^2\right]\left[ny^2 - (y)^2\right]}} \tag{4}$$

where x is observation, y is modelled value, $\overline{x}$ is average observation, $\overline{y}$ is average modelled value, S_x is the standard deviation of the observed data, S_y is the standard deviation of the modelled data, n is the number of data sets.

Contingency tables evaluate the consistency by using two main indicators i.e., Probability of Detection (POD) (Eq. 5) and False Alarm Rate (FAR) (Eq. 6).

$$\mathrm{POD} = \frac{a}{a+c} \tag{5}$$

$$\mathrm{FAR} = \frac{b}{a+b} \tag{6}$$

where "a" (detected by both ground and GMC), "b" (detected by only GCM), "c" (detected by only ground) and "d" (no detection in both).

3 Results

3.1 Hydrological Modelling Performances

The conceptual daily hydrological models are calibrated and validated from 1980–2011. Tables 1 and 2 show that the HBV model has a higher model efficiency for both basins and periods. The performance of HBV is slightly better than the HEC-HMS model in terms of NSE, mainly associated with peak flow accuracy. KGE accounts for various components and both models show similar performances in Karasu.

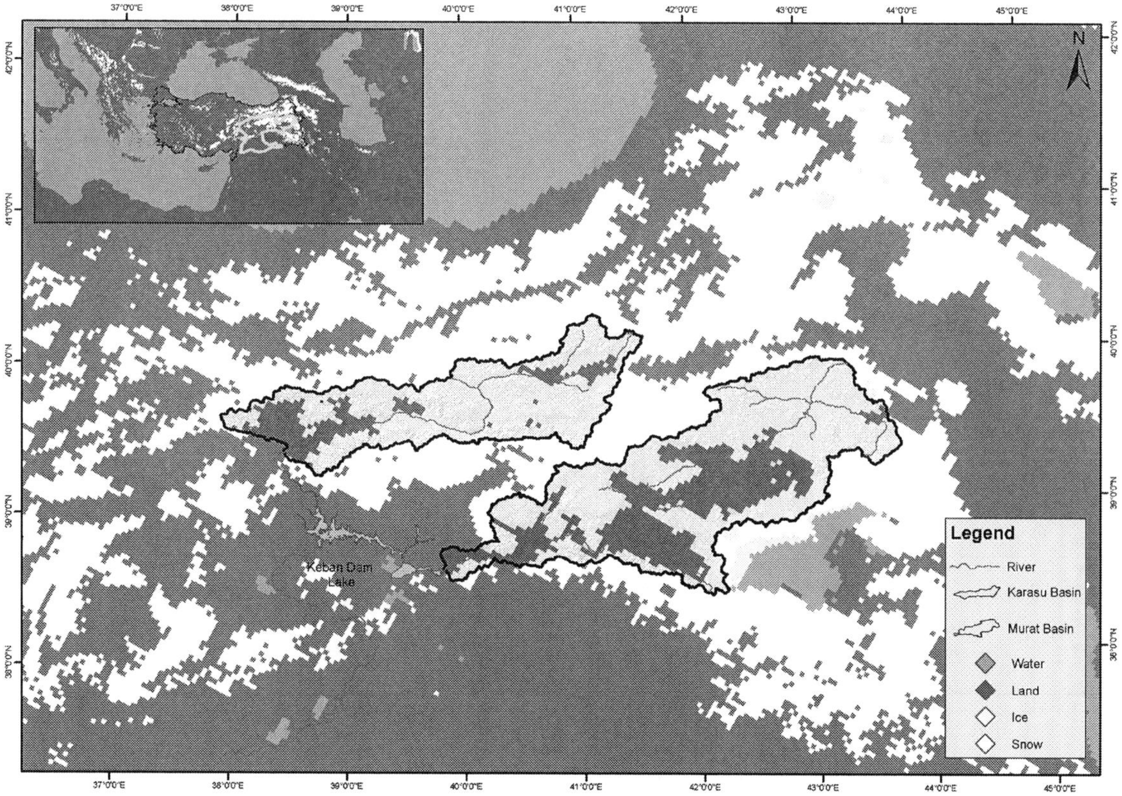

Figure 1. Snow cover across Türkiye from IMS satellite (30 March 2009) and the selected catchments on the Euphrates River.

Table 1. Model performances for Karasu basin.

	HBV*		HEC-HMS*	
	Cal	Val	Cal	Val
NSE	0.80	0.83	0.77	0.77
KGE	0.86	0.86	0.88	0.86
Pearson-R	0.91	0.93	0.88	0.88

*Cal: 1980–1999, Val: 1991–2011.

Table 2. Model performances for Murat basin.

	HBV*		HEC-HMS*	
	Cal	Val	Cal	Val
NSE	0.90	0.83	0.79	0.72
KGE	0.90	0.90	0.88	0.84
Pearson-R	0.95	0.91	0.89	0.85

*Cal: 1980–1999, Val: 1991–2011.

3.2 Raw and Bias-Corrected Climate Data Performances

Bias correction is applied for temperature and precipitation GCM data sets. Performances of the raw and bias-corrected GCMs data sets are compared with gauge measurements for the reference period. Pearson-R values of monthly average bias-corrected temperature against gauge measurement are around 0.96 for MPI-ESM-LR and CNRM-CM5 in both basins. Figure 2 depicts both basins' POD and FAR values with monthly (mo) threshold precipitations. POD values increased in both products after the bias correction application. CNRM-CM5 performs slightly better in comparison with MPI-ESM-LR.

3.3 Future Projections

The performances of the hydrological models, GMCs data are assessed and the change in future projections (runoff and SWE) are demonstrated against the reference period. Future projections are assessed for RCP8.5 as the worst-case scenario (Fig. 3). Both models demonstrate a possible increase in the runoff for two mountainous basins (more emphasized for Karasu Basin) for the first two projection periods and a significant decrease for the last projection period according

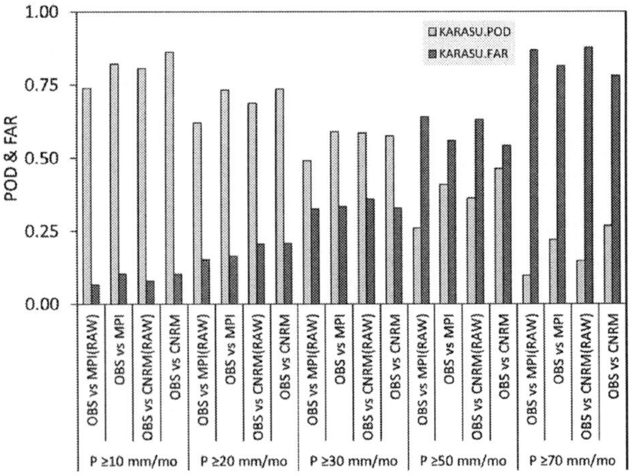

(a)

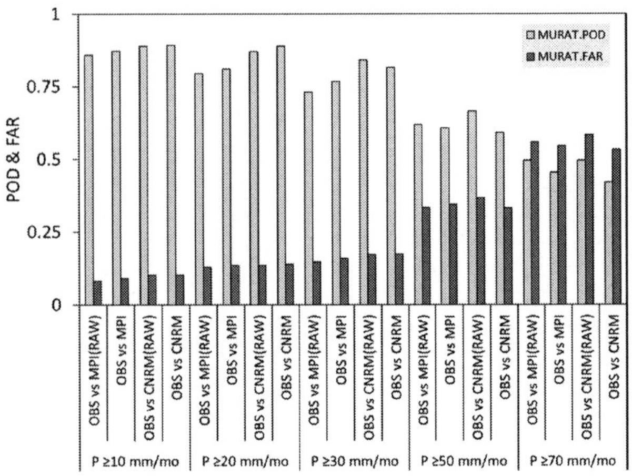

(b)

Figure 2. Assessment of the GCM precipitation **(a)** Karasu Basin, **(b)** Murat Basin.

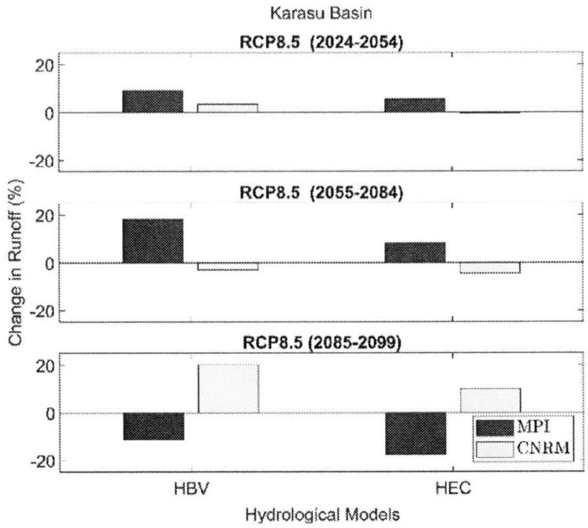

(a)

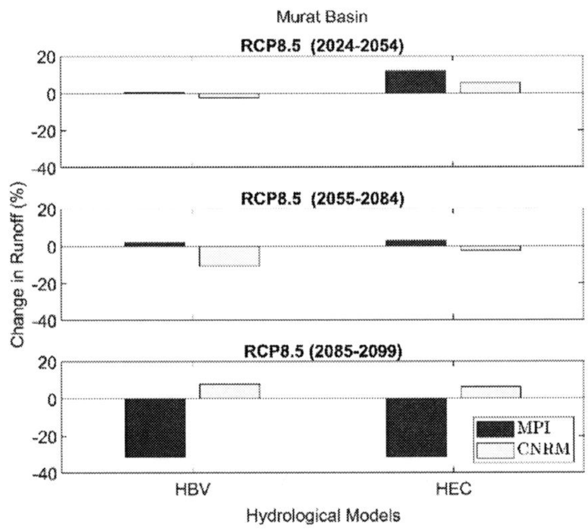

(b)

Figure 3. The change in runoff for three future projections **(a)** Karasu Basin, **(b)** Murat basin.

to MPI-ESM-LR based GCM. CNRM-CM5 based hydrological model results indicate a variational change with low percentages for the first two projection periods, however, an increase in runoff for the last projection period for both basins. These mismatches might be assigned to variations of precipitation in each GCM in each period.

Figure 4 depicts the change in snow water equivalent (SWE) in terms of percentages in the future projections. Regardless of the hydrological model and GCM, the reduction is expected in future SWE values. Fewer reduction is expected for the 2024–2054 period. The reductions are exacerbated along with a longer lead time period. This might be attributed to the increase in temperature in the future.

3.4 Discussion of the Results

A bias correction is required for GCM data sets, however the improvement for precipitation data is limited compared to the temperature data. Both hydrological models (HBV and HEC-HMS) provide satisfactory calibration and validation performances. The change in runoff is variable depending on GCM data set. For example, MPI-ESM-LR based runoff present no change or slight increase for change in percent runoff between 2024–2054 and 2055–2084, however an increase is expected for 2085–2099. On the other hand, the change for CNRM-CM5 is either vice versa or less. The future projection of the SWE is decreasing related to the increase in temperature data. Climate change will alter snowfall to rainfall,

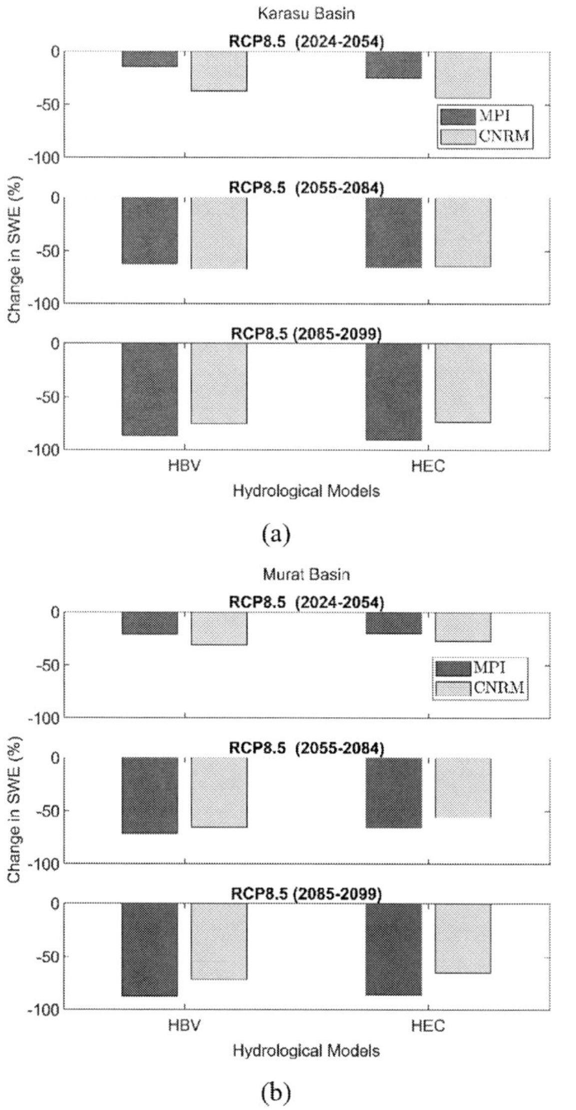

Figure 4. The change in snow water equivalent (SWE) for three future projections **(a)** Karasu Basin, **(b)** Murat basin.

thus will have various impacts on surface runoff and snow extent.

4 Conclusions and Recommendations

The study investigates the future implications of snowmelt runoff against the changing climate for the upstream part of the Euphrates River. EU-CORDEX GCM data sets MPI-ESM-LR and CNRM-CM5 are accounted to estimate future projections of discharge and SWE. The study accounts different hydrological models (HBV & HEC) in the assessment. Both hydrological models provide satisfactory validation NSE values (0.84–0.90). The bias correction is required for GCM data sets before using them in the models. A possible increase in the runoff for the first two projection period, and a significant decrease for the last projection period especially according to MPI-ESM-LR are expected. These results indicate a drastic and continuous decrease in SWE due to changes in the temperature. These changes will negatively affect the operation of reservoirs, the agriculture policy, and tourism management. Nonetheless of model selection, change in precipitation holds more uncertainty which might cause the variation in streamflow runoff. Changes in snow line elevation and lag time are being assessed in future studies. Considering the variation in GCMs, increasing the number of climate models would provide more information to interpret the results.

Data availability. Publicly available free climate datasets (https://esg-dn1.nsc.liu.se/search/cordex/, ESGF, 2024) are analyzed in this study. The hydro-meteorological data set is not public or open to third parties.

Author contributions. GU analyzed the data, wrote the draft; YOD and HSC performed the model simulations; AS designed the research; GU, YOD, HSC, AS, AAS reviewed and edited the manuscript.

Competing interests. The contact author has declared that none of the authors has any competing interests.

Disclaimer. Publisher's note: Copernicus Publications remains neutral with regard to jurisdictional claims made in the text, published maps, institutional affiliations, or any other geographical representation in this paper. While Copernicus Publications makes every effort to include appropriate place names, the final responsibility lies with the authors.

Special issue statement. This article is part of the special issue "IAHS2022 – Hydrological sciences in the Anthropocene: Variability and change across space, time, extremes, and interfaces". It is a result of the XIth Scientific Assembly of the International Association of Hydrological Sciences (IAHS 2022), Montpellier, France, 29 May–3 June 2022.

Acknowledgements. Special thanks are extended to the State Hydraulics Works of Türkiye (DSI) as well as the Turkish State Meteorological Service (MGM).

Financial support. This research has been supported by Eskisehir Technical University (Scientific Research Projects 22DRP190 and 22ADP051).

Review statement. This paper was edited by Christophe Cudennec and reviewed by two anonymous referees.

References

Bergström, S.: The HBV model, in: Computer models of watershed hydrology, edited by: Singh, V. P., Water Resources Publications, Highlands Ranch, Colorado, USA, 443–476, ISBN 9780918334916, 1995.

Bozkurt, D. and Sen, O. L.: Climate change impacts in the Euphrates–Tigris Basin based on different model and scenario simulations, J. Hydrol., 480, 149–161, https://doi.org/10.1016/j.jhydrol.2012.12.021, 2013.

ESGF: ESGF@LiU/CORDEX, MPI-ESM-LR (Max Planck Institute Earth System Model running on low resolution grid) and CNRM-CM5 (Centre National de Recherches Météorologiques-climate model version 5), Earth System Grid Federation (ESGF) [data set], https://esg-dn1.nsc.liu.se/search/cordex/, last access: 5 February 2024.

Fujihara, Y., Tanaka, K., Watanabe, T., Nagano, T., and Kojiri, T.: Assessing the impacts of climate change on the water resources of the Seyhan River Basin in Turkey: Use of dynamically downscaled data for hydrologic simulations, J. Hydrol., 353, 33–48, https://doi.org/10.1016/j.jhydrol.2008.01.024, 2008.

Gleick, P. H.: Climate change, hydrology, and water resources, Rev. Geophys., 27, 329–344, 1989.

Scharffenberg, W. A. and Fleming, M. J.: Hydrologic Modeling System-HEC-HMS-User's Manual-Version 2.0, ASIN B0006RUUPC, 2000.

Seibert, J. and Vis, M. J. P.: Teaching hydrological modeling with a user-friendly catchment-runoff-model software package, Hydrol. Earth Syst. Sci., 16, 3315–3325, https://doi.org/10.5194/hess-16-3315-2012, 2012.

Şensoy, A. and Uysal, G.: The value of snow depletion forecasting methods towards operational snowmelt runoff estimation using MODIS and Numerical Weather Prediction Data, Water Resour. Manag., 26, 3415–3440, https://doi.org/10.1007/s11269-012-0079-0, 2012.

Şorman, A. A., Uysal, G., and Şensoy, A.: Probabilistic snow cover and ensemble streamflow estimations in the Upper Euphrates Basin, J. Hydrol. Hydromech., 67, 82–92, https://doi.org/10.2478/johh-2018-0025, 2019.

Tabari, H.: Climate change impact on flood and extreme precipitation increases with water availability, Sci. Rep., 10, 13768, https://doi.org/10.1038/s41598-020-70816-2, 2020.

URL-1: https://climate-adapt.eea.europa.eu/, last access: 25 March 2022.

URL-2: https://www.euro-cordex.net/, last access: 1 December 2021.

Uysal, G. and Şorman, A. Ü.: Evaluation of PERSIANN family remote sensing precipitation products for snowmelt runoff estimation in a mountainous basin, Hydrolog. Sci. J., 66, 1790–1807, https://doi.org/10.1080/02626667.2021.1954651, 2021.

Uysal, G., Şensoy, A., and Şorman, A. A.: Improving daily streamflow forecasts in mountainous Upper Euphrates basin by multilayer perceptron model with satellite snow products, J. Hydrol., 543, 630–650, https://doi.org/10.1016/j.jhydrol.2016.10.037, 2016.

Proc. IAHS, 385, 471–476, 2024
https://doi.org/10.5194/piahs-385-471-2024

Open Access PIAHS

Comparison of Actual Evapotranspiration assessment by satellite-based model SEBS and hydrological model BBH in northern Tunisia

Nesrine Abid[1], **Zoubeida Bargaoui**[1], **Aymen Ben Jaafar**[1], **and Chris M. Mannaerts**[2]

[1]Ecole Nationale d'Ingénieurs de Tunis ENIT, Université de Tunis El Manar, Tunis, Tunisia
[2]Faculty of Geo-Information Sciences and Earth Observation (ITC), University of Twente, Enschede, the Netherlands

Correspondence: Nesrine Abid (nesrine_abid@ymail.com)

Received: 29 May 2022 – Revised: 1 February 2024 – Accepted: 1 February 2024 – Published: 19 April 2024

Abstract. Estimating actual evapotranspiration (AET) in agricultural semi-arid regions is important for crop yield and drought assessment. The Surface Energy Balance System (SEBS) model, a physically based energy balance model using satellite information is used to estimate AET at the 10 d scale, with a 3 km resolution. The bucket bottom hole (BBH) model, a conceptual daily water balance model is calibrated using the equifinality approach and run for simulating daily AET. Five watersheds located in northern Tunisia with areas varying between 56 and 448 km^2 were calibrated using daily rainfall and potential evapotranspiration data as entry and river discharge as output data. Sets of model parameters fulfilling both absolute relative errors of simulated discharge less than 20 % and Nash–Sutcliffe coefficients greater than 0.75 were selected. Three years were selected for the comparison (2010, 2017, and 2018). For every year, six subperiods of 10 d are considered belonging to January, March, April, May, July, and September. Boxplots of AET-BBH estimations are plotted to achieve a comparison with AET-SEBS estimates. It is found that AET comparisons are well favorable for January, March, and April while less satisfactory for May and September. They do not match for July. AET-SEBS are much higher in comparison with AET-BBH estimates with an RMSE and MAE equal respectively to 17 and 19 mm 10 d^{-1}. These results may help stakeholders to assess AET coming from different data sources and models.

Keywords. SDG 2; Modelling; Actual Evapotranspiration; Bucket Bottom Hole model (BBH); Surface Energy Balance model (SEBS); Northern Tunisia

1 Introduction

According to climate change studies, Tunisia experienced extreme phenomena over the past decade (drought and flooding) which would have a direct impact on agriculture and therefore on the country's food security (Verner, 2013). In this context, it is important to evaluate evapotranspiration, as a proxy for crop yield (Eagleson, 1994). Assessment of actual evapotranspiration is generally achieved using three different approaches (i) surface energy balance modeling (Bastiaanssen et al., 1998; Roerink et al., 2000; Su, 2002), (ii) Hydrological modeling (Kite, 1997; Kobayashi et al., 2001), and (iii) land surface modeling (Wang and Dickinson, 2012; Liou and Kumar Kar, 2014). Because there is a lack of published in situ evapotranspiration data in northern Tunisia, we aim to compare (i) and (ii) methods. We select the Surface Energy Balance (SEBS) model (Su, 2002) which is well-used all over the world except in Tunisia. It is also proposed to consider the Bucket Bottom Hole model (BBH) (Kobayashi et al., 2001) which was previously analyzed for northern Tunisia (Ben Jaafar and Bargaoui, 2020).

2 Study area

The study area is the northern part of Tunisia (North Africa) (Fig. 1). It is generally known for its Mediterranean climate. The main agricultural productions are cereals (wheat and barley) and olives. This area experienced several droughts in the

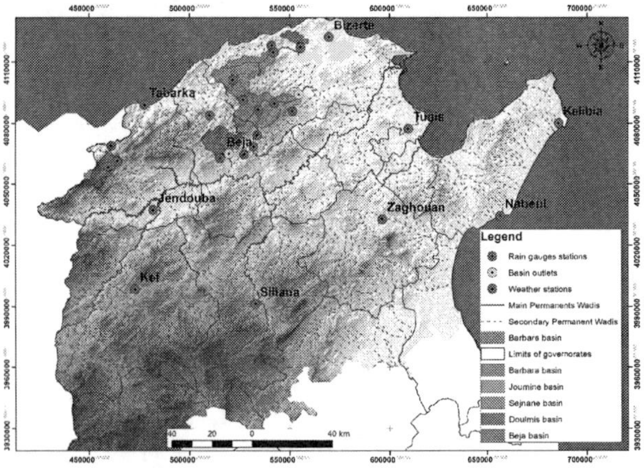

Figure 1. Map of the study area with the localization of the five watersheds (Barbara, Beja, Douimis, Joumine, and Sejnane) and the distribution of the meteorological and runoff stations.

last years. According to the National Journal of the Republic of Tunisia, 14 years from 1993 are identified as dry years for cereal crops resulting in damages for farmers and loss of revenues from cereal yields. Five watersheds are considered: Barbara, Beja, Douimis, Joumine, and Sejnane. The basin's areas vary from 56 to 448 km^2 (Fig. 1).

2.1 In situ data

2.1.1 Meteorological data

In situ meteorological data are obtained from the National Institute of Meteorology (INM) (https://www.meteo.tn/, last access: 14 February 2024) at the level of ten weather stations: Beja, Bizerte, Jendouba, Kef, Kelibia, Nabeul, Siliana, Tabarka, Tunis and Zaghouan (Fig. 1). Meteorological variables are mean daily air temperature in °C, mean daily air pressure in Pascal, daily duration of sunshine in hours, mean daily wind speed in $m\,s^{-1}$, and mean daily specific humidity ($kg\,kg^{-1}$). The period from January to September corresponding to the different stages of crop growth for the civil years 2010, 2017, and 2018 are studied.

2.1.2 Hydrological and rainfall data time series

Daily precipitation and runoff data are provided by the Hydrological National Service. The basin precipitation is estimated using the Thiessen method based on a network of available rain gauges (Fig. 1). Observations from the period 1950 to 1990 are used to calibrate the hydrological model (Ben Jaafar and Bargaoui, 2020). Observations from 2010 to 2018 are used also to run the model with calibrated parameters.

2.1.3 PET estimation using meteorological data

Monthly averages of minimum and maximum temperature in °C are used to compute monthly PET using Riou (1980) model. It is an entry variable to run the hydrological model BBH. The nearest meteorological station is considered for each basin (Fig. 1).

2.1.4 Soil map data

Soil texture data are used to estimate soil texture parameters in the BBH model. Soil texture maps are available from the digital agricultural map of Tunisia obtained from the local water and soil divisions.

2.2 Remote sensing data

Remote sensing data consists of The Land Surface Analysis Satellite Application Facility (LSA SAF) product: the Downward Surface Shortwave Flux (DSSF) ($W\,m^{-2}$), the Land Surface Temperature (LST) (°C), the Albedo parameter, the fraction of vegetation cover (FVC), and leaf area index (LAI) with a 3 km spatial resolution and instantaneous to daily temporal resolutions. In addition, we use the SPOT-Vegetation data Normalized Difference Vegetation Index (NDVI) product with 10 d and 1 km temporal and spatial resolutions respectively. Moreover, the topographic information (Digital Elevation Model DEM) was produced originally by NASA with a spatial resolution equal to 90 m. We use all data as input into the SEBS model.

ILWIS software environment is used to preprocess the data files. Data were downloaded and then imported and pre-processed using the GEONETcast Toolbox (Maathuis et al., 2011). The resampling of imported products was achieved into a geographic coordinate system datum WGS84 with a common 3 km spatial resolution.

3 Models and methodology

The Surface Energy Balance System SEBS is applied for the study area at 3 km spatial resolution. The hydrological balance model is applied at the scale of the watershed (lumped model). SEBS estimations are summed over the basin's pixels and then averaged to be compared to BBH estimations.

3.1 Bucket Bottom Hole (BBH) model

The BBH model is a water balance model developed by Kobayashi et al. (2001). Model inputs are daily precipitation and potential evapotranspiration. Outputs are mean daily river discharge data, actual evapotranspiration, percolation, and capillary rise. The model has seven parameters which are parameter (a) linked to the field capacity, parameter (b) linked to the recession of soil humidity, (c) representing capillary rise pressure, (p) representing soil porosity, (D) for the thickness of the active soil layer (mm), (η) the inverse of efficient porosity, and (σ) the resistance of vegetation to evapotranspiration. Bargaoui and Houcine (2010) reformulated parameters (a, b, c) using soil texture parameters: field capacity (SFC), saturated hydraulic conductivity (K_s), effective soil porosity (p), and the soil water retention curve shape parameter (B) as well as the thickness of the active soil layer (D). Rawls et al. (1982) and Cosby et al. (1984) models are adopted to estimate K_s and SFC respectively. The weights for computing spatial values of SFC and porosity are obtained considering the percent of basin area occupied by every soil texture type and the arithmetic mean. The geometric mean is used for K_s because it is known to be log-normally distributed.

The three remaining parameters (D, η, σ) are fitted using the observed rainfall and runoff time series. We divide the observations into a training subset (calibration) and a validation subset of 3 years in a systematic way (Ben Jaafar and Bargaoui, 2020). The model is running on the scale of the daily time step and the fitting evaluation adopts the 10 d and monthly time steps. Two criteria are considered to fit the parameters: (i) The relative error in absolute value (AARE) and (ii) the Nash–Sutcliffe coefficient (NSC) (Nash and Sutcliffe, 1970). All solutions with NSC > 0.75 are selected if they perform AARE < 0.2. The ranges of research are respectively $0 < D \leq 1500$ mm, $0 < \eta \leq 1$, $0 < \sigma \leq 2.5$. The generalized fractional sample test (GSST) proposed by Coron et al. (2012) is adopted as well as equifinality (Beven, 2006) to identify model parameters. Finally, the selected sets (D, η, σ) are used to run the model from 2010 to 2018 continuously.

3.2 Surface Energy Balance System SEBS

The SEBS model developed by Su (2002) combines remote sensing products and in situ meteorological data to estimate turbulent surface energy fluxes and the evaporative fraction. AET is estimated using the energy balance equation. The SEBS model considers a dry limit (λE flux $= 0$) and a wet limit controlling the sensible heat flux. The relative evaporation (Λ_r) is calculated from the sensible heat flux at wet (H_{wet}) and dry (H_{dry}) limits (Su, 2002).

The pre-packaged SEBS model under ILWIS is applied in this work. Considering crop evolution, essentially cereal crops only six 10 d periods are considered: the first 10 d of January (dec1), March (dec7), April (dec10), May (dec13), July (dec19), and September (dec25).

3.3 Comparison of AET-BBH and AET-SEBS estimations

Eighteen 10 d periods (six periods times 3 years) were considered to achieve comparison for each basin. Daily AET-SEBS are summed over 10 d periods and spatially aggregated at the level of the basin to be compared to BBH estimates.

Table 1. BBH model parameters (SFC, K_s, p) and observed ratios of mean annual AET to rainfall and runoff to rainfall.

Basin	Area	AET (%)	Runoff (%)	SFC	K_s (mm j^{-1})	Porosity, p	Obs. per
Barbara	108	49	51	0.43	410.2	0.43	1982 to 1989
Beja	207	59	41	0.45	199	0.46	1950 to 1961
Douimis	56	84	16	0.43	290.7	0.43	1963 to 1974
Joumine	448	56	44	0.44	243.8	0.44	1966 to 1980
Sejnane	376	72	28	0.44	352.7	0.43	1961 to 1975

Table 2. Range of BBH model parameters (D, η, σ) resulting from the calibrated period for each basin.

Basin	No. solution	D (mm)	η	Σ
Barbara	6	50–350	0.37–0.93	0.07–0.81
Beja	10	750–1500	0.27–0.46	0.01–0.76
Douimis	7	150–1500	0.32–0.51	0.13–2.45
Joumine	9	1200–1500	0.29–0.44	0.01–0.95
Sejnane	13	250–1500	0.28–0.46	0.01–0.78

The agreement between the actual evapotranspiration estimates is based on regression and box-plot analysis.

4 Results and discussion

Observation periods for BBH calibration are shown in Table 1. The long-term annual water balance is written as

$$P + R + E = 0, \tag{1}$$

where P, R, and E respectively average long-term precipitation, runoff, and AET. It is found that AET contribution to water balance varies from 49 % to 84 % which highlights its importance in this region. The results of estimating soil texture parameters SFC (capacity field), p (effective soil porosity), and K_s (saturated hydraulic conductivity) are shown in Table 1. The final number of solutions varies from 6 to 13 depending on the basin. The range of selected parameters under the equifinality framework is shown in Table 2.

Figure 2 shows the scatterplot per basin using the best solution. For Beja and Douimis basins, significant coefficients of determination are obtained ($R^2 = 0.6$) and ($R^2 = 0.5$) respectively. Contrary to Barbara, Joumine, and Sejnane with

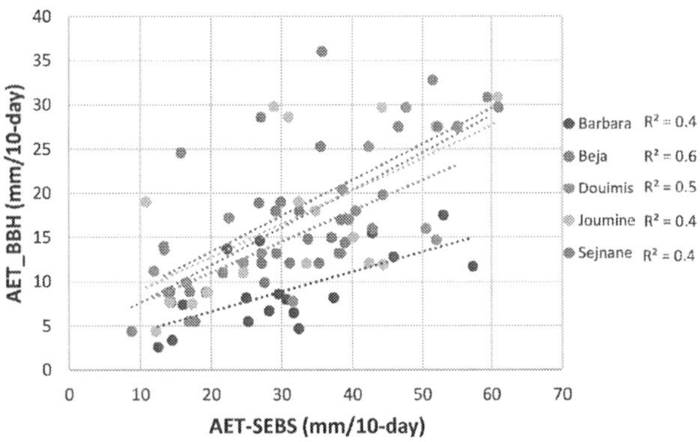

Figure 2. Scatter plot of actual evapotranspiration simulated by BBH model (AET-BBH) and derived by the SEBS model (AET-SEBS) at the level of five basin and 10 d periods of 2010, 2017, and 2018.

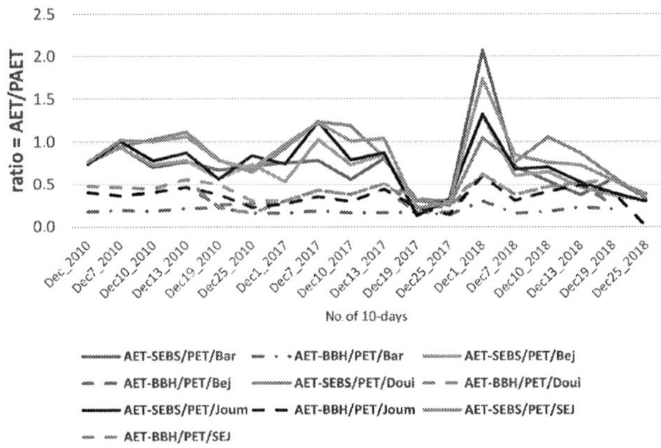

Figure 4. Time evolution of AET and PET ratio for both SEBS and BBH results.

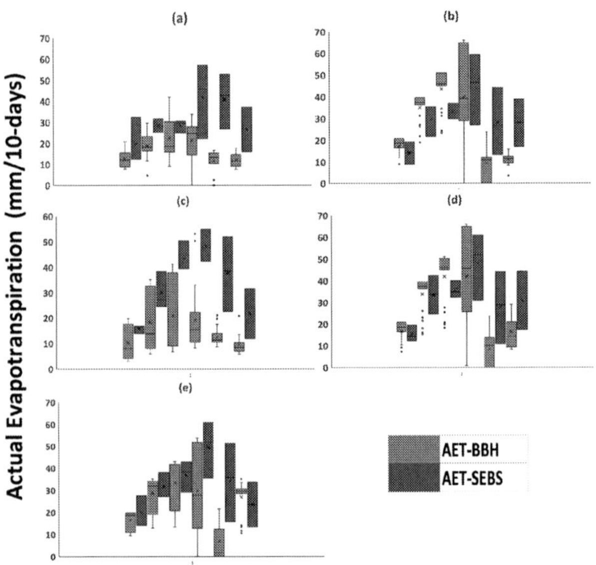

Figure 3. Boxplots of AET-BBH set data and AET-SEBS estimates for 2010, 2017, and 2018.

$R^2 \leq 0.4$. Merging results of all basins and periods, the MAE is found at $19\,\mathrm{mm}\,10\,\mathrm{d}^{-1}$ and the RMSE at $17\,\mathrm{mm}\,10\,\mathrm{d}^{-1}$. Boxplots of AET-BBH and AET-SEBS estimations per period (6 periods) and basin (5 basins) are presented in Fig. 3. It highlights that AET-SEBS values go well with the set of AET-BBH values described in boxes for the wet season which correspond to the first 10 d of January, March, April, and May, and are not satisfactory for the dry season corresponding to the first 10 d of July and September unless for Joumine and Sejnane basins.

AET estimations are constrained by PET with AET $\leq$ PET meaning that the ratio should verify the condition AET/PET ≤ 1. Figure 4 shows the time evolution of this ratio for both SEBS and BBH results. The condition is naturally met for BBH. However, it is seen that it is not obvious using SEBS estimations. The pre-packaged SEBS model seems to overestimate AET in northern Tunisia. The same result is found by Lu et al. (2012) and Gibson et al. (2011) over a semi-arid heterogeneous surface. They explain this discrepancy (overestimation) due to the complexity of the SEBS model which uses several parameters. These researchers indicate that the analysis of the remote sensing product and the interpolation of meteorological data over a semi-arid heterogeneous surface can be the potential source of errors that can be spread through the processing sequence. Likewise, studies by Gokmen et al. (2012), and Lu et al. (2012) indicate that in varied landscapes where classes of different land cover are combined into a single pixel, the SEBS model underestimates sensible heat flux leading to an overestimation of AET. As approved by Lu et al. (2012), the SEBS model overestimates evaporative fraction below complex characteristics of the surface cover. As highlighted by Huang et al. (2015), AET overestimation is high in arid and semi-arid areas due to the dominant water stress condition which is approved by our case study as the 3 chosen years are declared as stressed years.

5 Conclusions

Two models are selected to assess the actual evapotranspiration in northern Tunisia at the basin scale. The SEBS model adopts surface energy balance modeling and Satellite data (LST, DSSF, LAI, NDVI, Albedo) as well as ground meteorological data. The lumped hydrological model (BBH) is based on water balance modeling and ground data. Parameters of the BBH model are calibrated concerning two performance criteria using two thresholds: absolute relative bias < 0.2 and the Nash coefficient > 0.75 and adopting an equifinality framework. The set of calibrated parameters (D, η, σ) with the best value of Nash criteria is adopted to calculate the AET. Values vary from 10 to 60 mm 10 d^{-1} depending on the periods. We underline an overestimation of AET-SEBS compared to AET-BBH with a mean absolute error (MAE) and a root means square error RMSE of about 17 and 19 mm 10 d^{-1}. Therefore, other models and products as MOD 16 and LSA SAF AET should be investigated for estimating actual evapotranspiration in northern Tunisia.

Code availability. The SEBS model is included under Ilwis environment which is free at https://www.itc.nl/ilwis/download/ilwis33/ (University of Twente, 2024). The code of the BBH model was developed The code of the BBH model was developed under the programming language Fortran by Zoubeida Bargaoui and we have no permission to publish it.

Data availability. The hydrometeorological data are not publicly accessible. These data are purchased from the National Institute of Meteorology of Tunisia, and we have not been allowed to publish the observed data.

Author contributions. NA: satellite data, data processing, SEBS model running, writing. ZB: calculation code, ground data, supervision, writing. ABJ: run of the BBH model. CMM: know-how on using satellite data and the SEBS model.

Competing interests. The contact author has declared that none of the authors has any competing interests.

Disclaimer. Publisher's note: Copernicus Publications remains neutral with regard to jurisdictional claims made in the text, published maps, institutional affiliations, or any other geographical representation in this paper. While Copernicus Publications makes every effort to include appropriate place names, the final responsibility lies with the authors.

Special issue statement. This article is part of the special issue "IAHS2022 – Hydrological sciences in the Anthropocene: Past and future of open, inclusive, innovative, and society-interfacing approaches". It is a result of the XIth Scientific Assembly of the International Association of Hydrological Sciences (IAHS 2022), Montpellier, France, 29 May–3 June 2022.

Review statement. This paper was edited by Christophe Cudennec and reviewed by two anonymous referees.

References

Bargaoui, Z. and Houcine, A.: Sensitivity to calibration data of simulated soil moisture related to drought indices, Science et changements planétaires / Sécheresse, 21, 294–300, https://doi.org/10.1684/sec.2010.0263, 2010.

Bastiaanssen, W. G. M., Menenti, M., Feddes, R. A., and Holtslag, A. A. M.: A remote sensing surface energy balance algorithm for land (SEBAL). 1. Formulation, J. Hydrol., 212/213, 198–212, https://doi.org/10.1016/S0022-1694(98)00253-4, 1998.

Ben Jaafar, A. and Bargaoui, Z.: Generalized Split-Sample Test Interpretation Using Rainfall-Runoff Information Gain, J. Hydrol. Eng., 25, 04019057, https://doi.org/10.1061/(ASCE)HE.1943-5584.0001868, 2020.

Beven, K. J.: A manifesto for the equifinality thesis, J. Hydrol., 320, 18–36, https://doi.org/10.1016/j.jhydrol.2005.07.007, 2006.

Coron, L., Andréassian, V., Perrin, C., Lerat, J., Vaze, J., Bourqui, M., and Hendrickx, F.: Crash testing hydrological models in contrasted climate conditions: An experiment on 216 Australian basins, Water Resour. Res., 48, W05552, https://doi.org/10.1029/2011WR011721, 2012.

Cosby, B. J., Hornberger, G. M., Clapp, R. B., and Ginn, T. R. A.: Statistical Exploration of the Relationships of Soil Moisture Characteristics to the Physical Properties of Soils, Water Resour. Res., 20, 682–690, https://doi.org/10.1029/WR020i006p00682, 1984.

Eagleson, P.: The evolution of modern hydrology, Adv. Water Resour., 17, 3–18, https://doi.org/10.1016/0309-1708(94)90019-1, 1994.

Gibson, L. A., Münch, Z., and Engelbrecht, J.: Particular uncertainties encountered in using a pre-packaged SEBS model to derive evapotranspiration in a heterogeneous study area in South Africa, Hydrol. Earth Syst. Sci., 15, 295–310, https://doi.org/10.5194/hess-15-295-2011, 2011.

Gokmen, M., Vekerdy, Z., Verhoef, A., Verhoef, W., Batelaan, O., and Tol, C.: Integration of Soil Moisture in SEBS for Improving Evapotranspiration Estimation under Water Stress Conditions, Remote Sens. Environ., 121, 261–274, https://doi.org/10.1016/j.rse.2012.02.003, 2012.

Huang, C., Li, Y., Gu, J., Lu, L., and Li, X.: Improving Estimation of Evapotranspiration under Water-Limited Conditions Based on SEBS and MODIS Data in Arid Regions, Rem. Sens.-Basel, 7, 16795–16814, https://doi.org/10.3390/rs71215854, 2015.

Kite, G. W.: Manual for the SLURP hydrological model, Natl. Hydrol. Res. Inst., Saskatoon, SK, Canada, 10, 120 pp., https://api.semanticscholar.org/CorpusID:128060145 (last access: 14 February 2024), 1997.

Kobayashi, T., Matsuda, S., Nagai, H., and Teshima, J.: A bucket with a bottom hole (BBH) model of soil hydrology, IAHS-AISH publication, 270, 41–45, https://api.semanticscholar.org/CorpusID:140579764 (last access: 14 February 2024), 2001.

Lu, J., Li, Z. L., Tang, R., Tang, B. H., Wu, H., Yang, F., Labed, F., and Zhou, G.: Evaluating the SEBS-estimated evaporative fraction from MODIS data for a complex underlying surface, Hydrol. Process., 27, 3139–3149, https://doi.org/10.1002/hyp.9440, 2012.

Liou, Y. A. and Kumar Kar, S.: Evapotranspiration estimation with remote sensing and various surface energy balance algorithms – a review, Energies, 7, 2821–2849, https://doi.org/10.3390/en7052821, 2014.

Maathuis, B., Mannaerts, C., Retsios, B., Schouwenburg, M., and Lemmens, R.: Geonetcast toolbox: installation, configuration and user guide of the Geonetcast Toolbox plug-in for ILWIS

3.7.2: XML version 1.3: e-book, University of Twente, Faculty of Geo-Information Science and Earth Observation (ITC), http://www.itc.nl/library/Papers_2012/general/GNC_Toolbox.pdf (last access: 14 February 2024), 2011.

Nash, J. E. and Sutcliffe, J. V.: River flow forecasting through conceptual model's part I – A discussion of principles, J. Hydrol., 10, 282–290, https://doi.org/10.1016/0022-1694(70)90255-6, 1970.

Rawls, W., Brakensiek, D., and Saxton, K.: Estimation of soil water properties, T. ASAE, 25, 1316–1320, https://doi.org/10.13031/2013.33720, 1982.

Riou, Ch.: Une formule empirique simple pour estimer l'évapotranspiration potentielle moyenne en Tunisie, Les cahiers de l'ORSTOM, Série Hydrologie Bondy, 17, 129–137, http://geoprodig.cnrs.fr/items/show/152353 (last access: 14 February 2024), 1980.

Roerink, G., Su, Z., and Menenti, M.: S-SEBI: A simple remote sensing algorithm to estimate the surface energy balance, Phys. Chem. Earth Pt. B, 25, 147–157, https://doi.org/10.1016/S1464-1909(99)00128-8, 2000.

Su, Z.: The Surface Energy Balance System (SEBS) for estimation of turbulent heat fluxes, Hydrol. Earth Syst. Sci., 6, 85–100, https://doi.org/10.5194/hess-6-85-2002, 2002.

University of Twente: ILWIS 3.3, https://www.itc.nl/ilwis/download/ilwis33/, last access: 14 February 2024.

Verner, D.: Tunisia in a Changing Climate: Assessment and Actions for Increased Resilience and Development, Word Bank Studies, https://doi.org/10.1596/978-0-8213-9857-9, ISBN: 978-0-8213-9857-9, e-ISBN: 978-0-8213-9858-6, 2013.

Wang, K. and Dickinson, R. E.: A review of global terrestrial evapotranspiration: observation, modeling, climatology, and climatic variability, Rev. Geophys., 50, RG2005, https://doi.org/10.1029/2011RG000373, 2012.

Proc. IAHS, 385, 477–484, 2024
https://doi.org/10.5194/piahs-385-477-2024

Open Access PIAHS

How can altimetry data be used for water resources management (SDG 6.5.1)? Development of a method using altimetry data from the Envisat, Jason, Jason 2 and Sentinel 3A satellites

Thomas Legay[1], Yoann Aubert[1,8], Julien Verdonck[1], Jérémy Guilhen[2,3], Adrien Paris[6,7,10], Jean-Michel Martinez[4], Sabine Sauvage[3], Pankyes Datok[9], Vanessa Dos Santos[3], José Miquel Sanchez-Perez[3], Stéphane Bruxelles[5], Emeric Lavergne[2], and Franck Mercier[2]

[1]BRL Ingénierie. 1105, av. Pierre Mendès-France BP 94001, 30001 Nîmes cedex 5, France
[2]CLS Group – Pôle Terre et Hydrologie, 11 rue Hermès, Parc Technologique du Canal 31520 Ramonville Saint-Agne, France
[3]Laboratoire Ecologie Fonctionnelle et Environnement, Université de Toulouse, CNRS, INPT, UPS, Toulouse, France
[4]IRD – Institut de recherche pour le Développement, 14 Avenue Edouard Belin, 31400 Toulouse, France
[5]CENEAU, 265 avenue de l'Industrie 34 820 Teyran, France
[6]Hydro Matters, 1 Chemin de la Pousaraque, 31460 Le Faget, France
[7]LEGOS, Université de Toulouse, CNES, CNRS, IRD, UPS (Toulouse), France
[8]EDF Hydro – DTG, Département Eau – Environnement, Service Prévision, 4, rue Claude-Marie PERROUD – Bat A, 31100 TOULOUSE CEDEX, France
[9]AE-FUNAI EBONYI STATE NIGERIA, 44JR+PG7, IKWO IKWO LGA, Abakaliki, Nigeria
[10]Hydro Matters, 1 Ch. De la Pousaraque, 31460 Le Faget, France

Correspondence: Thomas Legay (thomas.legay@brl.fr)

Received: 31 May 2022 – Revised: 4 March 2024 – Accepted: 11 March 2024 – Published: 19 April 2024

Abstract. Water resources management relies on the use of hydrometric data collected mainly from in situ stations. Despite the efforts made in setting up and maintaining a network of hydrometric and meteorological stations, water resource managers face many problems, such as equipment degradation during floods, incidents of vandalism and issues related to inaccessibility. Satellite data can improve water resources monitoring, regardless of its objectives (strategic management of water resources, management of structures, forecasting of floods and low flows, etc.). These data have the advantage of covering the entire world at regular intervals, including difficult-to-access areas. Through several projects, various scientific and industrial partners (IRD, CNES, CLS, CNRS, CENEAU, etc.) have contributed to assess and enhance the use of satellite data for the management of water resources at the scale of a catchment (e.g. the Amazon) or a territory (Uganda). The objective is to deliver a method based on the combination of hydrological models, in-situ data and satellite data. This article presents the current possibilities and limitations of using satellite data to optimize the monitoring of water resources.

Keywords. Coupling; spatial altimetry; hydrological modelling; GR5J; regionalization of parameters; comparison

1 Introduction

The management of water resources, at local and regional scales, requires the use of hydrometric and meteorological data. These data usually come from in situ stations, whose location on a hydrographic network conditions the knowledge of flows. Despite the efforts made to set up and maintain a network of stations, water resource managers face numerous problems:

- The location of the stations is not systematically optimal, especially in southern countries (geopolitical context, access difficulties, etc.);
- The number of functional stations is insufficient, especially in the southern countries;
- Data are not available for a period of time that is long enough to analyze the historical context and understand the impacts of climate change.

Therefore, in situ hydrometric data alone are often insufficient for the design of hydraulic structures (for extreme events management) or the management of water resources (Vörösmarty et al., 2001).

Regarding in situ meteorological data, they are often too few, both historically and spatially, which limits their use (e.g. construction of a spatial rainfall, frequency analysis) (Lebel et al., 1997; Li and Heap, 2008).

Over the past two decades, two areas of spatial applications, particularly important for hydrology have been developed:

- Spatial altimetry: satellites equipped with altimeters are commonly used to measure the height of the world's water bodies and/or rivers (from the precursor Geosat, through Topex/Poseidon, ENVISAT, Saral/AltiKa, Jason 2 to Jason-3, Sentinel 3A and 3B, and even the future SWOT mission, etc.) (Paris, 2015);
- Spatial meteorology: various research centers are working on integrating satellite information, producing spatialized data (rainfall, ETP, temperature) and comparing them with in situ data (Satge et al., 2019).

At the same time, the field of hydrology is expanding and developing many tools for analysis, geostatistical processing and modeling. These tools can be used to generate information at variable spatial resolutions (pixel of 1 km^2 to several hundred km^2) and temporal resolutions (from 30 min to a month), even in regions without measurements.

The combination of spatial data with in situ data and hydrological tools can partially address the instrumentation problem (Paiva et al., 2013; Paris et al., 2016; Garambois et al., 2017; Bogning et al., 2018; Huang et al., 2020; Andriambeloson et al., 2020). The consortium composed of CLS, IRD, BRL Ingénierie, CENEAU, and CNRS developed an approach mixing altimetry, in situ data, and modeling in two projects:

- The HydroSim project studies the Amazon catchment; it is funded by the Occitanie region (France) and led by CLS (France);
- The SmartBasin project focuses on the Ugandan hydrographic network; it is financed by the General Directorate of the Public Treasury (France) and led by BRL Ingénierie

2 Objectives and method

2.1 Objectives

The objective of the HydroSim and SmartBasin projects is to develop a method for integrating and exploiting spatial data, in particular altimeter data, for the management and monitoring of water resources. This method is based on a combination of spatial data, in situ data and hydrological modeling tools. The objective of this manuscript is to evaluate the relevance of altimetry data for watercourse monitoring and to evaluate the method developed to integrate them into the management and monitoring of water resources.

2.2 Method

2.2.1 General approach

Satellites equipped with an altimeter can measure the water level of a river. Like any satellite, they are characterized by:

- Their orbit: the trajectory of the satellite around the Earth is fixed;
- Their frequency of passage over the same point: depending on the satellite, it varies from a few days (7 d) to over a month.

The ground track of a satellite is the projection of its orbit on the Earth's surface.

The satellites measure the water level of a river at the intersection between the ground track and the river. Those sites are called virtual stations. The virtual stations can be located on watercourses, or sections of watercourses where no information is available.

The approach consists of complementing water level data measured by satellites with flow data. This enables the construction of virtual rating curves, i.e., a relation between the water level measured by the satellite and the flow of the river. Flow data is generated using in situ hydrometric data and hydrological modeling tools.

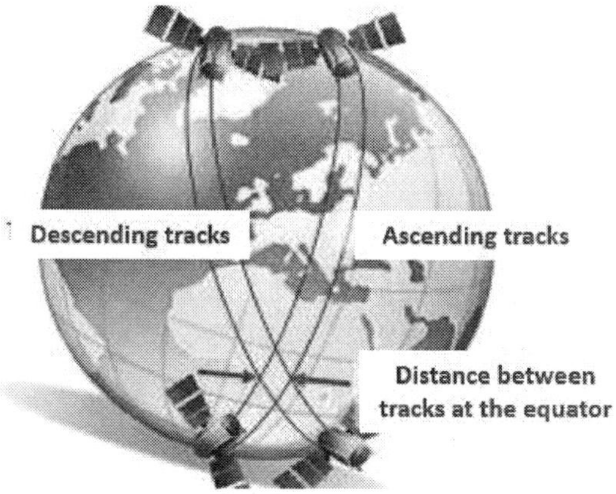

Figure 1. Trajectory of a satellite around the globe. Notions of tracks and inter-tracks.

The method is as follows:

1. Identification of virtual stations

   Over the study area, this step aims to identify the sites where satellites equipped with altimeters correctly measure the water height.

2. Modeling of a flow chronicle at the location of the virtual stations

   This step aims to model the flow at the virtual stations.

   This step relies on:

   - The collection and analysis of hydrometric and meteorological in situ data,
   - The calibration of a rainfall-runoff model at the hydrometric stations (the criteria used for calibration is the Kling Gupta Efficiency (Gupta et al., 2009)),
   - The flow modelling at the virtual stations,

3. Comparison of modeled flows with spatial altimetry data

   This step aims to relate model flow with water levels measured by spatial altimetry. It allows virtual rating curves to be built.

4. Verification of the modeled flows at the virtual stations

   This step consists in calculating the flow using the altimeter data and the associated virtual rating curve, then verifying that the modeled flow is consistent with the closest in situ data.

This methodology is frequently used in the scientific community and is proven to be useful for satellite monitoring of many basins around the world (Paris, et al., 2016; Bogning, et al., 2020; Andriambeloson et al., 2020). It is also the source of the deployment of free altimetry flow estimates on the Theia/Hydroweb website (http://hydroweb.theia-land.fr/, last access: 20 June 2019).

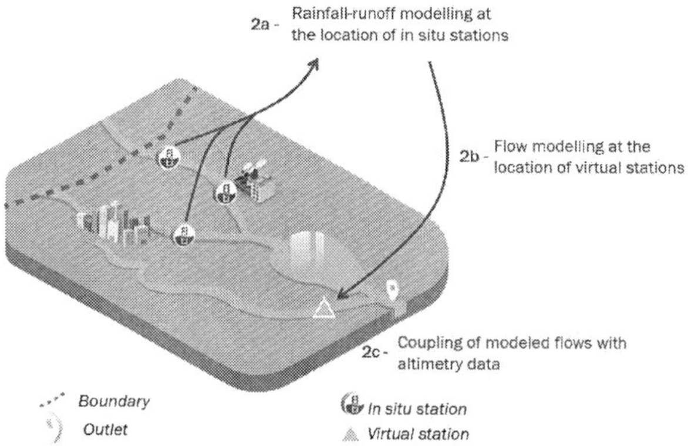

Figure 2. Diagram of the general approach applied for the modelling of the flow at the virtual stations (BRL Ingénierie).

2.2.2 Specificity of each project

Each project has its own specificities concerning the data used, the models implemented, the method to simulate the flow at virtual stations and the method of verification of the modelled flows.

Virtual stations

In the scope of the HydroSim project, the virtual stations available on HydroWeb were used. HydroWeb provides data from 981 virtual stations distributed among 3 satellite missions: 97 virtual stations for the Jason and Jason 2 satellite missions, 143 virtual stations for Sentinel-3A and 741 virtual stations for Envisat. The altimetry data used cover a period from September 2002 to June 2019. Depending on the satellite mission, data are available for frequencies of 10 d (Jason and Jason 2), 27 d (Sentinel 3A) and 35 d (Envisat).

For the SmartBasin project, 68 virtual stations have been identified by CLS. 34 virtual stations have been identified for the Jason2/Jason3 constellation and 34 virtual stations have been identified for the Sentinel 3A satellite. The collected data cover the period from July 2008 to February 2019. The frequency of available data is 10 d for Jason and Jason 2 satellites and 27 d for Sentinel 3A. After reanalysis of the data, 39 virtual stations were selected.

In situ data

For the HYDROSIM project, data from more than 1200 rainfall stations were collected from different countries (Brazil, Peru, Bolivia), and from different sources (ANA/IRD, SENAMHI Peru and WMO). For the Smart Basin project, data from 17 stations were collected from the Ministry of Water and Environment of Uganda and from the WMO. After data analysis with the KRITIKOR tool [EDF-DTG], only 413 and 17 rainfall stations were selected as relevant (minimum of 10 years of data, threshold on gaps in series, etc.)

respectively over the Amazon basin and the Ugandan hydrographic network.

For the Amazon catchment, flow records were collected from the IRD and their HYBAM database; for Uganda, they were collected from the Ministry of Water and Environment (MoWE). All stations were analyzed and criticized. At the end, 119 hydrometric stations were selected for the Amazon catchment. For Uganda, most of the collected hydrometric data were considered questionable: at the end, only 8 hydrometric stations were selected.

Choice of the rainfall-runoff model

Different classifications of models can be found in literature. One major criteria divide models into:

- Physically based models, such as SWAT. They represent the processes of the rainfall-runoff relationship using the physical laws governing these processes;

- Conceptual models, such as GR models (Le Moine, 2008). These models aim to represent the main hydrological processes in a simplified way without resorting to parameterizations of the physical laws that govern them.

Both types of models have advantages and disadvantages. Physically based models, such as SWAT (Arnold et al., 2007), can provide distributed information over the entire study area. Once the hydrological model is calibrated on observed flows at in situ stations (either over an entire catchment, or over parts of the catchment called sub-catchments), the model provides a series of modeled flows over the entire river system including the virtual stations.

The GR5J conceptual model is a global model. This model is used to model the flow of one catchment defined by its outlet. This outlet corresponds to the location of the hydrometric station. Thus, the GR5J model must be calibrated for each hydrometric station. In order to model the flow at the virtual stations, it is necessary to go through a parameter regionalization step.

The following models have been used:

- HydroSim: CNRS and BRL Ingenierie were in charge of carrying out the flow modelling of the Amazon catchment. CNRS modelled the flow using the SWAT model and BRL Ingenierie used the GR5J model;

- SmartBasin: the flow modelling work was carried out only by BRL Ingenierie, the flow has been modelled using GR5J.

Flow simulation at the location of the virtual stations

Depending on the rainfall-runoff model used, the method to simulate the flow at the location of the virtual stations differs.

With the SWAT model, as explained previously, once the hydrological model is calibrated on observed flows at in situ stations, the model provides a chronicle of modeled flows over the entire river system including the virtual stations.

With the GR5J model, simulation of the flow at the location of the virtual station implies the use of regionalization of parameters method.

In the context of hydrology, regionalization is defined as a transfer of information from a gauged site where this information is known to an ungauged site where this information is unknown. Here, the ungauged site is the virtual station and the information to be transferred are the parameters of the GR5J model.

The method to simulate the flow at the location of the virtual stations consists in:

- Calibrating the GR5J model for every hydrometric station,

- Applying regionalization of parameters methods to calculate the parameters for the catchments of virtual stations.

Different methods are available for regionalizing parameters. A preliminary step consisted in identifying the most robust method. For this, the "leave one out" method was applied with the in situ hydrometric stations. This method is deployed in several stages:

- We consider all the in-situ stations for which we have calibrated a GR5J model. We remove an in-situ station from this set.

- We calculate a new set of parameters by regionalization for the excluded station from the parameters of the calibrated models.

- We simulate the flow of the excluded station. We check the consistency of the simulated flow with the flow observed at this station.

This work is carried out for all the in-situ stations and for all the desired regionalization methods. The selected regionalization method is the one which provides the best simulations of the flow at the in-situ stations. The different regionalization method tested are:

- Geographical regionalization (Poncelet et al., 2017),

- Physiographical regionalization (Poncelet et al., 2017),

- Machine learning methods (Random Forest (Saadi et al., 2019), neuronal network (Riedmiller and Braun, 1993; Intrator and Intrator, 1993; Anastasiadis et al., 2005)).

- Simulating the flow at the location of the virtual stations using the retained regionalization method and the GR5J model.

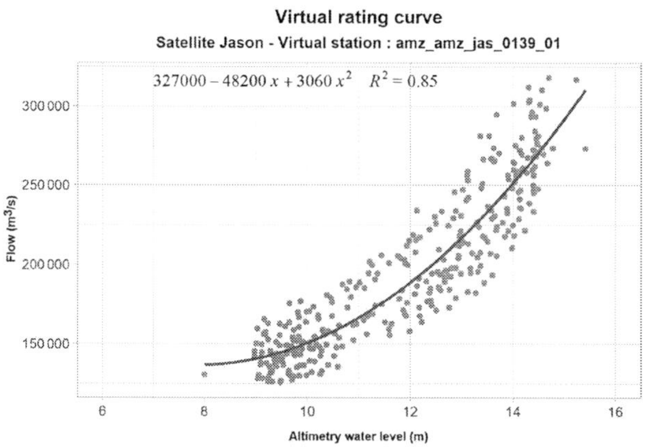

Figure 3. Correlation between modelled flows and satellite water levels – Station amz-amz-jas-0139-01 ($S = 4\,696\,804\,\mathrm{km}^2$).

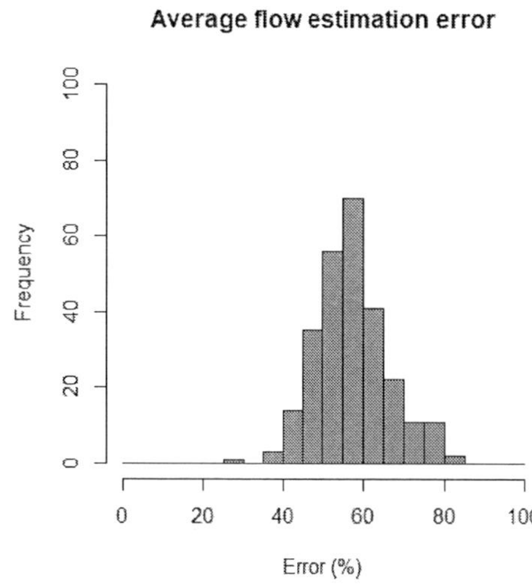

Figure 4. Distribution of average flow estimation error for the Madeira and Tapajos catchments.

Performance of the implemented models

Depending on the project, different verification methods were applied to assess the relevance of the flow modeled at the virtual stations.

For the HydroSim project, a comparison of the flows modeled by the SWAT and GR5J models at the virtual stations was performed.

For the SmartBasin project, no SWAT model was implemented because the CNRS did not take part of the project. The project includes the installation of 3 hydrometric stations at the virtual stations' location. These hydrometric stations will enable the verification processes of the modeled flow (its consistency with the observed flow) and the consistency between the water level measured by the station the one measured by spatial altimetry.

3 Results and discussion

We present hereafter the results obtained for 2 pilot sub-catchments of the Amazon catchment: the Madeira and the Tapajos sub-catchments. These sub-catchments have been modelled with 2 rainfall-runoff models: SWAT and GR5J. In these sub-catchments, 266 virtual stations were studied.

Figure 3 shows an example of correlation between the water level measured by the satellite Jason at virtual station amz-amz-jas-0139-01 and the regionalized flow of the GR model. The good correspondence between water level and discharge enable the construction of a virtual rating curve: a variation in water level measured by spatial altimetry is correlated with a variation in discharge. Here, the correlation coefficient is significant: 0.85. However, Fig. 3 shows a discrepancy between the regression curve and the observed flow. This gap varies between 0 % and 60 % compared to the regression curve, with an average of 37 %.

Figure 4 shows the distribution of the average flow estimation error for the 266 virtual stations of the Madeira and the Tapajos catchments. 34 % (90 virtual stations) present an error between 50 % and 60 % and 66 % of the virtual station (175 virtual stations) present an error between 45 % and 65 %.

Such discrepancies have several origins. They include:

- The imprecision linked to the measurement of water level by satellite. Even though technologies are advancing, measurement uncertainties remain.
- Uncertainties linked to flow modeling by regionalization of parameters.

However, such results provide discharge information at places where this knowledge is unavailable, or even inaccessible. Such results can be useful for estimating available resources and carrying out water needs-resources assessments.

Figure 5 shows the spatial distribution of the R^2 correlation coefficients between the modeled flows (GR5J at the top and SWAT at the bottom) and the satellite water level. We assume that any variation in elevation is associated with a variation in discharge. We also assume that the flows are natural.

The two figures clearly show that the flows modelled by the SWAT and GR5J models present a good correlation with the water level of the virtual stations: on average the R^2 are 0.79 with GR5J and 0.77 with SWAT. On the upstream parts of the sub-catchments, the correlation coefficients are lower. This can be explained by:

- the small size of the rivers making the interpretation of radar echoes more complex,
- a worse modeling of flows (lack of hydrometric data in these upstream parts, complex representation of rainfall in the Andean zone, etc.).

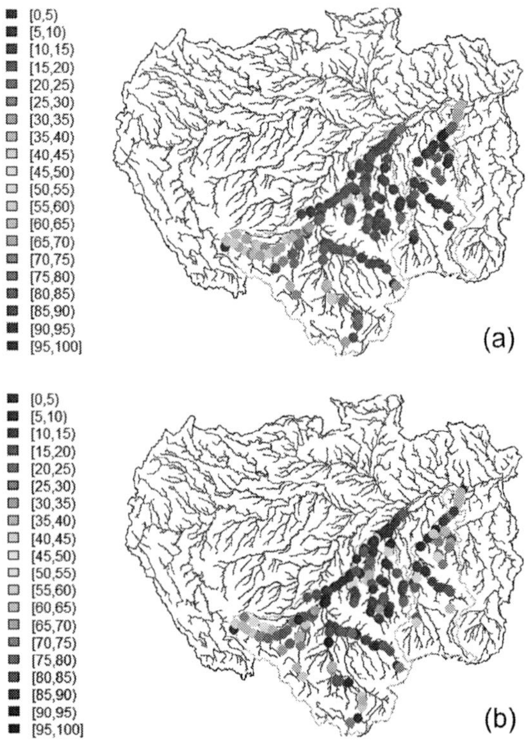

Figure 5. Spatial distribution of the correlation coefficients R^2 (satellite water level – modelled flows) on the Madeira and Tapajos sub-catchments. Correlation coefficients between satellite water level and modelled flows at daily time step and at dates for which altimetric and modelled data are available with GR5J (top picture) and with SWAT (bottom picture).

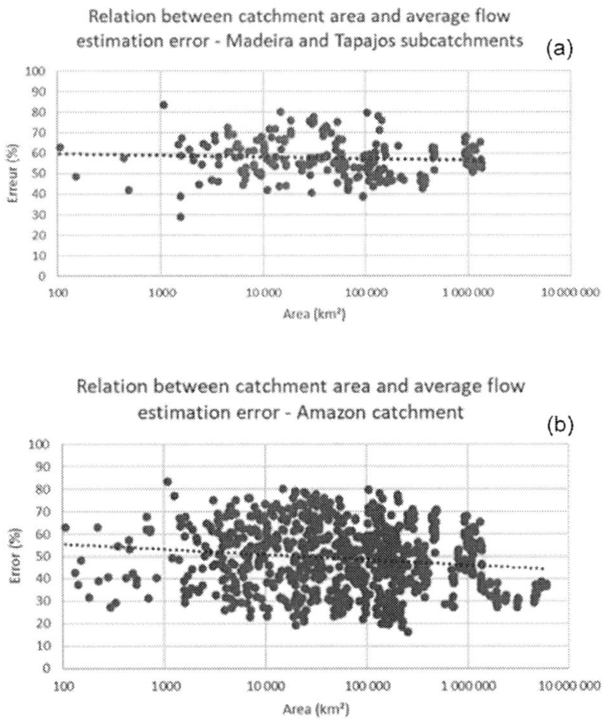

Figure 6. Relation between catchment area and average flow estimation error. For the Madeira and Tapajos subcatchments **(a)** and for the Amazon catchment **(b)**.

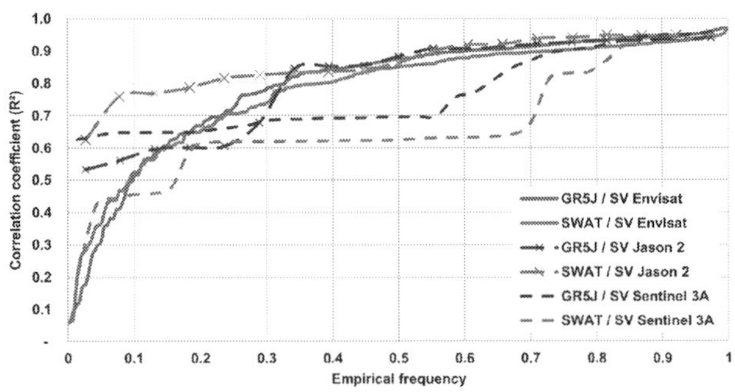

Figure 7. Frequency distribution of correlation coefficients as a function of model type and satellite data type.

We examined the relation between the mean error in the flow estimate and the surface area of the catchment (see Fig. 6). For the Madeira and Tapajos catchment, the mean error in the flow estimate seems to decrease as the area of the catchment is higher. This observation is clearer considering the whole Amazon catchment.

We examined the correlation coefficients according to the type of satellite (Sentinel 3A, Jason2 or Envisat). For this, the correlation coefficients were analyzed in frequency (ranking of R^2 values).

The following graph shows the frequency distributions of the correlation coefficients according to the type of model (GR5J and SWAT) and the type of satellite data:

- JASON 2: 2008–2012,
- ENVISAT: 2002–2010
- SENTINEL 3A: 2016–2019

In general, it is important to note that both models give good correlations between modeled discharge and satellite water level regardless of the satellite product (given the assumptions for each model and satellite data). However, the correlation coefficients calculated with the Sentinel 3A product are lower than with the other data. At the outlet of the Tapajos catchment, the Sentinel 3A satellite track is parallel to the river. The satellite measures the water level of the river at 17 geographically close virtual stations. The correlation coefficient of these 16 stations (43 % of Sentinel 3A virtual stations) is 0.65 to 0.70 for the GR5J model and 0.62 to 0.63 for the SWAT model. This influences the shape of the curve. By retaining only one of these 16 virtual stations (and removing the other 15 stations), the shape of the curve is corrected for GR5J, not for the SWAT model (see Fig. 8). This is due

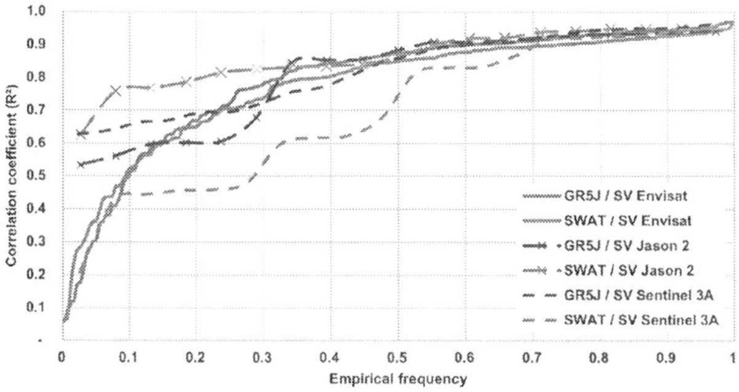

Figure 8. Frequency distribution of correlation coefficients as a function of model type and satellite data type after correction.

to the number of data in common: the SWAT model stops in 2016. The R^2 is not statistically significant.

On the Uganda river network, 39 virtual stations were studied. The figure hereafter shows the spatial distribution of the R^2 correlation coefficients between the satellite water level and the modeled discharge on the Ugandan river network.

For Uganda, the results are worse than those of the Amazon catchment. This is explained by:

- The low number of hydrometric stations available for calibrating the GR5J model. Data from 8 hydrometric stations were made available to us;
- The quality of hydrometric data. The analysis of the data highlighted numerous uncertainties regarding their quality. The calibration of the GR5J models provided Nash-Sutcliffe values between 19 and 70;
- Rainfall and ETP data used as input to the GR5J model. We did not have any climatic data on the ground. Therefore, we had to use spatial dataset. The choice of spatial data could not be validated by a comparison with in situ data.

4 Conclusions and perspectives

Satellite data have the advantage of covering the globe and provide regular and homogeneous altimetric information (same measurement methods) on regions with few or no in situ measurements. The HydroSim and SmartBasin projects confirm the interest of this spatial information, in particular spatial altimetry, for monitoring continental water resources.

The idea of spatial altimetry is to efficiently complement in situ data collection networks, and the coupling with these data and modeling is growing today. Within the framework of the HydroSim and SmartBasin projects, it is planned in the near future to enhance the use of spatial altimetry data to confirm the results of real-time hydrological modelling at many points of the network. It will be possible, in a few months, to use additional satellite information following the launch of the Surface Water Ocean Topography (SWOT, Biancamaria et al., 2016) mission with a much more precise temporal and spatial resolution than previous satellites. The analysis tools produced in these two projects will probably be used by future hydrologists.

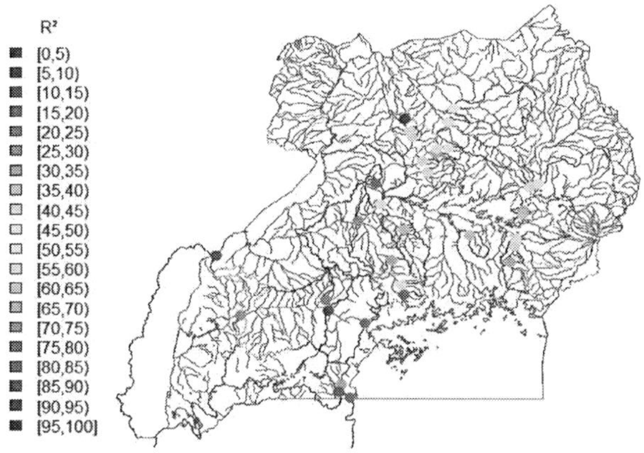

Figure 9. Spatial distribution of R^2 correlation coefficients (satellite water head – modelled flows) over the Uganda river system.

Data availability. The data used in the two projects (SmartBasin and HydroSim) are the property of institutions who do not wish their data to be publicly accessible.

Author contributions. JV and EL were respectively in charge of the Smart Basin and the HydroSim project administration and supervision. SB was in charge of the project administration and supervision for CENEAU for both projects. SS and JMSP were in charge of the HydroSim project administration and supervision on CNRS' side. JMM oversaw the relation with the beneficiary of the HydroSim project and ensured data collection. TL and YA analysed the data, calibrated the models at the location of the in-situ stations and modelled the flow at virtual stations location for both projects on BRLi's side. JG, PD and VDS analysed the data, calibrated the models at the location of the in-situ stations and modelled the flow at virtual stations location on CNRS's side. AP provided support for the critique of the altimetry data and the relationships between altimetric water level and simulated flow. FM identified the virtual stations, analysed and corrected the data for the Smart Basin project. TL and YA were in charge of writing the initial draft, making corrections and writing the final draft of this manuscript.

Competing interests. The contact author has declared that none of the authors has any competing interests.

Disclaimer. Publisher's note: Copernicus Publications remains neutral with regard to jurisdictional claims made in the text, published maps, institutional affiliations, or any other geographical representation in this paper. While Copernicus Publications makes every effort to include appropriate place names, the final responsibility lies with the authors.

Special issue statement. This article is part of the special issue "IAHS2022 – Hydrological sciences in the Anthropocene: Past and future of open, inclusive, innovative, and society-interfacing approaches". It is a result of the XIth Scientific Assembly of the International Association of Hydrological Sciences (IAHS 2022), Montpellier, France, 29 May–3 June 2022.

Acknowledgements. The results presented in this article are the result of the work of several partners: BRLi, CENEAU, CLS, CNES, CNRS, DG-Treasury, IRD and the Occitanie Region. Without them, we would not have been able to complete this process.

Review statement. This paper was edited by Christophe Cudennec and reviewed by two anonymous referees.

References

Anastasiadis, A., Magoulas Goerge, D., and Vrahatis Michael, N.: New globally convergent training scheme based on the resilient propagation algorithm, Neurocomputing, 64, 253–270, 2005.

Andriambeloson, J. A., Paris, A., Calmant, S., and Rakotondraompiana, S.: Reinitiating depth-discharge monitoring in small-sized ungauged watersheds by combining remote sensing and hydrological modelling: a case study in Madagascar, Hydrolog. Sci. J., 65, 2709–2728, 2020.

Arnold, J. G., Srinivasan, R., Muttiah, R. S., and Williams, J. R.: Large area hydrologic modeling and assessment part I: model development, J. Am. Water Resour. As., 34, 73–89, https://doi.org/10.1111/j.1752-1688.1998.tb05961.x, 2007.

Biancamaria, S., Lettenmaier, D. P., and Pavelsky, T. M.: The SWOT mission and its capabilities for land hydrology, Surv. Geophys., 37, 307–337, https://doi.org/10.1002/2014GL062764, 2016.

Bogning, S., Frappart, F., Blarel, F., Niño, F., Mahé, G., Bricquet, J.-P., Seyler, F., Onguéné, R., Etamé, J., Paiz, M.-C., and Braun, J.-J: Monitoring Water Levels and Discharges Using Radar Altimetry in an Ungauged River Basin: the case of the Ogooué, Remote Sens.-Basel, 10, 350, https://doi.org/10.3390/rs10020350, 2018.

Bogning, S., Frappart, F., Paris, A., Blarel, F., Niño, F., Picart, S. S., Lanet, P., Seyler, F., Mahé, G., Onguene, R., Bricquet, J.-P., Etame, J., Paiz, M.-C., and Braun, J.-J.: Hydro-climatology study of the Ogooué River basin using hydrological modeling and satellite altimetry, Adv. Space Res., 68, 672–690, https://doi.org/10.1016/j.asr.2020.03.045, 2020.

Garambois, P. A., Calmant, S., Roux, H., Paris, A., Monnier, J., Finaud-Guyot, P., Montazem, A. S., and Santos da Silva, J.: Hydraulic visibility: Using satellite altimetry to parameterize a hydraulic model of an ungauged reach of a braided river, Hydrol. Process., 31, 756–767, https://doi.org/10.1002/hyp.11033, 2017.

Gupta, H. V., Kling, H., Yilmaz, K. K., and Martinez, G. F.: Decomposition of the mean squared error and NSE performance criteria: Implications for improving hydrological modelling, J. Hydrol., 377, 80–91, 2009.

Huang, Q., Long, D., Du, M., Han, Z., and Han, P.: Daily continuous river discharge estimation for ungauged basins using a hydrological model calibrated by satellite altimetry: implications for the SWOT Mission, Water Resour. Res., 56, 537–546, https://doi.org/10.1029/2020WR027309, 2020.

Intrator, O. and Intrator, N.: Using Neural Nets for Interpretation of Nonlinear Models, Proceedings of the Statistical Computing Section, 244–249, 1993.

Lebel, T., Taupin, J., and D'Amato, N.: Rainfall monitoring during HAPEX-Sahel. 1. General rainfall conditions and climatology, J. Hydrol., 189, 74–96, https://doi.org/10.1016/S0022-1694(96)03155-1, 1997.

Le Moine, N.: Le bassin versant vu par le souterrain: une voie d'amélioration des performances et du réalisme des modèles pluie-débit ?, PhD, Université Pierre et Marie Curie, École Doctorale 398 Géosciences et Ressources Naturelles, France, 348 pp., https://hal.inrae.fr/view/index/identifiant/tel-02591478, 2008.

Li, J. and Heap, A.: A review of Spatial Interpolation Methods for environmental scientists, Geoscience Australia, GPO Box 378, Canberra, ACT 2601, Australia, 137 pp., 2008.

Paiva, R., Buarque, D., Collischonn, W., Bonnet, M.-P., and Frappart, F.: Large-scale hydrological and hydrodynamic modelling of the Amazon River basin, Water Resour. Res., 49, 1226–1243, 2013.

Paris, A.: Utilisation conjointe de données d'altimétrie satellitaire et de modélisation pour le calcul des débits distribués dans le bassin amazonien, PhD, Université Paul Sabatier – Toulouse III, 289 pp., https://theses.hal.science/tel-01357059, 2015.

Paris, A., Dias de Paiva, R., Santos da Silva, J., Medeiros Moreira, D., Calmant, S., Garambois, P. A., and Seyler, F.: Stage-discharge rating curves based on satellite altimetry and modeled discharge on the Amazon basin, Water Resour. Res., 52, 3787–3814, 2016.

Poncelet, C., Andréassian, V., Oudin, L., and Perrin, C.: The Quantile Solidarity approach for the parsimonious regionalization of flow duration curves, Hydrolog. Sci. J., 62, 1364–1380, 2017.

Riedmiller, M. and Braun, H.: A direct adaptive method for faster backpropagation learning: The RPROP algorithm, ProSceedings of the IEEE International Conference on Neural Networks (ICNN), San Francisco, CA, USA, 586–591, https://doi.org/10.1109/ICNN.1993.298623, 1993.

Saadi, M., Oudin, L., and Ribstein, P.: Random Forest Ability in Regionalizing Hourly Hydrological Model Parameters, Water, 11, 1540, https://doi.org/10.3390/w11081540, 2019.

Satge, A., Defrance, D., Sultan, B., Bonnet, M.-P., Seyler, F., Rouché, N., Pierron, F., and Paturel, J.: Evaluation of 23 gridded precipitation datasets across West Africa, J. Hydrol., 581, 124412, https://doi.org/10.1016/j.jhydrol.2019.124412, 2020.

Vörösmarty, C., Askew, A., Grabs, W., Barry, R. G., Birkett, C., Döll, P., Goodison, B., Hall, A., Jenne, R., Kitaev, L., Landwehr, J., Keeler., M., Leavesley, G., Schaake, J., Strzepek, K., Sundarvel, S. S., Takeuchi, K., and Webster, F.: Global water data: A newly endangered species, Eos, 82, 54–58, https://doi.org/10.1029/01EO00031, 2001.

Proc. IAHS, 385, 485–487, 2024
https://doi.org/10.5194/piahs-385-485-2024

Return periods in current and future climate

Dan Rosbjerg

Department of Environmental and Resource Engineering, Technical University of Denmark, Kongens Lyngby, 2800, Denmark

Correspondence: Dan Rosbjerg (daro@dtu.dk)

Published: 19 April 2024

Abstract. For Danish conditions, discrete rain intensity climate factors for a 100-year time horizon are generalized to continuous curves as functions of the return period. Assuming the tails of the distribution functions for T-year events to be approximately exponential, a general formula for projecting current return periods into future return periods is developed. Moreover, the uncertainty of the future T-values due to uncertainty in the climate factor is approximately assessed using first order analysis.

Keywords. UPH 21; SDG 13; Modelling; Urban drainage design; Prediction uncertainty

1 Introduction

In Denmark, recommended climate factors for design rain intensities looking 100 years ahead are given as discrete values for return periods T equal to, respectively, 2, 10 and 100 years (SVK, 2014). Multiplication of the current T-year event with the climate factor results in an estimate of the future T-year event. The recommended Danish climate factor values are based on simulations with climate models (Christensen et al., 1998; Gregersen et al., 2013) and therefore not exact values, but nevertheless useful for design of urban drainage structures. To account for the uncertainty, they are provided both as standard climate factors and as high climate factors. To facilitate design efforts, continuous curves are presented below. Climate factors are telling how much a given T_c-year event in current climate will increase in the future climate. Another, but related question is which future return period, T_f, corresponds to a T_c-year event in the current climate. Below a convenient, general formula for the future return period as function of the climate factor and the current return period is developed. In addition, by means of first analysis the uncertainty of T_f is assessed talking into account the uncertainty in the climate factor.

Table 1. Climate factors as function of return periods.

Return period, T_c	Standard climate factor	High climate factor
2	1.2	1.45
10	1.3	1.70
100	1.4	2.00

2 Continuous climate factor curves

If $x_{T,c}$ denotes the T-year event in the current climate, and $x_{T,f}$ denotes the T-year event in the future climate, the climate factor, k, is defined as the ratio between the events. Thus

$$k = \frac{x_{T,f}}{x_{T,c}} \quad \Rightarrow \quad x_{T,f} = k\, x_{T,c} \tag{1}$$

The climate factor depends on several different other factors, where the return period T by far is the most important one. The dependence of T is shown in Table 1 both for standard climate factors and for high climate factors.

A generalisation of the discrete values into a continuous curve might be done using a simple log-linear approximation. Here, however, it is chosen to use a second order polynomium approximation, which exactly corresponds to the three given climate factor values in Table 1. The relation for the Danish standard climate factor is found to be

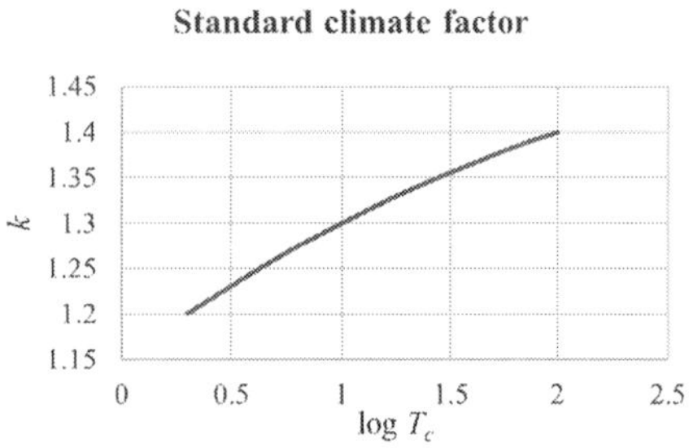

Figure 1. Standard climate factor as function of return period.

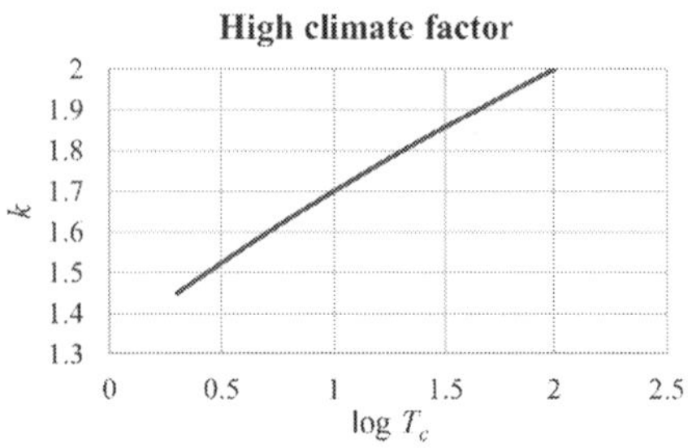

Figure 2. High climate factor as function of return period.

$$k = -0.0253\,(\log T_\mathrm{c})^2 + 0.175 \log T_\mathrm{c} + 1.150 \tag{2}$$

which is shown in Fig. 1. The corresponding formula for the high climate factor is

$$k = -0.0341\,(\log T_\mathrm{c})^2 + 0.402 \log T_\mathrm{c} + 1.332 \tag{3}$$

which is shown in Fig. 2. Both the relations are developed for $2 \leq T_\mathrm{c} \leq 100$, but can be assumed valid also for values of T_c greater than 100.

3 Projection of the current return period into a future return period

A wide class of distribution functions possesses exponential tails, and it is here assumed plausible that this is also the case for the distributions of $x_{T,\mathrm{c}}$ and $x_{T,\mathrm{f}}$. Thus for the extreme value regions we get for $x_{T,\mathrm{c}}$

$$F(x) = 1 - \exp\left(-\frac{x}{a}\right) = 1 - \frac{1}{T_\mathrm{c}} \quad \Rightarrow \quad T_\mathrm{c} = \exp\left(\frac{x}{a}\right) \tag{4}$$

and for $x_{T,\mathrm{f}}$

$$G(x) = 1 - \exp\left(-\frac{x}{b}\right) = 1 - \frac{1}{T_\mathrm{f}} \quad \Rightarrow \quad T_\mathrm{f} = \exp\left(\frac{x}{b}\right) \tag{5}$$

From Eq. (1) we have

$$P\left\{x_{T,\mathrm{f}} \leq x\right\} = P\left\{k\,x_{T,\mathrm{c}} \leq x\right\} = P\left\{x_{T,\mathrm{c}} \leq \frac{x}{k}\right\}$$

$$\Rightarrow \quad G(x) = F\left(\frac{x}{k}\right) \tag{6}$$

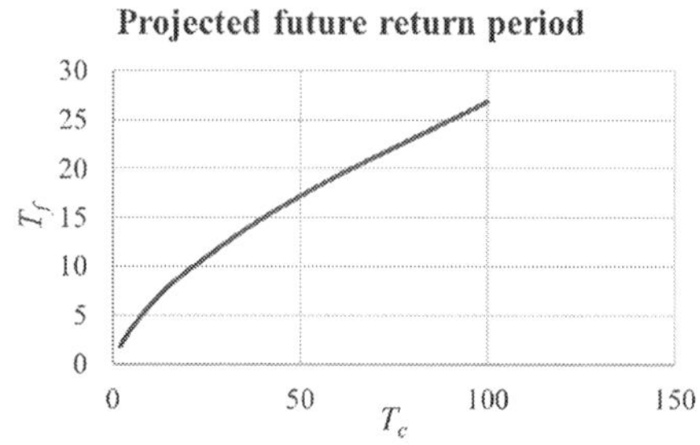

Figure 3. Projected future return period as function of current return period using standard climate factors.

Table 2. Current return periods, corresponding climate factors and projected future return periods.

T_{2017} (MST, 2021)	1	3	68	220	515
T_{2100} (MST, 2021)	1.2	1.6	20	50	100
k_standard Eq. (2)	1.15	1.23	1.39	1.42	1.44
T_{2100} Eq. (9)	1	2.5	20.1	44.3	76.3
k_high Eq. (3)	1.3	1.5	2.0	2.1	2.2
T_{2100} Eq. (9)	1	2.1	8.7	13.3	17.7

Combining with Eqs. (4) and (5) we get

$$\exp\left(-\frac{x}{b}\right) = \exp\left(-\frac{x}{k\,a}\right) \quad \Rightarrow \quad b = k\,a \tag{7}$$

By insertion into Eq. (5) we then obtain

$$T_\mathrm{f} = \exp\left(\frac{x}{b}\right) = \exp\left(\frac{x}{k\,a}\right) = \left[\exp\left(\frac{x}{a}\right)\right]^{\frac{1}{k}} \tag{8}$$

leading to the below general relation between T_f and T_c

$$T_\mathrm{f} = T_\mathrm{c}^{\frac{1}{k}} \tag{9}$$

The relation is exemplified in Fig. 3 using Danish standard climate factors. It is, e.g., seen that a 100-year value in the current Danish climate becomes a 27-year event in the future climate.

In Table 2, the values obtained using Eq. (9) are compared to values published in MST (2021) obtained by using results from climate models. The correspondence is found reasonable. Moreover, it can be seen from the table that the use of high climate factors gives rise to a dramatic change in the assessment of future return periods.

4 Uncertainty of projected future return periods

Due to uncertainty in the climate factor, the projection of a return period into the future will become uncertain as well. For fixed values of T_c, the uncertainty of T_f is assessed in terms of coefficient of variations. Interpreting Eq. (9) as $T_\mathrm{f} = f(k)$ and applying first order analysis lead to the general equation

$$\mathrm{CV}\{T_\mathrm{f}\} \cong \frac{k}{f(k)}\left|\frac{\mathrm{d}f(k)}{\mathrm{d}k}\right|\mathrm{CV}\{k\} = \frac{\ln T_\mathrm{c}}{k}\mathrm{CV}\{k\} \tag{10}$$

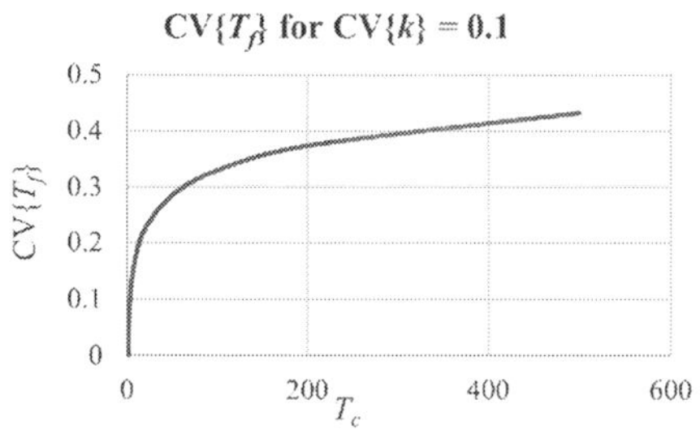

Figure 4. Coefficient of variation of the future return period as function of the current return period assuming the coefficient of variation of the climate factor equal to 10 %.

The relation is shown in Fig. 4. It can, e.g., be seen that a 10 % uncertainty in the climate factor implies a 33 % uncertainty in the future projection of a 100-year return period.

5 Conclusions

Discrete climate factors are for Danish conditions generalized to continuous functions of the return period for both standard and high climate factors. Using exponential tail approximations for the distribution function of current and future T-year events, a general analytical expression for projecting the current return period into the future return period as function of the climate factor is developed and evaluated. Finally, the uncertainty of the projected future return period due to uncertainty in the climate factor is assessed revealing a notable uncertainty even in case of a moderate uncertainty in the climate factor.

Code availability. The code is not publicly available.

Data availability. The applied data are available in MST (2021, https://www2.mst.dk/Udgiv/publikationer/2018/manual.pdf).

Competing interests. The author has declared that there are no competing interests.

Disclaimer. Publisher's note: Copernicus Publications remains neutral with regard to jurisdictional claims made in the text, published maps, institutional affiliations, or any other geographical representation in this paper. While Copernicus Publications makes every effort to include appropriate place names, the final responsibility lies with the authors.

Special issue statement. This article is part of the special issue "IAHS2022 – Hydrological sciences in the Anthropocene: Past and future of open, inclusive, innovative, and society-interfacing approaches". It is a result of the XIth Scientific Assembly of the International Association of Hydrological Sciences (IAHS 2022), Montpellier, France, 29 May–3 June 2022.

Acknowledgements. The editor Christophe Cudennec is acknowledged for his great efforts in handling the paper.

Review statement. This paper was edited by Christophe Cudennec.

References

Christensen, O. B, Christensen, J. H., Machenauer, B., and Botzet, M.: Very-high resolution regional climate simulations over Scandinavia – present climate, J. Climate, 11, 3204–3229, https://doi.org/10.1175/1520-0442(1998)011<3204:VHRRCS>2.0.CO;2, 1998.

Gregersen, I. B., Sørup, H. J. D., Madsen, H., Rosbjerg, D., Mikkelsen, P. S., and Arnbjerg-Nielsen, K.: Assessing future climatic changes of precipitation extremes at small spatio-temporal scales, Clim. Change, 118, 783–797, https://doi.org/10.1007/s10584-012-0669-0, 2013.

MST: Manual for application of Report no. 31 by IDA, Spildevandskomitéen, Ramboell to Miljoestyrelsen, https://www2.mst.dk/Udgiv/publikationer/2018/manual.pdf (last access: 21 January 2023), 2021 (in Danish).

SVK: Updated climate factors and design rain intensities, Report no. 30, IDA Spildevandskomitéen, https://ida.dk/media/2994/svk_skrift30_0.pdf (last access: 21 January 2023), 2014 (in Danish).

Proc. IAHS, 385, 489–497, 2024
https://doi.org/10.5194/piahs-385-489-2024

The International Commission on Continental Erosion (ICCE): a brief overview of its scientific focus and example outputs

Adrian L. Collins[1], **Des E. Walling**[2], **Valentin Golosov**[3,4], **Paolo Porto**[5,6], **Allen C. Gellis**[7], **Yuri Jaques da Silva**[8], **and Sergey Chalov**[3]

[1]Net Zero and Resilient Farming, Rothamsted Research, Okehampton, EX20 2SB, UK
[2]Geography, Faculty of Environment, Science & Economy, University of Exeter, Exeter, EX4 4RJ, UK
[3]Faculty of Geography, Lomonosov Moscow State University, Moscow, 119991, Russia
[4]Institute of Geography, Russian Academy of Science, Moscow, 119017, Russia
[5]Department of Agraria, University *Mediterranea* of Reggio Calabria, Feo di Vito, 89122 Reggio Calabria, Italy
[6]Faculty of Geographical Sciences, Kazimierz Wielki University, 85-033 Bydgoszcz, Poland
[7]U.S. Geological Survey, Baltimore, MD, USA
[8]Agronomy Department, Federal University of Piauí (UFPI),
Planalto Horizonte/Bom Jesus – PI, 64900-000, Brazil

Correspondence: Adrian L. Collins (adrian.collins@rothamsted.ac.uk)

Published: 13 May 2024

Abstract. Erosion and sediment-related problems are well documented globally and continue to warrant further scientific investigation, as well as improved policies and management strategies to protect soil and water resources. The International Commission on Continental Erosion (ICCE) has long been a very active commission of the International Association of Hydrological Sciences (IAHS) focussing on progressing scientific understanding of erosion and sediment systems. This paper provides a brief overview of its main scientific foci, examples of previous contributions to scientific conferences and finally, concludes by paying tribute to two key former members of the international scientific community engaged with its remit.

Keywords. Erosion; sedimentation; sediment budget; sediment source fingerprinting; river basin; connectivity

1 Introduction

Environmental problems associated with accelerated erosion (land and river channel), sediment redistribution, storage and transport, sediment-associated transfers of nutrients and contaminants, and the role of fine-grained sediment in degrading aquatic habitats have attracted increasing attention over the past 30 years (e.g., Hinderer, 2012). Whereas in the past, interest in erosion and sediment delivery focused primarily on those areas of the world with documented high rates of soil erosion and high sediment yields, sediment-associated problems have now frequently been shown to be equally important in other regions where erosion rates and sediment yields are relatively low. These regions are often very sensitive to human impact, since, for example, small shifts in sediment response can result in significant impacts on aquatic habitats. Furthermore, through advances in sediment budget techniques, including sediment fingerprinting and remote-sensed imagery (e.g., Lidar), the relative importance of channel sources versus soil erosion has been established. Globally, land use and climate change, along with commensurate channel readjustments and the construction of large dams, are exerting an important influence on erosion and sediment yields of the world's rivers (e.g., Verstraeten et al., 2009). Rivers are a key source of sediment supplied to deltas and the coastal seas and changes in their sediment flux can have wide-ranging implications, such as the alteration of the global carbon and nutrient cycles (e.g., Horowitz, 2008). More generally, human impact on erosion and sediment transport is clearly evident across a range of scales, and the role of fine sediment in diffuse or distributed source pollution has resulted in it being described as the world's most pervasive pollutant both currently and historically. These important environmental and societal problems are central to the interests of the International Commission on Continental Erosion (ICCE) and the work of the Commission has therefore gained a growing relevance in recent years. The Commission has responded by directing its attention to these important issues which are forecast to become more severe under projected near- and far-future climates.

2 Focus themes for the ICCE

The work of the ICCE has focused on a number of themes, aimed at emphasizing the sensitivity of sediment cascades and systems to human impact and global change (including changing climate), which, in turn, is directed at improving our understanding of sediment processes with a view to supporting targeted management strategies.

3 Sediment budgets

The first area reflects the need to develop an improved understanding of the mobilization, transfer and redistribution of sediment within the landscape across different spatial and temporal scales. This has, in turn, highlighted the importance of establishing sediment budgets, which link erosion and sediment mobilization to sediment transfer and storage and ultimately sediment output at the outlet of a river basin. Here, particular emphasis is commonly placed on identifying key sediment sources and sinks. Sediment sources are a key influence on sediment quality with associated biological and infrastructure impacts, whereas sediment sinks can exert a very important control on the transit and residence time of sediment fluxes, since much of the sediment mobilized from the surface and channel sources may reach the basin outlet over differing time periods. Sediment budgets are therefore adopted widely as a basic framework for understanding sediment systems and for providing system-wide data for supporting sediment management. ICCE can be seen as being ahead of its time in directing attention to this research theme, by organizing a symposium on "Sediment Budgets" in 1988 in Porto Alegre, Brazil. This was followed up by the Moscow symposium on "Sediment Transfer through the Fluvial System" in 2004 and a second symposium on "Sediment Budgets" held in Foz do Iguacu, Brazil in 2005. The symposium on "Erosion and Sediment Yields in the Changing Environment" in Chengdu, China, in October 2012, also featured work on sediment budgeting (Minella et al., 2012). Even when sediment budgets have not been an explicit focus of ICCE meetings, attendees have always presented on various aspects of the sediment cascade including at the symposium on "Sediment Dynamics from the Summit to the Sea" in New Orleans, USA, in December 2014, where papers were presented on hillslope erosion, channel erosion (e.g., Kuhnle et al., 2015), stream and river sediment transport, catchment and river basin sediment yield (e.g., Zhang et al., 2015) and sediment dynamics in coastal systems.

4 Development and application of measurement and monitoring technique

The development of improved measurement and monitoring techniques has also been an important focus for ICCE activity for many years. This reflects both the need to obtain more reliable information on river sediment loads and the need to develop the new techniques required to quantify the various components of sediment budgets. A major symposium on "Sediment Transport Monitoring in River Basins" was organized by ICCE in Oslo, Norway in 1992, and this was followed up by the Oslo Workshop on "Erosion and Sediment Transport in Rivers – Technical and Methodological Advances" in 2002. The Moscow Symposium on "Sediment Transfer through the Fluvial System" in 2004 emphasized the progress achieved in this research field by the application of different methods and techniques, providing detailed quantitative information on sediment behaviour and sediment transfer in small first order catchments (Porto et al., 2004), as well as in large river basins and their individual components (Walling and Zhang, 2004).

A specialized workshop focused on "Sediment Tracing", held in Melbourne in 2011, provided further thoughts on the need to assemble reliable information to characterize the sources in sediment budget investigations. In that workshop, ICCE scientists focused on the opportunity to combine direct observations of soil erosion/deposition with results obtained using sediment tracing techniques as an essentially unique strategy of assembling the required information on sediment sources, sediment sinks and spatially distributed sediment redistribution rates (Porto et al., 2013; Walling, 2006). Given the increasing drive towards integration of research methods and techniques, the symposium at Okehampton in England in July 2016, focused on "Integrating Monitoring and Modelling for Understanding, Predicting and Managing Sediment Dynamics". Regardless of any specific focus of individual meetings, ICCE scientists have always presented studies using a range of individual or integrated methods for understanding various components of sediment systems, including remote sensing for soil erosion by water (e.g., Fitrzyk, 2012), terrestrial laser scanning for gravel-bed morphology (Moretto et al., 2012), fallout radionuclide measurements combined with sediment load monitoring for catchment sediment response (Porto et al., 2015), fallout radionuclides for estimating overbank sedimentation rates (Golosov and Walling, 2015), acoustic sensor deployment for bedload transport monitoring (Rickenmann et al., 2017), the use of bed sediment sampling for water quality studies and monitoring programmes (Horowitz and Elrick, 2017) and the combined use of field walking and digital elevation model information to assess field-to-river connectivity (Favis-Mortlock et al., 2022).

5 Sediment source fingerprinting

One method which has featured strongly in many ICCE meetings over the years is sediment source fingerprinting. This direct method for quantifying the relative contributions from individual sediment sources to various types of target sediment including suspended, bed, floodplain or lake sediment, has experienced growing uptake globally over the past 25 years (Collins et al., 2020) and several ICCE scientists are now seen as being established experts in this specific field. This latter fact is evidenced by their co-authorship of a recent review of the approach and remaining challenges of sediment source fingerprinting, published in the Journal of Soils and Sediments (Collins et al., 2020). ICCE papers on this topic have covered a wide range of issues associated with source fingerprinting procedures and their applications, including the use of frequentist un-mixing models (Walling et al., 1993) and uncertainty analysis (Martinez-Carreras et al., 2008), combining source fingerprinting with artificial tracers to improve the resolution of source data (Collins et al., 2010) and demonstrating applications of different types of tracers including colour (Martinez-Carreras et al., 2008). The focus on sediment sourcing methods was again a feature of one session convened by the ICCE at the 27th IUGG meeting in Montreal, Canada, in 2019. This very successful session resulted in a special issue in the international journal Catena entitled Using Fingerprinting, Monitoring and Modelling to Explore Erosion, Connectivity and Transfers in Catchment Sediment Systems. This special issue includes papers on sediment fingerprinting using elemental geochemistry (Li et al., 2020; Laceby et al., 2021; Shi et al., 2021; Tsyplenkov et al., 2021; Nosrati et al., 2021a), spectroscopy, magnetism and geochemistry (Ramon et al., 2020), magnetic and colour fingerprints (Nosrati et al., 2021b) and optical properties (Amorim et al., 2021).

6 Human impacts on sediment systems

The importance of human impact on erosion and sediment transport has also been a recurring theme for ICCE, as illustrated by the symposia on "The Impact of Environmental Change on Sediment Sources and Sediment Delivery" held in Perugia, Italy in 2007, "Sediment Dynamics in a Changing Environment" held in Christchurch, New Zealand in 2008, and on "Sediment Dynamics for a Changing Future" held in Warsaw, Poland in 2010, as well as the symposium on "Erosion and Sediment Yields in the Changing Environment" held in Chengdu, China in October 2012. In those symposia, the potential impact of environmental change, including both land use change and climate change on sediment fluxes in catchments and river basins, was emphasized (Porto et al., 2009; Bogen, 2009; Zhang et al., 2009).

Wildfires are seen as an increasing threat to many of the world's forests and such fires can cause greatly increased sediment mobilisation and transport, causing both extensive damage within the area and major problems in rivers and reservoirs downstream due to increased sediment loads. Given the growing wildfire threat globally, ICCE organized a Symposium on "Wildfire and Water Quality: Process Impacts and Challenges", in Banff, Alberta, Canada in June 2012. This meeting included papers on a broad range of wildfire related topics including impacts on sediment yields (Jordan, 2012) and sediment accumulation (Southwell and Thoms, 2012).

The recent Fukushima disaster in Japan has clear parallels with the Chernobyl disaster of 1986 and has again directed attention to the fundamental role of erosion and sediment transport in controlling the redistribution and fate of human-induced radionuclide contamination. Attention was directed to this topic by a workshop held at the International Association of Hydrological Sciences (IAHS) Scientific Assembly in Gothenburg, Sweden, 2014. The important role played by sediment in river channel morphology and development and the many detrimental impacts of sediment within river basins, which include degradation of water quality and aquatic habitats and siltation of reservoirs and other infrastructure for water resource development, have necessarily highlighted the need for sediment management strategies. ICCE has addressed this theme through the symposium on "The Structure, Function and Management of Fluvial Sedimentary Systems" held in Alice Springs, Australia in 2002, the symposium on "Sediment Dynamics and the Hydromorphology of Fluvial Systems" held in Dundee, Scotland in 2006 and the workshop on "Sediment Problems and Sediment Management in Asian River Basins", which formed part of the Hyderabad Scientific Assembly in 2009. Together, these various ICCE symposia and workshops have generated a valuable set of IAHS Publications which are an important resource for those working in the field of managing the global sediment problem. The full archive of the publications and/or special issues is available at: https://iahs.info/Commissions--W-Groups/ICCE-Continental-Erosion/ICCE-Publications.do (last access: 7 May 2024).

Another widely discussed topic is related to research on mining-affected rivers, which continues to pose challenges because plans exist in many regions worldwide to develop opencast mines (e.g., Bartley et al., 2004; Chalov, 2014). Metal mining and mineral processing operations often produce substantial quantities of solid waste which are delivered to river systems due to erosion processes, and this material often results in substantial detrimental impacts on fluvial habitats. Responsible catchment management and mining impact control requires an understanding of contempo-

Figure 1. Case studies for erosion research on mining-impacted sites: 19th century abandoned gold-mining site (Sherlock creek, Montana, USA) (left) and ongoing platinum placer mining site (Kamchatka peninsula, Russia) (right) (photos by Sergey Chalov).

raneous catchment baseline conditions of sediment redistribution. Environments, which have historically suffered from mining operations (e.g., the Rocky Mountains in the USA, Fig. 1), or such regions as the Kamchatka Peninsula and the island of Madagascar, which are new frontiers for the mining industry, have been the key focus of the recent research activities related to erosion and sediment transport.

7 River hydromorphology and sediment yield

Another special issue which also included topics associated with human impacts but also comprised studies on interactions between river processes and sediment transport was published as a follow-up to the stand-alone conference hosted by the ICCE which was held in August 2018, in Moscow (Chalov et al., 2019). This particular special issue emphasized that links between sediment yield and channel patterns have been used as a key tool for channel management in many regional studies (e.g., Golovlyov et al., 2019). Estimating the total global suspended sediment flux from the land to the oceans for establishing the global pattern of sediment yields and its major controls has been another key focus of the ICCE (e.g., Walling and Webb, 1996). At the 2004 ICCE symposium "Sediment Transfer through the Fluvial System" a contribution by Dedkov (2004) argued that deposition very often prevails over erosion in the lower reaches of rivers. Analysis of available long-term records and sediment archives has provided an important means of assessing the sensitivity of sediment yields to environmental change; for example, in relation to the contribution of catchment and in-channel erosion to sediment transport.

8 Modelling of sediment dynamics

Whilst the ICCE has undoubtedly attracted more empirical scientists over the years, as opposed to modellers, modelling studies have frequently featured at ICCE symposia in recognition that statistical or process-based tools and frameworks can scale out and up the results generated by the more resource intensive methods used by most of the ICCE community. Examples include the MOSES (Modular soil Erosion System) model for predicting soil erosion at small scales (de Figueiredo and Davi, 2006), the Silsoe model for hillslope soil erosion (Matthews and Neave, 2008), the SWAT (Soil and Water Assessment Tool) model for predicting the impact of different land use scenarios on discharge and sediment transport (Dos Santos et al., 2010), the ADAS (referring to the former UK Agricultural Development and Advisory Service) Pollutant Transport (APT) model for sediment delivery to rivers (Collins et al., 2012), and an integrated sediment production and supply model (Fujita et al., 2015). In other cases, modellers have made a comparison between field observations and Digital Elevation Model (DEM)-based field-to-river routing, using for example, a steepest-descent-with-overtopping model (Favis-Mortlock et al., 2022).

9 Wider contributions of ICCE to international sediment initiatives and programmes

ICCE is active in promoting the contribution of IAHS to international activity in the field of erosion and sediment transport, through involvement in international activities and collaboration with other international bodies. These include the European SedNet initiative, UNESCO International Sediment Initiative (ISI), the World Association for Sedimentation and Erosion Research (WASER), the International Research and Training Centre in Erosion and Sedimentation (IRTCES), the International Geosphere-Biosphere Programme (IGBP), the International Geographical Union (IGU), the International Coordinating Committee on Reservoir Sedimentation (ICCORES), the International Commission on Large Dams (ICOLD), the International Association of Sediment–Water Science (IASWS), the International Soil Conservation Organization (ISCO) and the International Atomic Energy Agency (IAEA). ICCE has been an active participant in UNESCO's International Hydrological Programme for many years and this involvement has increased through its contributions to the UNESCO ISI. Several ICCE members have played a very active role in a number of Collaborative Research Projects organized by IAEA, where the application of new isotopic and radionuclide tracer techniques for documenting erosion and sediment redistribution rates and fingerprinting sediment sources has been progressed. ICCE has also been a major contributor to the Global Geochemical Mapping programme, coordinated by Norwegian scientists, which aims to exploit the sediment archives provided by river floodplains and deltas to document both global patterns of sediment geochemistry and recent changes associated with human impacts in upstream river basins.

During the COVID-19 pandemic, ICCE organised online educational activities. Here, the most wide ranging online seminar series was held online for Young Scientists under the topic "Pollutant and Sediment Mobility in River Systems: Monitoring Studies to Identify Human Impacts" (25–27 November 2020). The programme included lectures by leading experts from Russia, USA, Germany, France, Belgium, Netherlands, Brazil and the UK on the problems of sediment and water quality and geochemistry of river systems: Adrian Collins (Rothamsted Research, North Wyke, Okehampton, United Kingdom); Caroline Clason (University of Plymouth, Plymouth, United Kingdom); Sagy Cohen (Department of Geography, University of Alabama, Tuscaloosa, Alabama, United States); Sergey Chalov (Lomonosov Moscow State University, Moscow, Russia); Bjoern Helm (Dresden University of Technology, Dresden, Germany); Jeff Nittrouer (Department of Earth, Environmental and Planetary Sciences, Rice University, Houston, United States); Edgardo Latrubesse (Federal University of Goiás-Brazil, Goiânia, Brazil); Martina Flörke (Institute of Engineering Hydrology and Water Resources Management, Ruhr University Bochum, Bochum, Germany); Matthias Vanmaercke (Liège University, Liège, Belgium); Salomon Kroonenberg (Delft University of Technology, Delft, the Netherlands), and Daniel Karthe (German-Mongolian Institute for Resources and Technology, Ulaanbaatar city, Mongolia). Almost 400 participants attended the online lectures from almost 50 countries. The full archive of the lectures is available at: https://www.youtube.com/channel/UCCXZNFdglOzGczmnAisb1vg (last access: 7 May 2024).

Figure 2. Wayne Erskine (© IAHS, https://iahs.info/Commissions--W-Groups/ICCE-Continental-Erosion/Past-Information/, last access: 7 May 2024).

10 Notable former members of the ICCE community

10.1 Wayne Erskine

Professor Wayne Erskine (Fig. 2) passed away in Darwin in late July 2017. Wayne graduated with a Doctor of Philosophy in Fluvial Geomorphology (1987) from the University of New South Wales, Australia. He held academic positions at the University of New South Wales, University of Newcastle – Ourimbah Campus and Charles Darwin University and, also, research scientist positions with New South Wales Department of Water Resources, New South Wales State Forest, Office of the Supervising Scientist (OSS), and the Environmental Institute of the Supervising Scientist (ERISS). He had recently been appointed as the sub-editor of Geographical Research. Wayne was a keen fisherman and spoke often of his obsession for everything related to freshwater science.

Wayne was an exceptionally productive and influential fluvial geomorphologist publishing over 230 peer reviewed international and national journal manuscripts, conference papers and book chapters. He supervised 35 post graduate students. Wayne had been a strong supporter of, and contributor to, ICCE activities since the mid-1980s and will be warmly remembered for his in depth and wide readership of the scientific literature which permitted him to engage in detail with all scientists on their work, his ability to produce high quality papers for the Red Book series, his extremely engaging and word perfect presentations on diverse subjects and increasingly on work bridging pure and applied science, and his capacity to ask incisive questions at conferences backed up by his broad, but in-depth, knowledge and understanding. He had recently been discussing the possibility of hosting a future ICCE meeting with the current committee.

10.2 Arthur J. Horowitz

Art Horowitz (Fig. 3) passed away on 26 February 2021, after a relatively short battle with cancer. Art graduated in 1967 from Queens College, New York, with a BA in Chemistry and Geology. Between 1967–1971 he studied for a Master of Science in Marine Geology at the University of Miami, Florida, USA. Thereafter, he studied for a PhD in Geology (1971–1974) at the University of London, UK and

Figure 3. Art Horowitz (© IAHS, https://iahs.info/Commissions--W-Groups/ICCE-Continental-Erosion/Past-Information/).

a DIC in Applied Geochemistry (1971–1974) at Imperial College of Science and Technology, London, UK. His PhD furthered his interests in marine geology and involved research on the geochemistry of sediments associated with the mid-Atlantic Ridge. From 1975–1976 he worked as Assistant Professor in the Department of Earth and Environmental Sciences, Queens College, New York. He then worked as an oceanographer in the Environmental Studies Section, New York Outer Continental Shelf Office, where he served as a Contracting Officer's Representative for developing and monitoring the performance of both government and non-government contractors tasked with establishing baseline environmental levels at potential offshore oil and gas leasing sites.

Art's career changed direction towards things more hydrological when he joined the US Department of the Interior in 1976; first, with the Bureau of Land Management and then with the USGS Water Resources Division in 1978 as a Supervisory Chemist at the Atlanta National Water Quality Lab (ANWQL) in Doraville, Georgia. Between 1978–1983, he was Chief of the Automated Wet-chemistry Section for the ANWQL, responsible for analyses of filtered water, whole-water and bed sediment samples as well as the development of novel analytical techniques for use with these sample matrices. When the Atlanta NWQL closed in 1983, he relocated to the Georgia District Office, Atlanta (USGS-WRD) and began his long tenure as Chief Research Chemist of the Sediment Partitioning Research Project. Over the next 33 years, Art pioneered many of the laboratory analytical techniques used to partition sediment into its chemical components, with a primary focus on indicators of anthropogenic contamination. He developed and tested many of the field-sampling and processing techniques used daily across the US and internationally by the hydrological community. He undertook important work on evaluating de-watering devices for suspended sediment samples in preparation for chemical analyses and was responsible for the development of protocols for the collection and processing of water samples for subsequent trace element analyses. Other key work included testing the effects of membrane filtration on dissolved trace element concentrations, and an evaluation of short-term spatial and temporal variability in suspended sediment and associated trace element concentrations. From 1997 until his retirement, Art continued as Chief for the Sediment Partitioning Project, within the Georgia Water Science Center, Atlanta, where he was responsible for the sediment and sediment-associated chemistry work for the City of Atlanta Water Quality Monitoring Program and for the separation, trace element analyses, and data interpretation for all suspended and bed sediment samples collected for the national NASQAN program.

As well as work in the laboratory and field-testing sampling equipment and sampling protocols and strategies linked to sediment geochemistry, Art was also involved in a number of important field investigations of contaminant transport and fate and water-sediment interactions. These included investigations of mining-related impacts on the Belle Fourche, Cheyenne and Whitewood Creek rivers in South Dakota and high profile geochemical investigations of Lake Coeur d'Alene, the floodplains of the Coeur d'Alene River, and Terrace Reservoir in Idaho. Another nationwide field-based study undertaken by Art in the years before his retirement merits mention as indicative of his dedication to his research. In this, he undertook a nationwide sampling campaign, personally visiting sites as far downstream as practicable on 131 sizeable US rivers located around the coast of the entire country, to collect samples of fine bed sediment to provide ground truth for a national assessment of annual land-sea fluxes of sediment-associated trace and major elements and nutrients that he was undertaking. The sampling campaigns extended over a whole year during 2010–2011. This was followed up in late 2011 by revisiting and resampling 41 of these rivers located between the North Carolina / South Carolina border and the US Canadian border, to investigate the impact of extreme events Hurricane Irene and Tropical Storm Lee, which hit the Atlantic coast of the USA in late August and early September 2011, on the bed sediment chemistry of US Atlantic Coast rivers. Contrary to what might have been expected, these changes were limited. Art also had a long-standing research interest in the changing sediment load of the Mississippi River and demonstrated the importance of the disastrous 1993 flood in influencing such changes. His interests also extended beyond the US; he had a long-term collaboration with French scientists working on the River Seine basin, contributing to their work on sediment geochemistry and sediment-associated contaminants. He also collaborated with Brazilian hydrologists in exploring the problems of obtaining reliable estimates of suspended sediment loads based on infrequent samples and on designing effective sampling strategies.

Art was strongly committed to the need for rigour in the design of hydrological investigations, in their effective implementation and in the interpretation and communication of their results and many will have benefited from this guidance and constructive criticism. He was always available for discussions and enjoyed being a mentor on sediment chemistry and fluvial geomorphology. As a result of his long and focussed career on sediment and contaminant transport in river systems, Art leaves the international science community with a substantial body of more than 100 published outputs. These include the widely cited "Primer on Sediment-Trace Element Chemistry", which has been an important influence on sediment fingerprinting science. This was originally published as USGS Water Supply Paper 2277 in 1985 and a second edition was published by Lewis Publishers in 1991. Several of his key papers on fluvial sediment and sediment-associated contaminant fluxes can be found in the journal Hydrological Processes. His most recent co-authored publication appeared in December 2020 (Collins et al., 2020). Art was an active member of IAHS from 1992 and will be remembered for his fun presence, challenging questions and raucous laugh at many IAHS meetings including the 2005 IAHS General Assembly in Foz do Iguacu, Brazil, for which Art and Des Walling co-edited the notable twin volume Red Book on Sediment Budgets, containing a total of 80 papers. Throughout his IAHS membership, Art was a strong supporter of ICCE stand-alone meetings and during this phase developed friendships with many sediment researchers globally on the basis of his expert knowledge and advice and enthusiasm for high quality sediment and contaminant science. Art was Vice President of ICCE between 2003–2007. In addition to IAHS, Art was an active member of several additional professional societies including the Geological Society of America, the American Geophysical Union, the American Society for Testing and Materials and the International Association of Geochemistry and Cosmochemistry.

Data availability. No data sets were used in this article.

Author contributions. ALC: conceptualization, writing – original draft; all authors: writing – review & editing.

Competing interests. The contact author has declared that none of the authors has any competing interests.

Disclaimer. Any use of trade, firm, or product names is for descriptive purposes only and does not imply endorsement by the U.S. Government.

Special issue statement. This article is part of the special issue "IAHS2022 – Hydrological sciences in the Anthropocene: Past and future of open, inclusive, innovative, and society-interfacing approaches". It is a result of the XIth Scientific Assembly of the International Association of Hydrological Sciences (IAHS 2022), Montpellier, France, 29 May–3 June 2022.

Review statement. This paper was edited by Christophe Cudennec.

References

Amorim, F. F., da Silva, Y. J. A. B., Nascimento, R. C., da Silva, Y. G. A. B., Tiecher, T., do Nascimento, C. W. A., Minella, J. P. G., Zhang, Y., Upadhayay, H. R., Pulley, S., and Collins, A. L.: Sediment source apportionment using optical property composite signatures in a rural catchment, Brazil, Catena, 202, 105208, https://doi.org/10.1016/j.catena.2021.105208, 2021.

Bartley, R., Olley, J., and Henderson, A.: A sediment budget for the Herbert River catchment, North Queenland, Australia, in: Sediment transfer through the fluvial system, edited by: Golosov, V., Belyaev, V., and Walling, D. E., IAHS Press, Wallingford, IAHS Publ., 288, 147–154, 2004.

Bogen, J.: The impact of environmental changes on the sediment loads of Norwegian rivers, Catena, 79, 251–256, https://doi.org/10.1016/j.catena.2009.07.003, 2009.

Chalov, S., Golosov, V., Collins, A., and Stone, M.: Preface: Land use and climate change impacts on erosion and sediment transport, Proc. IAHS, 381, 1–1, https://doi.org/10.5194/piahs-381-1-2019, 2019.

Chalov, S. R.: Effects of placer mining on suspended sediment budget: case study of north of Russia's Kamchatka Peninsula, Hydrolog. Sci. J., 59, 1081–1094, https://doi.org/10.1080/02626667.2014.903330, 2014.

Collins, A. L., Zhang, Y., Walling, D. E., and Black, K.: Apportioning sediment sources in a grassland dominated agricultural catchment in the UK using a new tracing framework, in: Sediment dynamics for a changing future, Proc. IAHS, Wallingford, UK, 337, 68–75, 2010.

Collins, A. L., Foster, I., Zhang, Y., Gooday, R., Lee, D., Sear, D., Naden, P., and Jones, I.: Assessing "modern background sediment delivery to rivers" across England and Wales and its use for catchment management, in: Erosion and Sediment Yields in the Changing Environment, Proc. IAHS, 358, 125–131, 2012.

Collins, A. L., Blackwell, M., Boerckx, P., Chivers, C.-A., Emelko, E., Evrard, O., Foster, I., Gellis, A., Gholami, H., Granger, S., Harris, P., Horowitz, A. J., Laceby, J. P., Martinez-Carreras, N., Minella, J., Mol, L., Nosrati, K., Pulley, S.,, Silins, U., Jacques da Silva, Y., Stone, M., Tiecher, T., Upadhayay, H. R., and Zhang, Y.: Sediment source fingerprinting: benchmarking recent outputs, remaining challenges and emerging themes, J. Soil. Sediment., 20, 4160–4193, https://doi.org/10.1007/s11368-020-02755-4, 2020.

Dedkov, A.: The relationship between sediment yield and drainage basin area, Sediment Transfer through the Fluvial System, Proceedings of a symposium held in Moscow, Russia, August 2004, IAHS Publ., 288, 197–204, ISBN 1-901502-67-8, 2004.

de Figueiredo, E. and Davi, H.: MOSESS: a model for soil erosion prediction at small scales, in: Sediment Dynamics and the Hydromorphology of Fluvial Systems, Proc. IAHS, 306, 556–563, 2006.

Dos Santos, I., Andriolo, M. F., Gibertoni, R. C., and Kobiyama, M.: Use of the SWAT model to evaluate the impact of different land use scenarios on discharge and sediment transport in the Apucaraninha River watershed, southern Brazil, in: Sediment Dynamics in a Changing Future, Proc. IAHS, 337, 322–328, 2010.

Favis-Mortlock, D., Boardman, J., Foster, I., and Shepheard, M.: Comparison of observed and DEM-driven field-to-river routing of flow from eroding fields in an arable lowland catchment, Catena, 208, 105737, https://doi.org/10.1016/j.catena.2021.105737, 2022.

Fitrzyk, M.: Determining soil erosion by water using high resolution remotely-sensed data, in: Erosion and Sediment Yields in the Changing Environment, Proc. IAHS, 356, 319–326, 2012.

Fujita, M., Yamanoi, K., and Izumiyama, H.: A combined model of sediment production, supply and transport, Proc. IAHS, 367, 357–365, https://doi.org/10.5194/piahs-367-357-2015, 2015.

Golosov, V. and Walling, D. E.: Using fallout radionuclides to investigate recent overbank sedimentation rates on river floodplains: an overview, Proc. IAHS, 367, 228–234, https://doi.org/10.5194/piahs-367-228-2015, 2015.

Golovlyov, P., Kornilova, E., Krylenko, I., Belikov, V., Zavadskii, A., Fingert, E., Borisova, N., and Morozova, E.: Numerical modeling and forecast of channel changes on the river Lena near city Yakutsk, Proc. IAHS, 381, 65–71, https://doi.org/10.5194/piahs-381-65-2019, 2019.

Hinderer, M.: From gullies to mountain belts: A review of sediment budgets at various scales, Sediment. Geol., 280, 21–59, https://doi.org/10.1016/j.sedgeo.2012.03.009, 2012

Horowitz, A. J.: Determining annual suspended sediment and sediment-associated trace element and nutrient fluxes, Sci. Total Environ., 400, 315–343, https://doi.org/10.1016/j.scitotenv.2008.04.022, 2008.

Horowitz, A. J. and Elrick, K. A.: The use of bed sediments in water quality studies and monitoring programs, Proc. IAHS, 375, 11–17, https://doi.org/10.5194/piahs-375-11-2017, 2017.

Jordan, P.: Sediment yields and water quality effects of severe wildfires in southern British Columbia, in: Wildfire and Water Quality: Processes, Impacts and Challenges, Proc. IAHS, 354, 25–35, 2012.

Kuhnle, R., Wren, D., and Langendoen, E.: Erosion depth of sand from an immobile gravel bed, Proc. IAHS, 367, 117–121, https://doi.org/10.5194/piahs-367-117-2015, 2015.

Laceby, J. P., Batista, P. V. G., Taube, N., Kruk, M. K., Chung, C., Evrard, O., Orwin, J. F., and Kerr, J. G.: Tracing total and dissolved material in a western Canadian basin using quality control samples to guide the selection of fingerprinting parameters for modelling, Catena, 200, 105095, https://doi.org/10.1016/j.catena.2020.105095, 2021.

Li, Y., Gholami, H., Song, Y., Fathabadi, A., Malakooti, H., and Collins, A.L.: Source fingerprinting loess deposits in Central Asia using elemental geochemistry with Bayesian and GLUE models, Catena, 194, 104808, https://doi.org/10.1016/j.catena.2020.104808, 2020.

Martinez-Carreras, N., Gallart, F., Iffly, J., Pfister, L., Walling, D., and Krein, A.: Uncertainty assessment in suspended sediment fingerprinting based on tracer mixing models: a case study from Luxembourg, in: Sediment Dynamics in Changing Environments, Proc. IAHS, 325, 94–105, 2008.

Matthews, S. and Neave, M. N.: Modelling hillslope soil erosion at ANZAC Cove, Turkey, in: Sediment Dynamics in Changing Environments, Proc. IAHS, 325, 616–619, 2008.

Minella, J. P. G., Merten, G. H., Walling, D. E., and Moro, M.: Using ^{137}Cs measurements and sediment yield monitoring to document catchment scale sediment dynamics and budgets, in: Erosion and Sediment Yields in the Changing Environment, Proc. IAHS, 356, 338–344, 2012.

Moretto, J., Rigon, E., Mao, L., Deali, F., Picco, L., and Lenzi, M. A.: Using a terrestrial laser scanner to assess the morphological dynamics of a gravel-bed river, in: Erosion and Sediment Yields in the Changing Environment, Proc. IAHS, 356, 428–437, 2012.

Nosrati, K., Akbari-Mahdiabad, M., Fiener, P., and Collins, A. L.: Using different size fractions to source fingerprint fine-grained channel bed sediment in a large drainage basin in Iran, Catena, 200, 105173, https://doi.org/10.1016/j.catena.2021.105173, 2021a.

Nosrati, K., Akbari-Mahdiabad, M., Ayoubi, S., and Collins, A. L.: An exploratory study on the use of different composite magnetic and colour fingerprints in aeolian sediment provenance fingerprinting, Catena, 200, 105182, https://doi.org/10.1016/j.catena.2021.105182, 2021b.

Porto, P., Walling, D. E., and Callegari, G.: Validating the use of caesium-137 measurements to estimate erosion rates in three small catchments in Southern Italy, IAHS Publ., 288, 75–83, 2004.

Porto, P., Walling, D. E., and Callegari, G.: Investigating the effects of afforestation on soil erosion and sediment mobilisation in two small catchments in Southern Italy, Catena, 79, 181–188, https://doi.org/10.1016/j.catena.2009.01.007, 2009.

Porto, P., Walling, D. E., and Callegari, G.: Using ^{137}Cs and ^{210}Pbex measurements to investigate the sediment budget of a small forested catchment in Southern Italy, Hydrol. Process., 27, 795–806, https://doi.org/10.1002/hyp.9471, 2013.

Porto, P., Walling, D. E., La Spada, C., and Mallimo, N.: Combining caesium-137 measurements and suspended sediment load data to investigate the sediment response of a small catchment in southern Italy, Proc. IAHS, 367, 220–227, https://doi.org/10.5194/piahs-367-220-2015, 2015.

Ramon, R., Evrard, O., Laceby, J. P., Caner L., Inda, A. V., de Barros, C. A. P., Minella, J. P. G., and Tiecher, T.: Combining spectroscopy and magnetism with geochemical tracers to improve the discrimination of sediment sources in a homogeneous subtropical catchment, Catena, 195, 104800, https://doi.org/10.1016/j.catena.2020.104800, 2020.

Rickenmann, D., Antoniazza, G., Wyss, C. R., Fritschi, B., and Boss, S.: Bedload transport monitoring with acoustic sensors in the Swiss Albula mountain river, Proc. IAHS, 375, 5–10, https://doi.org/10.5194/piahs-375-5-2017, 2017.

Shi, Z., Blake, W. H., Wen, A., Chen, J., Yan, D., and Long, Y.: Channel erosion dominates sediment sources in an agricultural catchment in the upper Yangtze basin of China: evidence from geochemical fingerprints, Catena, 199, 105111, https://doi.org/10.1016/j.catena.2020.105111, 2021.

Southwell, M. and Thoms, M.: Double trouble: the influence of wildfire and flow regulation on fine sediment accumulation in the Cottee River, Australia, Proc. IAHS, 354, 90–98, 2012.

Tsyplenkov, A., Vanmaercke, M., Collins, A. L., Kharchenko, S., and Golosov, V.: Elucidating suspended sediment dynamics in a glacierized catchment after an exceptional erosion event: the Djankuat catchment, Caucasus Mountains, Russia, Catena, 203, 105285, https://doi.org/10.1016/j.catena.2021.105285, 2021.

Verstraeten, G., Lang, A., and Houben, P.: Human impact on sediment dynamics – quantification and timing, Catena, 77, 77-80, https://doi.org/10.1016/j.catena.2009.01.005, 2009.

Walling, D. E.: Tracing versus monitoring: New challenges and opportunities in erosion and sediment delivery research, in: Soil Erosion and Sediment Redistribution in River Catchments, edited by: Owens, P. N. and Collins, A. J., CABI, Wallingford, 13–27, https://doi.org/10.1079/9780851990507.0013, 2006.

Walling, D. E. and Webb, B. W.: Erosion and sediment yield: a global overview, in: Erosion and Sediment Yield: Global and Regional Perspectives, Proceedings of the Exeter Symposium, July 1996, Exeter, UK, IAHS Publ. no. 236, ISBN 0-947571-89-2, 1996.

Walling, D. E. and Zhang, Y.: Predicting slope-channel connectivity: a national-scale approach, IAHS Publ., 288, 107–114, 2004.

Walling, D. E., Woodward, J. C., and Nicholas, A. P.: A multi-parameter approach to fingerprinting suspended sediment sources, in: Tracers in hydrology, edited by: Peters, N. E., Hoehn, E., Leibundgut, C., Tase, N., and Walling, D. E., IAHS Publication No. 215, IAHS, Wallingford, 329–338, ISBN 0-947571-68-X, 1993.

Zhang, X., Tang, Q., Long, Y., He, X., and Wen, A.: Recent changes of suspended sediment yields in the Upper Yangtze River and its headwater tributaries, Proc. IAHS, 367, 297–303, https://doi.org/10.5194/piahs-367-297-2015, 2015.

Zhang, X. C., Liu, W. Z., Li, Z., and Zheng, F. L.: Simulating site-specific impacts of climate change on soil erosion and surface hydrology in southern Loess Plateau of China, Catena, 79, 237–242, https://doi.org/10.1016/j.catena.2009.01.006, 2009.

AUTHOR INDEX